Marine
Geology

Marine Geology

James P. Kennett

Graduate School of Oceanography
University of Rhode Island

PRENTICE–HALL, INC., Englewood Cliffs, N. J. 07632

Library of Congress Cataloging in Publication Data

KENNETT, JAMES P. (DATE)
  Marine geology.

  Bibliography: p. 752
  Includes index.
  1. Submarine geology. I.–Title.
QE39.K46   551.46′08   81–10726
ISBN 0-13-556936-2   AACR2

Cover photograph Deep-space view of the water planet, Earth, from Apollo 17 showing four of its major elements: the oceans, continents, atmosphere, and cryosphere. Clearly visible are the East Antarctic ice sheet, Africa, and the Arabian Peninsula. Oceanic areas include the Southeast Atlantic, western Indian Ocean, Southern Ocean, and Red Sea. Courtesy U. S. National Aeronautics and Space Administration.

Printed in the United States of America

10 9 8 7 6 5

Editorial/production supervision
and interior design by Ellen W. Caughey
Cover design by Diane Saxe
Manufacturing buyer: John Hall

ISBN 0-13-556936-2

Prentice-Hall International, Inc., *London*
Prentice-Hall of Australia Pty. Limited, *Sydney*
Prentice-Hall of Canada, Ltd., *Toronto*
Prentice-Hall of India Private Limited, *New Delhi*
Prentice-Hall of Japan, Inc., *Tokyo*
Prentice-Hall of Southeast Asia Pte. Ltd., *Singapore*
Whitehall Books Limited, *Wellington, New Zealand*

To all those workers
who painted the new view of the earth
and
to Diana, Douglas, and Mary

Contents

III Oceanic Sediments
and Microfossils   *395*

Contents   **ix**

Preface

Marine Geology was written because of the need for a single, modern, comprehensive text about the geology of the oceans, the lack of which has been keenly felt by both instructors and students. The field of marine geology has become rigorous and sophisticated. Our increasing use and misuse of marine resources have required a greater expansion of our knowledge about the geology and geophysics of the oceans. Enormous advances have occurred in our knowledge of marine geology leading to our new view of the earth. The revolution in earth sciences has resulted directly from scientific research in the marine realm. Yet, no book has brought together the many profound ideas developed during the last 15 to 20 years.

My goal has been to provide a text covering the broad spectrum of marine geology in a largely nonquantitative way: the rocks, sediments, geophysics, structure, microfossils, and stratigraphy and history of the ocean basins and their margins. This text has brought together a substantial body of information not otherwise easily accessible. A synthesis of this type is needed now following the revolutionary concepts of global plate tectonics and the enormous amount of new information that has been gathered by a diversity of programs including the Deep-Sea Drilling Project (DSDP). For the first time, it has been possible to examine the history of the oceans within a global evolutionary framework; paradoxically, the vast increase in knowledge is resulting in an increase in specialization. So vast is the subject that a researcher cannot hope to be expert in but a few branches, but yet cannot afford to be ignorant of the multitude of other branches. Syntheses, such as provided in this text, thus take on an increasingly important role.

This text, intended for a first course in marine geology at the advanced undergraduate or graduate level, assumes little background in oceanography. However, it should be of value to professional workers in geology and related fields who seek a modern treatment of the subject. I also hope that it educates about the intimacy of the several major earth elements, one to another: the lithosphere, hydrosphere, atmosphere, cryosphere, and biosphere. A change in one of these elements can ultimately affect one or more of the others.

The text provides a global view, concentrating on the history of the oceans. The historical approach is stressed—the recognition that the oceans are the product of their history. The historical nature of geology sets it apart from the familiar nonhistorical sciences. For this reason, the text includes a major section on *paleoceanography*—the evolution of the ocean system based on the analysis of marine sediments—the newest of geological disciplines. Paleoceanography has generated intensive interest and is just one of several marine-geology disciplines that provides a reorientation in perspective in a rapidly developing field of endeavor. In paleoceanography, perhaps more than in other subjects, a holistic approach is necessary. The solution of any paleoceanographic problem requires the integration of several investigatory avenues. Thus, this text has attempted to demonstrate the scope and diversity of studies that can be pursued within the marine geological arena and of global histories that can be unwoven from such investigations. For instance, much of what we have learned and will continue to learn about the earth's climatic history is derived from the study of marine sediments.

A major purpose of *Marine Geology* is to integrate modern and classical concepts of geology in an understanding of the history of the oceans. Geologists have traditionally viewed the earth from land. But the earth is predominantly a water planet: Enormous oceans surround island landmasses. Tectonism in the oceans ultimately controls the geography of the earth, and oceanic paleocirculation changes through geologic time exert enormous influences upon the character of the global environment. For these reasons, a course in marine geology is a necessary addition to any undergraduate geology curriculum. The development of such curricula, however, has often been frustrated by the lack of a suitable text.

The text is organized within a framework of plate tectonics and the evolution of the lithospheric plates. The 19 chapters are arranged into four sections: the structural and oceanographic setting; the ocean margins; ocean sediments and microfossils; and ocean history. The chapters are organized within a natural progression of ideas. The student is first introduced to fundamental concepts of marine geophysics, stratigraphy, geology, and oceanography that are necessary for an understanding of the chapters that follow about nearshore processes, the continental margins, marine sediments, and paleoceanography. The final chapter is a global synthesis about the evolution of the oceans during the last 200 million years. My approach has been to emphasize mechanisms and processes rather than to provide detailed descriptions.

In collating for the reader the main results and insights of marine geology, one is confronted with an enormous range of interacting subjects. It is difficult for one individual to be versed in all these areas, and I urge my colleagues and interested workers to provide critical comments and better examples for possible future revisions. Certain areas of marine geology are advancing at a rapid rate. Surely, the next few years will bear witness to enormous advances in our knowledge of global paleoceanographic evolution. Hence, the new view of the earth summarized here will be rapidly modified with continuing research.

Text citations to published work have been kept to a minimum. Many papers have not been cited because of a conscious effort to preserve a readable style, and I hope that such omissions will be considered within this perspective.

I am much indebted to colleagues and graduate students for their criticisms and suggestions which have added much to this work. Tj. H. van Andel and K. K. Turekian read the entire draft and offered innumerable suggestions. Many other colleagues critically read entire chapters. These include R. S. Detrick, E. P. Laine, T. S. Loutit, R. L. McMaster, T. C. Moore, Jr., H. Sigurdsson, M. S. Srinivasan, and M. Wimbush. Sections of the chapter on oceanic microfossils were critically read by B. H. Corliss, W. Dudley, P. E. Hargraves, Johanna Resig, W. R. Riedel, and F. Theyer. Others who assisted with various sections were J. P. Dauphin, J. D. Devine, R. L. Larson, Margaret Leinen, and Karen Romine. The interest and support of S. V. Margolis and J. H. Lipps have also been of great value. I am indebted to many workers who have assisted by way of discussion and provision of both published and unpublished information. D. F. Heinrichs, B. Lettau, B. T. Malfait, and R. E. Wall of the National Science Foundation (U.S.A.) have been supportive of my research over many years.

I also recognize with sincere appreciation those individuals who have helped see the book through to completion. From Prentice-Hall, Inc., they are Logan Campbell, who encouraged me to expand the scope of an earlier book I was writing, and my production editor, Ellen Caughey, for all her conscientious efforts toward successful publication. My secretary, Nancy Meader, typed the manuscript and assisted with other tasks; Kristin Elmstrom, D. Scales, Dolores Smith, and Michelle Emery assisted me in a variety of ways. Jane Carter edited for proper English, making numerous changes throughout; Nancy Penrose, likewise, improved several chapters.

This preface permits me to express my appreciation to some who have influenced my thoughts and work over the years. One's knowledge and biases result from constant interaction with instructors and colleagues. A number of scientists at the New Zealand Geological Survey were encouraging to a 12-year-old school boy curious about geology. One of these, N. de B. Hornibrook, has continued his encouragement over the years. Professors P. Vella, W. H. Wellman, R. H. Clark, and J. Bradley were especially influential during formative years. H. Pantin introduced me to marine geology and J. Brodie, of the New Zealand Oceanographic Institute, provided further encouragement. More recent colleagues who have been especially influential are the late O. L. Bandy and N. D. Watkins, as well as N. D. Shackleton, B. Malmgren, M. S. Srinivasan, M. L. Bender, and T. C. Moore, Jr.

I am especially grateful to my wife, Diana, for her encouragement during my two decades of continued research and writing.

JAMES P. KENNETT

one
Introduction

Difficulties are just things to overcome.

Sir Ernest Shackleton

Marine geology is concerned with the character and history of that part of the earth covered by seawater. The importance of marine geology is evident when we consider that three-fourths of the earth's surface is covered by water. Areas of concern range from the beach to marine marshes and lagoons, across the continental shelf, and down to the deepest parts of the ocean. Marine geologists and geophysicists rarely restrict their research to areas below sea level because a considerable body of important information about the history of the earth and the oceans is gained from rocks exposed above sea level. Marine stratigraphers and paleontologists often examine uplifted marine sediments. Likewise, those interested in the evolution of the oceanic crust often visit oceanic islands to understand better the processes of formation of the oceanic crust. The title of one of the papers by the marine geologist Philip Kuenen [1958] was "No Geology without Marine Geology."

The primary aim of marine geology and geophysics is to develop an understanding of the structure of the earth beneath the oceans, the history and character of processes that have shaped the earth beneath the sea, and oceanic and global history itself. The roles that geophysicists and geologists play in developing this knowledge overlap considerably, although geophysicists traditionally have been more concerned with the earth's structure, while geologists have been concerned mainly with the earth's history. Nevertheless, the discoveries by geophysicists about oceanic structure have directly led to some of our more fundamental advances in knowledge of the history of the earth. Marine geologists differ from land-based investigators mainly because they use different tools. Because marine geologists are unable to walk over the outcrops directly and sample the ocean floor (except from submersibles), special methods have been developed for submarine sampling. Virtually all marine research requires a vessel of some sort. Because the methods employed by marine geologists are so different from those of land geologists and because the ocean tends to act as one major geochemical system, the lines along which marine geologists reason tend to be different from those of land geologists. Marine geologists also differ from marine geophysicists in that they are concerned primarily with the study of rocks and sediments. Geophysicists, on the other hand, work mainly with data related to the earth's gravity, heat flow, magnetism, earthquakes, and with artificially generated sound waves transmitted through sediment and rock sequences.

Almost everything we know about the geology of the oceans has been discovered during the last 30 years. During the first half of the twentieth century, considerable intuition existed concerning the potential of the oceans toward providing critical information about the character and history of the earth. The dominant feature of the surface of the earth is its oceans, covering about 72 percent of the surface. The distribution of this water and land is not even, since about 81 percent of the Southern Hemisphere is covered by ocean, compared with 61 percent of the Northern Hemisphere. A hemisphere with a pole in New Zealand contains 89 percent sea and only 11 percent land surface. Continental areas of the earth are almost always (95 percent) antipodal to oceanic areas. The total volume of water in the oceans is 1350 million cubic kilometers (km³) [Menard and Smith, 1966], while the average depth of the ocean is about 3700 meters (m). This is in contrast with the average elevation of the land surface, which is only 850 m. However, both continents and ocean form only a veneer upon the vast globe. An examination of a globe of the earth also shows that the ocean is broken up by large continental areas into a few major oceans and numerous smaller oceans or seas. The longest uninterrupted oceanic stretch lies south of about 50°S—the Southern or Antarctic Ocean. Further north the oceans are divided by three large landmasses—Eurasia-Africa, North and South America, and Australia. This configuration of the earth's surface is only the most recent, since the relative positions of continents and land have been in a state of dynamic change through all of geologic time. To a large extent this book is the story of this dynamic oceanic evolution: the story of tectonic and of geographic change and of the interactions of the global environmental system, especially marine sediments and life.

HISTORY OF MARINE GEOLOGY

Most of the earth's surface has been explored only during the last few decades as marine geology developed in conjunction with oceanography. This period represents one of the greatest ages of geographic and scientific exploration. The foundations for this era of oceanic exploration were begun about 150 years ago by a few investigators involved in some of the earliest scientific expeditions. During the voyage of the H.M.S. *Beagle* from 1831 to 1836, Charles Darwin's observations about the origin of species by evolution and its implications concerning antiquity of life established a scientific base for the further exploration of the earth's history. Darwin also provided some of the earliest speculations concerning movement of the ocean floor during his attempts to explain the origin of coral atolls.

Development of knowledge about the bathymetry of the oceans began in the middle 1800s. Before that time there was little concept of oceanic depths because of the short lengths of rope used in most soundings. The first accurate deep-sea soundings seem to have been made by Sir James Ross in 1840 during a voyage to the Antarctic with H.M.S. *Erebus* and *Terror*. During this expedition, Ross discovered bottom at 14,550 ft at one location, demonstrating the considerable depths of the ocean basins. In response to the needs of shipping, extensive bathymetric surveys began in the nearshore areas of the eastern United States. By 1843 about 30,000 km^2 of the inner part of the continental shelf had been sounded by the "Survey of the Coast," which was later renamed the U.S. Coast and Geodetic Survey. In 1830 a parallel bureau was organized within the U.S. Navy Department to compile charts of deeper waters, under the direction of Lt. Charles Wilkes. In 1842 Lt. Matthew Fontain Maury succeeded Lt. Wilkes and expanded operations. This depot became the U.S. Navy Hydrographic Office in 1866. The first deep-sea bathymetric chart was published by Maury. Because of his influence within the Navy, Maury had the naval ships equipped with 10,000-fathom reels of baling twine and 64-1b cannonballs for use as sinkers on the sounding lines. He used the data generated from these surveys (180 deep soundings) to construct the first deep-sea bathymetric chart of the oceans in the Atlantic from 52°N to 10°S. This map served as a basis for laying the first trans-Atlantic telegraph cables. Because of these efforts, Maury can be regarded as the first marine geologist.

As the need increased to lay trans-Atlantic and other submarine telegraph cables, there was interest in the character of the deep-sea floor and whether life could exist at such depths. An influential British biologist of the time, Edward Forbes, claimed that no life existed in the oceans below about 600 m. He had developed these ideas during an expedition on H.M.S. *Beacon* in the Mediterranean in 1841 even though bottom-living organisms had been dredged to depths as great as 1380 ft. He gave the name **azoic** zone to the deep parts of the ocean inferred to contain no life. The azoic zone was thought to have been formed by the depletion of oxygen because of reduced circulation and stagnation. The azoic theory was not refuted until the 1860s, when ocean cables, on being raised to the surface for repairs, were found to have living organisms attached to them. This discovery of life in the deepsea in turn stimulated interest in the search for primitive life (living fossils) that might have survived in what was then considered to be the relatively unchanging deep-sea environment. Another major concept related to the deep sea resulted from the study of deep-sea calcareous oozes by Thomas Huxley, a close friend of Charles Darwin. Huxley had noticed that the chalk cliffs on land were composed of tiny calcareous shells of planktonic organisms like those preserved in deep-sea sediments and suggested that the chalks could be uplifted deep-sea sediments.

By the 1860s, a number of questions had been formulated about the deep ocean and a climate had developed in which an extensive expedition could be mounted to answer these questions. One of the most influential advocates of such an expedition was Charles Wyville Thomson, the successor of the Edward Forbes's chair in natural history at Edinburgh University. Thomson had not accepted the theory of evolution and believed that the vertical range of various groups of organisms in the ocean corresponded closely with their vertical range in strata. In other words, he believed more primitive creatures become dominant with increase in oceanic depths. The Royal Society of London was persuaded to sponsor the most ambitious and innovative scientific project ever attempted—a global survey of the deep ocean. This was the *Challenger* expedition (1872–1876), under the direction of Thomson. The *Challenger* was a corvette of 2300 tons with auxiliary steam power, and her assignment was to determine "the conditions of the deep-sea throughout all the great oceanic basins." The expedition, still the longest, covered nearly 70,000 nautical miles in 4 years, carried out almost 500 deep soundings and 133 dredgings and obtained various data from 362 stations (one every 200 miles). As a result of the expedition, 715 new genera and about 4500 new species were described. The introduction of steel cable in 1870 greatly facilitated deep-sea sampling operations. A deep-sea sounding of 8180 m in the Mariana Trench was the deepest up to that time.

The *Challenger* expedition provided a solid foundation for marine geology. Samples of bottom sediment provided the basis for the recognition of the major types of marine sediments, their classifications, and their general distribution patterns throughout the oceans. The formation of deep-sea oozes was extensively studied by Sir John Murray. He developed accurate ideas about the role of planktonic organisms in determining the character of the deep-sea sediments at different depths and latitudes throughout the oceans. He was also able to differentiate various planktonic microfossil assemblages from different latitudes, thus providing a foundation for the biogeography of oceanic plankton. Murray, because of the breadth and brilliance of his geological work, is generally considered the father of modern submarine geology. He took major responsibility for the publication of the *Challenger* reports, volumes of results of the expedition, which remained the main source of knowledge of the ocean floor until the 1930s. Murray capped his career by jointly publishing *The Depth of the Ocean* with J. Hjort in 1912. For many years, it was one of the most widely read books on oceanography. Although the *Challenger* expedition added much knowledge about the deep ocean, it did little to alter the false concept that had reigned for years before the expedition—that the deep sea is a tranquil environment with uninterrupted sediment deposition. This myth was not destroyed for almost another 100 years.

For the 70 years following the *Challenger* expedition, few new data were gathered about the geology and geophysics of the ocean floor. The oceans were largely left to the biologists. In America, charts of sediment patterns had been drawn by Delesse in the 1860s and Pourtales in the 1870s from samples collected along the continental margin from New England to Florida, but little else was done for many years. In 1912 the theory of continental drift was promoted by the German meteorologist, Alfred Wegener. But, after early heated debate, the theory fell into disrepute due to a lack of essential information about the geology and geophysics of the oceans. One of the major advances during the time period before World War II was the development of the electronic echo sounder to measure ocean depths, replacing the time-consuming and often incorrect wire soundings. The development of the echo sounder resulted from efforts at submarine detection and was first employed during the German *Meteor* expedition in the South Atlantic in the 1930s. Thus topographic mapping of the oceans made a rapid leap forward. The *Meteor* expedition demonstrated, through a number of survey lines across the South Atlantic, that a ridge occurred throughout the length of the mid-Atlantic region. The ridge had been discovered in the North Atlantic during cable-laying operations, when it was called the Telegraph Plateau, but its length was unknown. It has been renamed the mid-Atlantic ridge and has been found to be one part of an ocean-wide ridge system.

A second major development between the world wars was the development of techniques to measure gravity at sea. This was pioneered by the Dutch physicist F. A. Vening Meinesz and employed from submarines, representing relatively stable platforms. Vening Meinesz's most important discovery was large negative gravity anomalies associated with the deep oceanic trenches, such as the one off Indonesia. This suggested that tectonic activity prevents the earth's crust from reaching isostatic equilibrium in these regions. His observations played an important role in the development of theories of global tectonism. In 1932 Vening Meinesz carried out investigations using a U.S. submarine in the trench off Puerto Rico in association with the U.S. Navy–Princeton Gravity expedition. One of the members of the scientific party was a graduate student, Harry H. Hess, who later played a principal role in the development of the theory of sea-floor spreading (see Chapter 4).

Surprisingly little additional work was conducted on deep-sea sediments during this time. The German South Polar expedition of 1901–1903 had collected short sediment cores, which were studied by Philippi in 1912. In 1929–1930 during the *Snellius* expedition, Dutch investigators employed an explosive coring device called a Piggot gun, which obtained sediment cores up to 2 m long. Cores such as these allowed geological history to be determined from changes in the sediments and fossils. Similar cores taken from the North Atlantic in the 1930s showed that glaciations can be distinguished from interglacial episodes based on changes in planktonic microfossil assemblages and sediment type [Schott, 1935; Stetson, 1939; Phleger, 1939; Bramlette and Bradley, 1940]. Unfortunately, too few sediment studies had been carried out to disclose the environment of the deep oceans. Until World War II it was still believed that the deep-sea floor was monotonous, uninteresting, and a place of virtually no water movement. In 1942 it was stated in the volume *The Oceans,* by H. Sverdrup, N. Johnson, and R. Fleming [1942], that, "From the oceanographic point of view the chief interest in the topography of the sea floor is that it forms the lower and lateral boundaries of the water."

World War II had a profound effect on the development of oceanography and its branch, marine geology. Oceanographic research is expensive and the major advances must be sponsored by governments or wealthy individuals. Before the war, funds from governments were meager. Some solid financial support was provided by nonprofit institutions established by wealthy individuals. These set the stage for later developments that accelerated because of the war effort. World War II radically reordered the priorities of the United States and other nations; the importance of science grew and research appropriations soared. In relation to antisubmarine warfare, research was expanded on sound transmission through water, helping to lay a foundation for seismic studies of marine sediments. Numerous studies of the sea floor were initiated because they influenced the behavior of sound, and work on sediment charts began in early 1943. But for scientists, the most significant outcome of the war was an increased confidence in their potential for contributing to the nation's welfare [Schlee, 1973]. After the war, oceanography emerged as a modern science with a well-established base for growth in all disciplines. The concentrated exploration of the oceans began. This also coincided with the completion of geographic exploration of the landmasses other than Antarctica. A redistribution of interests occurred within the field; the preeminent position of marine biology was replaced by a more balanced approach emphasizing the physical sciences. Oceanography became tied to government support and government policy. In 1946, the Office of Naval Research (ONR) was established to sponsor long-term research including oceanography. The ONR played the leading role in the rapid development of the oceanographic institutions from the middle 1940s through the late 1960s. In 1950 the U.S. National Science Foundation was established and has taken on increasing responsibility for the support of marine geological research in the United States. Other governments, including the USSR, Britain, France, Canada, New Zealand and, slightly later, West Germany and Japan, also began to support marine geological research on a larger scale. The Swedish deep-sea expedition (1947–48) provided valuable additions to marine geology.

In the postwar environment marine geology developed rapidly. Marine geology and geophysics of the oceans could be studied systematically because of greatly increased funding levels. Most of the world was being explored for the first time. Expansion occurred in many of the large U.S. marine laboratories. Scripps Institution of Oceanography in La Jolla, California, expanded to become the largest oceanographic institution in the United States. It was established for marine research at the turn of the century and incorporated in the University of California in 1912. Roger Revelle, its director from 1948 to 1964, enhanced its emphasis on ocean-floor research. Woods Hole Oceanographic Institution in Woods Hole, Massachusetts, another large institution to grow rapidly during World War II, was chartered in 1930 as a private, nonprofit research institution. Ocean mapping was developed on a large scale by Soviet scientists, who have produced detailed atlases of ocean-floor features and sediments.

By the 1950s the echo-sounding technique had been highly refined, and it was possible to measure ocean depths throughout the world accurately and cheaply to a depth of 10,000 m. The resolving power of the modern **precision depth recorder**, is better than one part in 5000, so that a change as small as 1 m can be detected at depths of 5000 m. The oceanographic fleet increased at this time, making the expansion of bathymetric mapping of the oceans possible. After initial work by Maurice Ewing (Fig. 1-1) and Bruce Heezen at the Lamont Geological Observatory of Columbia University (now the Lamont-Doherty Geological Observatory) the mapping project was led by Bruce Heezen and Marie Tharp and led to the publication of the well-known maps of the Pacific, Atlantic, Indian, and Arctic oceans in *National Geographic* of the National Geographic Society, Washington D.C. These maps are so detailed that they have provided many scientists with stimulation in their individual areas of interest. Heezen concentrated his efforts on exploring the

Figure 1–1 Maurice Ewing and R/V *Vema*. (Courtesy Lamont-Doherty Geological Observatory)

oceanic-ridge systems. The mid-ocean ridge system, more than 3000 m in height, practically encircles the earth. It surpasses the Alps and the Himalayas in scale, yet was discovered in its entirety only 25 years ago. It was Ewing and Heezen who first realized that the earth is encircled by this ridge system [Ewing and Heezen, 1956]. They also discovered that at the crest of the ridges is a narrow trough or rift valley [Ewing and Heezen, 1956]. This central rift valley was interpreted by Carey [1958] to be a narrow block, sinking under tension as the sea floor on either side of the valley moves apart. H. W. Menard of Scripps Institution of Oceanography was concurrently exploring the crests of the East Pacific rise but found no rift valleys.

The Lamont-Doherty Geological Observatory, which was built by Maurice Ewing, played an important postwar role in the growth of knowledge of the geology and geophysics of the oceans. These activities "changed the subject from a polite academic backwater into one of the most exciting fields of inquiry being pursued today; that the floors of the oceans are no longer terra incognita is largely because of Maurice Ewing's curiosity, which was relentless" [Wertenbaker, 1974]. Very early in his career, Ewing recognized the fundamental differences between oceanic and continental crust, a difference which Ewing called a brutal fact. To develop an understanding about the character of the oceanic lithosphere, including its sediments, Ewing and his colleagues developed techniques using seismic waves artificially created by underwater explosions near the surface of the oceans. This area of investigation is called **marine-explosion seismology** and provided data on sediment thickness in the ocean and on the suboceanic crustal structure. Later, the **air-gun method** was developed; it has become the most widely used method for studying sediment character. Sound waves are generated by shooting an air gun into surface waters. Continuous reflections of sound waves off buried layers of sediment show the detailed configurations of oceanic sediments. After the war, large quantities of explosives were available for such studies at sea. Ewing began his seismic investigations at Woods Hole in the years immediately preceding World War II and continued them at Lamont after the war. In 1953 he purchased the *Vema*, the first research vessel for the observatory, which enabled the scientists at Lamont to develop seismic techniques more rapidly. **Seismic-reflection** techniques eventually became a routine procedure on most research vessels and provided a basis for the understanding of sediment distributions in the ocean basins and deep oceanic processes. Ewing made sure that the Lamont ships were always engaged in data collection of a wide variety including magnetic and gravity data, bottom photographs, and piston cores. Huge libraries of data and sediments accumulated for current and future investigations. David Ericson and Goesta Wollin carried out crucial pioneering micropaleontological studies of the ocean basins. At the same time, Emiliani pioneered the use of ocean isotopes for paleotemperature analysis of deep-sea sediments. Other major structures of the ocean floor were discovered and mapped. Among the most important were the extensive, linear fracture zones discovered by Dietz [1952] and Menard [1953] in the Pacific. These have been found to be widespread in all of the ocean basins and have played a critical role in tectonic theory. The mapping of the ocean basins during the 1950s produced rather simple patterns of sea-floor topography, providing a basis for the development of theories of sea-floor spreading and plate tectonics during the 1960s. The history of this phase of the exploration of the oceans is described in Chapters 2 and 4.

Many recent developments have occurred, including the development of methods for navigation using fixes from satellites, which provide accuracy for an oceanographic station to within 100 m. Deep-sea drilling has been continuing since 1968 and is discussed in Chapter 3. More recently, deep-diving submersibles have been used to observe the features of the ocean floor, especially on the oceanic ridges. Pioneering work on deep-diving developments has been carried out by Jacques-Yves Cousteau, who also initiated the use of scuba diving. Important centers of marine geological research outside the United States include the National Oceanographic Institute, Wormley, United Kingdom; Keil University, West Germany; Centre National pour l'Exploration des Océans (CNEXO), Brest, France; Institute of Oceanology in Moscow, USSR; Bedford Institute and Dalhousie University, Nova Scotia, Canada; New Zealand Oceanographic Institute, Wellington, New Zealand; University of Cape Town, South Africa; and the University of Tokyo, Japan. A large number of institutions within the United States carry out marine geological and geophysical research.

The Structural
and Oceanographic Setting

two
Geophysics
and Ocean Morphology

It is the business of the future to be dangerous. . . . The major advances in civilization are processes that all but wreck the societies in which they occur.

Alfred North Whitehead

GEOPHYSICS

Shape and Rotation of the Earth

Studies of the earth's shape (geodesy) have shown that it is a sphere that is flattened at the poles and bulges at the equator, that is an oblate spheroid. The earth has a circumference of about 40,000 km. The radius of the earth at the equator is 6378 km; it is 6356 km through the poles, a difference of 22 km. The earth is not a perfectly shaped oblate spheroid, but exhibits irregularities as determined from the motion of orbiting satellites. The flattening and bulging of the earth's shape results from the centrifugal forces of its rotation. Forces resulting from rotation have other effects, including wind patterns in the atmosphere, current directions in the oceans, and the flow of hot, viscous material inside the earth.

Internal Structure

Knowledge of the earth's density provides a basis of our understanding of the internal structure and composition of the earth. The planet's size, shape, mass, and moments of inertia give a mean density of 5.5 grams per cubic centimeter (g/cm^3). Since most rocks on the earth's surface have a density between 2.7 and 3.3 g/cm^3, there must be a substantial increase in internal density. This increase in density is not due primarily to an increase in pressure, but rather to compositional changes.

13

Almost all information about the interior of the earth at depths greater than a few kilometers has been obtained from seismological studies. Most of these studies have involved the analysis of seismic waves generated by large-scale seismic events—namely earthquakes and nuclear-test explosions. Since the early 1960s, seismology has advanced rapidly because of projects studying the upper mantle and new technology developed to detect underground nuclear explosions. A network of more than 100 sensitive seismographic stations was established in the 1960s. These new observatories have enabled seismologists to confirm a number of older hypotheses about the earth's interior, as well as providing much new, detailed information.

The seismic waves generated by earthquakes or explosions are bent, speeded up, slowed down, or even reflected as they pass through the earth (Fig. 2–1). Two principal types of seismic waves are transmitted through the earth: **Primary, compressional** or **P-waves,** have an initial velocity near 8 km/sec at the source and are compressional in character. The wave progresses by alternately compressing (push-pulling) and dilating the medium, and their direction of vibration is in the direction of propagation. These waves can pass through regions that are liquid. **Secondary, shear** or **S-waves,** have velocities approximately half that of P-waves and are transverse body waves. The S-waves cause particles to vibrate perpendicular to the direction of propagation. These lower velocity waves cannot pass through liquids because liquids flow too easily and cannot sustain shear forces.

Seismic waves are detected by seismographs and are usually taken to originate from a single point within the earth called the **focus.** The point of the earth's sur-

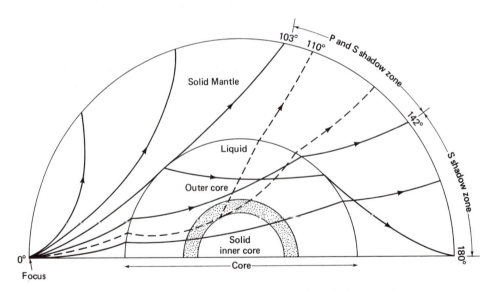

Figure 2–1 Selected ray paths for P-waves passing through the earth. Such rays provide the principal evidence for the internal structure as shown. The P and S shadow zones are marked on the right. The dashed ray paths represent weak P waves within the shadow zone which provide the principal evidence for the narrow inner high P velocity zone in the core. (After B. Gutenberg, 1959)

peridotite in composition. This conclusion is derived from a number of indirect observations and thus remains controversial. The compositional range is suggested by the following:

1. Inferred density characteristics from seismic velocities.

2. Observations of possible mantle material within **xenoliths,** which are blocks of foreign rock caught up in some volcanic eruptions. The xenoliths are ultramafic and of plutonic appearance. Unfortunately, it cannot be firmly proven that any of the surface rocks of the earth have been derived from the mantle without change.

3. Observations of dense materials at the earth's surface known to be unstable under surface conditions.

4. Experiments that subject typical earth materials to conditions which may exist in the interior to the earth. By reproducing mantle pressures and temperatures, it is possible to test whether the likely materials of the earth's mantle have the velocity, density, or other characteristics known or inferred for the earth's mantle.

5. Solar-system abundances as determined from analysis of meteorites.

Many of the ideas about the chemical and physical state of the earth's interior are associated with the concept of polymorphic transition in rocks and minerals. The concept is that an assemblage of minerals will become unstable and will recrystallize to form new minerals as pressure and temperature change. These changes are called **phase transitions.** For instance, minerals that are not in their most closely packed crystallographic orientation may be transformed into higher density–lower volume phases in a higher pressure-temperature environment. Many of the silicates found in crystallographic forms permit substitution of one metallic ion for another, such as magnesium for iron. Thus a complete substitution series may exist between minerals, such as forsterite (Mg_2SiO_4) and fayalite (Fe_2SiO_4) to form $(Mg,Fe)_2SiO_4$ [Knopoff, 1969]. Also at pressures equivalent to a depth of about 400 km, enstatite $(MgSiO_3)$ (density 3.1 gm/cm³) undergoes a rearrangement of its crystal structure to become olivine and stishovite (density 3.4–3.5 gm/cm³). Eclogites (pyroxene-garnet rocks) also undergo a small increase in density at this depth. At a depth of about 300 km, forsterite transforms into a spinel-form crystal structure (density 3.5–3.6 gm/cm³) [Knopoff, 1969].

It is possible that some seismic velocity transitions in the mantle are related to phase transitions between polymorphs of the same mineral or are transitions between two different chemical components. Many silicates of aluminum, calcium, and the alkali metals, the **sial** of classical geologic lore, undergo phase transitions at pressures of about 10 kbars and at temperatures of several hundred degrees [Knopoff, 1969]. These materials include the feldspars, of which the high-pressure phases are the garnets. Gabbro, a composite of such minerals, may undergo a transition to eclogite, its high-pressure form, at about 10 kbars or at depths of about 30 km. It has been suggested that the M-discontinuity represents such a phase change.

face vertically above the focus is the **epicenter.** Most earthquake foci are located in the upper 100 km of the earth, but earthquakes are known to occur to depths of 700 km. The location of a focus is found by a geometric calculation from several seismological stations. As the P- and S-waves progress deeper into the earth's interior, velocities for both waves increase (Fig. 2–2). The velocity of seismic waves depends on the density and flow properties of the rocks through which they pass—velocity is high in rigid, dense rocks and low in less rigid, less dense rocks. Waves that travel the greatest distances travel faster and become more refracted as they pass through the denser parts of the earth's mantle (Fig. 2–1). Refraction causes them to follow paths that are convex (Fig. 2–1). Waves continue to be refracted at higher angles at the successive boundaries (levels where changes occur in the physical properties of the medium) in the interior of the earth (Fig. 2–1). Such studies provide information on the depth of boundaries and the character of the materials that make up the interior of the earth.

Concentric layers

Seismic studies have shown that the earth's interior (Fig. 2–3) is comprised of three main concentric layers.

1. An outer **crust** ranging in thickness from 5 to 10 km beneath the oceans to more than 40 km beneath the continents. Crustal rocks are rich in feldspars, silicate minerals made up of elements with large ionic radii (Na, K, Ca, Al), and iron-magnesium silicate minerals such as pyroxene and olivine. The oceanic crust is discussed in Chapter 7.

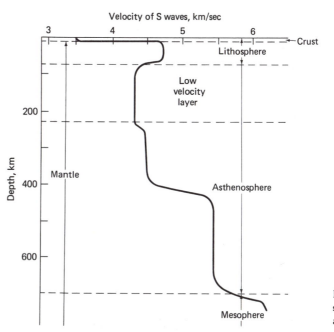

Figure 2-2 Approximate velocities of shear waves in the lithosphere and asthenosphere. (After J. Dewey, 1972)

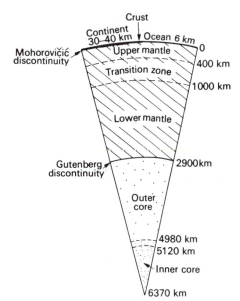

Figure 2-3 Layers of the earth and major discontinuities. (From M. H. P. Bott, 1971)

2. The **mantle**, a layer of thickness 2900 km whose upper limit is marked by the **Mohorovičić discontinuity** (or Moho or M. discontinuity), a seismic discontinuity between overlying cool crust, with compressional wave velocities less than about 7.4 km/sec, and underlying dense hot rock with a median velocity of 8.1 km/sec. The mantle contains two-thirds of the earth's mass and 83 percent of its volume. Indirect evidence strongly suggests that the mantle is composed dominantly of olivine, pyroxenes, and an aluminum-bearing mineral such as feldspar, spinel, or garnet, depending on the confining pressure (depth). The composition of the bulk mantle mineral assemblage is close to **peridotite.**

3. A liquid **outer core** at the earth's center, probably of iron-nickel composition, with a thickness of about 2000 km surrounding a solid **inner core** about 1200 km thick. Although P-waves can travel through the liquid outer core.

The crust, mantle, and core are distinguished from each other by differing seismic velocities. Another set of terms defining the concentric layering of the earth is based on strength and viscosity. These are the **lithosphere, asthenosphere,** and **mesosphere.** The lithosphere is the outermost shell of the earth (about 100 km thick) and includes the crust and uppermost mantle (Fig. 2-2). It is distinguished by its ability to support large surface loads, like volcanoes, without yielding. It is cool and therefore rigid [Wilson, 1963]. The lithosphere is underlain (to an approximate depth of 700 km) by the asthenosphere (*asthenos* is Greek for soft). The asthenosphere is near its melting point and because it has little strength, it flows when stress is applied over time. The upper asthenosphere is the zone of compensation for geodesists, the low velocity layer for seismologists, and the zone of magma

generation for most petrologists [Anderson, 1967; Wilson, 1963]. The next layer is the mesosphere. The mesosphere is more rigid than the asthenosphere, but more viscous than the lithosphere. The mesosphere extends to the core and thus incorporates most of the mantle.

The mantle

The mantle is mostly solid. Seismic waves are transmitted at increasing velocities with depth as density increases from 3.3 to 5.5 g/cm³. This increase in density occurs progressively as discrete steps. The mantle is complex and shows variations in structure both horizontally and vertically. A transition zone between about 400 and 1000 km (Fig. 2-3) divides the mantle into upper and lower parts, and almost all interest is in processes associated with the upper mantle. The most important vertical variation within the upper mantle is the decrease in S-wave velocities from 4.7 to 4.3 km/sec between 75 and 150 km (Fig. 2-2). This low velocity layer forms a natural discontinuity between the lithosphere and the asthenosphere (Fig. 2-2). The low-velocity layer probably represents a zone of partial melting in the upper mantle where a small amount of melt (1 percent) forms an intergranular film between silicate grains, similar to water-soaked sand, but at temperatures exceeding 1200°C. This layer is probably the source of melts, or **magma,** that rise to the surface and form igneous rocks. Magma is either intruded into the crust as **intrusive** or **plutonic rocks** or extruded onto the surface as **extrusive** or **volcanic rocks.** The mantle also gives off volatiles that rise to the earth's surface to contribute to two other major layers of the earth, the **hydrosphere** and the **atmosphere.** These layers are the outer parts of the decreasing density gradient of the earth.

Below the low-velocity zone are distinct steplike increases in seismic velocities at depths near 400 and 700 km. Gutenberg computed the average annual seismic energy release as a function of depth and found maxima centered at 350 and 600 km. Thus the depth distribution of deep earthquakes may be controlled by stress release in these seismic transition zones. One possibility is that the most stress is in those zones where the elastic properties are changing most rapidly. Another possibility is that the seismicity and velocity increase due to changes in mineralogical composition related to changing pressure-temperature relations.

Temperature distribution within the mantle is not well known. However, we know the lithosphere is cold and the underlying asthenosphere and upper mantle are hot (1000–1500°C). Seismic evidence shows that the mantle is solid, yet volcanism demonstrates that magma (1200°C) is formed locally within the mantle. The earth's heat is generated by input from (1) gravitational energy converted to heat during formation and accretion of the earth, (2) frictional energy from the earth's tides, and (3) radioactive decay. In turn the earth loses heat by (1) conduction, (2) radiation, and (3) convection. The process of convection may occur on a large scale in the earth's mantle, but this theory is still controversial. Convection is the most efficient means of heat transfer, as well as a very attractive driving force for global tectonics.

As discussed earlier, the mantle is probably dominated by material clo

However, Ringwood and Green [1966] question this hypothesis because of the difficulty of explaining a phase change at the very different depths and pressures exhibited by the M-discontinuity beneath the continents on one hand and the oceans on the other.

The core

The existence of a central core within the earth was theorized by Oldham [1906] because P-waves recorded near the angular distance of 180° from the earthquake epicenter arrived much later than expected. In 1913 at the University of Gottingen, Beno Gutenberg calculated that at a depth of about 2900 km there is a 40 percent drop in the velocity of P-waves. This is the Gutenberg discontinuity, which marks the boundary between the mantle and the core (Fig. 2–3). The core is made of material that cannot maintain a shear stress and therefore is regarded as liquid. A small increase in P-wave seismic velocities indicates the presence of a solid inner core from about 5120 to 6370 km (Fig. 2–3). It is assumed that the core consists of an iron-nickel-sulfur mixture. The boundary at 2900 km between the mantle and core marks a sharp change in density from about 5.5×10^3 kg/m^3 in the mantle to about 10^4 kg/m^3 in the core. So the core is more than twice as dense as the mantle, and although it is only 16 percent of the earth's volume, it is about 32 percent of the mass [Harris, 1972].

The Earth's Gravity Field

The universal law of gravitation defines the force of attraction that draws all objects toward the center of the earth. This force is directly proportional to the masses of the particles and inversely proportional to the square of the distance between them. The earth has an external gravity field that arises from its distribution of mass. Deviations, or *gravity anomalies,* provide important information on the distribution of density in the earth's outer layers. The measurement of gravity is affected by distance from the center of the earth and the density of the rocks in and beneath the area where the measurement is being conducted. Materials of greater density exert a larger gravitational force. Weight differs numerically from mass by a constant, called the *acceleration of gravity,* or *g*. At sea level (45° latitude) the accepted standard value of *g* is 9.8 m/sec^2. The unit of measure for differences in gravity is a *gal* (named for Galileo). One gal represents an acceleration of 1 cm/sec for each second of fall. At the earth's surface, the acceleration due to gravity is about 981 gal, but gravity anomalies need to be measured in a much finer unit, the *milligal* (1 mgal $= 0.001$ gal, or $10^{-6}g$). Modern instruments allow measurements as precise as 0.01 mgal, or $10^{-8}g$. This corresponds to a change equivalent to raising an object 4 cm from the earth's surface.

In the 1770s, P. Bouguer discovered that gravitational attraction is not equal around the earth, but differs with latitude. There are several factors that modify the average gravity values near the earth's surface. These include latitude, altitude, and density of rocks in the immediate area. The rotation of the earth creates a centrifugal force that accelerates matter partly in opposition to the gravitational force.

This force is greatest at the equator and decreases toward the poles, resulting in a systematic decrease in gravity from the poles to the equator. A gravity anomaly is the difference between the observed gravity and the expected gravity at that location based on a model for the earth's gravity field. In order to compare gravity anomalies measured at different locations, it is necessary to correct them to some common datum—sea level. This correction is called the free-air, or altitude, correction and is about 0.3087 mgal/m. After the altitude correction is made, any remaining difference or anomaly is called the **free-air anomaly.** A further correction, the Bouguer correction, takes into account the nature (density) of the material between the surface of equal gravity or **geoid** and the measuring station, using the average density of the inferred material. Any anomaly remaining after all these corrections have been made is called the **Bouguer anomaly** and represents the effect of rocks with densities different from the average or structural controlled rock masses not in isostatic equilibrium. As a result, elevated rocks on the continents are marked by large negative Bouguer anomalies, while over ocean basins the Bouguer anomalies are large and positive.

In the middle 1800s, J. H. Pratt and G. B. Airy discovered that the mass of mountains creates gravity anomalies that require correction and that the correction was not so large as that predicted from the mass of the visible mountain. This suggested the concept of **isostasy** to Pratt. Isostasy is the idea that the crust is in gravitational equilibrium through a buoyancy mechanism. In order to attain equal pressure everywhere, Airy suggested, low-density roots beneath the mountains compensate for the material in the mountains themselves. These early ideas were the beginning of the concept of the geoid. Gravity measurements over the earth show that, despite large mass differences at the surface, the earth is nearly in isostatic equilibrium, with various crustal blocks acting as if they floated on a dense viscous substratum. Deviations from the geoid show undulations of only about 80 m. If a body is floating in isostatic equilibrium, it is considered to be **compensated.** If the root beneath a mountain is too small, the mountain is **undercompensated** and thus will sink. If a root is too large for the body above it, the crust is overcompensated and will rise. The shape of the M-discontinuity beneath the continents is like that of the surface of the continent, but exaggerated. The bottom of a continental block changes its level in response to change in volume of the continent above. This maintains a constant exposed-to-submerged ratio, just like that of a melting iceberg. The depth to the M-discontinuity is about 11 km below the surface of the ocean and 30 to 50 km below the surface of the continents. These M-discontinuity depths are in agreement with the principle of isostasy in that the light crustal rocks are in buoyant equilibrium to the dense mantle [Knopoff, 1969].

Under ideal conditions of isostatic equilibrium, all crustal columns exert equal pressure at some depth (possibly 100 km) and the column extending above sea level has a mass equal to that of the compensating mass at depth. However, large **isostatic anomalies** do exist, indicating that portions of the earth's crust are not in isostatic equilibrium. In Scandinavia, for example, the accumulation of ice caps during the ice ages depressed the land several hundred meters. During the 10,000 years following the last ice age and the removal of the ice cap, the Scandinavian shield has rebounded 300 m, as shown by raised beaches and tide-gauge records.

Isostatic adjustment is still occurring. The extremely rapid isostatic uplift in Scandinavia indicates relatively rapid return flow of mantle under the crust and, in turn, a low viscosity.

Interpretations of gravity variations are insufficient by themselves to provide a unique solution of subsurface structure. As a result, the gravity data are best used in conjunction with seismic refraction and other geophysical data to provide an explanation for the anomalies. For instance, gravity anomalies over Atlantic-type continental margins have been used with other geophysical data to help delineate the boundary separating continental and oceanic crust. There is a positive free-air anomaly over the continental shelf break and slope and a negative anomaly over the upper part of the continental rise. This *edge-effect* anomaly is caused by the oceanward thinning of the continental crust [Watts, 1975].

The earliest measurements of marine gravity were made in the 1920s by Vening Meinesz, who conducted gravity surveys in the Indonesian archipelago. Vening Meinesz developed a gravity meter based on a pendulum mechanism. The period of a pendulum is directly related to the acceleration of gravity providing a means for its measurement. But, gravity measurement requires precise leveling of the gravity meter. This poses a major problem at sea, which was circumvented by Vening Meinesz by conducting surveys from a submarine. Measuring gravity at sea has been greatly improved by the development of spring-type surface-ship gravimeters, which can be operated under a wide variety of sea conditions [Watts, 1975]. Such spring balances are based on the pull of gravity against a delicately balanced spring. The balances are then mounted in devices that keep them motionless despite the ship's motion. Since about 1960, surface-ship measurements of gravity have been obtained on a continuous basis over the world's oceans.

The Earth's Magnetic Field

The earth's magnetic field bears some resemblance to the field of a dipole magnet thrust through the earth's center and inclined at a slight angle to the earth's axis of rotation (Fig. 2–4). A dipole field has two magnetic poles, where a magnetized needle stands vertically, and a magnetic equator, where the needle lies horizontally (Figs. 2–4 and 2–5). The north magnetic pole is in the vicinity of Northwest Greenland, 1200 miles from the geographic North Pole (79°N, 70°W); the south magnetic pole is nearly opposite in Antarctica near the Ross Sea, about 1000 miles from the geographic South Pole (79°S, 110°E). The earth's magnetic field at a particular location varies in strength and direction and is especially unsteady during magnetic storms. The positions of the magnetic poles have changed little during the last 100 years, although the strength of the magnetic field has been decreasing. However, the **dipolar axis** is believed to "wobble" about every 10,000 years so that the average location of the magnetic pole coincides with the geographic poles.

At any point on the surface, the magnetic field is defined by its strength and direction. Three directions are required to describe the magnetic force at any single location and a large number of directions are available for use. The force may be

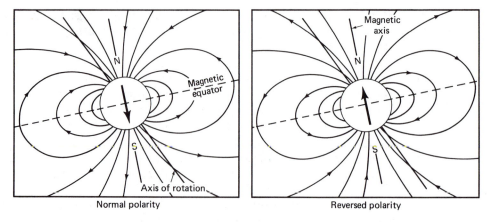

Normal polarity

Reversed polarity

Figure 2-4 Lines of force of the earth's dipole magnetic field. Normal polarity is shown at left and reversed polarity to the right. The magnetic axis is inclined with respect to the axis of the earth's rotation.

defined by determining the downward, northward, and eastward directions (Fig. 2-5). In Fig. 2-5 the vector **T** represents the direction and magnitude of the magnetic field, called the **total force.** Northward (**N**), eastward (**E**), vertical (**Z**), and horizontal (**H**) forces are indicated. The angle of dip of the force from the horizontal is called the **inclination.** The **declination** (δ) represents the angle between the magnetic north pole and the geographic North Pole. Figure 2-4 shows the **lines of force** of a dipole field outside a sphere. Tangents to the curvature of these lines reflect the direction of the field at points along it, and the components vary with the location. The study of paleomagnetism is complex because the components have varied through time. Also, the actual magnetic field measured at any one location will normally vary by a few percent from that predicted by the dipole field. These anomalies are caused by an irregular, or nondipole, field.

During the past 350 years a detailed picture has been developed of the magnetic field and its changes with time. Remarkably, this has been accomplished without an understanding of the mechanisms that have created the field. Mathematical geophysicists have been working to understand the mechanisms that

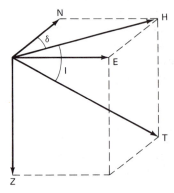

Figure 2-5 Components of the earth's magnetic field. N is the north component; E the east component; Z the vertical component; H the horizontal force; I the angle of dip or inclination; and δ the declination. The vector T is the direction and magnitude of the magnetic field and is called the total force. (After Harris, 1972)

generate the field; progress has been made, but the origin of the earth's magnetic field still remains a major scientific problem. One of the earliest theories was that the earth acts as a large permanent magnet; however the interior of the earth is too hot to retain magnetism. It is now generally accepted that the main features of the earth's field—the dipole field, the nondipole field, and their changes through geologic time—are all produced inside the earth. Of the various theories proposed, only the theory that the earth is an electromagnet has survived. This theory was developed during the 1940s and 1950s by W. M. Elsasser in the United States and Sir Edward Bullard in the United Kingdom and is known as the **dynamo theory.**

Paleomagnetism, the study of fossil, or remanent, magnetism acquired by rocks at the time of their formation and locked into the magnetic grains, is used to determine long-period variations in the earth's magnetic field. There are several forms of remanent magnetism, including **thermal remanent magnetism** (TRM) and **detrital remanent magnetism** (DRM). In TRM molten magnetic minerals, including magnetite, cool. Magnetite has a **Curie point** of 575°C. This means that at this temperature the material begins to become magnetic and the electrons are forced into alignment with the existing magnetic field of the earth. In DRM loose magnetic particles become aligned like iron filings in a magnetic field. The majority of paleomagnetic studies have been concerned only with the directions of magnetization.

One of the most important changes is that the earth's dipole field has vanished often in the geologic past, during which time the earth's magnetism has undergone a complete reversal. These are called *geomagnetic* or *paleomagnetic reversals* and constitute changes in **magnetic polarity.** These reversals are recorded in newly forming oceanic crust and sedimentary deposits. As we shall see later, the earth's magnetic polarity record has been fundamental in the formation of the hypotheses of sea-floor spreading and plate tectonics and in the stratigraphic correlation of marine sediments. The role of paleomagnetic reversals in stratigraphy is discussed in detail in Chapter 3 and its role in sea-floor spreading in Chapter 4.

MORPHOLOGY OF THE OCEANS

Hypsometry

Elevation changes between land and ocean depths can be shown as hypsographic curves (Fig. 2–6). These curves exhibit the cumulative percentage of the world ocean basin at different depths. The ocean floor can be divided into two major regions: the **continental margins** and the **ocean basins.** The continental margins include the continental slope and continental shelf and form only a small percentage (21 percent) of oceanic area. The earth's surface is dominated by two levels, one close to sea level (mean elevation 0.8 km, the continents); the other centered at 3.8 km, (the mean depth of the oceans). Therefore the average level of the continents is 4.6 km above the average level of the ocean floor. A steep transition zone, the **continental slope,** occurs between these two elevations. The deepest part of the oceans is the Mariana Trench, which is 11.04 km deep.

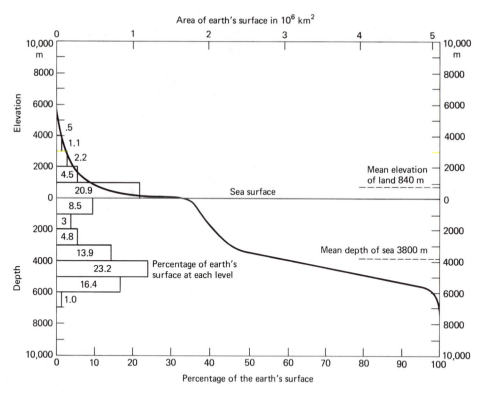

Figure 2-6 The distribution of levels on the earth's surface. The hypsographic curve, based on the histogram showing frequency distribution, indicates the percentages of the earth's surface that lie above, below, or between any levels. (Data from Sverdrup, Johnson, and Fleming, 1966; after W. A. Anikouchine and R. W. Sternberg, 1973)

The difference in elevation between the continents and the ocean reflects the fundamental difference between continental and oceanic crust in relation to isostasy and composition.

The boundary between continental and oceanic crust does not occur at the coastlines. Sea level shows virtually no relation with the major structural differences that occur between continental and oceanic crust because, for a number of reasons, it has constantly changed throughout geologic time. Instead, the transition between the oceanic and continental areas occurs at the continental slope. As a result, a high proportion (25 percent) of continental crust presently lies beneath sea level. Table 2-1 shows the areas of continents and oceans and mean ocean depths.

There are three major oceans: the Pacific, Atlantic, and Indian. The largest ocean is the Pacific (181×10^6 km²), which occupies more than one-third of the earth's surface (Table 2-1). The Pacific Ocean is also the deepest, with a mean depth about 200 m deeper than the oceanic average of about 3700 m (Table 2-1). Each of the oceans has its distinctive features. The Pacific Ocean is surrounded mainly by linear mountain chains, trenches, and island arc systems that, in most

TABLE 2-1   AREA, VOLUME, AND MEAN DEPTH OF THE OCEANS[a]

Ocean and adjacent seas	Area 10^6 km²	Volume 10^6 km³	Mean depth (m)
Pacific	181	714	3940
Atlantic	94	337	3575
Indian	74	284	3840
Arctic	12	14	1117
Total of all oceans	361	1349	3729

[a] Data from Menard and Smith [1966]. Area of world surface is 510×10^6 km².

areas, effectively isolate the deep-sea basins from the influences of continental, or **terrigenous,** sedimentation. The continental margins of the Pacific Ocean are narrow, forming a small proportion of the total area of the Pacific compared with the other oceans. Two other distinct features of the Pacific include large numbers of volcanic islands, especially in the central and western parts, and the presence of extensive marginal basins. **Marginal basins** of variable size lie close to the margins of the western Pacific and are separated from the deep-ocean basins by trenches or island arcs. They are particularly well developed adjacent to Indonesia, the Philippines, and eastern Australia. The depth of these marginal basins is usually greater than 2 km and, like trenches, they serve as sediment traps, resulting in thick accumulations of sediment.

The Atlantic Ocean is the second largest (94×10^6 km²), extending as a relatively narrow (about 5000 km wide) S-shaped basin lying between the Arctic and subantarctic regions. The Atlantic Ocean is marked by the largest north-south extension of any ocean, and because of its link between the two polar oceans, it serves as an important avenue for cold bottom waters produced in the polar oceans and ultimately transported to the world ocean. The Atlantic Ocean is slightly shallower (about 200 m) than the world ocean (Table 2-1). This results from large areas of continental shelves and margins. The Atlantic contains relatively few volcanic islands, and marginal seas are restricted to its most southern part (the Scotia Sea), the Caribbean, and the northern polar seas, such as the Norwegian Sea. Part of its northern boundary is Greenland, which is an eastward extension of the North American continent. Another distinctive feature of the Atlantic is that it receives the greatest amount of freshwater discharge from rivers. The Amazon and Congo rivers alone account for a total of one-quarter of the earth's river supply. Other significant rivers flowing into the Atlantic and adjacent areas include the Mississippi, St. Lawrence, Niger, and Parana rivers, and the Nile, Rhone, and Rhine. One of the major consequences of this is that terrigenous sediment input is much higher in the Atlantic.

The Indian Ocean is the third largest ocean (74×10^6 km²). Most of its area lies in the Southern Hemisphere. The boundary between the Indian and Atlantic oceans lies south of South Africa, while its boundary with the Pacific Ocean follows the Indonesian Islands to eastern and southern Australia and south of Tasmania to Antarctica. The average depth of the Indian Ocean is 3840 m, very close to the world average, and is marked by a small percentage (9 percent) of continental shelves. Although the Indian Ocean has few islands, it is marked by numerous submarine plateaus and rises. Nearly all river discharge occurs in the northern part adjacent to Asia.

The Arctic Ocean is a shallow (average 1117 m), circular, landlocked polar ocean of relatively small size (12×10^6 km²) centered over the North Pole. A high proportion of its total area (68 percent) is made up of continental shelves and slopes (Table 2-1). The Arctic Ocean is ice covered for much of the year with sea-ice thicknesses up to 3 to 4 m. It is linked with the world ocean only by narrow passages between Greenland and Iceland and between Iceland and northern Europe. Otherwise it is landlocked. Its average salinity is much lower than the world ocean as a result of its landlocked character and of high river discharge from the surrounding continents.

In addition to the earth's major oceans, there are also a number of much smaller, largely landlocked seas lying between continental blocks. These include the Mediterranean Sea, Black Sea, Gulf of Mexico, Red Sea, Bering Sea, Baffin Bay, and the North Sea. Each is distinctive, and each has physical characteristics strongly influenced by the climatic regime and topography of the surrounding continents, latitudinal position, proximity to ice sheets, amount of river discharge, nature of circulation with the open ocean, and several other factors. Their sedimentary characteristics thus vary widely.

Principal Topographic Features

If all the waters were removed from the ocean basins, there would be revealed a pattern of topographic features dominated by a system of ridges and rises encircling the globe with intervening deep-sea basins between the ridges and the continents. The pattern shows that the deepest parts of the oceans do not occur in the middle as one might expect, but close to the land. The middle of the ocean is shallower because of the mid-oceanic ridges. This is similar to the patterns of major mountain chains around the earth which, except for the Himalayas and a few other chains, are not located toward the middle of continental masses but near the edges facing deep oceanic trenches. Thus, both continental and oceanic areas exhibit greatest vertical change in narrow zones of the earth's crust. In this section we shall describe the principal topographic features of the ocean basins. The origin of these features within a global tectonic framework will be discussed in later chapters. Knowledge of the character of the sea floor emerged only during the past few decades following the invention of the continuously recording echo sounder. The mapping, investigating, and exploring of these broad physiographic features has occupied a whole generation of oceanographers [Heezen and Hollister, 1964], but much work remains to be done. Rock dredging and core sampling have provided data on ocean-floor rocks and sediments.

We have observed from the hypsographic curve (Fig. 2-6) that the earth's surface consists of two dominant physiographic provinces: the continental platforms and the ocean basins. These are linked by the continental margins (Fig. 2-7).

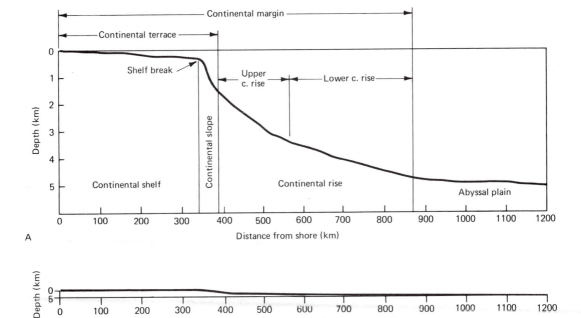

A

Figure 2-7 Principal features of the Atlantic continental margin. (A) vertical exaggeration 1/50; (B) no vertical exaggeration (Based on B. Heezen and others, 1959, and B. Heezen, 1962; after W. A. Anikouchine and R. W. Sternberg, 1973)

Seismic methods are not refined enough yet to show in detail how the oceanic and continental crusts merge, although the transition is known to occur over a distance of 200 km or less. The oceans are divided into three major topographic provinces (Fig. 2-8): the continental margins, the ocean basin floors, and the major oceanic-ridge systems. In addition to these major features, there are secondary topographic rises and plateaus that tend to break up the oceans into separate basins. Continental shelves and slopes, although covered by water, are part of the continental crust. If

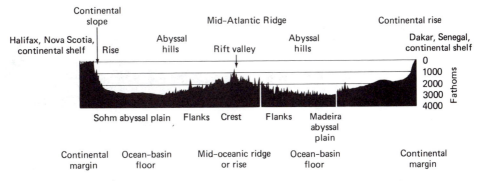

Figure 2-8 Principal morphologic features along profile across the North Atlantic between North America and Africa. (After Holcombe, 1977)

the areas of the ocean represented by continental shelves and slopes are excluded, it is found that only three oceanic topographic provinces occupy more than 2 percent of the ocean basin [Wyllie, 1971].

1. Ocean basin floors, 41 percent.

2. Ridges and rises, 33 percent.

3. Continental rises, 5 percent.

Profiles across the North Atlantic and North Pacific are shown in Fig. 2–8 and 2–9 to illustrate the major topographic features of the oceans and their relationships. Holcombe [1977] provides a useful review of oceanic topographic features.

The Continental Margins

Continental margins lie between the continents and ocean basins, and include those physiographic provinces of the continents and the ocean basins associated with the boundary between these two first-order morphologic features of the earth (Fig. 2–7). The margins underlie 20 percent of the total area of the oceans. Although continental margins vary considerably from region to region, there are two main forms. The first consists of three components with increasing depth: *continental shelf, continental slope,* and *continental rise* (Figs. 2–7 and 2–8). This has

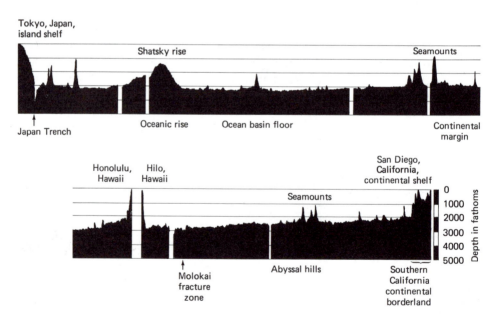

Figure 2-9 Principal morphologic features along profile across the North Pacific between Japan and California. Vertical exaggeration is 100×. (After Holcombe, 1977)

28 Geophysics and Ocean Morphology

been called the **Atlantic-type** continental margin. The second grades from shelf to slope to a deep trench (Fig. 2–9). This type of margin is widespread around the Pacific and has been called the **Pacific-type** margin. The Pacific-type margin has been subdivided into the **Chilean-type,** as characterized by a narrow shelf with a trench below the slope, and the **island-arc** or **Mariana-type,** which exhibits a shallow marginal basin separating the continent from the island-arc and trench system. In some cases the marginal basins are so broad (up to 2000 km) that they cannot be considered to represent part of a single continental margin.

Atlantic-type continental margins are often called **aseismic,** or **passive, margins** because they are seismically inactive, having developed when continents rifted apart to form new oceans. Pacific-type continental margins are often called **seismic,** or **active, margins** because they are seismically active. The tectonics and genesis of these two types of margins are discussed in more detail in Chapters 10 and 11.

Continental shelf

The continental shelf forms the seaward extension of the adjacent continent from the shoreline to a line called the *shelf break* or *shelf edge,* along which there is usually a marked increase of slope at the outer margin. Geophysical surveys have shown that the crust beneath the shelf is continental in character but thinner than the adjacent continental crust. The continental crust begins to thin abruptly near the shelf break. The average depth of the shelf break is uniform, averaging about 130 m over most of the world ocean. Slopes on the shelf are very gentle (less than 1:1000) and relief is subdued (less than 20 m). The width of the shelf ranges from a few kilometers to greater than 400 km (average 78 km). A series of benches or terraces occur on the continental shelf. The widest shelves occur in the Arctic Ocean and the deepest (about 350 m) occur around Antarctica. The present level and gross topography of the continental shelf is the cumulative effect of erosion and sedimentation related to numerous large-scale sea-level oscillations during the last 1 million years (late Quaternary). The present shelf break is believed to have formed about 18,000 years ago when sea level stood at that level due to the removal of water from the oceans as large icesheets. The position of the shoreline at that time coincided with or was located slightly inland of the present shelf break.

Continental slope

At the edge of the continental shelf, the depth falls off rapidly from 100 or 200 m to 1500–3500 m. The regional slope is steep (greater than 1:40, average of about 4°) and the boundaries with relatively flat continental-shelf and continental-rise provinces are abrupt. These are the most conspicuous boundaries on the earth's surface. The relief is very steep for the oceans and generally not precipitous, as gradients are normally low compared with those of the land. However, slopes are precipitous (35–90°) in some areas. This zone is narrow, with a width usually less than 200 km. The continental slope and shelf are together known as the *continental terrace* (Fig. 2–7). The continental slope may be divided up by faults into a series of escarpments that eventually become filled with sediment obscuring the original structure of the margin. Where the continental terrace is constructed of a series of basins and ridges, such as off southern California, it is called the *continental borderland* (Fig. 2–9). Very thick sediment accumulations cover many of the earth's continental margins. Around much of the Pacific, the margins extend down into deep trenches such as the Peru-Chile Trench.

Continental rise

The physiographical province between the slope and ocean basins is the continental rise (Figs. 2–7 and 2–8). The rise is a zone 100–1000 km wide, marked by a gentle seaward gradient (1:100 to 1:700) and low local relief (less than 40 m). Seaward boundaries are abrupt, occurring where the regional gradient decreases to that of the abyssal plains. Breaks in surface gradient on the rise allow subdivision into further sections in certain areas. The continental rise is formed of sediment accumulations several kilometers thick, transported from the continents and deposited at the base of the continental slope. **Submarine canyons** and **deep-sea channels** may cut through the continental rise, acting as channels for the seaward transport of sediments. The canyons may continue upwards, incising the continental slope and shelf. Because the continental rise is a depositional feature, it is intimately associated with **deep-sea fans,** where a reduction in current speed, because of decrease in surface gradients, causes sediment settling. Each submarine canyon has a fan at its mouth, and these fans may merge, forming broader, fanlike features. Thick sequences of continentally derived sediments can accumulate in these regions. Because most sediment eroded from the continents is deposited in the Atlantic and Indian oceans, the continental rises are more highly developed in these oceans. (Major rivers from China flow into a marginal sea.) The smooth surface of the continental rise owes its origin to the same processes that produced the abyssal plains. However, the rises are morphologically distinct from the featureless topography of the abyssal plains.

The Oceanic Ridges

Ridges

The map of the ocean basins (Figs. 2–10 and 2–11) shows that the most conspicuous topographic features are the **mid-oceanic ridges,** or rises. These ridges extend through all the oceans, with a total length of 80,000 km and an average depth of about 2500 m. They occur in the middle part of the oceans, except in the North Pacific where the ridge is confined to the far eastern region before it intersects with North America (Fig. 2–10). The mid-oceanic ridge essentially represents a broad undersea cordillera or mountain range that rises to its highest elevation at the axis and slopes away on either flank (Fig. 2–12). The ridges are gently arched uplifts, but locally the topography can be very rugged. Local relief consists of ridge-and-valley terrain paralleling the axis, cross-cut by fracture zones. Local relief ranges from 100 to 2000 m. The ridge crest is 1000 to 3000 m above the adjacent ocean basin floor, and the width of the ridge is greater than 1000 km. The **flanks** of the ridges lie between the crest and the ocean-basin floor. The two flanks are sym-

Features of the Ocean Floor

Figure 2-10 Major features of the ocean floor showing mid-ocean ridge system, basins, fracture zones, plateaus, trenches, and rises. (From *The Face of the Deep* by Bruce C. Heezen and Charles D. Hollister. Copyright © 1971 by Oxford University Press, Inc. Reprinted by permission).

Geophysics and Ocean Morphology **31**

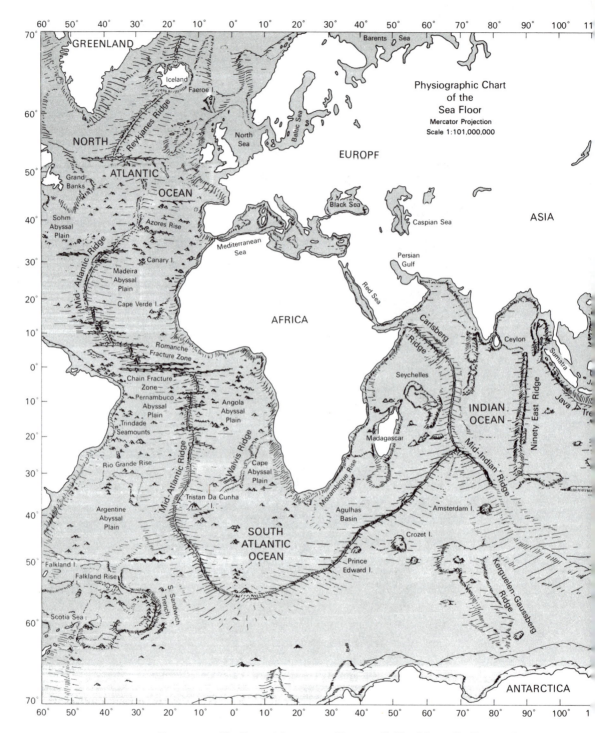

Figure 2–11 The floor of the oceans. (Courtesy Hubbard Scientific Company)

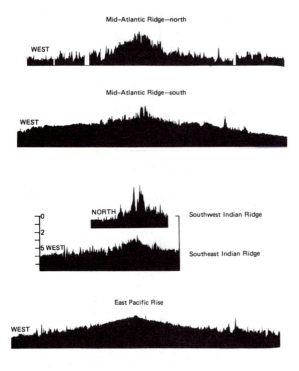

Mid-Atlantic Ridge—north

WEST

Mid-Atlantic Ridge—south

WEST

NORTH   Southwest Indian Ridge

0
2
5 WEST   Southeast Indian Ridge

East Pacific Rise

WEST

Figure 2-12 Selected profiles across the mid-oceanic ridge. Profiles in Atlantic and Indian oceans exhibit a well-defined rift valley set within very rugged relief. The South Pacific profile of the East Pacific rise lacks a rift valley. Baseline is 6500 m for all profiles. (After B. Heezen and M. Ewing, 1963)

metrical in their topography (Fig. 2-12). They are marked by moderate-to-rugged basement relief (100–1000 m) and variable sediment ponding and cover. The sediment cover usually thickens with increasing distance from the ridge crest.

The topography of the mid-ocean ridges varies throughout their length. The greatest contrast exists between the East Pacific Rise and ridges in the other oceans. The East Pacific rise is much broader and less rugged than the other ridges and has no prominent valley along its crest (Fig. 2-12). This character probably reflects different rates of formation to be discussed later. The East Pacific rise stands about 2 to 4 km above the adjacent ocean bottom and varies from 2000 to 4000 km in width. The rise passes into North America at the Gulf of California and reappears as a much less conspicuous feature off the Oregon coast, extending into the Gulf of Alaska.

The mid-Atlantic Ridge, that portion of the ridge system lying in the Atlantic Ocean, is best known. One profile across the mid-Atlantic ridge resembles many others (Fig. 2-12). The crest consists of extremely rugged relief with broad, rugged flanks. A central **rift valley,** 1 to 2 km deep and a few tens of kilometers wide, lies along the axis of the ridge (Fig. 2-13). This feature was discovered by B. Heezen, at Lamont, during the early stages of his oceanic bathymetric mapping. Rugged rift mountains formed by block faults run parallel to the rift valley on either side (Fig. 2-13). The crestal area of the mid-Atlantic ridge, with its central rift, is nearly identical to the rift valleys in the high plateau of East Africa (Fig. 2-13). In fact, the mid-ocean ridge system merges with the African rift valley in the Gulf of Aden; thus it is usually inferred that they have formed from a similar tectonic process.

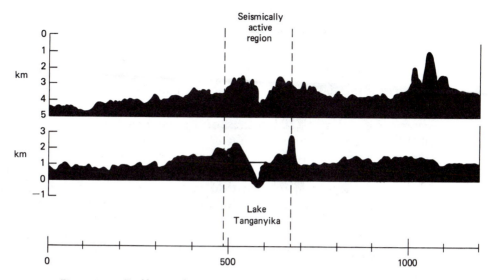

Figure 2-13 Profiles on the same scale of the mid-Atlantic ridge and African Plateau with their median valleys. (After M. Ewing and B. Heezen, 1960)

Detailed mapping from surface ships, deep-towed instruments, and observations from submersibles on the mid-Atlantic ridge during the FAMOUS (French-American Mid-Ocean Undersea Study) expedition south of the Azores Islands, have revealed that the central rift valley contains another even narrower inner-rift valley a few kilometers wide. The freshest volcanic rocks of the mid-ocean ridge have been observed and collected from this region, including closely associated central volcanic hills. On a small scale, fissures a few meters wide and up to 10 m deep parallel the axis of the rift for a few hundred meters [Heirtzler, 1979]. Basalt is the important rock type; gabbro, serpentinite, and other igneous rocks are also present. Volcanoes and volcanic islands, such as the Azores, Iceland, Ascension, and Tristan da Cunha, most of which are basaltic, occur at a few locations along the ridge.

The world-encircling belt of oceanic ridges can be traced from the Siberian continental shelf near the Lena River delta, across the Arctic Ocean (represented by the Nansen ridge), and through the Norwegian Sea to Iceland, where it continues through the Atlantic, around the south of Africa, and into the Indian Ocean. Between Madagascar and India, it splits. One branch runs northwest into the Gulf of Aden. Here it divides again, one branch continuing up the Red Sea, the other passing into the African rift-valley system. The other Indian Ocean branch extends to the south of Australia and New Zealand, across the South Pacific and northward into the Gulf of California.

A rare terrestrial expression of the mid-oceanic ridge is volcanically active Iceland; its central graben is the extension of the mid-Atlantic ridge. Geologically recent volcanism in Iceland is restricted to the central graben, with rocks becoming progressively older toward the east and west of it. The floor of the graben is marked by active fissures (gja) parallel to the bounding graben faults. Tensional tectonics prevail. The structure and origin of the mid-oceanic ridges are fundamental to the origin of oceans, as we shall see later.

Fracture zones

The ocean floor is cut by hundreds of fracture zones (10–100 km wide) that, on a sea-floor map, form a pattern of semiparallel stripes cutting across the mid-oceanic ridges (Fig. 2–11). The crest of the mid-oceanic ridge is frequently offset along fracture zones; the part of fracture zones lying between the offset limbs of the axis is seismically active. Fracture zones are extensive linear zones of irregular topography marked by troughs, escarpments, and other features such as large sea-mounts or steep-sided, asymmetrical ridges. They normally extend for long distances across the flanks of the ridge, in some cases extending as surface or sub-surface features across the ocean-basin floor to the continental margins. Fracture zones may reach a length of 3500 km and consist of several linear ridges and valleys. In many areas, the fracture zones separate basement surfaces lying at different levels, with total relief ranging from 4000 m or more to 100 m.

The mid-oceanic ridge crest may be displaced by lateral fracture zones for great distances (Fig. 2–11). For instance, in the equatorial Atlantic, a series of semiparallel fracture zones displaces the mid-Atlantic ridge by about 4000 km. The ridge, therefore, maintains its median position during its S-shaped sweep through the Atlantic. Figures 2–10 and 2–11 show the distribution of a few of the more important fracture zones [Menard and Chase, 1970].

The first fracture zone to be studied was the Gorda escarpment [Murray, 1939], a major west-east scarp offsetting the continental slope of northern California. Menard and Dietz, in 1952, were the first deliberately to map the nature and extent of a fracture zone in detail. They initially turned their attention to the Mendocino fracture zone (Fig. 2–11) which they found to have a length of at least 2000 km, a relief of 3 km, and slopes to 24°. This was soon followed by the discovery of a system of semiparallel fracture zones in the Northeast and Southeast Pacific. Menard [1954] defined fracture zones as long, narrow bands of grossly irregular topography, linear ridges, and scarps, typically separating distinctive topographic provinces with different regional depths. This definition remains accurate.

The linearity of fracture zones (Fig. 2–11) is diagnostic of their origin and the trend is critical in determining the direction of crustal block movement. Although fracture zones are straight normally, more detailed studies show that they change direction along their length (Fig. 2–11). For instance, Menard [1967] determined that several major fracture zones in the Northeast Pacific follow great circles for almost 10,000 km. In 1970 he further recognized eight trend changes each in the Clipperton, Clarion, and Mendocino fracture zones. The length of trends ranges from 100 to 1400 km and the mode of the lengths is about 500 km. The changes in trend range from 3° to 102° and the modal change is 5°. Like oceanic ridges, several fracture zones abut the continents. The Mendocino fracture zone displaces the northern Californian continental shelf break, and the continental extension of the Murray fracture zone is the Transverse Mountain range of southern California. Fracture-zone mapping advanced rapidly with the advent of extensive magnetic surveys of the ocean floor. The patterns of magnetic anomalies defined by these surveys proved the existence of large lateral offsets by fracture zones of the mid-oceanic ridge system and allowed mapping of new sets of fracture zones throughout the oceans. This mapping clearly demonstrated the close relations existing between mid-oceanic ridges and fracture zones. This relationship was initially clouded because the best-known fracture zones were in the Northeast Pacific, while the best-understood mid-oceanic ridge system was in the North Atlantic. The origin of fracture zones and their role in global tectonics are discussed later.

The Ocean-Basin Floor

The ocean-basin floor lies between the continental margins and the mid-oceanic ridges. There are three major subdivisions of the ocean-basin floor.

1. The abyssal floor, in turn divided into abyssal plains and abyssal hill provinces.

2. Oceanic rises.

3. Seamounts and seamount groups.

Abyssal plains

One of the most striking discoveries of the 1947 mid-Atlantic ridge expedition was the vast areas of flat, nearly level plains of the deep-ocean floor in the North Atlantic [Heezen and Laughton, 1963]. In 1948 the Swedish deep-sea expedition discovered a similar plain in the Indian Ocean south of the Bay of Bengal. Since then, several abyssal plains have been mapped throughout the world (Fig. 2–11). Abyssal plains are areas of the ocean-basin floor that are flat, with a slope of less than 1:1000 (1 m/km). They are among the flattest portions of the earth's surface. They occur adjacent to the outer margins of the continental rise, between 3000 and 6000 m deep. The horizontal extent of an abyssal plain ranges from less than 200 km to more than 2000 km. Some have a flat surface with a continuous gradient in one direction, whereas others are marked by broad irregularities. They are widespread in the Atlantic and Indian oceans and occur in marginal seas such as the western Mediterranean, Gulf of Mexico, and the Caribbean.

Seismic evidence indicates that abyssal plains are deposits of great thicknesses of sediment over the original topographic irregularities. The flattest abyssal plains occur in areas, such as the Atlantic, where abundant sediment is derived from the continents. Since relatively few rivers drain into the Pacific Ocean and because it is rimmed by a trench system that traps sediment, the ocean basin receives less sediment and fewer plains form (Fig. 2–11).

The landward edge of the abyssal plains is generally marked by an abrupt change in slope to values from 1:100 to 1:700—typical of the continental rise. In some cases the deep-sea fans of the rise merge with the abyssal plains, while in others the abyssal plains are cut by deep-sea channels at the seaward ends of submarine canyons.

Abyssal-hills province

Abyssal hills are small, sharply defined hills that rise from the abyssal plains to elevations no higher than about 1000 m (Fig. 2–8). Horizontal dimensions are generally from 1 to 10 km, but may range as high as 50 km or more. Hillside slopes range from 1 to 15°. Abyssal hills usually occur in groups, between the abyssal plain and the flanks of the mid-ocean ridges, in which case they are called the **abyssal-hills province.** The hills usually take their shape from the basement surface, although the same term may be applied to hills of sedimentary origin. Although the abyssal hills are abundant in the Atlantic and Indian oceans, they are important features in the Pacific where they cover 80 to 85 percent of the ocean floor. Abyssal hills are an integral part of the mid-oceanic ridge (Fig. 2–8), but are morphologically separate because they occur at a deeper level, away from the regional slopes of the ridges and because of their lower relief due to burial by sediment.

Seamounts

Volcanoes and volcanic ridges that form from the ocean basin are a topographic province distinct from volcanoes associated with the oceanic ridges. Volcanoes that rise more than 1000 m above the ocean floor are called *seamounts* (Fig. 2–9). Seamounts are distributed throughout the provinces of the deep-ocean floor. They do occur randomly, but more typically are clustered in groups or rows (see Chapter 7). Seamount slopes are typically from 5 to 15° and most are conical. Thousands of seamounts are distributed throughout the Pacific. **Volcanic ridges** or **seamount chains,** like the Hawaiian volcanic chain, are formed by overlapping volcanoes. Seamounts rise abruptly out of the abyssal plain, suggesting that their bases have been buried.

Once formed, submarine volcanoes exist for a long time (for example, volcanoes of the Hawaiian-Emperor Chain). But if the summit breaks the sea surface and a volcanic island is formed, the part above sea level is eventually eroded. Subsequent subsidence of the seamount produces a flat-topped feature, which occurs up to 2 km below the sea surface (see Chapter 7). These are called **guyots** or **tablemounts.** In tropical areas coral often grows, forming reefs around seamounts. After the seamount subsides, only the coral atoll remains.

Guyots are abundant in the ocean basins, especially in the Pacific. The formation of the flat tops had long puzzled geologists, and several hypotheses have been advanced to explain this feature. Drilling and dredging of the truncated surface has in all cases revealed shallow-water sediments and fossils immediately overlying the volcanic basement rocks. These sediments also commonly exhibit evidence of wave action. It is now accepted that guyots represent submarine volcanoes that have had their tops truncated by subaerial or shallow-water erosion. A large group of guyots extends northwestward across the central Pacific.

Marginal Trenches

Among the most spectacular features of the ocean floor are the marginal trenches. They are narrow, steep-sided troughs roughly parallel to continental margins at the seaward base of a continental platform, such as off western South America or at an island arc such as the Bonin-Mariana Islands (Figs. 2–9 and 2–11). Nearly all occur near the margin of the Pacific (Table 2–2), except the South Sandwich and Puerto Rico trenches in the Atlantic and the Java Trench adjacent to Indonesia in the Indian Ocean.

Trenches are remarkable for their length and continuity [Bott, 1971]. The Peru–Chile Trench is 5900 km long; the Mariana Trench is 2500 km long. Considering their length and depth, trenches are remarkably narrow; on the average, they are less than 100 km wide, like curved gashes across the seascape. They reach depths of up to 11 km and from 2 to 4 km below the adjacent ocean floor (Table 2–2). The greatest ocean basin depths occur in the trenches. The deepest is Challenger Deep, a part of the Mariana Trench, which reaches a depth of 11,034 m. This was visited in 1960 by J. Picard during a dive by the bathyscaphe *Trieste*. The bottom of the trench is a narrow, flat, sediment-filled base. The trench sides are steep-sided steps resulting from faulting. Trenches are separated from the ocean floor by an outer ridge, which rises 200 to 1000 m above the level of the adjacent ocean floor. In parts of the Indonesian region, trenches occur as double, parallel features. Island arcs and trenches, including the whole system of low swells subparallel to the trenches are considered a single province, covering only 1.2 percent of the earth's surface. Nevertheless, like the ocean ridges, they play a critical role in global tectonics.

TABLE 2-2 TRENCH CHARACTERISTICS[a]

Trench	Depth (km)	Length (km)	Average width (km)
Pacific Ocean			
Kurile–Kamchatka Trench	10.5	2200	120
Japan Trench	8.4	800	100
Bonin Trench	9.8	800	90
Mariana Trench	11.0	2550	70
Philippine Trench	10.5	1400	60
Tonga Trench	10.8	1400	55
Kermadec Trench	10.0	1500	40
Aleutian Trench	7.7	3700	50
Middle America Trench	6.7	2800	40
Peru–Chile Trench	8.1	5900	100
Atlantic Ocean			
Puerto Rico Trench	8.4	1550	120
South Sandwich Trench	8.4	1450	90
Indian Ocean			
Java Trench	7.5	4500	80

[a] From M. G. Gross, 1977; data after R. W. Fairbridge, 1966, pp. 932–933.

Ocean Seismic Reflection and Refraction

Knowledge of the thickness and layering of oceanic sediments comes largely from seismic reflection methods, while information on the oceanic crust and upper mantle (beneath the sediments) comes from seismic refraction work. These are by far the most widely used geophysical methods. In the 1950s and early 1960s, they were pioneered by M. Ewing and his colleagues at Lamont, M. N. Hill at Cambridge University, and R. Raitt and G. Shor at Scripps Institution. They were later employed and improved by the petroleum industry for hydrocarbon exploration both at sea and on the continents.

The technique basically involves measuring the time required for a seismic wave (or pulse) generated by an explosion, a mechanical impact, or vibration to return to the surface after reflection and refraction from subsurface interfaces possessing different physical properties (Fig. 2-14). The physical principles upon which reflection and refraction methods are based have been summarized thoroughly by Dobrin [1976]. In the reflection method, acoustic energy is reflected off an interface; in refraction, the energy is refracted so as to propagate along an interface. The principal difference between the geometries of the refraction and the reflection methods is in the interaction that takes place between the seismic waves and the lithological boundaries they encounter in the course of their propagation [Dobrin, 1976].

The waves reflected by the boundaries travel along paths that are quite easy to visualize (Fig. 2-14). An interface between two layers may be detected if the acoustic impedance (product of acoustic velocity and density) of the layers is different [Ewing and Ewing, 1970]. The ratio of reflected to incident energy (called the **Rayleigh reflection coefficient**) for the normal incidence of a plane wave is

$$\frac{P_2\,C_2 - P_1\,C_1}{P_2\,C_2 - P_1\,C_1}$$

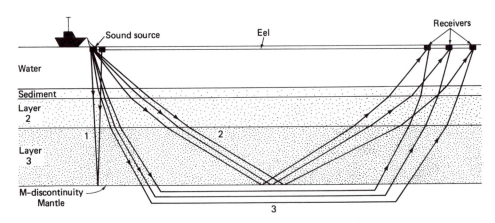

Figure 2-14 Reflected and refracted waves through earth layers between sound source and receiver. 1 = vertical incidence reflector; 2 = wide-angle reflectors; and 3 = refracted waves.

where P is bulk density and C is compressional velocity of the respective layers. The impedance contrast at an interface can be estimated by the amount of energy reflected from the interface. Most recorders show an increase in printing density with increasing amplitude of the reflected wave. Thin layers with small impedance contrast often escape detection. The reflection records, then, yield the two-way travel time to the reflecting horizons. In order to convert this time into depth or thickness, it is necessary to know the velocity at which the waves travel in the layers penetrated. Unfortunately, the study of deeper layers can be a problem because the wave velocities may not be known in the overlying layers. Nevertheless, the form of the reflecting interface is still determined.

Refracted waves follow more complex paths (Fig. 2–14), crossing boundaries between materials with different velocities in such a way that energy travels from source to receiver in the shortest possible time [Dobrin, 1976]. Most refraction work uses waves with paths along the tops of layers with speeds much higher than those in overlying beds. The seismic wave's speeds and depths of these beds are determined from the times required for the refracted waves to travel between the sources and receivers at the surface.

The *multichannel reflection method,* widely used in the petroleum industry, has probably become the most sophisticated, technologically advanced branch of geophysics. In the academic community, single-channel seismic reflection profiling and wide-angle reflection and refraction techniques have been widely used since World War II. More recently, with the development of ocean-bottom seismometers, new and more sophisticated analytical techniques, and the use of multichannel methods, seismology has become one of the more rapidly changing areas of marine geophysical research.

Seismic Reflection

Echo sounding

The **precision depth recorder** (PDR) was developed during World War II as part of antisubmarine warfare and has been widely used on oceanographic research ships since the early 1950s. The method is based on the simple principle of measuring the time for sound to propagate through the water column, reflect off the sea floor, and return to the surface ship. Records over areas of flat terrain are simple to interpret but over slopes or rough terrain, side echoes become a problem and the echo sounder will not record the true depth beneath the ship. This problem is partly alleviated by use of **narrow-beam echo sounders** that employ an electronically stabilized beam of width 2° to 3°.

Low-frequency echo sounders are also useful tools in marine geology. These are equipped with transducers and transceivers tuned for frequencies of 3.5 kilohertz (kHz), compared with the 12 kHz for echo sounders and the few hertz to few hundred hertz for most seismic profiling systems. The operation at intermediate frequencies is a good compromise between deep penetration and high resolution for studying the *upper part* of the sedimentary section. Penetration into sediment can often be 100 m with greater penetrations (about 200 m) in some areas. Profiling at

3.5 kHz provides an excellent view of the upper sediment layers, when used in conjunction with piston coring (Fig. 2–15).

Reflection profiling

The purpose of reflection profiling is the mapping of subbottom acoustic horizons. Frequencies generally employed range from 20 Hz to 200 Hz. This enables some of the waves to penetrate below the ocean floor and to be reflected from subsurface interfaces. Low frequency pulses are less attenuated (absorbed) when passing through sediment or rock.

Various types of sound sources have been employed. In earlier experiments explosives were the only sound source available, as they were in great abundance following World War II. A half-pound block of TNT was dropped over the ship's side every 3 minutes to provide a continuous seismic record. At Lamont this procedure was carried out continuously for 10 years, providing much valuable data. This tedious, potentially dangerous method has now been replaced, as other acoustic sources such as air guns and electric sparks have been developed. Air guns

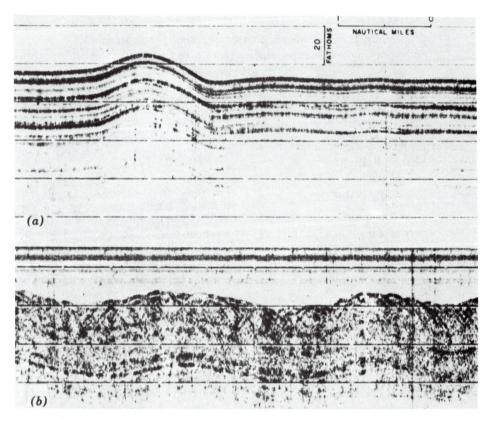

Figure 2–15 Seismic profiles of 3.5 kHz near the Blake Bahama Outer Ridge at about 4400 m. (From M. Ewing and J. Ewing, *The Sea, Ideas and Observations on Progress in the Study of the Seas*, Vol. 4, Part I, John Wiley & Sons, Inc., 1970)

release air at high pressure into the water and are fired every 10 or 12 seconds. The return waves are received by a towed hydrophone array. The receivers, or hydrophones, convert sound energy to electrical energy, thus making it possible to record the sound graphically. Often a string of hydrophones is contained within plastic tubes filled with liquid so that the tubes are nearly neutrally buoyant, eliminating "drag noise." These tubes are called **streamers,** or **eels.** The hydrophones can be used to receive weak subbottom reflectors while the ship is moving. Ships can be operated at near cruising speeds, which enables continuous seismic records to be made more economically. This is called **continuous seismic profiling.** Such systems are as common on research ships today as precision echo sounders were in the 1950s.

Multichannel seismic reflection profiling

During the past 20 years exploration seismology has been revolutionized by the development and application of increasingly sophisticated **multichannel seismic systems.** Until recently these techniques have not been used in the academic research community, primarily because of their great expense. However, during the past few years there has been increasing use of multichannel techniques, and it appears that they will be increasingly important during the 1980s.

Multichannel systems differ from the conventional single-channel systems in that very long streamers sometimes several kilometers in length, are employed. The streamers contain large numbers of hydrophones and are arranged in groups. Each group of hydrophones in the streamer acts as a separate recorder. These sets of recorders enable a large number of signals to be obtained from a single reflection point on the ocean floor (the **common depth point**). By **stacking,** or adding the signals together, the signal-to-noise ratio can be increased, thus enhancing the quality of the record. Another advantage is that velocities (and hence thicknesses) can be accurately determined from the variety of propagation paths recorded for each common depth point. The long streamers also allow the reception of the vertical incidence reflections, wide-angle reflections, and refracted arrivals (Fig. 2–14). The wide-angle reflection and refraction signals can be used to determine the speed of sound in the upper layers and accurately record the thickness and depth of the rock layers [Stoffa and others, 1980]. Energy sources are larger than with single-channel systems and use high-pressure air guns with 2000 lb/inch^2 (psi) (equal to 2 lb TNT), which allow deeper penetration. This new technology made wide-ranging geophysical surveys possible; multichannel, digitally processed seismic data for the study of crustal evolution has now become widely available. A comprehensive review of multichannel seismic reflection profiling is provided in Dobrin [1976] (Chapter 8).

In the interpretation of seismic profiles (Figs. 2–16, 2–17, and 2–18), the following considerations need to be kept in mind.

1. Lateral or vertical variations in velocity of the medium through which the sound has been transmitted produce the characters of a seismic profile. Profiles, therefore, are not geological cross sections but merely a picture created by the velocity variations of sound waves through the sequence.

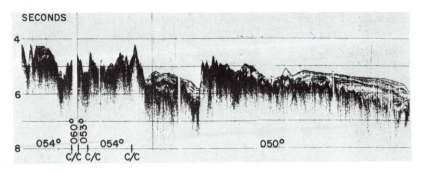

Figure 2-16 Seismic profile across the east flank of the mid-Atlantic ridge near 45°S. (From M. Ewing and J. Ewing, *The Sea, Ideas and Observations on Progress in the Study of the Seas*, Vol. 4, Part I, John Wiley & Sons, Inc., 1970)

2. Stratigraphic resolution is limited by the frequency or wavelength of the seismic energy. For instance, at 25 to 50 Hz and at velocities of 5 km/sec, seismic wavelengths are 100 to 200 m. Thus any feature smaller than these wavelengths cannot be resolved.

3. The seismic section is only as good as the processing that was used to produce it. Errors introduced during filtering, velocity analysis, or in other procedures can dramatically change the appearance of the record.

4. A seismic profile is a two-dimensional record of a three-dimensional system.

Seismic Reflection Features

Seismic profiles show that the most prominent reflection below the seabed is the interface between the ocean sediments and basement rocks. The layer of ocean sediments is called **Layer 1,** while the immediately underlying oceanic basement

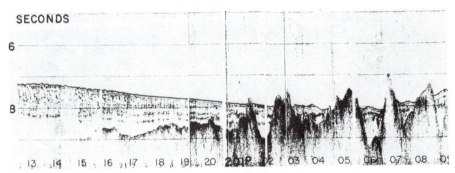

Figure 2-17 Seismic profile across the southern part of the Argentine Basin toward the Falkland Escarpment north of South Georgia. (From M. Ewing and J. Ewing, *The Sea, Ideas and Observations on Progress in the Study of the Seas*, Vol. 4, Part I, John Wiley & Sons, Inc., 1970).

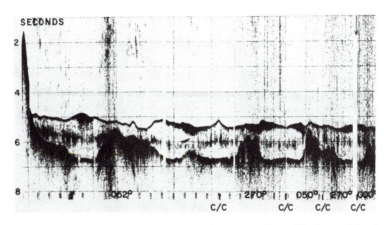

Figure 2–18 Seismic profile across the North American Basin (Northwest Atlantic northeast of Bermuda). Horizon *A* is exhibited midway within the sediment column. The peak at the left is a seamount nearby 300 km east northeast of Bermuda. (From M. Ewing and J. Ewing, *The Sea, Ideas and Observations on Progress in the Study of the Seas*, Vol. 4, Part I, John Wiley & Sons, Inc., 1970)

rocks belong to **Layer 2.** This interface is rougher than the seabed, with an amplitude of up to 1 km and an average wavelength of undulations of 10 to 20 km. This roughness continues under the abyssal plains.

Reflection data, supplemented by deep-sea drilling, have provided information on the thickness and velocity distributions of ocean-basin sediments. Seismic observations in the Atlantic show three major sedimentary sequences [Ewing and others, 1966; Houtz and others, 1968].

1. Unconsolidated sediments in which there is a uniform increase in velocity with depth of burial (velocity between 1.6 and 2.2 km/sec).

2. Semiconsolidated sediments which exhibit a discontinuous increase in velocity with increasing depth, called Layer *A* (velocity between 1.7 and 2.9 km/sec).

3. Consolidated sediments exhibiting a rather constant velocity of between 2.7 and 3.7 km/sec, called Layer *B* (Ludwig and others, 1970).

The thinness of the sedimentary layer was a great surprise in early seismic studies (Fig. 2–19). This suggested that either the rates of accumulation were much lower than previously suspected or that the oceans were younger than predicted. The reflection studies also showed that oceanic sediment distribution is not uniform (Fig. 2–19). Sediment thickness varies from several kilometers in some areas to zero in others. Sediment thickness depends on several factors, including the age of the ocean floor, proximity of the ocean basin to the continents and to areas of large river and terrigenous sediment outflow, relative productivity of biologically produced components, chemical dissolution of biogenic components, and the importance of bottom currents.

One of the most conspicuous features is the absence of an appreciable cover in

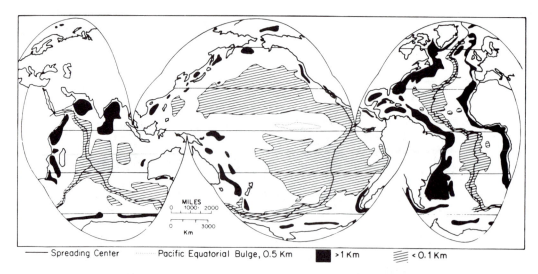

Figure 2-19 Sediment thickness to acoustic basement in the world ocean. (After W. H. Berger, 1974)

a belt along the crest of the mid-oceanic ridges (Figs. 2–16 and 2–17). Thicknesses increase steadily down the flanks of the ridge. In middle-latitude regions, especially in the Pacific, even the flanks are marked by thin sediment cover because of low biological productivity and little terrigenous sediment input. Fracture zones often contain thick sequences of carbonate due to slumping of the steep slopes.

The major accumulations of oceanic sediment are adjacent to well-drained continental masses (Fig. 2–19). Large rivers carry enormous quantities of sediment to the ocean, the thickest sections occurring just beyond the foot of the continental slopes (Fig. 2–18). This is particularly marked in the Atlantic Ocean, which is rimmed by large rivers, and in the northern and western Indian Ocean. The Pacific has thin sections over much of its area (Fig. 2–19). Sediments in the Pacific marginal areas are confined to pockets at the foot of the continental slopes and in small abyssal plains at the bottom of the marginal trenches.

Sediment thickness in the ocean basins distant from land is controlled by biological productivity, preservational character of biologically produced sediment, and bottom current redistribution. A thicker belt of Pacific sediments associated with the equator and a zone around Antarctica is related to high biological productivity in these regions (Fig. 2–19). Seismic-reflection work provided much information on sediment thickness distributions in the ocean basins before it was possible to drill into these layers to determine their compositions.

One of the early discoveries resulting from seismic-reflection profiling was the presence of distinct reflection interfaces that could be traced over millions of square kilometers of the ocean basins. **Horizon A,** the most extensive, earliest detected, and best-known of such acoustic reflectors, is continuous over most of the western North Atlantic (Figs. 2–16 and 2–18) between the Lesser Antilles and the Grand Banks, westward to the lower continental rise of North America, and eastward to the lower flank of the mid-Atlantic ridge (Fig. 2–16). The horizon defines the top of a stratified zone in the sediments, identified seismically as a closely spaced, mutually conformable group of reflectors. The sediment both above and below the stratified zone is acoustically transparent but may contain several weak reflectors. Horizon A is covered by an average of 300 to 500 m of sediment with the greatest measured thicknesses of 3 to 3.5 km at the southern New England continental rise and the Blake-Bahama outer ridge. Reflectors that resemble Horizon A acoustically and morphologically have been observed in other basins of the Atlantic as well. At first Ewing and others (1964) believed Horizon A was a fossil abyssal plain with a local concentration of turbidite layers. Piston cores collected near outcrops of Horizon A suggested that it was Cretaceous in age. The problem of the nature of Horizon A, still inadequately explained in the middle 1960s, was one of the many reasons for the development of the Deep-Sea Drilling Project (DSDP). One of the earliest deep-drilling cruises in the North Atlantic in 1969 was partly devoted to this problem. This drilling revealed that the strong reflectors were caused by chert layers of middle Eocene age. Older, distinct reflectors occurring below Horizon A, usually referred to as **Horizon β** and **B**, have also been shown by deep-sea drilling to be chert layers, though of Cretaceous age.

Seismic Refraction

It has been possible to sample only the uppermost 500 m of the oceanic crust. Knowledge about the character of the rest of the crust has been based largely on the results of marine refraction work and correlation of the seismic velocities determined in these experiments with possible rock types. Seismic refraction experiments were first carried out by M. Ewing and by R. W. Raitt in the late 1940s and early 1950s, resulting in two significant discoveries: The ocean crust is much thinner than continental crust, and it is made up of a few well-defined layers that appear to be remarkably homogeneous throughout the ocean basins [Ewing and others, 1954; Raitt, 1956].

By the mid 1960s a remarkably coherent picture of oceanic crustal structure had emerged from these studies using simple techniques. Recently, however, with the development of new instrumentation and interpretation techniques, it has become possible to obtain much more detailed information on oceanic crustal structure. The results from these studies indicate that the oceanic crust is much more heterogeneous than previously thought, and this work has developed into one of the most active areas of marine geophysical research.

When refraction methods were first conducted and when little was known about the structure and genesis of the ocean basins, one of the most important questions to be answered was the composition of Layer 2. Seismic reflection experiments supplemented by piston cores has shown that Layer 1 consisted of sediments, but the seismic velocities of Layer 2 were consistent with volcanic rocks of basaltic composition and with highly consolidated sediments. If Layer 2 consisted of consolidated sediments, they could be ancient and would support the theory of permanency of the ocean basins; if they were volcanic rocks overlain by relatively young unconsolidated rocks, then the ocean basins must be young. Thus the distribution of the layers as defined by the refraction experiments became a major problem.

The refraction method is based on refraction of sound waves at the interface of layers with different densities. As they continue, the waves are returned to the ocean surface until the energy is dissipated (Fig. 2–14). At any fixed distance from the explosion, the first large signal to arrive will be the sound that has traveled along the fastest path. Subsequent returns represent successively slower paths. To uncover the crustal layers controlling the refractive behavior of the waves, sound returns must be recorded at several locations. Dobrin [1976] summarized the basic principles of sound refraction as it applies to marine geophysics.

In seismic refraction studies the basic datum is the travel-time curve. This is obtained by plotting the recorded times of arrival of distinctive waves against distance from the source. Fig. 2–20 shows the ray paths for direct, refracted, and reflected waves for a three-layer model and their relationships to each other on a travel-time plot [Ewing, 1963]. The direct wave (D) in the water (G_1) and the refracted waves (G_1 and G_2) in layers G_2 and G_3 follow straight-line segments. The reflected waves (R_l and R'_l) follow hyperbolic curves. A critical distance occurs at which a refracted wave from the deeper layers can be returned to the surface. The distance is dependent upon the velocity contrast between the layer and seawater and on the velocity and thickness of the intermediate layers. In the travel-time curve in Fig. 2–20, the refraction curves are tangent to the reflection curve associated with the top of that layer. Tangency develops between the reflection and refraction at a critical distance. The determination of these relationships is fundamental for interpreting sound-wave arrivals, [Ewing, 1963]. The travel-time curve (Fig. 2–20) illustrates sound-wave behavior for direct, reflected, and refracted waves. The wave (G_2), refracted at the boundary of Layer 1 and Layer 2, occurs only as a first arrival for a short distance. With further thinning of Layer 2, this wave may not occur as a first arrival, and reflection waves (R_l) would be required to define the position of the boundary of Layer 1 and Layer 2 [Bott, 1971].

The main survey methods used in the early refraction experiments are described by Hill [1963] and Shor [1963]. Basically two approaches were used—two-ship and single-ship refraction experiments. In the two-ship refraction experiments, the ships start in a position close to each other. One ship remains in this position deploying a receiver, while the other steams away detonating explosions at various times. As the intervening distance increases, the size of the explosive charge increases. Most refraction experiments extend to a ship separation of about 100 km. Energy returns from deeper layers are received first because of higher velocities. Because of the expense and logistical difficulties of two-ship refraction work, a single-ship method involving self-contained **radio sonobuoys** were developed for seismic work, with considerable success. The sonobuoys consist of a single, disposable hydrophone suspended at a depth of 20 to 40 m below the sea surface. The sonobuoy, powered by a seawater-activated battery, transmits the output of this hydrophone to the ship by radio as the ship steams away from the buoy, firing an air gun. The buoy is located at distances far enough from the sound generating ship to receive large angles of incidence and refracted rays; thus the sonobuoy acts

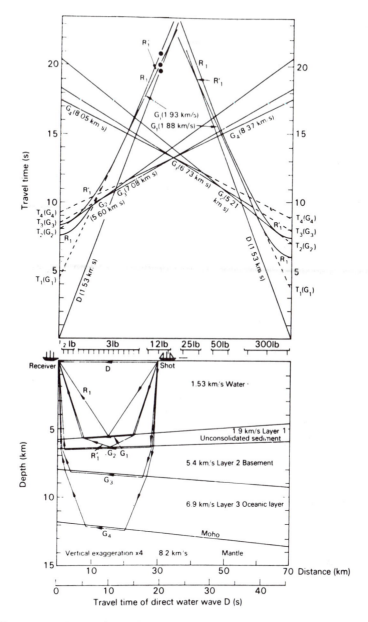

Figure 2-20 A typical time-distance graph for refractions—reflections and the direct water wave. The wave paths are shown for 30-km separation between acoustic source and receiver. (After Talwani, 1964; redrawn by M. H. P. Bott, 1971)

as a receiving ship. This allows calculation of the velocity-depth structure in the sediment column, refraction velocities, and data on velocity gradients.

During the late 1960s and early 1970s two approaches were introduced that led to further refinements in knowledge of the ocean crust—the use of repetitive sound sources (air guns) and the development of **ocean-bottom seismometers** (OBS).

One of the major limitations of the early refraction experiments was the relatively large shot spacing (often several kilometers) necessitated by the use of explosives as a sound source. The introduction of large-volume air guns (500 and 1000 in³) and arrays of air guns in the late 1960s made shot spacing of less than 1 km feasible for the first time. Another major advance was the development of ocean-bottom receivers of various types used in conjunction with a shooting ship. All of these instruments sink via free fall to the sea floor and have a timed release, an acoustic release, or both. All employ internal recording and magnetic tape. Major advantages include the fixed position of the receiver and the quiet environment.

The most important result of oceanic refraction work has been the information it has provided on the composition of the oceanic crust. By the early 1960s a coherent picture had emerged on the character of the crustal structure. The oceanic crust consists of three principal layers above the mantle with the following velocities [Raitt, 1963]:

Layer 1	sediments	< approximately 500 m
Layer 2	5.07 ± 0.63	1.71 ± 0.75 km
Layer 3	6.69 ± 0.26	4.86 ± 1.42 km
Mantle	8.13 ± 0.24	

The mantle is separated from Layer 3 by the Mohorovičić discontinuity. This interface is only 6 to 7 km below the ocean-basin floor, compared with 40 km beneath the continents (taken at sea level). The observed refraction velocities have been compared with the velocities measured on core samples from DSDP holes [Hyndman and Drury, 1976], rocks dredged from the sea floor [Fox and Opdyke, 1973], and ophiolites [Peterson and others, 1974] to infer the composition of the lower crust. This is discussed in more detail in Chapter 7.

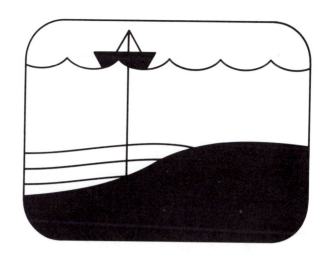

three
Marine Stratigraphy,
Correlation, and Chronology

What seest thou else
In the dark backward and abysm of time?

William Shakespeare

STRATIGRAPHY

Primary Objectives

The sediments and other stratified rocks that mantle the earth's surface and ocean floor record information about the history of the earth's environment and life. Sedimentary sequences provide most of our knowledge about the history of life on the earth and its changing environments, including paleoclimatic history, the history of oceanic circulation, sea-level change, oceanic geochemical changes, and the history of the earth's magnetic field. Sedimentary strata cover the ocean floor, except in certain very young areas at the mid-ocean ridges; they occur in great thicknesses at the continental margins and as uplifted strata on the continents themselves. **Stratigraphy** is concerned with the age relations of rock strata, their form, distribution, lithologic and fossil composition, paleoenvironmental interpretations, and geologic history. Stratigraphy also embodies the relations between rock strata and nonstratiform rock, including certain igneous bodies. **Stratigraphic correlation** is the determination of age equivalence or relative position between different rock strata. A few years ago nearly all of our knowledge of pre-Quaternary geological history came from sedimentary sections exposed or drilled on the continents, which represented only a small portion of the earth's surface. As a result of the DSDP, stratigraphy of Mesozoic and Cenozoic rocks has become global in scope. More data about the evolution of the earth are now being derived from stratigraphy below the sea's surface. Paleozoic rocks are largely absent from the oceans, except beneath certain continental platforms; conventional terrestrial geology will continue to provide all of our information about the earlier evolution of the earth. In this book, we are concerned only with stratigraphic sequences beneath the ocean surface. This information is gained by drilling and other coring, supplemented by direct observations from submersibles. Large areas of the sea floor are suitable for direct studies and sampling, especially the well-exposed, truncated strata at continental margins. However, expensive field expeditions using submersibles are required.

Stratigraphy provides a basis for the understanding of the history of geological and paleobiological events and processes of the earth. The economic geologist, primarily concerned with the discovery of hydrocarbons, also uses the full range of stratigraphic approaches and is required to analyze geological sequences from a historical perspective. The stratigrapher is essentially a historian working within a framework of long periods of time. Geological samples contain primarily two kinds of information necessary for the analysis of the earth's history: age (relative and absolute) and paleoenvironmental (including physical and biological) history. For well over a century, there has been a well-established stratigraphic system that has provided the framework to extract information about the history of the earth. In general this involves the determination of **sequence, relative age, correlation,** and **absolute age** of a sedimentary section or historical event. It has usually been necessary to proceed utilizing these steps because obtaining the true ages of rocks is very difficult due to the severe limitations of radiometric dating methods. Nevertheless, the ultimate aim of stratigraphy is to understand earth history, which requires accurate dating of rocks because only an absolute chronology can furnish *rates* of processes. The resulting temporal framework provides a foundation for understanding geological processes and their possible interrelationships. Therefore solid stratigraphic work is essential for paleoenvironmental research. Other than the difficulty of access to the rocks, marine stratigraphy is essentially not different from land-based stratigraphy, which has been practiced for over a century.

In geology we are most interested in that interval of time of the earth's history when life was most abundant and diverse, as reflected in the animal and plant fossils. This time period is called the *Phanerozoic*. This interval of time since the pre-Cambrian (less than 570 Ma*) includes the Paleozoic, Mesozoic, and Cenozoic eras. Marine geologists concentrate on the two youngest eras of the Phanerozoic.

Stratigraphic Subdivision and Nomenclature

The three most widely used elements of stratigraphy are **lithostratigraphy, biostratigraphy,** and **chronostratigraphy** (Table 3–1). Lithostratigraphy is con-

---

*The abbreviation Ma (megannum) refers to units of yr × 10⁶ measured from the present (A.D. 1950 by international agreement) pastward. It means the same as the cumbersome *millions of years before present* and is a fixed chronology analogous to the calendars tied to historical events. The abbreviation m.y. (million years) is used to express simple duration in units of yr × 10⁶ in any given past interval.

TABLE 3-1 UNITS USED IN STRATIGRAPHIC CLASSIFICATION

Stratigraphic categories	Principal stratigraphic units	Equivalent geochronologic units
Lithostratigraphic	Group formation member bed	
Biostratigraphic	Biozones: Assemblage-zones Range-zones Acme-zones Interval-zones Other kinds of biozones	
Chronostratigraphic	Erathem System Series Stage Chronozone	Era Period Epoch Age Chron

cerned with the classification, description, and lateral tracing of rock units based on their lithologic character. Principles of lithostratigraphy were among the first to be developed in geology and are well understood. Biostratigraphy concerns the organization of strata into units based on their fossil content. Chronostratigraphy is the organization of strata into units based on age relations. Two other major approaches that are also now becoming widely used are **magnetostratigraphy** and **stable-isotopic stratigraphy**. Magnetostratigraphy bases the organization of strata on units defined on paleomagnetic reversal patterns in combination with age. Stable-isotopic stratigraphy is based on the relative abundances of stable isotopes, especially those of oxygen and carbon. Stratigraphic terminology in these fields is rigidly defined in formal stratigraphic codes approved by national and international bodies; the most important of these is the International Stratigraphic Guide, produced under the aegis of Hollis D. Hedberg. A summary of categories and unit terms in stratigraphic correlation is shown in Table 3-1. The purposes of the guide are to promote international agreement on principles of stratigraphic classification and to develop common, internationally acceptable stratigraphic terminologies and rules of stratigraphic procedure to improve effectiveness in stratigraphy. Different types of units have been established to express the variations exhibited within rock strata, but the units are concerned only with different aspects of the same rocks and are not meant to distract from the general unity that exists in stratigraphy.

Lithostratigraphic, biostratigraphic, and similar kinds of stratigraphic units are restricted by the areal extent of the features chosen to distinguish them. Few, if any, of these features are both distinctive and present on a global basis. Stratigraphic correlation is a simple matter when confined to a small area where the rock and fossil sequence is constant, but becomes increasingly uncertain with increasing distance between the sequences being correlated. The traditional solution to this problem, which forms the basis for most stratigraphic codes, is the designation of **type sections,** or **stratotypes,** which serve as reference standards or stratigraphic anchors. Correlations are carried out with these stratotypes using the approach that works best.

Chronostratigraphy and the Time-Rock Concept

The stratotype is central to chronostratigraphic correlation. Because chronostratigraphic units are based on **time** of deposition or formation, a universal property, the stratotype defines the time in a particular unit. Employment of chronostratigraphic units therefore automatically implies global correlation, for which interest has increased during the last decade as a result of deep-sea drilling. However, because the principles of chronostratigraphy were developed in discontinuous, incomplete sections with significant differences in sediment character, considerable disagreement developed as to the best methods for establishing age equivalence, or **synchroneity.** The conventionally used hierarchy of chronostratigraphic and geochronologic terms is shown in Table 3-1. This nomenclature shows that one category (the chronostratigraphic or time-rock category) is characterized by the rocks deposited during a **defined** geologic time unit; the second category is the equivalent time or geochronologic unit itself, which is independent of any rock record. Some stratigraphers believe that parallel sets of time and time-stratigraphic units are unnecessary, and that it is only necessary to work with time units. However both schemes represent the concept of the flow of geologic time, subdivided in an almost arbitrary way by earth scientists into units which carry the names of our present *periods*. The Tertiary period has been further subdivided into five epochs, such as the Eocene and Pliocene epochs. The Cenozoic has also been grouped into two broad subdivisions of time. The older part is known as the **Paleogene** and consists of the Paleocene, Eocene, and Oligocene. The younger part, called the **Neogene,** consists of the Miocene, Pliocene, and Quaternary. These divisions have found increasing use in the literature mainly because the earlier and later parts of the Cenozoic represent natural divisions of the evolutionary development of marine life. The terms *Paleogene* and *Neogene* are extremely valuable and widely used because they represent intermediate, and as yet unnamed, chronostratigraphic categories between a period and an epoch.

These subdivisions of time constitute the **International Time Scale,** or the **Standard Stratigraphic Scale** (Fig. 3-1) used daily by earth scientists. Any stratigrapher will easily understand a colleague who discusses the Cretaceous, the Miocene, or the Aptian, because these are globally based, time-dependent subdivisions. For instance the Cretaceous is the interval of time (an intangible property) between 135 and 65 Ma. When used within a time context, we refer to the *Cretaceous period,* but we speak of the *Cretaceous system* for those rocks laid down during the Cretaceous.

Thus the purpose of modern chronostratigraphy is the adequate definition of

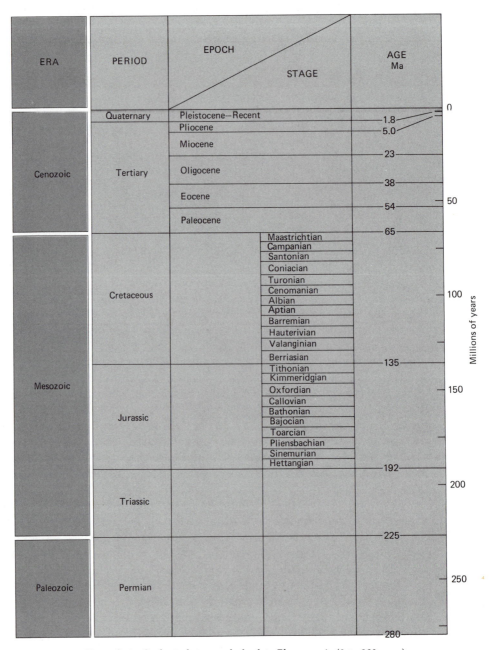

ERA	PERIOD	EPOCH / STAGE		AGE Ma
Cenozoic	Quaternary	Pleistocene—Recent		1.8
		Pliocene		5.0
	Tertiary	Miocene		23
		Oligocene		38
		Eocene		54
		Paleocene		65
Mesozoic	Cretaceous		Maastrichtian	
			Campanian	
			Santonian	
			Coniacian	
			Turonian	
			Cenomanian	
			Albian	
			Aptian	
			Barremian	
			Hauterivian	
			Valanginian	
			Berriasian	135
	Jurassic		Tithonian	
			Kimmeridgian	
			Oxfordian	
			Callovian	
			Bathonian	
			Bajocian	
			Toarcian	
			Pliensbachian	
			Sinemurian	
			Hettangian	192
	Triassic			225
Paleozoic	Permian			280

Figure 3–1 Geological time scale for late Phanerozoic (0 to 280 m.y.)

a standard stratigraphic scale for worldwide application (Fig. 3–1). Its purpose is not to provide a uniform set of correlation units, but an operational time-scale to which all students of the earth's history can compare their various data. Standardized correlation units have been established as a nomenclatural tool for enhancing communication, but their essence is the time framework of the units. The basic chronostratigraphic working unit is the **stage,** or **age,** which represents small intervals of geologic time, and is suited in scope and rank to the needs and purposes of interregional classification (Table 3–1). The stratotypes of the international stages are all located in Europe, where the earliest work was carried out.

Because the original definitions of boundaries between stages were often imprecise, it is sometimes unclear which stage stratotype is to be used to define the boundary. And although numerous stages have been clearly defined, many have partially or completely overlapping time values and are redundant. Firmer placement of boundaries is resulting, however, from more detailed work in the stratotype sections. Most stratigraphers believe that the upper and lower boundaries of each of the stratotypes represent fixed stratigraphic reference points or "golden pegs." These pegs fix the top and bottom of each chronostratigraphic unit in time, and the essence of stratigraphic correlation is to correlate geological sequences or rocks with these time levels. This approach enhances stratigraphic stability and provides an objective basis for correlations.

Correlations with the stratotypes of the Standard Stratigraphic Scale can be difficult, however, because most occur in the Mediterranean, which is isolated from the mainstream of the world ocean. Also, the sequences were laid down in shallow water and, because of biogeographic isolation, generally contain atypical fossil assemblages compared with the open ocean and thus complicate correlation attempts. Fortunately when units such as the Miocene and Pliocene were defined, the boundaries were chosen at levels exhibiting greatest paleontologic change. In these cases the boundaries represent the largest changes in the scheme of natural geological events, which tend to be widespread and even global in extent. These boundaries serve as excellent markers in our attempts to establish the earth's history.

Having established the Standard Stratigraphic Scale as the chronologic basis, it is now important to describe the *methods* used to tie rock sequences into this scale and establish true ages of the units, forming a totality of time. Several methods exist for correlation within lithostratigraphy, biostratigraphy, magnetostratigraphy, and stable-isotope stratigraphy. Ages are established using a wide range of geochronological techniques, which together represent the basic tools used in dating geological events.

Lithostratigraphy

Rock sequences change in physical character both vertically and laterally. This reflects changes in the depositional environment. The purpose of lithostratigraphy is to describe and organize these changes systematically into distinctive units. Lithologic units are differentiated based on certain lithologic types such as limestone, sand, tuff, biogenic ooze or chert, or other distinctive and unifying features (Fig. 3–2). The material may be consolidated or unconsolidated. The units are defined by observed physical features and not by interpretations about their formation. However, their lithologic character normally provides a fingerprint of their development, including sources of rock material and environment of deposition.

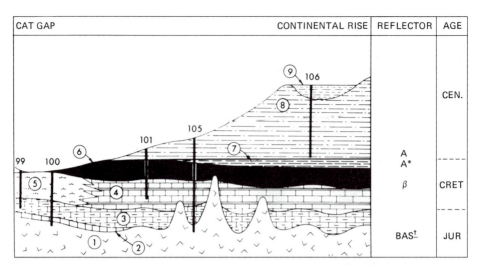

Figure 3-2 Lithologic units and seismic reflector horizons summarized by Lancelot and others [1972] for the North American Basin of the Northwest Atlantic. Location is also shown of DSDP sites. 1. Basalt; 2. Callovian-?–Oxfordian greenish gray limestone; 3. Late Jurassic red clayey limestone; 4. Tithonian-Neocomian white and gray limestone; 5. Tithonian-Neocomian calcareous ooze and chalk; 6. Cretaceous black clay; 7. Late Cretaceous-?–early Tertiary multicolored clay; 8. Tertiary hemipelagic mud; 9. Quaternary terrigenous sand and clay.

The geographic or lateral extent of lithologic units is restricted by the original limits of the environment in which they formed. Boundaries of lithological units are based only on physical characteristics of the sediment or rock and are independent of time relations. Often these boundaries cross time lines (that is, they are **diachronous**), since the lithological units are related to paleoenvironments that change in location regardless of time. However, when lithologic and chronostratigraphic units are superimposed, they complement each other; Lithological units provide information on what happened during geological history, and chronostratigraphic units define the age.

A sediment's environment of deposition controls its **aspect,** or **facies.** Rock strata exhibit differences in physical aspect, or **lithofacies,** and in paleontological aspect, or **biofacies.** The term *facies* is useful in describing lateral gradations that exist between different sediments that reflect differences in environment of deposition, for example, the gradation between sediments deposited in estuaries and those deposited on the continental shelf. In the deep sea, numerous and continous gradations occur in response to lateral and vertical changes in oceanographic conditions, biological productivity, terrigenous input, and other factors. Changes in facies through time result from sedimentation occurring with sea-floor spreading from an oceanic ridge (see Chapter 16). As the sea-floor sedimentary environment changes with increasing depth (distance) down a mid-ocean ridge, the sediments respond by changing laterally and also vertically in sequence because any point on the ocean floor is conveyed with time through different sedimentary environments.

In traditional land-based geology, it has been normal practice to name each lithologic unit formally. The primary unit of lithostratigraphic classification is the **formation.** However, in marine stratigraphic work, there has been little support for naming formations. Units are recognized, and in the description of drilled sequences, are numbered but not named, other than in the simple, informal terminology shown in Fig. 3–2. Many parameters are used for the subdivision and correlation of marine lithological units and include basic sediment types (biogenic, terrigenous, or volcanically derived material), the type and dominance of biogenic components, grain size of terrigenous component, color, degree of lithification, and other factors.

There are several reasons for the absence of deep-sea formation names compared with land sections, fewer sediment types and building components, less structural complexity over larger areas, large distances between the drill sites to provide stratigraphic control, and obvious difficulties to carry out detailed, local mapping. The mapping and differentiation of lithologic units in deep-drilled sites are assisted by various logging techniques down the drill hole. For instance, changes in the response of strata to electrical conductivity and sonic velocity provide continuous logs for a drilled sequence. The value of logging is that it provides information on the physical character of sediments even when the sediments have not been sampled. Incomplete stratigraphic records normally result from conventional rotary drilling, and logging helps to complete the stratigraphic record.

Marine stratigraphic sequences, in the deep parts of the ocean, are often disrupted by intervals of erosion or nondeposition. These intervals of time not represented by sediment are known as **unconformities** (or **hiatuses**). Most unconformities in the ocean result from active periods of bottom current erosion. Periods of nondeposition may occur in areas where bottom currents are active enough to prevent sediment deposition but not of high enough velocity to erode. Unconformities may also result from almost complete dissolution of biogenic sediments with few terrigenous components. Unconformities represent *negative* geological features; that is, they are not represented by any rock sequences containing age or paleoenvironmental information. Marine geologists normally attempt to date unconformities because they often result from important oceanic processes such as intensified intervals of bottom-water activity. Individual unconformities can be traced in sequences over vast areas of the oceans, and understanding their stratigraphic relations has received much attention.

Tephrochronology

Most lithologic units have only limited use in long-range correlations required in most marine geological studies because of restricted geographic extent. One exception is the use of volcanic ash or glass layers to aid in correlation and dating of marine sequences. Because the volcanogenic sediments that form individual ash layers are blasted out of volcanoes during just a few days or weeks and can be traced over distances as great as 4000 km from the source (Fig. 3–3), they have been called **key beds.** Key beds are deposited virtually instantaneously over wide areas and differ from other lithological units because they represent time planes over the extent of their distribution. The study of ash layers in marine sequences, including

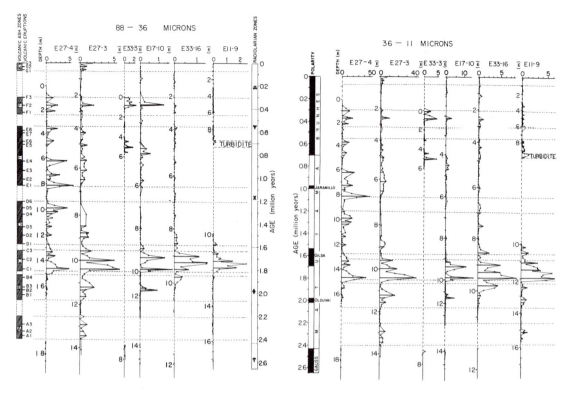

Figure 3-3 Tephrochronology of a suite of piston cores in the Antarctic Ocean (Southern Ocean) between the Ross Sea sector and the Southeast Pacific. Periods of maxima on volcanic-glass accumulation rates (in two size fractions at left and right) are labeled A to G, with subscripts 1 to 8 indicating possible single volcanic eruptions or closely spaced series of eruptions. Accumulation rates are in mg/1000 yr/cm². Ages of cores based on paleomagnetic stratigraphy. (From T. C. Huang and others, *Geol. Soc. Amer. Bull.* Vol. 86, p. 1307, 1975, courtesy The Geological Society of America)

their age, distribution, and geochemistry, is called **marine tephrochronology**. A review of marine tephrochronology is provided by Kennett [1981].

Ash layers are usually lithologically distinctive. The glass in each layer may be distinctive, or possess a geochemical fingerprint, because it represents a single major eruption or series of eruptions over a brief time span, from a single eruptive center with characteristic magma mix. The magma may not necessarily have been uniform, and single tephra layers may exhibit distinct compositions resulting from mixed magma eruptions. This is easily determined by analysis of only a few grains of volcanic glass from the ash layers using the electron microprobe. This technique is extremely valuable in the correlation of individual ash layers.

Ash layers can also provide suitable material for radiometric dating, and these dates establish ages of the rocks and fossils in which the layers are interbedded. Deriving ages is only one of the many valuable contributions of tephrochronology. Others are as follows:

1. Information on individual or series of volcanic explosions. The general size of eruptions, volumes of material produced, and perhaps even the duration can now be estimated within certain limits.

2. Information on the geochemistry of volcanic sources and changes in the geochemical character of a source region through time.

3. Insight into the tectonic history of source regions from tephra sequences. Attempts have been made to relate the history of volcanism to local or regional tectonics. Episodicity of volcanism has been observed in volcanic sequences and volcanic activity may have occurred throughout the Cenozoic over large areas irrespective of local setting (see Chapter 12). The initiation of volcanism is also important for heralding the beginning of active tectonism in a region.

4. The best available sequences for studying temporal relations between paleoclimatic and volcanic explosive history from tephra deposits in the deep sea.

Individual ash layers can often be distinguished using mineralogical composition, including heavy minerals (such as titanomagnetite) and the mix of phenocrysts, in addition to volcanic glass, which is dominant. Plagioclase feldspars are often the principal mineral, followed by hornblende. Geochemical analyses also include the silica content of glass and the trace and minor element composition. Tephrochronologic correlations are also enhanced by quantitative analysis of the grain-size distributions (Fig 3–3), including the fraction finer than sand (< 63 μm). This approach provides more information than distributions of megascopically distinctive ash layers. Many levels of maxima of volcanic-glass accumulation occur in zones that are mixed in with other sediment and are not necessarily identifiable as distinct layers. An example of the enhanced stratigraphic resolution that results from this approach is shown in Fig. 3–3.

The stratigraphic distribution of tephra layers has provided important information about Cenozoic volcanism. For instance, the chronology of volcanism can be difficult to study on land near the volcanic centers themselves. Volcanic deposits close to the source are either terrestrial and unfossiliferous, or sparsely fossiliferous and complex. In shallow-marine sequences, sediments are thick and monotonous and correlations are difficult. In areas of high volcanic activity, the older deposits are buried and unavailable for analysis. Therefore a powerful approach to determine the history of pre-Quaternary explosive volcanicity is studying ash layers of deep-sea sequences close to source regions. The DSDP has made such sequences available, and they now form a basis for the study of Cenozoic explosive volcanicity.

The two most important complicating factors in the interpretation of explosive volcanic history from Tertiary ash deposits are diagenetic alteration and plate motions that effect the long-term global distribution of deep-sea ash deposits. Volcanic glass may convert rapidly to zeolites and clays in deep-sea sequences. Such conversions may have an effect on ash-distribution studies. The distribution of clay resulting from alteration of glass layers is sometimes used to evaluate explosive volcanic history. Frequencies of other volcanic minerals such as plagioclase feldspars, which are more resistant to dissolution than glass, are also used to interpret volcanic history.

The second complicating factor affecting tephra distribution is plate motions. These are inconsequential immediately after any eruption and not important for Quaternary tephrochronology, but can produce considerable displacement of Tertiary tephra deposits. This is a significant problem when dealing with tephra deposits moving toward, or away from, source regions. On plates migrating toward centers of explosive volcanism, ash deposits increase in volume as source areas are approached. Such relations have been demonstrated in sites southeast of Japan [Ninkovitch and Donn, 1975].

Biostratigraphy

The purpose of biostratigraphy is to organize and correlate rocks into units based on a wide range of fossil characteristics. Biostratigraphic units have been used for over 150 years to subdivide and correlate rock strata and provide basic information used by marine geologists for determining rock age and environment of formation. Organisms are particularly sensitive to environmental conditions and may change in response to subtle changes compared to inorganic material such as sediments. Hence fossils provide much detailed paleoenvironmental information once these relationships are determined.

Fossils provide biologists with historical information for better understanding present-day organisms, including their evolution and distribution. However, in general geology, one of the major contributions of micropaleontology is the assignment, through biostratigraphic approaches, of the relative ages of sedimentary strata. Changes in fossil faunas and floras are the essential markers in biostratigraphy (Fig. 3–4). Irreversible changes have been caused by evolution, extinction, and large-scale migration. These may or may not approximate time planes over large areas, but they form the boundaries of the basic biostratigraphic unit, which is called a **biozone** or a **biostratigraphic zone** (Table 3–1). Reversible faunal changes have been caused by temporary changes in the environment and are less useful in biostratigraphy.

Basic biostratigraphic methods were developed on land-based marine and terrestrial sequences before studies began on the oceans. The fossil record of past marine life was fundamental in establishing the geologic time scale. The fossil sequences established on land were the standard references for earlier biochronology, but at present the most useful marine biostratigraphic sequences are those cored in the oceans.

Approaches used in biostratigraphy are basically no different when used on land-based or oceanic sedimentary sequences. However, there are several differences between land-based and oceanic sedimentary sections that have made biostratigraphy of greater relative value to the marine geologist. First, in land-based sections, fossils may form only a minor part of the strata. Even in fossiliferous

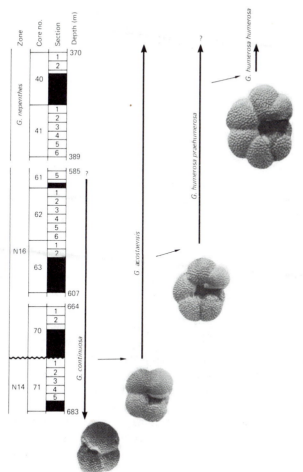

Figure 3-4 Biostratigraphic ranges of four planktonic foraminiferal species (in the genus *Neogloboquadrina*) within an evolutionary lineage through the middle and late Miocene in DSDP Site 397 (Northeast Atlantic). Principal morphologic change in the evolutionary bioseries is the successive upward increase in number of chambers in final whorl. Planktonic foraminiferal zones shown at left. (From Salvatorini and Cita, 1979)

sequences, they are rarely found in every bed or formation. Much land-based biostratigraphic work uses megafossils such as mollusks or brachiopods. In contrast, marine sediments contain rich assemblages of microfossils and, over much of the ocean, the dominant sediment types (oozes) are those made up almost exclusively of microfossils. Most of these are planktonic in character and are widely dispersed over large areas. Biostratigraphy benefits from these rich microfossil deposits. Second, many sections of sedimentary rocks in the oceans have relatively continuous records compared with epicontinental deposits. As a result, organic evolution on land is difficult to follow with the same detail found in many oceanic sections where the continuity of morphologic change, variation, and abundance is often clearly shown through time.

Planktonic microfossils play a major role in the process of correlation and determinations of age equivalence due to their diversity, abundance, rapid evolution, wide distribution, and abundance in cores. The principal groups employed are foraminifera, calcareous nannofossils, radiolaria, diatoms, and silicoflagellates (see

Chapter 16). Although megafossils have been extensively applied to stratigraphic problems of shallow-marine deposits exposed on land, they are of virtually no use in oceanic paleontology because of their rareness in deep-sea samples.

The high potential of planktonic microfossils for their use in biostratigraphy is exemplified by the diversity found in the modern, tropical, open-ocean environment. This environment contains an assemblage of forms with calcareous and siliceous skeletons consisting of about 200 species of radiolaria, 50 species each of calcareous nannofossils and diatoms, 30 species of planktonic foraminifera, and about 15 species of silicoflagellates. Diversity is highest in tropical water in all groups except the diatoms and silicoflagellates and conspicuously decreases toward high latitudes. Reduction in diversity of microfossils leads to fewer biostratigraphic events in the geologic record and thus to a reduction in biostratigraphic resolution.

Biostratigraphic correlation evaluates the similarities of fossil assemblages from different sedimentary sequences and is based on paleontological events arranged in order of superposition. Biostratigraphic resolution is higher when the ancestry, or **phylogenetic sequence,** of species within a lineage is known [Srinivasan and Kennett, 1976].

There are several kinds of biostratigraphic zones, depending on the paleontological feature considered (Fig. 3-5). These may be based on a natural association of fossils, on the range of a fossil taxon or taxa, on frequency and abundance of fossils, on morphological features of fossils, on stages of evolutionary development, or on any other variations related to the fossil content of strata. Biostratigraphic zones, like lithostratigraphic units, are relatively objective products of classifications because they are based on directly observable features in the rock strata. Biostratigraphic zones are based on the appearance or disappearance of the fossil groups used to define their limits, except for **assemblage zones** and **acme zones.** The assemblage zone is a group of strata marked by a distinctive natural assemblage of all forms regardless of their ranges. Such a natural assemblage is the product of environmental conditions, rather than a reflection of any precise stratigraphic position or age. The acme zone is based on the abundance of certain forms, independent of association or range. The **range zone** is a group of strata

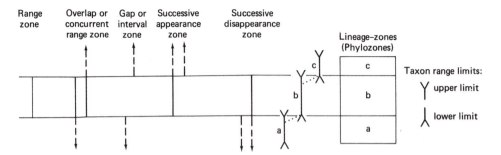

Figure 3-5 Different types of biostratigraphic zones. Vertical lines indicate stratigraphic occurrence of species. Arrow end indicates direction of increase age, either up or down. Lineage zones or phylozones are shown at right and are based on the range zones (a,b, and c) of an evolutionary succession of species.

representing the stratigraphic range of some selected fossil element (Fig. 3–5). Each species exhibits a restricted stratigraphic range. The **overlap zone, or concurrent range zone,** is marked by overlapping ranges of specified fossils. The **interval zone, or gap zone,** is not recognized by the range of any particular element within the interval but only from its position relative to the underlying and overlying zones. Continuous changes in microfossil sequences can also lead to the recognition of **successive appearance zones** and **successive disappearance zones** (Fig. 3–5).

The marine sedimentary record contains much potential for studying continuous evolutionary sequences. In some lineages, morphologic change appears to have occurred continuously, not stepwise, and consequently the dividing line between species has been placed almost arbitrarily [Malmgren and Kennett, 1981]. These divisions form **lineage zones** (Fig. 3–5) representing a segment of an evolutionary trend. Theoretically it should be possible to subdivide these lineages into a greater number of short lineage zones, provided that the gradational morphological changes are described precisely. Most modern biostratigraphic zonations are based on first and last appearances of fossil groups and employ combinations of concurrent range zones, range zones, successive appearance zones, and successive disappearance zones. The extent of geographic use of a zone is based on the distribution of the taxa used to define it.

The resolution and geographic range of biostratigraphic correlations is limited by the inadequacies of the fossil record and numerous other factors. The most useful zonal schemes are often based on relatively conservative elements in the assemblages, those elements which are common, easily identifiable, and dissolution-resistant. Resolution can be increased by more precise definition of the morphologic limits of taxa and greater subdivision of evolutionary lineages. Complicating factors are largely related to the local environmental conditions where the organisms lived, as well as reworking, dissolution, and alteration by burial. The accuracy of correlation is controlled by the geographic range of the taxa and the morphologic consistency of the taxa and of its stratigraphic range over its geographic extent.

There are two important factors to be emphasized in using planktonic species in biostratigraphic correlation. First, all planktonic species exhibit restricted geographic ranges, so zonal schemes are also restricted in regional extent. No species distribution is so widespread as to have occurred in all oceanic areas at one time [Srinivasan and Kennett, 1981]. Second, a species does not normally appear or disappear everywhere in the oceans at the same instant of time. Most exhibit diachronous ranges throughout their geographic extent. Planktonic organisms take time to adapt to different water masses when extending their geographic ranges. The appearance or disappearance of certain species is also time-transgressive in response to changing oceanographic conditions.

In order to test the integrity of any zonal scheme, it should be correlated to some reliable chronostratigraphic framework. The most reliable framework available is the record of paleomagnetic reversals and of oxygen isotopic change. Without these, a zonal scheme for another microfossil group should be used that has been correlated to the paleomagnetic reversal record or oxygen isotopic stratigraphy. Conservative species persist for 20 m.y. or more; most species have a range of about 5 m.y., and rapidly evolving forms may have a range of 1 m.y. or less. Rates of evolution in assemblages are not constant through time; planktonic foraminifera seem to have varied in their rate of evolution by a factor of 10 through the Cenozoic.

In recent years the zonal concept in biostratigraphy has been supplemented by the use of **datum levels.** Datum levels are based mostly on abrupt evolutionary first appearances (**FAD, first appearance datum,** or **LAD, last appearance datum**) of microfossils and are the most widespread and distinctive *events* in paleontologic history. The value of datum levels was well illustrated for the first time by Hornibrook and Edwards [1971] when they applied it to a sequence of integrated Cenozoic planktonic foraminifera and calcareous nannofossils in New Zealand. Such events play a central role in correlation over long distances.

Not all microfossil species are equally useful for correlation purposes in practice and can be grouped subjectively into three orders of reliability, as follows:

1. *First order.* Easily identifiable, common, and persistent; either first rapid evolutionary appearance or highly consistent range.

2. *Second order.* Easily identifiable, fairly consistently present, but not necessarily common; or first gradual evolutionary appearance, fairly consistent range.

3. *Third order.* Consistency of identification difficult, rare; or not well defined ecologically, local geographical restrictions, or known in very few sections; or limits of range uncertain due to sampling gaps or unconformities.

Rapid progress in biostratigraphy over the past two decades has standardized low-latitude zonal schemes for calcareous and siliceous microfossils. Two major low-latitude Cenozoic planktonic foraminiferal zonations are now widely used. The first was established by Bolli [1957] using Caribbean sections on land, the second by Banner and Blow between 1962 and 1965 for circumequatorial areas of the world. The zonation of Banner and Blow [1965] (Fig. 3–6) has found widespread use not only because of its wide applicability in the warmer region of the world, but also because of the shorthand nomenclatural method that numbered zones sequentially. Twenty-two zones exist for the Paleogene, numbered P1 (Paleocene) through P22 (latest Oligocene) and 20 zones for the Neogene, numbered N4 (early Miocene) through N23 (late Quaternary; see Fig. 3–6). The value of a numbered zonal scheme is that any marine geologist, even a nonpaleontologist, can readily grasp the stratigraphic position of a zone when the discussion is not closely linked with superposition. Thus it is easier to remember that zone N19 is of early Pliocene age than it is to recall its paleontological basis, which is the *Sphaeroidinella dehiscens–Globoquadrina altispira* concurrent range zone. Another widely used scheme was also developed by Martini and Worsley [1970] for the calcareous nannofossils of the Neogene, consisting of 21 zones numbered NN1 (early Miocene) to NN21 (late Quaternary), where NN is nannofossil Neogene zone, and for the Paleogene, consisting of 25 NP zones, where NP is nannofossil Paleogene zone. Figure 3–6 shows the sequence of these zones, and their correlation with one another and with the

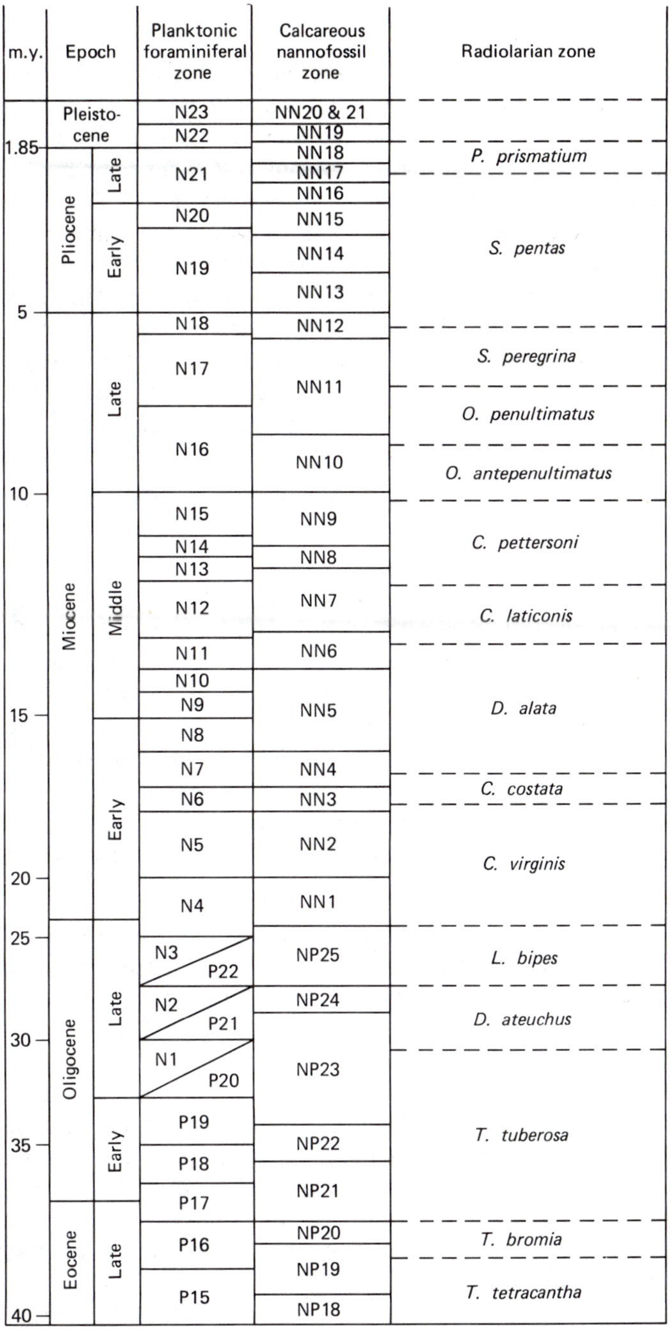

m.y.	Epoch			Planktonic foraminiferal zone	Calcareous nannofossil zone	Radiolarian zone
	Pleistocene			N23	NN20 & 21	
				N22	NN19	
1.85	Pliocene	Late		N21	NN18	P. prismatium
					NN17	
					NN16	
		Early		N20	NN15	S. pentas
				N19	NN14	
					NN13	
5	Miocene	Late		N18	NN12	
				N17	NN11	S. peregrina
						O. penultimatus
				N16	NN10	O. antepenultimatus
10		Middle		N15	NN9	C. pettersoni
				N14	NN8	
				N13		
				N12	NN7	C. laticonis
				N11	NN6	
				N10		
15				N9	NN5	D. alata
				N8		
		Early		N7	NN4	C. costata
				N6	NN3	
				N5	NN2	C. virginis
20				N4	NN1	
25	Oligocene	Late		N3 / P22	NP25	L. bipes
				N2 / P21	NP24	D. ateuchus
30				N1 / P20	NP23	
		Early		P19	NP22	T. tuberosa
35				P18		
				P17	NP21	
	Eocene	Late		P16	NP20	T. bromia
					NP19	T. tetracantha
40				P15	NP18	

Figure 3-6. Planktonic foraminiferal zones, calcareous nannofossil zones, and radiolarian zones for last 43 m.y. plotted against geologic time scale. Foraminiferal zones are largely those of Blow [1969]; calcareous nannofossil zones mostly of Martini and Worsley [1970], and the radiolarian zones are from Riedel and Sanfilippo [1971]. The chronology is after Hardenbol and Berggren [1978] and Berggren and van Couvering [1978].

standard stratigraphic scale. These zonal schemes work well for tropical to warm subtropical areas of the oceans, but are only partially applicable to sediments of the same age in temperate or colder regions. These areas have their own well-developed zonal schemes, such as that of Jenkins [1967] for temperate areas of the Southern Hemisphere. During the last decade the siliceous radiolarians and diatoms have also become increasingly important for Cenozoic biostratigraphy, especially in high-latitude areas, where they are more important than the calcareous microfossil forms, which exhibit low diversity in those regions. The Cretaceous also has well-developed zonal schemes based on a variety of planktonic microfossils, but the early Mesozoic largely is lacking in schemes other than radiolarians, because three major planktonic microfossil groups of the modern ocean, namely the foraminifera, diatoms, and calcareous nannofossils, had not evolved until the late Jurassic to early Cretaceous.

Geochronology

To understand the history of the earth, including the oceans, it is vital to determine the actual age of rocks, sequences, and events. The science of dating rocks is called **geochronology.** Even without geochronology, it has still been possible, using fossils and a wide range of other geological criteria, to place geological events in sequential order. Geochronology has provided a time framework in which to place this sequence and therefore is important for two main reasons.

1. Knowledge of age enables the determination of the rates of geological processes, including biological evolution. This provides a better perspective on the character of the processes.

2. Age data on rocks and sequences provide another dimension to correlation between different rock types, especially over long distances, and help to determine the correct sequence of events.

Geochronology has played an important role in marine geological studies. Most of the methods were initially used on continental and oceanic island rocks before application to oceanographic problems. All the methods of dating are based on radioactive decay. Each radioactive isotope has a specific decay constant, or half-life. Decays with short half-lives may be useful only for dating younger rocks, whereas under suitable conditions, isotope decays with long half-lives can be used over a wide range of ages. Ages determined using radioisotopes are known as **radiometric ages.**

A few other dating methods are also used, such as the counting of **varves,** or annually deposited sediment layers, an approach that is restricted to very young sedimentary deposits. However, even with these, radioactive dating is required to confirm that they are annual. **Concordant ages** are similar ages obtained by more than one dating method. **Discordant ages** are frequently obtained using geochronology and provide important information, such as about the effects of tectonism upon radiometric dates.

A *relative* time scale has been progressively developed and refined since earliest geological studies were carried out. Correlation is conducted throughout the earth using fossil sequences. Calibration of this relative time scale in time units (years) has also been conducted for many years. Between 1955 and 1965, a physical time scale was established; it consisted of the intercalibration of time with various recognizable stratigraphic units throughout the earth. This time scale is used daily by earth scientists. The physical time scale was developed using all available dating methods, but application of the potassium-argon method dominated.

Intercalibration of age with stratigraphic data is carried out on sequences containing material (especially volcanic material) suitable for radiometric dating, interbedded with sediments containing a diagnostic fossil assemblage. This enables the position of the relative time scale to be precisely determined. Such sequences are not common and, as a result, intercalibration is often based on meager stratigraphic information, and ages applied to the relative time scale require continuous modification as new data becomes available. For instance, during the early 1960s, the age of the Miocene-Pliocene boundary was considered to be 9 Ma, but it is now generally considered to be about 5 Ma.

In marine geology, several methods have been used to date both the sedimentary sequences and volcanic rocks found in the oceans. In addition, ages determined for marine sediments on land have also been widely used and intercalibrated with those in the ocean basins. There are eight direct geochronological approaches that have been used successfully to date marine sediments and rocks. Other methods widely used in dating rocks on land are not applicable to marine geological problems. The following table shows six of the methods that are important in directly or indirectly dating marine sediments and rocks. Shown also is the applicable age range for each.

1. Potassium-argon; $> 0.5 \times 10^6$.

2. Magnetostratigraphy; 0–200×10^6.

3. Fission-track; 1–100×10^6.

4. Thorium 230; 2–25×10^4.

5. Carbon 14; 1×10^2–4×10^4.

6. Varve chronology; $<$ approximately 7×10^3.

The range of ages of rocks spanned by these various methods is extremely large. At least one method applies to any particular interval of geological time.

Fundamental conditions which must be met for radiometric dating include the following:

1. The isotopic composition must not have changed by fractionation or other processes, but only by decay of the parent isotope.

2. The decay constants of the isotope must be accurately known.

Potassium-argon isotopic dating method

The potassium-argon (K-Ar) dating method is a widely applied technique to determine isotopic age using potassium-bearing rocks and minerals. Potassium is an abundant element in many important rock-forming minerals such as micas, feldspars, and clay minerals. The method is based on the decay of naturally occurring ^{40}K to stable ^{40}Ar, which has a half-life of about 1250 my. The isotope ^{40}K has a dual decay; about 90 percent decays to produce ^{40}Ca, and the remaining 10 percent decays to produce ^{40}Ar. The K-Ar method is based on argon, a noble gas that is normally lost during the formation of igneous rocks and begins to accumulate from the decay of ^{40}K as new radiogenic argon. The amount of ^{40}Ar present is therefore a measure of the time the isotopic clock has run. The method has been applied to rocks ranging in age from Precambrian through Pleistocene. Summaries of the method include those of Faure [1977] and McDougall [1977]. Because of its wide applicability to time ranges and mineral and rock types, the K-Ar method has been very significant in developing the Mesozoic and Cenozoic time scale. It has had an enormous influence on marine geology and geophysics. By the early 1960s advances in the technique enabled young rocks of Pliocene and Pleistocene age to be dated with rather high precision, and this was crucial in the development of the paleomagnetic time scale.

As with all dating methods, certain assumptions must be met if a K-Ar age measurement is to give a valid estimate of age. Two of the most important follow:

1. At the time of crystallization of the rock, all pre-existing radiogenic argon must have been lost. Any argon retained by the time crystallization has terminated, will increase the measured age relative to the true age. Most extrusive or shallow-intrusive rocks do not retain argon at crystallization.

2. The analyzed rock or minerals must have remained a closed system since crystallization. No loss or gain of Ar or K should have occurred since crystallization except for that related to the radiogenic process.

The latter causes many incorrect K-Ar ages. Argon loss commonly occurs as a result of chemical weathering, or metamorphism, resulting in younger ages. Subaqueous alteration by fluids also creates changes in the potassium content of minerals. Only the freshest mineral and whole-rock samples should be used for K-Ar dating, and altered samples or those that contain devitrified glass, secondary minerals, and xenoliths should be avoided.

Argon retention in terrestrial basaltic glass is generally good, and fresh, tuffaceous deposits interbedded with fossil-bearing sediments have been very useful in stratigraphic studies. Entire rock, or **whole-rock analyses** are commonly employed on rocks dominated by these minerals. Plagioclase and pyroxene retain argon satisfactorily.

Decay constants of the isotopes must be accurately known to provide realistic ages. Much of the Cenozoic and Mesozoic time scale is based on K-Ar data calculated using the decay constants for ^{40}K proposed by Aldrich and Wetherill [1958]. However, later work by Beckinsale and Gale [1969] provided a more accurate set of decay constants that require about a 2.6 percent change in ages. This is often within the precision quoted for many K-Ar ages, although the correction is unidirectional. These values have little effect on age assignments of most geological boundaries but are significant in the geochronology of the late Cenozoic (less than 5 Ma) where greater detail is possible.

Potassium-argon dating was developed and used on many land-based rocks and sediments. Extension of the method to **direct** dating of rocks and sediments in the ocean has been disappointing for two reasons: First, basalts that are extruded on the ocean floor under high hydrostatic pressure contain excess ^{40}Ar that is concentrated in the glassy crusts of individual "pillows," apparently because argon in solution in the lava is trapped there by rapid quenching. Excess argon decreases inward from the glassy rind of the pillows because a slower rate of cooling allows the argon to migrate toward the rind. Excess argon in basaltic glass increases with water depth, suggesting that it is controlled primarily by hydrostatic pressure. Second, widespread alteration of oceanic rocks occurs because of extensive seawater circulation through these rocks. This increases the amount of potassium. Thus, K-Ar ages cannot usually be obtained from oceanic crustal rocks, and the method has little direct influence on our understanding of the ocean crust. Instead, ages are assigned to ocean crust by indirect approaches including the following:

1. Mapping of magnetic anomalies, where ages are based on assumed rates of sea-floor spreading, and correlation of the magnetic anomalies at sea with dated paleomagnetic reversals on land sections. These ages, in turn, have been obtained by direct application of the K-Ar method.

2. Micropaleontological defined ages of sediment immediately overlying oceanic crust.

Magnetostratigraphy

Magnetostratigraphy is one of the latest stratigraphic approaches developed to aid historical geology and stratigraphy and has been very important in marine geological problems. During sediment deposition or cooling of molten rock, magnetic iron oxide minerals align themselves with whatever magnetic field exists. This preferred alignment of grains, the remanent magnetism, acts as a geomagnetic field recorder. Magnetic stratigraphy, or **magnetostratigraphy,** is based on geologically frequent reversals of the earth's north and south magnetic poles. For instance, our present polarity (which is called **normal polarity**) began about 7×10^5 years ago and for about 1 m.y. before that, the poles were reversed (**reversed polarity**) except for several short episodes of normal polarity. In 1906 the French physicist B. Brunhes found that some old volcanic rocks were magnetized in a direction exactly opposite to the present magnetic field. Further work has shown that the earth's magnetic field has two stable states: either toward the North Pole, as it does today, or toward the South Pole. It has repeatedly alternated between the two.

In 1963 it was still debated whether the earth's magnetic field had reversed at all and the techniques for dating relatively young rocks were inadequate to state when the earth's magnetic field reversed. This changed when refinements in the

K-Ar dating technique made it possible to accurately date lava flows extruded on land.

Development of the polarity time scale   Magnetostratigraphy was developed as a new discipline in the early 1960s by a small group of scientists working in northern California (A. Cox, R. Doell, B. Dalrymple) and in Australia (I. McDougall, D. Tarling and F. Chamalaun). In only 5 years this group of workers developed a **magnetic polarity sequence** as a result of their combined and competitive research on the paleomagnetism and chronology of young sequences of volcanic rocks. This development has been summarized by Cox [1973], Watkins [1972], and McDougall [1977]. Magnetostratigraphy was then extended to deep-sea sedimentary sequences, and this has had a major impact on marine geology. Development of the polarity time scale (Fig. 3–7) has relied on the application of two different techniques: determination of the remanent magnetism of rocks to establish paleomagnetic polarity and radiometric dating using the K-Ar method. This approach is stratigraphically powerful for two reasons.

1. Magnetic reversals are synchronous, worldwide phenomena and are thus different from most other diachronous stratigraphic criteria used in correlation. Other synchronous events, such as wind-deposited volcanic ash layers also represent powerful correlation tools, but are too geographically restricted for widescale correlations.

2. Magnetostratigraphy covering at least the last 6 m.y. was established using radiometric dating of terrestrial lava sequences. It has provided stratigraphers with an accurate chronological framework. The magnetostratigraphy provides a source of dates to establish age relationships of sediment and fossil sequences. The chronology, in turn, has provided the basis to accurately determine rates of geologic change. This approach continues to be applied to the entire Cenozoic.

Magnetostratigraphy organizes strata according to their magnetic properties acquired at the time of deposition. Magnetostratigraphy of sediments differs from the chronologic techniques which provided the widely used polarity time-scale [Cox, 1969] (Fig. 3–7). The latter was constructed by organizing radiometrically dated samples of known polarity according to their isotopic age, often without knowledge of stratigraphic relations between individual samples or groups of samples. Magnetostratigraphy of sediments, in comparison, is employed in the absence of any direct radiometric data on the examined sedimentary sections. Because of the repetitive nature of reversals, identification of particular polarity events within incomplete sequences is possible only through comparisons with other stratigraphic or radiometric data. Such independent criteria are usually well-established biostratigraphic schemes.

Nomenclature   Magnetostratigraphy, like any branch of stratigraphy, has required a clear set of definitions to reduce potential ambiguities that arise from studies carried out by many investigators. There now exists a well-established

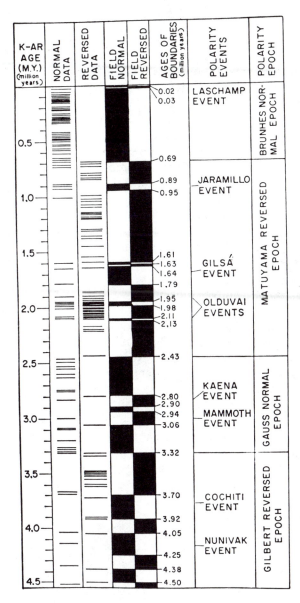

Figure 3-7 Principal components of the geomagnetic polarity scale for the last 4.5 m.y. Normal polarity is shown in black; reversed polarity in white. Since this figure was published by Cox [1969], the geomagnetic polarity scale has continued to be updated as new events are recognized and the ages of polarity boundaries adjusted as a result of new data. (From Allan Cox, *Science*, vol. 163, 1969, pp. 237–245; copyright © 1969 by the American Association for the Advancement of Science)

nomenclatural convention in magnetic polarity scale studies. Periods of one polarity of the order of 1 m.y. are called *epochs* and those lasting about 50,000 to 100,000 years are termed *events*. The four epochs recognized for the Pleistocene and Pliocene (**Brunhes, Matuyama, Gauss,** and **Gilbert;** Fig. 3–7) are named after past workers in terrestrial magnetism, while paleomagnetic events have been named after the type locations where they were first recognized. Older paleomagnetic epochs have been numbered. This is similar to the numerical system developed by the Lamont group during the 1960s for sea-floor magnetic anomalies, but the two systems do not coincide.

By the early 1970s it had become apparent that inevitable inconsistencies and nomenclatural conflicts in magnetostratigraphy required resolution. Groups were formed to examine nomenclatural problems and the Subcommission on Stratigraphic Classification made recommendations published as a supplement to the *International Stratigraphic Guide.** The most significant change in terminology adopted in this document is the replacement of the word *epoch,* as in Brunhes epoch, with the word *chron,* as in Brunhes chron. Furthermore, the word *event,* as in Jaramillo event, is replaced by the word *subchron,* as in Jaramillo subchron. These are chronological units. The chronostratigraphic equivalents for these terms are *chronozone* for chron and *subchronozone* for subchron. The recommended terminology for magnetostratigraphic polarity units and their geochronologic and chronostratigraphic equivalents is summarized in Table 3–2. These new terms are not used in this book because they only began to be used in the literature in 1980 [Liddicoat and others, 1980].

Understanding the nature of the polarity time scale has continued to evolve slowly, and the latest versions are not widely accepted although changes in the ^{40}K decay constants have required recalculation of the age of the boundaries, and later versions of the polarity scale are **required** to be adopted.

Magnetostratigraphy in deep-sea sediments   Deep-sea sediments are multichannel recorders of the earth's history particularly those related to ocean paleoenvironmental biotic evolution. By the middle 1960s, studies of deep-sea cores were rapidly accelerating. Although the sequence of events was often arranged using stratigraphic superposition, chronology of events was still poorly known and correlation between high- and low-latitude sequences was still a new idea. With this background, the potential for applying the polarity time scale to stratigraphy was rapidly perceived by a few workers, and magnetostratigraphy was attempted on deep-sea sediments. Iron detrital minerals are dispersed in most deep-sea sediments which are thus capable of recording a magnetic signature, such as those in volcanic rocks. A thorough review of the paleomagnetism of deep-sea sediments until 1971 has been provided by Opdyke [1972].

TABLE 3–2

Magnetostratigraphic polarity units	Geochronologic equivalent	Chronostratigraphic equivalent
Polarity superzone	chron (or superchron)	chronozone (or superchronozone)
Polarity zone	chron	chronozone
Polarity subzone	chron (or subchron)	chronozone (or subchronozone)

*"Magnetostratigraphic Polarity Units," a supplementary chapter of the ISSC *International Stratigraphic Guide,* was published by the IUGS International Subcommission on Stratigraphic Classification and the IUGS/IAGA Subcommission on a Magnetic Polarity Time Scale. See *Geology* 7: 578–583.

The first workers to recognize paleomagnetic reversals in deep-sea sediments were C. Harrison, a geophysicist, and B. Funnell, a paleontologist, fellow students at Cambridge University, England, who collaborated in a joint study at Scripps Institution of Oceanography, California, in 1964. They recognized that deep-sea sediments possess a relatively stable natural remanent magnetism and that cores could provide an excellent opportunity for the study of rather complete sequences laid down over several million years. These initial pioneering efforts were followed by more intensive efforts by other groups, especially the one led by N. Opdyke at Lamont Observatory. This group decided to study the paleomagnetism of several deep-sea cores from the Antarctic Ocean because the cores were long (about 5–12 m) and contained good potential for a relatively long paleomagnetic record. Also, these cores had a radiolarian biostratigraphy established by J. Hays that allowed independent verification of a magnetostratigraphy and of intercore correlations. Of primary importance in the selection of these cores, however, was their geographic location at high latitudes, where the inclination of the magnetic field is high. Hence a magnetic vector pointing upward indicates an opposite polarity to the magnetic vector pointing downward in the same core. It is therefore necessary only to know which part of the core being measured is at the top and which is down. In equatorial areas the inclination of the magnetic field approaches zero and was more difficult to interpret in unoriented cores during the early days of less-sensitive instrumentation.

The collaborative research at Lamont paid off and presented the first example of detailed correlations between several deep-sea cores up to 4 m.y. old using magnetostratigraphy. This work also enabled detailed dating of each of the cores, including a previously established radiolarian zonation. Intensified work resulted in magnetostratigraphy in all oceans of the world, and this began to play a role in the reinforcement of the polarity time-scale as established on terrestrial volcanic sequences. The magnetic stratigraphy of deep-sea sediments and shallow marine and terrestrial sequences on land is a powerful tool when applied to stratigraphic and historical geological problems. Results of these applications to marine geology include the following:

1. Dating of ash layers and correlation refinement. This provides a firm chronology for the major explosive volcanic activity in various regions.

2. Dating of biostratigraphic schemes based on a number of microfossil groups and determination of isochroneity, or diachronism, of microfossil events, such as appearances or extinctions, especially between different water masses. Using fossils to correlate over long distances has been difficult even when planktonic groups are used. This is because different ocean water masses are marked by distinct planktonic assemblages. Also, species that range over wide areas often have changing stratigraphic ranges. Because of these longstanding problems, paleontologists and stratigraphers were enthusiastic about magnetostratigraphy as a new correlation tool especially between the tropics and polar regions. Figure 3–8 shows the paleomagnetic correlation of cores from the Arctic, Pacific, Indian, and Atlantic oceans.

Berggren and others [1967] were among the first to attempt magnetostratigraphic correlation between high- and low-latitude planktonic foraminiferal zonations on the boundary between the Pliocene and Pleistocene. Since then, much progress has been made on the dating and correlation of the Plio-Pleistocene boundary, as shown in Fig. 3–8, where radiolarian stratigraphic ranges in different regions and different lithologies were correlated with both paleomagnetic stratigraphy and foraminiferal zones.

3. Dating of sedimentary history associated with ice-rafting and biogenic productivity.

4. Rapid age-mapping of sedimentary deposits over large areas of the ocean floor and development of isopach maps based on isochrons provided by the paleomagnetic reversals.

5. Changing concepts of geological processes. While developing a firm chronological framework for late Cenozoic marine sequences, it was discovered that fossils previously used for the differentiation of the Miocene-Pliocene boundary were only about 5 m.y. old and that the changes occurred near the base of the Gilbert epoch. Until magnetostratigraphy, radiometric evidence had suggested a boundary of 9 m.y. ago. The Pliocene epoch was suddenly discovered to be only about half as long as previously believed. Magnetostratigraphy substantially transformed our concepts of the rates of late Cenozoic geological processes.

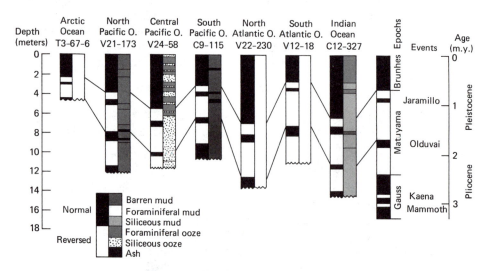

Figure 3–8 Paleomagnetic correlation of cores from the Arctic, Pacific, Indian, and Atlantic oceans, all of which have different lithologies and fossil assemblages. (From N. D. Opdyke, *Rev. Geophysics and Space Physics*, Vol. 10, pp. 213–219, 1972, copyrighted by the American Geophysical Union)

Pre-Pliocene extension of polarity time scale Throughout the Cenozoic and late Cretaceous, paleomagnetic reversals occurred frequently. During the Mesozoic and Paleozoic, reversal frequency was much more variable, and at times there were periods of constant polarity. The polarity time-scale has been extended to sequences older than the last 6 m.y. by both direct and indirect approaches.

1. Direct measurements in older volcanic and sedimentary sequences. Decreasing resolution of the K-Ar dating method as rocks become older than about 6 m.y. has made it difficult to extend the polarity scale. Some success has resulted using areas where thick exposed sequences of lavas represent significant amounts of time and where stratigraphic relationships between successive lavas are clear. Careful work in Iceland has resulted in the polarity scale extension to almost 14 m.y. (middle Miocene). Because of the paucity of sections of sufficient age and stratigraphic resolution exposed on land, further extension using this approach is unlikely.

2. Indirect use of the sea-floor magnetic anomaly pattern. This forms a critical component of dating methods for marine rocks and sediments and is discussed in Chapter 4 in relation to sea-floor spreading.

Extension of magnetostratigraphic studies to cores older than 6 m.y. has been progressing more slowly. Piston cores older than 6 m.y. are uncommon and the long sedimentary sequences provided by deep-sea rotary drilling are usually too disturbed to provide paleomagnetic data. Two different approaches have been used; both require studying piston cores. Piston cores usually represent limited amounts of time because they are relatively short, but long piston cores marked by low sedimentation rates have solved part of this problem. Such sequences often extend well into the Miocene. Using this approach, magnetostratigraphy has been extended in deep-sea sediments to about 12 Ma.

The second approach using piston cores combines biostratigraphic datums and magnetostratigraphy in cores of overlapping age (Fig. 3–9). Paleomagnetic intervals are assigned by correlation with the polarity-scale based on sea-floor anomaly patterns as described in Chapter 4. The most valuable approach is the use of the hydraulic piston corer, which can provide long, undisturbed sequences as old as the Paleogene.

Magnetostratigraphy was received with a great deal of excitement in the middle 1960s because of the limitations of other methods and the need for a practical way to extend the chronology to sequences older than the Quaternary. Some workers initially regarded it as a panacea to late Cenozoic stratigraphic problems. However like all approaches, magnetostratigraphy possesses characteristic limitations, such as very low magnetic intensity in some sediments, magnetic overprinting of younger polarity signals upon older signals, difficult interpretations of polarity records due to missing sections and hiatuses, bioturbation blurring the record, and errors in technique and polarity recording lags resulting from delays between original deposition and final consolidation of sediments. Also, because the paleomagnetic polarity record consists of a large set of reversals, with each reversal

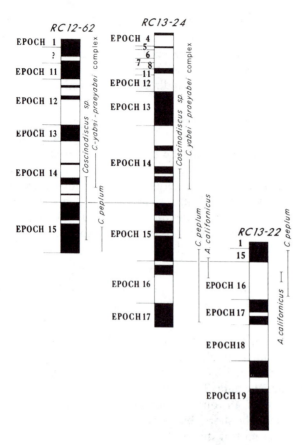

Figure 3-9 Geomagnetic polarity correlations in a set of overlapping piston cores assisted by the ranges of selected diatom species. (From N. D. Opdyke, L. H. Burckle, and A. Todd, *Earth and Planetary Science Letters*, vol. 22, pp. 300–306, 1974)

being similar to another, independent evidence is required of stratigraphic position. However despite these limitations, magnetostratigraphy is a powerful tool and will continue to be so in the future.

Fission-track method

Fission-track dating is a widely used radiometric method different from others because it measures the physical effect of isotopic decay rather than ratios of parent isotopes to daughter isotopes or the product of decay. The method is normally employed as an alternative to K-Ar dating of tephra. Charged particles resulting from the spontaneous fission of radioisotopes leave a trail of damage when traveling through a solid. These tracks, which are up to 10 μm in length, are due mainly to the fission of ^{238}U atoms and are visible at high magnifications. The number of tracks per unit area is a function of the age of the specimen and its uranium concentration. The method has been successfully employed in the dating of micas, tektites, and glass and is useful for young rocks or sediments where tracks have not been reheated, healing the damage [Macdougall, 1971]. This method has been used on many volcanic ash layers, which has assisted in establishing the chronology of interbedded, fossiliferous sediments. Compared with other radiometric methods, fission-track dating is based on a very small number of decays that can be identified even in small samples. The uranium concentration has to be normalized because it also contributes to the number of fission tracks. Concentrations are determined after irradiation in a nuclear reactor using thermal neutrons by counting tracks created by fission of ^{235}U.

Thorium-230 method

One of the most useful radioactive isotopic methods for determining deep-sea sedimentation rates is the **Thorium-230 (^{230}Th) method.** This method is based on the decay of ^{230}Th from ^{238}U, directly from ^{234}U. Although uranium is highly soluble in sea water, thorium is rapidly removed from seawater by adsorption of sediments or incorporation into certain authigenic minerals. After separation from ^{234}U, ^{230}Th decays at a rate determined by its half-life (7.52×10^4 years) to ^{226}Ra and progressively down the decay chain to ^{206}Pb. Thus the concentration of ^{230}Th in deep-sea sediments should decrease exponentially with increasing depth in cores if the rate of accumulation of sediment and ^{230}Th were constant. If it is assumed that both the rates of sedimentation and of the precipitation of ^{230}Th have been constant, the age of the sediments in the sequence can be determined from the concentration of ^{230}Th relative to that in the surface sediments. The method depends on relative amounts of ^{230}Th remaining in core samples. The following assumptions must be satisfied for the ^{230}Th method to be of value in geochronology:

1. Any supported ^{230}Th present in the mineral phases of the sediment is accounted for during analysis.

2. The thorium does not migrate in the sediment, nor does ^{238}U migrate downward with seawater through the sediment (often a problem), thus supporting the decay of ^{230}Th.

An age is determined by establishing the ratio of ^{230}Th to ^{232}Th. Because of the relatively short half-life of ^{230}Th, the maximum determinable age is about 300,000 years. Ages obtained by this method have been useful, though, because they transcend a gap between maximum ^{14}C and minimum K-Ar ages, especially in age determinations of middle Quaternary paleoclimatic cycles.

Carbon-14 method

The *carbon-14 method* has been used extensively for dating young deep-sea sediments. Carbon-14 is a radioisotope with a half-life of 5730 years so the method is limited to the dating of sediments younger than 40,000 years, but may be extended by a few tens of thousands of years using special enrichment techniques. This method of dating was developed about 1946 by W. F. Libby; he later received a Nobel Prize for it. Carbon-14 is produced in the atmosphere by interreactions of neutrons produced by cosmic rays with ^{14}N. The ^{14}C atoms rapidly mix with the other carbon isotopes throughout the atmosphere and hydrosphere and reach a steady-state concentration. The ^{14}C atoms are incorporated into carbon dioxide molecules in proportion to this equilibrium concentration. The carbon dioxide containing ^{14}C is incorporated into plant tissue by photosynthesis and eventually into the tissues of plant-eating animals. Upon their death, incorporation of ^{14}C ceases and the ^{14}C activity declines by radioactive decay. This activity, in relation to the total amount of carbon present, is used to determine the amount of elapsed time since death and is called the *carbon-14 date*.

A major assumption of the ^{14}C dating method is that ^{14}C activity has been constant during the last 60,000 years. Detailed work has shown that past systematic variations of the radiocarbon content have produced inaccuracies in radiocarbon dates (± 10 percent). For instance, if the initial activity (and thus radiocarbon content) of a sample was less than that of the nineteenth century, the resulting radiocarbon age will be too old. However, corrections have been made for this temporal variation in ^{14}C for the last 8000 years by analysis of wood samples (especially bristle-cone pine) whose age is established by counting tree rings (dendrochronology) and by the chronology of varves that contain sufficient amounts of organic debris for radiocarbon analyses.

The ^{14}C method was first used to date marine sediments by Arrhenius, Kjellberg, and Libby in 1951, based on an *Albatross* expedition core from the Pacific Ocean. In 1958, Broecker, Turekian, and Heezen did the first ^{14}C rate study in deep-sea cores based on multiple data. The ^{14}C method has been important in providing accurate ages for climatic changes about 10,000 years ago when the earth underwent its latest change from glacial to the interglacial conditions typical of the present day. This boundary is the **Holocene-Pleistocene boundary,** marked in most deep-sea cores by changes in planktonic microfossil assemblages. A postglacial rise in sea level occurred at the boundary and has been dated, in shallow-water sequences, using ^{14}C. Carbon-14 dating of marine sediments is based mainly on the analysis of calcium carbonate microfossils. Recently, atomic accelerators have been employed as mass spectrometers for ^{14}C dating. The importance of this approach is that it enables an age assignment using smaller amounts of carbon, which is often important in geochronology because of the small size of samples available for analyses.

Varve chronology

Marine varves are the tree rings of the oceans; they form as a result of climatically induced seasonal changes in clastic sediment deposition or seasonally cyclic biogenic productivity. Examples are diatomaceous varves in the Guyana Basin, Gulf of California, and clastic varves in the Santa Barbara Basin, California, resulting from increased sedimentation during the winter associated with continental precipitation. Varves and other laminated sequences are preserved in environments marked by little or no benthonic sedimentary burrowing activity. These environments are invariably anaerobic (depleted in oxygen).

Varve counts are valuable to chronological studies because they provide a direct measure of elapsed years. Individual varves may be very thin (less than 1 mm) and difficult to differentiate. In such cases, X-ray photographs are of great assistance. Independent radiometric datings, such as ^{14}C, are usually intercalibrated with varve counts. In the Santa Barbara Basin, California, varve counts have demonstrated that ^{14}C ages are consistently too old by about 2000 years, probably due to contamination of continentally derived organic debris. Varve sequences provide unique opportunities to conduct climatic and biostratigraphic work on a detailed basis. A good example is the detailed paleoclimatic history for the last 7000 years established for the Santa Barbara Basin, California, by N. Pisias [1978].

Oxygen Isotopic Stratigraphy

Oxygen isotopic stratigraphy has become a major tool for correlating marine sediments (Fig. 3-10). This approach is now one of the most common and powerful tools in correlating Quaternary marine sequences. Attention is now being focused on the Tertiary, as the technique is becoming potentially useful for this time period. Oxygen isotopic changes occur in sequences of carbonate marine microfossils and are synchronous throughout the oceans. Changes in the oxygen isotopic record can

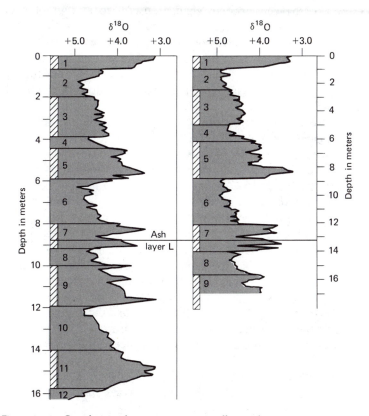

Figure 3-10 Correlation of oxygen isotopic oscillations between two piston cores from the eastern equatorial Pacific. The depth in the cores is scaled according to ash layer L in both cores. Isotopic stages 1 through 12 are indicated. (After D. Ninkovitch and N. Shackleton, 1975)

be precisely matched and correlated between the Indian, Pacific, and Atlantic oceans. Other than oxygen isotopes and paleomagnetic reversals, few global phenomena change in unison over such wide areas.

The study of marine oxygen isotopic stratigraphy was pioneered by C. Emiliani in several classic papers on foraminifera in deep-sea cores beginning in 1954. Emiliani had been a student at the University of Chicago under H. Urey, who in 1947 had determined that oxygen isotopes fractionate at different temperatures during precipitation and evaporation of water. Isotopes of oxygen and several other elements are incorporated into invertebrate skeletons of marine organisms. Atoms of different isotopes have different numbers of neutrons in their nuclei and hence different atomic weights. These differing weights can produce fractionation, or differential concentration, of isotopes. Oxygen isotopes are *stable;* in other words they are not subject to radioactive decay. The common oxygen isotope ^{16}O occurs in association with the rare, heavy isotope ^{18}O. Although these two isotopes have identical chemical properties, the lighter molecules (^{16}O) exhibit higher vapor pressure. Thus during evaporation, higher concentrations of the lighter isotope (^{16}O) are fractionated in vapor, and seawater becomes enriched in the heavier isotope. This process makes oxygen isotopic changes useful to stratigraphic studies of the late Cenozoic when large ice sheets occurred in the polar regions of the earth and fluctuated in volume. Ice sheets lock up huge amounts of freshwater and prevent normal recirculation of this water back into the oceans. Freshwater is enriched in ^{16}O, and increased growth of ice caps draws more water from the oceans, leading to a drop in sea level and relatively higher concentrations of ^{18}O (the heavier isotope) in the remaining seawater. Thus marine organisms, such as foraminifera, which precipitate calcium carbonate ($CaCO_3$), record the isotopic changes of seawater. These changes are measured as $^{18}O/^{16}O$ ratios using a sensitive mass spectrometer. The isotopic composition of seawater becomes uniform during the 1000-year mixing time of the oceans, a short interval in geologic time. Thus oxygen isotopic changes are recorded instantly by organisms living in the oceans and provide the isochronous levels (Fig. 3–10). However, there are several important limitations to the method. First, several organisms have been found to fractionate oxygen isotopes biologically during metabolism and secrete $CaCO_3$. These perturbations are called *vital effects.* The amount of fractionation is not always consistent within a group or even within the same genus. Second, local differences in salinity or precipitation may create large oxygen isotopic differences. Highly saline marginal marine waters have high evaporation rates and are enriched in ^{18}O. Usually, salinity and ^{18}O concentration vary directly. Third, and most important, is the isotopic fractionation due to preferred growth temperature of the organism. As temperature decreases, there is an increase in the ^{18}O concentration in relation to the ^{16}O concentration in the fossil shell.

Shackleton [1967] used the following equation, modified after Craig [1957], to estimate paleotemperatures from the oxygen isotope data.

$$T(°C) = 16.9 - 4.38(\delta_c - \delta_w) + 0.10(\delta_c - \delta_w)^2 \qquad (1)$$

This equation expresses isotopic equilibrium between water (*w*) and calcite (*c*). In this equation, the δ notation defines the deviation in per mil or parts per thousand (‰) of the isotopic ratio of the oxygen of the sample analyzed from that of an arbitrary standard.

$$\delta^{18}O = \left[\frac{(^{18}O/^{16}O) \text{ sample} - (^{18}O/^{16}O) \text{ standard}}{(^{18}O/^{16}O) \text{ standard}}\right] \times 1000$$

The most common standard in oxygen isotopic studies is the PDB (PeeDee belemnite) standard from a Cretaceous belemnite from the PeeDee formation of South Carolina. A foraminiferal test that has a δ value of 2 per mil to PDB means that the carbon dioxide (CO_2) derived from that test is 2 parts per mil enriched in relation to the CO_2 derived from PDB.

When Emiliani pioneered the use of oxygen isotopes for studies of deep-sea cores, he recognized that the isotopic signal is due to two main features: growth temperature and isotopic composition of seawater. Observed increases in $^{18}O/^{16}O$ ratio in planktonic foraminifera during times of global glaciation (polar-ice buildup) reflect both an increase in the $^{18}O/^{16}O$ ratio of seawater as discussed above and an additional enrichment in the ^{18}O content of the shell due to temperature decrease (see Chapter 17). Emiliani argued that about two-thirds of the signal was due to temperature change and one-third was due to change in the $^{18}O/^{16}O$ ratio of seawater. After much debate, it is now accepted that the dominant signal (about two-thirds) is due to change in seawater composition (Fig. 3–10). Most workers now agree that variations in isotopic compositions of Quaternary foraminifera reflect climatic events, but the emphasis has shifted from using oxygen isotopic ratios for direct quantitative temperature variations to using them as a "continental ice-volume" signal.

Paleotemperatures from the isotopic composition of fossils can be determined only if the difference between the isotopic composition of the fossil's ancient seawater environment and average seawater today is known. This is determined by measuring the magnitude of ice effect, a two-part problem: (1) What is the change in oxygen isotopic composition of the oceans from a Pleistocene full-interglacial regime to a full-glacial regime? (2) What would the oxygen isotopic composition of the oceans be if all the existing continental ice melted into the ocean?

To correct for changing isotopic composition during each Quaternary glacial-interglacial cycle, Emiliani originally estimated that seawater composition changed by 0.4‰. This correction was criticized by Olausson [1965], who estimated the isotopic composition of Pleistocene polar ice sheets and evaluated the effect at about 1.1‰. The isotopic composition of seawater is closely reflected in changes in the isotopic composition of benthonic foraminifera from Quaternary deep-sea sediments because the temperature of bottom water is held constant by ice in the polar regions where bottom water forms. Shackleton realized this in 1967 and also noted that because the record of ocean isotopic composition through the Quaternary can be used in interocean correlation, it is more valuable than the temperature record. Most researchers today agree that the ocean was slightly more than 1‰ isotopically positive at the last glacial maximum than it is now.

A comparison done by Shackleton and Opdyke [1973] strongly supported the theory that most oxygen isotopic variation in deep-sea sediments is due to isotopic composition changes. They compared the oxygen isotopic records of planktonic and benthonic forms showing similar amounts of fluctuation and timing (Fig. 3–11). These results have often been misinterpreted to indicate that the Quaternary planktonic foraminiferal isotopic records reflect only changes in ice volume. This is incorrect because temperature fractionation of oxygen isotopes remains a significant factor in surface waters of the ocean, which fluctuate widely in temperature compared to bottom waters. However, isotopic variations in planktonic foraminifera do reflect variations in ice volume. Temperature and ice signals are often in unison and in the same direction, even in regions where the isotopic record mostly reflects temperature change. Glacial maxima and minima can be identified as stratigraphic markers in correlating ocean sediment cores (Fig. 3–10).

The oxygen isotopic composition of foraminiferal tests has oscillated in response to glacial-interglacial fluctuations between maximum and minimum values through most of the Quaternary. Intercalibration of this sequence of oscillations with the paleomagnetic reversal scale (Fig. 3–12) has shown that the amplitude and cyclicity of the oscillations are fairly regular over about the last 1 m.y. (since about the Jaramillo normal paleomagnetic event). Oxygen isotopic oscillations older than 1.0 m.y. are less pronounced (Fig. 3–12) than recent ones. Their decreased amplitude is due to smaller volumes of polar-region ice production. Each distinct isotopic event over the last 1 m.y. has been assigned a stage number by Emiliani (Fig. 3–12). These **isotopic stages** represent alternating interglacial and glacial episodes. They are valuable in correlating Quaternary deep-sea cores. In this scheme, odd-numbered stages (starting from Stage 1, the unfinished stage containing recent sediments) are characterized by lighter values of the oxygen isotope (less continental ice) and even-numbered stages by heavier values of the oxygen isotope (more continental ice). Thus the most important stratigraphic aspects of oxygen isotopic stratigraphy are that the events are isochronous, widely occurring, and ideally preserved to make possible intercalibrations with other scales, such as magnetostratigraphy, and with biostratigraphic zonal schemes. The paleoenvironmental significance of the oxygen isotopic oscillations through the late Cenozoic are discussed in more detail in Chapter 17.

We have established that oxygen isotopic isochrons of great stratigraphic value are caused by the melting and freezing of high-latitude ice sheets and that the magnitude of these isotopic oscillations is proportional to the volume of ice

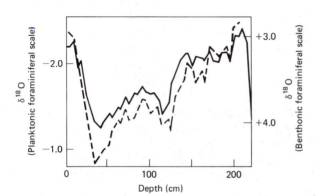

Figure 3–11 Oxygen isotopic records of planktonic versus benthonic foraminifera in piston core V28–238. Note scale differences at left and right figure margins. (After N. J. Shackleton and N. D. Opdyke, 1973)

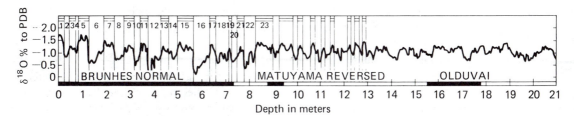

Figure 3-12 Oxygen isotopic oscillations in planktonic foraminifera expressed as deviation from Emiliani B1 standard for last 2 m.y. scaled against paleomagnetic record in piston core V28-239. Stages in oxygen isotopic record (1 through 23) are numbered after Emiliani [1955; 1966] and Shackleton and Opdyke [1973]. The last 0.7 Ma (Brunhes) contains glacial stages represented by amplitudes in excess of 1 per mil and with periodicities of about 100,000 years. Isotopic minima are approximately the same in the different "core stages." The mid-Matuyama (~1.4-0.7 Ma) interval contains isotopic fluctuations with approximately 40,000-year periodicities. Amplitudes are lower (0.7 per mil) than during the Brunhes. (After N. J. Shackleton and N. D. Opdyke, 1976, courtesy The Geological Society of America)

involved in this process. At some time in the past, no major ice sheets existed anywhere on earth and, as a result, the initiation of ice-sheet formation should provide additional correlation surfaces. Thus a second major consideration concerns the isotopic effect resulting from the melting of all continental ice. This is controlled by the amount of ice stored on Antarctica, which has the most of any continent. Today, the overall average isotopic composition of ocean water is about −0.28 per mil. However, prior to the accumulation of the Antarctic ice sheet, the value was isotopically more negative, since the ice now stored in Antarctica is isotopically more negative. Shackleton [1967] estimated the Antarctic ice sheet to have a $\delta^{18}O$ composition of −50 per mil. Upon complete melting of the continental ice sheets, seawater would become depleted in $\delta^{18}O$ by 0.92 per mil. Later we shall see that calculations indicate the beginning of ice-sheet development on Antarctica about 14 m.y. ago in the middle Miocene. Since then, ice volume has strongly affected the isotopic composition of the oceans. Prior to that, oxygen isotopic changes were the result of oceanic temperature changes. Such changes vary widely from area to area and provide a less valuable method of long-distance correlation.

Carbon Isotopic Stratigraphy

One of the latest methods in stratigraphic correlation of marine sediments is the use of carbon isotopes. This is known as **carbon isotopic stratigraphy** and employs two carbon isotopes: the common ^{12}C and the much rarer ^{13}C, which are stable isotopes (unlike radioactive ^{14}C). Carbon isotopic stratigraphy employs many of the same methods used with oxygen isotopes, and analyses are normally conducted in parallel with the oxygen isotopic measurements on the mass spectrometer [Duplessy, 1972; Douglas and Savin, 1973; Shackleton and Kennett, 1975]. As with oxygen isotopes, variations in $\delta^{13}C$ (that is, the difference between $^{13}C/^{12}C$ in a sample compared with a standard) result from fractionation of the heavy and light isotopes during physical or chemical processes. This fractionation is recorded in the calcium carbonate tests of microfossils. If any change occurs in the ^{13}C composition of seawater where the shell-secreting organisms live, there is potential for using the geochemical change for correlation purposes.

In general, the $\delta^{13}C$ in shells changes as the proportions of carbon from different sources change. Low ^{13}C content occurs in various organic materials, such as detritus from marine plankton. As a result, CO_2 derived from the decay of organic material is poor in ^{13}C in relation to PDB-1. Increased supply of carbon from this source further depletes the $\delta^{13}C$ in the water and in the microfossil shells. Deep waters of the oceans typically exhibit higher concentrations of nutrients and CO_2 compared with surface waters as a result of organic processes involving photosynthesis, sinking of organic debris, and respiration by bottom organisms. As deep water remains near the ocean floor for longer periods, nutrients and CO_2 continue to accumulate and oxygen content decreases. Because of the intimate relation between accumulation of $\delta^{13}C$ and CO_2, the $\delta^{13}C$ can represent a valuable tracer of nutrient flow through the world ocean. The $\delta^{13}C$ of marine organic (and hence respired CO_2) material is about $-25‰$ compared with CO_2 in the atmosphere, which is about $0‰$. Respiration causes $\delta^{13}C$ to decrease as the concentrations of nutrients and CO_2 rise. In other words, the $\delta^{13}C$ of deep-water CO_2 is lower than in surface water due to the accumulation of metabolic carbon. Therefore, any major change in the cycling of nutrients and CO_2 in the deep ocean is reflected in $\delta^{13}C$ and thus recorded in microfossils.

A distinct and permanent shift (decrease by $0.7‰$) in $\delta^{13}C$ values has been recorded in benthonic foraminifera in deep-sea sediment sequences in the Indo-Pacific region in the late Miocene (Fig. 3–13) [Keigwin, 1979]. This has been paleomagnetically dated at 6.2 Ma. Integration with paleomagnetic stratigraphy and biostratigraphy in the sequences indicate that the $\delta^{13}C$ shift is isochronous. Ap-

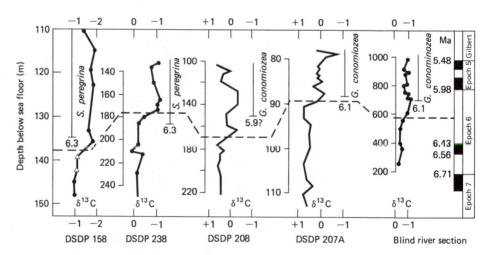

Figure 3-13 Carbon isotopic ($\delta^{13}C$) oscillations in a suite of late Miocene marine sections plotted against depth. Dashed line represents inferred isochron based on carbon isotopic shift in each record which is paleomagnetically dated in the Blind River section (at right) as 6.1 Ma.

parently, ancient oceans changed at that time from a previous steady-state $\delta^{13}C$ balance into the modern ocean state. It is not clear which of several oceanic mechanisms created the shift, but Bender and Keigwin [1979] favor a mechanism that involves a rapid change in CO_2 cycling in the ocean, resulting from a decrease in upwelling rate or an increase in the fraction of phosphate (PO_4) reaching deep-ocean waters as particulate organic matter. This shift may thus reflect some change in deep-ocean circulation and upwelling patterns. Such changes tend to affect large areas of the ocean basins simultaneously and the $\delta^{13}C$ shift represents a valuable tool for correlation of late Miocene sequences (Fig. 3–13).

Stratigraphic Integration: Toward A Single Time Scale

Quantification in geology requires a single numerical time scale resulting from the integration of stage stratotype information, biostratigraphy, paleomagnetic stratigraphy, and radiometric dating. Dating a sediment or rock sequence normally requires complex correlations with a known time scale because of limitations of the dating methods. Intercalibration of a number of parameters may be involved. Thus an age assignment for a geological event may be incorrect due to indirectness of correlation between radiometrically dated sequences. A reasonably accurate chronological framework was required by the DSDP for use in dating geological events in the drilled cores. Because of rotary drilling disturbance, magnetostratigraphy could not be applied directly to most sequences, and thus the intercalibration of paleomagnetic stratigraphy with the biostratigraphies established in the deep-sea sections had to be carried out indirectly using radiometrically dated microfossil zonations in land-based sequences for comparison.

Three different stratigraphic categories are available for the development of a time scale in marine sequences. These are (1) **dates,** relative to the present day; (2) **scales,** comprising a sequence of changing conditions; and (3) **datum levels,** representing unique events in progressive faunal and floral evolution.

Examples of scales are sequences of oxygen isotopic oscillations and paleomagnetic reversals. Correlations can be accurately carried out between different scales when they occur in the same cores. Scales are dated using radiometric approaches, although it is often difficult to obtain high-quality radiometric dates from sequences with well-preserved scales, and the required indirect correlations may be quite inaccurate. Nonetheless, accurate correlations must be developed between biostratigraphy and geochronology.

The two aspects of geochronology that are widely used are **radiochronology,** based on isotopic decay rates, and **biochronology,** based on organic evolution. Biochronology is the organization of geologic time according to the irreversible process of evolution in the organic continuum. Nevertheless, the absolute ages used in biochronology still come from radiochronology. Biochronology is an ordinal framework that measures all but youngest Phanerozoic time with greater resolution (about 1 m.y., the average age range of species in rapidly evolving lineages), but less accuracy, than radiochronology. The basis of biochronologic correlations are notable singular occurrences, or "datum levels," in the fossil record (for example, the *Orbulina* datum with a FAD at 15 Ma) that have geographic ranges overlapping those of coeval but distinct biostratigraphic zones. Long-distance correlations are primarily biochronological in character.

Biochronological correlations are based on three procedures: (1) recognizing widespread and distinctive events in paleontological history; (2) placing these events within local biostratigraphic schemes and determining their age with respect to as many reinforcing criteria as possible; and (3) relating these events to other biochronological datum levels and to radiometrically dated levels, such as a tuff bed or a paleomagnetic reversal. As observations such as these are synthesized, a time scale of dated events is developed and integrated with microfossil zonal sequences representative of different areas. Confidence in the application of fossil datums as isochronous levels is gained when a sequence of events appears in a section in the same order and with the same relative spacing as it does in stratotype sections. Differences commonly exist in the order of events in observed sequences. These are reconciled by the evaluation of environmental conditions and preservation of microfossils. When large numbers of taxa in several sequences are involved, statistical procedures can assist in objectively handling the data. Shaw [1964] developed a graphical method that helps in the placement of successive paleontologic events through time (Fig. 3–14). The first and last appearances of species are plotted against depth for one stratigraphic section on the horizontal graphical axis and another section on the vertical axis. Those points that represent an identical biochronological sequence for each section tend to lie along a straight line on the graph. Points that plot away from this line represent diachronous events in the two sections, and objective evaluations can be made regarding the isochroneity of a series of individual microfossil or other events. The soundest stratigraphic assignments, of course, are those built upon multiple datums comprising different fossil groups.

Zonal sequences and datum levels are intercalibrated where possible with the magnetic reversal time scale. Such biochronological scales have been constantly updated by Berggren since 1968. Deep-sea cores of strata deposited during the last 5 m.y. are correlated to the land-based K-Ar calibration of the paleomagnetic polarity intervals with little ambiguity. This part of the time scale seems accurate and reliable. For the period older than 5 m.y., the deep-sea record is not as well calibrated for several reasons: First, the K-Ar method loses its resolution with increasing age, so that pre-Pliocene sediment sequences are more difficult to date and, in turn, to correlate with the polarity record. Second, discrepancies can occur between biostratigraphic ages assigned to the oldest sediment overlying basement and to the predicted ages from the magnetic anomaly time scale of Heirtzler and colleagues [1968] (see Chapter 4).

A third factor that can introduce systematic distortions into biochronological scales are uncertainties originating from a small number of radiometric age determinations in certain parts of the scale. Examples of the interrelationships developed between Cenozoic planktonic zones, standard European ages, and the paleomagnetic time scale are shown in Figs. 3–15 and 3–16. Similar intercalibrations have been presented for the Mesozoic, such as the Cretaceous time scale of Van Hinte [1976].

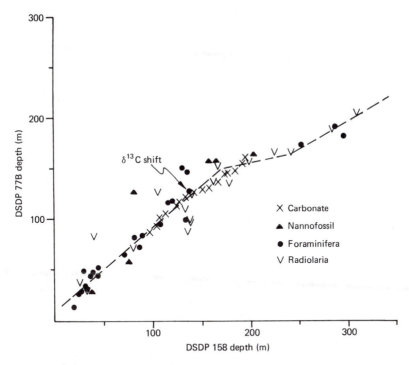

Figure 3-14 Graphic plot showing the correlation line between two DSDP sites based on first or last appearances of a number of planktonic microfossils or an oscillation in percent CaCO₃. Each data point represents the level in each of the two cores of a particular stratigraphic event. The position of the late Miocene carbon isotopic shift (shown in Fig. 14-13) is also shown. (Courtesy D. Dunn and T. C. Moore, Jr.)

Epoch and age boundaries

Because stratotypes provide the global chronostratigraphic framework, attention has recently focused on the precise definition and age of the boundaries between the various epochs. Three of these are discussed in the following section. Two boundaries—the Cretaceous-Tertiary and Eocene-Oligocene—are so significant paleoceanographically that they are discussed later in Chapter 19. Multiple criteria have been developed to facilitate identification of these boundaries in the stratotype sections and in other sections around the world. Central to these efforts has been the development and utilization of the paleomagnetic reversal record and the use of multiple micropaleontological criteria. The three examples that follow reflect problems that remain in international correlation.

Pliocene-Pleistocene boundary   An old problem for micropaleontologists is the relationship of the deep-sea sedimentary record to the standard sections of the Pliocene-Pleistocene of Italy. The base of the Quaternary is philosophically linked with the development of ice sheets in the Northern Hemisphere, and for this reason

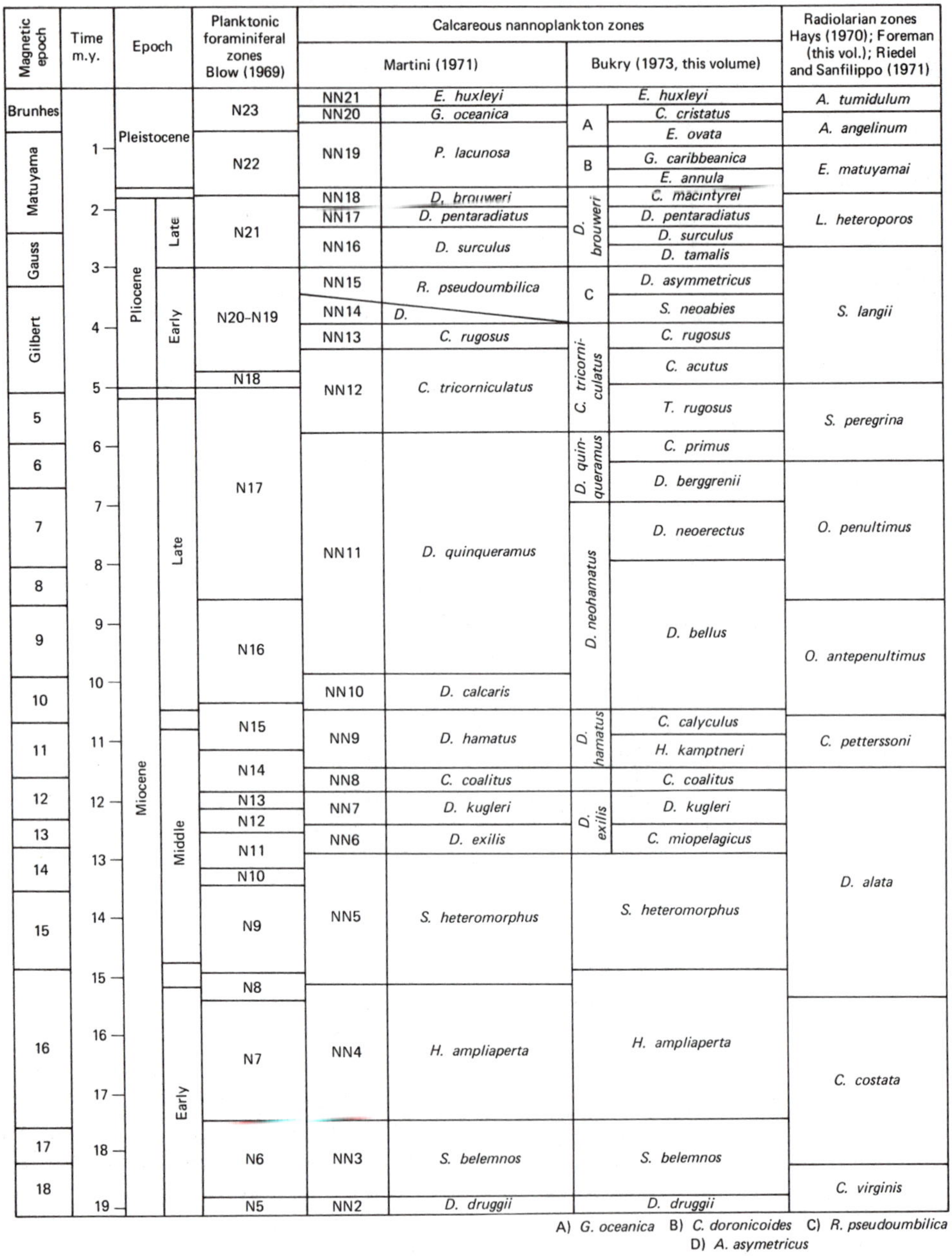

Magnetic epoch	Time m.y.	Epoch	Planktonic foraminiferal zones Blow (1969)	Calcareous nannoplankton zones — Martini (1971)		Calcareous nannoplankton zones — Bukry (1973, this volume)		Radiolarian zones Hays (1970); Foreman (this vol.); Riedel and Sanfilippo (1971)
Brunhes		Pleistocene	N23	NN21	E. huxleyi		E. huxleyi	A. tumidulum
				NN20	G. oceanica	A	C. cristatus	A. angelinum
	1		N22	NN19	P. lacunosa		E. ovata	
Matuyama						B	G. caribbeanica	E. matuyamai
							E. annula	
	2			NN18	D. brouweri		C. macintyrei	
		Late	N21	NN17	D. pentaradiatus	D. brouweri	D. pentaradiatus	L. heteroporos
				NN16	D. surculus		D. surculus	
Gauss	3						D. tamalis	
		Early	N20–N19	NN15	R. pseudoumbilica	C	D. asymmetricus	
	4			NN14	D.		S. neoabies	S. langii
Gilbert				NN13	C. rugosus		C. rugosus	
			N18			C. tricorniculatus	C. acutus	
	5			NN12	C. tricorniculatus		T. rugosus	
5	6					D. quinqueramus	C. primus	S. peregrina
6		Late	N17				D. berggrenii	
7	7			NN11	D. quinqueramus		D. neoerectus	O. penultimus
8	8					D. neohamatus	D. bellus	
9	9		N16					O. antepenultimus
10	10			NN10	D. calcaris			
11	11		N15	NN9	D. hamatus	D. hamatus	C. calyculus	C. petterssoni
			N14				H. kamptneri	
12	12	Middle	N13	NN8	C. coalitus		C. coalitus	
13			N12	NN7	D. kugleri	D. exilis	D. kugleri	
14	13		N11	NN6	D. exilis		C. miopelagicus	
			N10					D. alata
15	14		N9	NN5	S. heteromorphus		S. heteromorphus	
	15		N8					
16	16	Early	N7	NN4	H. ampliaperta		H. ampliaperta	C. costata
	17							
17	18		N6	NN3	S. belemnos		S. belemnos	
18	19		N5	NN2	D. druggii		D. druggii	C. virginis

A) G. oceanica   B) C. doronicoides   C) R. pseudoumbilica
D) A. asymetricus

Figure 3–15 Correlations of microfossil zones, paleomagnetic epochs, and the time scale for northern Indian Ocean DSDP sites of Neogene age (early Miocene to Present Day). This is an example of intercalibrations between a diverse set of biostratigraphic zones. (From E. Vincent, 1976)

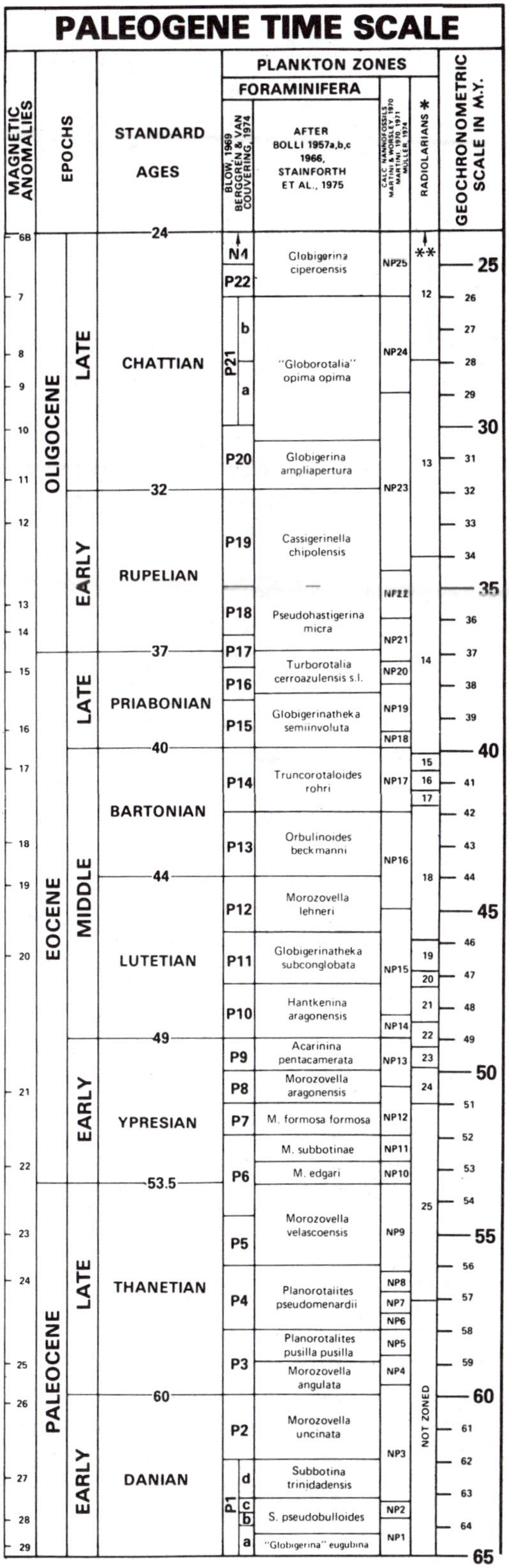

PALEOGENE TIME SCALE

MAGNETIC ANOMALIES	EPOCHS	STANDARD AGES	FORAMINIFERA (BLOW 1969; BERGGREN & VAN COUVERING, 1974)	FORAMINIFERA (AFTER BOLLI 1957a,b,c 1966, STAINFORTH ET AL., 1975)	Nannofossils (C.C.C. NANNOFOSSILS; MARTINI & WORSLEY 1970, 1971; MULLER 1974)	RADIOLARIANS *	GEOCHRONOMETRIC SCALE IN M.Y.
6B	OLIGOCENE / LATE	CHATTIAN	N4	Globigerina ciperoensis	NP25	**	—24— / 25
7			P22			12	26
8			P21 b	"Globorotalia" opima opima	NP24		27 / 28
9			P21 a				29
10							30
11			P20	Globigerina ampliapertura		13	31
12	OLIGOCENE / EARLY	RUPELIAN			NP23		—32— / 33
			P19	Cassigerinella chipolensis			34
13				—	NP22		35
14			P18	Pseudohastigerina micra	NP21		36
15	EOCENE / LATE	PRIABONIAN	—37— P17	Turborotalia cerroazulensis s.l.	NP20	14	37
			P16		NP19		38
16			P15	Globigerinatheka semiinvoluta	NP18		39
17	EOCENE / MIDDLE	BARTONIAN	—40— P14	Truncorotaloides rohri	NP17	15 / 16 / 17	—40— / 41
18			P13	Orbulinoides beckmanni	NP16	18	42 / 43
19			—44— P12	Morozovella lehneri			44 / 45
20		LUTETIAN	P11	Globigerinatheka subconglobata	NP15	19 / 20	46 / 47
			P10	Hantkenina aragonensis	NP14	21 / 22	48
	EOCENE / EARLY	—49—	P9	Acarinina pentacamerata	NP13	23	49 / 50
21		YPRESIAN	P8	Morozovella aragonensis		24	51
			P7	M. formosa formosa	NP12		52
22				M. subbotinae	NP11		53
		—53.5—	P6	M. edgari	NP10	25	54
23	PALEOCENE / LATE	THANETIAN	P5	Morozovella velascoensis	NP9		55 / 56
24			P4	Planorotalites pseudomenardii	NP8 / NP7 / NP6		57 / 58
			P3	Planorotalites pusilla pusilla	NP5		59
25	PALEOCENE / EARLY			Morozovella angulata	NP4		—60—
26		—60—	P2	Morozovella uncinata		NOT ZONED	61 / 62
27		DANIAN	P1 d	Subbotina trinidadensis	NP3		63
28			P1 c / b	S. pseudobulloides	NP2		64
29			P1 a	"Globigerina" eugubina	NP1		65

Figure 3-16 Paleogene biostratigraphic units correlated with European stages ("standard ages"), magnetic anomalies, and the geologic time scale. Details of the radiolarian zones are provided in following reference source. (After J. Hardenbol and W. A. Berggren, 1978, with permission of the American Association of Petroleum Geologists)

arguments have often been made to set its lower boundary at the first indication of major marine cooling. However, the timing of this event has been controversial because effects of cooling vary according to the criteria used, and large ice sheets did not develop rapidly. Thus a climatically defined base of the Quaternary has not been stable enough for correlation and dating. Proposals that the Pleistocene epoch should be defined by paleoclimatic, paleontologic, or anthropologic evidence were rejected by the resolution adopted at the 1968 International Geological Congress, which concluded that the concept of the base of the marine Pleistocene should be permanently placed and defined at the base of the Calabrian stage, whose stratotype is located in southern Italy. This decision marked the beginning of the golden-spike concept, important in global correlations. In tropical-subtropical deep-sea sediments, Berggren and others [1967] dated the Pliocene-Pleistocene boundary at about 1.8 m.y. because of this boundary's association with the Olduvai Paleomagnetic event (Gilsa of some workers). The Pliocene-Pleistocene boundary in the deep-sea sediments was based on the evolution of the planktonic foraminifera *Globorotalia truncatulinoides,* (from its ancestor *G. tosaensis*). This species also first appears within the stratotype section in Italy, but is rare and a poor stratigraphic tool. A reevaluation of the Calabrian biostratigraphy, principally of the calcareous nannofossils, indicates that the boundary stratotype at Le Castella in southern Italy should be correlated with the top of the Olduvai event at about 1.6 Ma. The new data agree with the age of a marked climatic cooling observed in deep-sea cores. The possibility exists for a reconciliation between a biostratigraphically and paleoclimatically defined Plio-Pleistocene boundary.

Miocene-Pliocene boundary   Traditionally, the boundary between the Miocene and Pliocene epochs has been placed at the top of the Messinian stage evaporitic sediment sequences in the Mediterranean, and the base of the early Pliocene-Zanclean stage. Until recently, no boundary stratotype section had been defined, but in 1975, M. Cita formally proposed the first definition of the Miocene-Pliocene boundary in terms of a fixed reference point. The "golden spike" (in actuality, an iron bar) was pounded into the seacliff at Capo Rossello in southern Sicily, at the stratigraphic plane at the base of the Zanclean stratotype, where Messinian shallow-water, anoxic "Arenazzolo" beds are conformably overlain by deep-water planktonic chalk of the Zanclean "Trubi marls." By indirect correlation, this level seems to be associated with the lowermost, reversed event of the Gilbert Paleomagnetic epoch with an age of about 5.2 Ma. The latest part of the Miocene is linked with major global paleoceanographic changes discussed in Chapter 19.

Oligocene-Miocene boundary   The definition of the Oligocene-Miocene boundary is, by international agreement, at the base of the Aquitanian stratotype, in the bank of the Saucets River at Moulin de Bernachon, France. In retrospect, the golden spike might have been more conveniently located since an unconformity lies at the base of the Aquitanian stratotype. Below this, upper Oligocene rocks as well as the older part of the Aquitanian formation are missing. The boundary is placed between the Aquitanian and Chattian stages, but there is a short interval of time unrepresented by sediments between the Chattian (upper Oligocene) and Aquitanian (lower Miocene) stages. The Oligocene-Miocene boundary is *approximately* equal to the *Globigerinoides* datum, represented by the first evolutionary appearance of this planktonic foraminiferal genus. Intercalibration of the boundary with paleomagnetic stratigraphy in deep-sea cores suggests an age of about 23 Ma.

It should be clear, therefore, that global correlation is very difficult and is successful only when a wide range of approaches are employed. The ultimate aim of stratigraphy, however, is to date rock sequences in order to understand the history of the earth and its life. This process often requires a series of integrated steps, but with radiometric dating always forming the fundamental base.

SEDIMENT SAMPLING METHODS

Historical Review

Our understanding of deep-sea sediments and of paleoceanographic history has grown with the availability of suitable vessels and the development of appropriate sampling equipment. The development of deep-sea drilling and the Kullenberg piston-corer demonstrate this.

During the 1860s, the nature of deep-sea sediments was virtually unknown. The *Challenger* expedition (1872–1876) marked the beginning of the systematic study of deep-sea sediments. The responsibility of sediment mapping was largely taken over by John Murray [Murray and Renard, 1891]. Although the work of Murray and his co-workers provided a foundation for all later marine geological studies, the expedition lacked coring equipment so it was not able to examine the geological record of deep-sea sediments. This research was not possible for another half century.

The first expeditions to take cores and examine the deep-sea sedimentary record were in the late 1920s and 1930s, beginning with the Dutch *Snellius* and the German *Meteor* expeditions. It was Schott [1935] who first demonstrated that changing planktonic foraminiferal assemblages record the glacial phases of the Quaternary. These efforts were later greatly expanded by workers of the Swedish deep-sea expedition; Kullenberg's [1947] invention of the piston corer provided even longer Quaternary sequences that opened up investigations of Quaternary paleoclimatic, sedimentological, and volcanic history.

More extensive investigations using piston cores began in the 1950s by an increasing number of investigators. For instance, at Lamont, M. Ewing, B. Heezen, D. Ericson, and G. Wollin carried out critical studies on the late Quaternary paleoclimatic and sedimentary record, while C. Emiliani (University of Miami) pioneered the use of oxygen isotopes for Quaternary paleoclimatic history. Studies such as these formed a solid foundation for more comprehensive examinations of the Quaternary sedimentary record during the 1970s. These were based on extensive piston-core collections assembled by several institutions during the 1960s and 1970s, including collections from the *Eltanin* expedition's circum-Antarctic surveys. These investigations were mostly limited to the study of the Quaternary record because the coring device rarely penetrated deeper than 20 m. Thus the long record of global history believed to exist in the deep-sea basins could only be reached by drill string. Pilot drilling of the Mohole Project was the first attempt to obtain this record. The project ended in a fiasco, but not before a useful drilled section was obtained off Baja California. The *Submarex* expedition also obtained a drilled sequence in the Caribbean. Although these expeditions recovered only modest amounts of material by today's standards, they were important in demonstrating the potential of deep-sea drilling, showing the uses of conventional petroleum industry technology, and for laying a foundation for later developments in deep-sea drilling.

This potential came to fruition on July 28, 1968, when the newly built drilling vessel *Glomar Challenger* sailed from Orange, Texas, for its first deep-sea drilling expedition. This program ushered in a new era in marine geology, improving our understanding of the evolution of the earth and its biota. It came at an appropriate time, immediately following the discovery of sea-floor spreading and plate tectonics in the middle 1960s, which provided a framework for quantitative and predictive investigation of the earth. Indeed, early drilling results from the DSDP itself [Peterson and others, 1970; Maxwell and others, 1970] persuaded all but a few diehards of the reality of a spreading ocean and of drifting and colliding continents. The failure of the Mohole Program created a climate for the development of a program based on national and international cooperation which has been truly successful. The DSDP was funded exclusively by the U.S. National Science Foundation during its first 8 years, but later funding was provided by a consortium of six countries, including the United States, under the International Program for Ocean Drilling (IPOD). Scripps Institution of Oceanography has been managing this program under the scientific guidance of a set of internationally staffed advisory panels (JOIDES—Joint Oceanographic Institute for Deep-Earth Sampling). The early stages of the program focused on reconnoitering, with drilling directed toward broader-scale dating of a wide range of submarine features. The program later turned to answering a wide range of oceanographic questions. The primary advisory panels established reflect these questions—the Ocean Crust Panel, the Passive Margin Panel, the Active Margin Panel, and the Paleoenvironmental Panel. The IPOD has revolutionized our understanding of ocean history. Most data generated by the program are published in the Initial Reports of the DSDP (Washington, D.C.: U.S. Government Printing Office). Brief summaries of each expedition were published in the journal *Geotimes*.

Piston Coring

The Kullenberg piston corer consists of a tight-fitting piston inside a core barrel. The piston is held at or near the sediment-water interface during penetration (Fig. 3-17), creating a suction to hold the sediment column in place. This suction reduces the effect of internal wall friction, thus increasing the length of recovered material (generally 7–20 m in length). The length of the core is the principal advantage that this approach has over simple gravity coring (generally recovering less than 5 m). The penetration energy is increased by weights of 3000 to 5000 lb that are fitted at the top of the rig (Fig. 3-17). The system requires a triggering action to

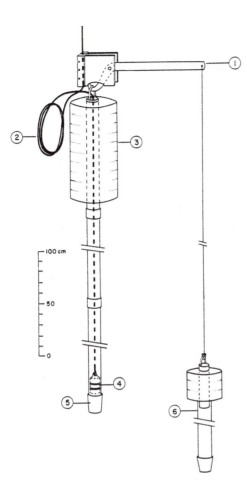

Figure 3-17 Standard piston core. 1 = trigger arm; 2 = wire loop (scope) for free-fall mode; 3 = bomb; 4 = piston; 5 = core cutter; 6 = trigger core. (From T. C. Moore, Jr. and R. Heath, *Chemical Oceanography*, vol. 7, pp. 75-126, 1978. Copyright by Academic Press, Inc. (London) Ltd.)

allow for free fall several meters above the ocean floor. A triggering device (usually a short gravity core) is positioned to hang from a trigger (release) arm several meters below the base of the piston core. Upon impact this triggering mechanism activates the free fall (Fig. 3-18). The wire lengths are calculated to guarantee that the piston begins to move up the core barrel just ahead of the sediment. Details of individual rigs may vary.

The length of cores obtained varies depending on the nature of the sediment samples. Most cores are 20 m in length: Foraminiferal oozes are generally no longer than 10 m, terrigenous sands are greater than 8 m, siliceous oozes about 15 m, and abyssal clays 15 to 25 m.

Deep-Sea Drilling

Using the technology developed by the petroleum industry, very long sections (up to 1500 m) can be drilled in water depths of up to 5500 m. Between August, 1968, and December, 1980, *Glomar Challenger* sailed on 84 legs, each about 2 months long, for a total of 250,000 nautical miles. About 500 sites have been drilled

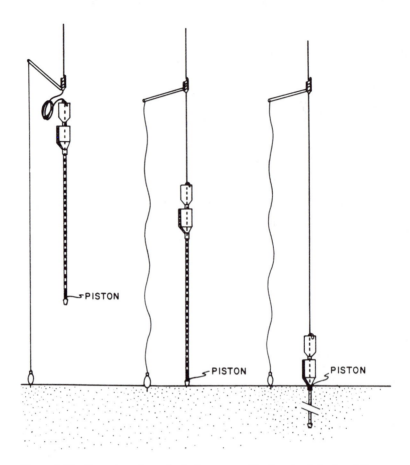

Figure 3-18 Operation of standard piston core. The length of the free-fall wire loop (scope) is calculated to insure that the piston begins to move up the core barrel at the instant when the core barrel begins to penetrate the sediment.

throughout the oceans except the Arctic. The *Challenger,* which is approximately 121 m in length and 20 m in beam and has a 6 m draft, was the first vessel ever equipped for full dynamic positioning and has supplies for up to 90 days at sea. It was the first vessel ever to drill in water depths greater than 6000 m. This vessel is capable of handling 6860 m of drill string and has a maximum water depth limit of 6000 m, with a 762 m penetration into the sea floor. Dynamic positioning is performed manually or automatically (fully computerized) by positioning the vessel in relation to a sonar beacon placed on the sea floor (Fig. 3-19). Positioning is accomplished by the main screws and by tunnel thrusters at the bow and stern for lateral motion.

The most conspicuous aspect of the vessel is its 43-m drilling derrick; an automatic pipe racker stores and handles up to 7620 m of drill pipe. The entire drill stem is rotated for the drilling operation. The base of the drill stem consists of a massive bottom-hole assembly, the weight of which ensures continued penetration.

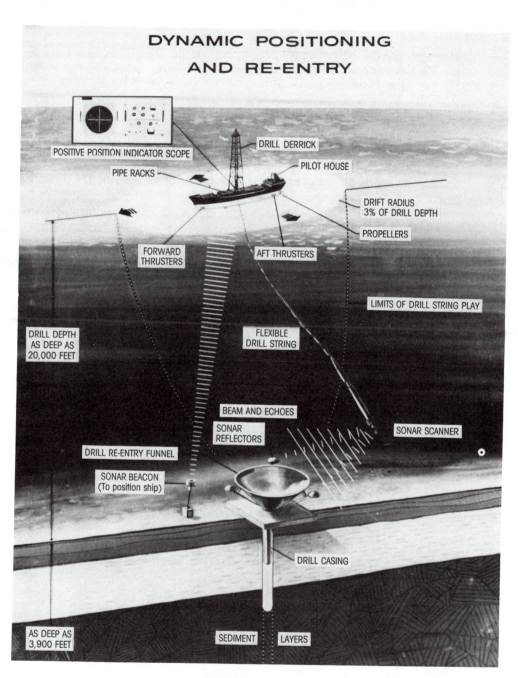

DYNAMIC POSITIONING AND RE-ENTRY

POSITIVE POSITION INDICATOR SCOPE

PIPE RACKS

DRILL DERRICK

PILOT HOUSE

DRIFT RADIUS 3% OF DRILL DEPTH

PROPELLERS

FORWARD THRUSTERS

AFT THRUSTERS

LIMITS OF DRILL STRING PLAY

DRILL DEPTH AS DEEP AS 20,000 FEET

FLEXIBLE DRILL STRING

BEAM AND ECHOES

SONAR REFLECTORS

SONAR SCANNER

DRILL RE-ENTRY FUNNEL

SONAR BEACON (To position ship)

DRILL CASING

AS DEEP AS 3,900 FEET

SEDIMENT   LAYERS

Figure 3–19 Dynamic positioning and re-entry of *D/V Glomar Challenger* for deep-sea drilling. When a drill bit is worn out, it is possible to retract the drill string, change the bit, and return to the same bore hole through a re-entry funnel placed on the ocean floor. High resolution scanning sonar is used to locate the funnel and to guide the drill string over it. The figure shows a sonar beacon used for dynamic positioning and a sonar scanner at the end of the drill string searching for the three sonar reflectors on the re-entry core. (Courtesy the Deep Sea Drilling Project, Scripps Institution of Oceanography)

However, the drill stem is not under tension during the drilling operation because the ship's motion is compensated within the bottom-hole assembly by telescoping bumper subs. Cores are retrieved by a coring device which moves vertically within the drill stem by wire worked from a separate winch. When a core is to be taken (up to 9.5 m of sediment), the corer is locked into place just above the bit and sediment fills the core as the drill string continues downward.

Re-entry into the same drill hole, which is now a fairly routine procedure, is nevertheless a remarkable feat when considering the scale (Fig. 3–20). Re-entry is assisted by a bell on the ocean floor that emits sound beacons for homing (Fig. 3–19).

Each *Challenger* cruise carries a party of geophysicists, sedimentologists, and paleontologists. The ship has excellent onboard laboratory facilities and much scientific investigation is carried out during the cruise. The investigative procedure involves measurement of physical properties, core splitting, lithological description, smear-slide analysis, photography, and paleontological investigations. These onboard investigations are followed by 1 to 2 years of more detailed analyses.

Hydraulic Piston Corer

One of the major disadvantages of using conventional rotary drilling techniques to obtain sedimentary cores from the deep sea is that the sediments are

Figure 3-20 Scale of deep-sea drilling at a depth of 5500 m. Length of *Glomar Challenger* is 120 m. Drill string width should hardly be detected at this scale.

highly disturbed. This is not a problem with the hydraulic piston corer because it is rapidly punched into the sediment without rotation (Fig. 3–21). Even finely laminated sequences are obtained using this method without discernible disturbance. The piston corer is lowered and retrieved by the drill string hanging from the drill ship. Thus, by repetitive operation in the same hole, high-quality cores may be taken at about 4.5-m increments or more through unlithified intervals. The rapid punch of the cores is activated by the buildup of hydraulic pressures in special chambers that exceed the strength of a shear pin (Fig. 3–21). Penetration takes only 1 to 2 seconds. Sequences of at least up to 200 m can be cored in this manner. These sequences are of the same quality as conventional piston cores; thus they provide superb opportunities for high-resolution stratigraphic and sedimentological studies over much longer intervals of geological time than conventional piston cores, which are very restricted in length.

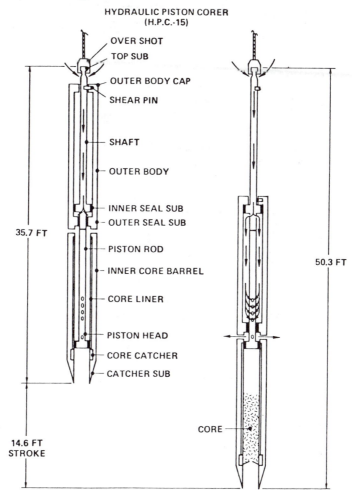

Figure 3–21 Hydraulic piston core mechanism used at end of drill string in deep-sea drilling. The mode at right is that of penetrating the sediment column. (Courtesy the Deep Sea Drilling Project, Scripps Institution of Oceanography)

Additional Sampling Techniques

Kasten corer

The Kasten corer is a large-diameter, rectangular gravity corer (about 23 cm), generally 3 m long. It was designed to obtain large samples at any stratigraphic level. As with gravity cores, penetration energy is provided by heavy weights; a core catcher at the base prevents the sample from falling out. These cores are useful for providing sufficient material at narrow intervals for ^{14}C dating rather than the usual wide sampling over stratigraphic intervals in narrow-diameter cores.

Gravity corer

Gravity corers are used where only short cores are required. The method is used extensively on the continental shelf and for obtaining surface-sediment samples but is not commonly used in the deep sea because piston cores are much more useful sedimentary sequences obtained with little additional time or effort.

Giant piston corer

This instrument was designed by Silva and Hollister [1973] to obtain long piston cores (30–40 m) in water depths of approximately 5000 m. This device is basically a much larger version of the conventional piston corer, with slight modifications: 14-cm diameter (inside diameter 11.5 cm); rigged for barrel lengths of 20 to 40 m; weights are up to 11,000 lb, and a parachute controls penetration speeds. Unfortunately, these have only limited use because large ships with heavy-duty winches and lifting equipment are required. The great advantages of such a system are that structural disturbance is reduced compared with conventional piston cores, which often exhibit disturbance near the margins, core length may be doubled, large-volume samples are obtained, and cores of good length (up to 20 m) can be obtained from shallow depths such as the continental shelf.

Box corer

This device is used to obtain samples of large volume or large surface area with shallow penetration. Most box cores have been used to study sedimentary structure in relatively shallow water. More recently they have been used in deep-sea sedimentary studies to collect samples that retain their in situ characteristics such as the orientation of manganese nodules and their detailed relationships with surrounding sediment. The device must be arranged to have complete closure of water-flow passages after sampling and before leaving the seabed, thereby minimizing sample washing during ascent.

Vibratory corer

Gravity and piston corers are unsatisfactory for use on the continental shelf because of the difficulty in penetrating sand layers and shell beds. Vibratory corers are instead successfully used for this purpose. The corers are driven into the substrate by vibration. A vibratory mechanism at the core head is activated using a variety of power sources, including compressed air or electricity. Cores up to 15 m in length can be obtained and have been important in providing sedimentary sequences necessary in understanding latest Quaternary sea-level history and shallow water sedimentary structures.

Grab samplers

Various grabs can provide a quick method for obtaining large surface-sediment samples at various water depths. The resulting surface-sediment material is normally mixed, although at times the sediment-water interface may be distinguished. Grabs have been used successfully to provide samples for obtaining large assemblages of scarce microfossils.

Relative Merits of Various Sampling Methods

Each approach is different and must be considered in terms of cost, that is, ship-station time versus the type of information being sought. Table 3–3 sum-

TABLE 3–3   SEA-FLOOR SAMPLING DEVICES[a]

Core type	Corer I.D. (cm)	Max. length (cm)	Disturbance	Use and remarks
Grab samplers	—	—	Negligible in center	Surface samples primarily for lithological and petrographical analyses. Layering usually preserved.
Free-fall grab	—	—	—	Samplers fitted with position-finding transmitter.
Gravity				
Open-barrel	4•75–8•1	300	Shortening of sediment	Upper few meters of sediment and surface samples.
Phleger	2•5–3•8	36	Slight deformation	Surface sample and pilot core.
Hydroplastic	8•1–10•1	400	Negligible	Undisturbed surface samples for engineering and mass physical properties.
Multiple (3–5 barrels)	2•5–3•8	48	Negligible	Surface samples to test reliability of very small samples to represent comparatively large areas.
Free-fall	4•0–6•5	122	Negligible	Short cores, working from small boats. Hydrodynamic release with glass sphere floats for recovery.

cont.

TABLE 3-3   SEA-FLOOR SAMPLING DEVICES[a] *(cont.)*

Core type	Corer I.D. (cm)	Max. length (cm)	Disturbance	Use and remarks
Piston	4·75–8·1	3400	Dependence upon piston set-up	Down-core studies.
Large volume Sphincter	12	1200	Negligible	Wide diameter piston corer.
Kasten	15 × 15	1500	Negligible	Undisturbed down-core studies.
Box	20 × 30 (50 × 50)	60	Negligible	Large rectangular sample at water-sediment interface.
Large diameter	15	600	Negligible	Enlargement for large volume samples.
Giant piston	12·7	3000	Negligible	Long cores in abyssal depths.
Vibro-corer	5·0–8·9	1000	—	Disturbs unconsolidated formations. Suitable for semiconsolidated formations.
Hydrostatic/Gas	3·8	600	—	Dependence upon sediment type, layering distorted. Operating depth less than 90 m.
Deep-sea drilling (rotary)	3·5–5·0	—	Much	Drilling tower on board ship. Discontinuous coring. Hard formation rotary bits with drilling fluid.
Deep-sea drilling (hydraulic piston coring)	—	—	Negligible	Undisturbed sections.

[a]From Moore and Heath [1978]

marizes the various characteristics of a large number of different deep-sea sampling methods. For most sedimentological and micropaleontological investigations, the concern is primarily with the length and continuation of the sequence and the amount of mechanical disturbance of the sediments. The latter factors largely control the quality of the data obtained from the sequence and the feasibility of the magnetostratigraphic investigations. Except for deep-drilled sequences, all methods retrieve late Quaternary material (0–1.0 Ma). The samples and sequences obtained by conventional deep-sea drilling, though this is the only approach that provides sequences of considerable age, are usually disturbed and incomplete. But the detailed investigations performed on Quaternary piston or gravity cores are now possible by using hydraulic piston cores activated at the bottom of the drill string. This applies only to unlithified sequences.

Another problem concerns the recovery of suitable sections of interbedded cherts, semilithified sediments, or alternating lithified and soft sediments. In these sediments, water must be circulated to keep the bit clear and the drill string free. Unfortunately, the water circulation washes away the soft sediments in the sections [Moore, 1972]. Often the drilling operation itself causes severe mechanical disturbance of the sediment, especially in younger, softer sediments. Caution needs to be exercised in the common problem of down-hole contamination of sediment and microfossils.

There is also strong evidence indicating that even when the core barrels are returned full, the actual sampling is incomplete. There are a disproportionate number of biostratigraphic boundaries within core barrels near the base of each cored section or between continuous cores [Moore, 1972]. As much as 35 percent of the core section may be lost during drilling. Deep-drilled sequences have played a major role in our understanding of oceanic history, but better tools are still required.

four

Continental Drift
and Sea-Floor Spreading:
Prelude to Plate Tectonics

*. . . the most prominent component is the assumption of great
horizontal drifting movements which the continental blocks underwent
in the course of geologic time and presumably continue even today.*

Alfred Wegener, 1924

INTRODUCTION

The earth is an active planet. It exhibits dynamic interaction between its inner and
outer parts, which puts the earth's surface layer into a state of lateral and vertical
motion. This layer involving both oceans and continents is in motion, with new
material in the oceans created as older crust descends back into the interior of the
earth. Earth scientists, in general, began to understand the earth as a mobile,
dynamic body only during the 1960s. This change represented a major scientific
revolution, which, like the Copernican and Darwinian revolutions, should be
named for its champion, German meteorologist Alfred Wegener. He was the first to
collect much evidence in support of the theory of continental drift, though this data
did little to convince his stoical colleagues. It has long been observed that mountains, volcanoes, and earthquakes are not randomly distributed, but are found in
narrow, distinct belts. Several theories have been proposed during the twentieth
century to account for these various types of evidence for instability in the earth's
crust, including global contraction, global expansion, the foundering of large areas
of the earth's crust (sunken continents), and lateral motion; lateral motion in its
modern concept is called *global plate tectonics*. In this chapter and in Chapter 5, we
shall examine the principal steps in the conceptual evolution from a largely static to
a largely mobile lithosphere involving enormous lateral continental motions. During the middle part of the nineteenth century, the Darwinian revolution transformed thought about the permanence of life on earth. It required another century
of scientific work to transform thought about the permanence of the oceans and
continents.

105

CONTINENTAL DRIFT

Early Studies

The seeds for the concept of continental drift were planted hundreds of years ago. As long ago as 1620, Francis Bacon was impressed by the parallelism of the shores of the Atlantic and discussed the possibility of the drifting of the two opposite continental masses. Several others followed with similar suggestions during the 1800s, including Antonio Snider-Pellegrini who, in "La Création et ses mystères dévoilés" (1858), linked the fracturing and pulling apart of the Americas from the Old World with Noah's flood. As modern evolutionary biology began to develop, biogeographers began to examine the distribution of plants and animals and formulate concepts about their phylogeny. For instance, Edward Forbes quickly noticed relationships between organisms on distant continents that were impossible to explain without the existence of connecting "land bridges." During the remainder of the nineteenth and much of the twentieth centuries, biologists continued to be impressed by the distributions of living organisms, such as earthworms and marsupials, that provided evidence of former continental connections. Biological evidence led to speculation about large-scale foundering of previously interconnecting continental masses or land bridges. Although favored by many biologists, this concept was supported by few geologists and geophysicists because isostasy makes it unlikely that continental masses will sink. Earth scientists either favored other theories involving lateral motion of continents or simply ignored the important biological evidence.

At the turn of the century, one of the most favored theories of the earth's genesis was that of the Austrian synthesizer Eduard Suess—the **global contraction hypothesis.** It was formulated to explain **orogenesis,** or the formation of the great linear mountain ranges around the globe. Some mechanism was required to explain the great fold belts. Suess invoked global contraction; he believed that the earth's surface wrinkled and cracked like drying fruit. According to this theory, the earth is progressively solidifying and contracting from a molten mass. Differences in rock densities lead to the formation of a superficial granitic and metamorphic layer rich in Al, Na, and K silicates underlain by a more dense layer richer in Fe, Ca, and Mn silicates. Large troughs of sediments up to 10,000 m thick formed in the wrinkles, or **geosynclines,** which were uplifted into mountain belts. Suess proposed that all the continents had been at one time joined in a single immense land mass called *Gondwanaland,* named after a key geological province in India (the Gonds is a tribe of central India). He also noted the close correspondence of geological formations in the continents of the Southern Hemisphere and ascribed their later dispersion as a result of differential contraction. The contraction theory has lost ground steadily since Suess's work because the large amount of total folding in mountain ranges required too much cooling and contraction and because of the different ages of mountain chains and the relative youth of many of them.

Although Wegener is considered the father of continental drift, an American, Frank B. Taylor [1910], independently suggested a mechanism that could account for large lateral displacements of the earth's crust. Taylor used the distribution of mountain belts to indicate a creeping movement of the earth's crust toward the southern periphery of Eurasia and of the western Americas, too far to be counted for by the contraction hypothesis. He speculated that continental movement resulted from large tidal forces and a slowing of the earth's rotation resulting from the capture of the moon, which he also speculated occurred during the Cretaceous.

Major lateral movements of the earth's crust had also been suggested by the Austrian geologist Otto Ampferer [1906], to account for the large fold mountains. As early as 1925, he further proposed convection currents in the interior of the earth as the tectonic driving force. He also anticipated substantial ideas incorporated within modern plate tectonics, including the subduction of ocean crust beneath the fold mountains at the leading edge of drifting continents and the formation of new ocean crust in the center of the ocean basins. About this same time, his colleague Robert Schwinner also attempted to explain the formation of oceans by convection currents in the earth's interior. Nevertheless, it is Wegener who is largely remembered because of the comprehensiveness of his research (beginning in 1912) on continental drift, during which he assembled evidence of close affinities to rocks, fossils, and structures on opposite sides of the Atlantic Ocean, akin to reading lines of type across a piece of torn newspaper. Between his first paper of 1912 and his untimely death during a Greenland expedition in 1930, he continued to publish papers and books which were translated into five other languages, increasing the impact of his discoveries. His most important contribution, however, was his book first published in 1915, *Die Entstehung der Kontinente und Ozeane.* Wegener presented evidence for all the continents' having previously been joined in a single supercontinent called **Pangaea** about 200 Ma (Fig. 4–1). In this configuration, the continents of the Western Hemisphere butted against Africa and Europe, and those of the Southern Hemisphere assembled in the southern part of this supercontinent. Wegener postulated that about 180 Ma during the Mesozoic and continuing up to the present, Pangaea fragmented, first in the Southern Hemisphere and later in the north, with northern Europe and North America connected until the Quaternary.

In addition to collecting information about parallelism of opposite sides of Atlantic shores, similarities in geological features in areas previously joined, and the biogeographic arguments, Wegener also amassed a considerable amount of paleoclimatic evidence. He mapped the distribution of coal measures and glacial deposits to determine previous climatic belts. Special importance was attached to the distribution of deposits indicative of extensive glaciations. Glacial deposits and other data indicate that in the Devonian a polar ice cap covered the Sahara, while eastern North America lay near the equator. Such drastic shifts in climate suggested not only continental drift, but polar wandering. Polar wandering can result from either a change in the spin-axis of the earth or the displacement of the entire crust over the earth's mantle. Wegener conducted tests to differentiate between these two processes. Also of importance are the glacial deposits (**tillite**) of Carboniferous to Permian age distributed in the Gondwanaland continents. These deposits left a record of a succession of ice caps during the Permocarboniferous. Reassembly of Gondwanaland brings these dispersed glaciated regions back as a single unit. The glacial deposits of India, which now lie far to the north of the equator, were ex-

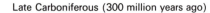
Late Carboniferous (300 million years ago)

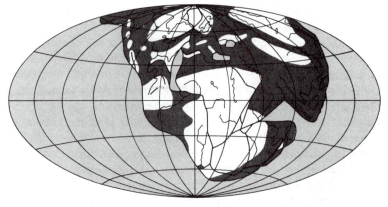

Eocene (50 million years ago)

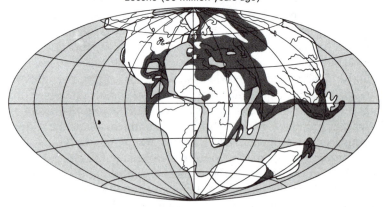

Early Pleistocene (1.5 million years ago)

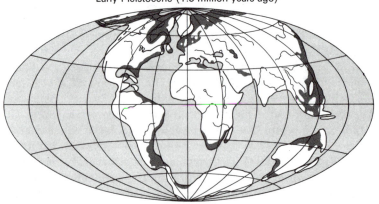

Figure 4-1 Paleogeographic reconstructions for three periods according to Wegener. Africa is shown in its present-day position for reference. Shallow seas are indicated by heavy shading. (After A. Wegener, 1924)

plained by extensive northward drift of the Indian subcontinent from Gond-wanaland.

Wegener unfortunately chose to describe drift of continents as rigid bodies moving through a yielding sea floor. He believed that in spite of the apparent high velocity of subcrustal materials, small forces acting over long periods of time could cause the material to yield and allow the continents to flow slowly through the mantle. These forces were derived from the differential gravitational attraction of the equatorial bulge (**Polflucht force**) and of westward drift due to the tidal forces of the sun and moon. Geophysicists correctly objected that the sea floor was rigid and not viscous and that the forces involved in Wegener's mechanism of drift were insufficient to cause continental drift. Thus his arguments converted few geologists, who were as fixed as the continents in their traditional beliefs. Wegener's idea remained in disfavor until the 1950s; just about everything that could be argued for or against continental drift had been written and the debate faded for lack of additional evidence. Further development was stalled because of a critical lack of information from the oceanic areas. Thirty years passed before oceanography could provide the crucial information about the ocean floor.

Mantle Convection

One of the most important developments that occurred between Wegener's work and the early 1950s was the formulation of an alternative theory to explain continental drift. Arthur Holmes, of the University of Edinburgh, reactivated the convection theory as the primary mechanism for earth's dynamic processes. In 1931 Holmes suggested that the interior of the earth is in a state of extremely sluggish thermal convection, forming large cells with the mean flow vectors changing only over distances on the order of the earth's radius and during periods of time comparable to the geological periods. He was the first to suggest, specifically, that the basaltic layer acts as a conveyor belt on top of which a continent is transported to the location of the downward sinking of the convection cell. New oceanic crust is generated by the intrusion of basaltic magma into the oceanic crust. Of particular importance is that in proposing this mechanism, Holmes alleviated the major problem in Wegener's model by suggesting that the continents were carried along with the adjacent ocean floor rather than plowing through it. Vening Meinesz added that ocean trenches were formed by the downdragging by this same type of convection mantle flow. These large downwarps, or tectogenes, came close to plate tectonic theory but subduction was not invoked. They also showed that convection currents were necessary to account for the transfer of heat flowing from the earth's interior through the poorly conducting mantle region. The theory of mantle convection replaced that of earth contraction as an explanation of the compressional features, as well as the hypothesis of an expanding earth [Carey, 1958], which had been advanced to explain the more recently discovered tensional features. Although Holmes's theory had little immediate impact, it formed the basis for interpretation of vast amounts of data obtained 20 to 30 years later.

Polar Wandering

Wegener's attempt to change earth science failed because he used criteria, such as paleoglacial evidence, which was considered too uncertain by geophysicists. Beginning about 1956, new discoveries—paleomagnetism of ancient continental rocks, broad-scale paleomagnetic patterns on the ocean floor (relating to the youth of the sea floors and ocean sediments), and seismology—began to change this. No field has contributed more to continental drift and related tectonic interpretations than paleomagnetism.

The first important use of paleomagnetism was to test polar wandering and continental drift. The idea of polar wandering was invoked early by Wegener in his discussion of Permocarboniferous glaciation in support of continental drift. Paleomagnetism was to provide the first quantitative and geophysical test for these processes. By measuring the paleomagnetism of rocks from different continents, it has been possible to reconstruct the position of the ancient magnetic poles. By 1956 several workers, especially K. Creer, K. Runcorn, and E. Irving, were able to demonstrate that the paleomagnetic evidence from single continents showed change of polar position, but that polar wandering curves charted from different continents were increasingly incongruent over time. This incongruence was eliminated when the continents were reassembled. At present, continental migration is favored over polar migration to explain these paleomagnetic changes because it is unlikely that several poles existed simultaneously. Also, it is unlikely that the magnetic pole has shifted far from the axis of the earth's rotation or that the axis of rotation has changed radically in position from the principal mass of the earth.

About the same time, E. Irving [1956], then at the Australian National University, determined the paleolatitudes of North America, Europe, and Australia from paleomagnetic data and compared these with the paleoclimatic records on these continents. He demonstrated that except for the Precambrian, the two data sets were in remarkably good agreement: Coral reefs and rocks formed in ancient deserts lay at low paleomagnetic latitudes, whereas ancient glacial deposits lay at high paleomagnetic latitudes. Using paleomagnetic data, Irving calculated pole paths for North America and Europe and noted that the path for North America lay to the west of the path for Europe. He concluded that during the Mesozoic and Paleozoic, North American lay closer to Europe and that India, Australia, North America, and Europe had all drifted.

Continental Arrangements

Many different continental arrangements have been suggested over the years. Some configurations are widely accepted, while others remain speculative. One of the reconstructions is the way in which the continents bordering the Atlantic Ocean are joined together. On the basis of computerized geometrical fit of the continental margins, the predrift reconstruction of South America to Africa was shown by the British geophysicist Sir Edward Bullard and others [1965], to be so precise that few doubt its validity. These workers also demonstrated that equally good fits result from matching bathymetric contours between 100 and 1000 m. Similar good fits were determined for North America and Africa by the marine geologists Dietz and Holden [1970]. The critical test for such reconstructions is the degree to which overlaps or gaps occur.

Two different reconstructions have been proposed for the original configuration of the groups of continents in the Northern and Southern Hemisphere. Wegener believed that all the continents were assembled as a single mass (Pangaea) at the beginning of the Mesozoic. This was modified by the South African geologist Du Toit [1937]; he suggested the existence of two large land masses: *Gondwanaland* in the Southern Hemisphere and *Laurasia* in the Northern Hemisphere. These two enormous land masses were separated by a large body of water between Africa and Eurasia called the *Tethys Sea,* which was located north of Arabia and extended from the former location of the Atlas Mountains to the east of the Himalayas. Great thicknesses of sediments were deposited in the Tethys Sea over the past 200 m.y. When Gondwanaland broke up and moved northward against Eurasia, the motion closed the Tethys Sea and buckled these thick accumulations of sediments creating the mountain chain that extends from the western Atlas range through the Mediterranean, the western Alps, the Caucasus, and the Himalayas. The most spectacular thrusting occurred in the Himalayas as India rammed into Asia.

Gondwanaland Configurations and Separation History

Geologic evidence for the existence of Gondwanaland from the Devonian to the Jurassic is overwhelming. It is supported by evidence used in continental reconstructions, namely, paleomagnetism, geometrical fit of continental outlines, the age and character of the separating oceans, geological correlations, and the geological history of continental margins. Various reconstructions have been proposed, but precise reconstruction and the limits of Gondwanaland remain undefined. A recent Gondwanaland reconstruction by Barron, Harrison, and Hay [1978] (Fig. 4–2) uses previously determined features and includes data for several critical areas. By preventing overlap and satisfying geological constraints, this reconstruction places the Antarctic Peninsula to the west of South America, adjoining the Andean province, leaving Antarctica as a single unit. This aligns the geosynclinal belts of South America, Africa, the Falkland Plateau, and Antarctica. Madagascar is adjacent to Mozambique, and the Seychelle Islands in the Northwest Indian Ocean are a continental fragment. Australia is rotated 250 km to the west with respect to Antarctica to obtain a better geometric fit. The remarkable fit of these two continents is shown by the computerized fit by Sproll and Dietz [1969] in Fig. 4–3. The fit of Australia and India indicates that they were originally one continuous continental mass (Fig. 4–2).

The timing of southern continental separation has been determined by a variety of geological evidence. Continental separation is associated with the first marine sedimentation or volcanic activity and the time of first generated oceanic crust. It is important to realize, however, that first evidence of rifting may have preceded separation by a significant amount of time, and be associated with uplift, extrusion

Figure 4-2 A revised reconstruction of the southern continents. Africa is shown in its present-day position. Lambert equal area projection centered at −30° latitude, 20° longitude. (From E. J. Barron, C. G. A. Harrison, and W. W. Hay, *EOS*, vol. 59, p. 446, 1978, copyrighted by the American Geophysical Union)

and intrusion of basic and alkaline igneous rocks, and rift formation. Barron and others [1978] provide the following summary of the breakup sequence of Gondwanaland: The initial separation of the southern continents occurred between western Gondwanaland (Africa and South America) and eastern Gondwanaland (Madagascar, Antarctica, India, and Australia) during the late Jurassic (140–160 Ma). This is associated with Middle to late Jurassic volcanic activity in the southern Andes and the Antarctic Peninsula, followed by the deposition of marine sediments. Early to late Jurassic marine sediments of the Cape Province also indicate this separation. The rifting of eastern and western Gondwanaland is associated with the widespread Ferrar (Jurassic) dolerites present in the Transantarctic Mountains. The paleomagnetic poles of eastern and western Gondwanaland began to diverge as early as the Permotriassic.

The rifted western margin of Australia consists of Triassic continental deposits with Jurassic marine incursions from the north. A continental separation during the late Jurassic is also indicated by the age of the ocean floor west of Australia. The separation of India from Antarctica is associated with the Rajmahal and Sylhet traps of Jurassic and Cretaceous age in India and Jurassic volcanic activity in coastal Antarctica. Barron and others [1978] concluded that Madagascar-Seychelles-India (Fig. 4-2) separated from Antarctica-Australia approximately 140 Ma. India and the Seychelles parted from Madagascar in the late Cretaceous (100 Ma), a conclusion based on the age of the sea floor south of the Seychelles and paleomagnetic pole determinations. The separation of India and the Seychelle

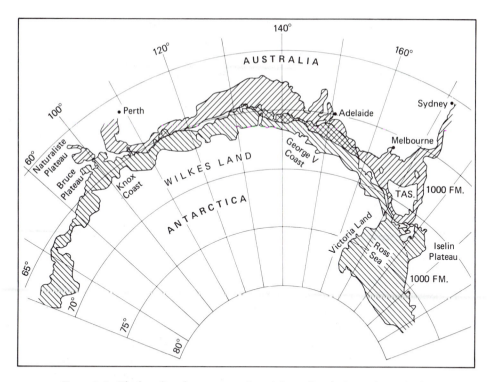

Figure 4–3 The best fit paleoreconstruction of Australia relative to Antarctica by computerized matching of the 1000–fathom isobath (outer contour). Overlap areas are cross ruled and underlap areas are blank. Australia began to separate from Antarctica at 53 Ma (middle Eocene). (From W. P. Sproll and R. S. Dietz, 1969, reprinted by permission from *Nature*, vol. 222, pp. 345–348, copyright © 1969 Macmillan Journals, Ltd.)

Islands followed during the earliest Tertiary associated with the formation of dyke swarms in the Seychelles and the Deccan Traps of India.

Uplift, fracturing, and volcanic activity occurred during the separation of Africa and South America during the early Cretaceous with the first marine sediments accumulating during the Aptian. The age of the earliest ocean floor in the South Atlantic suggests that separation began about 127 Ma.

Because the separation of eastern and western Gondwanaland took place at least 20 m.y. before the opening of the South Atlantic, the reconstruction of the Antarctic Peninsula is no obstruction to the westward movement of South America from South Africa (Fig. 4–2). This is because the Antarctic Peninsula had already moved sufficiently south along the west coast of South America before the opening of the South Atlantic so as not to interfere with the separation of South America and Africa.

The age of the ocean floor in the Southwest Pacific shows that the Campbell Plateau (south of New Zealand) separated from Antarctica during the late Cretaceous about 80 Ma. The oldest sea floor in the Tasman Sea (between New

Zealand and Australia) is 80 Ma and is associated with marine transgressions and volcanism on the east coast of Australia and in New Zealand.

The latest separation is Australia from Antarctica (Fig. 4-3). Marine sediments on the southern coast of Australia and Ferrar dolerites in East Antarctica indicate a prior rifting phase, but the oldest sea floor between these two continents is of middle Eocene age (55 Ma), heralding the northward movement of Australia from Antarctica. The northward drift was recognized by Wegener [1912] who stated: "The drift from the poles is also very clear in the case of Australia, for this continent is moving towards the north-west, as is consistently shown by the deformation of the series of islets forming the Sunda Archipelago, by the high and youthful mountains of New Guinea . . ."

SEA-FLOOR SPREADING

Development of the Concept

Until the 1960s, most geologists believed in the permanence of continents and ocean basins. Although the theory of continental drift has long been established and was known to most earth scientists, it still did not form a theory to which researchers applied their data. Paleomagnetic studies of continental rocks were beginning to make a mobile crustal theory more acceptable, but much more data were needed to convert the majority of earth scientists. This revolution came in the mid-1960s, and the critical data were obtained from the oceans. A new, detailed model that provided a tenable alternative to Wegener's ideas was provided in the classic paper, "History of the Ocean Basins," by H. H. Hess [1962] of Princeton University (Fig. 4-4), in which he incorporated some of the concepts of A. Holmes. In contrast to the earlier theories of continental drift, which required the continents to plow through the ocean crust, the mechanism suggested by Hess conveys them passively by lateral movement of the crust from the source of a convection current to its sink. New oceanic crust is generated at the mid-oceanic rises, where hot mantle material raises the topographic elevation of the rises due to lower density of hotter rocks. At the rises, the surface rocks are broken by tension and pulled apart, and the rift is filled by new volcanic material from the mantle. The ocean floor thus spreads laterally as if on a conveyor belt. Where the convection cells converge, the ocean crust is dragged downward to form trenches. These descending currents are the sites of compression, marked by mountain ranges and volcanic arcs associated with the trenches. The continents, having been built up by the accumulation of lighter silica-rich rocks, are not dragged downward at the trenches but accumulate or are thrust upward as mountains. Earthquakes occur at depths beneath the trenches as descending cool material fractures. For this general process R. Dietz, then of the U.S. Coast and Geodetic Survey [1961], coined the term **sea-floor spreading** to describe a theory primarily concerned with the process by which the ocean floor is created and destroyed. The beauty of this theory is that it satisfied most of the results available from the fields of marine geology and marine

Figure 4–4 Harry Hess as a navigation officer aboard a World War II troop ship. During the war he took echo-sounding records across parts of the Pacific and discovered several flat-topped volcanoes which he later called guyots. Such early observations were to help convince him of the lateral mobility of the ocean floor. (Courtesy Department of Geology, Princeton University)

geophysics. Its mechanism is in harmony with physical theory and much of geological and geophysical observation, the directions and rates of motion of both sea-floor spreading and continental drift are entirely compatible, and the model provides a mechanism for disrupting and moving continents. Since the theory of sea-floor spreading was formulated in detail by Hess and Dietz, it has created a profound understanding of the ocean basins and their margins, and new data are nearly always compatible with this unifying concept. In effect, the formulation of these new theories created the well-known revolution in earth science, the **Wegenerian revolution** [Wilson, 1973]. A true scientific revolution causes a convergence of scientific disciplines, with results from one discipline affecting the course of research in another.

Like all revolutionary concepts, sea-floor spreading was based upon and influenced by a number of ideas including the following:

1. The paleoreconstruction fits of the continents and other evidence for continental drift.

2. The height and topography of ocean ridges.

3. Systematic thickening of sediments away from ridges.

4. Volcanism on ridge axes.

5. Seismic zones, island arcs, and volcanism at or near certain oceanic margins.

Furthermore, it had been shown by J. Tuzo Wilson [1963] of the University of Toronto that a systematic increase in age of islands in the Atlantic occurs with increasing distance from the mid-ocean ridge. By this time topographic surveys by B. Heezen and others at Lamont had revealed the basic symmetry of the flanks of mid-ocean ridges and a large number of seismic reflection profiles had been obtained from the ocean basins by M. Ewing and associates at Lamont. These profiles showed that the thickness of sediments increases from essentially zero on the crest of the mid-ocean ridge to several kilometers at some continental margins. The distribution pattern is symmetrical about the ridge axis. These trends also suggested more time for sediment to accumulate farther from the spreading ridge.

Oceanic magnetic stripes

Although various lines of evidence pointed toward a concept of sea-floor spreading, the most crucial evidence was provided by magnetometer surveys of large segments of the ocean floor. By 1958, sensitive magnetometers had been developed to measure the earth's magnetic field to 1 part in 10^5. Using these tools, Vacquier and others [1961] and Mason and Raff (1961) of the Scripps Institution of Oceanography uncovered a pattern of linear magnetic anomalies in the Northwest Pacific (Fig. 4–5) orientated approximately north-south and bounded by steep gradients between highs and lows. This discovery aroused considerable interest, and they postulated that the linear anomalies resulted from an alternation of lava flows and sediments, forming a sequence of troughs. Also of significance was clear evidence of major offset (up to 1400 km) of the anomaly pattern across major fracture zones. These offsets were initially interpreted to indicate extreme lateral strike-slip motion of the ocean crust.

Another discovery that also required adequate explanation was a positive magnetic anomaly associated with the median rift valley of the mid-Atlantic ridge. This anomaly persists where the median rift valley is absent or poorly developed. More surprising was the direction of the anomaly, as it was opposite that which was expected because of the decrease in volume of crustal material at the median rift valley.

During the early 1960s the polarity time scale was being developed independently of the activities of marine geologists (see Chapter 3). About that time F. Vine and D. Matthews [1963] of Cambridge University offered a simple explanation to account for the enigmatic central magnetic anomaly observed over ridge crests and the puzzling linear magnetic anomalies (**magnetic stripes**) that are re-

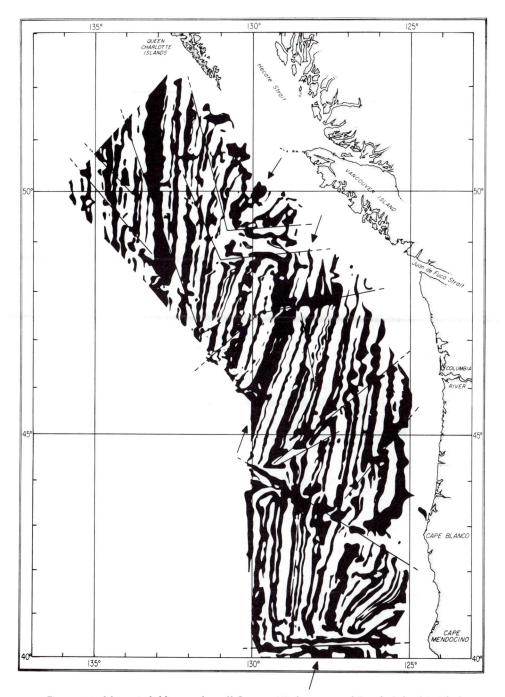

Figure 4-5 Magnetic field anomalies off Oregon, Washington, and British Columbia. Black areas represent positive anomalies; white represent negative anomalies. Arrows show the axes of three short ridges—from north to south, Explorer, Juan de Fuca, and Gordon ridges. The magnetic anomaly patterns are offset at a series of fracture zones. (From R. G. Mason and A. D. Raff, *Geol. Soc. Amer. Bull.,* vol. 72, pp. 1267–1270, 1961, courtesy The Geological Society of America)

corded in the Northeast Pacific. They combined Hess's sea-floor spreading theory with the new data on the history of reversals of the earth's magnetic field and suggested that as new sea floor forms at the mid-ocean ridges through extrusion of submarine lava, it becomes magnetized in the earth's magnetic field prevailing at the time it cooled (Fig. 4-6). When the newly formed material was subsequently pushed laterally from the ridge axis by new sea-floor material, it would form stripes of alternating normal and reversed magnetization. The magnetic anomalies revealed by the shipboard magnetometer at the ocean surface detected these stripes as positive and negative anomalies superimposed on the earth's smooth field (Fig. 4-6). When the stripe is magnetized in the direction of the earth's present field, the effect is additive, and a strong magnetic intensity results. When the stripe is magnetized in the opposite direction (reverse polarity), it subtracts from the present magnetic intensity, leaving a low value. By the middle 1960s, when the younger part of the polarity time scale was becoming well established, Vine and Wilson [1965] and Vine [1966] began to correlate the linear magnetic anomaly patterns with the polarity time scale. J. Tuzo Wilson, for instance, had been stimulated by data presented by N. Opdyke on the magnetic stratigraphy of sediments at the International Gondwana Conference in 1965 in Montevideo. From these data he formulated his developing concepts of sea-floor spreading. The excellent correlations between linear marine magnetic anomaly patterns and the polarity time scale thus led to the convincing demonstration of the relationships between marine magnetic anomaly patterns and the sea-floor spreading hypothesis of Hess [1962] and of Vine and Matthews [1963]. Because the oceanic ridges are long, linear features, the overall effect was one of long, linear stripes of normal and reversed polarity (Fig. 4-5). Furthermore, bilateral symmetry of the anomaly patterns along the ridge axis resulted from the separation of formerly adjoining segments of ocean crust exhibiting the same magnetic polarity (Fig. 4-7). The central positive magnetic anomaly over the ridge crest (Fig. 4-7) represents formation of oceanic crust during

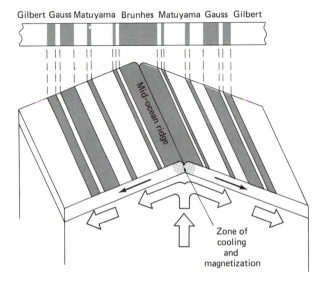

Figure 4-6 Production of magnetic anomaly patterns (magnetic polarity time scale shown at top) due to accretion of new volcanic rock at the ridge crest followed by bilaterally symmetrical sea-floor spreading away from the ridge axis. (Modified after A. Cox and others, 1967)

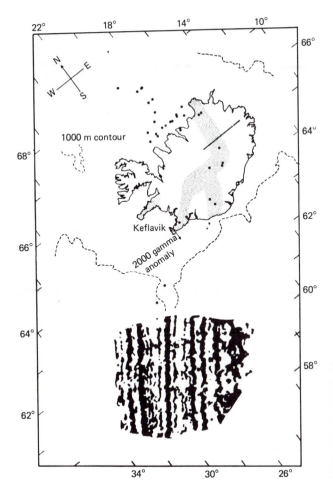

Figure 4-7 Magnetic anomalies observed over Reykjanes ridge south of Iceland. The axis of the ridge is marked by a central positive anomaly. (Reprinted with permission from *Deep Sea Research*, vol. 13, Heirtzler et al., "Magnetic Anomalies over the Reykjanes Ridge," 1966, Pergamon Press, Ltd.)

the current normal polarity of the earth's magnetic field. Ironically, a similar model had been formulated by the Canadian geophysicist L. W. Morley and submitted for publication in 1963, but it was rejected for being too speculative [Cox, 1973].

There is at present no general agreement on the thickness and susceptibility of the layer of basalt that produces the magnetic anomalies. Vine and Matthews [1963] suggested that it was the entire thickness of the oceanic crust, but measurements of the susceptibility of pillow lavas dredged from the axis of the mid-Atlantic Ridge and uplifted onto land suggests that only the pillows have a stable remnant magnetism, and their susceptibility is sufficient to produce the observed anomalies if the total thickness is 400 m.

The ideas of Vine and Matthews and other closely related studies were presented in 1966 at the annual meeting of the American Geophysical Union in Washington, D.C., and at the annual meeting of The Geological Society of America in San Francisco. The combined effect was to promulgate the new discoveries, effectively destroy remaining effective opposition to the theories, and usher in a conceptual revolution about how the earth operates.

The paleomagnetic time scale

The symmetry of the magnetic anomalies along ridge crests persists beyond the crest and upper flanks of the oceanic ridges and extends across the lower flanks into the deep-ocean basins for distances as great as 2000 km. This shows that virtually all of the modern ocean has been created by sea-floor spreading. If the ocean floor magnetic anomalies were correlated to the geological time scale for the total duration of the magnetic anomaly record in the ocean basins, the age of the crust could be related directly to the distance from axis of the ridge, and *rates* of sea-floor spreading determined for each of the ocean ridges. This step required magnetic profiles of high quality and great length, run at right angles to the ridge axis from which the anomalies were generated. Such profiles showed symmetry along the ridge axis, as had been shown by Heirtzler and others in 1966 for the Reykjanes ridge southwest of Iceland (Fig. 4–7). Fortunately, a large body of magnetic profile data was immediately available at Lamont, having been routinely collected for several years. Thus this phase of the work was rapidly completed by J. R. Heirtzler, W. C. Pitman, G. O. Dickson, and X. Le Pichon. They showed in a series of papers published in the *Journal of Geophysical Research* [1968] that the records of magnetic anomalies in different parts of the world ocean were very similar (Fig. 4–8) and that the records could be correlated in detail using the anomaly patterns. Furthermore, symmetry was established for ridges in the South Pacific by Pitman and others, in the South Atlantic by Dickson and others, and the Southeast Indian Ocean by Le Pichon and Heirtzler. In 1966 a profile was obtained for the Southeast Pacific across the East Pacific rise south of Easter Island. The highest rate of spreading known is in this region. Consequently, the reversal time scale is reproduced in remarkable detail (Fig. 4–8).

The reversals of the earth's magnetic field are not uniformly periodic (see Chapter 3). If they were, the magnetic anomalies would be consistent, thereby making them difficult to use to date the ocean crust. However, just as the variation in tree rings reflect growth differences because of seasonal changes, the magnetic anomalies produced at certain times are distinctive and can often be immediately recognized. The "wiggles" of the anomalies exhibit shape differences (Fig. 4–8), and the shapes of several anomalies are particularly distinctive. These are key isochrons for correlation and dating (Fig. 4–8). The sequence of anomalies for the Cenozoic are identified numerically from 1 to 33 in order of increasing age (Fig. 4–8). Anomalies of Mesozoic age are provided with a prefix M and are numbered from M0 (108 /± 2 m.y.) to M25 (153 m.y.). The quality of magnetic profiles varies considerably; those employed for construction of the magnetic time scale exhibit particularly clear records. Others are much less clear, however, and identification of individual anomalies is ambiguous. Anomalies can be poorly developed or obscured in areas of structural complexity—for instance, where the ridge axis is offset by numerous fracture zones.

Based on the study of magnetic lineations in normal to active spreading centers, a time scale of geomagnetic reversals was predicted by Heirtzler and others [1968] for the Cenozoic and late Cretaceous (0–79 Ma; Fig. 4–9) using the polarity time scale then known for the last 3.5 m.y. (see Chapter 3). This predicted time scale is based on two main assumptions: Magnetic anomaly profiles above the

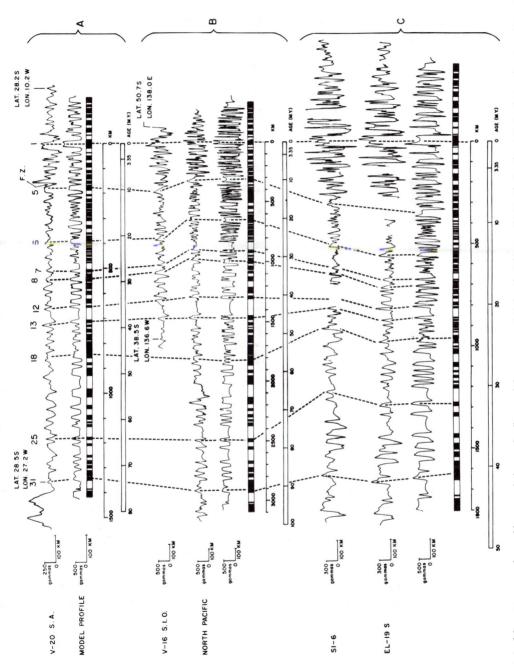

Figure 4-8 Magnetic anomaly profiles recorded over the floors of three oceans: (A) South Atlantic; (B) North Pacific; (C) South Pacific. Correlation lines are shown between the individual profiles of particular anomalies identified by numbers at the top of the figure (1 through 31). Note the similarity of the profiles. The horizontal bars of dark and light shapes represent the succession of normally and reversed magnetized bodies of oceanic crust running parallel to the oceanic ridge. (From J. R. Heirtzler and others, *J. Geophys. Res.*, vol. 73, pp. 2119–2136, 1968, copyrighted by the American Geophysical Union)

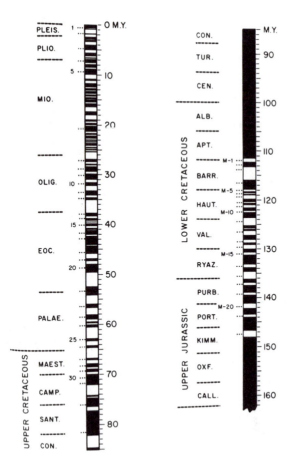

Figure 4-9 Geomagnetic polarity time scale for last 162 m.y. (early Jurassic to Present Day). Black = normal polarity; white = reversed polarity. Abbreviations for Cretaceous and late Jurassic stages are shown in full in Fig. 3–1. Numbered magnetic anomalies are shown to left of each column. (From R. L. Larson and W. C. Pitman, III, *Geol. Soc. Amer. Bull.*, vol. 83, p. 3651, 1972, courtesy The Geological Society of America)

ocean ridges and basins were manifestations of earlier reversals in the polarity of the earth's magnetic field, as hypothesized by Vine and Matthews [1963]; and that sea-floor spreading rates have been constant. Resolution was derived from magnetic anomaly profiles over the rapid sea-floor spreading system in the North Pacific. Such a time scale extrapolates almost 20 times the length of the baseline, and, as emphasized by Heirtzler and others [1968], there is a possibility of progressively larger systematic error with increasing age. Nevertheless, their resulting magnetic time scale was assigned an age only slightly younger than the radiometrically determined age of the Cretaceous-Tertiary boundary.

Since development of the polarity time scale by Heirtzler and others, several attempts (for example, Fig. 4–9) have been made to improve resolution of the reversal pattern, using new magnetic anomaly profiles, new data on the ages of critical polarity boundaries in land-based sections, and especially comparison of assigned oceanic basement ages with biostratigraphic ages of basal sediments in deep-drilled holes. The ages of several anomalies have now been directly determined by drilling through the deep-sea sediment column to the ocean crust. These studies have shown that all anomalies younger than Anomaly 32 were formed during the last 76 m.y.

(Fig. 4–9). Good first-order agreement exists between predicted ages and paleontological ages for the last 60 m.y. The time scale based on spreading rates was confirmed in general and has also shown that there is little sediment buried beneath the acoustic basement at most sites in which the basement has been drilled. However, much of the work has shown that the original assumption of a constant sea-floor spreading rate is often slightly and occasionally greatly in error. For instance, Anomaly 24 has been assigned values as different as 60 Ma in the extrapolation of Heirtzler and others [1968] and 49 Ma by Tarling and Mitchell [1976]. Also Schlich [1975] and Larson and Pitman [1975] documented a consistent offset between Cretaceous basement ages determined from magnetic anomaly patterns and biostratigraphic ages of overlying deep-drilled sediments (Fig. 4–10). This offset has since been reconciled by La Brecque and others [1977] through modification of several biostratigraphic ages. These discrepancies or differences in opinion seem to

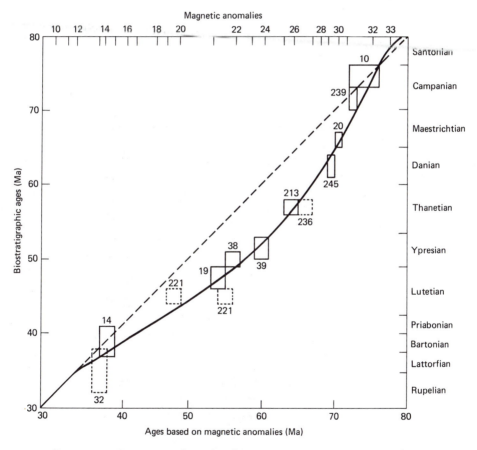

Figure 4-10 Comparison of ages based on magnetic anomaly patterns of the sea floor and those based on fossils in sediments immediately overlying basement rocks in numbered DSDP sites. (After R. Schlich, 1975, with permission of the Societé Géologique de France)

increase with increasing age, although attempts to rectify these by assuming different rates of sea-floor spreading have been successful in many cases. Usually, the agreement between predicted and biostratigraphic ages is surprisingly close, considering the problems that erosion, nondeposition of sediment, renewed volcanism, and uncertainties in absolute dating of geologic epochs give to the biostratigraphic dating of the underlying basaltic rocks. The results from the DSDP not only have been consistent with the concept of sea-floor spreading, but have provided impressive support for the hypothesis. Future revisions of the polarity time scale will allow for changes in the rate of sea-floor spreading and use chronological information from several approaches.

The time scale has also been extended further into the Mesozoic using sea-floor magnetic anomaly patterns. Larson and Pitman [1972] extended the geomagnetic reversal time scale back to the beginning of the late Jurassic (162 Ma). This shows (Fig. 4–9) that a period of reversals occurred from between 150 and 100 m.y. ago, and these reversals are bracketed by long periods of dominantly normal polarity known as the Cretaceous and Jurassic **magnetic quiet zones.** These are marked by subdued magnetic signature over areas of the ocean floor.

Age of Oceanic Crust

Once the magnetic anomaly sequence and the ages of individual anomalies had been established, an efficient tool became available to map the age of large sections of oceanic crust. The mapping revealed sequences of magnetic lineations, separated by fracture zones forming a zebralike mosaic of varying complexity throughout almost all of the deep ocean floor (Fig. 4–11). By 1974, Pitman, Larson, and Herman had produced the first global map of the age of the oceanic crust (Fig. 4–12) using anomaly patterns from different sources. This map shows the broad sweep of the youthful oceanic ridges and increasing age of oceanic crust down both flanks of the ridge system, as predicted in the sea-floor spreading model.

One of the most striking discoveries made during the exploration of the oceans is the relative geologic youth of all of the ocean basins. The early piston-coring work throughout the ocean basins had not collected any deep-sea sediments older than the Cretaceous. Before the development of the polarity time scale this had suggested that the oceans were young. However, the evidence was inconclusive without deep-sea drilling, and it was again the pattern of oceanic magnetic anomalies that provided the convincing evidence.

That part of the oceans for which Anomalies 1 through 32 have been mapped is over half the area of deep-ocean floor (Fig. 4–12), implying that over half of the ocean floor was formed less than 76 Ma. The inferred young age of the oceanic crust was proved by direct drilling in the DSDP. The oldest deep oceanic crust dated either geophysically or by direct paleontological dating is only of middle to late Jurassic age, or about 170 m.y. old. Crust of this age occurs at the margins of the North Atlantic, in the Northwest Pacific, and to the west of Australia. Not a fragment of pre-middle Jurassic oceanic crust has been identified. Thus the present area of the oceans has been created by sea-floor spreading during only about 5 percent of the recorded geologic history of the earth. Rocks much older than this have been

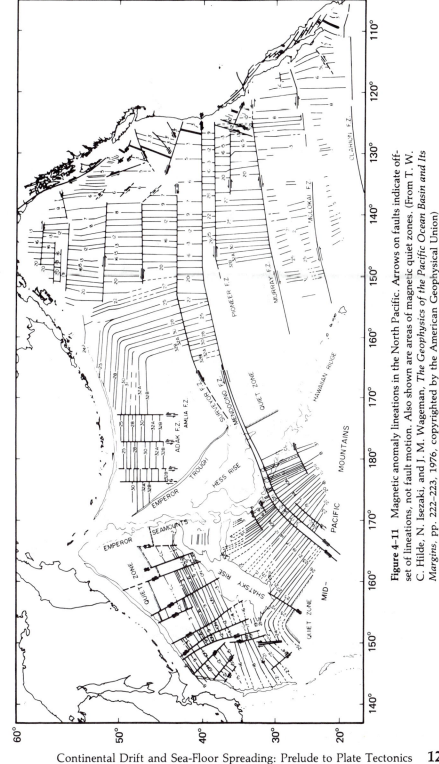

Figure 4-11 Magnetic anomaly lineations in the North Pacific. Arrows on faults indicate off-set of lineations, not fault motion. Also shown are areas of magnetic quiet zones. (From T. W. C. Hilde, N. Isezaki, and J. M. Wageman, *The Geophysics of the Pacific Ocean Basin and Its Margins*, pp. 222–223, 1976, copyrighted by the American Geophysical Union)

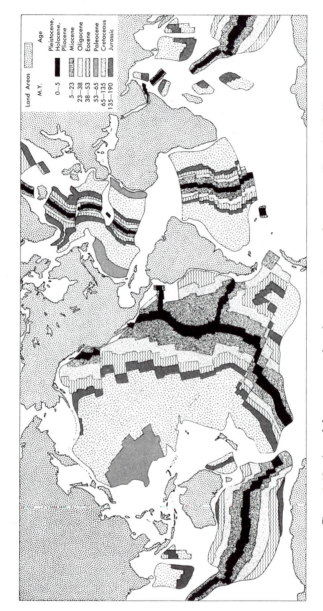

Figure 4-12 Age of the ocean crust based on magnetic anomaly patterns. (From W. C. Pitman, III, R. L. Larson, and E. M. Herron, *Geol. Soc. Amer. Bull.,* 1974, courtesy The Geological Society of America)

Legend:
Land Areas

M.Y.	Age
0—5	Pleistocene, Holocene, Pliocene
5—23	Miocene
23—38	Oligocene
38—53	Eocene
53—65	Paleocene
65—135	Cretaceous
135—190	Jurassic

identified on certain continental margins, but these form part of a continental block. Many continental rocks are much older than the late Mesozoic. The oldest known rocks are about 3.7 billion years old. All known rock older than the late Mesozoic is found on the continents. The preservation of continental rocks over long periods of time is due to their low density, which prevents them from being subducted into the mantle beneath the ocean trenches even when a continent (for example, India) is transported into such a zone. The absence of pre-Jurassic oceanic crust implies a global system of ocean crustal renewal and destruction. Figure 4–13 is a histogram of oceanic crustal age versus area of the modern ocean. This shows a decrease in the area of ocean crust as it becomes older, especially older than 80 m.y. (late Cretaceous). Geologically the ocean floors are temporary features.

The map of crustal ages of the ocean shows important broad features; the implications of these patterns will be discussed in more detail later.

1. The North Atlantic contains rocks of late Jurassic age near the margins of North America and North Africa, while the South Atlantic lacks these. This indicates a younger age of opening of the South Atlantic.

2. Much of the area of the present Pacific Ocean has been created by northwest spreading from the East Pacific Rise; thus large sections of the central and western Pacific are relatively old (Cretaceous and late Jurassic), while most of the Southeast Pacific is of Cenozoic age.

3. The oldest known rocks in the ocean are of Jurassic age and occur in the Northwest Pacific.

4. All of the sea floor between Australia and Antarctica was formed within the last 55 m.y., implying that the separation of these two continents was the last stage in the fragmentation of Gondwanaland.

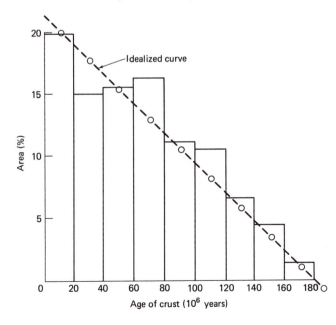

Figure 4-13 Histogram of oceanic crustal age (in 20 m.y. units) versus percent area for today's ocean, derived from measuring areas on crustal age map (Fig. 4-12). The idealized curve is an average of the steps in the histogram. (After W. H. Berger and E. L. Winterer, 1974)

5. Almost all of the Indian Ocean is younger than the late Cretaceous, indicating recent development of this ocean.

Magnetic Quiet Zone

The magnetic quiet zones are found beyond the oldest correlated anomalies and are essentially devoid of the linear anomalies characteristic of oceanic crust of Cenozoic age. In the Atlantic Ocean, these quiet zones lie near the margins of the continents. The origin of these quiet zones has been controversial. Paleomagnetic studies on land have shown that much of the Cretaceous is marked by few magnetic reversals and that, throughout most of this period, the earth's magnetic field exhibited normal polarity (Fig. 4–9). Also, the late Triassic to early Jurassic was marked by predominantly normal magnetic polarity. This contrasts with the Cenozoic, which exhibits frequent polarity reversals (Fig. 4–9). Thus the magnetic signature of oceanic crust found during periods of more or less constant polarity should exhibit a subdued or constant magnetic signature. This has been confirmed by Larson and his colleagues and accepted by most workers to explain the origin of the quiet zones. The change from constant polarity to frequently reversing polarity occurred during the late Cretaceous. It has been suggested by Leyden and others [1971] and La Brecque and Rabinowitz [1977] that the seaward edges of these quiet zones in the Atlantic must be an isochron, since the African boundary can be fitted against the North American boundary. However, not all quiet zones may result from crustal formation during times of infrequent reversals. Other theories summarized by Emery and others [1970] include (1) erasure of the original magnetic signature by metamorphism, (2) poor paleomagnetic record due to complexity of rifting during early stages of the opening of oceans, and (3) formation of the oceanic crust at low magnetic latitudes.

Spreading-Rate Differences and Temporal Oscillations

The pattern of magnetic anomalies shows that rates of spreading are not constant throughout the length of the mid-ocean ridge system, but vary from region to region. Spreading rate is normally used as a half-rate of separation from an axis which is assumed to be bilaterally symmetrical. Rates of spreading vary from a few millimeters per year on the Nansen Ridge, Arctic Ocean, to 1 cm per year on the mid-Atlantic ridge near Iceland, to about 6 cm per year on the East Pacific rise in the equatorial Pacific. Differences in spreading rate appear to influence ocean-ridge topography. For instance, ridges exhibiting a median valley at ridge crests and high relief in general seem to be associated with slow rates of spreading. Ridge crests in the Atlantic and Northwest Indian oceans, which are marked by spreading rates of less than 2.5 cm per year, exhibit a median valley and high relief. In contrast, ridges in the Pacific, which are marked by faster spreading rates, typically lack a median valley and exhibit less relief.

It is also clear that spreading rates have varied through time. As early as 1967, J. Ewing and M. Ewing suggested that spreading rate differences may have occurred in the past and that the present rate of spreading may have only been operative during the late 10 m.y. They suggested that a widespread decrease in spreading rate may have occurred before Anomaly 5 time (9 Ma), based on a sudden increase in sediment thickness at that time. A decrease in the spreading rate at this time is not clear but could have lasted more than a few million years.

Variations in spreading rates imply several things. First, if there are significant changes in spreading rates at all ridge crests and the earth is not expanding, then there must be corresponding changes in the worldwide rate of crustal subduction. This might be reflected in increased volcanic activity in the island arcs and continental areas adjacent to the subduction zones (a phenomenon that is discussed in greater detail in Chapter 12). Larson and Pitman [1972] have called for very rapid spreading (up to 18 cm per year) at all spreading centers in the Atlantic and Pacific oceans in the Cretaceous from 110 to 85 Ma. Also, they related the rapid spreading to increased circum-Pacific intrusive and extrusive activity and orogenesis during this period. For instance, it appears that extensive plutonism occurred in eastern Asia, West Antarctica, New Zealand, the southern Andes, and western North America during the early to middle late Cretaceous. This is best documented in western North America, where more than 50 percent of the exposed batholiths range in age from 115 to 85 Ma. If the granodiorites and granites that make up these batholiths were related to underthrusting of oceanic lithosphere, then they represent enormous amounts of lithospheric subduction, which would be a consequence of rapid spreading at this time [Larson and Pitman, 1972].

The second major consequence of widespread changes in spreading rates is the effect this should have in changing the total volume of the mid-ocean ridge system, which in turn, should create major changes in sea level. Valentine and Moores [1970] suggested that such a change might result from continental collision when mid-oceanic ridges would be consumed in the subduction zones and thus create a reduction in total ridge volume. This would cause a lowering of sea level and a marine regression. In contrast, the rifting of continents and the development of new oceans would create new ridges, causing higher sea levels and transgressions. In 1963, Hallam suggested that ridge volume should be altered by changing rates of sea-floor spreading, with fast spreading yielding wide ridges and slow spreading yielding narrow ridges. Since then, specific correlations between global transgressive and regressive episodes and rates of sea-floor spreading have been shown. Hays and Pitman [1973] used the spreading-rate data of Larson and Pitman (1972) calculate to volume changes of the mid-oceanic ridges for each 10 m.y. from 100 Ma to the present. They were able to show that rapid spreading in the late Cretaceous accounted for a large contemporaneous transgression, which was followed by regression during the early Tertiary. Thus changes in the rates of sea-floor spreading, if occurring over wide areas, can produce large changes in the environment and in sedimentation through sea-level changes (see Chapter 9).

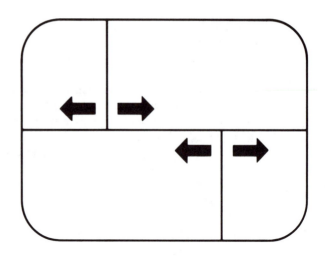

five

Plate Tectonics

. . . in times to come, these twenty years when men gained a new level of understanding of their planetary home will be thought of as one of the great ages of exploration. . . . The simplicity and grandeur of this hypothesis of sea-floor spreading has caught the imagination of scientists throughout the world.

Roger Revelle

THE BASIC CONCEPT

A major extension of the conceptual revolution about the earth is the unifying theory of global plate tectonics, which embraces continental drift and sea-floor spreading. The theory of plate tectonics was formulated by J. T. Wilson [1965] with the description of a continuous and integrated network of ridges, transforms, and subduction zones bounding large, rigid *plates,* and, by W. J. Morgan [1968] who quantified the geometry of the plates.

The essential idea of plate tectonics is that the entire surface of the earth is composed of a series of internally rigid, undeformable, but relatively thin (100–150 km) plates. Although the size of the plates is variable, most of the earth's surface is covered by seven major plates (Fig. 5–1), such as the plate that occupies almost all of the Pacific Ocean, supplemented by several small, sometimes insignificant plates, such as that which corresponds to most of Turkey. These plates are continuously in motion, both in relation to each other and to the earth's rotational axis. The plates, which are essentially aseismic, are bounded by active ridges, trenches, rifts, and great fractures, or *megashears,* associated with extensive seismicity. Plate motions and their interactions are responsible for the present positions of the continents, for the formation of the earth's mountain ranges, and for most of the earth's topographic characteristics and its major earthquakes. This explains why earthquakes and volcanoes are for the most part concentrated in very narrow zones and why some earthquakes are shallow and others are deep. Plate tectonics, as the in-

131

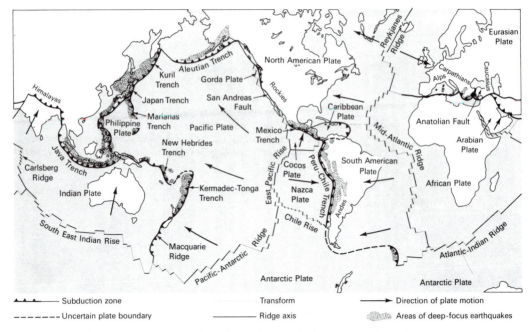

tegrator of continental drift and sea-floor spreading, has been very successful in explaining many of the earth's features and in providing information about processes in the earth's interior. Wilson has compared the central idea of plate tectonics to the theory of the Bohr atom in its simplicity, its elegance, and its ability to explain a wide range of observations.

Plates are constantly added to at the active mid-ocean ridges where hot volcanic material is exuded. This material becomes part of the plate where it has cooled enough to develop mechanical strength. Plates are, therefore, thinner at the area of accretion near the ridges and gradually thicken with increasing distance from the accreting boundary. Once formed, the new crust becomes part of a rigid plate. However, the oceanic crustal plates are transient features that will eventually be destroyed in the **subduction zones**, beneath ocean trenches (Fig. 5–1). This destruction occurs at the same rate as new crust is added at the mid-ocean ridges. During the subduction process, the cool crustal slab descends into the hot mantle material. Because plates are poor conductors of heat, they must descend several hundred kilometers into the mantle before being brought to the temperature of the mantle. These are zones of deep seismicity called the **Benioff zones,** after their discoverer, Hugo Benioff, which dip toward the continental areas from the ocean trenches and island arcs.

THE GEOMETRY OF PLATE TECTONICS

Plate Boundaries

Plates interact in three ways: They diverge, converge, or they slide past each other. These three plate interactions form three types of plate boundaries. **Ridges** are the centers of divergence, or spreading apart, of two plates. These are the principal areas for formation of oceanic basalts and represent **constructive plate margins.** Much more is known about the structure and evolution of ridges than of trenches. Ridges are often perpendicular to the motion between the two plates on each side.

Trenches, or sinks, are the areas of convergence of two plates, where one is thrust under the other and where older crust is discarded. Unlike ridges, trenches show no tendency to lie perpendicular to the motion between two plates. These boundaries are often curved, with the plate that is not being destroyed exhibiting a convex plate boundary. Normally only one plate is consumed at these boundaries, so trenches are asymmetric, although there is no geometric reason why the edges of both plates are not consumed at a subduction zone. These are **destructive plate margins.** Three types of plate convergence have been recognized in the present day: (1) ocean-to-ocean plates, like the Mariana Arc or Tonga Arc, (2) continent-to-ocean plates, as in the Peru-Chile Trench and the adjacent Andean Cordillera, and (3) continent-to-continent plates, such as India into Asia and the adjacent Himalayas-Tibetan Plateau (Fig. 5-1).

Transverse fractures or transform faults, representing most of the oceanic fracture zones, exist where two plates move parallel to each other but at different rates or in opposite directions. The azimuth of these fracture zones indicates the direction of movement of the two plates. No plate is either formed or destroyed along these boundaries; they are slip lines and have been called **conservative plate margins.**

It is important to recognize that the relative motions of all types of plate boundaries are motions on a sphere and therefore have a pole of rotation. Because of this, the rate of convergence, divergence, or slip at a margin will depend upon its distance from the pole of rotation.

Transform Faults

Major geometric aspects of plate tectonics are the transform faults and associated great fracture zones that enable the plates to rotate relative to each other. The mid-ocean ridge does not wander unbroken throughout the oceans but is constantly offset by transform faults (Fig. 5-1), which may create high submarine cliffs. These structures are normally marked by narrow, linear slots running across the ocean floor. These slots may be as deep as 1500 m below the crest of the mid-ocean ridge and become closer to average ocean depths with increasing age from the ridge crest. In the Pacific Ocean, the slots become filled with volcanic rock-flows away from the area of the ridge crest. In the Atlantic, there is less filling with volcanic rocks, and a slot often remains down the flanks of the oceanic ridge. At 4000 m the slot can still be 500 m deeper than average ocean depth. A transform fault is the offset between the axes of the mid-ocean ridge, where the plates slip past each other to accommodate their relative motions (Figs. 5-2 and 5-3). These offsets are connected with, and continue for great distances as, fracture zones beyond the offset ends of the ridge axis (Fig. 5-2), forming topographic irregularities on the sea floor. Originally, transform faults were believed to represent lines of offset of the mid-ocean ridge, but J. T. Wilson, who coined the term in 1965, suggested that they are offsets of the spreading axis and that they transform motion between two segments of the ridge. This was confirmed by seismic first-motion studies by Isacks, Oliver, and Sykes [1968], who showed that only that part of the fault joining the ridge ends is marked by earthquakes, while the fracture-zone extensions are inactive traces of transform faults. The motion across transform faults is lateral and its rate is proportional to the rate of spreading at the two ridge axes. The direction of motion of the two plates is exactly parallel to the fault (Figs. 5-2 and 5-3) and should be described by small circles around the center of rotation of the plates.

It was the tendency of transform faults and their associated fracture zones to lie along small circles and to define parts of circles of rotation that first suggested the idea of plate tectonics geometry to W. J. Morgan [1968] and X. Le Pichon [1968]. The inactive continuation of transform faults as fracture zones defines circles of rotation for the previous history of transform motion. These represent the geological evidence of earlier plate rotation. Inactive fracture zones, which pass

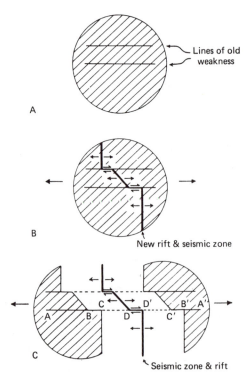

Figure 5-2 Three stages of rifting and separation of a continent into two parts, such as South America and Africa. Changes in relative motion along the rift are accomplished along transform faults, such as D–D¹. Seismic activity occurs along the heavy lines only. The seismically inactive lateral extensions of the transform faults are fracture zones which may extend into the continent along the lines of old weaknesses. (From J. T. Wilson, 1965, reprinted by permission from *Nature*, vol. 207, pp. 343–347, copyright © Macmillan Journals, Ltd.)

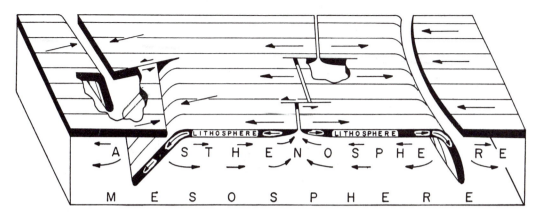

Figure 5-3 Lithospheric plate motions in three dimensions shows plate generation along the mid-ocean ridges; relative movement of adjoining blocks away from the ridges and the subduction of the cold sinking slabs to great depths beneath the trenches is shown. One arc-to-arc transform fault appears at left between oppositely facing zones of convergence (island arcs). (After B. Isacks, J. Oliver, and L. R. Sykes, 1968)

into transform faults following the same circle of rotation, provide evidence of long-term similar plate motion. Fracture zones that abruptly change direction are evidence of ancient changes in plate motions. Transform faults are conservative plate margins because there is normally no addition or subtraction of the lithosphere along them. However, if transform faults do not exactly correspond with small circles of rotation, they must undergo a small amount of subduction, or spreading, along their length. A transform fault exhibiting a spreading component is called a **leaky transform fault.** These can form accessory ridges nearly at right angles to the axis of the principal ridge.

The Gulf of California and the San Andreas fault are related to a system of transform faults (Fig. 5-4). The East Pacific rise extends northward from the equatorial Pacific and disappears near the mouth of the Gulf of California (Fig. 5-4). From the north, the San Andreas fault and a related system of parallel faults run through California and disappear under the Salton Sea Trough. Wilson proposed that the San Andreas fault, which runs through much of California and upon which San Francisco sits, is a ridge-to-ridge transform fault with clockwise, or dextral, relative motion. As such it forms a connecting link, and transforms motion between the East Pacific rise at the mouth of the Gulf of California and the reappearance of the same rise as the Gorda rise and the Juan de Fuca ridge in the North Pacific off the coast of Washington (Fig. 5-4). Since the Jurassic, offset on the San Andreas fault has been 560 km. The motion along the San Andreas fault has resulted from sea-floor spreading from the crest of the rises at the northern and southern ends (Fig. 5-4). Larson and others [1968] have modified this model to show that the Gulf of California and the San Andreas fault are not associated with a single transform fault, but a family of parallel transform faults that run slightly oblique to the Gulf (Fig. 5-4). At the northern end of the Gulf, this series of parallel offset lineaments extends into southern California along the San Andreas

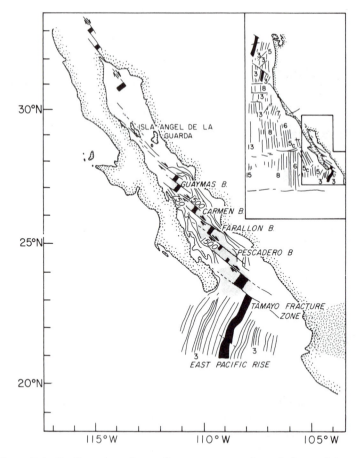

Figure 5-4 Configuration of spreading centers, transform faults, and fracture zones in the Gulf of California. The heavy, black lines represent possible spreading centers. (From L. A. Lawver and D. Williams, *Journal of Geophysical Research*, vol. 84, pp. 5465–78, 1979, copyrighted by American Geophysical Union)

fault system. Strike separation along each of the transform faults and sea-floor spreading off the ridge segments have caused the opening of the Gulf of California. Larson and others [1968] have shown that southern Baja California has separated from mainland Mexico by 260 km during the last 4 m.y.

Geometry of Plate Motions

The new global tectonics deals with the motion of seven major lithospheric plates: Eurasian, Indian, Pacific, Antarctic, North American, South American, and African plates. Plate tectonics has gone beyond the concepts of sea-floor spreading by the analysis of the motions of the various parts of the lithosphere in terms of a spherical shell rather than a plane (Fig. 5–5). The key to this geometry was provided largely by Morgan [1968] and Le Pichon [1968], who discovered that transform

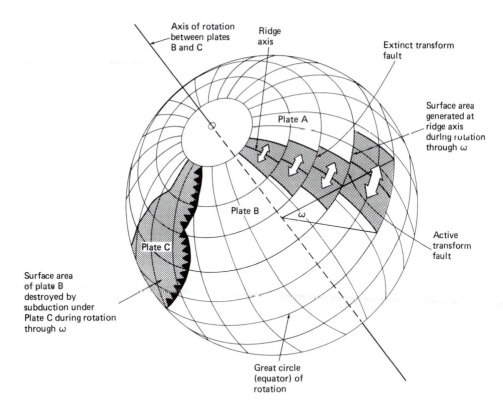

Figure 5-5 Example of the relative motion of three rigid plates on the surface of a sphere. Shown is the axis of rotation between plates B and C. The relative motion increases between adjoining blocks with increasing distance from the pole of rotation to the great circle of rotation. The relative differences in motion are taken up along transform faults. As the plate B rotates through angle omega (ω), new surface area is added symmetrically to both plates A and B about the ridge axis, and surface area of plate B is destroyed by subduction under plate C. (Modified after J. Dewey, 1972)

faults and fracture zones could be accurately fitted by a set of small circles about a pole of relative motion between two plates. Rotational poles can be located by determining the intersection of great circles perpendicular to transform faults. If two plates on the earth's surface are spreading on either side of the ridge, their relative motion must pivot around a point called the *pole of rotation* (Fig. 5-5). This is a unique point that does not move relative to either of the two plates. The **axis of rotation** is perpendicular to the earth's surface, passing through the pole of rotation and the center of the earth (Fig. 5-5). Plate tectonics follows Euler's geometrical theorem, which states that a layer on a sphere can be moved to any other orientation by a single rotation about a suitably chosen axis through the center of the sphere. This theorem permits any rigid motion between two plates to be described by three parameters: the latitude and longitude of the intersection of the rotation axis with the earth's surface and the angle through which the plate must be rotated [McKenzie, 1972]. Most poles of rotation are at high latitudes, but do not necessarily fall near the geographic poles of the earth, to which they are unrelated. The poles of rotation for the Pacific and Atlantic oceans are close to the earth's magnetic poles: near Greenland and in the Antarctic south of Australia. The pole of rotation for the Indian Ocean is located less certainly, but it is in North Africa or the Pacific north of New Zealand.

The relative motion between two plates is described as the **angular velocity** (w or *omega*, ω), which is independent of the latitude of rotation (Fig. 5–5). The velocity of relative motion across a constructive or destructive boundary is proportional to both the angular velocity about the axis of rotation for the motion of the plates and the angular distance of the point of the boundary from the axis of rotation (Fig. 5–5). Thus the relative motion will vary continuously along all such boundaries, being smallest at high latitudes of rotation and greatest at low latitudes of rotation.

The point where three plate boundaries intersect is known as a **triple junction** (Fig. 5–1). There are two main types.

1. Stable triple junctions between three ridges. In this case the three plate boundaries are all accretionary and relatively stable.

2. Unstable triple junctions at the intersection of three subduction zones.

The motions of all plates are to some extent interdependent, so a change in the velocity or direction of motion of one plate must be reflected by changes in motions elsewhere. All motion is *relative,* since there is no known method for defining absolute motion of plates. The relative motion between two plates can be established by determining their movement at two points along a common boundary. Plate motion may or may not involve rotation. Wilson discussed many of the consequences of relative motion of rigid plates, such as the changing geometry of transforms between ridges and subduction zones, evolving mosaics of ridges, transforms, and subduction zones, and the changing area of plates. In order to determine the relative motion of a plate mosaic, a method of finite rotations can be employed. At any particular time, the relative motion of a plate pair, or the motion of the plates relative to a fixed reference coordinate system, can be determined by a set of **instantaneous rotation vectors** [Morgan, 1968]. Chase [1972] and Minster and others [1974] have developed refined instantaneous plate motions for the global plate mosaic. The approach of Minster and his colleagues was to establish a zero summation of instantaneous angular velocity vectors to describe the relative motion between 11 major plates by continuously juggling the positions of rotational axes and using angular velocities to determine those that best fitted observational data, including a set of transform fault trends [Dewey, 1972]. Models of instantaneous plate motions have been successful in predicting both magnitudes and directions of relative plate motions and have been tested in many ways, leading to a good understanding of present-day tectonic configurations. However, attempts to understand the details of the evolution of the present system through geologic time is much more difficult. Normally this is done by choosing a pole about which the past few million years of crust (for instance, 10 m.y.) can be rotated back to the ridge axis. Then a second pole is chosen to rotate back the next time segment (for example, 10 to 20 m.y.), and so on throughout the extent of time encompassed by the evolutionary scheme. This approximation of how the tectonic system evolved is called a **series of finite rotation poles.** It is only an approximation to reality because it is likely that infinitely small adjustments in pole location occurred throughout the evolution of the system.

Plate motion can undergo changes through geologic time for various reasons. Foremost are the interplate motion changes that result from the contact of a continental fragment with a subduction zone. This causes either the motions at the boundary to change or relative motion between the two plates to cease. If the latter occurs, a single plate forms, and adjustment in motion is necessary at other plate boundaries.

Seismicity and Plate Tectonics

Seismology has provided important evidence in support of plate tectonism. Earthquakes result from the release of stresses associated with rock breakage. Their global distribution patterns and depths of origin had been studied for years before the ideas of sea-floor spreading and plate tectonics were formulated. However, once the theory was formulated, a large body of seismological data neatly fell into place.

The basic patterns of global earthquake activity had been determined by seismologists by the middle 1950s. Gutenberg and Richter [1954] had determined that earthquake activity is largely confined to the young fold mountains and ocean trench systems of the Alpine-Himalayan and circum-Pacific belts and to the crests of the mid-ocean ridges (Fig. 5-6). Furthermore, by the middle 1950s M. Ewing and B. Heezen had discovered that numerous earthquake epicenters were not only associated with mid-ocean ridges, but lay along the median valleys of these ridges. Seismological research rapidly advanced during the following decade as a result of the establishment of 125 seismological stations, constituting the Worldwide Standardized Seismograph Network (WWSSN), which was initially established to differentiate between natural earthquakes and nuclear tests [Cox, 1973]. The work of the WWSSN proved that the earthquake distribution pattern was a narrow, sinuous zone encircling the globe and surrounding large areas of infrequent earthquake activity. Figure 5-6 shows a seismicity map of the earth for the period 1961 through 1967, prepared by Barazangi and Dorman [1969] using data from about 29,000 earthquakes. The next major conceptual development occurred during the middle to late 1960s by B. Isacks, J. Oliver, and L. Sykes of Lamont, who related the distribution and behavior of earthquakes to plate tectonic theory. They demonstrated that seismicity is concentrated along plate boundaries. The patterns of earthquakes represent one of the major criteria used in defining the positions of plate boundaries. They showed the following:

1. The seismic belts are narrow; in some places, particularly along the ocean ridges, they are very narrow, supporting the earlier observation by Ewing and Heezen of association with ridge crests.

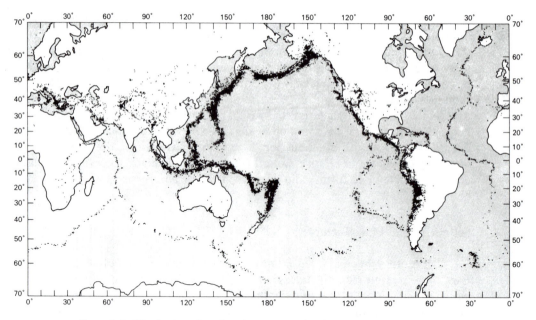

Figure 5-6 Distribution of earthquake epicenters for the period 1961 through 1967. Those in the oceans are concentrated in a narrow, often stepped zone along the mid-oceanic ridges. Broader, more concentrated zones mark the areas of convergence. (From B. Isacks, J. Oliver, and L. R. Sykes, *Journal of Geophysical Research*, vol. 73, p. 588, 1968, copyrighted by American Geophysical Union)

2. The belts are continuous, with few gaps in the major belts. Few free belt ends are observed and belts do not cross one another.

3. Seismic belts divide the earths surface into a small number of stable blocks of irregular configuration.

Seismology indicates that the earth can be divided into a series of rigid plates with little seismic activity occurring within their areas. The seismically active boundaries are areas where the oceanic crust is created (ridges) or destroyed (trench-arc systems), where continental crust is compressed (orogenic belts) or extended (rifts and fracture zones), and where different sections of oceanic crust slip past each other (transform faults). (See Fig. 5-3).

Seismic activity is of two types: shallow and deep earthquakes. Shallow earthquakes are concentrated at ocean ridges and in the transform faults joining the ridge crests. The earthquakes occur at depths less than 30 km. Sykes [1963] observed that on fracture zones, earthquakes are limited to the transform fault sector (that is, only the segment offsetting the ridge crest). Activity there is much weaker than at the major zone of convergence. The largest earthquake known in the ocean-ridge system had a magnitude of about 7 [Gutenberg and Richter, 1954], about one-eighth the energy of the largest recorded shock in the arc systems.

Deep earthquakes occur in zones beneath the oceanic trenches, the regions where plates descend into the mantle. Seismic belts associated with zones of convergence form a much broader band of epicenters than those associated with the ocean ridges (Fig. 5–6). This is because intermediate-to-deep earthquake foci are located along the surface, which dips at a fairly steep angle from 30° to 80° (usually about 45°). (See Fig. 5–7.) These zones plunge from the trench floor to depths of 200 to 250 km (Fig. 5–7) and are known to extend downward as deep as 700 km. Zones of deep earthquake activity—the Benioff zones—are restricted to all active island-arc systems. These zones have the greatest concentration, and generate the largest earthquakes. Benioff-zone seismicity is associated with the underthrusting of the lithosphere.

Thus the areal and depth distribution patterns and the magnitude of earthquakes have provided vital information on the nature of lithospheric plates and the process of plate tectonism. Seismology has also provided a technique by which motion on fault planes associated with earthquakes can be determined, thus giving information on the orientation of faults and the slip directions of earthquakes. This is called **fault-plane solution,** or **first-motion study,** and provides needed information on fault movement, most of which occurs at depths inaccessible to observers. Earthquakes with magnitudes greater than about 5.5 are large enough to be re-

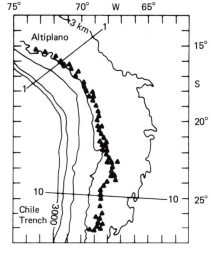

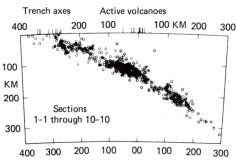

Figure 5-7 Southern Peru and Chile showing area (between cross sections 1-1 and 10-10) examined to obtain composite section at bottom exhibiting earthquake foci along the dipping Benioff zone. Also indicated on the top figure are volcanoes (solid triangles = historic volcanoes; open triangles = Quaternary volcanoes) and the projection on the lowest figure of the volcanic zone and trench axis. (From M. Barazangi and B. Isacks, *Geology,* vol. 4, pp. 686–692, 1976, courtesy The Geological Society of America)

corded by seismographs over the entire earth. The first motion of the ground at seismic stations can either be **compressional,** in which the first motion of the ground is away from the source, or **dilatational,** in which the first motion is toward the source. Elastic waves generated by an earthquake can be mixture of both types. When a cavity explodes, the first waves are compressional; an imploding cavity initially produces dilation waves. When faulting occurs, a pattern of compressional (extension) and dilation (contraction) waves are generated; the pattern defines the movement along the fault plane (Fig. 5-8). A network of seismological stations allows the changing character of waves associated with an earthquake to be determined. From first-motion studies the seismologist can determine the orientation of faults and the slip directions of earthquakes anywhere on the globe.

In 1966 Sykes realized that such focal mechanism studies (Fig. 5-8) could be employed to examine Wilson's [1965] hypothesis of transform faults, in which exactly the opposite direction of motion had been predicted on the two sides of a fracture zone as that predicted using traditional geological models. The earlier studies

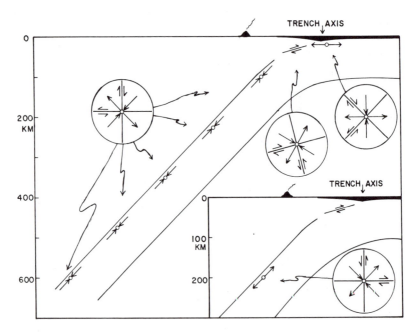

Figure 5-8 Vertical sections at right angles to an island arc schematically show typical orientations of double-couple focal mechanisms. The horizontal scale is the same as the vertical scale. The axis of compression is represented by a converging pair of arrows; the axis of tension is represented by a diverging pair; the null axis is perpendicular to the section. In the circular blowups, the sense of motion is shown for both of the two possible slip planes. The features shown in the main part of the figure are based on results from the Tonga Arc and the arcs of North Pacific. The insert shows the orientation of a focal mechanism that could indicate extension instead of compression parallel to the dip of the zone. (From B. Isacks and others, *Journal of Geophysical Research*, vol. 73, p. 5874, 1968, copyrighted by American Geophysical Union)

suggested that the ridges had been offset by transcurrent faulting. Sykes confirmed Wilson's hypothesis and provided much impetus to global tectonic theory [Sykes, 1967]. His studies also showed that many of the strike-slip faults on land, such as the San Andreas fault, may be interpreted as transform faults.

Focal mechanism studies support plate-tectonic models. They show that the ridge axes are in tension, that there is lateral movement along transforms, and that island arc regions are dominated by compressional tectonics. It has also been shown that mechanisms can either be compressionally or tensionally parallel with the dip of the Benioff zone (Fig. 5–8). First-motion studies at island arcs indicate that two types of earthquakes occur at consuming plate boundaries. They may be either tensional or compressional. Many tensional events are due to a bending plate. Tensional earthquakes, associated with a plate sinking through a low density substratum, can occur only to depths of 300 to 350 km, where a phase change in the mantle takes place (Fig. 5–9). At this level the orthorhombic mineral olivine is transformed into a dense, cubic form—spinel—creating a 10 percent density increase. Below this level the stress on the subducted plate is compressional (Fig. 5–9), causing the plate to bend, contort, or disintegrate due to differential motions in the underlying mantle. At the zones of divergence two kinds of focal mechanisms are found [Sykes, 1967]. For shocks associated with the central rift of the ridges, the mechanisms are those of normal faulting, in agreement with the concept of a freshly formed thin lithosphere being pulled apart. Along the transform faults the mechanisms are predominantly strike-slip along a near-vertical surface. The seismic events associated with trenches and island arcs are mainly thrust-fault related and compressional, associated with the underthrusting of the plates.

Asymmetric Spreading and Migrating Ridge Axes

The sea-floor spreading theory, as originally formulated, states that midocean ridges spread symmetrically, adding to flanks on either side at the same rate. This approximation, although not demanded in the models, is constantly assumed and has proven to be so successful throughout the oceans that exceptions to the rule are of major interest. Asymmetric spreading results in the migration of the ridge axis. The first good example of a migrating ridge axis was given by Weissel and Hayes [1974]. The mid-ocean ridge between Australia and Antarctica spread more rapidly to the south of the ridge before 10 Ma. The asymmetry in the spreading of a ridge can result from two processes.

1. An asymmetric axial process that continuously adds more crustal material through volcanism to one side of a ridge relative to another (called *true* asymmetry).

2. A symmetrical axial process with subsequent discontinuous shifts in discrete jumps in the rise axis (called *apparent* asymmetry).

Jumps of a ridge axis have been described by several workers; a good example is the East Pacific rise between the equator and 20°N. Here seamount lineations west of

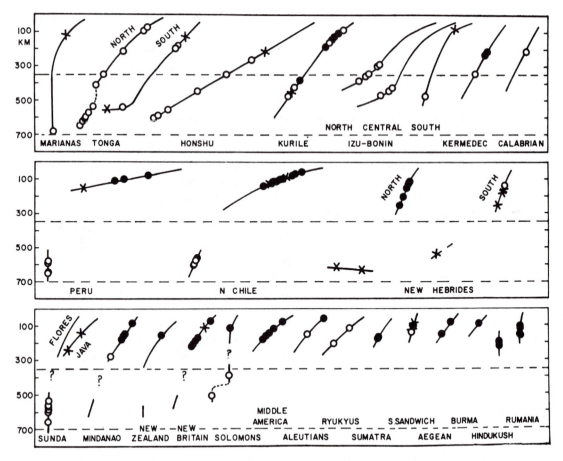

Figure 5-9 Distribution of down-dip stresses in inclined seismic zones. Open circles show mechanisms with the compressional axis parallel to the dip of the zone. Solid circles show mechanisms with the tensional axis parallel to the dip of the zone. Crosses indicate mechanisms with other orientations. Solid lines show the approximate configuration of the seismic zones. (From B. Isacks and P. Molnar, *Rev. Geophys. Space Phys.*, vol. 9, p. 117, 1971, copyrighted by American Geophysical Union)

the present ridge axis mark the former position of the ridge axis. Also in the Galapagos region, Hey and others [1977] presented evidence for a sequence of jumps in the rise axis along the Cocos-Nazca spreading center. These began only 3 Ma and have caused a westward migration of the ridge crest of 150 km. When the distance of a jump is large, anomalies will be repeated on one side and not recorded on the other. However, if there are many rapid, short jumps compared to the narrowest anomaly widths, only parts of the anomaly sequence will be either missed or repeated. In this case the resulting magnetic anomaly pattern will be indistinguishable from that produced by true asymmetric spreading. Theoretically, it should be possible to distinguish between true and apparent asymmetric spreading when a rise axis jumps so far that an anomaly is missing on one ridge flank and repeated on the other.

HEAT DISTRIBUTION AND AGE
OF THE OCEANIC CRUST

Crustal Heat Flow

The interior of the earth is hot, and heat constantly flows to the exterior with a resulting depth-related thermal gradient. In the 1950s, measurements were first made through the ocean floor of the heat flow from the earth's interior. These were made using a thermistor probe to measure the temperature gradient through deep-sea sediments and by study of thermal conductivity in cores obtained from the same vicinity. These measurements showed that the heat flow through most areas of the ocean floor is generally similar to that through the continents, with an average of about 1.0 microcalories (μ cal)/cm²/sec. However, it was also discovered that heat flow through the ocean ridges was several times greater (Fig. 5-10). The temperature anomaly over the ridge was interpreted by Hess as evidence that the oceanic crust is created by intrusion of hot material at the ridge crest. This material cools by conduction, then solidifies and moves away from the ridge crest with additional intrusion. The concept of upwelling hot mantle material under the mid-ocean ridges is supported by gravity measurements. Before sea-floor spreading was understood, the lack of a gravity anomaly despite excessive mass was not understood. Anomalous mantle densities were invoked under the ridges. The absence of a gravity anomaly over the ridge suggested that the excess mass must be

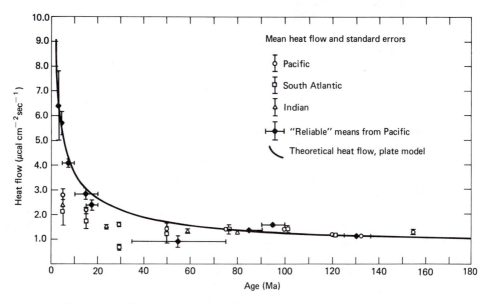

Figure 5-10 Changes in mean heat flow with increasing age away from the mid-oceanic ridge. (From B. Parsons and J. G. Sclater, *Journal of Geophysical Research*, vol. 82, p. 818, 1977, copyrighted by American Geophysical Union)

compensated for by a corresponding deficiency of mass at depth. The thermal expansion of material below the ridge resulting from anomalously high heat flow readily accounts for this. Thus the topographic expression of the ocean ridges is caused by thermal expansion and, possibly, a small amount of partial melting.

Created at high temperatures at the center of the mid-Atlantic ridge, the plate cools first at its upper surface, where it is in contact with seawater at a temperature of about 0°C. As heat flows out of the upper surface into the ocean, the plate cools and thickens (see Chapter 7). Sclater and Tapscott's [1979] theoretical description of thermal conduction gives a simple equation to predict heat flow through the plate. The equation yields the expression $q = 11.3 / \sqrt{t}$. The flow of heat (q) out of the plate into the ocean is a function of the age (t) of the ocean floor. The dimension of q is the heat-flow unit, or 10^{-6} cal/cm²/sec, where cal stands for calorie; the dimension of t is 1 m.y. [Sclater and Tapscott, 1979]. On the axis of the mid-ocean ridges, the intrusion of magma into seawater causes extensive fractures to form and the crustal rocks to become highly permeable to seawater. As a result, most loss of heat from the earth's interior in these areas is due to the advection of water through the rock. Heat loss through conduction is less important. The importance of water in transporting heat out of the ocean ridges is now also supported by observations from submersibles of hydrothermal activity, in some cases as plumes of hot water rising as submarine geysers. On the flanks of the ridge and deeper, where a sediment layer has formed, water flow through the rock is reduced and heat loss is mostly through conduction. When heat-flow values are plotted against age of oceanic crust (Fig. 5–10), it is apparent that highest heat-flow values through the ocean crust are associated with the crest of the mid-ocean ridge and that they show a systematic logarithmic decrease down the flanks of the ridge. Ocean crust younger than 15 to 20 m.y. exhibits highly elevated heat-flow values in relation to the ocean-floor average.

High heat-flow values are also associated with the areas behind the island arcs. All island-arc areas are marked by volcanic activity. The sea-floor spreading hypothesis requires that island-arc areas be loci of descending mass transport and destruction of oceanic crust. However, the downward sinking of cool, dense matter in the subduction zone should not necessarily be associated with the generation of high temperatures. Instead, it might be expected that the descending mass transport in the trench areas would result in a strong depression of the isotherms and, therefore, a lower than average surface-heat flux. This is not the case. Several authors, such as McKenzie and Sclater [1968] and Oxburgh and Turcotte [1968], have suggested that a feasible mechanism for generating the heat in these areas is by frictional dissipation along the seismically active Benioff zone. A major problem with this model is that once frictional heating contributes enough heat to cause partial melting, the stress should be dissipated and the mechanism of heat generation could not be maintained. Other models have invoked secondary convective flow in the mantle beneath backarc basinal areas as a source of the higher heat-flow values (see Chapter 12). The distribution of changing heat-flow values throughout the ocean basins has been fundamental in the development of models to explain bathymetric relationships as follows.

Age and Depth Relations of the Ocean Basins

Ridges are elevated above the surrounding sea floor because they consist of rock that is hotter and less dense than the older, colder plate. They are, therefore, somewhat analogous to floating icebergs. As hot rock moves away from the spreading center, it cools and contracts; as the crust contracts, the water depth increases. A simple relation thus exists between depth and age of the oceanic crust [Sclater and Francheteau, 1970; Sclater and Detrick, 1973; Parsons and Sclater, 1977] (Fig. 5-11). This relation is explained by the cooling of slabs approximately 100 km in thickness and of underlying mantle material as it moves away from the spreading center. It is also possible that the depth increase is, in part, due to a thickening of the oceanic crust. When the ocean crust is formed over the ridge, it is thin (5–10 km) but thickness increases by accretion; material is added from below and, as the material cools, it becomes welded onto the lower boundary of the plate. As the dense lithosphere thickens, it subsides into the less-dense mantle beneath, causing a depth increase away from the ridge axis.

Studies in the ocean basins have shown that the variation in mean depth of the ocean away from the ridge follows a linear relationship with the square root of the age of the ocean crust (Fig. 5-11). Subsidence of the sea floor is proportional to the the square root of the age, determined either by magnetic anomalies or by the dating of deep-drilled sediments, just as heat flow shows a systematic (logarithmic) variation as a function of distance from the ridge. Depths and basement ages of deep-drilled cores agree with curves showing empirical depth versus age, determined from magnetic anomalies and topographic profiles.

The rate of vertical thermal contraction can be calculated from the heat flow. This contraction, in conjunction with the gravitational-loading effect of the ocean water, is responsible for the depth of the ocean floor and can be represented by a simple mathematical expression: $D = 2500 + 350\sqrt{t}$, where D is the depth in meters and t is the age of the floor in millions of years. Empirical data indicates that this expression, like the one for heat flow, yields correct values for lithosphere no older than 60 m.y. (Fig. 5-11) [Sclater and Tapscott, 1979]. A mean rate of subsidence of about 40 m/m.y. (the rates decrease exponentially) indicates contraction of 3 percent of plate thickness. Theoretical calculations suggest a thermal expansion coefficient of 3.2×10^{-5}°C.

In a very general way these relationships were recognized by Wegener [1924], who pointed out that ". . . the most ancient oceanic floors have the greatest depths" and ". . . for that reason the differences of depth are also explained by the temperature relationships. The old ocean floors became more strongly cooled, and therefore of greater density than the younger."

In general, the depth of the ocean floor increases from 2700 ±300 m at a spreading center to 5800 ± 300 in oceanic crust of early Cretaceous age (Fig. 5-11). For instance the axis of the mid-Atlantic ridge is 2500 m below sea level. The 3000-m isobath (contour of equal depth) lies on ocean crust that is 2 m.y. old, the 4000-m isobath on crust that is 20 m.y. old, and the 5000-m isobath on crust that is 50 m.y. old (Fig. 5-11) [Sclater and Tapscott, 1979].

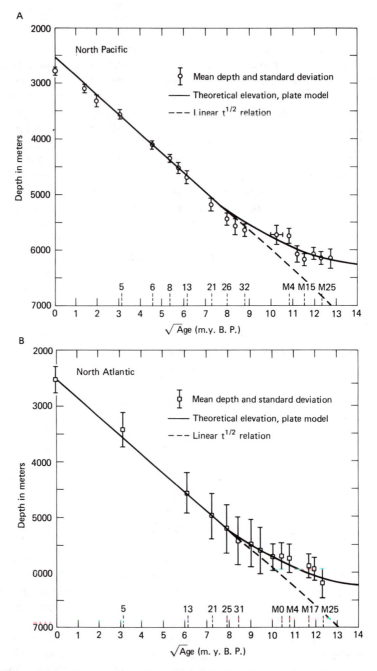

Figure 5-11 Plots of mean depths and standard deviations of oceanic crust versus the square root of its age for (A) the North Pacific and (B) the North Atlantic. (From B. Parsons and J. G. Sclater, *Journal of Geophysical Research*, vol. 82, p. 819, 1977, copyrighted by American Geophysical Union)

The East Pacific rise in the North Pacific shows a uniform decrease from a crustal elevation of about 2700 m to a depth of 5500 m for crust that is 75 m.y. old (Fig. 5–11). All other mid-ocean ridges, with the exception of the mid-Atlantic ridge south of Iceland, show the same uniform increase of depth with age. These ridges have a shallower crestal elevation than the East Pacific rise, but the differences are small, in most cases within 400 m. The great similarity between the ridges of the Pacific, Atlantic, and Indian oceans is evidence of a fundamental principle underlying the relation between depth and age of the oceanic crust.

It is apparent that the relation between depth and age is compatible with plate theory and, with certain limitations, can be used within this framework to predict the age of the oceanic crust.

One consequence of this topographic dependence on time is a variation in spreading rates. Since subsidence is proportional to the square root of age, a fast-spreading ridge will have a greater volume than a slow-spreading ridge. During global episodes of fast spreading, sea level will rise as a result of the bulging lithosphere. It is possible that fast-spreading maxima may coincide with periods of high heat flow and increased mantle upwelling. The estimated turnover time of the upper mantle is estimated at 300 m.y., an extremely slow process. Although they appear to have little surface expression, convective motions within the mantle below the plates must exist to transport material in closed circulation cells between the ocean ridges and island arcs.

Because the depth-versus-age relationship is a consequence of plate creation, there is no reason to believe that plate creation was any different during the late Mesozoic (oldest known oceanic crust) than during the present. With this relationship and a knowledge of previous positions of continents and of tectonic history, it is possible to construct paleobathymetric charts of the ocean floor at particular times in the past. The first such charts were constructed by Sclater and McKenzie [1973] for the Atlantic Ocean. Since then, Sclater and his colleagues have constructed paleobathymetric charts of the Indian Ocean and additional charts of the Atlantic Ocean.

ISLAND ARCS, TRENCHES, AND BACKARC BASINS

Island arcs and trenches are among the most spectacular tectonic features of the earth (Fig. 5–12). They occur mainly along the margins of the Pacific and represent the sites where the oceanic plate is subducted under another plate to form the most complex type of plate boundary (Fig. 5–3). At this point we need to discuss briefly some aspects of the geology and geophysics of the island-arc system as they relate to plate tectonics. They are discussed in greater detail in Chapter 12. These regions are collectively referred to as **magmatic arcs**, being characterized by an arcuate shape and by intensive volcanism and seismicity. The magmatic arcs are of two types: *island arcs,* such as the Aleutians, Mariana, Izu-Bonin, Tonga-Kermadec (Fig. 5–12), Lesser Antilles, and South Sandwich, where both plates are veneered by oceanic crust; or *continental margin arcs,* such as the western margin of North and

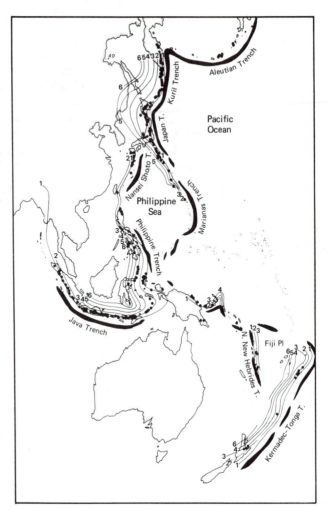

Figure 5-12 Distributions of trenches, volcanoes, and isopachs of dipping Benioff zones (in hundreds of kilometers below the sea floor) for the western Pacific. (From R. Oxburgh and D. L. Turcotte, *Geol. Soc. Amer. Bull.*, Vol. 81, p. 1668, 1970, courtesy The Geological Society of America)

South America, Indonesia, Kamchatka, and Japan (Fig. 5-12), where the overriding plate is constructed of continental crust.

It was Eduard Suess [1904] who, in his classic work *The Face of the Earth*, focused attention on the arcuate structure of the great mountain systems of the earth. However, the importance of island arcs in terms of large-scale tectonics was not appreciated until the pioneering marine gravity work of Vening Meinesz in the early 1930s. He demonstrated that island arcs and their deep trenches, which usually lie on their oceanward side, are associated with distinct anomalies in the earth's gravitational field running parallel to the trench system centered near the trench and the arc. He attributed the large negative anomalies to a downbuckling of the crust, which he called a **tectogene,** and considered it to be held down either by compressional forces or by the drag of converging convection currents. The trenches are also associated with widespread seismicity. In 1948, Hess introduced important new observations—that deep-sea trenches tend to be continuous along the various interconnecting arcs and that there is a close spatial association of seismic activity, volcanism, and gravity anomalies along the arcs. These observations, combined with accurately located earthquakes beneath the arcs, assisted in the formulation of the concepts of sea-floor spreading and plate tectonics, with the ocean trenches being the zones of crustal destruction.

The island-arc regions exhibit the following characteristics [Sugimura and Uyeda, 1973]:

1. Arcuate line of islands.

2. Prominent volcanic activity.

3. Deep trench on the ocean side and shallow seas on the continent side.

4. Distinct linear gravity anomaly indicating large departures from isostasy.

5. Active tectonism.

6. Coincidence of arcs with recent orogenic belts.

7. High heat flow on the continent side of the arc.

If we consider that island arcs are marked by recent volcanic activity, trenches deeper than 6000 m, and earthquake foci deeper than 70 km, the following features are island arcs (Fig. 5-12): (1) New Zealand to Tonga; (2) Melanesia; (3) Indonesia; (4) Philippines; (5) Formosa (Taiwan) and western Japan; (6) Mariana and eastern Japan; (7) Kurile and Kamchatka; (8) Aleutian and Alaska; (9) Central America; (10) West Indies (Lesser Antilles); (11) South America; and (12) Scotia Arc and West Antarctica.

Island arcs are conspicuous features in the circum-Pacific, especially in the western sector (Fig. 5-12), but occur in the Atlantic only as the Lesser Antilles Arc and Scotia Arc (Fig. 5-1). The island-arc system does not completely girdle the Pacific Ocean. Notable gaps occur adjacent to Antarctica and in western North America. Each island arc is a thousand to several thousand kilometers in length with a narrow width of only 200 to 300 km, including the trench. Island arcs can occur as single or double arcs, as in the Indonesian Arc, in which case the active volcanoes are associated with the inner arc. Although most arcs are bordered by a relatively shallow sea on the continent side, there are some, such as the Mariana Arc, that are more or less isolated by another trench (Fig. 5-12). Others continue into the continent, forming continental counterparts; for example, the Indonesian Arc trends into Burma (Fig. 5-12) and the Aleutian Arc trends into Alaska.

Most island-arc systems also contain backarc or marginal basins between the intraoceanic arcs and the continents. These basins, such as the Philippine and Japan seas, constitute a high proportion of the western Pacific ocean floor and also include the Caribbean and Scotia seas. They are discussed in detail in Chapter 12.

Trenches are complicated features and are still poorly understood compared with the ocean rises. This is in part because of their great depth, but also because they can consume the sea floor, leaving little evidence of the material consumed. Even where sediments are accreted, the tectonics are complex; geological history is thus also difficult to interpret. The oceanic trenches are the most important feature of the island arcs (Fig. 5–12). The evolution of the continents is closely connected to that of the trenches, because volcanism in the island arcs produces rocks similar to those which form continents. Trenches are the deepest features of the ocean basins, ranging from about 7000 to 11,000 m at their axes. The deepest trenches are the Mariana and Tonga trenches, which are 10,860 and 10,800 m deep, respectively. Cross sections at right angles to most trenches exhibit bilateral asymmetry with shallow-water depths (to 1000 m) on one side and average oceanic depths (about 4000–6000 m) on the other.

Island arcs are marked by a high negative gravity anomaly over the trench of the order of -300 milligal (mgal) and a positive anomaly over the island arc itself (Fig. 5–13). These anomalies show that isostatic equilibrium is absent in these areas. Specifially, these features indicate mass deficiency in the trench region and mass excess in or under the arc. Furthermore, minor positive anomalies occur seaward of the trench (Fig. 5–13) and are attributed to flexuring of the oceanic lithospheric plate as it approaches the trench. This flexuring forms minor elevations on the sea floor that correspond to minor positive anomalies. The major positive gravity anomaly over the volcanic arc (Fig. 5–13) is related to a mass excess related to the subducting plate. The cold, dense (3.4 g/cm^3) plate is inserted into the hot asthenosphere (3.35 g/cm^3) with a temperature contrast of about $400°$C.

There is a remarkable parallelism between the seismic zones and the trenches, even in areas of major directional changes in the trench. For instance, a sharp bend at the northern end of the Tonga Trench is also reflected in the Benioff zone. Not only do distinct relations occur between diverse geological and geophysical features of island arcs, but these are arranged in zones with a definite order [Sugimura and Uyeda, 1973]. From the ocean side toward the continents, they are the following: trench, negative gravity anomaly, positive gravity anomaly, axis of islands, volcanic belt, and deep earthquake zone (Fig. 5–13). This systematic arrangement indicates that all the island arc-trench systems are caused by a common mechanism—crustal subduction. This process and its control on the tectonic and sedimentary character of the active margins are discussed in Chapter 12.

VOLCANISM AND PLATE TECTONICS

General Patterns of Volcanism

Since most tectonism takes place at the boundaries of plates, it is not surprising that most volcanism also occurs in these regions (Fig. 5–14). Volcanism occurs in three physiographic and tectonic settings: volcanism at convergent plate boundaries (zones of crustal convergence); volcanism on the mid-ocean ridges (zones of

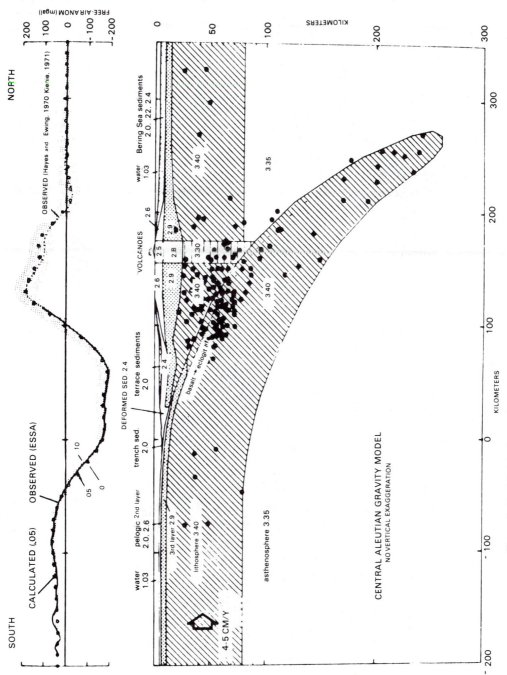

Figure 5-13 Model of crustal structure based on gravity survey across the central Aleutian Arc, North Pacific. Free-air gravity anomalies shown at top. Calculated gravity values are plotted as circles. Density of various layers shown in g cm⁻³. (From J. A. Grow, *Geol. Soc. Amer. Bull.*, Vol. 84, p. 2181, 1973, courtesy The Geological Society of America)

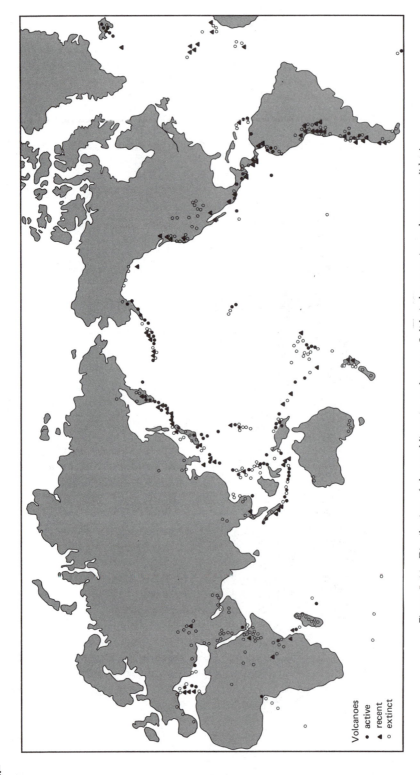

Figure 5-14 Distribution of the world's active volcanoes. Solid circles = active volcanoes; solid triangles = recently active volcanoes; open circles = extinct volcanoes.

Volcanoes
- active
- recent
- extinct

154

tension in the oceanic crust); and volcanism at particular locations in the interior of the plates. Although there are thousands of known volcanoes scattered over the earth's surface, only about 800 are either active or known to have been active historically (Fig. 5-14). Only a minor fraction of these volcanoes (about 70) are known to have undergone submarine eruptions. Volcanoes are highly visible events, though the high visibility of terrestrial volcanism is quite unrelated to the total volumes of volcanic rocks produced. Schilling [1973] estimated that 80 percent of volcanic rocks are produced at the mid-ocean ridges, while the remainder are formed at convergent plate boundaries and as intraplate volcanoes, such as Hawaii and the Society Islands. The tremendous dominance of constructive plate boundary volcanism is not surprising, since about two-thirds of the earth's crust is formed in these regions (see Chapter 7). Divergent boundaries produce tholeiitic basalts formed by partial melting of the upper mantle at relatively shallow depths, under conditions of comparatively low pressure and high temperature under the oceanic ridges. More than 75 percent of the active and recently extinct volcanoes are located in the circum-Pacific belt (Fig. 5-14). This belt is commonly known as the *ring of fire*.

Volcanism at Convergent Margins

The volcanoes of the circum-Pacific region are associated with young mountain chains and island arcs that extend as narrow, linear, arcuate zones (Fig. 5-14). The belt can be clearly traced northward from New Zealand through Melanesia, into Indonesia, the Philippines, Japan, the Kuriles, Kamchatka, and eastward through the Aleutian Islands to southern Alaska (Fig. 5-14). The belt continues southward along the western edge of North America with a gap through California and Mexico, continuing through Central America and the western edge of South America to the Antarctic Peninsula. The southern part of the girdle is partially completed by ridge systems in the Southern Ocean. Mountains that are not associated with trenches, such as the North America cordillera, are marked by relatively little volcanism. Two broad, yet distinctive, features marked by compressional tectonics occur in the circum-Pacific volcanic belt—island arcs and continental margins not of the island-arc type. The distribution of these features is different throughout the circum-Pacific. The island arcs in the western Pacific exhibit the convex side of the arc facing toward the Pacific Ocean (Fig. 5-14). Volcanically active continental margins, in contrast, occur on the eastern side of the Pacific and include the Cascade, Central American, and the Andes ranges. These features exhibit little curvature and are also associated with trench systems [Williams and McBirney, 1979].

The margins of the Atlantic Ocean are largely free of volcanoes, seismicity, and orogeny. The two major loops of volcanoes associated with island arcs in the Atlantic, namely the Lesser Antilles and the Scotia arcs, are considered extensions of the circum-Pacific belt. Curiously, both of these arcs—which occur at the eastern edge of the Pacific—occur at gaps between continental blocks, as between Antarctica and South America and in young continental crust between North and South

America. Williams and McBirney [1979] point out that the asymmetry in east-facing island arcs and west-facing active continental margins on opposite sides of spreading axes suggests that their orientations are influenced by fundamental directional forces such as the earth's rotation. On the other hand Wilson [1973] stated that the formation of arcs results from particular plate interactions. When two plates approach each other at a subduction zone, the continental plate overrides an oceanic one; Wilson [1973] suggested that the plate which is more nearly stationary over the mantle determines the character of the continental margin. When a continental plate advances and slides under it, island arcs will form on the coast facing the advancing plate, such as in the Northwest Pacific. If two continents collide, mountain ranges of the Himalayan type will form.

Most of the continental margins of the Indian Ocean are nonconvergent, thereby lacking widespread volcanism. However, a zone of particularly intensive volcanism occurs in the Indonesian Arc, which represents a spur of the circum-Pacific belt (Figs. 5–12 and 5–14). The Indonesian Arc contains about 14 percent of the earth's active volcanoes. A second belt of recent orogeny associated with active volcanism is the Mediterranean-Asian orogenic belt, but volcanism in this system (Fig. 5–14) is relatively sparse compared with the circum-Pacific belt.

The association of active volcanoes with subduction zones and their trenches is so clear-cut (Fig. 5–12) that the character of the tectonism that occurs in these regions is critical for the understanding of the genesis of volcanism. Circum-Pacific andesites with associated basalts, dacites, and other rock types are erupted from island arcs and continental margins, which generally coincide with belts of intermediate to deep seismicity paralleling the adjacent trenches. Benioff himself, in 1949, pointed out that seismic zones delineate tectonic dislocations, which serve as preferred loci for the production of magmas by heat generated by stresses in the mantle. Sykes [1966] demonstrated that an association exists between orogenic volcanism and the distribution of intermediate and deep earthquakes. All regions of active andesitic volcanism are subject to intermediate to deep seismicity (Fig. 5–12). For instance, the active volcanoes in Japan are restricted to areas above intermediate depth seismicity in the Benioff zone. Volcanoes of the Aleutian Range also occur above the zones of epicenters of intermediate-depth events in Alaska. Geochemical data suggest that these magmas rise to the surface without undergoing contamination sufficient to erase the petrochemical fingerprint imposed along the Benioff zone. The detailed relations between volcanism and underlying seismicity varies from area to area. Beneath Japan there tend to be fewer earthquakes at intermediate depths in the Benioff zone (beneath the volcanoes), which suggests that stresses may be relatively reduced in this zone. Typically, active volcanoes mark the boundary between shallow and intermediate seismicity (Fig. 5–13), though active volcanoes occur farther from the trench in some arcs. In the Andes region, Barazangi and Isacks [1976] determined that volcanism is most active above steeply inclined Benioff zones and above the zone of low velocity at the intersection between the Benioff zone and the lithosphere. Whatever the process, it is clear that the subduction of oceanic crust creates stress to melt the volcanic magmas. It is not clear, however, whether the volcanic rocks are derived only from mantle sources or whether subducted oceanic crust melts as well.

The circum-Pacific suite of volcanic rocks, displaying a broad spectrum of petrologic variation, is variously known as the *calcalkaline, orogenic,* or *high-alumina* suite (see Chapter 7). These rocks play a major role in continental accretion, and it may not be coincidental that the composition of andesite is close to that of the crustal average. Although some areas of subduction are marked by an abundance of andesites of the calc-alkalic series, such as New Zealand and Oregon (western United States), there are others—such as in the Kermadec Islands—which lack andesitic rocks and the calc-alkalic series. Therefore it is inappropriate to characterize the subduction zone volcanic rocks as calc-alkalic andesites.

The composition of volcanic suites may differ between island arcs and continental margins. Although siliceous rocks, such as rhyolite and ignimbrites, are normally small in comparison with more basic rocks such as basalts and andesites, at times their volumes are enormous in continental margin areas such as the central volcanic complex of New Zealand, which is marked by massive outpourings of Quaternary ignimbrites. However, this type of volcanism is almost unknown in island-arc regions constructed on oceanic crust [Williams and McBirney, 1979]. Volcanism that occurs in the subduction zones is explosive compared to that of the intraplate and ocean-ridge areas. This leads to the production of a large pyroclastic (airborne) fraction that is carried by wind vast distances from the source. The high explosiveness is probably due to the high water content of the magmas. It is possible that the souce of this water is from the descending oceanic crust.

Commonly proposed mechanisms for the generation of andesitic magma include partial melting of continental material; partial melting of upper mantle material containing small amounts of water; partial melting of upper mantle material with subsequent contamination by sialic material; or melting of subducted oceanic crust, either of deep lithospheric material only or of the entire oceanic crust, including pelagic sediments. Because many of the island arcs with widespread andesitic volcanism lie on oceanic crust quite distant from areas of continental rocks, the processes requiring continental crust cannot be considered essential. Thus andesite production seems restricted to either partial melting of mantle material or of subducted oceanic crust. Because of the relationships discussed before between magma generation and the Benioff zone, it is clear that the andesitic suite results from partial melting of the subducted oceanic crust. Opinions differ as to the roles of crustal material and their overlying sediments in this process. Experimental work suggests that the water in sediments overlying the oceanic crust provides a hydrous zone in which andesite melts are produced.

Iceland Volcanism Astride the Mid-Ocean Ridge

The volcanism of Iceland is particularly important because it takes place on the only large island that lies astride a spreading ridge (Fig. 5–15). Iceland is the largest single area above sea level which is entirely of volcanic origin and is the most active volcanic region in the world. It consists of fissure-erupted lava plateaus and large conical volcanoes and is bounded to the north by the Tjornes fracture zone and to the south by the Reykjanes fracture zone. About one-fourth of terrestrial lavas of the world formed during the last 400 years have been erupted in Iceland.

Because of Iceland's position on the mid-ocean ridge, the island is continu-

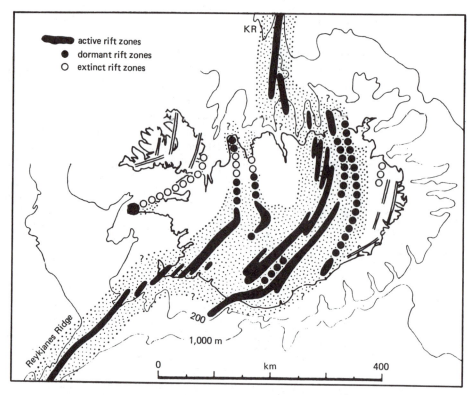

Figure 5-15 Map of Iceland showing its position astride the mid-Atlantic ridge (as represented by the Reykjanes and Kolbeinsey ridges) and the distribution of rift zones. The position of the spreading axis (shown by rift zones) has changed with time. (After G. P. L. Walker, reprinted by permission from *Nature*, vol. 255, no. 5508, pp. 448–471, 1975, copyright © 1975 Macmillan Journals, Ltd.)

ously in a state of tension as the east and western portions of the island are moved apart by sea-floor spreading. Thus processes operating there should be similar to those processes operating on the remote mid-ocean ridges.

The island spreading creates fractures, which are intruded with magmatic material, forming dike swarms. Material erupted at the surface has formed thick sequences of late Cenozoic lava flows. Each dike extends the crust, with estimated extension totaling 400 km since the island began to form about 15 m.y. ago. Iceland is not only growing wider but thicker. Almost all present-day volcanism is confined to fracture zones cutting through the central part of the island (Fig. 5-15). Because of lateral spreading away from this central zone, the oldest rocks occur in the eastern and western sectors of Iceland. Detailed volcanological investigations show that active fracture zones jump to different locations across the island (Fig. 5-15). These reflect jumps of the spreading axis; thus increments of volcanic rocks are complex. There is no evidence to suggest that volcanism has ceased for any geologically significant period of time during the last 15 m.y., although Vogt [1972b] has presented evidence for pulses of more intense volcanism. If future volcanism is reduced to rates typical of other parts of the mid-Atlantic ridge, Iceland would be bisected, forming two isolated ridges on either side of the mid-ocean ridge crest like the Rio Grande–Walvis ridges in the South Atlantic.

Because of its position over the mid-Atlantic ridge, the rocks of Iceland are dominated by basalt close to tholeiite in composition. However, large volumes of more silicic and alkalic rocks, such as rhyolite, are probably the result of differentiation of basaltic magmas.

Intraplate Volcanism

Although about 90 percent of volcanic activity on the earth is concentrated within, or adjacent to, the zones of plate divergence or convergence, other volcanism does occur at locations remote from the plate margins. These include large numbers of seamounts and volcanic islands, some of which occur in chains like those of the Hawaiian Islands, Line Islands, and Society Islands (Fig. 5–16). Although seamounts are widespread throughout all ocean basins, they are most abundant in the Pacific basin. About 1000 islands occur far from the continental margins in the Pacific. However, only a small proportion of volcanic mountains are high enough to break sea level and form oceanic islands. The others remain below sea level as seamounts, which are volcanoes that either never grew to sea level or stopped erupting above sea level and have subsequently been eroded and subsided [Hess, 1946]. Seamounts and volcanic islands are not distributed uniformly in the Pacific. Large volcanoes (islands and large seamounts) are more abundant in the southwestern and western Pacific, while smaller seamounts are more abundant in the northeastern sector. Only a few oceanic islands are known to have been historically active, including two or three of the Hawaiian Islands and Rocard, Moua Pihaa, and Macdonald islands in the South Pacific.

Seamounts are morphologically similar to subaerial shield volcanoes, being broadly rounded with slopes ranging from 5° to 25° and with circular, ovoid, or lobate shapes. Although all are of volcanic origin, those that emerged close to sea level in the tropics are usually capped with coraline sediments or fossil coral reefs. These subsequently subside systematically as guyots, as the associated sea floor moves away from the ocean ridges. In some cases, where subsidence has been slow enough, growth of coral reefs on top of the guyot has been fast enough to keep up, maintaining a coral biota near sea level and ultimately forming thicknesses of at least 1000 m of fossil reefs.

Intraplate volcanoes are immense features that may rise to about 10,000 m above the ocean floor, dimensions much greater than the largest mountains on the continents. Little is known about the composition of rocks that form the lower part of a seamount, although it is probably a tholeiitic basalt similar to that of the ocean crust on which the seamounts are built. The suite of volcanic rocks that make up the higher parts of seamounts and islands, such as Hawaii, Iceland, the Galapagos, and Reunion, consists of alkali basalts marked by less silica and higher alkali content than tholeiites. Furthermore, the diversity of rock types is greater than that of typical ocean basalt and may be rich in silica and alkalis and poorer in calcium, iron, and magnesium. These rocks include trachytes, phonolites, and rhyolites.

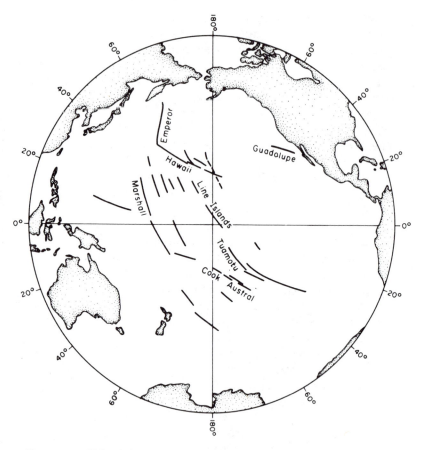

Figure 5-16 Volcanic lineations in the Pacific that may represent hot-spot trails. (From T. H. van Andel, G. R. Heath, and T. C. Moore, Jr. (simplified after Winterer, 1973), *Geol. Soc. Amer. Memoir*, 143, 1975, courtesy The Geological Society of America)

Island chains

Striking features of the deep-ocean basins, particularly the Pacific basin, are the long linear chains of islands and seamounts far from the spreading axes, which are much younger than the underlying oceanic crust (Fig. 5-16). The chains tend to be parallel to one another (Fig. 5-16) and are nearly perpendicular to the magnetic anomaly patterns. Most chains trend in a northwesterly direction, although the larger ones change slightly to west of north. The importance of the linearity and parallelism of the Pacific island chains was anticipated in some earlier studies, such as the one by Chubb [1957], who thought that the geometry of the island chains was tectonically controlled, either by ancient crustal fractures or by compression into fold belts.

Another major aspect of the linear island chains is that some of and possibly all these chains, are marked by a progressive increase in age of the volcanoes away from the mid-ocean ridges. The first naturalist [Darwin, 1842] and geologist [Dana, 1885] to visit these island chains recognized a geomorphic progression from high volcanic cores in the southeast to coral atolls in the northwest, and they believed that this resulted from increased age toward the north. Several authors have since suggested that such linear volcanic chains result from volcanism along a propagating fracture zone or along transform faults. Even the ancient Hawaiians seemed to recognize this age progression, as illustrated in their legend of Pele, fire goddess of Hawaiian volcanoes. The most common account [Bullard, 1976] relates how Madam Pele, in search of a home, first visited one island and then another looking for a suitable home. Her visit began in the older islands to the northwest in the chain and continued through the progressively younger islands to the southeast. Prominent landmarks on the eastern end of the island of Oahu, such as Koko Head, represent Pele's attempts to build a home on Oahu. She finally came to Kilauea, the most active volcano on the island of Hawaii, where she finally established her home. Carey [1958] suggested that many of the aseismic ridges observed on the ocean floor were formed from localized intensive volcanic activity and assumed, as Dietz and Holden [1970] did later, that these aseismic ridges could be used in reconstructions of continents.

Hot spots and plumes

On the basis of chain geometry and fossil age data, J. T. Wilson [1963] proposed a mechanism for the remarkable pattern of migration exhibited by the island chains, which now seems compatible with the plate-tectonic model. Wilson suggested that the magmas that built the volcanoes are derived from a relatively fixed magma source in the upper mantle, which he called a **hot spot** (Fig. 5–17). Because the magma source lies beneath the crustal plate and because the plate is moving laterally, the active volcanoes are eventually separated from the hot spot, causing a cessation of volcanism (Fig. 5–17). Ultimately this process creates a chain of extinct volcanoes moving away from the hot spot in the direction of sea-floor spreading and thus becoming progressively older. The hot spot that has built the Hawaiian chain is therefore Pele, the "hot-spot goddess," who resides under the big island of Hawaii. Extensive radiometric age determinations of the rocks that form the various islands of the chain support this concept. For instance, the basalts underlying Midway Island are 16 m.y. old, while those of Necker are about 11 m.y. old.

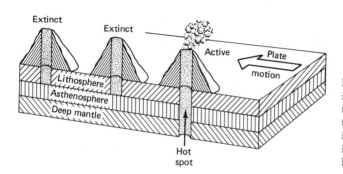

Figure 5-17 Formation of a volcanic-island chain by the lateral movement of the oceanic plate moving over a stationary hot spot. The age of the islands increases toward the left, and new islands will continue to form over the hot spot. (After N. Calder, 1974)

Calculations using these data show that the average rate of migration of the center of active volcanism in the Hawaiian Chain is about 15 cm/year.

In 1972, Morgan expanded Wilson's model to include the chains in the Pacific trending in other directions. Topographic maps of the Pacific show that several of the island chains shift direction from southeast to northwest to slightly west of north (Fig. 5-16). He proposed that the change in trend between the Hawaiian, Emperor, Gambier-Tuamotu, Line, Austal-Cook, and Gilbert-Marshall chains reflect a dramatic change in the direction of Pacific plate relative to the underlying mantle. Morgan also realized that large volumes of rocks are needed for the partial melting of the seamount material and suggested that the melting spots supplied parent material as well as heat. He proposed that magmatic material necessary for hot-spot volcanism was supplied from the upper mantle by **plumes** of a few hundred kilometers in diameter and moving at velocities of a few meters per year (Fig. 5-17; see Chapter 7). Several dozen hot spots have been identified throughout the oceans and the continents (Fig. 5-18). Each has had a life ranging from tens of millions to even 100 m.y. Wilson [1973] identified the various characteristics of hot spots.

1. Each hot spot is an uplift, marked by elevated basement rocks on land and shallow water at sea.

2. The uplifts are capped by active volcanoes, characteristically producing alkaline basalts and rhyolite as well as tholeiitic basalts. These lavas have distinctive isotopic ratios and geochemical patterns.

3. The hot spots represent areas of high heat flow.

4. Gravity highs accompany at least some of the hot spots.

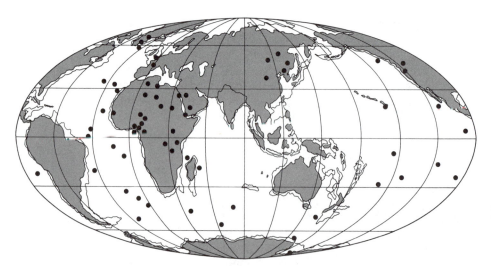

Figure 5-18 Location of possible hot spots rising as plumes deep in the mantle. Most occur on either plates that are believed to be nearly fixed (Africa, Nasca, and Southeast Asia) or near mid-ocean ridges. (After J. T. Wilson, 1973)

5. In the oceans and in some continental areas, one (or sometimes two) lateral aseismic ridges stretch away from the hot spot.

6. When a hot spot is located on an actively spreading ridge, like Iceland, chains of volcanic islands or seamounts are formed on both sides of the ridge because the plates on both sides are spreading away from each other.

There are numerous volcanic sources regarded as hot spots. These include the Hawaiian, Society, Marquesas, and Galapagos islands in the Pacific Ocean; Iceland, Tristan da Cunha, and Bouvet islands in the Atlantic Ocean; and Prince Edward–Marion and St. Pauls–Amsterdam islands in the Indian Ocean.

Schilling and others [1978] calculated annual production rates of lava from hot spots. Assuming a minimum production rate of 0.01 km^3/yr for a total of 122 hot spots [Burke and Wilson, 1972] provides an annual lava production rate of 3.5 × 10^{12} kg/yr. If, on the other hand, a more conservative total of 36 hot spots is assumed [Morgan, 1972], but with a minimum weighted average production rate of 1.5 km^3/yr, the total production rate of rock is 4.35 × 10^{12} kg/yr assuming an average density of 2.9 × 10^3 kg/m^3. Both estimates show that hot spot magma production rate represents about 10 percent of oceanic crust generation rate.

The Hawaiian-Emperor chain

The classic example of a systematic age progression along a linear chain is the Hawaiian-Emperor Chain. The Hawaiian Islands lie on the Hawaiian Ridge, which runs for about 2600 km from the island of Hawaii to the coral atolls of Midway and Kure (Fig. 5-19). Northwest of Kauai, the ridge continues for another 2000 km, mainly as a submarine ridge. Nihoa and Necker represent eroded stumps of volcanoes just above sea level. After the Hawaiian-Emperor Chain bends, it strikes north-northwest toward the Meiji Guyot (Fig. 5-19). Extensive radiometric dating of over 18 volcanoes along this chain [McDougall, 1979; McDougall and Duncan, 1980] shows age progression (Fig. 5-19 and 5-20), though with some complications. Situated on the island of Hawaii at the southeastern limit of the chain are several active volcanoes. Kilauea is the most active, while Mauna Loa and Hualalei are less active. Mauna Kea is either dormant or extinct. Seismic data below Hawaii indicate that the cause of the volcanism lies in the asthenosphere at a depth of about 60 to 70 km.

The present spreading direction of the Pacific crustal plate, toward the Japan and Aleutian trench systems from the East Pacific rise, is closely parallel to the strike of the Hawaiian Chain, supporting Wilson's hot-spot concept. The isotopic-age data on the Hawaiian volcanoes allow the rate of migration of the center of volcanism to be calculated and compared with the velocity of sea-floor spreading for the East Pacific rise (Fig. 5-19). The K-Ar results of McDougall [1979] and of McDougall and Duncan [1980] indicate that the construction of each of the large Hawaiian shield volcanoes above sea level took place in about 0.5 m.y. The data on the ages of the volcanoes in the Hawaiian-Emperor Chain are plotted as a function of distance from Kilauea in Fig. 5-20. This shows that age increase to the northwest is almost linear. However, simultaneous eruptions do occur throughout as much as

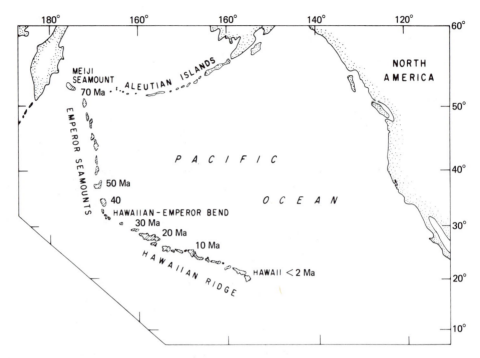

Figure 5-19 The Hawaiian-Emperor volcanic chain showing ages (Ma) of particular islands or seamounts. (After D. A. Claque and others, 1975)

300 km of the chain, and thus some overlap in ages of basalts occurs on adjacent islands. However, the simplest model [McDougall, 1979], preferred by some, shows the relationship between island age and distance as a linear one. Possible departures from linearity are observed in the section of the chain between Kauai and Hawaii. However, if one assumes a linear rate of migration of volcanism from Hawaii to French Frigate Shoals, this provides a rate of migration of nearly 10 ±0.25 cm/yr. This rate is very close to those determined for four other chains (Marquesas, Pitcairn-Gambier, Society, and Austral). The migration of the volcanoes along the Hawaiian Chain is much faster (10 cm/yr) than the spreading rate of the Pacific plate away from the East Pacific rise (6–7 cm/yr). Thus either the hot spot is moving toward the East Pacific rise or the rise is migrating toward Hawaii.

A major corollary of the melting-spot hypothesis [Morgan, 1972] is that the sharp bend in the chain between the Hawaiian and Emperor chains (near the Kanmu and Yuryaku seamounts; Fig. 5–19) represents an abrupt change in the direction of the Pacific plate over the Hawaiian hot spot. This direction change seems to have occurred between 40 and 50 Ma (Eocene epoch). The extrapolated age, using an average migration rate of volcanism of 9.66 (±0.27) cm/yr over the last 27 m.y., provides an age of nearly 38 m.y. for the intersection of the Hawaiian and Emperor chains [McDougall, 1979]. The oldest ages yielded for the Emperor Chain are at its northern limit, where it intersects the Aleutian-Kamchatka cusp and is about 72 Ma (late Cretaceous). The rate of migration changed at the same time as

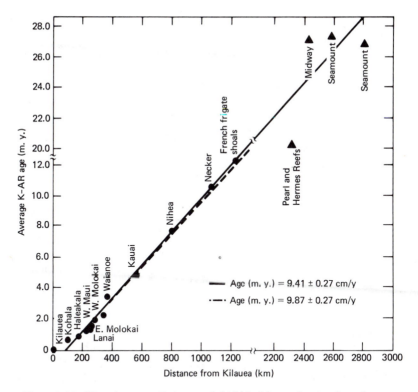

Figure 5-20 Plot of average K–Ar age of shield-building volcanism for volcanoes of the Hawaiian Island chain versus distance from Kilauea. Filled circles indicate data from tholeiitic basalts of the main shield; filled triangles show results from alkalic lavas that postdate the main shield-building phase. (From I. McDougall and R. A. Duncan, *Tectonophysics*, vol. 63, p. 280, 1980)

the direction. The average rate of migration along the Hawaiian Chain (10 cm/yr) compares to only about 6 cm/yr along the Emperor Chain. There are 18 major volcanoes per 1000 km along the Hawaiian ridge. Despite the lower rates of migration along the Emperor ridge, there are only 13 major volcanoes per 1000 km [Williams and McBirney, 1979]. Because more volcanoes should occur along a given distance from a hot spot on a more slowly moving plate, the intensity or frequency of volcanism was probably lower during the formation of the Emperor ridge.

The concept of fixed plumes to explain hot spots or melting anomalies is not universally accepted. One of the main appeals of the plume hypothesis has been that the plumes are fixed relative to the lower mantle, thereby furnishing a fixed reference system for plate motions. However, evidence has since emerged that there may be a gradual drift of melting anomalies [Molnar and Atwater, 1973]. Other theories, which allow for the drift of the hot spot relative to the lower mantle have been formulated. One of these, **the gravitational anchor theory,** was proposed by Shaw and Jackson [1973]. Their theory states that the volcanic activity at the ends of the Hawaiian, Tuamoto, and Austral chains is the result of shear melting caused by plate motion. Once such melting begins, a dense residuum is formed and sinks. This downwelling forms gravitational anchors that stabilize the anomalies and cause inflow of fresh, partially fractionated, parental materials into the source areas of the basalts. These anchors act as pinning points for melting anomalies within the asthenosphere. Another alternative theory is that linear chains are formed by volcanism along propagating fractures. If forces causing the fractures are broad-scale features, this would explain the broad parallelism of the chains. However, with the **propagating crack hypothesis,** there is nothing to explain the narrowness of volcanism along a crack, the similar rates of propagation in different chains, and the fact that cracks propagate to the east-southeast, never to the west-northwest [Claque and Jarrard, 1973]. These observations are, however, consistent with the hot-spot hypothesis.

Aseismic ridges

Structural highs are prominent features throughout the ocean basins, standing 2 to 3 km shallower than the surrounding sea floor. Some of these features are long, linear ridges while others are plateaulike regions of broad elevation. Because these features are free from earthquake activity, they are known as **aseismic ridges,** a term coined by Laughton and others [1970]. Aseismic ridges differ from the surrounding ocean floor, not only by their morphology, but by distinct geophysical characteristics. Some of the ridges are contiguous with an adjacent continent or a mid-ocean ridge, while others are isolated features. The best studied examples are in the Atlantic and Indian oceans. One of the dominant features of the Indian Ocean which differentiates it from the central and southern Atlantic and the central and eastern Pacific is the large number of aseismic ridges. These include the Broken, Ninetyeast, and Chagos-Laccadive ridges, the Mascarene Plateau, the Madagascar ridge, and the Crozet and Kerguelen plateaus.

The several hypotheses that have been proposed to account for the origin of various aseismic ridges include: (1) isolated fragments of continental crust (*microcontinents*); (2) uplifted oceanic crust; and (3) linear volcanic features.

Microcontinents Some aseismic ridges are continental in origin, having been separated during rifting, sea-floor spreading, or both; these ridges include Orphan knoll, Rockall Plateau, and Jan Mayen ridge in the North Atlantic, the Agulhas Plateau in the South Atlantic, and the Mascarene Plateau in the western Indian Ocean. Changes in plate motions and the development of new spreading centers has, in some instances, led to the isolation of small segments of continental crust and lithosphere surrounded by oceanic crust. These isolated continental fragments have been called **microcontinents.** Since most continental segments that protrude into the oceanic areas remain in contact with the continents, microcontinents are unusual features. There are several ways in which small continental masses become isolated from their parental continental blocks. This includes ridge jumping and the formation of a new spreading center landward of a previous spreading center, thus bisecting a continental fragment and carrying it oceanward by subsequent sea-floor spreading. A second mechanism is by ridge jumping near a pole of rotation, which can slice off small pieces of continents.

Volcanic features The most prominent examples of aseismic ridges, however, are those composed not of continental rock but of oceanic crustal basalt; these include the Hawaiian-Emperor volcanic chain in the North Pacific, Ninetyeast ridge in the Indian Ocean and Walvis ridge and Rio Grande rise in the South Atlantic. Some rises previously considered to be microcontinents or uplifted oceanic blocks are now accepted as linear volcanic features. Aseismic ridges intercept the linear pattern of oceanic magnetic anomalies parallel to ocean ridges. This makes it difficult to estimate the age of the rises without direct sampling of the underlying volcanic rocks.

One of the best known aseismic ridges and the longest straight ridge to develop in the oceans is Ninetyeast ridge, which extends linearly for 4500 km in the eastern Indian Ocean from a latitude equivalent to southern Australia to the Bay of Bengal. It takes its name from its parallelism to the 90°E meridian. Sedimentary evidence indicates that this feature has undergone considerable subsidence during its northward movement across the Indian Ocean. Lignite and lagoonal sediments were recovered at water depths now over 1600 m. Other shallow-water indicators include oyster beds recovered at depths of 3000 m. Various evidence shows that this feature was formed at sea level, close to an active spreading center, and that it has subsided with age. In fact, this depth increase is close to that predicted by assuming that the ridge has subsided at the same rate as the oceanic crust on the Indian plate to which it is attached. The fauna and flora recovered from sediment cores drilled at various locations throughout its length not only reflect this history of subsidence but also reflect the concomitant increase in warmer-water species as the ridge moved northward into lower latitudes.

The deep crustal structure of aseismic ridges is poorly known. Francis and Shor [1966] determined that the Mohorovičić discontinuity is located about 17 km beneath the Chagos-Laccadive ridge and this boundary is reported at 16 to 18 km beneath the Iceland-Faeroe ridge. Thus the crust has thickened by 8 to 10 km beneath aseismic ridges, since oceanic crust is normally 5 to 7 km thick. Crustal thickening may not, however, have occurred beneath Ninetyeast ridge. Free-air gravity anomalies over aseismic ridges are small in amplitude, and the ridges appear to be in isostatic equilibrium. Isostatic adjustments of the thickened crust associated with aseismic ridges must have occurred with the vertical movement of large crustal blocks, because the lithosphere at ridge crests is weak and cannot carry the loads produced by the vast volcanic rock accumulations typical of aseismic ridges. These adjustments occurred concurrently with volcanism, explaining the large scarps and block-fault morphology of many aseismic ridges. Moats around Hawaii are caused by isostatic adjustment of the lithosphere due to the weight of the islands.

Evidence shows that aseismic ridges have undergone significant long-term subsidence since their formation (Fig. 5–21). Detrick and others [1977] have found that the aseismic ridges have the same subsidence rate, though not the same absolute depth, as normal ocean floor. Thus, using their present offset in depth and allowing them to subside at the same rate as normal ocean floor, estimates can be made of their depth changes through time. The most powerful evidence for such subsidence is from sediments that cap the ridges. Deep-sea drilling on several ridges

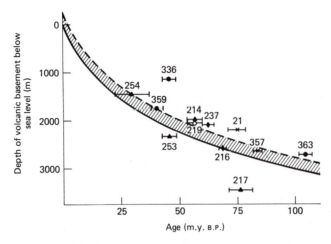

Figure 5-21 Plot of depth versus age for DSDP sites on aseismic ridges inferred to have formed near sea level. The heavy black line is an empirical curve of Parsons and Sclater [1977] showing depth range expected on the assumption of normal subsidence. The dashed line is the same curve displaced upward by 300 m. (From R. S. Detrick, J. G. Sclater, and J. Thiede, *Earth and Planetary Science Letters*, vol. 34, pp. 185–196, 1977)

has shown that sedimentation began in shallow water, even subaerial conditions, followed by progressively deeper water and more open marine depositional environments. Some have subsided several thousand meters below present sea level. For instance, the Orphan knoll in the Labrador Sea is a small continental fragment that received coastal plain deposits rich in redeposited anthracite during the Jurassic. During the late Jurassic and early Cretaceous, the knoll was uplifted and partly eroded followed by a minor subsidence to a depth of 200 m during the late Cretaceous. During the Paleocene, impressive subsidence occurred of up to 115 cm per 1000 years, at which time the knoll sank to its present depth of 1800 m.

Twin aseismic ridges

In the South Atlantic, pairs of aseismic ridges lie on opposing sides of the ocean. These are the Ceará rise–Sierra Leone rise pair and the Rio Grande rise–Walvis ridge pair (Fig. 5–22). The origins of these pairs of topographic features are related to early opening phases of the South Atlantic. In the equatorial Atlantic, the Ceará and Sierra Leone rises lie on opposing sides of the mid-ocean ridge and are equidistant from its axis (Fig. 5–22). The northern and southern boundaries of these two rises are bounded by the same fracture zones. The area of shallowest acoustic basement under the Ceará rise coincides with a seismic layer 1 to 2 km thick (velocity, 3.5 km/sec) overlying Layer 2. This layer is a pile of volcanic rocks from eruptions which began about 80 Ma (late Cretaceous) when the two now-opposing ridges were sited on the mid-Atlantic ridge. When the high rate of volcanism ceased, the originally contiguous pile began to separate into the distinct topographic features observed today.

The Rio Grande rise and Walvis ridge also appear to be demarcated at their

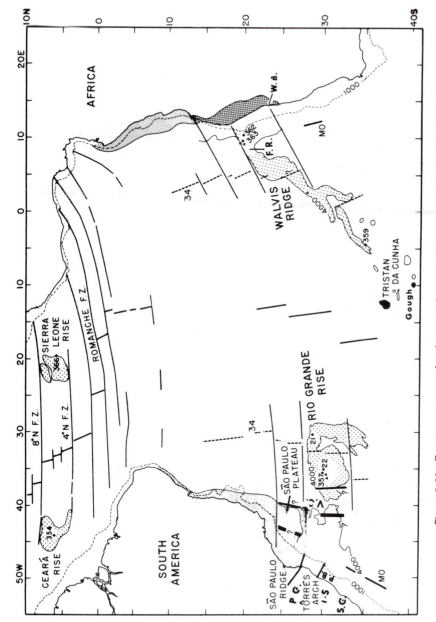

Figure 5–22 Prominent structural and tectonic features of the equatorial and South Atlantic. Note especially the paired aseismic ridges: Rio Grande Rise–Walvis Ridge and Ceará Rise–Sierra Leone Rise each demarcated to the north and south by their associated fracture zones. Zones of salt diapirs are diagonally shaded; area of overlap upon predrift reconstruction is shown by cross hachuring. Magnetic anomalies 34 (80 Ma) and MO are shown. DSDP sites also shown. P.G. = Ponta Grossa Arch; T.S. = Tôrres Syncline; S.G. = Sao Gabriel Arch; V.C. = Vema Channel; P.B. = Petolas Basin; F.R. = Frio Ridge; W.B. = Walvis Bay. (From N. Kumar, *Marine Geology,* vol. 30, p. 176, 1979)

northern and southern boundaries by equivalent fracture zones (Fig. 5-22). Both are made up of basement rocks, consisting of alkali basalt suites typical of oceanic volcanic islands, and both have been subsiding since their creation 100 to 80 Ma (middle to late Cretaceous). At that time, volcanism was abnormally active, giving rise to the ridges. Since then volcanism has decreased and the ridges have separated, due to sea-floor spreading [Wilson, 1963] (Fig. 5-22). The Rio Grande rise, in the middle to late Cretaceous, was a volcanic island 2 km high. Its highest peaks subsided below sea level about 30 Ma (middle Oligocene) and continued to subside at rates comparable with the surrounding oceanic crust. This picture is complicated by a shift, about 80 Ma, of the volcanic center that originally formed the eastern sector of the Walvis ridge. This hot spot has remained beneath the African plate throughout the Cenozoic and is presently located at Tristan da Cunha (Fig. 5-18), about 300 km east of the mid-Atlantic ridge. Following the shift in the hot-spot position, volcanism became more intermittent, forming a series of thin seamounts to the southwest of the Walvis ridge instead of a continuous ridge system linking Tristan da Cunha and the Walvis ridge (Fig. 5-22). Gravity measurements along the ridge-seamount complex support this hypothesis and suggest that the western part of the Walvis ridge was thin enough to flexurally load the thinner oceanic crust but not to produce major isostatic adjustments. This load created isostatic differences between the eastern and western sectors. Thus the western Walvis ridge is similar topographically to the Hawaiian-Emperor seamount chain.

Subduction of aseismic ridges Aseismic ridges on underthrusting oceanic plates often create cusps, or irregular indentations, in the trace of the subduction zone. In general, subduction zones resemble a scalloped alternation of arcs and cusps (Fig. 5-1). Vogt [1973] believes that the complexities of consuming plate boundaries are caused by the buoyancy of aseismic ridges on the downgoing plate. This increases buoyancy of the plate, causing greater resistance to sinking. This, in turn, inhibits backarc or interarc extension by preventing an island arc from freely migrating toward the consumed plate, as proposed by Karig [1971], resulting in the production of a notch in the subduction zone. Island arcs, therefore, may acquire their curvatures not only by the earth's curvature but from colliding aseismic ridges.

The best example of an aseismic ridge moving into a subduction zone is that of the Hawaiian-Emperor Chain moving into the Kuril-Aleutian cusp, at the intersection of the Kuril-Kamchatka and the Aleutian arcs (Fig. 5-19). The Aleutian Arc is one of the major arclike subduction zones, along which the Pacific plate underthrusts the American plate. The eastern junction of the Aleutian Arc lies near south central Alaska, where another cusp forms from several lesser seamount chains moving into the arc. Thus the main, southward convex bulge of the arc lies between two groups of aseismic ridges on the downgoing plate.

DRIVING MECHANISMS OF PLATE TECTONICS

Although the geometry of the plates and the kinematics of their motions are now quite well understood, there is still little known about the driving mechanism of plate tectonics. Geophysicists are now conducting research on the driving motion of the plates. In the meantime, the absence of a well-developed dynamic theory has in no way affected the use of the kinematic theory of plate tectonics to decribe the evolution of the earth's surface. It is widely accepted that an energy source which accounts for the earth's heat loss, seismicity, volcanism, and tectonism lies in the heat production due to the radioactive decay of uranium, thorium, and potassium. This heat source was discovered in the early 1900s when R. J. Strutt, later Lord Rayleigh, discovered the presence of radioactive materials in crustal rocks which have since been found to be extremely widespread. Alternative energy sources are derived from the process of differentiation in the mantle and core and tidal energy within the earth.

Several mechanisms have been proposed as the *driving force* of plate tectonics (Fig. 5-23). These include the following:

1. Thermal convection in the upper mantle, with ascending limbs under oceanic ridges and descending limbs near subduction zones.

2. Sliding of lithospheric plates due to thermally induced elevation of oceanic ridges.

3. Sinking by gravity of cold, rigid lithosphere into the less dense upper mantle of trenches.

4. Sinking or sliding of the lithosphere on a sloping surface, due to the wedge shape of the lithosphere.

The mechanism of plate tectonism requires an enormous amount of energy, at least 10^{26} ergs/yr, to cause movement of the large plates. The earth acts as an inefficient heat engine and a small fraction of the escaping heat, amounting to $\frac{1}{1000}$ to $\frac{1}{100}$,

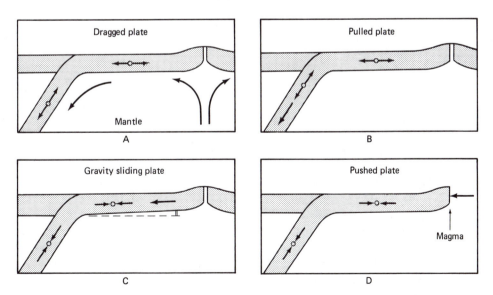

Figure 5-23 Possible mechanisms for moving lithospheric plates. States of compression and tension in the plates are indicated. (From P. J. Wyllie, *The Dynamic Earth: Textbook in Geosciences*, p. 354, 1971, courtesy John Wiley & Sons, Inc.)

is converted into strain energy in the lithosphere; this is largely released by tectonic activity concentrated along mobile belts.

The Convection Hypothesis

The only known process by which the lithosphere can be continually strained is by some type of thermal convection within the mantle. The hypothesis of thermal convection as a possible mechanism for tectonics is not new, having been proposed in the early 1800s by Hopkins and used in 1889 by Fisher. However, the idea was largely ignored until 1929, when Arthur Holmes revived and elaborated the idea of convection in the mantle as the cause of global tectonism (see Chapter 4). Convection occurs in fluids when the density distribution deviates from stable equilibrium. The resulting buoyancy forces create flow until equilibrium has been established. A convection system is established when density anomalies such as those caused by thermal and by chemical disequilibrium are produced as rapidly as they are dissipated. Modern knowledge of the rheology of the mantle has suggested that the convection is probably restricted to the upper 500 km of the mantle. There still remains much uncertainty about scale and character of convection, how it might cause strain in the lithosphere, and even if it is large enough to be a primary driving force for plate movement. This uncertainty reflects ignorance about the composition and properties of the mantle, for which there is little available quantitative evidence. Oxburgh [1972] believes the following characteristics should apply to mantle convection.

1. Convection can be expected to occur in the upper mantle and to be an effective mechanism for heat transfer.

2. Convection in the lower mantle may be inhibited by inferred high viscosity.

3. The earth's rotation probably has little effect on the pattern of convection.

4. Turbulent flow is unlikely to occur.

An important question concerning mantle convection is the depth of the counterflow from trench to ridge. Some argue that the counterflow occurs in the asthenosphere above a depth of about 300 km while others have proposed mantle-wide convection.

Oxburgh [1972] demonstrated that the mantle has a viscosity highly dependent upon temperature and pressure and, in zones of ascending flow, warm material from below rises at a rate that is high by comparison with the rate at which it can lose heat by conduction. The ascending material thus undergoes reduction in pressure but little change in temperature, and its viscosity diminishes with decreasing depth. With the upward decrease in viscosity, the velocity of upward flow increases. The hot material reaching the surface at the ridges diverges to form horizontal surface flows which extend down to about 200 km. The thermal gradient is very steep in these horizontal flows and is the main factor in controlling viscosity. In the upper 100 km, the temperature is so low that it creates a rigid, elastic body, while the lower hotter part of the uppermost mantle behaves as a viscous fluid. At some distance from the source of hot upwelling mantle, the density excess associated with the cold boundary overcomes the ability of the plate to resist downward deflection and the currents descend, dragging the crustal plate and destroying it by heat. The flow geometry is complex and related to the global distribution of sinks and sources, their relative activity, and their direction of migration. Ascending convection is thus associated with ocean ridges (Fig. 5-3).

Oceanic trenches are associated with the descending convection currents (Fig. 5-3). Seismic evidence indicates that the upper part of the descending slab is usually in tension, indicating a pull on the surface plate, but that the deeper part of the slab is in compression. There is also evidence that, in some cases, the descending slab breaks into two or more pieces.

Forces of Plate Motion

A number of possible forces have been proposed for driving plates. These include pushing from the mid-ocean ridges, pulling by downgoing slabs, suction toward trenches, and coupling of the plates to mantle flow. Forsyth and Uyeda [1975] have attempted to determine the relative magnitude of the various possible forces that drive the plates (Fig. 5-24). Before considering each of these, we need to examine some aspects concerning the character of plate motions.

1. Plate velocity correlates negatively with total continental area of each plate. For example, the angular velocity of the Eurasian plate is $0.038°/m.y.$ and the Antarctic plate is $0.054°/m.y.$ Both of these plates are dominantly continental. These rates are slow compared with the dominantly oceanic Cocos

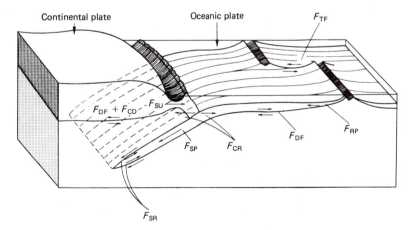

Figure 5-24 Possible forces acting on the lithospheric plates. The abbreviations of forces are as follows: F_{RP} = ridge push, F_{DF} = drag force, F_{CR} = colliding resistance, F_{SP} = slab-pull, F_{SR} = slab resistance, F_{SU} = suction, F_{CD} = continental drag, and F_{TF} = transform fault resistance. (From D. Forsyth and S. Uyeda, *Geophys. J. R. Astr. Soc.*, vol. 43, p. 165, 1975)

plate, which is moving at 1.422°/m.y. and the Pacific plate, moving at 0.967°/m.y. Thus drag on the convecting mantle is weak under oceanic crust but strong beneath continental plates [Forsyth and Uyeda, 1975].

2. The velocity of plates correlates with the fraction of a plate boundary being consumed. The presence of a slab being subducted is a dominant factor.

3. There is no obvious relation between ridge length and plate velocity, so plate motion is not simply the result of pushing from the ridge as new oceanic crust is added.

4. Plate velocities correlate with geographic latitude. Plates move more rapidly near the equator and more slowly near the poles.

5 As exhibited by the Philippine Sea plate, plates can move even when not bounded by oceanic ridges at any edge. Therefore ridge push cannot be the only motivating force.

6. The plates on both sides of the mid-Atlantic ridge are moving apart, despite the absence of downgoing slabs attached to them.

Apparently some combination of forces is responsible for maintaining plate motions. The forces acting on plates are classified into two groups [Forsyth and Uyeda, 1975]: the forces acting at the bottom surface of plates and those acting at plate boundaries.

Mantle drag force (F_{DF})

The force acting at the bottom surface of plates is due to the viscous coupling of plates and the underlying asthenosphere (Fig. 5–24). This would be an important driving force if mantle convection were an important mechanism. Conversely, F_{DF} may be a resistive force if the asthenosphere is passive with respect to plate motion. Since plate velocity is independent of the area of the plate, F_{DF} is probably a resistive force which increases with the area of the plate. This implies that oceanic plates are moving faster than the underlying mantle. Modern studies of thermodynamics suggest that the *classical* theory of mantle convection is probably unacceptable for several reasons. First, the viscosity of the upper mantle implies that cell scale is small. Small cells exert only a periodic stress on the base of the plate, which makes little net contribution when integrated over the entire area of the plate. These small cells, which may control the heat flow in ocean basins away from ridges, cannot drive the plates for long distances although they may exert enough stress to initiate breakup of large continental masses. Furthermore, the supposition that mid-oceanic ridges are the sites of upwelling and trenches are sinkings of the large-scale convective flow (Fig. 5–3) may be invalid, because it is now established that actively spreading, oceanic ridges can migrate and collide with trenches [Atwater, 1970; Larson and Chase, 1970]. Although convection of mantle material at depth is necessary to conserve mass during plate subduction, thermal convection as the primary driving force is insufficient [Forsyth and Uyeda, 1975].

This is an additional resistive force acting on the plate due to viscous drag. The force F_{SR} is proportional to the viscosity of the asthenosphere and velocity of subduction (Fig. 5-24).

Colliding resistance (F_{CR}) *and transform fault resistance* (F_{TF})

Additional resistance forces between plates include F_{CR} at converging plate boundaries and F_{TF}. Most shallow earthquakes are caused by interplate resistances (Fig. 5-24), and Forsyth and Uyeda [1975] believe that the magnitude of these forces is independent of the relative velocity. As the plates move past each other, strain energy accumulates at the locked plate boundary. When the stress reaches a certain level, slip on the fault occurs, releasing strain energy as an earthquake. Seismicity increases with increasing velocity, but since earthquakes are a reflection of stress release, no greater stress should result.

Driving forces

There are various possible driving forces for plate tectonics. These include *ridge push* (F_{RP}), *slab pull* (F_{SP}), and *suction* (F_{SU}) (Fig. 5-24).

Ridge push is due to gravitational sliding which pushes plates apart at the boundary. Inherent in the elevated topography of ridges is a potential energy that forces the ridge to spread out to obtain a lower energy state. This drive is created by the rising convection beneath the ridge. Gravitational sliding off the ridge is a significant force in driving the plates. Within plate interiors, compressional stresses dominate over tensile stresses, suggesting that plates are more strongly pushed from ridges than pulled from trenches. However, other important intraplate stresses must be considered.

The lithosphere-asthenosphere boundary is a sloping surface shallow at the ridge and deeper near the distal plate margins. In cross section, the lithosphere is wedge shaped, ranging in thickness from a few kilometers at the thin end to 100 to 150 km at the thick end. As the lithosphere moves away from the ridge, it becomes thicker because cooling reaches deeper into the mantle, and new material is added to the base of the dense lithosphere. This thickening leads to horizontal density differences across the plate and to a sloping surface at the base of the lithosphere (Fig. 5-24). The principal factors in this process are the horizontal mass difference along the length of the lithospheric slab and the slope of the surface due to the wedge shape of the lithosphere, causing sinking or sliding.

Because the depth dependence of the ocean ridge on the square root of age is not valid beyond about 80 m.y. (130-km lithosphere thickness), it seems the lithosphere does not thicken appreciably beyond that point. Thus, under older ocean floors, the lithosphere-asthenosphere boundary is horizontal. Consequently, the ridge-push driving force only implies to younger (less than 80 m.y. old) parts of the plate. As the elevation of the ridge is maintained by the continual advection of heat in the rising mantle, plate motions generated by the ridge force may be driven by thermal convection.

Slab-pull is a negative buoyancy force that acts on the dense, downgoing slab (Fig. 5–24). Though some claim that the downward pull of the cold, dense slab may be transmitted to the horizontal lithosphere to drive its motion, there are two limiting considerations here. First, the pull of the slab requires a density contrast that is due to a temperature contrast. The temperature contrast is time dependent; hence it is dependent on the rate of subduction: the faster the subduction rate, the higher the contrast. Second, increased viscosity and density in the mantle at depths of 500 to 600 km precludes penetration of the slab to very great depths. Pulling from the trenches, like pushing at the ridges, is a form of thermal convection in which the driving forces are supplied by gravity acting on the density contrasts induced by the cooling of the upper mantle.

Suction, as Elsasser [1971] suggests, may draw the American and Eurasian plates toward the trenches surrounding the Pacific. The plates surrounding the Atlantic are moving apart, overthrusting the plates in the Pacific. Because the radius of the earth probably remains constant, the Pacific must be growing smaller and the trenches migrating seaward. However, the physical nature of the suction force is not clear. Elsasser [1971] visualizes this force as a downwarping of the oceanic plates at trenches, creating an empty space which is continually filled by the seaward movement of the continental plate.

The correlation between velocity and length of trench is a significant clue to plate driving forces and an indication that F_{SP} is very large. However, the motion of plates without subduction zones requires that F_{RP} (ridge push) is also an important mechanism. The importance of F_{SP} and F_{RP} has been debated, the importance of the two varying from plate to plate [Forsyth and Uyeda, 1975].

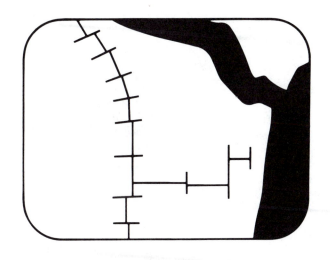

six

Tectonic History
of the Oceans

Geologists . . . inhabit scenes that no one ever saw, scenes of global sweep, gone and gone again, including seas, mountains, rivers, forests, and archipelagoes of aching beauty rising in volcanic violence to settle down quietly and then forever disappear . . . .

John McPhee

GLOBAL PATTERNS

The Ocean Life Cycle

In the early stages of the theory of plate tectonics, Wilson [1965] and Dewey and Bird [1970] recognized that ocean basins have a life cycle of their own. Patterns of ocean development change through geologic time and a succession of ocean basins are born, grow, diminish, and close again (Table 6–1). Spreading rates of a few centimeters per year, when applied to the ocean of a few thousand kilometers, imply a life span of a few hundred million years.

The Atlantic and Indian oceans are growing oceans approaching maturity (Table 6–1). Their mid-ocean ridges lead into narrow seas in the subarctic region, the Gulf of Aden, and the Red Sea, all of which exhibit features indicating a younger stage of ocean development. These, in turn, lead into the East African rift valley, which is an embryonic ocean (Table 6–1).

The Pacific Ocean, though still the largest ocean, is much smaller than when the continents were assembled as Gondwanaland and Laurasia and continues to decline (Table 6–1). The history of the Tethys Ocean suggests that the Mediterranean Sea is a former ocean in a terminal stage. The Himalayan Mountain chain shows the final phase of oceanic destruction, being a scar (**geosuture**; Table 6–1) of the eastern sector of the former Tethys Sea. The nature of the *oceanic life cycle*

TABLE 6-1 STAGES IN THE LIFE CYCLE OF OCEAN BASINS AND THEIR CHARACTERS[a]

Stage	Example	Mountains	Motions	Sediments	Igneous rocks
Embryonic	East African rift valleys	Block uplifts	Uplift	Negligible	Tholeiitic flood basalts, alkalic basalt centers
Young	Red Sea and Gulf of Aden	Block uplifts	Uplift and spreading	Small shelves, evaporites	Tholeiitic sea floor, basaltic islands
Mature	Atlantic Ocean	Mid-ocean ridges	Spreading	Great shelves (miogeosynclinal type)	Tholeiitic ocean floor, alkali basalt islands
Declining	Pacific Ocean	Island arcs	Compression	Island arcs (eugeosynclinal type)	Andesitic volcanics, granodiorite-gneiss plutonics
Terminal	Mediterranean Sea	Young mountains	Compression and uplift	Evaporites, red beds, clastic wedges	Andesitic volcanics, granodiorite-gneiss plutonics
Relic scar (geosuture)	Indus line, Himalayas	Young mountains	Compression and uplift	Red beds	Negligible

[a] After Wilson [1968].

Tectonic History of the Oceans 179

[Wilson, 1965; Dewey and Bird, 1970] is such that mountain belts resulting from continental collision preserve only the record relating to initial continental breakup and terminal continental collision, since subduction essentially destroys the intervening oceanic historical record.

The present episode of continental drift and sea-floor spreading began about 200 Ma with the opening of the Atlantic and Indian oceans. This episode cannot continue longer than another 200 m.y. because the Pacific Ocean, which is decreasing in size now, will be completely closed 200 m.y. from now. Thus the life span of the present phase of spreading and drift is expected to be 500 m.y. This is considered the last in a series of discrete episodes of spreading that have occurred during the earth's geological history, possibly since the original formation of the crust. The oceanic crust previously separating continents has been largely destroyed by the subduction process. Linear mountain chains contain a record of tectonism since at least 2000 Ma, and paleomagnetic evidence also shows that the continents were moving during the Precambrian. It is possible that continents with very thin crust covered most of the earth when they were originally formed. Since then, the continental crust has been shortened across mountain belts and by accretion at convergent margins. It is possible that a progressive reduction has occurred in the continental area with a reciprocal increase in the oceanic area, though the volume of continental crust has possibly increased with time through continued accretion at the convergent margins. Volcanism associated with the subduction of plates produces volcanic rocks with the same bulk composition as continental crust. The volcanic chains are, therefore, embryonic crust adding to the volume of the continents. Because there is no effective means of destroying continental crust once formed, it follows that the total volume of the continental crust has been increasing during the past two billion years.

During the Cenozoic, oceanic crust has been doubled by creation at ridge crests; as much has been destroyed in the subduction areas. About one-third of the present earth's crust has been formed during only the last 1½ percent of geologic time, the Cenozoic. The rest of the oceanic crust is from the middle to late Mesozoic. Any history of the ocean basins prior to the middle Mesozoic (Jurassic) can be found only in their old convergent margins, but such information is difficult to obtain since such margins are extremely complex due to compressional and transcurrent deformation, metamorphism, uplift, and erosion. Ranges like the Alps and the Andes are relatively young and are associated with currently active plate boundaries, and the sequences of rocks in older, nonactive mountain ranges such as the Appalachians and Urals are similar to the presently active mountain chains. The Appalachian Mountains were in part formed of sediments deposited in a "proto-Atlantic" ocean during the late Precambrian and compressed by an oceanic closure during the late Paleozoic (about 300 Ma). To determine the history of the ocean basins, a number of techniques must be employed. Dating magnetic anomalies up to the continental margins is useful for determining the initiation of continental drift, but it cannot be employed in all areas. The geologic record on the trailing margin of continents and the paleomagnetism of continental rocks is also used.

Changes in Plate Geometry

Sea-floor spreading and orogeny are essentially continuous, although certainly not steady-state processes. A particular pattern of crustal movements, such as that currently occurring, may well continue for some time until stresses set up between or within blocks force a geometric change. The poles of relative motion between plates are not fixed for very long periods of geologic time. Colliding continents and island arcs create changes in the poles of rotation. Le Pichon and Hayes [1971] have shown that at least two different pole positions are required to account for the directions of fracture zones in the South Atlantic. The motion of India relative to Antarctica has been even more varied; at least five pole positions are required to describe India's motion during the last 75 m.y. Thus changing geometric patterns in the earth's plates are discontinuous, and because the plates are essentially rigid, the change from one geometry to another is rapid [Menard and Atwater, 1968]. It seems probable that the last change of this type occurred about 5 Ma; the geometry of spreading in the Northeast Pacific and at the mouth of the Gulf of California clearly changed then from a dominantly east-west to the northwest-southeast direction of today [Vine, 1968; Larson and others, 1968].

In this chapter brief summaries are provided of the three major oceans. Their paleoenvironmental evolution is summarized in Chapter 18.

INDIVIDUAL OCEANIC HISTORY

Tectonic History of the Pacific Ocean

The modern Pacific Ocean, still the largest, consists of a number of plates; the largest of these is the Pacific plate, encompassing about 22 percent of the earth and the majority of the Pacific basin (Fig. 5-1). Other plates that make up the Pacific Ocean are the Antarctic plate to the south, the Nazca and Cocos plates to the east, and the Indian and Philippine plates to the west (Fig. 5-1). The eastern margin of the Pacific is bounded by plate boundaries near the continental border of North and South America. Western South America is bordered by the Peru-Chile Trench, a destructive boundary to the east of the Nazca and Cocos plates. The North American boundary is more complex and is believed to be underlain by a constructive boundary—the northward extension of the East Pacific rise (Fig. 5-1). The Pacific plate is bounded to the east and south by constructive boundaries at the Pacific-Antarctic ridge and East Pacific rise. To the west and north, the Pacific plate is bounded by a trench system extending from the Kermadec-Tonga Trench in the South Pacific to the New Hebrides Trench, Mariana Trench, and the Japan Trench, and then northeast through the Kurile and Aleutian trenches (Fig. 5-1).

The Pacific Ocean is the oldest ocean. It was a *superocean* in the early Mesozoic when the continents were clustered as Gondwanaland and Laurasia. The

Pacific Ocean has since steadily decreased in size as the Americas moved westward while the Atlantic grew, and Australia drifted northward, constricting the southwest and western sectors of the Pacific basin. The Pacific Ocean will continue to decrease until, in about 200 m.y., it is eventually destroyed. It is estimated that since the early Cretaceous, an area of oceanic crust equal to most of the Pacific basin has been subducted and destroyed beneath northeastern Asia, North America, South America, and probably Antarctica. This loss of crust has eliminated much of the evidence that could have been used to interpret the older history of the Pacific Basin.

The Northwest Pacific: ancient Mesozoic plates

One of the oldest and best preserved records of sea-floor spreading in the world's oceans is in the Northwest Pacific. R. Larson and his colleagues have examined the late Jurassic and early Cretaceous magnetic anomaly patterns in the Northwest Pacific and reconstructed the history of the plates and their boundaries. Larson and Chase [1972] discovered three sets of magnetic lineations: the Japanese, the Hawaiian, and the Phoenix (Fig. 6-1). These lineations are of early Cretaceous to late Jurassic age, and no older magnetic anomalies are known anywhere in the ocean basins. The set of lineations include M1 through M22 and range in age from about 110 to 150 Ma (Fig. 6-1). We have already determined that the pattern of magnetic anomalies represents one of the most powerful tools for establishing paleoreconstructions of the ocean floor. However, the presence of the magnetic quiet zone in the middle Cretaceous makes Mesozoic paleoreconstructions much more difficult (Fig. 4-9). Nevertheless, Larson and Pitman [1972] and Larson and Chase [1972] have used the geometry of the Mesozoic lineations in the Northwest Pacific to provide a possible plate configuration of the Pacific Ocean about 110 m.y. ago. They proposed the presence of at least four plates, five oceanic ridges, and two triple junctions in the Pacific during the early Cretaceous (Fig. 6-2). Three of these plates have since been entirely or partially destroyed by subduction. From the time the Pacific plate was being generated in the Southern Hemisphere, it has subsequently rotated slightly in a clockwise direction. The Pacific plate moved north while the Kula plate was being destroyed by subduction under Japan and the Kurile and Aleutian trenches. Likewise, the Farallon and Phoenix plates were destroyed at the subduction zones adjacent to the Americas (Fig. 6-2). The most important implication of this study is that sea-floor spreading is a process that has continued to occur over a long period of time and that previous plates and plate boundaries were formed, destroyed, and replaced by younger features.

The Pacific plate: Cenozoic northward movement

There are several aspects of the history of the Pacific plate which are particularly important [van Andel and others, 1975]:

1. During the Cenozoic, the plate continuously changed its position relative to the equator.

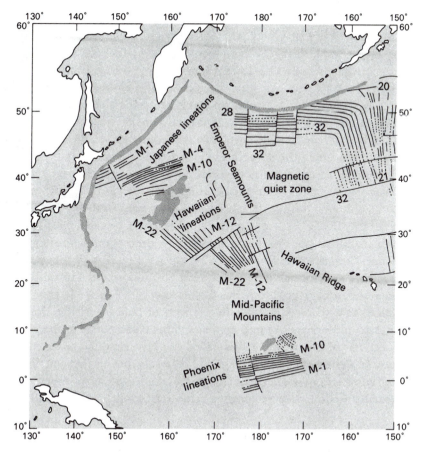

Figure 6-1 Mesozoic and Cenozoic magnetic lineations in the Northwest Pacific. Mesozoic lineations are prefixed by M with M-1 the youngest. (After R. L. Larson and C. G. Chase, 1972)

2. Spreading from the East Pacific rise and its precursors has resulted in a northward or northwestward increase in age and subsidence of the sea floor.

3. The location of the East Pacific rise has not remained fixed throughout the Cenozoic.

4. The East Pacific rise is a relatively young feature. Before the formation of the East Pacific rise, the eastern boundary of the Pacific plate in the Southern Hemisphere (south of 5°S) was located at the Galapagos rise, further to the east. North of this position, the previous spreading center is represented by a fossil ridge. It appears that several large crestal jumps occurred in the position of this spreading center through geologic time. The last such jump seems to have occurred about 12 Ma [Herron, 1972].

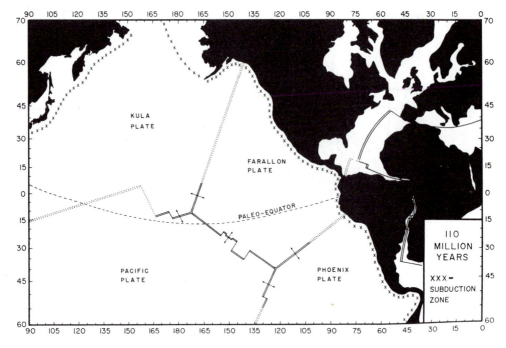

Figure 6-2 Possible configuration of lithospheric plates in the Pacific Ocean 110 m.y. ago (middle Cretaceous). (From R. L. Larson and W. C. Pitman III, *Geol. Soc. Amer. Bull.* vol. 83, p. 3654, 1972, courtesy The Geological Society of America)

A northward movement of the Pacific plate was originally postulated by Francheteau and others [1970] on the basis of paleomagnetic data from seamounts. Since then, northward drift of the plate has been postulated to explain the progressive northward displacement of equatorial sedimentary deposits in stratigraphic sections in the equatorial region [van Andel and others, 1975]. At the same time, Morgan [1972] and others have shown that linear volcanic trends, such as the Hawaiian, Emperor, and Line island chains, are trails resulting from drift of the plate over hot spots probably fixed deep in the mantle. Based on this geometry, the evidence from the Hawaiian Chain suggests that during the Cenozoic, the Pacific plate rotated around two successive absolute poles, the rotations being represented by the Emperor and Hawaiian volcanic chains (Fig. 5–16). The first pole of rotation, the Emperor, has a position at about 17°N, 107°W [Claque and Jarrard, 1973]. The second and most recent rotation is determined by the Hawaiian pole. Its position, based on the Hawaiian, Cook-Austral, and Guadelupe chains, has been calculated at 72°N, 83°W [Claque and Jarrard, 1973], but the circle of confidence is large and includes poles at 67°N, 73°W [Morgan, 1972] and 67°N, 45°W [Winterer, 1973; van Andel and others, 1975]. The change in rotation between the Emperor and Hawaiian poles occurred about 40 Ma near the Eocene-Oligocene boundary. Other approaches to paleoreconstruction and sedimentary implications are discussed in Chapter 17.

The Southwest Pacific: New Zealand and Australian migrations

Beginning in the late Cretaceous about 80 Ma, sea-floor spreading began in the Southwest Pacific and caused the northward movement of New Zealand, followed by Australia, beginning in the early Eocene (53 Ma). These motions played the principal role in the construction of the western part of the Pacific Ocean and represented an important element in the destruction of the Tethys Sea, which had previously extended between the Pacific and the Atlantic via the northern Indian Ocean. Molnar and others [1975] have presented a set of reconstructions of the Southwest Pacific Ocean (Fig. 6–3) which exhibits the principal elements in plate-tectonic history. The paleoreconstructions, in part, are based on histories previously established for the Southeast Indian Ocean [Weissel and Hayes, 1972], the Tasman Sea [Hayes and Ringis, 1973], and the area to the southeast of New Zealand [Christoffel and Falconer, 1972]. These reconstructions describe the rigid plate motions of the Campbell Plateau, the Chatham rise, and New Zealand southeast of the Alpine fault with respect to West Antarctica. The motion of Australia to East Antarctica is that described by Weissel and Hayes [1972]. Hayes and Ringis (1973) demonstrated that the Tasman Sea was created by sea-floor spreading between 70 and 60 Ma. It seems that no relative motion has occurred between Australia, the Lord Howe rise, and Northwest New Zealand since 60 Ma. At no time did more

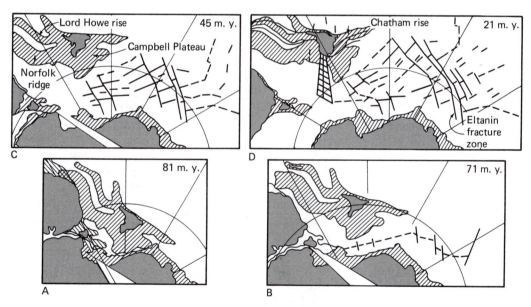

Figure 6–3 Possible configuration of continental fragments (hachured) and plate boundaries in the South Pacific at successive times in the past. (A) 81 Ma—late Cretaceous; (B) 71 Ma—late Cretaceous; (C) 45 Ma—late Eocene; (D) 21 Ma—early Miocene. Plate boundary in New Zealand shown to be active for the first time in 21 Ma paleoreconstruction. The 81 Ma reconstruction shows the grouping of Australia, Antarctica, New Zealand, and associated ridges. An additional, unknown, plate boundary is probably needed to avoid the overlap of Australia and the Norfolk ridge. (After P. Molnar and others, 1974)

than three major plates exist in the Southwest Pacific. The most important interactions occurred between the Antarctic, Pacific, and Indian plates. Several important segments of the Southwest Pacific are continental fragments, including the Campbell Plateau, Chatham rise, Lord Howe rise, and Norfolk ridge. These are shown as continental extensions of the New Zealand Plateau in the reconstructions (Fig. 6–3). The South Tasman rise south of Tasmania is also of continental origin.

Bathymetric and magnetic anomaly data indicate that the floor of the South Pacific Ocean between New Zealand and Antarctica formed since the late Cretaceous (since 81 Ma; Fig. 6–3(a)) by sea-floor spreading at the Pacific-Antarctic rise and the southern portion of the East Pacific rise. This system of rises appears to have been the plate boundary between the Pacific and West Antarctica plates since the late Cretaceous [Christoffel and Falconer, 1972; Pitman and others, 1968]. During the late Cretaceous, the Campbell Plateau, Chatham rise, and New Zealand split away from West Antarctica (Fig. 6–3(b)); between about 81 and 63 Ma, the rate was about 6 cm/yr. South of New Zealand, a few fracture zones with left-lateral offsets were formed, but further northeast, four major fracture zones with right-lateral offsets also developed. The beginning of spreading in the South Pacific coincides with that in the Tasman Sea [Hayes and Ringis, 1973]; both may have begun as interarc spreading (Fig. 6–3(b)). It appears that the directions of relative motion in these regions were different, indicating that East and West Antarctica were being rifted apart (Fig. 6–3(b)).

Following the initial breakup of the Campbell Plateau and Antarctica, the Pacific plate rapidly (about 9 cm/yr) moved away from the Antarctic plate. Between 81 and 63 Ma, East and West Antarctica separated from one another, and a triple junction existed where the spreading centers in the Tasman Sea, the South Pacific, and between East and West Antarctica meet. Between 63 and 38 Ma, the rate of spreading on the Pacific-Antarctic rise slowed and the direction changed (Fig. 6–3(c)). The fracture zones south of New Zealand, with left-lateral offsets, appear to have ceased; the Eltanin system of fracture zones changed orientation and spacing. The change in rate corresponds with the end of spreading in the Tasman Sea about 60 Ma [Hayes and Ringis, 1973] and the beginning of spreading between Australia and East Antarctica about 53 Ma [Weissel and Hayes, 1972]. The data require deformation in Antarctica until as recently as about 40 Ma.

Bending of the "New Zealand Geosyncline" occurred more recently than 40 Ma, at the same time the southern portion of the Tasman sea floor probably formed by slow spreading near the present Macquarie rise, between the Pacific and Indian plates. Deformation in New Zealand began near the end of the Eocene with an oblique separation of different parts of New Zealand; then beginning in the Miocene, one part rotated with respect to the other, obliquely closing the separation to form the present configuration (Fig. 6–3(d)).

Setting the late Eocene age as the beginning of this motion is consistent with H. W. Wellman's inference that the Kaikoura orogeny occurred during this period. Wellman [1971] states that during the Eocene and Oligocene, normal faulting occurred in New Zealand near the Alpine fault, but the direction of motion reversed in the Miocene. In addition, Lillie and Brothers [1970] observed that prior to the Oligocene, the history of New Caledonia and New Zealand were similar, but they began to diverge afterward. Suggate [1963] concluded that a belt of early Tertiary sediments was deposited along the location of the Alpine fault and were uplifted during the change in tectonic conditions in the late Oligocene or early Miocene. Since then, spreading rates have been increasing. The present relative velocity of the plates along the Alpine fault ranges from about 6 cm/yr in the southwest to about 10 cm/yr in the northeast.

During the middle Eocene, Australia began to move northward and the volcanic arcs and marginal seas in the Southwest Pacific began to form a few million years later. At this time, the Coral Sea and New Hebrides basin, as well as the western part of the Fiji basin (Packham and Terrill, 1975) and the Norfolk basin, began forming. The basins show a general age progression eastward from the western province, as demonstrated by late Cretaceous volcanoes of the Lord Howe rise, the middle Eocene to Oligocene basement age of the South Fiji basin, and the late Miocene age of the initiation of the Lau basin. Arc migration has usually been associated with marginal sea development, and seems to be an effect of this development rather than a cause [Packham and Terrill, 1975]. Further to the northwest, new sea floor in the western Philippine basin began to form from about 45 to 37 Ma [Karig, 1975]. The eastern and western Caroline basin immediately to the north of New Guinea developed slightly later, during the early to middle Oligocene. This basinal development north of Australia reflects increased tectonism related to the northward movement of Australia. The uplift of New Guinea, beginning in the early late Miocene [Burns, Andrews, and others, 1973], is related to even further tectonism of this region.

The Northeast Pacific: boundary between the North American and Pacific plates

One of the most important aspects of the tectonic history of the Northeast Pacific has been the changing relationships between the boundaries of the Pacific and North American plates. Atwater [1970] and Atwater and Molnar [1973] have calculated the relative positions of these two plates at several different times (Fig. 6-4) by examining global plate reconstruction. The plate tectonics of the region has been largely controlled by the eastward advance of the ocean ridge toward North America and its contact and relocation under the western sector of North America (Fig. 6-4).

The magnetic lineations to the west of North America are considered to be only one side of a spreading ridge. The East Pacific rise was once located in the Northeast Pacific, just as it is now located in the modern Southeast Pacific. At that time the spreading ocean floor advanced from the ridge crest toward North America and was consumed in a trench like the Peru–Chile Trench off South America (Fig. 6-5). Atwater and Molnar [1973] propose a plausible model summarized in Fig. 6-4. At 38 Ma the Farallon plate was the symmetrical counterpart of the main Pacific plate and thus was formed by the creation and eastward spreading of oceanic crust from the East Pacific rise (Fig. 6-4(A)). Between 38 and about 30 Ma, the Pacific plate moved slowly west with respect to North America, but the spreading center migrated east. By about 30 Ma initial contact was made

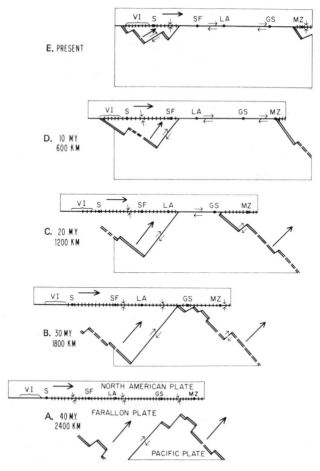

Figure 6-4 Schematic model of interactions between the North American, Pacific, and Farallon plates during last 40 m.y. based on the assumption that the North American and Pacific plates moved with a constant relative motion of 6 cm/yr parallel to the San Andreas Fault. The west coast of North America is almost parallel to the San Andreas Fault. The distances are shown for each time interval that the North American Plate must be displaced to reach its present position with respect to the Pacific Plate. Place name abbreviations are as follows: VI-Vancouver Island; S = Seattle; SF = San Francisco; LA = Los Angeles; GS = Guyamas; MZ = Mazatlan. (From T. Atwater, *Geol. Soc. Amer. Bull.*, vol. 81, p. 3516, 1970, courtesy The Geological Society of America)

between the East Pacific rise and North America (Fig. 6-4(B)). At that time, the San Andreas fault, which is a ridge-ridge transform fault, began its motion. At present, this fault connects the East Pacific rise at the mouth of the Gulf of California to the Juan de Fuca ridge in the North Pacific (Fig. 6-6).

Between about 30 and 20 Ma, the motion of the Pacific and North America plates was rapid (about 5 cm/yr). During this time, the Farallon plate was broken up between the Mendocino and Murray fracture zones and the fragments thrust

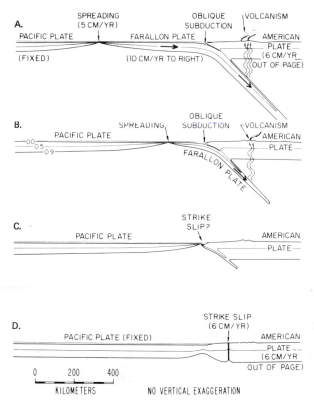

Figure 6-5 Collision of North American Plate (with its previous marginal trench) with the mid-oceanic ridge during the Cenozoic. The Pacific Plate is held fixed. The spreading center spreads at 5 cm/yr half rate, accreting material onto both plates so that it moves to the right at 5 cm/yr and the Farallon Plate moves to the right at 10 cm/yr. The American Plate moves out of the page at 6 cm/yr. A represents the plates in the early Cenozoic; C represents the configuration at the time of collision of the ridge and trench; B is an intermediate situation; and D is the present-day configuration. (From T. Atwater, *Geol. Soc. Amer. Bull.*, vol. 81, p. 3528, 1970, courtesy The Geological Society of America)

under western North America (Fig. 6-4(C)). The destruction of the Farallon plate continued, and by about 5 m.y. ago the crest of the East Pacific rise had reached the Gulf of California (Fig. 6-4(D and E); Fig. 6-5). This caused Baja California to dislocate from the North American plate and to move northward at a velocity that has not significantly altered since that time. The Gulf of California was produced as a result of strike separation along a series of en echelon transform faults spreading in the Gulf (Fig. 5-4). The relative motion between two oceanic plates that are spreading apart at an oceanic ridge can be determined if well-defined magnetic anomalies are generated from the spread. The only place along the North American–Pacific plate boundary exhibiting well-defined magnetic anomalies is in the Gulf of California, and these were formed only about 5 m.y. ago (Fig. 5-4). The recent rate of motion between these two plates is about 5.5 cm/yr [Minster and others, 1974]. Thus 5 m.y. ago, the Pacific plate was about 250 km southeast of its present position relative to the North American plate. At present, the East Pacific rise trends northward into the Gulf of California (Fig. 6-4(E) and 6-6) and probably continues north under western North America through the *basin and range province* of Utah, Nevada, and Arizona. Here, high heat flow and tensional tectonics suggest the possibility of an area underlain by an active rifting system. Continued motion in the future will eventually lead to the complete separation of the southern California–Baja area from the rest of North America. This area will become a separate microcontinent.

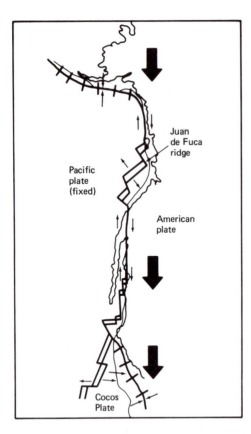

Figure 6-6 Present-day configuration of plates in the Northwest Pacific and in western North America. San Andreas Fault is long vertical line in center of figure. (After T. Atwater, 1970)

Tectonic History of the Atlantic Ocean

The tectonic history of the Atlantic Ocean has been summarized from several sources by many people, such as Phillips and Forsyth [1972], van Andel and others [1977], and, most recently, Sclater and Tapscott [1979] and Sclater and others [1977]. Finite rotations of the African, European, and North and South American plates about the poles described by Bullard and others [1965] have been employed to reconstruct the ancient configuration of the Atlantic Ocean. The rates and duration times for the motions have been estimated from marine magnetic profiles and deep-sea drilling results based on sea-floor spreading models. The following summary is largely from Sclater and Tapscott [1979] and Sclater and others [1977].

Early to middle Mesozoic (Fig. 6-7(A))

The central Atlantic, Caribbean, and Gulf of Mexico began to form in Triassic time about 200 Ma, with Africa and South America drifting away from North America. The opening of the North Atlantic began 165 Ma (middle Jurassic) with the formation of oceanic crust between North America and Africa. The North Atlantic basin remained closed to the north by the juncture of Spain and the Newfoundland fracture zone and to the south by the juncture of the Guinea nose and

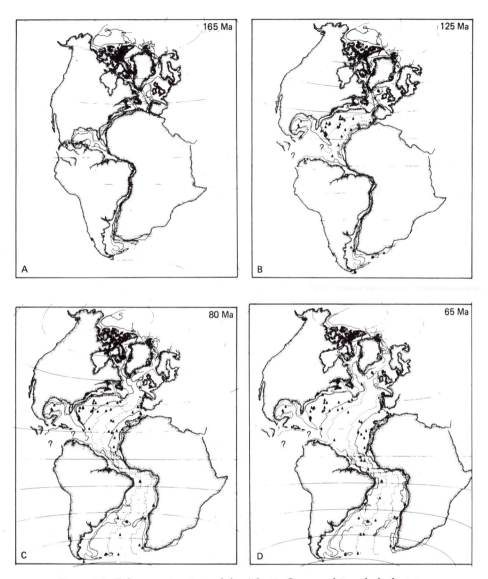

Figure 6–7 Paleoreconstructions of the Atlantic Ocean and its paleobathymetry between 165 Ma (middle Jurassic) and 65 Ma (earliest Cenozoic). DSDP sites are shown as triangles. (From J. G. Sclater, S. Hellinger, and C. Tapscott, *Journal of Geology*, vol. 85, pp. 500–522, 1977, reproduced by permission of the University of Chicago Press)

Bahama platform. Farther south, Africa and South America remained snugly fit together. The Falkland Plateau fit around the southern tip of Africa and against the Mozambique ridge. By 125 Ma the North Atlantic had developed an active mid-ocean ridge and had reached depths of up to 4000 m (Fig. 6–7(B)). By this time Africa had begun to separate from Iberia, causing a connection between the North Atlantic and the Tethys Sea.

Early to middle Cretaceous (Fig. 6–7(B))

The next major event was the opening of the South Atlantic. It is not certain when this began, but various geological constraints limit the event to between the earliest Cretaceous (125 Ma) and the boundary between the early and middle Cretaceous (110 Ma). This is based on the age of the magnetic anomaly pattern that lies in the Cape Basin directly adjacent to the continental edge of South Africa. The North and South Atlantic remained unconnected at the bulge of Africa. During its early history, the South Atlantic was divided into two basins by the mid-Atlantic ridge, and these were closed to the north by Africa and to the south by the Falkland Plateau. Each basin was divided in two by a volcanic ridge: the Rio Grande rise across the western basin and the Walvis ridge across the eastern (Fig. 6–7(C)). It is possible that during the very early stages of opening, the South Atlantic may have been a freshwater lake.

Middle Cretaceous

During the middle Cretaceous, Africa and North America continued to separate. About 95 m.y. ago, Europe began to separate from North America. Greenland started moving away from both Europe and North America, leaving narrow, shallow seas jutting northward from the North Atlantic.

In the South Atlantic, the bulge of Africa was separated from South America, possibly allowing shallow-water connections for the first time between the North and South Atlantic. However, deep-water communications were almost certainly prevented by topographic elevations associated with the fracture-zone system. In the South Atlantic, the Rio Grande rise–Walvis ridge complex began to subside, which would have allowed shallow-water connections between the two basins. Also, the ocean entered from the south as the Falkland Plateau moved away from Africa and subsided.

Late Cretaceous (Fig. 6–7(C))

By 80 Ma the North Atlantic was a full-fledged ocean. Parts of it were 5000 m deep and its water could flow into the oceans through the Caribbean and the Tethys. In the South Atlantic, four distinct basins were forming at depths greater than 4000 m (Brazil, Guinea, Argentine, and Cape), separated by the mid-Atlantic ridge and the Walvis–Rio Grande rise complex. At 80 Ma it is still unlikely that a deep-water connection had been established with the North Atlantic. Furthermore, deep-water flow was still probably blocked at the Falkland Plateau–Falkland fracture zone.

Early Tertiary (Fig. 6–7(D))

A major reorganization in the North Atlantic occurred between 65 and 53 Ma. Spreading ceased in the Bay of Biscay and Iberia became part of the European Plate. The Norwegian-Greenland Sea started opening between 60 and 63 Ma, and the relative motion between Europe and Greenland increased in velocity. Baffin

Bay began to open and the Labrador Sea developed significant sections deeper than 4000 m.

By this time, the South Atlantic was wide and deep, and it is likely that both shallow and intermediate waters penetrated the North Atlantic. By this time, the Rio Grande rise had subsided to close to its present level.

Middle Tertiary

Most of the major topographic features of the Atlantic had formed by 36 Ma. Both the North Atlantic and the South Atlantic were broad and deep, although the North Atlantic was deeper. The Walvis and Rio Grande rises assumed present-day morphology and depths. The Vema gap, in the Southeast Atlantic, was open to depths below 4000 m between the Rio Grande rise and South America, allowing deep water to penetrate into the North Atlantic. The Caribbean had acquired its present general shape. Spain and Africa were still significantly separated by a deep channel between the Tethys and the Atlantic. Over the past 36 m.y., Africa and Europe have slowly converged, almost completely closing the passage between the Atlantic and the Mediterranean. The major change occurring during this time was a change in motion between Greenland and North America. Greenland changed from a rotation about northern Baffin Bay to a rotation about a pole near the center of the Sahara. This movement opened Baffin Bay and the Labrador Sea.

Late Tertiary (Fig. 6–8)

The Iceland-Faeroes ridge, which had been above sea level since its initial opening, began to subside about 28 Ma. Other changes included the cessation of motion between Greenland and North America, a westward jump of the Iceland-Jan Mayan spreading center at about 26 Ma, and a change of relative motion between Europe and Africa; this brought Iberia to Morocco, adding to the isolation of the Mediterranean basin. The present-day morphology of the Atlantic had been established by about 21 Ma. At some time during the early to middle Miocene overflow water from the Norwegian Sea and Arctic began to flow in significant quantities into the North Atlantic, creating the basic circulation patterns existing in the modern Atlantic Ocean.

Tectonic History of the Indian Ocean

The Indian Ocean is the most complicated and least-understood of the earth's major oceans. The magnetic anomaly pattern and the bathymetry of the Indian Ocean shows it as being more complex than the evolution of the Atlantic. At least five different poles of rotation are necessary to describe the motion between India and Antarctica since the Cretaceous; the complete reconstruction process consists of a series of rotations, compared with the single rotation necessary to close the South Atlantic. The floor of the Indian Ocean has four large north-south ridges, two of which are still actively spreading (Figs. 2–10 and 2–11). The active mid-

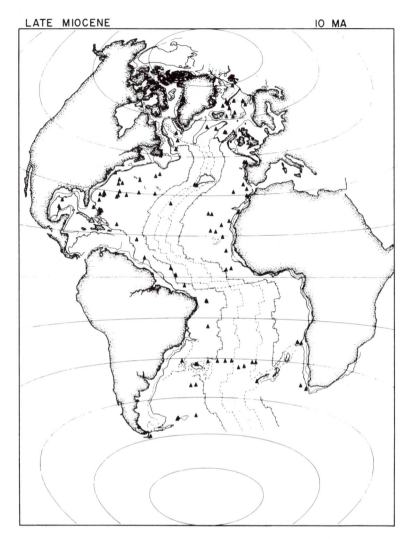

Figure 6-8 Paleoreconstruction of the Atlantic Ocean and its paleobathymetry at 10 Ma (late Miocene). DSDP sites are shown as triangles. (From J. G. Sclater, S. Hellinger, and C. Tapscott, *Journal of Geology*, vol. 85, pp. 500–522, 1977, reproduced by permission of the University of Chicago Press)

ocean ridge forms a broad inverted Y, with links extending northwest toward the Gulf of Aden, southwest toward the South Atlantic, and southeast toward the Southern Ocean south of Australia. Large fracture zones offset the ridge axis and the rough topography of the ridge contrasts with the long, linear, smooth-capped aseismic Chagos-Laccadive and Ninetyeast ridges (Figs. 2–10 and 2–11). The Indian Ocean has average thicknesses of oceanic sediments, except in the northern region, Arabian Sea, and Bay of Bengal. In these areas, vast thicknesses of terrigenous sediments are the result of Himalayan erosion during the late Cenozoic. Seismic studies in the Bay of Bengal [Curray and Moore, 1971; Curray and others, 1981] show that sediment may exceed 12 km and that denudation of the Himalayan Mountains may be occurring at an average rate of 70 cm per 1000 years. Other thick deposits of terrigenous sediments are found off East Africa. The Southern Ocean region south of the Antarctic convergence also has thick biogenic sedimentary deposits resulting from high biological productivity.

The understanding of the evolution of the Indian Ocean is of considerable importance because the development of its northern part is linked with the destruction of the Tethys Sea and the development of the Himalayan Mountains. The early evolution of the southern part of the Indian Ocean was a necessary step in the development of the circum-Antarctic Ocean. A number of extensive studies have been conducted on the tectonic history of the Indian Ocean, including McKenzie and Sclater [1971; 1973]; Sclater and Fisher [1974]; Pimm and others [1974]; Johnson and others [1976]; Sclater and others [1977]; and Curray and others [1981].

Paleoreconstructions of the Indian Ocean normally assume interactions between four major plates: the African plate, bounded by the Carlsberg, Central Indian and Southwest Indian ridges; the Indian plate, bounded by the Carlsberg, Central Indian, Southeast Indian, and Ninetyeast ridges; the Australian plate, bounded by the Ninetyeast and Australia-Antarctic ridges; and the Antarctic plate, bounded by the Southwest and Southeast Indian ridges and the Australia-Antarctic ridge.

The modern Indian Ocean involves active interactions between only three plates (Fig. 5–1), as the Ninetyeast ridge ceased to be a plate boundary in the late Cenozoic, and the Indian and Australian plates began to act as one plate.

Mesozoic evolution

During the Jurassic (about 170 Ma), Australia was part of the southern border of the Tethys Sea, which opened to the east and was several thousand kilometers wide at this longitude. The west coast of Australia does not seem to have been open to the ocean until the late Jurassic. Based on nonmarine sedimentary sections, a continental landmass abutted the coast of western Australia [Veevers and others, 1971] as part of Gondwanaland. India was adjacent to the Enderby land bulge of Antarctica, and, prior to the Jurassic, "Greater India" is inferred to have been contiguous with western Australia (Fig. 6–9). The initial separation of India from Gondwanaland began sometime prior to 127 Ma. Magnetic anomalies in the Bay of Bengal suggest that this separation occurred parallel to the eastern margin of India (Fig. 6–10(A) and (B); [Curray and others, 1981].

Prior to the rifting episode off northwestern Australia, the Tethyan spreading center was in a more central position to the north in the Tethys Sea. It was probably connected with spreading centers active in the Pacific Ocean at that time [Larson and Chase, 1972]. About 150 Ma the Tethyan spreading center moved southward and began to tear away fragments from northern Australia, which later became portions of the Asian continent. This event coincides with the initial opening of the central Atlantic by separation of North America and Africa. This motion continued until the early Cretaceous. About 125 Ma the South Atlantic and the Indian Ocean

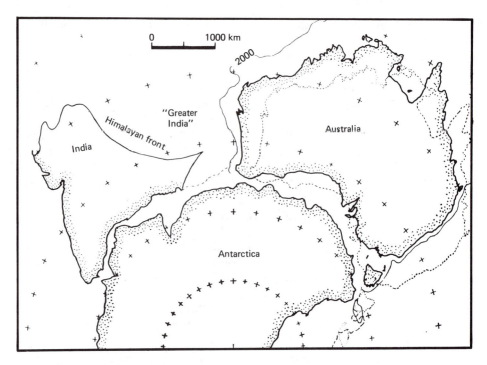

Figure 6-9 Computer fit of continental fragments of East Gondwanaland of Norton and Molnar [1977] with computer fit of Sproll and Dietz [1969] shown as dashed line. (From I. O. Norton and P. Molnar, reprinted by permission from *Nature*, vol. 267, pp. 338–339, 1977, copyright © 1977 Macmillan Journals, Ltd.)

began to open. These events marked the beginning of the present-day oceans and the demise of the Tethys Sea.

Separation of India from Antarctica probably continued in a northwesterly direction until about 90 Ma (Fig. 6–10(C)) at which time the plate edge was reorganized and spreading proceeded north-south, parallel to a transform fault which lay on the east side of the northern echelon section of the Ninetyeast ridge (Fig. 6–10(C)) [Curray and others, 1981]. This marks the beginning of the Indian Ocean as a significant ocean. The Ninetyeast ridge at that time was an active transform fault and facilitated the northward slippage of the Indian plate. During this time, Australia, Tasmania, and its continental southward extension remained joined to Antarctica. Thus, except for a small region between Broken ridge and the Naturaliste Plateau, the basins between Antarctica and Australia were totally closed to deep water from both east and west.

Cenozoic evolution

From 90 to 53 Ma India made its spectacular, northward journey (Fig. 6–10(D)). This occurred between two great transform faults: Ninetyeast ridge to the east and the parallel, but shorter, similarly aseismic Chagos-Laccadive ridge to the west (Fig. 6–11). Together, these meridional highs resemble train tracks upon which

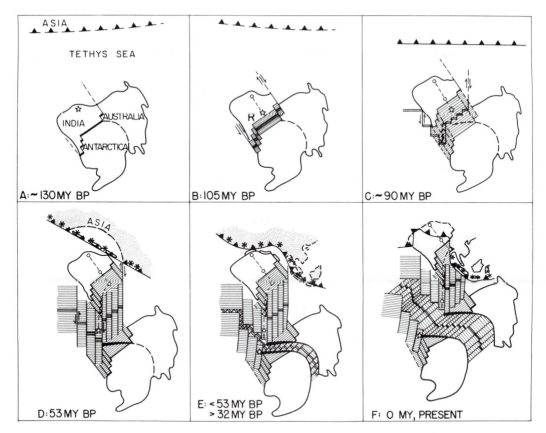

Figure 6-10 Schematic paleoreconstructions of the eastern Indian Ocean between 130 Ma (earliest Cretaceous) and the present day. Kerguelen hot spot indicated by a star. R indicates location of Rajmahal Traps. (From J. R. Curray and others, in *Ocean Basins and Margins*, vol. 6, *The Indian Ocean*, eds. A. E. M. Nairn and F. G. Stehli. New York: Plenum Press, in press, 1981)

India traveled toward Asia [Sclater and Fisher, 1974]. It was about 53 Ma (early Eocene) that India began colliding with Asia (Fig. 6-10(D)). A spreading center in the Wharton basin became active, forming the basins in the eastern Indian Ocean. In the western Indian Ocean, north-south spreading occurred from the mid-ocean ridge (Fig. 6-10(D)).

The end of this phase of spreading occurred about 53 Ma when Australia and Antarctica began to separate (Fig. 6-10(E)). For the next 20 m.y. Australia, Antarctica, and India were on separate plates and spreading continued on the east side of the Ninetyeast ridge (Fig. 6-11). Thus the Ninetyeast transform fault still represented the line of motion between India and Antarctica. Spreading of the eastern Indian Ocean occurred in a northeast-southwest direction along a southeastern Indian Ocean spreading ridge between Australia and Antarctica (Fig. 6-10(E)). Subduction continued into the Sunda subduction zone to the north of India (Fig. 6-10(E)). By this time spreading between the India and Antarctic plates

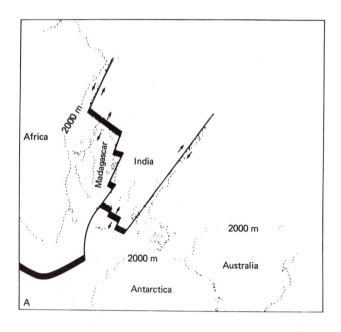

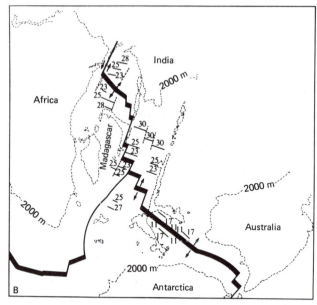

Figure 6-11 Paleoreconstructions of plates and their boundaries in the Indian Ocean at 75 Ma (late Cretaceous) and 36 Ma (early Oligocene). Numbers in (B) represent magnetic anomalies. (From D. P. McKenzie and J. G. Sclater, *Geophys. Jour. of Royal Astron. Sci.*, vol. 24, pp. 437–528, 1971)

was more rapid than between the Antarctic and Australian plates, resulting in increased distance between the Southeast Indian and Australian-Antarctic ridges.

The next major event occurred about 32 Ma (Anomaly 11), when relative motion on either side of the Ninetyeast transform fault ended and India and Australia became part of the same plate (Fig. 6-10(F)) [Sclater and Fisher, 1974].

The direction of motion between these two continents and the Antarctic plate changed from due north-south to northeast-southwest. The 6° north-south transform fault between the Southeast Indian ridge and the Antarctic-Australian ridge became an S-shaped series of ridges and transform faults, with the faults trending northeast-southwest. This new geometry and spreading direction has continued to the present (Fig. 6–10(F)). This is marked by the rapid northward movement of Australia and India from Antarctica, associated with spreading on the central Indian and Australia-Indian mid-ocean ridges. The distance between India and Australia remains about the same because they are on the same plate. The most recent events in the history of the Indian Ocean have been the formation of the Red Sea and the Gulf of Aden, which separated Arabia from Africa in the latest Cenozoic.

History of the Ninetyeast ridge

The submarine topography of the central Indian Ocean east of 70°E is dominated by the meridional Ninetyeast ridge (Fig. 6–12). This flat-topped ridge extends almost north-south as a topographic high for more than 4500 km from 31°S to 9°N, and it may be traced several hundred kilometers farther north until it disappears beneath sediments of the Bengal fan system in the Bay of Bengal. To the south it adjoins Broken ridge. Ninetyeast ridge separates the deep central Indian basin from the even deeper Wharton basin to the east (Fig. 6–12). Drilling at various locations on the ridge has shown several important aspects [Luyendyk, 1977].

1. The ridge has an extrusive volcanic basement.

2. Ages of basal sediments show that the ridge is older toward the north from the late Cretaceous (Campanian) or older at 9°N to Eocene-Oligocene at 31°S. The gradient in age is similar to that of the Indian basin immediately to the west, implying that the ridge is attached to and belongs tectonically to the Indian plate.

3. The ridge was formed in shallow water, sometimes subaerially, and has subsided with time in accordance with known age-depth curves. Several environmental facies have been recognized in the overlying sediments, including a subaerial facies consisting of low-grade coal beds (lignite). Other facies include lagoonal reef, shallow bank, deeper bank, and fully oceanic. The stratigraphic succession of these facies indicates an increase in depth in the overlying sediments.

4. The ridge was formed in more southerly latitudes and has moved northward. Two lines of evidence which support northward motion are cold-water microfossils from the older, northern part of the ridge now located in tropical waters and paleomagnetic studies [Klootwijk and Peirce, 1979]. Also, pollen assemblages from the northern tropical part of the ridge are typical of temperate vegetation of southern Australia and New Zealand [Kemp and Harris, 1975].

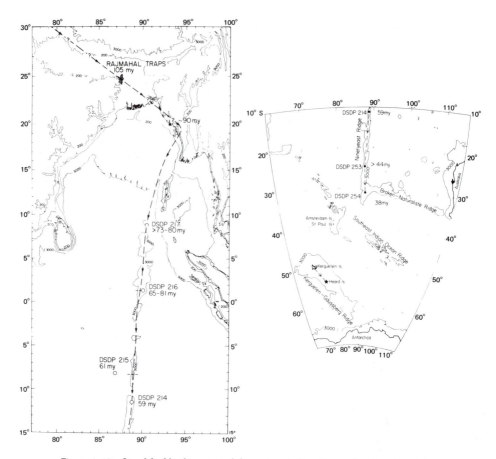

Figure 6-12 Simplified bathymetry of the eastern Indian Ocean showing the origin of the Ninetyeast ridge, as a trace of the Kerguelen hot spot. Buried continuation of the ridge to about 17°N is based on geophysical evidence. (From J. R. Curray and others, in *Ocean Basins and Margins*, vol. 6, *The Indian Ocean*, eds. A. E. M. Nairn and F. G. Stehli. New York: Plenum Press, in press, 1981)

These observations, in combination with geophysical data, show that the Ninetyeast ridge is primarily a sunken oceanic island and seamount chain (Fig. 6-12). Thus the ridge is an extrusive pile with a low-density shallow root rather than a horst or an uplift resulting from the convergence of plates. Various models have been proposed to explain the volcanic activity. The most favored hypothesis is that volcanism was produced from a fixed hot spot between 90 and 20 Ma [Morgan, 1972]. However, the current position of the hot spot is in doubt because there is a break in the volcanic chain between the southern part of the ridge and the possible present-day hot spot at the Amsterdam–St. Paul Islands to the south, probably requiring a jump in position of the original hot spot (Fig. 6-12). On the other hand, Peirce [1978] has presented paleomagnetic evidence which suggests that basement rocks, underlying basal sediments on the ridge and the Rajmahal Traps in northern

India (Fig. 6-12) all originated at the present latitude of the Kerguelen Island hot spot, which is considerably farther south than the Amsterdam-St. Paul Islands. If it is assumed that the hot spot has remained approximately fixed through geologic time (the last 130 m.y.), it is apparent that the Rajmahal Traps and the Ninetyeast ridge represent the trace of the hot spot on the overriding Indian and Antarctic plates (Fig. 6-12).

North-south spreading continued and the hot spot was carried to the spreading axis between the Indian and Antarctic plates during the early Tertiary [Curray and others, 1981]. If this model is correct, the Antarctic plate must have grown northward over the hot spot, before the next phase of spreading separated Australia from Antarctica across the southeastern Indian Ocean spreading axis after 53 Ma, because no trace of the hot spot is apparent in this younger sector of the southeastern Indian Ocean. Instead, the active or recently active volcanoes of Amsterdam and St. Paul Islands sit on the mid-ocean ridge where this trace would have occurred (Fig. 6-12) [Curray and others, 1981]. Thus, Broken ridge–Naturaliste Plateau and the Kerguelen Plateau represent a formerly continuous volcanic ridge which became separated during the episode of spreading beginning about 53 Ma.

Collision of India with Asia

The northward migration of India, in large part, created the Indian Ocean (Fig. 6-10); the Indian Ocean and the Himalayas have a common origin. During its northward migration, India crossed the Tethys Sea toward an inferred subduction zone along the southern margin of Asia. The first contact between India and Asia occurred in the early Eocene about 53 Ma (Anomaly 22). This event coincided with a change in relative plate motions in the Indian Ocean and a decrease in the rate of sea-floor spreading. It was at this time also that Australia began its northward drift away from Antarctica (Fig. 6-10). These changes in motion were probably caused by the resistance to plate motion after contact of India with Asia. Initially the contact was probably related to an attempt by the trench north of India to consume northern India. The initial collision seems to have created a succession of marine transgressions and regressions in the area during the Eocene and Oligocene. This suggests that the first collision was a "soft collision," or a contact between continental India and an island arc lying seaward of a marginal basin [Curray and others, 1981]. The marginal basin would have been steadily closed by the continued northward movement of India. "Hard collision," or a collision with the continent, occurred much later, during the Miocene, with the initial uplift of the Himalayan Mountains [Gansser, 1964]. The effect of the late Cenozoic collision was the development of the highest mountain chain on the earth. Considerable crustal shortening must have occurred with the development of the fold mountains derived from great thicknesses of sediments laid down in basins at the southern Asian margin. This could have happened by the underthrusting of continental crust under continental crust, by the thickening of continental crust by thrusting, or by a combination of both, resulting in the uplift of the Tibetan Plateau. The Himalayas are about 300 km wide and crustal shortening of about this amount is supported by geological and geophysical observations. The sum of these distances (600 km) is about the same as Greater India (Fig. 6-9) in the original reconstruction of Gondwanaland (Fig. 6-9). Thus all of Greater India could be absorbed in the Himalayas without the need for postulating the subduction of a large continental fragment beneath India [Norton and Molnar, 1977].

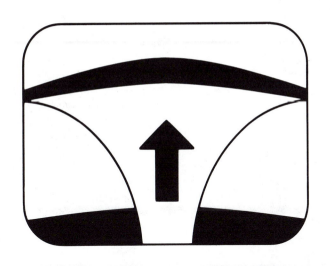

seven

The Oceanic Crust

INTRODUCTION

Igneous activity within and at the margins of the ocean basins can be divided into four broad categories according to composition, structural setting, and mode of emplacement. These include tholeiitic volcanism of the mid-ocean ridges and certain large oceanic islands, alkali basalts of seamounts and many oceanic islands, andesitic and acidic volcanism of the continental margins, and a complex of andesitic and tholeiitic and alkali basaltic volcanism that make up the island arcs and backarc basins. The greatest volume of lava produced at the margins of the Pacific is andesite, a rock somewhat low in alkali metals considering its high silica content (about 55 percent). Andesites are richer in Si, K, and large lithophile elements (that is, elements with R > 1Å) and do not occur in the oceanic areas away from the arcs and margins. So clear is their distribution that a line, called the **andesite line,** is drawn around the Pacific delineating andesitic occurrence. The andesitic rocks are described along with the associated active margins in Chapter 12. Here we shall be concerned only with the igneous rocks formed at the mid-ocean ridges and the oceanic hot spots.

According to concepts of sea-floor spreading and plate tectonics, most new oceanic crust is created at volcanically active, spreading mid-ocean ridges, with spreading rates varying from about 1 cm/yr per ridge flank on the mid-Atlantic ridge near Iceland to about 10 cm/yr per ridge flank on the East Pacific rise in the equatorial Pacific. The exuded magma which makes up the upper part of the ocean

204

crust is composed largely of basaltic rocks that are depleted, relative to most other basalts, of elements that are enriched in the continental crust. Activity at the spreading plate boundaries is quite accessible for observation because the asthenosphere penetrates to within a few kilometers of the earth's surface, and all the activity is concentrated in that shallow, narrow zone (less than 20 km wide and 10 km deep). The magmatic activity at the mid-ocean ridge is responsible for volcanic processes at the ridge crest. This volcanism is, however, only a fraction of all the magma emplaced at the ridge; most solidifies within dikes and layered intrusives at greater depths and is not extruded as volcanic rocks at or near the sea floor.

Although the lateral migration of new oceanic crust away from mid-ocean ridges is well documented, little is known about the mechanism and processes involved in the actual generation of new crust. It is likely that the intense volcanic activity at the ridge crest results from the lateral movements of plates caused by tectonic processes not directly related to volcanic intrusion beneath the ridges (see Chapter 5). Much remains to be learned about the nature and variability of the ocean crust and upper mantle, particularly the depths and character of magma generation, its emplacement at ridge crests, and subsequent history.

Magmatic activity at ocean ridges is relatively continuous over the lifetime of an ocean basin, which can be as high as several hundred million years. The apparent regularity of the crustal layer over hundreds of millions of square kilometers also indicates that the process that has produced this layer must be persistent. The great length of geologic time during which such submarine volcanism has occurred and its vast linear extent through the ocean basins have important ramifications beyond the generation of the ocean crust. The degassing of volatiles from the earth's interior as a result of this volcanism has produced the atmosphere and hydrosphere. Circulation of seawater through newly created oceanic crustal rocks may provide the main buffer for magnesium in the sea, generate hydrothermal deposits, and assist in the chemical alteration and even metamorphism of ocean-floor rocks. This metamorphism of ocean-floor rocks is relatively low grade, forming rocks of the green schist or the amphibolite facies. The minerals in these rocks (albite, epidote, chlorite, and hornblende) are different from the original basaltic rock (calcic plagioclase, pyroxenes, and olivine), but the chemical composition remains similar.

Information about the character and evolution of the ocean crust has been obtained from a number of sources, several of which have been linked to the development of sophisticated instrumentation. Approaches used to study the oceanic crust include the recovery of rocks by dredging and deep-sea drilling; marine geophysical measurements, especially from seismic surveys; direct observations of the ocean floor from deep-diving submersibles, and examination of specific types of rock sequences on land that are considered to represent uplifted ocean crust, called *ophiolite complexes*. Until recently most information on the oceanic crust was from seismic refraction studies and dredge hauls from the mid-ocean ridges and fracture zones.

The DSDP has involved over 300 holes drilled into the upper part of the ocean crust using *Glomar Challenger* to sample basement rocks at depths of less than 10 m beneath the oldest sediments. But only since 1974 have many cores been taken from basement depths of 60 to 600 m. These sequences have made examination of volcanic stratigraphy possible and have aided the study of the history of the ocean crust and the processes that formed it. The basement stratigraphy from such drill holes is defined by its lithology, petrography, chemistry (both bulk analyses and electron microprobe analyses of glasses), magnetic properties, and physical properties. Before deep crustal drilling it was not possible to study the character of deeper ocean crust except by using remote geophysical techniques.

Direct observation and sampling of the volcanic rocks at the mid-ocean ridge has been possible by use of submersibles. To carry out such an examination of a mid-ocean ridge, the FAMOUS project was organized in 1971. This involved a detailed survey of the mid-Atlantic ridge rift valley near the Azores at about 37°N using submersibles and remote instruments. New techniques of navigation allowed the mapping of topographic features to a scale of a few tens of meters, and the principal faults and scarps were identified. Sampling made possible the detailed mapping of rock types. The project was important in that it provided, for the first time, detailed data about the character of the youngest part of the ocean crust, that part where the most recent volcanic accretion has occurred. This was followed in the late 1970s by similar large-scale projects concentrating on the Galapagos spreading center and the East Pacific rise at 21°N. Detailed surveys employing submersibles and bottom photography allowed mapping of sheet flow and pillow lava distributions and bottom topography. The most spectacular of these expeditions, however, was the discovery of numerous active hydrothermal vents associated with the zones of active spreading at the ridge crest and their associated distinctive, benthonic biota (see pages 231 and 493).

STRUCTURE, PETROLOGY, AND SOURCES OF OCEANIC CRUST

Structure of the Oceanic Crust

Four approaches are used to determine the structure of the oceanic crust: seismic refraction and reflection, deep-sea drilling, dredging of fracture zones, and comparison with ophiolite complexes. Since deep-sea drilling has penetrated only as far as 600 m in the upper part of Layer 2, the structure below this is only speculated upon. Available data have shown that Layer 2 (the layer beneath the sediments) is formed largely of basaltic lavas rather than consolidated sediments. These are underlain by intrusive dikes and by gabbros. The gabbros are, in turn, underlain by the peridotites of the upper mantle. It is unclear where the boundary between Layers 2 and 3 is, but most evidence suggests that it is between the lavas and dikes.

Seismic refraction data

Seismic velocity values for Layer 2 range between 3.4 and 6.3 km/sec, but most are between 4.5 and 5.5 km/sec (see Chapter 2; [Ludwig, and others, 1970]; Table 7-1; Fig. 7-1). The range of seismic velocity has indicated that Layer 2 represents either a wide variety of rocks or differences in the amount of sealing of

TABLE 7-1 OCEANIC CRUSTAL LAYER CHARACTERISTICS[a]

Material	P-velocity (km/s)	Average thickness (km)	Approximate density (gm/cm³)
Water	1.5	4.5	1.0
Layer 1: Sediment	1.6–2.5	0.5	2.3
Layer 2: Basalt	4.0–6.0	1.75	2.7
Layer 3: Oceanic layer	6.7	4.7	3.0
	Moho		
Upper mantle	7.4–8.6		3.4

[a] Modified after Bott [1971].

cracks and fractures at depth. The average thickness of Layer 2 in the deep-ocean basins is about 1.7 km. In some areas, the seismic velocities measured for Layer 2 can represent either basalt or consolidated sediment. However, there is evidence showing it to be largely of basaltic composition.

1. Seismic profiling shows that Layer 2 emerges to the seabed on the flanks of mid-ocean ridges, where abundant dredge hauls of basaltic rocks have been recovered. Sediments older than the late Cenozoic have not been recovered from the area of exposure of Layer 2 on the ridges.

2. The irregular shape of the interface of Layer 1 and Layer 2 below the abyssal hills province is due to the volcanic origin for Layer 2.

3. Deep-sea drilling has been successful into Layer 2 many times and the contact between Layer 1 and Layer 2 coincides with the change between sediment and volcanic basement rocks. The deeper basement drilling has shown a continued dominance of the volcanic rocks below the interface, with only rare sediment pockets.

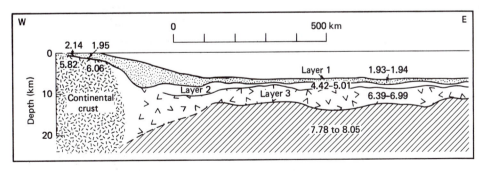

Figure 7-1 Oceanic crustal structure and its acoustic velocity as determined by seismic refraction in the Southwest Atlantic off Argentina. (After M. Ewing, 1965)

4. Oceanic magnetic anomalies show that highly magnetic rocks such as basalt (compared with sediments that are not highly magnetic) are a substantial part of Layer 2.

In refraction experiments, Layer 2 may be measured by the first arrivals over a short portion of a refraction profile, but the returns are often not clear because arrivals from the next deeper layer (Layer 3) often interfere with arrivals from basement. There is often little impedance contrast at the interface of Layer 1 and Layer 2, especially in older areas of the ocean, because consolidated sediments exhibit seismic velocities similar to basalt. Houtz and Ewing [1976] compiled the results from hundreds of air gun–sonobuoy profiles and divided Layer 2 into three distinct sublayers (Fig. 7-2):2A, 3.64 km/sec; 2B, 5.19 km/sec; and 2C, 6.09 km/sec.

Layer 2 is thickest on the ridge axis in the Atlantic and Pacific oceans, thinning with increasing age until it disappears at about 30 Ma in the Pacific and 60 Ma in the Atlantic (Fig. 7-2). Another change observed in Layer 2A is that its sonic

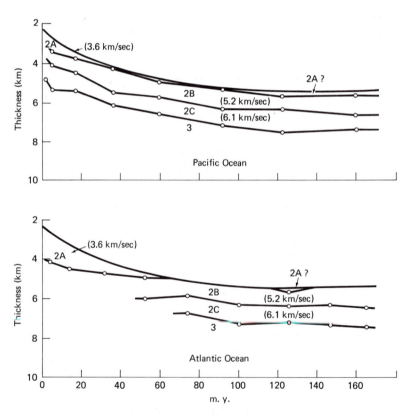

Figure 7-2 Acoustic velocity structure of the upper oceanic crust in the Pacific and Atlantic oceans as a function of age. The sediment layer has been removed. Average velocities are shown for layers 2A, 2B, and 2C. Note how layer 2A thins with increasing age. (After R. E. Houtz and J. Ewing, *Journal of Geophysical Research*, vol. 81, p. 2494, copyrighted by the American Geophysical Union)

velocity increases with age of the rock, from about 3.3 km/sec on the ridge axis to Layer 2B values (5.2 km/sec at more than 40 Ma). Velocity through fresh basalts average 6.0 km/sec. In situ downhole logging experiments also reveal low sonic velocities, much lower than for the basalt samples recovered from the holes. The low velocities for the sequence of rocks of the upper oceanic crust are believed to be due to porosity on a scale larger than the laboratory samples (core diameter is 6 cm) but less than about 1 m. Houtz and Ewing [1976] argued that Layer 2A does not really thin; what appears as a thinning may actually be the result of an increase of its refraction velocity with age, possibly caused by closing of cracks and fissures due to hydrothermal precipitation. Thus young crust produced at the axis of the ridges is highly porous, with porosity values near its upper surface of about 30 to 40 percent. Older parts of the upper oceanic crust do not seem completely sealed, but water circulation may be limited by the presence of massive basalts and the absence of shallow sources of heat.

Seismic reflection experiments show that the two most distinctive reflectors are the two major discontinuities at the top of Layer 2 and the Mohorovičić discontinuity (Fig. 7–3). Most reflection profiles do not reveal a sharp distinction between Layer 2 and Layer 3 (Fig. 7–3), suggesting that this interface probably represents a rather modest change and gradual transition in velocity with depth [Stoffa and Talwani, 1978]. The interface between Layers 2 and 3 is revealed in seismic refraction surveys.

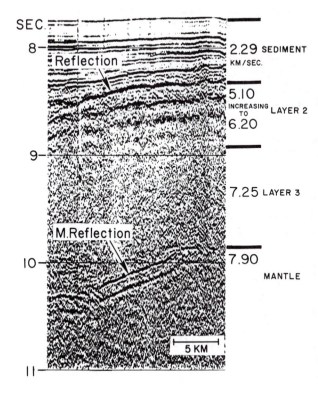

Figure 7–3 Seismic reflection profile of the oceanic crust. Distance is measured along the horizontal; the reflection time to the discontinuity is measured along the vertical. The reflection from the ocean bottom takes a time of just under 8 sec. There are several reflections within the sediment, and a prominent reflection at the base of the sediment is indicated at about 8.3 sec. Oceanic crustal layers 2 and 3 also marked. The boundary between layer 3 and the mantle (Moho, or M-reflection) is undulating. The speed of the travel of sound in various layers shown on the profile was obtained by separate experiments using a sonobuoy. Knowing the speed of sound in various layers, the depths to the various discontinuities can be obtained from the reflection times. This profile was made in the western Pacific just seaward of the Japan Trench. (From Stoffa and Talwani, 1978)

Layer 3 is the main oceanic crustal layer, more uniform in thickness and seismic velocity than Layer 2. When Layer 2 is not observed, the velocity values for Layer 3 are anomalously low: 6.65 ± 0.25 km/sec. Whenever Layer 2 is observed, the velocity in Layer 3 increases to 6.83 ± 0.31 km/sec [Ludwig and others, 1970]. The average thickness of Layer 3 is about 4.7 km (Table 7-1). The composition of the lower oceanic crust is still unknown. Layer 3 may consist mainly of gabbro and metagabbro rocks. Gabbros are marked by velocities of about 7.0 km/sec at ocean crustal pressures close to the lower average crustal velocity of 6.7 km/sec, determined from refraction experiments [Hall and Robinson, 1979]. Other models proposed by Christensen and Salisbury [1975] and Salisbury and others [1979] account for the boundary between Layer 2 and Layer 3 as separating unmetamorphosed or low-grade metamorphosed basalts (green-schist facies) above from more highly metamorphosed igneous rocks below (amphibolite facies). That Layer 3 may be largely of gabbroic composition is widely accepted and supported by observations and sampling from deep-diving submersibles in the mid-Cayman rise spreading center in the Caribbean Sea [Ballard and others, 1979]. Deep crustal drilling has yet to penetrate the gabbroic layer.

Deep-sea drilling data

Important information relating to the upper part of the oceanic crust has been collected as a result of deep-sea drilling. Most deep crustal drilling has taken place in the North Atlantic, and the results are compiled in a volume of collected contributions by Talwani and colleagues [1979] and in a paper by Hall and Robinson [1979]. Three types of units are commonly recognized in basement rock sequences: lithological, geochemical, and paleomagnetic. Lithologic units have been defined by macroscopic features of the rocks (such as presence of pillow fragments or breccias), the abundance and proportions of phenocrysts, and the occurrence of other distinctive rock types (such as gabbros and serpentinites). Deep crustal drilling has shown that Layer 2 is composed predominantly of pillow basalts with minor intercalated biogenic sediment to a depth of at least 600 m. Sediments are commonly interlayered with the extrusive basalts in the upper 200 to 300 m of crustal rock. The sediments are mostly fine-grained chalk or limestone of varying degrees of induration [Hall and Robinson, 1979]. Both massive and pillowed lava flows are interbedded; dikes and sills are rare (less than 2 percent of recovered basement) [Bryan and others, 1979; Hall and Robinson, 1979]. Until deeper penetrations can be made, the nature of rocks at greater depths will remain unknown. But because extrusive basalt persists to the greatest depths drilled, it seems likely that basalt persists to 1.5 km or more, with an increasing proportion of intrusive dikes and sills. The normal thickness of Layer 2 is not confirmed by existing drilling data, but it is inferred to be at least 1.5 km [Bryan and others, 1979].

Crustal construction seems episodic; distinct stratigraphic units typically make up the drilled rock sequence, each sharply distinguished from other units. The average thickness of individual units is about 45 to 60 m. Paleomagnetic evidence suggests that each individual unit formed in less than 100 years. Intervals of time between formation of the units has been calculated as about 5000 years [Hall and

Robinson, 1979]. These relatively long periods of quiescence are supported by altered or weathered zones caused by relatively long seawater contact. Lateral lithologic and stratigraphic continuity is lacking in the crust, even between holes drilled a few hundred meters apart. This indicates that sea-floor eruptions are local, with reduced lateral transport due to rapid chilling.

Evidence from ophiolite complexes

Ophiolite sequences represent sections of oceanic crust and upper mantle originally created at mid-oceanic ridges and later uplifted at convergent boundaries. Among the best preserved examples are the Bay of Islands complex, New-foundland, the Troodos complex, Cyprus, and the Oman ophiolites. Although it is difficult to determine exactly how this crustal material looked before it was uplifted on land, it is believed that they provide an otherwise unavailable view of the structure and rocks of the oceanic crust and upper mantle, even if this is not entirely typical. This is important because the oceanic data base is inadequate for the assignment of various lithologies to specific positions within the oceanic crust, except for the shallowest levels. Ophiolites are important guides for modeling the oceanic structure, by providing direct comparisons of ophiolite seismic stratigraphy with oceanic seismic refraction solutions.

In a well-developed ophiolite complex (Fig. 7–4), associated sediments are of deep-sea origin, cherts are present, and there are three principal units stratigraphically underlying the pelagic sediments. The upper unit is dominated by pillow basalts, beneath which is a sheeted-dike complex, in turn underlain by a gabbro layer (Fig. 7–4). In the Troodos complex of Cyprus, the pillow lavas are about 1000 m

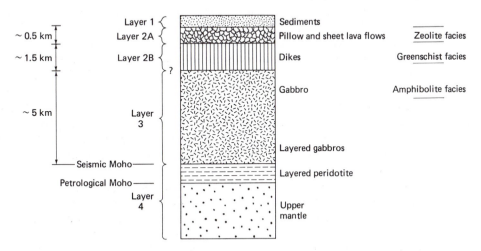

Figure 7–4 Cross section of the ocean crust exhibiting inferred layers. The approximate thickness of the layers is defined by seismic experiments; inferred igneous rock composition is based largely on observations and in fracture zones and with comparison with ophiolite sequences. The metamorphic sequence is produced by the reaction between seawater and igneous rock. (After J. Cann, 1974)

thick and show signs of extensive hydrothermal alteration. The upper part of the basalts do not have many dikes, but their quantity increases rapidly in the lower few hundred meters of pillow lavas. At 1000 m the pillow lavas give way completely to a 1000-m thick sheeted-dike complex. The average width of such dikes is about 1.5 m. Estimates of the width of this dike injection zone vary from 50 m to several kilometers. During dike injection, cooling is more rapid near the margins, creating more finely grained rock than in the middle. With such dike injections in a relatively narrow zone, it is not surprising that many dikes are multiple; that is, that successive injections occur up the center of still molten, older dikes. A complex picture of chilled margins results, depending on whether the edges of successive dikes are adjacent to material that has already been significantly cooled and how many single dikes have been split. The base of the dike complex grades into a plutonic unit dominated by gabbro (Fig. 7–4). At greater depths within the plutonic complex, the basic rocks grade into ultramafics (such as dunites) showing little evidence of hydrothermal alteration, where layering occurs from accumulation of dense magma differentiates, or **cumulates.**

Summary of oceanic crustal structure

By integrating the results from these various studies, the following picture of the oceanic crust has emerged (Fig. 7–4). A disrupted sequence of pillow lavas, massive basalt flows, intercalated sediment, and rubble zones at least 500 m thick grades downward into a sequence of sheeted dikes 1 to 1.5 km thick. It is assumed that the dikes are injected into fractures in the ocean crust caused by tensional stresses. This extrusive and shallow intrusive carapace is underlain by an assemblage of 3 to 5 km of gabbro. At shallow levels the gabbro is uniform; at deeper levels it grades into cumulate gabbro with interbedded ultramafic units (Fig. 7–4). Below lie upper mantle rocks characterized by tectonized harzburgite (olivine plus orthopyroxene). Layer 2 is characterized by steep velocity gradients of 1.5 to 2.0 sec. Velocities increase much more slowly in the gabbros (0.1–0.2 sec). The crust-mantle boundary appears to be spatially variable—sometimes there is a sharp velocity contrast between crust and mantle, while at other places the transition is much smoother, occurring over 1.0 to 2.0 km with a basal high-velocity layer overlying the Mohorovičić discontinuity. At other places it has been argued that there is a low-velocity zone above the Mohorovičić discontinuity. The nature and origin of these crustal structural variations are still not understood.

The rate or eruption of lavas onto the ocean floor seems related to the mechanism of dike intrusion into the spreading zone. On oceanic ridges exhibiting slow spreading rates, only thick dikes reach the surface. Thinner dikes are cooled before the lava has a chance to reach the surface. This causes a differential dilation of the crust which, in turn, is reflected in normally shallow faulting. In Iceland, a volcanically active spreading zone, more than half of the dikes failed to reach the surface and erupt as lavas. Instead, normal faulting occurred at the surface forming gaping fissures (gjás) and grabens. Bödvarsson and Walker [1964] have shown that for a typical cross section 53 m long, about 1000 dikes with a total lateral width of 3 km have produced lavas about 1500 m thick. With increasing spreading rates, wall rocks maintain higher temperatures and most dikes reach the surface and produce lava flows.

Rocks of the Oceanic Crust

Rocks are classified according to their texture and chemical-mineralogical composition. The **modal** mineralogical composition of a rock is the relative volume of its minerals. This can be determined by microscopic studies of thin sections of rocks, called **modal analysis.** More precise analyses of composition can be determined from the chemical composition, using a large range of techniques. The following minerals occur in rocks of the oceanic crust. Their variations are particularly important in understanding the formation of the oceanic crust.

Ferromagnesium minerals

OLIVINE:	$(Mg,Fe)_2SiO_4$
PYROXENES:	$Ca(Mg,Fe)Si_2O_6$ (augite)
	$(Mg,Fe)SiO_3$ (hypersthene)
AMPHIBOLE:	$Ca_2(Mg,Fe)_5Si_8O_{22}(OH)_2$ (hornblende)

Feldspars

PLAGIOCLASE SERIES:	$NaAlSi_3O_8$ (albite)
	$CaAl_2Si_2O_8$ (anorthite)

Oxides

SPINEL:	$(Mg,Fe)Al_2O_4$
MAGNETITE:	(Fe_3O_4)

Silica minerals

QUARTZ:	SiO_2

The texture of igneous rocks is controlled by the speed at which a melt cools. Rocks extruded onto the ocean floor have an outer coating of glass because of very rapid cooling by seawater. For example, a submarine basalt lava may be 70 percent glass, 25 percent feldspar, and 5 percent pyroxene. Rocks which cool more slowly in the crust are more coarsely crystalline and contain little or no glass. For instance, a basalt dike rock may be 60 percent feldspar, 30 percent pyroxene, and 10 percent olivine. Such rocks may have identical chemical compositions but may be differentiated by the textural differences, which provide information on their mode of emplacement.

The dominant type of basement rock found within the oceans is basalt. Basalt is a dark, fine-grained, extrusive igneous rock composed chiefly of plagioclase and clinopyroxene in a glassy, or fine-grained, groundmass. Basalt is in some places intrusive in the form of dikes; its intrusive equivalent, however, is generally called *dolerite* (diabase) or gabbro. Coarser grained basic and ultrabasic rocks, such as dolerites, gabbros, serpentinites, and peridotites, are frequently dredged from the fracture zones.

In the oceanic regions two major basaltic magma series are recognized: the

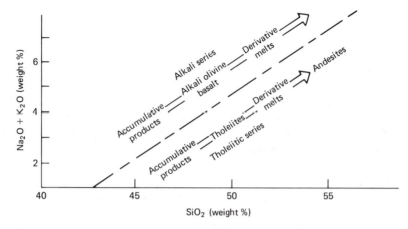

Figure 7-5 Plot of weight percent of Na_2O + K_2O versus that of SiO_2 showing relations among two major rock series in the Miocene to Recent of Japan, Korea, and Manchuria. The dashed line marks the boundary between the fields of the tholeiite series and the alkali rock series. (After H. Kuno, 1966)

tholeiitic (or subalkaline) and the alkali basalt series. The tholeiites are saturated with silica and are the dominant basalts of the ocean floors (Fig. 7-5). The alkali basalt series includes those basalts which are undersaturated in silica and which dominate many of the oceanic islands. Members of these series can be distinguished by mineralogical criteria, such as by calcium-poor pyroxenes or quartz in the tholeiitic series or by calcium-rich titanaugite or nepheline in the alkali basalt series, or by chemical criteria, principally the relative abundances of alkalies (Na_2O + K_2O) and silica (Table 7-2). The difference can be seen in the Na_2O + K_2O (weight percent) versus SiO_2 (weight percent) content of lavas (Fig. 7-5) [Kuno, 1966]. The line separating the two series can vary slightly depending on the particular volcanic province. Laboratory studies and studies of natural examples of fractional crystallization have shown that at low pressure these magma series are separated by a "thermal divide." Fractionation processes operating at low pressure cannot, in general, produce members of one magma series from a parent magma of the other series [Yoder and Tilley, 1962; Green and Ringwood, 1968]. This has led to the concept of at least two parent basaltic magma-types, both derived from the mantle but different in chemical composition and leading to divergent and distinctive fractionation series [Green and Ringwood, 1968].

The most widely used and valuable classification of basalts now is the **normative classification,** which is based on norms. The **norm** is the calculated mineral composition of a rock in terms of a hypothetical assemblage of minerals. The classification is entirely chemically based and groups together rocks of similar bulk composition irrespective of their mineralogy. The norm thus differs from the mode, which represents the mineral assemblage actually present. Nevertheless, the hypothetical (or calculated) minerals are those that are inferred to form a particular mineral assemblage (the mode) upon complete crystallization from the same chemical constituents at relatively low pressures in the absence of volatiles [Yoder,

TABLE 7-2 AVERAGE COMPOSITION AND VARIATION OF OCEANIC THO-LEIITIC BASALTS COMPARED WITH ALKALI BASALTS FROM ISLANDS AND SEAMOUNTS[a]

Composition	Oceanic tholeiitic basalt	Mean deviation	Alkali basalt	Mean deviation
	(parts per million)			
Ba	14	7	498	136
Co	32	3	25	5
Cr	297	73	67	57
Cu	77	6	36	13
Ga	17	2	22	2
La	<80	—	90?	—
Li	9	6	11	5
Nb	<30	—	72	9
Ni	97	19	51	33
Rb	<10	—	33	—
Sc	61	19	26	4
Sr	130	25	815	375
V	292	57	252	32
Y	43	10	54	7
Yb	5	1.5	4	1
Zr	95	35	333	48
	(weight percent)			
SiO_2	49.34	0.54	47.41	3.08
TiO_2	1.49	0.39	2.87	0.24
Al_2O_3	17.04	1.78	18.02	1.71
Fe_2O_3	1.99	0.65	4.17	1.16
FeO	6.82	1.50	5.80	1.17
MnO	0.17	0.03	0.16	0.03
MgO	7.19	0.67	4.79	1.35
CaO	11.72	0.69	8.65	0.91
Na_2O	2.73	0.20	3.99	0.41
K_2O	0.16	0.06	1.66	0.38
H_2O+	0.69	—	0.79	—
H_2O-	0.58	—	0.61	—
P_2O_5	0.16	0.05	0.92	0.22
Fe_2O_3/FeO	0.29		0.72	
K/Rb	1300		418	
Sr/Rb	130		25	
Na/K	16		2	
K/Zr	14		4	
K/Ba	121		28	
K/Cr	4		206	

[a] From Engel and others [1965].

1976]. However, the purpose of the norm is not to achieve correspondence with the mode, but to try to indicate affinities that would otherwise be masked by differences in grain size and mineralogy caused by different water content and cooling history [Cox and others, 1979]. Using a set of basic principles summarized by Cox, the various oxides are allotted to minerals in a particular sequence to calculate the normative mineral assemblage. The norm calculation is particularly useful for projecting the compositions of natural basalts into a phase diagram (Fig. 7-6). In the normative basalt tetrahedron (Fig. 7-6), the major basalt types occupy certain parts of the subtetrahedron as follows: Alkali basalts occupy the subtetrahedron Ne-Fo-Di-Ab; olivine-tholeiites occupy Fo-Di-Ab-En and quartz-tholeiites occupy Di-Ab-En-Q [Cox and others, 1979].

The oceanic tholeiites are remarkably uniform in composition (Table 7-3) and exhibit characteristic chemical fingerprints. They are marked by about 50 percent silica, low K_2O and Ti, and high alumina content. Furthermore, they have fewer lithophile elements, like K, U, Th, Pb, Rb, Zr, Ba, Sr, Cs, and La, than the basalts of either oceanic islands or continental crust. Although the oceanic tholeiitic basalts range widely in age from the Jurassic to the present; they have very similar compositions. Jurassic basalt taken from the Northwest Atlantic closely resembles modern basalt from the mid-Atlantic ridge. This indicates that magma of similar composition has been extruded in much the same manner throughout the opening of the Atlantic Ocean. This may imply continuous renewal of source material and great consistency in the temperature-pressure conditions at the site of partial melting [Bryan and others, 1979]. Although most oceanic tholeiitic basalts are erupted at the oceanic ridges, some basalts are known to have erupted at varying distances from the ridge axis. Bonatti and Fisher [1971] and Bryan and others [1979] have determined that there is no appreciable difference in the chemical parameters of ridge and nonridge basalts. This suggests that the conditions of magma generation are similar both beneath and away from the ridge axis, again confirming the uniformity of ocean-floor basaltic rocks.

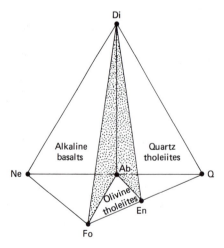

Figure 7-6 The normative basalt tetrahedron of Yoder and Tilley [1962] showing the compositional placement of alkali basalts, olivine tholeiites, and quartz tholeiites. Olivine tholeiites occupy the subtetrahedron Fo-Di-Ab-En.

TABLE 7-3 MICROPROBE ANALYSES (PERCENTAGES OF COMPONENTS) OF
BASALT GLASSES[a]

	b	c	d	e
SiO_2	49.6	49.6	49.8	50.3
Al_2O_3	14.7	14.1	15.7	14.7
TiO_2	1.02	1.41	0.84	1.89
FeO	8.99	10.4	7.13	9.62
MnO	0.18	0.21	0.15	0.19
MgO	8.71	7.32	8.85	7.32
CaO	11.89	11.7	12.9	10.7
Na_2O	1.92	2.30	2.05	2.94
K_2O	0.06	0.07	0.11	0.08
Cr_2O_3	0.09	0.07	0.08	0.06
Totals	97.16	97.18	97.61	97.80

[a] After Bryan and others [1979].
[b] JOIDES leg 11, site 105, western Atlantic margin.
[c] Chain cruise 100, core 15, Red Sea rift. Core catcher chips.
[d] Chain cruise 43, site 104, dredge sample 17, mid-Atlantic ridge at 45°N.
[e] Atlantis II cruise 60, site 2, dredge sample 5, mid-Atlantic ridge at 23°S.

Alkali basalts dominate the upper flanks and crests of most oceanic islands and submarine volcanoes. The main volcanic edifice extending beneath sea level may, however, be tholeiitic in composition, though this is still uncertain. On Reunion Island in the Indian Ocean, alkali basalts are underlain by tholeiitic basalt. In the Galapagos Islands, although both alkali and tholeiitic basalts are present, tholeiites underlie alkali basalts only on James Island [McBirney and Williams, 1969]. The *shield volcanoes* of the Hawaiian Islands and Iceland are almost completely dominated by tholeiites, though smaller, younger cones are constructed of alkali basalts. The alkali basalts are olivine-bearing and quartz-free. They are associated with a series of rock types that appear to originate by differentiation from the alkali basalts and include trachytes, phonolites, and alkali rhyolites. This series shows a progressive enrichment in SiO_2, Na_2O, and K_2O.

Compared with the oceans, continental volcanism produces a much wider range of rock compositions, with the extrusion of acid and intermediate volcanic rocks, such as andesites, dacites, and rhyolites, as well as the more highly fluid basaltic magmas. In the oceans, acid and intermediate volcanism are almost completely absent. Basaltic volcanism on the continents can be very widespread; the most spectacular are the **plateau** or **flood** basalts, which are of tholeiitic composition and can occur in enormous volumes and stratiform sheets, such as the Columbia Plateau basalts of the western United States and the Deccan Plateau basalt of western and central India. Although the composition of continental and oceanic tholeiitic basalts are quite similar, there are important differences. In most continental tholeiites the amounts of silica and potassium are greater than in the oceanic tholeiites. The only important exceptions are the parent lavas of the stratiform sheets, which can exhibit compositions remarkably similar to oceanic tholeiites. Most continental tholeiites also contain certain trace elements in greater abundance than in oceanic tholeiites, reflecting contamination of the continental tholeiites by these elements scavenged from sialic inclusions and the walls of crustal conduits and magma chambers.

Compositional differences between continental and oceanic lavas often cause subaerial volcanism to be more violently explosive than that in the deep ocean. Higher silica content of many continental lavas creates higher viscosity. The effusion of these lavas is also accompanied by larger emissions of gases such as water vapor, CO_2, H_2S, and SO_2. If the rate of gas escape is large, many bubbles form causing the lava to lose its coherence and explode. These factors produce explosiveness, leading to phenomena, such as caldera collapse, eolean ash, and ejected bombs. The critical water depth at which gas is reduced in continental lavas, preventing volcanic explosions, is about 500 m.

Oceanic magmas are not explosive by nature; however, in shallow-water eruptions the oceanic magma erupts explosively in contact with seawater. This often forms **hyaloclastites,** which are highly fragmented volcanic rocks. With increasing depth (greater than 300 m) the weight of the water column is sufficient to suppress all explosive water-magma interaction. Thus submarine lavas at depths greater than about 300 m effuse very quietly. Lavas erupted beneath water are also cooled more rapidly, because water has a high thermal conductivity and heat capacity.

Volcanism at Oceanic Ridges

Oceanic ridges probably form from the rapid upwelling of material from the upper mantle. The partial melting of this material results in the formation of basaltic liquids, which are then injected through tensional crustal fissures into a narrow zone of only a few kilometers at the ridge axis. The axis of active ridges is also a zone of relatively high heat flow, reflecting high temperatures at shallow depths due to mantle upwelling. Volcanism is concentrated near the axis of a crestal rift or graben (the central rift valley), which is contained by symmetrical hills.

Observations from submersibles have provided valuable data on the volcanic processes that occur at the sea floor within the mid-ocean rift valley. These have shown that **sheet flows** are important in this volcanic environment, just as they are in the subaerial environment of Hawaii, in which both pahoehoe and aa lavas are produced [van Andel and Ballard, 1979; Ballard and others, 1979]. As in Hawaii, collapsed blisters, pits, and tunnels are associated with the lava flows. In the Galapagos rift, the largest sheet flows are known to extend for 7 km from the eruptive center. Pillow lavas are also of widespread importance on the mid-ocean ridge, but far less prevalent than originally assumed, based on the analysis of bottom photographs (Fig. 7-7). Pillow lavas are marked by numerous bulbous protrusions, each up to a few meters in diameter and surrounded with a corrugated skin of glass (Fig. 7-7). Detailed mapping of the volcanic features within the Galapagos rift zone using submersibles has enabled processes of accretion to be better understood [van Andel and Ballard, 1979; Ballard and others, 1979; Ballard and others, in press]. From Hawaiian analogs Ballard and others [1979] concluded that sheet flows can be

Figure 7-7 Thick, wrinkled, pillow basalts at nearly 2000 m, Puna ridge, Hilo, Hawaii. (Courtesy B. Heezen)

considered to be a submarine equivalent of surface-fed pahoehoe, while the pillow basalts are analogous to subaerial tube-fed pahoehoe. They also suggested that the sheet flows represent early and brief but voluminous eruptions, followed by more sustained and slower but steadier eruptive phases that produce pillow basalts after an internal plumbing system is well established. Thus a stratigraphic sequence of alternating series of sheet and pillow flows tends to be built up. The position of the axis of volcanism either oscillates within the rift valley or changes position randomly. This distribution of volcanic features suggests that volcanism is episodic about every 10,000 years [Ballard and others, in press].

In the FAMOUS area on the mid-Atlantic ridge, the central rift valley is V-shaped from 2 to 5 km wide, has linear hills in its trough, and is bounded by walls 1200-m high that are broken into terraces by numerous faults. Within the central rift valley is an inner-rift valley. The linear hills are mounds of pillow basalts freshly extruded from beneath the ridge crest and forming low ridges about 200 m high and 500 to 1000 m wide [Moore and others, 1974]. Observations made during the FAMOUS project suggest that new material is emplaced along a linear zone a few hundred meters wide near the center of the inner rift valley [Ballard and van Andel, 1977]. Loading produced by many of the thicker flows from a newly formed edifice exceeds the strength of the crust, causing it to yield by normal faulting at the perimeter of the rift. Movement along such growth faults appears to cause tectonic disruption in the area [Hall and Robinson, 1979].

Paleomagnetic study of extrusive sequences in drill cores suggests that the formation of the linear hills in the FAMOUS area is brief (less than 100 years) compared with the interval between edifice formation (about 10^4 years) [Hall and Robinson, 1979]. Hence volcanism appears to be episodic, as in the Galapagos region. Larger but similar elongate structures have been described by Menard and Mammerickx [1967] and Luyendyk [1970] from the crest of the East Pacific rise. These elongated hills are major volcanic deposits up to 300 km long, 20 km wide, and 300 m high. The spacing of these hills relative to inferred spreading rates of the East Pacific rise indicates intervals of about 1 m.y. between major eruptive episodes [Williams and McBirney, 1979].

The new rock material is transported toward one of the outer walls of the inner median valley during the continuous process of sea-floor spreading. By this stage, the construction of the upper part of the ocean basement is essentially complete. The segments are then removed from the median valley to the adjacent plate via the rift mountains [Ballard and van Andel, 1977; Hall and Robinson, 1979]. A major problem, however, is the mechanism by which the oceanic crust in the rift valley is uplifted at the rift-valley walls. The lifting forces appear to break off blocks along the edge of the valley floor and raise them up to form terraces. Terrace formation on the valley walls maintains the median valley as a narrow feature. According to Sleep and Biehler [1970], the process that creates uplift of the valley walls may be linked to temperature-regulated buoyancy of upwelling magma. Consolidation of upwelling magma at depth would create buoyancy, because the magma is higher than the surrounding material.

High heat-flow values are associated with the part of the oceanic ridges within a few hundred kilometers of the axis. Ages taken from magnetic anomalies suggest that this part of the crust is less than 5 to 7 Ma [Langseth and Von Herzen, 1970]. The width of this heat-flow zone is much smaller in the slow-spreading Atlantic and Indian oceans than in the East Pacific. In contrast, those parts of the ridge flanks older than 10 Ma and up to about 50 Ma are associated with low heat-flow values.

Origin and Differentiation of Magmas

Molten rocks which erupt to form lavas on the ocean floor or intrude into the crust result from a complex history of melt generation and later geochemical changes within the upper mantle and crust. **Petrogenesis** is the study of this history. Magma will form whenever conditions are suitable for partial fusion of rocks. The slightly lower density of melts causes them to rise through the overlying lithosphere toward the surface of the earth. As they rise they undergo differentiation, which alters their composition. Thus each melt is affected by a long chain of factors, including the following:

1. Composition of the parent mantle.

2. Degree of partial melting of the mantle material and the depth at which magma segregation occurs from the residual crystals.

3. Speed of rise of the partial melt and the conditions and extent of fractiona-

tion of magma at various depths after segregation from the residual mantle material.

4. Fractionation at specific depths due to rise interruption.

5. Changes in the partial pressure of water during crystallization.

6. Mixing of magmas with different histories and compositions at any stage in the processes.

With so many factors controlling the final composition of a melt, it is not surprising that small but significant differences can occur in the geochemistry, mineralogy, or texture, or all three, between individually erupted cooling units.

Basaltic magmas are generated by partial melting of peridotite or eclogite in the upper mantle [Bowen, 1928]. After segregation, magma may migrate slowly to the surface or be held at various depths for slow cooling and crystal fractionation. New crystals are segregated by gravity, which changes the melt composition by **fractional crystallization.** It has been argued since Bowen's early work in 1915 that most melts can be derived from a parent peridotitic magma by different degrees of fractional crystallization. By this process, if the first-formed minerals, such as olivine, are precipitated out, their removal from the system changes the overall melt composition and, consequently, the geochemical interreactions of any later melt. In the olivines (Mg_2SiO_4 or Fe_2SiO_4), the magnesium-rich minerals crystallize first. If these precipitate out, the successive olivines are more iron-rich. With increased fractionation, rocks become increasingly richer in SiO_2. Low-pressure crystal fractionation (less than 10 to 15 km) is common, particularly among tholeiitic magmas; thus tholeiitic magmas tend to diverge toward quartz tholeiites and alkali basaltic magmas toward trachytes. Another important process in the generation of the composition of the final melt is the mixing of different magmas.

Several questions have risen related to the composition and evolution of those primary magmas that form basalts: What is the composition of upper mantle rock that forms primary melts or magmas? What is the composition of primary magmas? Do oceanic basalts closely approach the composition of the primary magma or are they differentiated to some degree?

Primary magma is produced in the remote upper mantle, which makes it difficult to determine the composition of the parent material (primary mantle composition) and the primary magma itself. It is theorized that the source rock is composed of either peridotite or eclogite. The *pyrolite* model of Clark and Ringwood [1964] and Green and Ringwood [1967] assumes that primary basalts are derived from a peridotitic source rock by partial melting and exhibit a specific, but hypothetical chemical composition of one-third basalt and two-thirds peridotite. If the mantle composition is similar to pyrolite, primary mantle magma must have a ratio of $100 \times Mg/Mg + Fe = 68$ to 73 [Green, 1970; Ito, 1973] because the model assumes that chemical equilibrium is maintained between the liquid and residual crystals of the parent pyrolite. In other words, the basaltic liquid is continually buffered by residual olivine and pyroxene.

Another major question concerns the composition of the primary magma. A primary magma is that liquid formed by equilibrium melting of the source rock, undifferentiated by crystal fractionation and capable of producing basalts by solidification directly or after crystal fractionation [Frey and others, 1974]. Obviously, in order to identify a primary magma from the mantle, it must be clearly demonstrated that such material has not undergone compositional changes at any stage between initial melting and final consolidation. However, it is still unknown whether the primary magma rises rapidly to the surface with little or no change, or whether there are significant compositional changes through fractionation. The very abundance of oceanic tholeiitic basalts and their relative uniformity through the ocean basins suggests to some that these must be close to the composition of a primary magma, but O'Hara [1965; 1968] considers basalts the residual liquid of advanced crystal fractionation, not primary magmas. It is generally believed that most oceanic basalts have been affected by fractional crystallization. At some depth, called the **depth of magma segregation,** the degree of partial melting is extensive enough (perhaps 20–40 percent) and the tectonic environment is such that the liquid segregates from residual crystals. After this, liquids and solids are out of equilibrium and fractionation proceeds.

In summary, the various models for the generation of basalts are that primary magma compositions are controlled by the depth of partial fusion of peridotite mantle [Kushiro, 1968], that tholeiitic basalts result from fractionation of olivine from a deeper seated parental picrite liquid [O'Hara, 1965; Yoder and Tilley, 1962; Ito and Kennedy, 1967], and that basaltic compositions are controlled by the degree and depth of partial melting of dry mantle peridotite and later fractionation [Green and Ringwood, 1968].

Single or multiple mantle sources?

So far we have discussed the origin of basalts as a partial fusion and differentiation product of mantle material. This introduces another major question, which proposes the possibility of more than one major mantle source of basaltic melt. During the last few years, the study of trace elements and the isotopic composition of oceanic basalts has shown that basalts cannot be derived from a single mantle source, but that the mantle must be heterogeneous. This produces basalt types with distinct geochemical fingerprints. We have already seen that basalts of spreading ridges and some oceanic islands, like Hawaii and Iceland, are dominated by low-K tholeiites, but despite their superficial compositional resemblance, the basalts of these two environments vary significantly in minor and trace-element chemistry and in isotopic composition. J. G. Schilling, S. R. Hart, and others have demonstrated that spreading ridge tholeiites (Figs. 7–8 and 7–9) are marked by low concentrations of large-ion lithophile (LIL) elements such as La, Rb, Cs, Ba, and Sr (and therefore with high ratios of K to Rb, K to Ba, and so on). There are also low concentrations of the minor elements K, P, and Ti and low ^{87}Sr to ^{86}Sr ratios (Fig. 7–9). These distinctions may reflect different mantle sources and depth of origin [Schilling, 1973]. Lead-isotope data suggest that these mantle sources have been distinct for long periods of time (1–2 billion years before present) [Sun and others, 1975]. However, because many oceanic islands are located close to spreading ridges, it is difficult to see how such distinct magma sources can remain so.

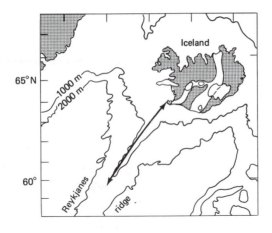

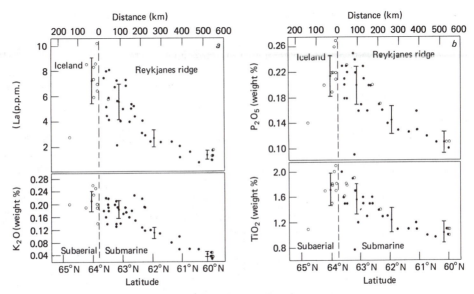

Figure 7–8 Variations in La, K_2O, P_2O_5, and TiO_2 concentrations in tholeiitic basalts sampled along the Reykjanes ridge southwest of Iceland. Regular gradients are observed. Error bars are marked by vertical lines. (After J. G. Schilling, reprinted by permission from *Nature*, vol. 242, pp. 565–571, 1973, copyright © 1973 Macmillan Journals, Ltd.)

Some of the proposed models reconcile this problem by invoking magma supply from different depths in the earth's mantle. The earliest and best-known work on compositional differences between oceanic island and mid-ocean ridge basalts is that of J. G. Schilling and his colleagues. This involved geochemical analyses of dredged basalts along the mid-ocean ridge south of Iceland (Reykjanes ridge) and north of the Azores Islands; sampling involved both oceanic-island and ridge-crest basalts. Striking LIL variations occur along the mid-Atlantic ridge axis north to

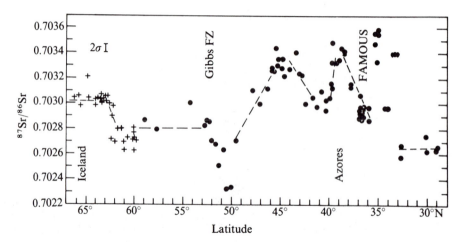

Figure 7-9 Variation of the ratio of strontium-87 to strontium-86 with latitude along the mid-Atlantic ridge. Relatively high values are found not only on or near Iceland and the Azores, the location of hot spots, but also at 45° and 35°N. The location marked by FAMOUS is the site of the French-American Mid-Ocean Study. Precision is marked by the error bar in the upper left-hand corner. (From J. G. Schilling and W. M. White, reprinted by permission from *Nature*, vol. 263, p. 661, 1976, copyright © 1976 Macmillan Journals, Ltd.)

Iceland and the Azores. These elements decrease regularly and progressively south of Iceland along the mid-Atlantic ridge and increase again in the vicinity of the Azores. The large-scale chemical gradients are interpreted by Schilling to result from the mixing of two major parts of the mantle. The reservoir that supplied the MORBs is considered the low-velocity seismic zone of the asthenosphere, with the partially molten layer of mantle rock lying at depths between 75 and 250 km. The source for oceanic-island basalts is considered to lie at mantle depths beneath the asthenosphere (greater than 250 km) and to be connected to the surface by rising plumes of magma (Fig. 7-10). These are the oceanic hot spots of Morgan and Wilson, discussed in Chapter 5. The **mantle plume** magmas are enriched in light rare earths (RE) and radiogenic Sr and Pb isotopes. Mantle plume basalts are three to four times richer in Cl and Br than the ridge basalts, which are depleted in light RE and low Sr and Pb isotopic ratios [Unni and Schilling, 1978]. The compositional gradients between the basalts of these two environments (Figs. 7-8 and 7-9) are caused by large-scale mixing of plume and deeper oceanic-ridge (abyssal) tholeiitic melt types. Such hybrid rare-earth patterns occur along the Reykjanes ridge and the Gulf of Aden [Schilling, 1973]. Therefore the geochemical evidence has suggested to some that there are at least two magma sources, but recent evidence suggests that compositional variations may be even more diverse, indicating the need for additional distinct mantle sources.

O'Hara [1973] and others have interpreted the LIL differences to be caused by fractional crystallization or variations in the degree of partial melting of the same primary magma source. They invoke a steady-state relation between intrusion into the magma chamber and fractional crystallization. The decoupling of major

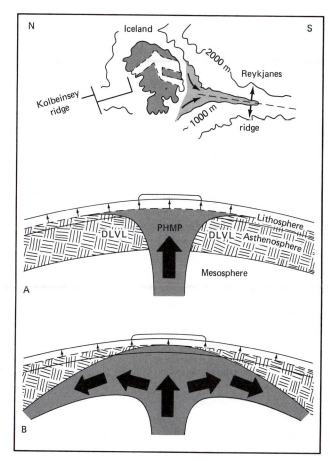

Figure 7-10 Two mantle plume models proposed by Schilling [1973] to account for observed chemical gradients along the Reykjanes ridge southwest of Iceland. Hachured zone represents influential depth range where upward injection can start. In model A, the density of PHMP (primordial hot mantle plume) and flow about Iceland is rapidly used up along the spreading axis. In model B the rising PHMP flow is shown to be progressively sinking about the Iceland plume as a thunderhead and the total plume discharge rate is larger than in A. DLVL = depleted low-velocity layer. (After J. G. Schilling, reprinted by permission from *Nature*, vol. 242, pp. 565–571, 1973, copyright © 1973 Macmillan Journals, Ltd.)

elements from LIL trace elements could arise because LIL elements may be strongly partitioned into liquid relative to mineral phases in the mantle, and the early liquid phases should contain most of these components. This was rejected by Schilling [1973] and Hart and others [1973] on the basis of the distinct Sr and Pb isotopic ratios in the two magma types. They consider it improbable that the Sr isotope ratios can be affected by fractional crystallization. O'Hara [1973], however, believes that the isotopic differences are not large enough to eliminate a fractional crystallization model, even though the mechanism that would create such changes must not cause major differences in element composition. Nevertheless, the isotopic evidence is compelling that these mantle sources are distinct and that the hot mantle plume source or sources have remained deep. The sources may have remained closed systems long enough to build up their radiogenic isotopes and heat content before rising as plumes and transporting primordial material to the earth's upper surface [Schilling, 1973]. During partial melting in the mantle, Pb and Sr isotopes are homogenized and subsequent fractional crystallization does not affect the isotopic ratios of such heavy elements if it occurs over just a few million years and without crustal contamination [Sun and others, 1975]. Since concentrations of LIL elements can vary during partial melting and fractional crystallization, isotopic ratios are probably more reliable fingerprints of primary mantle magmas. Such an interpretation is possible when the evidence is based on the ratio of ^{87}Sr to ^{86}Sr because different chemical composition can affect the isotopic composition. For example, the radioactive decay of ^{87}Rb produces ^{87}Sr, but ^{86}Sr has no radioactive precursor. Thus the higher ratios of ^{87}Sr to ^{86}Sr for oceanic islands suggest that, on average, the mantle source contains larger amounts of Rb relative to Sr than those magmas that supply the ocean ridges.

Isotopic analyses are being used to determine when the mantle separated into the chemically different reservoirs of isotopically distinct basalts. Initial estimates of 1.6 billion years are now being challenged by theories that the process may have been continuous for several billion years.

Thus, in summary, strong evidence exists for an isotopically depleted, low-velocity layer, a mantle source for normal ridge basalts that is remarkably uniform in space and time and possibly worldwide in extent. On the other hand, mantle magmas that feed oceanic islands are not only distinct from normal ridge basalt magmas in isotopic and rare-earth composition, but are also distinct from one another, are localized geographically, and seem to be derived from greater depths in the mantle.

Fractional crystallization of mid-Atlantic ridge basalts

Based on trace-element and isotopic composition, the magma that supplies normal mid-Atlantic ridge basalts seems remarkably uniform and widespread, but there are major element variations of some significance. The earlier studies suggested major element uniformity, but subsequent investigations involving dredged and drilled basalts show chemical variation among samples from the same portion of a ridge, as well as among samples from different spreading centers. This compositional variability results from fractional crystallization and magma mixing, rather than from different primary magma sources.

Basalts range from aphyric (fine-grained, no phenocrysts) to highly porphyritic, with most containing about 10 percent phenocrysts. Olivine, plagioclase, and clinopyroxene are the most common phenocryst phases; spinel is also present in some olivine-rich units. Rich plagioclase, phyric basalts with individual phenocrysts of up to 15 mm in diameter, are fairly common [Hall and Robinson, 1979]. Primitive types with a high ratio of MgO to MgO + FeO and high Ni have also been described (from the FAMOUS area) along with other basaltic types [Bryan and Moore, 1977] from drilled sections on the mid-Atlantic ridge [Frey and others, 1974]. These may be derived from the primary mantle, but are not important. Most basalts are too chemically evolved to be primary magmas. Olivine-phyric basalts have been sampled in the FAMOUS area, where they may be from the youngest eruptions along the rift axis [Bryan and Moore, 1977].

Leg 46 basalts on the mid-Atlantic ridge seem to be of two main chemical types: a low CaO and Al_2O_3 type, which corresponds to sparsely phyric basalts, and a high CaO and Al_2O_3 type, which corresponds to porphyritic basalts [Kirkpatrick and others, 1978]. Neither type deviates very much in composition from an average for the MORBs. Mineralogically typical MORBs exhibit a primary groundmass assemblage consisting of olivine, plagioclase, augitic clinopyroxene, titano-magnetite, and, in some cases, ilmenite. Phenocryst phases are olivine, plagioclase, and clinopyroxene. All of these basalts, as well as several others, vary in composition as a result of fractional crystallization.

Basalt variation can be displayed by plotting major elements or their ratios against each other, such as FeO/MgO versus TiO_2 (Fig. 7-11). North Atlantic data plotted in this manner tend to lie within the parallelogram-shaped area, with the longest diagonal being about a one-to-one variation of FeO/MgO as against TiO_2 [Bryan and others, 1979]. Thus major element variation can be considerable over short distances on the mid-Atlantic ridge, reflecting the effects of fractional crystallization operating over short distances. Likewise, major element variations among basalts at a given drilled site have been found to exceed those between different sites [Tarney and others, 1979].

Crustal drilling sites at varying distances from the mid-Atlantic ridge show that the observed chemical and mineralogical characteristics of Layer 2 seem to have been established very early in the history of the opening of the Atlantic and show no major variations as a function of time. Early work [Aumento, 1967; McBirney and Gass, 1967] suggested that regular compositional variations of basalts occur as a function of distance across a mid-ocean ridge, but such trends are now disproved by additional work, especially deep-sea drilling. On the other hand, detailed sampling in the FAMOUS area from the center of the rift valley to the walls of the rift valley shows regular compositional variation [Bryan and Moore, 1977]. The central lava samples show higher ratios of olivine relative to clinopyrox-

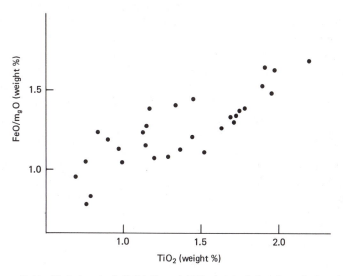

Figure 7-11 Variation in FeO/MgO and TiO_2 for fresh basalt and glass from drilled cores and dredged samples for the Atlantic. (After W. Bryan, G. Thompson, and F. Frey, 1979)

ene and plagioclase phenocrysts and contain chrome spinel. Glasses from flank lava samples are enriched in SiO_2, TiO_2, K_2O, H_2O, and FeO/MgO relative to central lava samples. These compositional trends are attributed to fractional crystallization. Flank lava glass can be derived by the removal of approximately 29 percent (by weight) of analyzed phenocrysts (in the ratio 5.7 plagioclase, 2.5 olivine, 1.8 clinopyroxene) from the central lava glass. In addition, other processes (possibly involving volatile transfer) must enrich the flank lavas in K_2O, TiO_2, and H_2O. Bryan and Moore [1977] proposed a model in which crystal fractionation occurs in a shallow, narrow magma chamber underlying the median valley. The chamber is compositionally zoned and central lavas feed off dikes which tap a hot axial zone, whereas flank lavas feed off cooler, differentiated margin melts.

Model for magma generation and modification

Several primitive magma types have been defined that cannot be accounted for by fractional crystallization in shallow crustal magma chambers. They must be formed from either different mantle composition or melting processes. This implies either several zones of melting and magma ascent in the asthenosphere or a compositionally heterogeneous mantle. For normal mid-ocean ridge basalts, Hall and Robinson [1979] have proposed a model for magma generation and modification (Fig. 7–12). Primary olivine tholeiite basalt magmas are formed by partial melting of a peridotite mantle at depths of about 30 to 35 km, which corresponds to a zone of low shear-wave velocity. Extensive melting of the mantle occurs only when its temperature and pressure are above the melting curve for dry peridotite. Variations in the degree and depth of partial melting create small differences in the primitive melts. The buoyant melts are then carried through the overlying mantle (Fig. 7–12). Olivine and plagioclase are fractionated and are phenocryst phases in many lavas. Fractionation in deep magma chambers results in the formation of cumulate, orthopyroxene-bearing gabbros and peridotites. While the magma chamber is located over the ridge crest, new magma continually enters it and the magma from high in the reservoir is erupted to form overlying lavas and dikes.

Primitive magma is repeatedly injected into a magma chamber and mixed with a more evolved magma that has fractionated from previous episodes of injection and mixing [Rhodes and Dungan, 1979]. This steady-state model reflects the generation of volumes of moderately evolved tholeiites, without the need of large magma chambers. Such a process successfully predicts that both primitive and highly differentiated basalts will be rare and that moderately evolved basalts should dominate Layer 2 of the oceanic crust [Rhodes and Dungan, 1979]. Distinct basalt types occur in units from 50 to 200 m thick, suggesting generation and fractionation of distinct batches of magmas. This implies that magma emplacement is an episodic, rather than a steady-state, process. Furthermore, the interbedding of basalts of distinct composition implies the coexistence of several magma chambers of restricted size, rather than a single large continuous magma chamber. The episodic eruptions from these subrift chambers onto the sea floor form the complex basalt stratigraphy in Layer 2 and probably lead to the formation of a sheeted dike complex at depths near the boundary of Layer 2 and Layer 3. Away from the ridge

Extrusion of pillow basalts onto rift floor, producing mixtures of basalt, basalt breccia, and minor sediment. Olivine-phyric basalt erupted in rift axis and plagioclase-phyric basalt on flanks. Lower part cut by sheeted dike complex. Minor quantitites of diapirically emplaced plutonic rock.

Magmas trapped in shallow magma chambers undergo low pressure crystal fractionation involving olivine + plagioclase ± clinopyroxene ± spinel. Small pockets of magma are isolated and undergo extreme fractionation to form plagioclase-phyric lavas. Other magma chambers receive periodic infusions of magma from depth resulting in mixing of magmas. Upper part of layer 3 cut by sheeted dike complex.

Zone of rising magmas in the upper mantle. Some magmas rise directly to higher levels, others are held temporarily in deep magma chambers. These undergo varying degrees of high-pressure fractionation involving olivine + plagioclase + clinopyroxene ± orthopyroxene to form gabbros, lherzolites, and harzburgites. Periodic influxes of primitive magma from depth results in varying degrees of magma mixing.

Zone of partial melting of mantle peridotite to form primitive olivine tholeiite magmas.

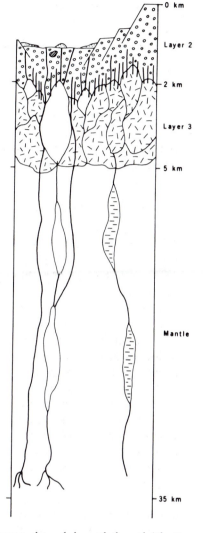

Figure 7–12 Cross section of the ocean crust and mantle beneath the mid-Atlantic ridge crest according to one possible model for magma generation and movement (not drawn to scale). In this model, magma is generated at a depth of about 35 km as mantle rock partially melts. Being lighter this magma rises, but can be held at several places. It may be held temporarily in deep mantle chambers, where its chemical composition can change due to the formation of new mineral crystals or the injection of a new batch of magma from the mantle. At higher levels, it can once again be trapped in magma chambers of varying sizes in the suspected layer of gabbro in the lower crust. This rock is produced by the slow cooling of magma in a chamber. From there, any magma remaining after the formation of gabbro can finally be extruded into sheet flows and pillow lavas, the volcanic rocks forming the upper layer of the crust. The boundary between this layer and the gabbro beneath it is cut by sheeted dikes, the remains of the conduits that carried the magma between the two layers. (From J. M. Hall and P. T. Robinson, *Science*, vol. 204, p. 581, 1979, copyright © 1979 by the American Association for the Advancement of Science)

crest, the magma chambers become filled, the accumulation of crystals on the floor meeting those build up on the roof.

CRUSTAL CHANGES AFTER FORMATION

Alteration and Weathering of Ocean Crust

Once formed, oceanic crustal rocks undergo widespread alteration as a result of weathering and metamorphism. As might be expected, weathered rocks, marked by alteration of glassy groundmass material and glassy pillow rinds, increase in abundance away from the ridge axis. Alteration is produced by the action of cold water on basalts and involves a substantial exchange of elements between rock and water. Deep-sea drilling has shown that the most important process operating to weather the ocean crust is the percolation of seawater through abundant fractures and voids. This process has been very important in regulating the chemical composition of seawater. Mineralogical evidence shows that the temperature at which this alteration takes place is less than 20°C. High thermal gradients at the ridge crest enhance convective circulation of seawater through the oceanic crust, raising water temperatures slightly to create the low temperature alteration. Unlike most metamorphism that takes place in continental rocks, the oceanic crustal metamorphism takes place in a tensional framework. Thus temperatures of metamorphism are much lower in oceanic than continental rocks. Little evidence exists of high-temperature, *hydrothermal circulation* even in the deepest levels penetrated, which is surprising, because most models of heat flow through oceanic crust postulate extensive high-temperature, hydrothermal circulation, and hot springs are probably quite abundant along mid-ocean ridge rifts [Edmond, 1980].

Low-temperature alteration is persistent throughout the upper part of the oceanic crust; this could not occur without a considerable network of fractures. Lister [1972] has calculated that the average space between cracks in young crust is about 30 cm. These fractures are original joints (for example, pillow and columnar jointing) that resulted from rapid cooling of the volcanic rocks. The rate of cooling, and hence fracturing, is greatest near the ridge crests. Hydrothermal circulation thus begins at the ridge crest immediately after the new crust is generated. Comparison with ophiolitic complexes suggests that penetration of seawater into the crust extends to depths of 2 to 5 km. Most low-temperature alteration takes place during the first 1 to 2 m.y. following eruption of the basalts. However, the persistence of easily altered minerals in old crust (50–100 Ma) indicates that after early stages of formation, the crust becomes sealed off from seawater contact [Hall and Robinson, 1979]. This results from increasing fracture constriction by secondary minerals precipitated from the circulating seawater.

In deep-drilled sequences two distinct phases of alteration are apparent. In the earliest phase, olivine and interstitial glass are replaced by the clay mineral smectite, resulting in widespread hydration of basalts. The second phase, the diffusion of pore water into the rock, creates alteration haloes often 20 to 30 times the width of the fractures [Hall and Robinson, 1979]; the new minerals formed are zeolites. During all these processes, immense volumes of seawater become fixed in the oceanic crust as hydrous minerals, such as clay, zeolite, and chlorite.

High-Temperature, Hydrothermal Activity

During the 1970s much evidence of widespread high-temperature hydrothermal acivity at the oceanic ridge crests accumulated. This has been summarized by Edmond [1980]. The first evidence discovered for such activity at the ridge crests are widespread iron-manganese-rich sediments on fast-spreading ridges and as basal sediments immediately overlying basement in deep-drilled sections (see Chapter 14). Decisive evidence came from heat-flow data of ocean-ridge crust, which indicated that a major proportion of the heat associated with the emplacement of new basaltic material must be removed by convection [Anderson and others, 1977]. Calculated heat loss for all of the ridge axes is 10^{19} cal/yr, implying that hydrothermal activity is an integral part of sea-floor spreading. During the middle 1970s, several high-temperature spikes were discovered in bottom waters over the Galapagos spreading ridge [Williams and others, 1974]. This was followed in 1977 by the discovery of active fields of hot springs in the same area using a submersible [Corliss and others, 1979]. These discoveries revolutionized the hydrothermal exploration program [Edmond, 1980].

The most spectacular features discovered are constructional features, composed of sulphides, sulphates, and oxides up to 10 m tall that belch out hot solutions (at least 350°C) which, on mixing with ambient waters, precipitate these various minerals and form black and white **smokers** [MacDonald and others, 1980]. Mounds of pyrite-chalcopyrite several meters high may cover the sea floor around the vents. The fluid jetting from the vents is clear, without any evidence of suspended particulates, but black plumes are produced upon contact with ambient waters. Extremely fine-grained clouds of sulphide minerals billow to heights of several tens of meters above the vents. Multiple vents occur at the top of massive constructional features like chimneys, several meters at the base and 6 to 9 m high. Sedimentary deposits resulting from the hydrothermal activity are apparently not extensive, but thinly coat the pillow lavas where present. The hot hydrothermal waters are very rich in H_2S, SiO_2 and Mn, and CO_2, H_2, and CH_4 [Edmond, 1980]. K, Ca, Li, Rb, and Ba are also enriched.

Because of the high concentrations of H_2S, very large biological populations are supported by sulphide-oxidizing bacteria, which form the base of the food chain for a diverse filter-feeding benthonic fauna, a self-contained biological community. In addition to this major biological discovery, the hydrothermal activity is of major importance in terms of oceanic geochemical cycles and budgets. For instance, fluxes into the oceanic ridges may represent important sinks for magnesium, sulphate, and dolomites. The conventional sinks are not large enough to balance the annual river input of these materials. Similarly, a calcium "excess" in the sedimentary column over that supplied by the continents may be accounted for by extra hydrothermal supplies. All of the manganese being deposited in the deep sea can be supplied by hydrothermal activity at the ridges [Edmond, 1980].

MAGNETIZATION OF THE OCEANIC CRUST

Linear marine magnetic anomalies are magnetization contrasts between normally and reversely magnetized oceanic crust [Vine and Matthews, 1963; see Chapter 5]. But is the signal derived from the magnetization of the entire lithosphere or only certain parts of it? Vine and Matthews [1963] assumed that the depth of magnetization of the ocean crust reversal patterns is related to the Curie point isotherm (the isotherm above which occurs spontaneous magnetic ordering) which lies well within the upper mantle. Subsequent models, however, have considered the bulk of the remanent magnetization to reside in basaltic Layer 2 or even the upper 200 to 500 m of Layer 2, where dredge hauls have shown a high intensity of magnetization.

Deep holes drilled in the Atlantic show the magnetic structure is much more complex than was expected. The simple model of uniformly magnetized crustal blocks of alternating polarity is not correct. The complexity is known from several features: clear reversals in polarity with depth in drill holes, large systematic deviations of inclinations from expected dipole values, large variations in intensity, and conspicuous lateral magnetic heterogeneity (including reversals) in the basalts where adjacent basement holes have been drilled a few hundred meters apart. Furthermore, basalt magnetic properties are strongly influenced by the degree of cold-water alteration, although stable remanence remains the dominant part of magnetization of the basalts. The bottom of the strongly magnetized layer was not reached in any of the deep-drilled holes.

Contrary to earlier beliefs, the upper few hundred meters of oceanic basement may be less significant in producing the linear marine magnetic anomaly patterns. There is little agreement between the magnetization in the drilled cores and the associated linear anomalies. The intensity of magnetization is now known to be too low in basalts drilled in the upper part of Layer 2 to account for the amplitudes of the marine magnetic anomalies. If the average remanent magnetization value obtained for drilled basalt is representative of the magnetic source layer, then this layer must be deeper than 1 km [Lowrie, 1974; Kent and others, 1978]. A significant fraction of anomaly amplitudes may originate from a magnetic source deeper in the oceanic crust. It now appears likely that the gabbroic Layer 3, although more weakly magnetized than Layer 2, may contribute significantly to the magnetic anomaly pattern of the ocean. The enormous volume of gabbroic material may well compensate for its weaker magnetization.

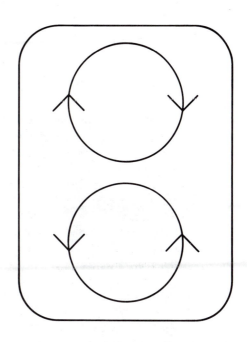

eight
Ocean Circulation

... my soul is full of longing
for the secrets of the sea,
And the heart of the great ocean
Sends a thrilling pulse through me.

Henry Wadsworth Longfellow

INTRODUCTION

The marine geologist must understand the characteristics of the circulation of oceanic waters and the forces that control this circulation. Patterns of sediments, organisms, many broad-scale geological features, and geochemical fractionation between the oceans are controlled chiefly by circulation patterns through the water column. The marine geologist is concerned mainly with parameters that are controlled by seawater characteristics and circulation.

The earth is heated differentially by the sun's energy, with low latitudes receiving a much greater amount than the polar regions. Convective currents in the atmosphere and oceans redistribute the sun's heat energy in a complex three-dimensional circulation pattern. Climatically the oceans are very important since they act as a vast heat reservoir, supplying heat to a cold atmosphere and receiving heat from a warm atmosphere, while retaining relatively constant temperatures. Heat is transferred by ocean currents and wind from low to high latitudes. The prevailing wind systems represent the primary driving force of ocean surface currents and of heat transport. Water is also set in motion because of density differences in seawater, resulting from variations in temperature (warmer waters less dense) and salinity (saltier water more dense). Because water is colder in the polar regions, it is more dense and sinks. A sinking water mass will tend to find its appropriate density level in the water column and spread out. Thus lateral changes in temperature and salinity are much smaller than vertical changes.

Density-driven circulation is called **thermohaline circulation.** As early as 1814

A. von Humboldt recognized that the cold water at depths in the low latitudes cannot be formed locally, but must flow toward the equator from high latitudes. These currents, which flow at greatest velocities at or near the ocean floor, are strongly controlled by bottom topography. Hence the topographic configuration of the ocean floor affects bottom current directions and velocities and, in turn, the currents have a major effect on the geology of the ocean floor.

The Thermocline and Pycnocline

Density differences through the water column are vertically stratified; currents at different levels can flow in different directions. The rate of change of the density with depth determines water-mass stability or unwillingness to move vertically. The water column can be divided into three zones, based on changes in temperature and density (Fig. 8–1). Almost everywhere in the oceans, temperatures decrease with depth; more rapid decreases occur in the upper part of the water column. In low latitudes, typical temperatures are 20°C at the surface, 8°C at 500 m, 5°C at 1000 m, and 2°C at 4000 m. There is a shallow lens of relatively warm water (to 200 m) floating on an immense volume of much colder and more saline water—the cold, deep-water zone. The surface zone comprises only about 2 percent of the ocean's volume. These cold- and warm-water spheres are separated by a zone of rapidly changing temperature, called the **thermocline** (Fig. 8–1), and density, called the **pycnocline.** Due to the density stratification, there is little transfer of horizontal momentum between the cold- and warm-water spheres. Low-density surface water cannot easily move downward through the pycnocline. The surface layer is largely a wind-driven system of currents, a turbulent zone which has little connection with deeper circulation dominated by thermohaline processes. Any interaction that takes

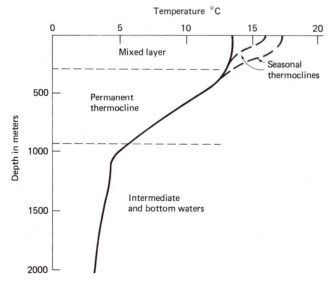

Figure 8–1 The permanent thermocline. The upper few hundred meters represent a mixed layer in which seasonal thermoclines can develop. (After B.W. Pipkin and others, 1977)

place is by upwelling of cold water and by downwelling of warm water in certain areas.

The surface layer of the oceans is well mixed by winds, waves, seasonal cooling, and salinity increase resulting from evaporation. For these reasons it is often called the **mixed layer.** Photosynthesis by marine plants occurs here and makes it the primary source zone of food in the oceans. At high latitudes, where surface temperatures are low and close to the temperature of deep water, there is no permanent thermocline, although a seasonal thermocline usually develops (Fig. 8–1). In middle latitudes, seasonal temperature changes at the surface cause the thermocline to show larger gradients during the summer than during the winter. The ocean waters below the thermocline are, for the most part, products of polar and subpolar processes. It is at high latitudes, the source areas for the cold deep-water zones, where contact is made between the cold-water sphere and the atmosphere. This leads to cold-water oxygenation, without which the deep-ocean environment would become anoxic.

The Oxygen Minimum Layer

The atmosphere is the primary source of oxygen in seawater, and surface water is close to being saturated. In most parts of the ocean, dissolved oxygen decreases from the surface to relatively low values in a zone at intermediate water depths (150–1000 m). This zone, called the **oxygen minimum layer,** is geologically important because it leads to the formation of organic-rich sediments through less oxidation (see Chapter 14). Oxygen from the atmosphere, as well as that produced by photosynthetic organisms in the euphotic zone of the ocean, is available for respiration by living organisms. At greater depths, the oxygen is consumed by living organisms and oxidation of detritus, but there is no replacement by photosynthesis or mixing processes since this zone occurs just below the thermocline. Above the thermocline, the mixing process ensures high oxygen content, but photosynthetic organisms deplete the water of phosphate and silicate; below the thermocline, in the oxygen-minimum layer there is enrichment in these nutrients. Below the thermocline there is reduced downward mixing, but zooplankton and nekton thrive because of the close proximity of food sources. Their respiration depletes the water of oxygen, but at greater depths reduction in the numbers of these organisms allows oxygen levels to rise again. Movement of these nutrient-rich waters through upwelling can cause high biological productivity, which in turn makes the oxygen minimum layer shallower because of increased oxygen consumption by animals.

Low oxygen values are also characteristic of older, deeper water masses. Bottom waters produced at or near the surface in the polar regions are rich in oxygen. As these bottom waters move through the deep-ocean basins, they become older and the oxygen becomes increasingly depleted by animal respiration and oxidation of detritus. Hence the oxygen content of deep water is a general indication of its age.

SURFACE CIRCULATION

Surface-Water Temperatures

Three of the most important properties of seawater are temperature, salinity, and density. Surface temperatures range from freezing point at high latitudes in winter to more than 28°C in low latitudes (Fig. 8-2). The distribution of temperatures is meridional; the lines of constant temperatures (**isotherms**) are in a general east-west direction (Fig. 8-2). Highest water temperatures occur slightly north of the equator. The mean annual surface water temperature for all oceans is about 17°C, being much higher in the Northern Hemisphere (19°C) than in the Southern Hemisphere (16°C). These differences result from different current patterns, which are controlled by the relative positions of sea to land. In the Northern Hemisphere, meridionally positioned landmasses deflect warm and cold currents northward and southward. However, in the Southern Hemisphere—where the land area is much smaller—a large body of cool water is retained close to Antarctica as the Antarctic Circumpolar Current. The isotherms do not extend east-west everywhere because some currents, which lie close to the margins of continents, carry warm water poleward (for example the Gulf Stream, Kuroshio or Japan Current, and the East Australia Current); others carry cool water equatorward (California, Peru, and Oyashio currents). Furthermore, upwelling of deep, cool waters near the eastern margins of some oceans, such as the Pacific, also affects temperature distributions. The highest temperatures (greater than 30°C) occur in restricted tropical basins such as the Persian Gulf and the Red Sea.

Surface Water Salinities

Salinity distribution in surface waters is roughly zonal, although more diffuse than the temperature pattern (Fig. 8-3). Evaporation, precipitation, and ice melting are the processes from which salinity patterns develop. Salinity values are related to decreasing evaporation and increasing precipitation, with salinity values lower toward both the high latitudes and toward the equator (Figs. 8-3 and 8-4). However, the normal range of salinities in the open ocean is small, usually from about 33‰ to 37‰, with average salinity being about 35‰. The Atlantic Ocean is, on the average, the most saline ocean (35.37‰), because it contains a tongue of high salinity water from the Mediterranean Sea.

Salinity at the surface of the sea is highest in the middle latitude subtropical gyres (Figs. 8-3 and 8-4) where evaporation exceeds precipitation. These gyres represent great lenses of highly saline water surrounded by lower salinity waters. Near the equator, the lowest values (less than 33.5‰) are found in the east. Particularly high salinity values occur in restricted regions of high evaporation, such as the eastern Mediterranean (39‰) and the Red Sea (41‰). Conversely, salinity decreases coastward, particularly near the mouths of large rivers. The salinity and temperature characteristics, in addition to numerous other parameters, define

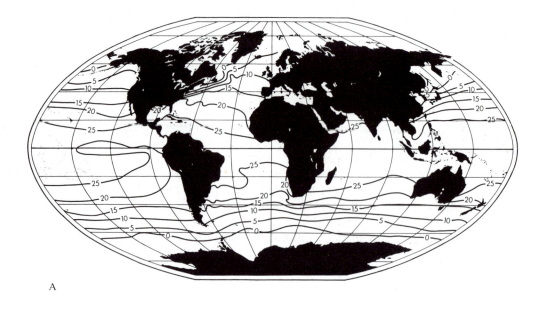

A

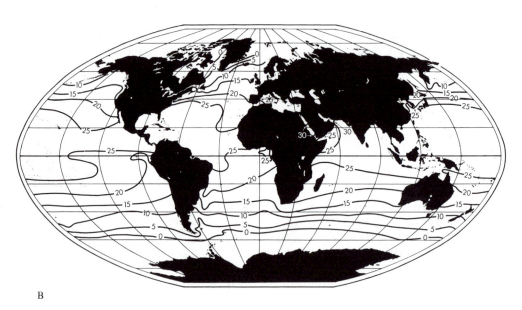

B

Figure 8-2 Surface water temperatures (in °C) of the world ocean in February (A) and August (B). (After Sverdrup, Johnson, and Fleming, 1942, with permission of the University of Chicago Press)

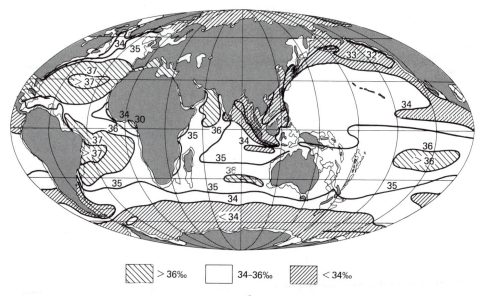

> 36‰   34–36‰   < 34‰

Figure 8–3   Surface water salinity (in ‰) in the world ocean. (After Sverdrup, Johnson, and Fleming, 1942)

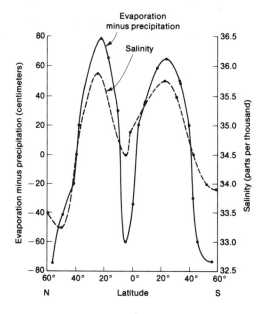

Figure 8–4   Changes in surface salinity (in ‰) with latitude plotted against evaporation minus precipitation (cm). (From S. Defant, *Physical Oceanography*, vol. 1, p. 163, 1961, reprinted by permission of Pergamon Press, Ltd.)

distinct water masses; environmental parameters may change markedly at the boundaries between different water masses.

Importance of Density

Density differences in ocean water lead to vertical circulation, which, in turn, controls the temperature and salinity distributions in the water column. The density of seawater ranges from about 1.02 to 1.07 g/cm^3 and depends on temperature, salinity, and pressure. The density of seawater can be calculated if these three variables are known. In general, density increases as salinity and pressure increase and temperature decreases. Density changes result primarily from evaporation, heating, precipitation, or sea-ice formation at the ocean's surface. As density increases, the seawater sinks, carrying relatively rich supplies of oxygen. Thus thermohaline circulation is largely driven by density differences. High temperatures at low latitudes cause low-density surface water (Fig. 8–5). Higher-density surface waters in middle latitudes (Fig. 8–5) are due to higher salinities created by high evaporation and low precipitation (Fig. 8–4). Waters of highest densities form near Antarctica (Fig. 8–5) through sea-ice production.

Importance of Wind

The primary force for surface currents of the world ocean are the winds in the lower atmosphere, which generate stress on the surface of the sea. Surface winds directly affect only the uppermost layers, although the depth to which surface circulation patterns penetrate is dependent upon stratification in the water column. It seems that the equatorial surface circulation pattern extends to only 300 to 500 m, but in the Antarctic and Arctic, surface circulation patterns extend to depths of several thousand meters.

It is easy to imagine how global wind patterns can establish the current patterns shown in Fig. 8–6. At high latitudes in the North Pacific and North Atlantic

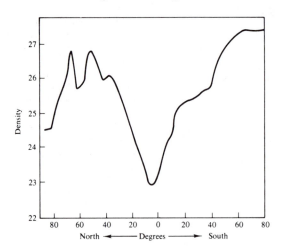

Figure 8–5 Changes in density (ϱ) of surface waters with latitude in the Atlantic. (After G. Wust and others, 1954)

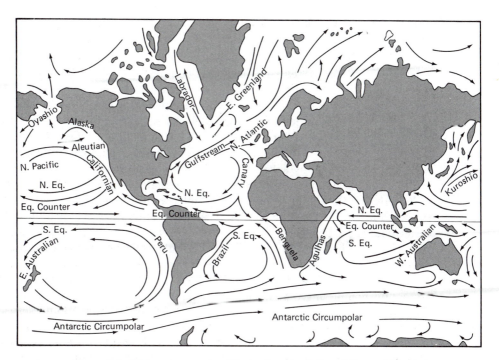

Figure 8-6 Major components of the surface circulation of the world ocean.

oceans, there is a belt of prevailing westerlies; south of this are the horse latitudes, and south of these latitudes is the northern trade-wind belt blowing from the east. These winds decrease near the equator as the **intertropical convergence** (ITC), or doldrums, are approached and then increase again in the southern trade-wind belt. The ITC, which separates the wind systems of the two hemispheres and can be considered the climatic equator, occurs between 3° and 10°N. Further south, the tradewinds give way to prevailing westerlies (the "roaring forties"), which are replaced by easterlies close to Antarctica. The winds of the Northern Hemisphere thus apply a clockwise torque to the ocean surface and the winds of the Southern Hemisphere apply a counterclockwise torque. This produces the major ocean current gyres in middle to low latitudes (Fig. 8–6). Because such large amounts of energy are stored in the open oceans, a shift in the wind field does not produce an immediate shift in ocean current directions. An ocean-wide balance occurs between surface currents and the climatic wind field. One region where drastic seasonal changes in wind patterns do cause reversals in current directions is the northern Indian Ocean. In this region, the summer monsoon winds blow from the western Indian Ocean onto India. Ocean currents (Southwest Monsoonal Drift) thus flow mainly from west to east. In contrast the cold, northern winter winds blow southward from the Asian continent over the Indian Ocean. Ocean currents (Northeast Monsoonal Drift) respond by flowing mainly from east to west or to the southwest. The circulation pattern further south in the Indian Ocean away from monsoonal influences is similar to the gross features in the South Atlantic and Pacific.

Primary Elements of Surface Circulation

The circulation patterns of surface waters of the oceans (Fig. 8–6) have gradually been discovered since the early compilations of drifts of sailing vessels were begun. The most important aspect of the pattern is the control exercised by the surface wind circulation. The most conspicuous features of the oceanic circulation are the large anticyclonic gyres (clockwise in the Northern Hemisphere; counterclockwise in Southern Hemisphere) in the tropical and subtropical regions of each ocean. The Northern Hemisphere gyres are separated from those in the Southern Hemisphere by a well-defined series of eastward and westward zonal currents, including an eastward-flowing undercurrent near the equator (Fig. 8–6). In each gyre, the flow is most narrow and intense on the western side of the ocean (such as the Gulf Stream and Kuroshio currents), and rather weak and diffuse in the eastern regions. The **western boundary currents,** like the Gulf Stream, Brazil, Agulhas, Kuroshio, and East Australian currents, flow in the direction of the poles. Conversely, the **eastern boundary currents,** such as the Canaries, Benguela, West Australia, California, and Peru currents, flow back toward the equator as cold currents, thus completing the system of circulation crossing a wide latitudinal range. The gyre in the North Indian Ocean is distinct in that its circulation reverses direction as a result of the seasonal changes in monsoonal winds.

Weak cyclonic gyres (counterclockwise in Northern Hemisphere; clockwise in Southern Hemisphere) occuring in the subpolar areas of the North Atlantic and Pacific oceans contribute to the general eastward flow at mid-latitudes. Components of these smaller gyres are the Alaska and Oyashio currents. Such subpolar gyres are lacking in the Southern Hemisphere because mid-latitude zonal flow is not obstructed by continental barriers. Instead, the Antarctic Circumpolar Current circles Antarctica and flows eastward at all depths (Fig. 8–6). This is the largest of all transport currents—over 200×10^6 m^3/sec compared to about 100×10^6 m^3/sec for the Gulf Stream and 15 to 50×10^6 m^3/sec for most other major currents.

As these major gyres circulate around the oceans, their limbs either receive or give up heat to the atmosphere or to adjacent currents. Surface-water temperatures thus change around the circuit. The westward-flowing limbs near the equator are heated as a result of higher insolation and heat transfer from the warm atmosphere and adjacent warm parts of the central gyres. Conversely, the eastward-flowing limbs at high latitudes lose heat. Thus the poleward-flowing western boundary currents are considerably warmer than the eastern boundary currents.

The Coriolis Effect

The general character of circulation of the oceans is the formation of large gyres rotating clockwise in the Northern Hemisphere and counterclockwise in the Southern Hemisphere. Likewise, high atmospheric pressure imparts a clockwise motion to cells in the Northern Hemisphere and a counterclockwise motion in the Southern Hemisphere. Furthermore, there is a marked east-west asymmetry in the circulation of ocean currents. This asymmetry was first explained by Stommel [1958] as due to forces related to the rotation of the earth. In fact, the fundamental effects of rotation of the earth on the motion of fluids was determined by G. C. de Coriolis in 1835. He determined that the effect of the earth's rotation is to deflect moving objects to the right in the Northern Hemisphere and to the left in the Southern Hemisphere. This is referred to as the **Coriolis effect,** or acceleration. This effect does not set wind or ocean currents in motion, but deflects them after they are in motion.

The earth rotates from west to east. If viewed from the North Pole, this rotation appears to be counterclockwise; but it is clockwise when viewed from the South Pole. Because the sense of the earth's rotation is different in both hemispheres, the direction of deflection of currents changes at the equator. At the equator, the Coriolis effect is zero (the centrifugal force is greatest) and the deflection increases, with latitude reaching a maximum at the poles (Fig. 8–7). An object at rest on the equator rotates at a velocity of 1000 miles per hour (mph), or 1670 km/h, because the circumference of the earth here is about 24,000 miles; this is reduced to only 1 mph at a latitude close to the pole, where the circumference is only 24 miles. In accordance with the law of conservation of momentum, a particle at the surface of the Northern Hemisphere that is moving toward the equator will lag increasingly further to the west of particles at rest on the surface at lower latitudes because the immobile particles have successively greater tangential velocities toward the east as the equator is approached. In contrast, a particle moving toward the North Pole will move progressively faster to the east compared with immobile bodies because of its greater initial tangential velocity. Thus particles moving across latitudes in the Northern Hemisphere are deflected increasingly to the right. The directions of movement are reversed in the Southern Hemisphere. The Coriolis acceleration is proportional to the sine of the latitude, and so increases from zero at the equator to $1.5 \times 10^4 \, v$ at the pole (where v = velocity): Coriolis acceleration is $1.5 \times 10^{-4}v$ sine ϕ cm/sec². Although this acceleration is very small compared with gravity, its atmospheric and oceanographic effects are large because of the cumulative effect over large distances and because it is horizontal, while g is vertical. Thus it is a major factor in controlling oceanic circulation.

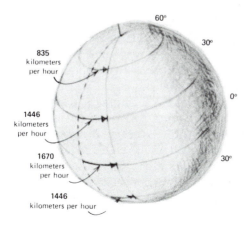

Figure 8–7 Change in speed of the earth's surface between the equator and the poles. An object moving rapidly from the equator to the north pole seems to be deflected to the right, illustrating the Coriolis effect. (From M. G. Gross, *Oceanography, A View of the Earth* © 1977, p. 190, reprinted by permission of Prentice-Hall, Inc.)

The above deflections relate to the motion of fluids across latitudes. We shall now see that equivalent deflective effects related to centrifugal forces act on particles moving east or west. A particle moving toward the east is subject to greater centrifugal acceleration because it is moving in the same direction as the rotation of the earth. Compared with a fixed object, the particle tends to be pulled away from the earth's axis of rotation. Because of the countering force of gravity, the particle is not pulled out vertically but, in the Northern Hemisphere, it slips to the right, toward the equator (which is further from the earth's axis of rotation). In contrast, a particle moving westward against the spin of the earth will experience a deficit in centrifugal acceleration compared with a particle at rest. Thus it moves poleward, or closer to the earth's axis of rotation (right in the Northern Hemisphere; left in the Southern Hemisphere).

Geostrophic Currents

Within the oceans, slight differences in water density produce forces strong enough to create water movement. Lateral differences in water density cause an acceleration of water particles. Once the water begins to move, it is influenced by the Coriolis effect and is deflected. Flows where the Coriolis effect is exactly matched by the pressure-gradient force are called **geostrophic currents;** most major currents, such as the Gulf Stream, are effectively geostrophic currents.

Persistent wind blowing over the surface of the ocean causes the water to pile up in the direction the wind is blowing. These winds create a sloping sea surface, which in turn creates pressure differences in the water column. Currents resulting from such pressure differences do not flow down the slope; they flow perpendicular to the slope along the isobaric surfaces because the Coriolis effect is at right angles to the pressure gradient. Thus air or water does not flow directly from high- to low-pressure areas but flows parallel to the isobars, clockwise around high-pressure cells and counterclockwise around low-pressure cells in the Northern Hemisphere. The effect is opposite in the Southern Hemisphere.

The accumulation of water occurs in the subtropical gyres as a result of wind movement toward these regions. Furthermore, the Coriolis effect deflects water toward the middle of these gyres until the surface slope is so great that gravity pulls the water down the slope. Gravity is balanced by the Coriolis deflection around the mound. Water is also stacked up against the western margins of the oceans owing to the strong, westward-flowing equatorial currents. This can cause a slope in the water surface of as much as 1 m per 100 km across the Gulf Stream. These sloping sea surfaces may contribute to the important eastward-flowing Equatorial Countercurrent. The gradients in topography that occur on the ocean's surface are referred to as **dynamic topographies** and elevations are called **dynamic heights.** Because the heights involve only a few tens of centimeters, the topography is computed only indirectly from density distributions. Strongest currents occur where the dynamic topography is steepest; the weakest occur where slopes are gentlest.

The Ekman Spiral and Upwelling

We have seen that the principal driving force of surface currents is the wind, which pushes water along via the force exerted upon ripples and waves and drags water along by friction. Water at depth is also dragged along by friction within the water column. Velocities decrease with increasing depth as the energy is dissipated. However, surface water is not usually driven directly before the wind. In the Northern Hemisphere, the drift of icebergs may deviate 20° to 40° to the right of the wind direction. In 1902, V. W. Ekman calculated that due to the Coriolis effect, a steady wind would drive surface waters at a maximum angle of 45° to the wind direction (Fig. 8-8). The deflection is to the right in the Northern Hemisphere and to the left in the Southern Hemisphere. The Coriolis force deflects each succeeding water layer further from the wind direction (Fig. 8-8). This continues to a depth at which the frictional forces and the water movement resulting from this process become insignificant. Thus current velocities are reduced in each layer with increasing depths. This spiral structure of the motion is called the **Ekman spiral** (Fig. 8-8) and the general effect, called the **Ekman transport,** is that the upper layer of the ocean (about 100 m in middle latitudes) may exhibit, under the influence of wind, a net motion of 90° to the wind direction. The most important geological significance of this phenomena is an *upwelling* effect in certain coastal regions, such as the west coast of North America. Upwelling is a term applied to the ascent of water from the thermocline or below to the ocean surface. Where the coast lies to the left of the wind direction, the light, warm, surface water is transported away from the coast and is replaced by upwelling colder, denser, subsurface water. Normally these upwelled waters contain higher concentrations of nutrients (such as phosphates, silicates, nitrates) than does the original surface water, which has been depleted as a

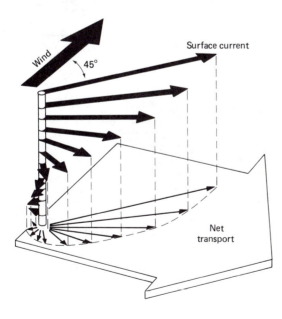

Figure 8-8 Schematic representation of a wind-driven current with depth in the Northern Hemisphere (the Ekman spiral). (After Sverdrup, Johnson, and Fleming, 1942)

result of biological demands. Thus upwelling replenishes the surface layers with the nutritional components necessary for biological productivity. Regions of upwelling are among the richest biological areas of the world and increased skeletal biogenic material is contributed to the bottom sediments. Thus the cool temperatures associated with the eastern boundary currents are not only due to the origin of these waters in high latitudes, but result also from the upwelling of cool, deeper waters.

Divergence and Convergence

Processes other than Ekman transport can cause upwelling. These can occur in either coastal areas or in the open ocean and are shown in Fig. 8-9. Wherever surface waters are blown or otherwise transported away from a certain region, a **divergence** occurs, and subsurface waters move upward to replace the water moving away. This results in higher biological productivity.

In wind-driven upwelling (Fig. 8-9(B)), offshore wind blows surface water away from the coastline. Wind and currents flowing past a headland will draw water away from the headland and create divergence called **obstruction upwelling**

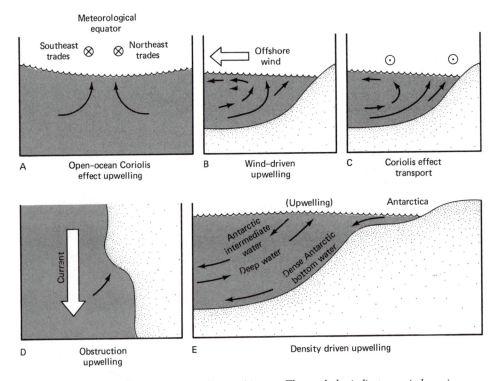

Figure 8-9 Upwelling processes as discussed in text. The symbol • indicates a wind moving toward the reader; the symbol x represents a wind moving away from the reader. (After Pipkin and others, 1977)

(Fig. 8-9(D)). In the open ocean the trade winds north and south of the equator blow at an angle away from the equator in a general westerly direction, creating the **equatorial divergence,** where important upwelling occurs (Fig. 8-9(A)). Especially pronounced upwelling associated with the **Antarctic divergence** also occurs around Antarctica. This divergence results largely from northward Ekman transport associated with the strong, persistent **west-wind drift** or prevailing circum-Antarctic westerlies. This upwelling is reinforced by two other processes. First, very strong *katabatic winds* blow northward over the Southern Ocean away from Antarctica, creating *wind-driven upwelling.* Cold, dense air constantly produced over the Antarctic ice cap flows outward and downward from the Antarctic Plateau as the katabatic winds. Second, upwelling intermediate North Atlantic deep water (NADW) or Circumpolar deep water (Fig. 8-10) compensates for the loss of dense bottom waters generated in shallow regions adjacent to Antarctica. This upwelling thus results from thermohaline circulation, in which denser Antarctic bottom water (AABW) sinks and replaces less dense water. This is called *density-driven upwelling.*

Downwelling of surface waters toward greater depths occurs at **convergences.** The best-known example is the *Antarctic convergence,* or *polar front,* which lies a few degrees north of the Antarctic divergence and represents the confluence of Antarctic and subantarctic surface waters tending to move toward each other (Fig. 8-10). Downwelling is powered from the momentum of currents and from gravity when surface waters become more dense. Within the subtropical gyres, a con-

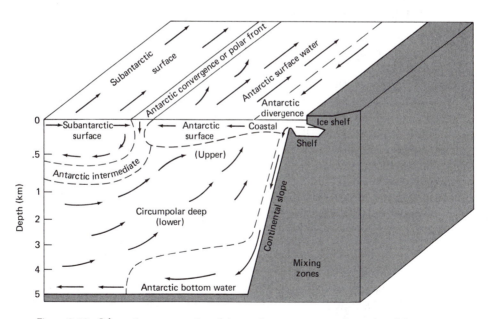

Figure 8-10 Schematic representation of Antarctic water masses and meridional flow. (After Sverdrup, Johnson, and Fleming, 1942, and Gordon, 1971)

vergence of surface water occurs and the accumulation of water causes a thickening of the mixed layer, especially in the western regions.

Equatorial Currents

The major ocean currents near the equator are a response to the wind system. The equatorial circulation consists of three primary components: the westward-flowing north and south equatorial currents which lie beneath the trade winds, a relatively narrow eastward-flowing countercurrent that occurs between the trade-wind belts at the zone of minimum wind stress, and the eastward flowing equatorial undercurrent, found directly on the equator beneath the sea surface. The equatorial wind system produces some upwelling along the equator and along the northern fringes of the surface countercurrent.

The North and South Equatorial currents increase in volume toward the west as a result of lateral influx. These currents are broad, but seldom exceed velocities of 20 cm/sec. The Pacific North Equatorial Current divides itself into two branches east of the Philippines; one turns south along Mindanao and the larger flows north and becomes the Kuroshio Current.

The Pacific Equatorial Undercurrent, or Cromwell Current, flows east at or near the equator. This current is narrow (300 km) and thin (200 m) and exhibits maximum velocities of up to 150 cm/sec. Its depth of flow is centered at about 200 m in the west to 50 m in the east. The origin of the Cromwell Current seems to be linked with the balancing of geostrophic flow near the equator as a result of an increasing pressure gradient toward the west.

Western Boundary Currents

The western boundary currents at the surface of the oceans, such as the Gulf Stream and Kuroshio currents, play an important role in global climates because they transport heat from tropical to polar regions. They also play an important role in the transfer of warm-water planktonic organisms toward higher latitudes, thus mixing gene pools within species. We have already seen that because of the latitudinal variation of the Coriolis effect, the western boundary currents are swift and narrow compared with the other limbs of the gyres. The Gulf Stream transports between 75 and 115 × 10^6 m³/sec of seawater northward. This current is not confined to the surface. It extends to the bottom over much of its path, perhaps including depths greater than 4000 m. In the shallow Florida Straits and along the Blake Plateau (500–1000 m), evidence exists of sediment erosion and transport at the ocean floor. The Gulf Stream originates in the North Equatorial Current, which flows westward between 10° and 20°N. This current passes through the Caribbean Sea, the Southeast Gulf of Mexico, and the Florida Straits. Additional water enters north of the Bahamas and flows as the Gulf Stream along North America; it is a narrow intense jet, about 100 km wide and with maximum surface velocities of 200 to 300 cm/sec. Near Cape Hatteras, the Gulf Stream turns eastward and passes the Grand Banks at 40°N, continuing its flow east toward Europe.

Antarctic Surface Circulation

The almost circular outline of Antarctica and the continuous ring of water favor the development of simple wind and current systems and free interocean circulation at high southern latitudes. A low air-pressure belt encircles the continent at about 65°S. Prevailing westerlies north of the low-pressure belt have a profound effect on oceanic circulation.

The oceanographic processes that occur in the Antarctic are among the most important in the world ocean. Much of the early work, in the 1930s, was carried out by Sir George Deacon and more lately by A. Gordon. Gordon [1975] summarizes the importance of Antarctic circulation as follows:

1. Maintenance of the intermediate, deep, and bottom water of the world ocean as an aerobic environment.

2. Removal of heat and addition of fresh water necessary for the steady-state character of the deep water.

3. Renewal of the warm-water sphere, which is depleted by the production of North Atlantic deep water.

4. Equalization of water characteristics of the three major oceans.

These processes lead to high biological productivity in Antarctic waters and a net heat flux from the ocean to the atmosphere. The driving mechanisms in Antarctic circulation are the wind and the thermohaline effect along the coast of Antarctica. Both cause upwelling of deep water, which can be replaced only by a southward migration of the circumpolar deep water (Fig. 8–10). Because of the importance of these processes and similar circumpolar climatic conditions, the ocean surrounding Antarctica is often referred to separately as the Southern Ocean, or the Antarctic Ocean. The circulation of Antarctic waters is responsible for massive interoceanic water exchange and for the oxygenation ("airing out") of ocean waters by exposing large quantities of these waters to the Antarctic atmosphere before they become deep waters (Fig. 8–10).

Southern Ocean circulation is dominated by the *Antarctic Circumpolar Current* (ACC), which flows clockwise around Antarctica under the influence of the prevailing west wind (Fig. 8–10). The eastward flow has a significant northward component in the surface and bottom layers and a southward component at intermediate depths. The northward movement at the surface is due partly to the effect of the earth's rotation, combined with that of the wind stress and the relatively high density of the Antarctic water compared with warmer surface waters further to the north.

The ACC is believed to be the most voluminous of ocean currents. Its velocity is usually less than 25 cm/sec (0.5 knots) and the motion extends to the bottom with little attenuation [Gordon, 1971B]. The total geostrophic volume through the Drake Passage south of South America has been calculated by Gordon [1971C] at about 240×10^6 m³/sec, while it has been estimated at 233×10^6 m³/sec between

Antarctica and Australia [Callahan, 1971]; this is over 1000 times the flow in the world's largest river, the Amazon.

As the ACC circulates around Antarctica, it loses water at the surface and bottom and is resupplied by intermediate waters originally formed in the North Atlantic and transported southward through the entire length of the Atlantic Ocean.

The position of the ACC is strongly influenced by oceanic and continental topography; the limiting factor in net transport is the Drake Passage. The flow is also confined south of New Zealand by the Campbell Plateau and the Macquarie ridge. As the ACC approaches a north-south orientated ridge, such as the Macquarie ridge, it intensifies and flows southward around the ridge. This deflection carries relatively warm water far to the south, accelerating sea-air exchange of heat in such areas. Close to the Antarctic continent, the prevailing east wind drives the water counterclockwise, but the Antarctic Peninsula blocks the current from being completely circumpolar. This flow is called the **east wind drift** and plays an important role in the circulation in the Weddell and Ross seas (Fig. 8-10).

Abrupt changes in the temperature gradient, which disturb the orderly spread of isotherms, occur in the Southern Ocean. The most southerly, at about 55° to 60°, is the Antarctic convergence, which forms one of the major boundary zones of the world's ocean (Fig. 8-10). Here within a distance of 200 to 400 km, a steep temperature gradient exists in the surface waters, from 4° to 8°C during the summer and 1° to 3°C in the winter. It is in this zone that the convergence takes place between the Antarctic surface water to the south and the subantarctic surface water to the north. An axis of flow of the Antarctic Circumpolar Current occurs slightly south of the Antarctic convergence.

About 10° north of the Antarctic convergence, there is another sharp surface temperature rise from 10° to 14°C in winter and 14° to 18°C in summer. This is the subtropical convergence (Fig. 8-10). North of this boundary, subtropical water occupies the surface layer of the ocean; south of it, some 4°C cooler, is the subantarctic surface water. The sharp change seems to reflect a convergence of these two surface water masses.

DEEP CIRCULATION

Formation of Bottom Waters

Coupled with the system of horizontal gyres and zonal flows predominantly near the surface is a deep-water convective circulation pattern in which air-sea interaction processes produce relatively dense water at the surface of the sea in high latitudes; this water sinks, fills the central ocean basins, and is replaced by poleward-moving water, which rises from intermediate depths. Bottom and intermediate water masses of the oceans differ markedly in their circulation patterns from surface water masses, because, once formed, they seek their respective levels in the ocean and spread widely, displacing the surrounding water [Warren, 1971].

Surface waters, on the other hand, have narrow latitudinal ranges closely matching climatic belts. The processes that provide deep and intermediate waters with their particular properties act almost exclusively at the surface, making it possible to trace the origin of these waters back to their region of formation at the surface. Since a water type is produced continuously, the oceans must continue to circulate to accommodate these new types of water supplies. Evidence that the deep ocean is active came first from observations in the early 1960s by Crease and Swallow, who discovered deep-sea currents of 5 to 10 cm/sec with complete changes in direction in about a month [Swallow, 1971].

The cold ocean waters that lie beneath the thermocline exhibit far smaller variations in physical properties than those at the ocean surface. Nevertheless, relatively small temperature and salinity differences are powerful markers of different areas or modes of formation. For instance, salinity is determined at the surface by evaporation, and precipitation and temperature are determined by variables that determine the ocean surface heat budget. Salinity and temperature exhibit little change once the water has sunk to the ocean floor, while, O_2, CO_2, and nutrient levels are altered by biological activity as the age of the bottom water increases. Bottom water must be dense to sink to ocean depths. This effectively limits the areas of formation of bottom waters to high latitudes and to certain enclosed basins with a high evaporation rate, such as the Mediterranean and Red seas, because it is only in these areas that high-density water is found at the surface.

The most important sources of bottom water of the oceans are in the Antarctic and the polar-subpolar North Atlantic (Norwegian and Greenland seas). The dense waters formed in the Southern Ocean compose 59 percent of the world ocean [Warren, 1971]. Circulation of AABW brings oxygenated, cold water as far north as 50°N in the Pacific and 45°N in the Atlantic (Figs. 8–11 and 8–12). In general, AABW includes all water in the Indian and Pacific oceans with potential temperatures less than 3°C and all water in the Atlantic with potential temperature less than 2°C (except in the Arctic and Greenland and Norwegian seas). **Potential temperature** is the temperature of a deep-water mass recalculated for sea-level pressure. On this basis, water from the Antarctic would comprise 24 percent of the Atlantic Ocean, 70 percent of the Indian Ocean, and 71 percent of the Pacific Ocean.

The formation of sea ice is particularly important in the formation of AABW and thus in thermohaline circulation. As ice freezes it incorporates only about 30 percent of the salt. The remaining salt is added to the nearly freezing water below, increasing salinity and density. This dense water mixes with warmer, intermediate waters of high salinity, resulting in the production of even more dense water, which sinks to the sea floor. This AABW then spreads north, influencing vast areas of the world ocean (Figs. 8–11 and 8–12). Seasonal sea-ice formation is considered the most important process in the formation of AABW. Northward transport of AABW is greatest during March and April, coinciding with the period of most rapid sea-ice production. During the summer, sea ice melts, returning fresher water back to the ocean surface and reducing salinities and densities. A possible secondary mechanism of AABW formation involves the freezing of sea water at the bottom of the large ice shelves in the Weddell and Ross seas. As in the formation of sea

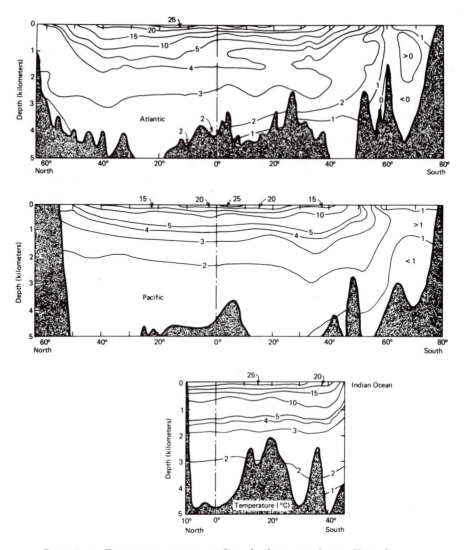

Figure 8–11 Temperature transects (0°C) in the three ocean basins. Vertical exaggeration is about 1000×. (After Dietrich, 1963)

ice, such freezing would increase the salinity and reduce the temperature of water remaining unfrozen. However, the bulk of the bottom water must result from sea-ice produced shelf water. It is significant that the bulk of AABW is produced in the large embayments of Antarctica—the Weddell and Ross seas near the world's two largest ice shelves (Filchner and Ross ice shelves). The classical value of −0.4°C temperature for Antarctic bottom water is the in situ value [Mosby, 1934]. It is more proper to use the potential temperature when determining the mixing of waters derived from vastly different depths because it eliminates the effects of changing pressure. If this is done, assuming an average bottom depth of 4000 m, the potential temperature of the classically defined Antarctic bottom water is about

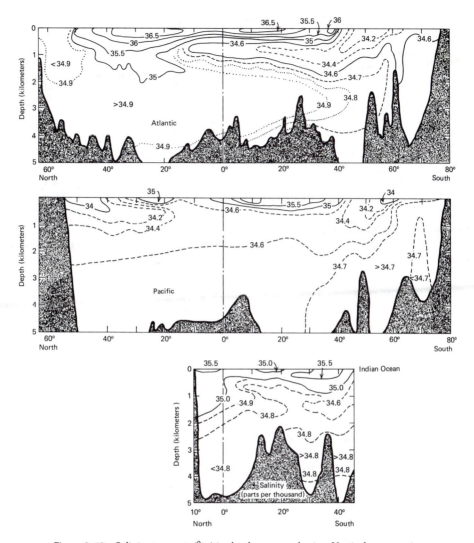

Figure 8-12 Salinity transects (‰) in the three ocean basins. Vertical exaggeration is about 1000×. (After Dietrich, 1963)

−0.7°C [Gordon, 1971C]. Salinities are typically about 34.65‰. Of the possible areas of formation of AABW around Antarctica, the Weddell Sea is considered the most important. The most significant outflow of bottom water takes place along a contour current along the periphery of the Weddell basin. The rate of production is not known, but estimates of 20 to 50 × 10⁶ m³/sec have been made [Gordon, 1971A].

The formation of bottom waters in the Northern Hemisphere is confined to the Arctic Ocean and to the Norwegian and Greenland seas. The northern Indian Ocean is in tropical latitudes so no deep or bottom water is formed. Likewise no important supply of deep water is formed in the North Pacific, even in the presence of sea-ice formation each winter in the Bering Sea. Passages are available for the southward transport into the Pacific of any bottom water that might be formed in the Bering Sea, but this does not occur. This is because surface salinities in the Bering Sea are too low to create a dense water, even after freezing.

In the Atlantic Ocean there are deep waters of various origins other than the Weddell Sea. Warren [1971] has distinguished three main subdivisions of NADW.

1. Upper part, derived from the Mediterranean Sea.

2. Middle part, from the areas near southern Greenland.

3. Lower part, from the overflow of the Arctic bottom water across ridges separating the Greenland and Norwegian basins from the Atlantic.

The formation of NADW involves the transfer of relatively warm, salty, North Atlantic central waters into the Norwegian and Greenland seas, where heat is lost and sea ice forms. The resulting denser water mixes with an overflow of cold Arctic waters and the NADW flows southward between Iceland and Great Britain into the North Atlantic (Fig. 8–12). Another deep flow from the Norwegian Sea has been measured in the Denmark Strait between Greenland and Iceland. The NADW is marked by a potential temperature of 2°C and a salinity of about 34.91‰ and is produced at an average rate of 10×10^6 m^3/sec.

The dense water follows contours through the western Atlantic; by the time it has reached the southern part of the North Atlantic, it begins to overlie the denser AABW, thus becoming a water mass at intermediate water depths (Fig. 8–12). By this process, the NADW moves southward (Fig. 8–12), is modified on the way, and ultimately spreads into most of the world's oceans. The details of its flow path are discussed in Chapter 15. In the Southern Ocean, it forms a southward-moving layer up to 2000 m thick, overlying the AABW. Heat is introduced into abyssal depths by NADW and Mediterranean and Red sea outflows (all of which are warmer than the average abyssal temperature), geothermal heat flux, and downward heat diffusion across the main thermocline. The North Atlantic intermediate water is important because, as it enters high latitudes of the South Atlantic and western Indian Ocean, it combines with circumpolar deep water (Fig. 8–10), which is relatively warm, salty, and low in oxygen; it then upwells at the surface near the Antarctic divergence (Fig. 8–10). Here it loses heat and is transformed into Antarctic surface water by sea-air interaction and vertical diffusion. Most of this surface water then flows northward (Fig. 8–10). Part of it becomes incorporated into the sub-Antarctic surface water and the rest eventually sinks at the Antarctic convergence, contributing to the formation of Antarctic or subantarctic intermediate water (Fig. 8–10). Some of the surface water flows southward to the shores of Antarctica, where the sea-ice formation creates AABW. The AABW then sinks to the ocean floor, thus completing the circuit system existing between the surface and deep sectors of the ocean (Fig. 8–10). During the northward transport of AABW through the ocean basins, it is warmed by heat flux from the surface and the sea floor and the oxygen is used up by biological activity (Fig. 8–13). The AABW eventually combines with intermediate waters such as North Atlantic intermediate water and flows southward

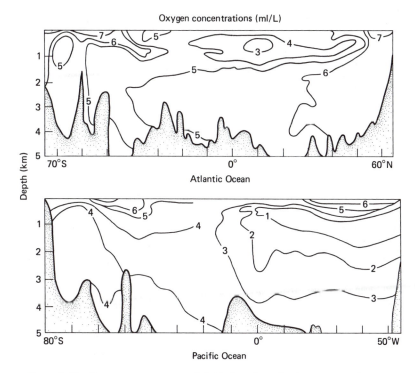

Oxygen concentrations (ml/L)

Atlantic Ocean

Pacific Ocean

Figure 8–13 Oxygen concentrations (ml/L) along latitudinal transects in the Atlantic and Pacific oceans. (After Sverdrup, Johnson, and Fleming, 1942)

to begin the cycle again. These represent the fundamental processes of oceanic vertical circulation, or overturning, which are vital for the "airing out" of the deep ocean. Geologically these processes are fundamental in controlling the geochemical processes of the ocean basin, the sedimentary patterns, and many geomorphic features on the ocean floor. More specifically, the character of bottom water plays an important role in (1) oxygenating the deep-ocean waters; (2) oxidizing organic material and sediments; (3) eroding bottom sediments and creating unconformities; (4) redistributing and redepositing sediments and creating particular bedforms; (5) dissolving calcium carbonate and silica; (6) distributing deep-sea benthonic organisms; and (7) creating rich manganese nodule fields in certain areas.

Bottom-Water Circulation

Bottom-water circulation is an integral part of global thermohaline circulation. The influence of bottom topography in ocean circulation, especially in high latitudes, is also well established. However, the primary topographic control exerted upon bottom currents is overridden in certain regions by secondary features, such as a funneling effect through narrow passages like the Romanche and Gibbs fracture zones in the Atlantic Ocean. Bottom currents often follow depth contours, and thus are often referred to as **contour currents.** Because of stratification in the water column, currents tend to flow along contours rather than up and over topographic features. They are best developed in areas of steep topography, especially at the continental margins where the bottom topography also extends through the greatest thickness of the stratified water column.

Because of the Coriolis deflection, bottom currents exhibit strong asymmetry within the basins. The scheme of deep circulation originally proposed by Stommel and Arons [1960] and Stommel [1958] postulated boundary currents only along the western sides of oceans because, for simplicity, they disregarded the ridge systems that divide oceans into multiple basins. If these ridges rise high enough above the ocean floor, they can prevent deep flow between the basins and require separate circulation systems. There are six known western boundary currents: one in each of the Southwest Pacific, the eastern Pacific and western Atlantic, and three in the Indian Ocean—the western, central, and eastern sectors. Three of these flow northward through basins separated by ridges from the western parts of the particular ocean. For instances, the West Australian basin is open to the Antarctic at depths greater than 5000 m near longitude 106°E. Thus when AABW enters this basin, it flows as a western boundary current northward along the Ninetyeast ridge system [Warren, 1971]. Current speeds can reach 20 cm/sec or more in the region of steeply sloping isotherms typical of the western boundary currents (Fig. 8–14), and these currents represent areas of particularly high geostrophic velocities (Fig. 8–15).

In the Pacific Ocean all the deep water comes from the Antarctic. As a result, deep Pacific water is much more homogeneous than that in the Atlantic (Figs. 8–11 and 8–12). Deep bottom water flows northward at depths greater than 2500 m and a

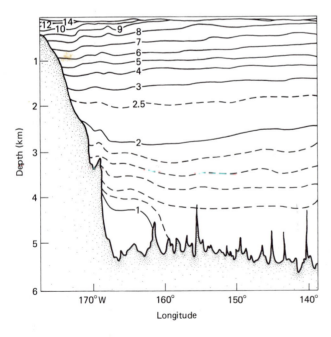

Figure 8–14 Temperature profile (°C) along 43° S near western boundary of deep Southwest Pacific. The core of water of less than 1.2°C occurs along the western edge of the basin. Flow of the bottom water is to the north. (After B. A. Warren, 1970)

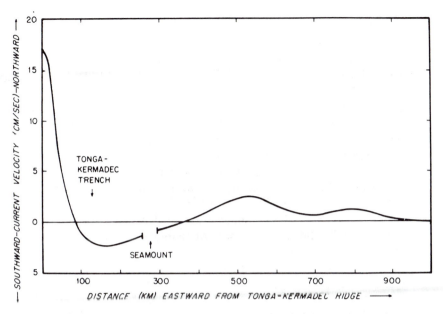

Figure 8-15 Calculated geostrophic velocity (cm/sec) at 4000 m relative to 2500 m along 28° S near the western boundary of the deep Southwest Pacific. (From B. A. Warren, *Research in the Antarctic*, ed. L. O. Quam, AAAS Pub. No. 93, p. 642, 1971, copyright © 1971 by the American Association for the Advancement of Science)

southward return flows above 2500 m. The relatively fresh bottom waters in the South Pacific exhibit temperatures between 0.5° and 2.5°C and salinities between 34.70 and 34.74‰ (Figs. 8–11 and 8–12). Because temperature and salinity represent rather conservative properties, these properties have changed only slightly by the time the waters reach intermediate depths (2000–2500 m) in the North Pacific (Figs. 8–11 and 8–12). It takes only about 1000 years for this water to travel from the South to North Pacific. Nevertheless, this represents enough time to significantly reduce oxygen levels (Fig. 8–13) and increase phosphate (nutrient) levels.

In the Pacific, the pattern is one of persistent northward migration through at least four principal basins, which are separated by relatively shallow sills that constrain the flow of water below 4000 m. The narrow Samoan Passage is perhaps the most important channel with depth sufficient to allow a significant flow of bottom water from the southern basin into the central basin. Bottom-current velocities of 5 to 16 cm/sec (mean 9.3 cm/sec) have been measured in the passage toward the north, and a sharp decrease in temperature at 4400 m records the upper boundary of AABW flowing northward through the passage [MacDonald, 1973]. AABW continues its northward flow, following the bathymetric contours of the nearly linear submarine volcanic chains. One main arm of the flow travels northward in the western boundary region and another flows into the Northeast Pacific through a deep passage in the Line Island Chain south of Hawaii.

CLASSIFICATION OF MARINE ENVIRONMENTS

The various physiographic features of the oceans are major marine environments for organisms. Assemblages of organisms (*biocoenosis*) often closely coincide with the physiographic provinces, forming distinctive marine environments, or **biotopes.** It is necessary to define the environmental terms because they are often used in marine geological studies. Figure 8-16 represents one such scheme for the classification of biotopes. Environments are subdivided into the **pelagic,** pertaining to the water mass, and **benthonic** or **benthic,** for the substrates. Pelagic environments may be subdivided into the **neritic realm,** for the water over the continental shelf, and the **oceanic realm,** which is associated with the deep, open ocean. In the benthonic environments, the **littoral zone** lies between high and low tide, the **sublittoral zone** is over the continental shelf (some workers call it the *shelf zone*), the **bathyal zone** over the continental slope, the **abyssal zone** over the abyssal plains, and the **hadal zone** over the trenches.

A more detailed classification (Fig. 8-16) describes four sets of terms for four separate categories [Edwards, 1979] (see Fig. 8-16).

1 . **Benthonic depth zones** are the vertical distributions of bottom-dwelling life.

2. **Pelagic depth zones** are the vertical distributions of floating (planktonic) and swimming (nektonic) life. In descending order, these are **epipelagic, mesopelagic, bathypelagic, abyssopelagic,** and **hadopelagic.**

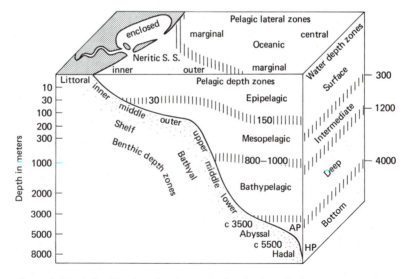

Figure 8-16 A classification of major marine benthonic and pelagic environments. Position of boundaries commonly varies between different investigations. AP = abyssopelagic; HP = hadopelagic. (After Edwards, 1979)

3. **Pelagic lateral zones** (or realms) are the horizontal distributions (generally distant from land) of pelagic organisms—the neritic and oceanic realms.

4. **Water depth zones** subdivide the water column according to physical and chemical characteristics (Fig. 8–16).

The boundaries are not fixed accurately and are debatable. Figure 8–16 shows the approximate depths usually assigned to these boundaries.

The Ocean Margins

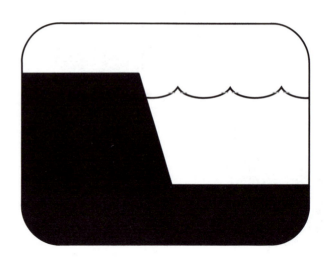

nine
Sea-Level History
and Seismic Stratigraphy

Now rolls the deep where grew the tree.

Alfred Lord Tennyson

SEA-LEVEL CHANGE AND THE COASTAL ZONE

The Coastal Zone

The **coastal zone** is the area at the edge of the oceans from the continental shelf to the coastal plains [Inman and Nordstrom, 1971; Gorsline and Swift, 1977] (Fig. 9-1). Because of its accessibility, the shoreline was the first part of the sea floor to be studied. The shoreline around the world is 440,000 km long [Inman and Nordstrom, 1971] and is marked by great environmental instability and by rapidly changing geologic character, over intervals of time ranging from minutes to several thousand years. This environmental instability results from two major processes: an intensive interaction between high-energy forces related to waves, tides, wind, and currents and constantly changing sea level over geologic time. These processes are considered to have operated throughout geologic time; thus the doctrine of uniformitarianism has formed the basis for the interpretation of ancient shallow-water geological features. However, the tectonic-geologic settings of today's coastal zones differ from some of those of the geologic past.

The continental shelf is the shallow, submerged platform bordering the continents (see Chapter 2) with a total area of 29×10^6 km^2, or about 8 percent of the area of oceans [Inman and Nordstrom, 1971]. Fundamental differences between shelves result from whether the edge of the continent is an active margin, a passive margin, or adjacent to a marginal sea (Fig. 9-1). Shelves associated with active

263

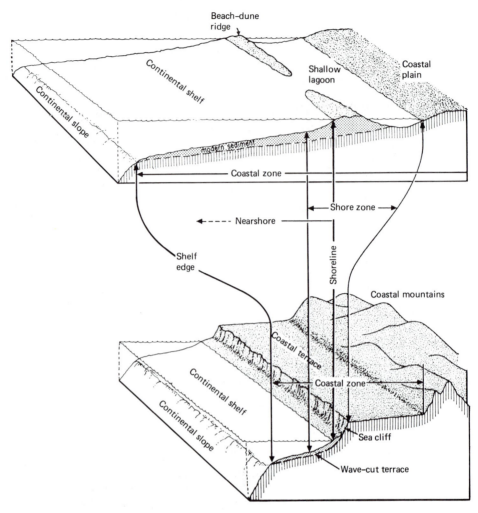

Figure 9-1 General coastal zone characteristics for trailing edge margins (top) and collision edge margins (bottom). Note the wide shelf plains coast of trailing edge margins and narrow shelf mountainous coast of collision edge margins. (From D. L. Inman and B. M. Brush, 1973)

margins are narrow and bordered by ocean trenches. Compressional tectonics give rise to rugged shorelines, often marked by irregular sea cliffs and backed by coastal mountains. The rising slopes may contain elevated sea terraces representing former wavecut platforms [Gorsline and Swift, 1977]. Typical examples are represented by the east coast of New Zealand's North Island and the west coasts of Central and South America. These have been called *collision coasts* by Inman and Nordstrom [1971] but are appropriately referred to as **leading edge coasts.** Active transform margins, such as the coast of California, exhibit similar features.

Coasts associated with passive margins have been called **trailing edge coasts** [Inman and Nordstrom, 1971] and occur on the edge of a continent that faces a spreading center, thus being located on the stable portion of a plate. The east coasts of the Americas are trailing edge coasts. The comparative tectonic stability of these coastlines produces broad continental shelves and slopes but no bordering trench. The coastal plain is wide and of low relief, allowing abundant and often extensive estuaries to intrude inland. Trailing edge coasts also often contain widespread lagoons and barrier islands (Fig. 9-1). Large deltas are also most commonly associated with trailing edge coasts or marginal seas because the configuration of mountains and plains that results from plate tectonics fundamentally affects drainage patterns. The development of large deltas at the coasts of marginal seas is also encouraged because of reduced intensity of wave action associated with smaller seas. Marginal sea shorelines often exhibit relatively wide shelves and hilly coastal areas (for example, the Gulf of Mexico and the China Sea).

Shoreline characteristics are also controlled by climate. In high latitude areas of both hemispheres, geologic processes associated with glaciers and ice caps have extensively modified the character of shorelines. The shorelines of the Northern Hemisphere show much more widespread evidence of glacial influences, including the entire Arctic Ocean margin, the North Atlantic, and the North Pacific, than do those of the Southern Hemisphere.

In subtropical and tropical areas, the warm oceanic temperatures encourage the growth of corals, calcareous algae, and other calcareous organisms in areas where terrigenous sedimentation is not extensive. This process dominates the shorelines of the west coast of Australia, the Caribbean, and the tropical Indo-Pacific.

The Importance of Sea-Level Change and its Causes

The process of overwhelming importance in dealing with the continental shelf and the geomorphology and sedimentary character of the coastal environments is the eustatic changes in world sea levels. Because most of the world's shorelines and continental shelves lie in shallow water near the peripheries of continental blocks, fluctuations in sea level cause the shorelines to migrate back and forth across the shelf. This was particularly important during the late Cenozoic ice ages, when continental ice sheets in the Northern Hemisphere repeatedly contracted and expanded, resulting in changes in sea level. Today's continental shelves are products of the Quaternary. The sea-level fluctuations created the morphology of shelves and the abrupt outer margin or shelf break. This break has a worldwide average depth of 100 to 130 m [Shepard, 1963], which most believe approximates the lowest sea level during the Quaternary [Curray, 1964; Milliman and Emery, 1968]. Estimates vary as to the maximum low of the sea level at the height of the last glacial episode, but it may have dropped to 130 m below the present level. If the remaining ice caps in Antarctica and Greenland melted, sea level would rise an additional 70 m [Russell, 1968].

There are a number of mechanisms which cause sea-level change, other than glacio-eustatic causes, which may be comparatively rare through geologic time. It may ordinarily not be possible to distinguish between glacio-eustasy and the following factors, although most are relatively slow processes as compared to glacio-eustasy.

1. **Tectono-eustatic** changes, which result from tectonic modification in the shape of ocean basins, a major control on the dimensions of the container.

2. **Sedimento-eustatic** changes, which are due to accumulation of sediment in ocean basins. These also affect the dimensions of the basins but are unidirectional positive shifts.

3. **Addition of juvenile water** derived from submarine volcanism. The rate of accumulation of juvenile water is unknown. However, if constant rates are assumed since the formation of the earth (about 4.5×10^9 years), accumulation rates are suggested of less than 1 m/m.y. If this is correct, sea level in middle Cretaceous (100 Ma) could have been up to 100 m lower than present, if other factors remained constant. Small amounts of water may also be removed due to hydrothermal alteration of oceanic crust.

4. **Glacial isostasy,** which results from ice loading and unloading. Glacial isostatic subsidence helps counteract a glacio-eustatic drop in sea level. The reverse is true when the ice sheets melt.

5. **Hydro-isostatic deformation,** which results from the loading effect of seawater over the continental shelf and nearshore regions. As sea level rises, subsidence occurs. This effect amplifies a transgression.

6. **Tectonic erosion,** which results because through the subduction process, a certain amount of deep-sea sediment is removed from the ocean basins to the earth's interior. Little is yet known about the relative effectiveness of this process compared with accretion at the active margins, which decreases the volumetric capacity of the ocean basins.

All of the above processes can effect changes in sea level through geologic time. Another process of potential geological importance is **geodetic sea-level changes,** which are undulations and of several meters amplitude due to variations in the geoid from region to region [Clark and Lingle, 1979]. The geoid is a function of the earth's gravity, which in turn is controlled by a number of variables (see Chapter 2). Mörner [1976; 1980] proposed that sea level will vary regionally due to rotational changes (in rate and tilt) of the earth and mass redistribution. This may occur rapidly as large ice sheets are built and destroyed at high latitudes. Mörner's model also provides a mechanism to explain different sea-level curves derived for different regions. Mörner [1980] argues that inherent in the theory of geodetic sea-level history is the corollary that all eustatic curves are valid only regionally, not globally as often assumed. Nevertheless, the history of geodetic change is still unknown, including the magnitude of the differences from region to region through geologic time. It is possible that the variations are small and of little consequence in our interpretations of the long-term global sea-level history.

It is generally believed that the most important *long-term* mechanism affecting sea level is tectono-eustatism. Plate tectonics has revolutionized ideas about the effects on sea level of the changing shape and volume of the ocean basins through geologic time. Hallam [1963], Russell [1968], Menard [1969], Valentine and Moores [1970], and Hays and Pitman [1973] have all suggested that changes in volume of the mid-ocean ridges have caused major sea-level changes in the past. Significant volume changes in the mid-ocean ridges have affected sea levels throughout the late Phanerozoic and even earlier.

Repeated sea-level advances and retreats, or **transgressions** and **regressions**, have been instrumental in shaping the continental shelf, especially during low sea-level stands, when the shelf was exposed to subaerial processes. These cyclical patterns have been very effective in producing the flat surfaces of the shelves and of the adjacent coastal plains, which are effectively emergent parts of the shelves. During subaerial exposure of the continental shelves, coinciding with late Cenozoic glacial maxima, alluvial deposition and erosion became dominant processes over the surface of the shelf. Major rivers cut their channels across the shelf and, in some cases, developed deltaic complexes at the outer shelf margin. During these same periods, drastic changes occurred in the depositional regime of the continental slopes and even in adjacent deep-sea basins [Curray, 1964]. Increased sediment deposition at the shelf edge or upper slope led to rapid transfer of terrigenous sediments to the slopes and some deep basins and trenches. In some areas, permanent progradation, or seaward building, of the shelf resulted from increased rates of sediment deposition at the shelf edge. Deposition of great thicknesses of sediments on the steeper slopes of the upper continental slope led to instability, slumping, sliding, and the generation of turbidity currents. These, in turn, increased the sediment supply to the deeper basins, which increased the rate of formation of abyssal plains [Curray, 1965].

As the continental glaciers receded, sea level rose, producing a rapid transgression across the shelves [Curray, 1965]. These transgressions are reflected in the shoreward migration of the shoreline environments. These transgressions have drowned all coasts except those with rapid glacial-isostatic or other tectonic uplift. However, the transgressions have been so rapid that sedimentary deposition has not kept pace with the shoreline migration, so **relict** subaerial and very shallow-water sediments remain exposed at the surface of the outer parts of the wider shelves [Curray, 1965]. The nearshore sediment characteristics remain at odds with the present environmental conditions; shoreline configuration has not yet adjusted to the prevailing wave climate [Komar, 1976].

Most rivers enter the sea either through estuaries or across wide deltas. Today, estuaries remain unfilled and act as a major sediment catch, starving the continental shelves of sediment. Headlands are being cut back and coasts straightened. Many coastal landforms are relict from the last interglacial period and have only recently been reoccupied by the sea [Bloom, 1978]. Such land forms clearly show that the present-day sea level is not typical of much of the late Cenozoic and that the sedimentary regimes of continental shelf areas and nearshore environments are out of phase with the present sea level.

SEA-LEVEL HISTORY

Quaternary Sea-Level History

Sea level is the most important reference surface on earth. Some have even implied, albeit naively, that it has been a relatively fixed reference surface, especially during pre-Quaternary times when the earth was assumed to be in a ice-free state. In contrast, others have posited constantly changing sea levels. Wheeler is known to have asked, "What is so sacrosanct about sea level?" [Fairbridge, 1960] and this has gone on for a long time. In the middle 1600s, Pepys, for instance, believed that England and Europe had once been connected [Curray, 1964]. Historically recorded sea-level changes are now well established. Gutenberg [1941] and later workers found from tide-gauge records that the world mean sea level was rising at about 1.1 mm/yr during the first half of the twentieth century [Fairbridge, 1966]. In Newport, Rhode Island, northeastern North America, sea level rose about 3 mm/yr between 1930 and 1970, although perhaps half of this is due to tectonic subsidence. Certain periods of time are marked by higher rates of sea-level rise than the average. For instance, records show that sea level rose by 70 mm between 1875 and 1877 [Fairbridge, 1966]. It is assumed that most of this change was glacio-eustatic.

It is now accepted that sea level has risen an incredible 100 to 130 m between 18,000 and 6000 years ago (the **Holocene transgression**), after which time sea level has remained close to that of the present day [Curray, 1964]. During the most rapid phase of deglaciation, from about 10,000 to 7000 years ago, sea level probably rose at a rate of 10 mm/yr. A rise of world sea level of 1 mm requires the melting of about 0.36×10^{12} m^3 of ice, or a layer 36 cm thick from ice surfaces covering 1×10^6 km^2. This rise in sea level is certainly one of the most important geologic events of the recent past and can be considered to be as important as the development of the Quaternary ice sheets themselves [Curray, 1964]. The full significance of this sea-level rise was slow in being recognized, mainly because the evidence was hidden beneath the sea. Glacio-eustatic variation is an efficient barometer of climatic change. According to Fairbridge [1966], the absolute range of sea level due to the ice effect is about 200 m, corresponding to an ice volume of 72×10^6 km^3. Present-day ice volume totals 30×10^6 km^3 and was about 40 to 45×10^6 km^3 greater during glacial episodes. The polar-ice volume today is thus about half that of the Quaternary glacial maximas. Almost all of this ice is stored on Antarctica.

Latest Quaternary sea-level history

The presence on the continental shelf of drowned beaches, mastodon teeth [Whitmore and others, 1967], submerged peat beds, and lagoonal mollusks [Emery and Garrison, 1967] all demonstrate that during the last glacial episode, sea level stood 100 m or more below that of the present day. The history of the most recent sea-level oscillation, during the last 30,000 to 40,000 years, has been studied using radiocarbon dates of materials known to have close relations with sea level, including salt-marsh and freshwater peat deposits, very shallow-water marine or brackish molluska, coralline algae, and beach rock. The presence of these materials in situ demonstrates that the sea was once at that level. It would seem on first analysis that such an approach would be very straightforward, as long as the inferred paleodepth of the material being dated is well established in relation to its contemporary sea level. However, many complications exist; as a result, published curves of late Quaternary sea level vary significantly. For instance, the most commonly used sea-level indicators are shallow-water mollusca, particularly the oyster, *Crassostrea virginica*. This species is restricted to very shallow, brackish waters with favorable salinity limits of 30‰ to 5‰ [MacIntyre and others, 1978]. However, it has now been established [MacIntyre and others, 1978] that significant postdepositional landward transport of these oysters has occurred in some areas, calling into question some previous interpretations of sea-level history. Sea-level curves must be established using materials which have not been transported, like articulated fossil oysters—which are unlikely to have moved far, if at all, from the environment in which they lived. Plots of radiocarbon dates show considerable scatter for other reasons, including dating errors, contamination of the carbon in the dated materials, errors in the assumed depth ranges of organisms, and depth of accumulation of peat and local crustal instability [Curray, 1964].

It is apparent that individual curves reflect local rather than global sea-level curves. There can be no single universally applicable curve because of the effects of tectonism in different regions and in the geodetic surface. No areas of the continental margins are completely stable, and the major problem posed in establishing eustatic sea-level history is to differentiate vertical changes due to sea level from those due to tectonism. Early studies on sea-level history were confined to relatively stable shelves and, until recently, shelves associated with active margins were generally avoided. Perhaps the most reliable approach to sea-level studies would be to study oceanic islands—considered by Bloom [1967] to represent "Pleistocene dip-sticks"—but, as we have seen (Chapter 5), even the intraplate crustal areas exhibit persistent subsidence with increasing age. High-latitude areas undergoing isostatic rebound following relatively recent removal of ice sheets are also difficult to interpret. Vertical changes in the lithosphere are even caused by the isostatic loading or unloading effects of water over the continental shelf [Bloom, 1967; Clark and Lingle, 1979]. As more data become available on the chronology of late Quaternary sea-level change, controversies about the timing of events and their relative levels will continue. Nevertheless, the work until now has exhibited a number of general trends which seem to be well established.

A chronology of sea-level changes over the last 35,000 years is shown in Fig. 9-2. This includes two frequently used curves proposed by Curray [1965], mainly from Gulf of Mexico data, and by Milliman and Emery [1968], based on data from continental shelves of the Atlantic and Gulf of Mexico. A modified version of the Milliman and Emery curve was developed by Dillon and Oldale [1978]. Although the various studies differ in detail, several trends are indicated (Fig. 9-2). The sea may have been close to its present level between about 35,000 and 25,000 years ago. It then began to recede as the last full glacial episode began. By about 20,000 to 15,000 years ago, sea level had dropped by about 120 to 130 m, though the extent of this drop has been debated. The chronology of sea-level change corresponds to the glacial chronology established for continental areas. Beginning at

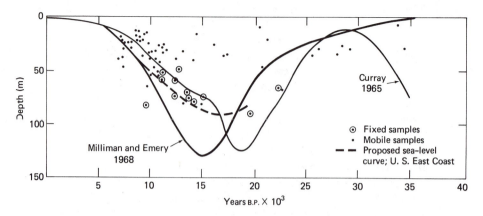

Figure 9-2 Age-depth relations of radiocarbon dated samples from east coast United States and sea-level curves for late Quaternary of Curray [1965], Milliman and Emery [1968], as well as Dillon and Oldale [1978] shown by heavy dashed line. Samples are classified into those considered to have undergone postdepositional transport and those that have remained at or close to their site of initial deposition. (After W. P. Dillon and R. N. Oldale, 1978)

about 15,000 to 17,000 years ago, the Holocene transgression began (Fig. 9–3). This is represented by an extremely rapid rise (about 8 mm/yr) in sea level of over 100 m, which lasted until about 7000 years ago, when sea level was about 10 m below its present level (Fig. 9–3). Since then, sea level has continued to rise, but at a much slower rate (about 1.4 mm/yr) and by 5000 years ago, sea level was within 5 m of its present level [Emery, 1969] (Fig. 9–3). By 2000 to 4000 years ago, sea level had reached its present level (Fig. 9–3), although historical records suggest that it has continued to rise slowly during the last 100 years. Although some workers have suggested that sea level was higher than at present between 2000 and 6000 years B.P. (Fig. 9–3), it is now generally believed that today's sea level is at its highest stand. The shape of curves for oxygen isotopic (δO^{18}) variation in deep-sea cores supports the chronology based on radiocarbon dating of nearshore materials.

Although the chronology of the lastest Quaternary sea-level changes is now fairly well established, much uncertainty still exists about the extent of the drop in sea level. Although much of the earlier work indicated a drop of as much as 130 m 18,000 years ago (Fig. 9–2) [Curray, 1965; Milliman and Emery, 1968], more recent work suggests that sea level may not have fallen more than 100 m during the late Quaternary. More detailed work by Oldale and O'Hara [1980] (Fig. 9–2) and Blackwelder and others [1979] have caused this change of opinion. Oldale and O'Hara [1980] estimated the extent of glacio-eustatic change after compensating for effects of isostatic rebound. In some areas, uplift of land is faster than the eustatic sea-level rise, thus producing a relative drop in sea level. Ratios of uplift are commonly 1 m per century and may be as much as 4 m per century on the west coast of Baffin Island. The effect of crustal rebound has been to create dipping terraces representing reference levels that were formerly horizontal. These provide a means of correction for the tectonic effects upon the samples previously used to construct sea-level rise curves. By making the correction, a sea-level curve that remains above

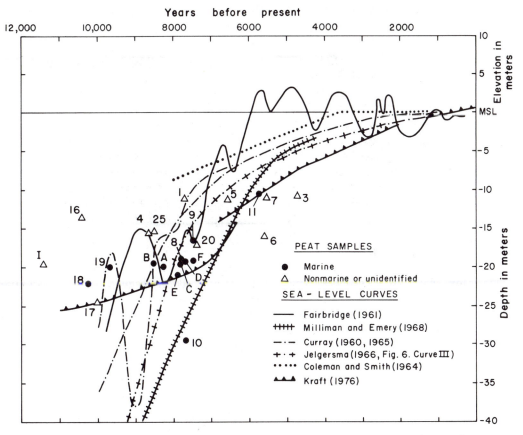

Figure 9–3 Age-depth relations of Holocene peat samples from east coast of the United States and family of sea-level curves for the last 10,000 years (modified by Field and others [1979] after Curray [1969]. (From M. E. Field and others, *Geol. Soc. Amer. Bull.*, Part I, vol. 90, p. 626, 1979, courtesy The Geological Society of America)

−100 m is produced (Fig. 9-2). Likewise, Blackwelder and others [1979] have reestimated sea-level change based on new dates from material considered to be in situ. They indicate an even lower total range of late Quaternary sea-level change, with a maximum lowering of only 60 m about 17,000 years B.P. If correct, these data would indicate that substantially less ice was stored on the continents during the latest Wisconsin glaciation than implied by estimates based upon earlier sea-level curves.

Earlier Quaternary sea-level history

Because radiocarbon dating is limited to materials younger than 40,000 years, an understanding of older sea-level history is obtained using ^{230}Th dating of fossil corals in uplifted terraces and oxygen isotopic studies. These studies indicate that a sequence of glacio-eustatic sea-level oscillations must have extended throughout the

Quaternary and at least into the late Pliocene about 3 Ma. These fluctuations probably did not remain constant because the average expansion of the Northern Hemisphere ice caps changed throughout this period. The oxygen isotopic evidence suggests that the largest glacio-eustatic oscillations occurred during the late Quaternary [Shackleton and Opdyke, 1977].

Quaternary interglacial high stands of sea level are preserved as marine terraces and beach ridges high above the present level in areas that are experiencing tectonic uplift, such as the Huon Peninsula of northern New Guinea. This peninsula is located near the boundary between the Pacific and Indian plates, and uplift along a major fault has created a spectacular sequence of coastal terraces consisting mainly of coral reefs and deltas [Chappell, 1974]. These have been dated using an assumed uniform rate of uplift between 120,000 and 400,000 years B.P. A detailed curve has been compiled by Bloom and others [1974] for the last 140,000 years based on the dated terraces of New Guinea and Barbados (Fig. 9-4). This shows sea level 6 m above the present level 125,000 years ago. Sea level fell in a series of pulses following the high stand 125,000 years B.P. (Fig. 9-4). These pulses were separated by brief intervals of high sea level, which were still well below present sea level. The low stands of sea level are associated with ice advances called **stadials**, while the higher stands are associated with retreats of the ice sheets called **interstadials**. The entire glacial interval between about 80,000 and 10,000 years B.P. is known as the *Wisconsin glaciation* in North America and the *Würm glaciation* in Europe. The maximum drop in sea level during the Wisconsin glaciation seems to have occurred

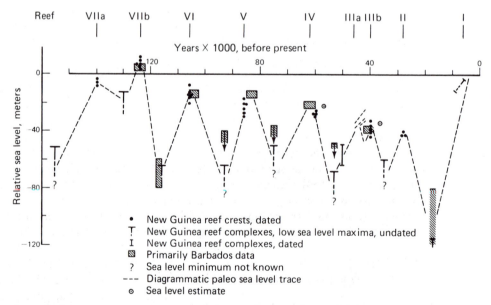

Figure 9-4 Sea-level fluctuations during the last 140,000 years determined from ages and altitudes of coral-reef terraces on tectonic coasts. The amplitudes of the oscillations were calculated using the assumption of constant rate of tectonic uplift. (From A. L. Bloom and others, *Quaternary Research*, vol. 4, no. 2, p. 203, 1974)

during its last stage about 18,000 years B.P. (Fig. 9-4). Three distinct high stands occurred at about 122,000, 103,000, and 82,000 years B.P. The data from New Guinea and Barbados indicate that most of the interglacial high stands of sea level were lower than the present level.

Changes in sea level determined from the study of uplifted wave-cut terraces are now subject to independent corroboration of oxygen isotopic curves, which directly reflect changing polar-ice volumes [Bowen, 1978]. Sea-level history inferred from oxygen isotopic change in deep-sea cores closely matches curves based on the dated uplifted terraces. Oxygen isotopic measurements indicate a slightly higher sea level than the present during the last interglacial episode. Oxygen isotopic curves are important for inferring sea-level history well back into the late Cenozoic [Shackleton and Opdyke, 1973; 1976]—much older than by using dated marine terraces. Little can be deduced about low stands of sea level using marine terraces because the present ocean covers nickpoints and terraces that were created during these low stands. These features appear near the outer shelf and upper continental slope and as submerged coral reefs. Persistence of such features at similar depths around the world indicate relationships with sea level rather than with tectonism. Some of these terraces have been found at surprisingly great depths. Several workers have discovered narrow terraces and coral reefs around Australia at depths between 175 and 238 m [van Andel and Veevers, 1967; Dill, 1969; Veeh and Veevers, 1970; Pratt and Dill, 1974]. Ages determined for these terraces had previously been considered to range from 17,000 to about 14,000 years B.P. and drops in sea level greater than 200 m, if correct, would imply much larger volumes of polar-ice buildup during a particular interval of the Quaternary than generally assumed. Such events are particularly significant because almost all continental shelves and the upper part of most continental slopes would have become exposed. Interregional differences in sea-level histories are apparent, and much work is still required to explain them.

Sedimentary Cycles

A large majority of the earth's continental shelves mark the upper surface of thick prisms of sedimentary strata deposited in relatively shallow water and under the constant influence of sea-level change. We have seen that sea level has oscillated in response to continental ice volume, ocean basin capacity due to tectonism, and sediment infilling. These sea-level oscillations cause a migration of the littoral (inshore) and other marine facies, which affect the three-dimensional structure of the sedimentary sequences being deposited. The succession of transgressions and regressions has created the broad, relatively flat coastal plain or continental-shelf complex. The subsurface geologic records of the continental shelves are the optimum place to study the history of shoreline migration in the past. As long as any large, local, tectonic influences can be differentiated, the subsurface record of these eustatic changes gives a solid basis for establishing a worldwide natural chronostratigraphic framework. The sediment record is now known to record advances and retreats of the sea over widely distributed geographic areas. Because of the synchroneity of these events, they have usually been attributed to eustatic sea-level fluctuations.

Oscillations in sea level create sedimentary cycles. Vail and others [1977] have called these cycles **depositional sequences,** which are stratigraphic units composed of genetically related strata. The upper and lower boundaries of depositional sequences are unconformities or correlative conformities. Because the distribution and type of facies of many shallow-water sedimentary sections are controlled by global changes of sea level, depositional sequences provide ideal bases for establishing a comprehensive stratigraphic framework on regional or global scales. As early as 1885, Suess [translated by Sollas, 1906] recognized three orders of sedimentary cycles that he attributed to sea-level change. Later, Stille [1924] identified unconformities that he considered to be synchronous throughout the earth and attempted to explain these by the hypothesis of **periodic diastrophism,** or a change in the tempo of tectonism and mountain building. The unconformities identified by Stille correspond to the boundaries of the sedimentary cycles recognized by Suess. Since these early studies, there have been many proponents of global sedimentary cycles and recently, before the work of Vail and his colleagues, these included Fairbridge [1961] and Vella [1965]. Both concluded that eustatic sea-level changes have taken place continuously and that they were caused partly by climatic changes and partly by tectonism and sedimentary processes, as Suess envisioned. As summarized by Vella [1965], sedimentary cycles are basic stratigraphic units, closely related to lithostratigraphic, biostratigraphic, and chronostratigraphic units. They can be traced over large areas, and each cycle corresponds to a sea-level rise and fall. A sedimentary cycle cuts across lateral facies changes and, like a stage, consists of numerous interdigitating lithofacies. Unlike a stage, though, it is defined by lithologic as well as paleontologic criteria. By using sedimentary cycles, biostratigraphic correlation can be reinforced by matching successions of sedimentary cycles using the unconformities or correlative conformities as the boundaries of each sedimentary cycle. Vella proposed that such units be called *cyclic-time-stratigraphic units* because they appear to be synchronous over much of the world. Mitchum and others [1977] called these units **sechrons** (from *sequence* and *chron,* or time) and defined them as the total interval of geologic time during which a sequence or sedimentary cycle is deposited. An advantage of sechrons is that many are linked with the type of chronostratigraphic units defined in Europe. Many of the geologic systems and stages were originally named for rocks lying between certain major unconformities, because these appeared to mark natural breaks in lithology and biostratigraphy. As a result, many of the chronostratigraphic units of Europe originated in this way. Thus, the concept of sedimentary cycles was linked with a classic concept of historical geology in which extensive periodic orogenies have furnished natural boundaries in earth history.

Suess, Haug, Stille, and Grabau paid special attention to the chronological and regional distribution of transgressions, and each concluded that a large number of major transgressions and regressions have occurred. During the late Cretaceous, epicontinental seas covered Europe and other areas, followed by a global regression toward the end of the Mesozoic and through much of the Cenozoic. The marine units represent high sea-level stands within natural eustatic cycles, and their boundaries are represented by unconformities and separate the traditional units on which European stratigraphy is based.

It should be pointed out that the view of synchronous global sea-level changes during the Mesozoic and Tertiary is still not universally accepted. The intensity of the debate can be measured by the following statement of Dott and Batten [1971, pages 69–70]:

"Until the middle 20th century there still lingered a faith among geologists that the stratigraphic record was naturally divided by the worldwide rhythms of mountain building reflected as very long-period, worldwide transgressive-regressive cycles, assumed to conform neatly with the system boundaries. This concept reflects a century-old influence of Hutton's and Lyell's cyclic view of the earth, and it provided a convenient rationale for a universal time scale. Modern stratigraphic studies have shown this scheme to be a fraud in its simple form; mountain building and unconformities have not been so perfectly uniform either in age or magnitude over large regions. Reaction set in and any widespread synchroneity was, for a time, denied categorically, but recently there has appeared evidence of certain long-term events which may, after all, prove to be more or less universal."

One problem adding to the debate is the identification of a mechanism to create large transgressions and regressions during times when there can be no change of sea level because of the melting and growth of ice sheets. Hays and Pitman [1973] believe that the worldwide transgressions and regressions which occurred during the middle and late Cretaceous may have been caused by contemporaneous oscillations in the spreading rates of mid-ocean ridges. Increase in spreading rate causes the ridges to expand, thus reducing the volumetric capacity of the oceans.

SEISMIC STRATIGRAPHY

A new and very important approach to global stratigraphic correlation and chronology is **seismic stratigraphy,** which is the stratigraphic interpretation of seismic reflection data. Primary seismic waves are reflected by physical surfaces in rock sequences, especially those bedding planes and unconformities that separate rocks by seismic velocity and density contrasts. Seismic reflections parallel these stratal surfaces (Fig. 9-5) and provide an effective means of mapping rock sequences over vast areas usually accessible only by expensive and time-consuming drilling. This method has been used for several years by petroleum-industry geophysicists in stratigraphic and structural studies of continental margins. The importance of the approach was highlighted in 1977 by the publication by the American Association of Petroleum Geologists of a comprehensive set of papers by P. R. Vail, R. M. Mitchum, and others on the techniques and implications of seismic stratigraphy [Payton, 1977].

In summary, several stratigraphic interpretations can be made using the geometry of seismic reflection patterns; these include geologic correlations, definition of genetic depositional units, thickness and depositional environment of genetic units, and relief and topography of unconformities. In addition, postdeposi-

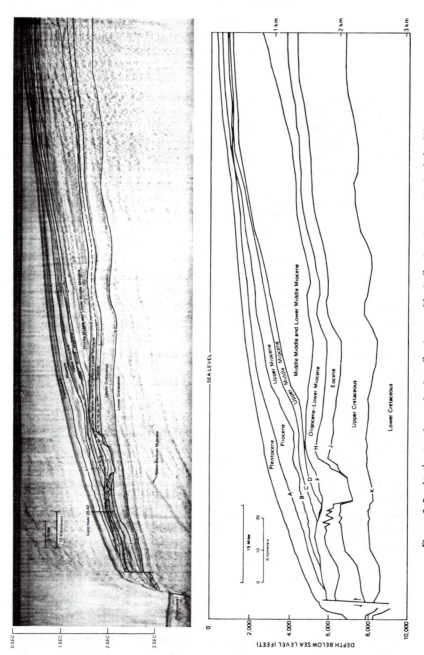

Figure 9-5 At the top is a seismic reflection profile (reflection-time section) of the West Florida continental shelf. Horizon F marks a dramatic change in sedimentary environment providing a conspicuous offlap of younger beds. Note fossil channel to right of core-hole 29-42 which is an ancestral part of the DeSoto Canyon system. At bottom, the same profile is converted to a depth scale in feet. Vertical exaggeration is about 20:1. (From R. M. Mitchum, Jr., *Framework, Facies and Oil-trapping Characteristics of the Upper Continental Margin*, AAPG Studies in Geology no. 7, p. 199, 1976, reproduced with permission of the American Association of Petroleum Geologists)

tional structural deformation can be determined. Four steps are necessary to make these interpretations.

1. Seismic sequence analysis.
2. Seismic facies analysis.
3. Analysis of relative sea-level changes.
4. Global correlations of relative sea-level changes.

The first step is the identification of individual sedimentary cycles or depositional sequences (Figs. 9–5 and 9–6(A)). To define and correlate a depositional sequence correctly, the sequence boundaries must be defined and traced accurately. Usually the boundaries are defined at unconformities and traced to their correlative conformities. A **conformity** is a surface that separates younger strata from older rocks, along which there is no physical evidence of erosion or nondeposition and no significant hiatus is indicated.

Analysis of seismic facies is based on interpretations of reflection geometry, continuity, and seismic velocities within a framework of depositional units. This provides the required information on sedimentary processes and paleoenvironments (Fig. 9–6(A)). Next it is necessary to analyze the relative changes of sea level through time using the geometry of marine sequences. A relative rise in sea level is indicated by coastal **onlap**, (Fig. 9–7(A)), which is the landward onlap of littoral or nonmarine coastal deposits, or both. The vertical component, coastal aggradation, can be used to measure a rise in sea level, but it should be adjusted for any thickening due to differential basinward subsidence. A **stillstand** of sea level is indicated by coastal toplap (Fig. 9–7(B)). A drop in sea level is indicated by a downward shift in coastal onlap from the highest position in a sequence to the lowest position in the overlying sequence (Fig. 9–7(C)). A typical eustatic cycle consists of a gradual rise, a period of stillstand, and a rapid fall of sea level. In Fig. 9–6(A), eustatic changes are determined by measuring the aggradation, the vertical component of coastal onlap. These are plotted against age (Fig. 9–6(B)) using available biostratigraphic data. This shows (Fig. 9–6(C)) a series of five asymmetric cycles, each exhibiting a slow eustatic rise to a stillstand, followed by a rapid fall of sea level of varying magnitude. Vail and others [1977] found strong similarities between the regional cycles of various continental margins. The relative magnitudes of the changes seem to be similar.

From the regional cycles, Vail and others [1977] also determined a global sea-level history, shown in Fig. 9–8, for the time since the late Triassic (200 Ma). Numerous unconformities seen on seismic data can be correlated globally and appear to be related to eustatic changes of sea level (Fig. 9–8). Thirteen appear to be related to major falls of sea level and are, therefore, considered major interregional unconformities. A marked asymmetry occurs in the global cycles, with a gradual rise and abrupt fall in sea level (Fig. 9–8). Although the ages and durations of the cycles are considered by Vail and others [1977] to be accurate, the amplitudes of the eustatic changes are only approximations. Sea level reached a **global high stand** in

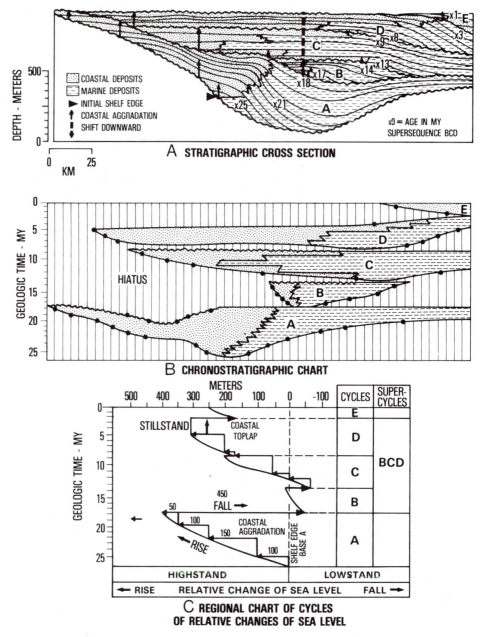

Figure 9-6 Three steps necessary in the construction of charts showing relative sea-level history. A represents stratigraphic cross section; B is the section plotted within a time framework; and C represents the conversion of the data represented in B to relative sea-level changes. (From P. R. Vail, R. M. Mitchum, Jr., and S. Thompson III, *Seismic Stratigraphy—Applications to Hydrocarbon Exploration*, Memoir 26, p. 85, 1977, reproduced with permission of the American Association of Petroleum Geologists)

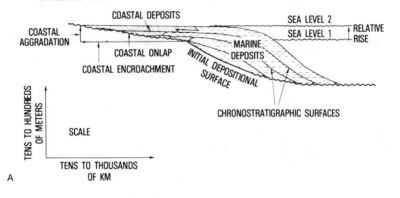

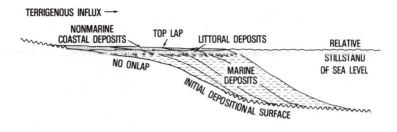

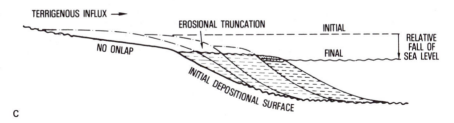

Figure 9-7 Modes of coastal deposition resulting from relative rise in sea level (A); Relative stillstand of sea level (B); Relative fall of sea level (C). A—coastal onlap indicates a relative rise of sea level. Relative rise of base level allows coastal deposits of a maritime sequence to aggrade and onlap initial depositional surface. B—coastal toplap indicates relative stillstand of sea level. With no relative rise of base level, nonmarine coastal and/or littoral deposits cannot aggrade, so no onlap is produced; instead, bypassing produces toplap. C—downward shift of coastal onlap indicates relative fall of sea level. With relative fall of base level, erosion is likely: Deposition is resumed with coastal onlap during subsequent rise. (After P. R. Vail, R. M. Mitchum, Jr., and S. Thompson III, *Seismic Stratigraphy–Applications to Hydrocarbon Exploration*, Memoir 26, p. 85, 1977, reproduced with permission of the American Association of Petroleum Geologists)

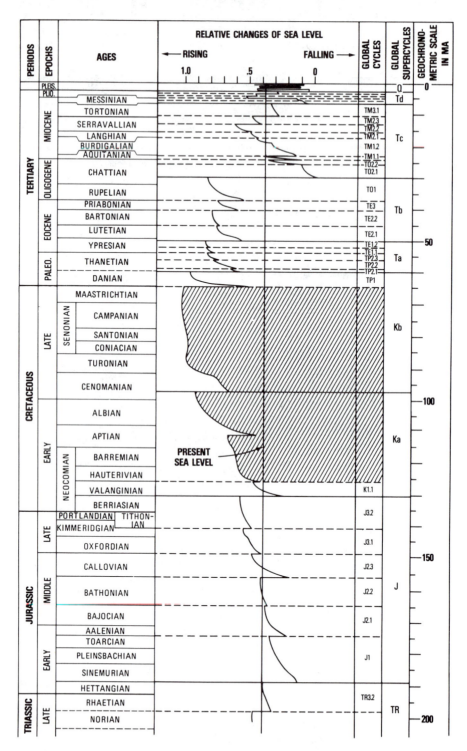

Figure 9–8 Global cycles of relative sea level during Jurassic-Tertiary time. Cretaceous cycles (hachured area) not yet published. (From P. R. Vail, R. M. Mitchum, Jr., and S. Thompson III, *Seismic Stratigraphy–Applications to Hydrocarbon Exploration*, Memoir 26, pp. 66, 70, 72, 78, 1977, reproduced with permission of the American Association of Petroleum Geologists)

the late Cretaceous, about 350 m above present sea level. In the Tertiary (Fig. 9–8), Vail and others [1977] show an eustatic curve with an overall trend of falling sea level. High sea levels mark the early Tertiary and low sea levels mark the late Tertiary. The largest drop, or low stand, was during the middle Oligocene (about 29 Ma) and was probably 250 m over an interval of only 1 to 2 m.y. Major falls occur about 60, 50, 40, 30, 22, 11, 7, and 4 Ma (Fig. 9–8). The amount of total sea-level change from the high stand during the Cretaceous to the lower present-day level is controversial, with estimates ranging from 300 to 100 m. Global highstands are marked by widespread shallow marine to nonmarine deposits over the continental shelves. If the supply of terrigenous sediment were abundant, deltaic lobes tended to extend over the edge of the continental shelf into deep water. Global low stands are marked by erosion and nondeposition on the shelves and deposition of deep marine fans in basins adjacent to the continental margins. Interregional unconformities are related to low stands when subaerial erosion and nondeposition occurred on the continental shelves and upper part of the continental slope.

The structure of Atlantic-type margins is that of a seaward-thickening mass of stratified sediment overlying a deeply subsided, basement platform. The sedimentary strata consist of seaward-thickening wedges separated, at least in the shallower sections, by remarkably undisturbed planar horizons [Pitman, 1979]. The thickness of the sediment that has accumulated at the shelf edge off the east coast of North America is often greater than 10 km. A program of shallow drilling [Hathaway and others, 1979] has shown that the entire sequence from the Mesozoic to the present was deposited at or within a few hundred meters of sea level [Pitman, 1978]. This indicates slow, persistent subsidence of the shelf matched by sedimentation at a rate sufficiently rapid to maintain paleodepths close to sea level. Distinct unconformities subdivide the sequence.

It is believed that transgressive or regressive events that have occurred simultaneously on geographically separated continental shelves have been caused by worldwide sea-level rise or fall (Fig. 9–9). Pitman [1978], however, has developed a model which suggests that this is not necessarily the case. He suggests instead that these events may be caused by changes in the *rates* of sea-level rise or fall. The shoreline tends to stabilize at that point on a margin where the rate of rise (or fall) of sea level is equal to the difference between the rate of subsidence of the shelf and the rate of sediment infill. Thus as sea level rises more rapidly or falls more slowly, a transgression will occur. Alternatively, if sea level is made to rise less rapidly or fall more rapidly, a regression will occur. Pitman's model follows the work of Sloss [1963], who indicated that transgressions and regressions are in part controlled by

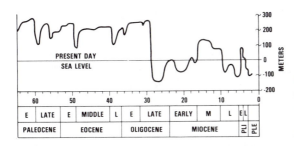

Figure 9-9 Tertiary eustatic changes of sea level. Absolute scale of changes in meters are tentative. (From P. R. Vail and J. Hardenbol, *Oceanus Int. Mag. of Mar. Sci.*, Vol. 22, no. 3, p. 71, 1979)

the relative magnitude of sea-level change and the rate of subsidence. A critical component of this model is the hypothesis that rates of sea-level change are always less than 1 cm per 1000 years (except with glacio-eustatic fluctuations) and are less than the subsidence rate at the edge of passive margins, which are often more than 2 cm per 1000 years. Subsidence of the shelf is assumed to be hingelike, with greatest subsidence at the shelf edge. The position of the shoreline is also affected by the sedimentation rate. The total rate of subsidence is generally slow and sediment influx is sufficient to keep pace. Thus if sea level is falling, the shoreline will move to that point on the shelf where the rate of sea-level fall is equal to the rate of subsidence minus the sedimentation rate. A major implication of Pitman's model is that even if sea level falls persistently for millions of years, it is not rapid enough to keep up with the subsidence of the shelf edge. In this model, the shoreline will always remain on the shelf rather than migrating onto the continental slope, except during glacial times. During glacial times of rapid eustatic fall, the shoreline is known to migrate to the upper continental slope.

We have already seen that present-day shelves are generally not in equilibrium. The rise in sea level during the past 10,000 years has been so rapid that sedimentation has not been able to keep up. We have also seen that the present day shelf break is at 135 m, which is considered by Pitman [1978] to be much deeper than the estimated average of 40 m for much of the Tertiary. The earlier Tertiary would have experienced less rapid sea-level rises and falls than those of the glaciated late Tertiary and Quaternary. Under these conditions, most of the sediment would have been deposited on the shelf. During times of rapid, glacially controlled sea-level falls, when much sediment bypassed the shelves, the rate of subsidence of shelves would have been reduced through reduced sediment load. Equilibrium conditions would not return until the shoreline returned to the shelf, allowing recommencement of terrigenous sedimentation.

Causes of Mesozoic-Tertiary Sea-Level Changes

We have seen that Quaternary sea-level changes were glacially controlled and modified locally only by tectonism. Of the various possible mechanisms that can cause long-term, sea-level changes, only tectonic mechanisms appear to be of sufficient duration and magnitude to account for the large changes throughout the Mesozoic and Cenozoic. It is, therefore, generally believed that global tectonic changes gave rise to major transgressions or regressions. Tectonic changes are unlikely to produce frequent, rapid sea-level changes like those that are glacially driven, and the rapid sea-level falls at times in the Tertiary are difficult to explain. In general, it is not possible to distinguish between the effects of glacial or tectonic causes, but tectonic effects are assumed to be dominant.

Pitman [1979] quantitatively considered the effect of abrupt changes in rates of sea-floor spreading on seismic-profile geometry on the continental margins. Except for glacial effects, volumetric change in the mid-ocean ridges related to changes in the rate of sea-floor spreading is potentially the fastest and volumetrically the most significant way to change sea level. Much more work is necessary to explain the causes of each of the smaller-scale eustatic changes, but, in general, high rates of sea-floor spreading should be associated with expansions in volume of the mid-ocean ridges and relatively shallower ocean floors. This should cause eustatic high stands. Eustatic low stands would result from slower rates of spreading. A hypothetical sea-level curve for the late Cretaceous to Miocene was calculated by Pitman [1979] from inferred changes in volume of the mid-oceanic ridge system. This model quantitatively relates the position of the shoreline to the rates of subsidence, rates of sea-level change, and rates of sedimentation.

The first component of the model has been well established in that age-depth relationships for all mid-oceanic ridges are the same regardless of the spreading history. The age-depth relationship follows a time-dependent exponential cooling curve (see Chapter 5). Thus the sea floor on a fast-spreading ridge (such as the East Pacific rise) is at the same depth as that of equivalent age of a slower-spreading ridge (such as the mid-Atlantic ridge). The second major component of this model is that the cross-sectional area or volume of the ridges will vary according to spreading rates. In Fig. 9–10 it can be observed that a ridge which has spread at 6

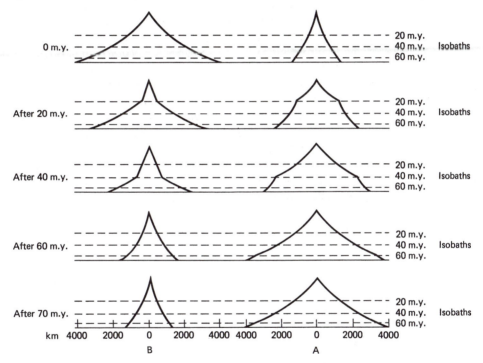

Figure 9–10   Differences that develop in cross-sectional areas of ridges spreading at different rates over a 70 m.y. period. (A) Top shows profile of ridge that has been spreading at 2 cm/yr for 70 m.y. At time 0 m.y., spreading rate is increased to 6 cm/yr. Sequential stages in consequent expansion of ridge profile are shown: first at 20 m.y. after spreading-rate change, at 40 m.y., then at 60 m.y., and finally at 70 m.y. after spreading-rate change. At 70 m.y., ridge will be at a new steady-state profile; cross-sectional area at this time will be three times what it was at 0 m.y. (B) Top is profile of ridge that has spread at 6 cm/yr for 70 m.y. At 0 m.y., spreading rate is reduced to 2 cm/yr. Sequential stages in subsequent contractions of ridge are shown. At 70 m.y., after change in spreading rate, ridge will be at a new steady-state profile; cross-sectional area will be one-third what it was at 0 m.y. (After Pitman, 1978)

cm/yr for 69 m.y. (Fig. 9–10(B), top) will have three times the cross-sectional area of one that has spread at 2 cm/yr for 69 m.y. (Fig. 9–10(A), top). It follows that if the spreading rate of 2 cm/yr were to be increased to 6 cm/yr, the cross-sectional area of the ridge with the lower rate would gradually increase (Fig. 9–10(A)). Conversely, a decrease in spreading rate from 6 cm/yr to 2 cm/yr would cause a decrease in cross-sectional area (Fig. 9–10(B)). A third factor to consider is that the various segments of the ocean-ridge system are not necessarily changing synchronously. Some may contract while others expand. Little is yet known, however, of the relationships among spreading rates between different segments of the global ocean ridge system. It is possible that large segments may exhibit similar patterns of expansion and contraction. Nevertheless it is the net effect of the total ridge system that is critical in controlling sea-level change. Pitman [1979] calculated that a maximum rate of sea-level change resulting from changes in geometry of the oceanic ridge system would be about 1 cm per 1000 years. Also, large sea-level changes can be caused by the inferred changes in the volume of the ridge system.

Computed sea-level change indicates that sea level may have been 350 m higher than the present during the late Cretaceous, which was a time of rapid sea-floor spreading (from about 110 to 85 Ma). Such a rise would be sufficient to cover 35 percent of the present continental (subaerial) surface, thus creating quite a different global ocean-to-landmass relationship than today. Since the Cretaceous, there has been a net contraction of the ridges, so the sea level has lowered (Fig. 9–11). Sea level is inferred to have fallen rapidly during the Oligocene and less rapidly during the early Miocene. Thus there was a minor regression in late Cretaceous time, a large Paleocene regression, a large Eocene transgression, an Oligocene regression, and an early Miocene transgression (Fig. 9–11). The computed sea-level curve of Pitman [1979] does not agree at all times with the geologically determined data (Fig. 9–11). The most important discrepancy is in the

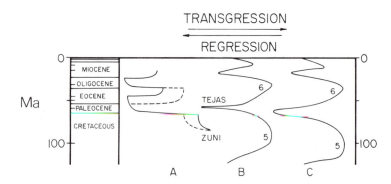

Figure 9-11 Sequence of late Mesozoic-Tertiary transgressive and regressive events as computed by Pitman [1978] and determined geologically; scale of magnitude of change is arbitrary. (A) Solid line shows transgressive and regressive sequence. Dashed line indicates possible change if DSDP results are used to recalibrate magnetic time scale. (B) North American transgressions and regressions. (C) African transgressions and regressions. (From W. C. Pitman III, *Geol. Soc. Amer. Bull*, vol. 89, p. 1401, 1978, courtesy The Geological Society of America)

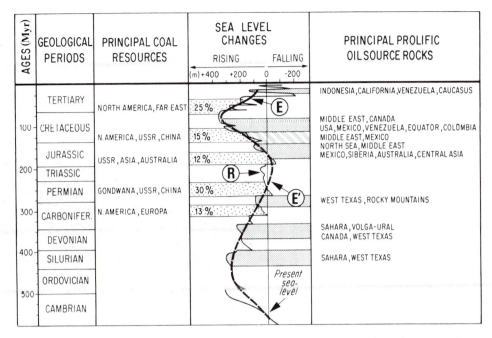

AGES (Myr)	GEOLOGICAL PERIODS	PRINCIPAL COAL RESOURCES	SEA LEVEL CHANGES	PRINCIPAL PROLIFIC OIL SOURCE ROCKS

Figure 9-12 columns (reconstructed):

- SEA LEVEL CHANGES: RISING / FALLING, (m) +400 +200 0 -200

AGES (Myr)	GEOLOGICAL PERIODS	PRINCIPAL COAL RESOURCES	PRINCIPAL PROLIFIC OIL SOURCE ROCKS
	TERTIARY	NORTH AMERICA, FAR EAST 25%	INDONESIA, CALIFORNIA, VENEZUELA, CAUCASUS
100	CRETACEOUS	N. AMERICA, USSR, CHINA 15%	MIDDLE EAST, CANADA / USA, MEXICO, VENEZUELA, EQUATOR, COLOMBIA / MIDDLE EAST, MEXICO
	JURASSIC	USSR, ASIA, AUSTRALIA 12%	NORTH SEA, MIDDLE EAST / MEXICO, SIBERIA, AUSTRALIA, CENTRAL ASIA
200	TRIASSIC		
	PERMIAN	GONDWANA, USSR, CHINA 30%	WEST TEXAS, ROCKY MOUNTAINS
300	CARBONIFER.	N. AMERICA, EUROPA 13%	SAHARA, VOLGA-URAL / CANADA, WEST TEXAS
	DEVONIAN		
400	SILURIAN		SAHARA, WEST TEXAS
	ORDOVICIAN		Present sea-level
500	CAMBRIAN		

Figure 9-12 Principal periods of deposition of prolific oil source rocks and major coal deposits are compared with worldwide transgressions and regressions as determined by Vail and others [1979] = R, and Pitman [1978] = E. The periods considered for petroleum source rocks are responsible for more than 10^8 tons of oil Myr^{-1} during the first cycle (570–200 Myr) and more than 10^9 tons Myr^{-1} during the second cycle (200–0 Myr). (From B. Tissot, reprinted by permission from *Nature*, vol. 277, p. 465, 1979, copyright © 1979 Macmillan Journals Ltd.)

Eocene, where the computed curve shows a continuous, although varying, rate of fall, while the geologic record consistently exhibits an Eocene transgression.

Global eustatic cycles also have major economic implications. Tissot [1979] showed the existence of alternate episodes of major coal and petroleum accumulations throughout the Phanerozoic (Fig. 9-12) that are associated with global sea-level cycles. Petroleum-rich sediments accumulate at times of high stands of sea level during the largest extension of epicontinental seas. These lead to greater shallow basinal sedimentation and sufficient accumulation of marine organic material. Particularly prolific oil deposits are typical of the Jurassic and Cretaceous (180 to 85 Ma), which make up 70 percent of resources in rocks but only 12 percent of Phanerozoic time. Considerable accumulations of coal occurred during global lowstands of sea level during the middle and late Carboniferous, middle and late Permian, early and middle Jurassic, early Cretaceous, and latest Cretaceous to Eocene (Fig. 9-12). These deposits represent 95 percent of coal resources deposited during only 30 percent of Phanerozoic time.

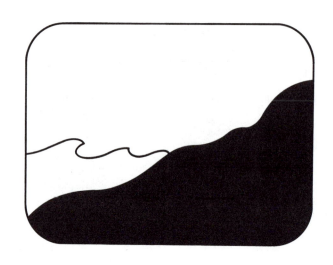

ten

Nearshore

Geological Processes

and the Continental Shelf

The shore is an ancient world, for as long as there has been an earth and sea there has been this place of the meeting of land and water.

Rachel Carson

INTRODUCTION

In this chapter we shall be concerned with certain geological processes that occur in the nearshore environments and on the continental shelf. As already discussed, these environments have been strongly affected by sea-level changes throughout geological history. Much of the shallow-water region is also strongly affected by waves, tides, and shallow currents. Thus the sedimentary history of the shelves and nearshore environments reflects the ever-changing relationship between tectonic subsidence, sea-level oscillations, and the dynamic processes acting at these water depths. These environments also lie within the photic zone, so they are influenced by biological processes and productivity that do not occur in most other parts of the ocean. Here we shall be concerned with a broader picture of the sedimentary history of shallow-marine environments, especially the modern distribution of terrigenous sediments in relation to sea level, past and present. The nearshore environments and the continental shelves have played a major role through time in the storage or redistribution (or both) of land-derived, or terrigenous, sediment to the deeper ocean. This is hardly surprising; however, it is not well known that 80 percent of all sediments of the earth are trapped in vast piles on certain continental margins. Most of the sediment has origin in very shallow water; it was deposited on the continental shelf and has since subsided. Subsidence is a key process in the preservation of these sequences. Rock weathering on land produces particulate matter like sand and clay, as well as dissolved material. This matter is then carried to the continental margins and some is deposited in the deep sea. Garrels and Mackenzie [1971] have estimated that each year some 250×10^{14} g of material is carried into the oceans. Rivers carry about 85 to 90 percent of this material, ice transportation accounts for about 7 percent, groundwater for 1 to 2 percent, and wind less than 1 percent. About 80 percent of the river load is particulate matter and the remainder is of dissolved form.

This vast amount of terrigenous sediment delivered to the marine environment is discharged by only a few rivers. Twenty-five percent of sediment transport is by four rivers: Huang Ho, Ganges, Brahmaputra, and the Yangtze. Of the 12 major rivers, all but the Amazon, Indus, and Ganges-Brahmaputra systems deposit their sediment loads in marginal seas. Almost all of the sediment is deposited near the shore in estuaries, deltas, or offshore basins. Little sediment is deposited directly beyond the continental shelf. The Congo River is one of the few rivers in which sediment completely bypasses the shelf by way of a submarine canyon.

In the modern ocean, ice transportation is important only in the polar regions, principally Antarctica, but conditions were markedly different during the Quaternary. During that time, ice-rafted deposition was important in the subpolar North Atlantic south of Iceland. Small changes in global climates can create large variations in the sediment supply to the oceans and drastically affect the sedimentary history of the deep-ocean basins.

Because of the diversity of oceanographic influences in shallow-marine areas, it is not surprising that there is so much diversity of sedimentary environments grading into each other. Sands are generally restricted to the shore zone at maximum depths of about 20 m, while muds are not distributed much beyond 30 km from the shoreline, except where sediment input is large [Curray, 1964; 1965; 1977]. The development of shoreline sedimentary environments is a function of numerous parameters, including rate of influx of terrigenous sediment, tidal regime, exposure to waves, current patterns, climate, and tectonism [Selley, 1970]. The character of modern shoreline deposits reflects the balance between the rate of injection of sediment and the ability of marine processes to redistribute the materials [Selley, 1970]. This balance, in turn, determines the chemical character of shelf sediments. Where terrigenous input is large, terrigenous sediments dominate (clastic shorelines); as terrigenous sediment input declines, carbonate sedimentation begins to become important and in some areas becomes dominant (carbonate shorelines).

Sediment Disequilibrium of Modern Coastal Zones

The character of the world's continental shelves varies considerably in structure, sediment type, and topographic evolution. The development of these shelves has ranged from the truncation of older rocks to the deposition of roughly conformable sediment sequences of Mesozoic to Recent age. The broad, flat character of the shelves has resulted from erosional activity related to a series of transgressions and regressions (see Chapter 9). Both erosion and sedimentation are concentrated in the nearshore areas, and the repeated sea-level migrations have leveled the margins forming the shelf surface.

Many marine geologists have been impressed by the strong influence that sea-level changes have had on the delivery of sediments to ocean basins and rises [Curray, 1965; Rona, 1973, A]. When sea level is high, shelf width is at its maximum; shelves act as storage areas for sediments or influence coastal processes that trap material in estuaries, coastal barriers, or inner-shelf shoals. This is the present state and has been amplified by the rapidity of the glacio-eustatic rise in sea level during the last 10,000 years. The morphology and sediment cover of most continental shelves is a relict of the previously lower sea level. This is because, since the Holocene transgression, insufficient time has elapsed for return of a typical shelf equilibrium with present-day environments. During this transgression, the shoreline migrated so rapidly that rates of sand supply were insufficient for appreciable beach deposits or ridges to develop over the shelf surface. The effect of the Holocene transgression is major transgressive, or drowned, coasts around the world, perhaps best noted along the Atlantic coastline of North and South America. This coastline exhibits extensive barrier-lagoon coasts, drowned estuaries, and other features created by rapid inundation.

Because the shallow-water environments are in a state of disequilibrium and because sediments are trapped in the nearshore areas, most continental shelves are covered with **relict sediments** and are still dominated by morphology related to the Quaternary low stands of sea level. Relict sediments are those that are related to past conditions even though they lie at the surface of the ocean floor. Some of the nearshore sediments, such as peats and nonmarine sands, were even deposited above sea level. Emery [1968] estimated that 70 percent of the continental shelf area is covered with relict sediments. Modern or equilibrium shelf-facies muds may extend over most of the surface of narrow shelves, but they are generally restricted to the inner sectors of wider shelves. The shelves of the east coast of North America are almost entirely covered with relict sediments because modern sediment is trapped in the estuaries [Curray, 1977]. Relict sediments are transported over the shelf surface, but their character is not representative of modern environmental conditions. However, since sea level has been close to its present level during the last 4000 to 7000 years, there has been much progress toward a return of equilibrium of sedimentation. For instance, littoral deposits began to develop about 7000 years ago in the form of widespread barrier islands [Curray, 1969; Kraft, 1978]. Also, coastal processes have been attempting to reestablish an equilibrium by the straightening of coastlines. This is accomplished by cutting promontories and building barrier bars and spits. However, the amount of sediment that has been able to reach the continental shelf is still limited because estuaries, which were formed by the drowning of river valleys during the Holocene transgression, have not yet been filled with sediment [Kraft, 1978].

THE NEARSHORE ZONE

Estuaries

An **estuary** has been defined by Pritchard [1967] as a semienclosed coastal body of water freely connected to the ocean and within which seawater is measurably diluted with freshwater from the land. Mixing of fresh and salty water within the estuary by tide, winds, and river influx produces density gradients that drive the characteristic estuarine circulation patterns. The particular characteristics of circulation are governed by the volume and rate of flow of freshwater entering the estuary, size and shape of the basin, and the effects of tides and winds. These factors create large differences; in the Laguna Madre, Texas, where there is virtually no freshwater input, high evaporation rates lead to salinities greater than the ocean.

Estuaries are dominant features along many modern coastlines as a result of the Holocene transgression [Nelson, 1972]. This transgression produced many more estuaries than would be expected from most other periods of geologic time [Curray, 1977]. Hence most estuaries fill not only ancestral river valleys, but fjord basins, bar-built basins, or semienclosed coastal basins produced by tectonism [Schubel and Pritchard, 1972]. Present-day estuaries are forming near tidal flats, barrier islands, and river deltas. Many estuaries are being filled in by sediment from associated river systems. Estuaries have been eliminated where sediment filling kept up with the transgression. In the geologic past, it is almost certain that estuaries were much less widespread, confined to areas of rapid subsidence with resultant drowning of the coastal topography.

Estuaries represent one of the most important modern geological features controlling the flow of terrigenous sediment from land to ocean. At present, rivers transport more than 8 billion tons of sediment annually to the sea. However, most of this material is trapped in estuaries or the immediate nearshore areas. This is particularly conspicuous on the east coast of North America, where rivers tend to transport relatively small sediment loads and empty into large estuaries. For instance, the Susquehanna River, which is a major river flowing into Chesapeake Bay, eastern United States, has a sediment discharge of 0.75×10^{12} g/yr. This compares with 312×10^{12} g/yr for the Mississippi River and 8.5×10^{12} g/yr for the Rio Grande [Holeman, 1968; Gibbs, 1977].

Thus most estuaries and associated coastal marshes represent sediment sinks or areas of entrapment. Little material is carried to the continental shelves. Exceptional areas include the Amazon River. Freshwater input there is much greater than the tidal volume, so freshwater extends outward over the continental shelf and a proportion of the river-transported sediment is deposited near the edge of the shelf, the remainder being carried in nearshore currents. Intermediate conditions exist in areas of strong seasonal river flow, such as the Mississippi.

Tidal processes are of the utmost importance in controlling estuarine sedimentary conditions [Elliott, 1978]. The term *estuary* is derived from the Latin word *aestus,* meaning boiling or tide. In a generalized estuary, low-salinity river water flows in at the head of the estuary and spreads out over the more dense salt water intruding from the ocean. Various circulation patterns develop within estuaries, depending on the manner in which the water bodies mix. The rate of mixing is strongly controlled by tidal flow in relation to basin geometry and freshwater inflow. In estuaries with minimal tides, freshwater extends over a saltwater wedge as a distinct layer, and the water column is highly vertically stratified. In estuaries exhibiting greater tides, turbulent mixing of the two water bodies occurs, leading to a more homogeneous water column. The highly stratified or salt-wedge estuaries have been called Type A estuaries, while those that are partially mixed are Type B estuaries [Schubel and Pritchard, 1972]. Other types have also been defined by Pritchard [1967] and Schubel and Pritchard [1972].

Type A estuary

A **salt-wedge,** or Type A, **estuary** forms when the estuary is essentially tideless and the floor of the estuary slopes down continuously toward the sea. The circulation pattern is dominated by river discharge. The salt wedge extends up the estuary (Fig. 10-1) at distances dependent on the rate of freshwater discharge. A sharp density contrast exists between the salt water and freshwater. Partial mixing is accomplished by internal waves, which carry saline water upward into the freshwater layer. Vertical advection of salty water causes the overlying fresher water to become more saline as it approaches the ocean. To replenish this water lost to the upper layer, there is a slow landward movement of salt water up the estuary [Biggs, 1978].

Type B estuary

In the **partially mixed,** or Type B, **estuary,** the tidal influence is increased to the extent that river discharge does not dominate circulation [Biggs, 1978]. The

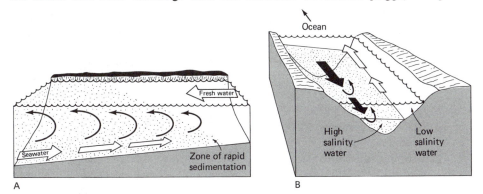

Partially mixed, type B estuary

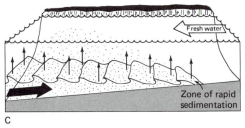

Highly stratified, type A estuary

Figure 10-1 Top figure shows schematic views of partially mixed, type B estuary. (A) side view; (B) view looking seaward in the Northern Hemisphere; (C) view of a highly stratified, type A estuary. (From J. R. Schubel and D. W. Pritchard, *Journal of Geol. Education,* vol. 20, pp. 60–68, 1972)

greater tidal influence increases turbulence and upward advection of salt water. Thus these estuaries exhibit greater mixing of salt water and freshwater, and the contact is more gradational between these layers (Fig. 10–1). Because of greater efficiency of mixing, the net seaward flow of the upper, lighter layer may be greater than the river discharge. Thus the upper layer requires higher rates of replenishment from the lower, denser layer. This causes a greater landward intrusion of the saltwater layer into the estuary. The circulation in these estuaries can also be influenced by the Coriolis effect. In the Northern Hemisphere the upper salt-water interface is raised slightly on the right side of estuaries extending north-south (Fig. 10–1), such as Narragansett Bay in New England. This estuary, which is a drowned river valley, is a good example of a partially mixed estuary; that is, it has little river-water input and is dominated by tidal flow.

Tidal range also plays a major role in the development of other features. Guilcher [1958] and Hayes [1976] consider that river deltas and barrier islands develop best in areas of minimal tidal range, while tidal flats and salt marshes are most abundant in areas of maximum tidal fluctuation.

Control of sedimentation in estuaries

Sources for estuarine sediment include rivers, the sea floor, and the adjacent shoreline, in addition to those generated by biological productivity. However, river sediment input is by far the most important source in most estuaries.

The settling of sediments is closely tied to the character of estuarine circulation and is a fundamental process in controlling the flow of sediment between continents and oceans. Particles the size of sand settle from suspension as flow velocities decrease. Finer material, particularly clays, tends to **flocculate** as freshwater mixes with salt water; that is, the particles aggregate as electrolytic forces bring them together. As these floccules increase in size, their settling velocity increases and they are deposited. Salinities of 2 to 5‰ are believed to be sufficient to cause such flocculation. Organic aggregation also seems to be an important process. The ingestion of suspended material by planktonic and benthonic animals leads to particle aggregation in the form of fecal pellets. This may also account for the considerable decrease in suspended loads in many nearshore waters. The importance of these processes cannot be overemphasized because the fine sediment material is deposited within the estuarine environment rather than being transported out of the estuary into the open ocean.

Flocculation in the estuary is closely tied to the area of contact between salt water and freshwater. This, in turn, determines the principal location of estuarine sedimentation. For instance, in salt-wedge estuaries, sediment deposition is concentrated at the *tip* of the salt wedge (Fig. 10–1). As the tip of the salt wedge migrates, so does the region of maximum sediment deposition [Schubel and Pritchard, 1972]. The Southwest Pass of the Mississippi River is a classic salt-wedge estuary. Major seasonal migrations of the tip of this wedge (up to about 250 km) create major migrations in the locus of sediment deposition, causing navigation problems to shipping [Biggs, 1978]. Partially mixed estuaries exhibit somewhat different sediment patterns. The increased mixing typical of these estuaries creates a *turbidity maximum* slightly downstream from the furthest salt-water intrusion. The locus of sedimentation is associated with this turbidity maximum.

It is apparent that flocculation of sediment in estuaries cannot be 100 percent efficient, and large rivers with huge suspended loads transport much sediment to the oceans. However, much sediment that escapes from the estuary is transported along the coast and deposited in other enclosed bodies of water [Curray, 1977].

Sediment concentration has been filling up the estuaries steadily during the late Holocene. Sedimentation rates are high where coarse sediment is deposited in estuarine deltas and at the tip of the salt wedge. In addition to the rate of sediment input, infilling is also controlled by the rate of subsidence or uplift of the area. If subsidence is persistent, the estuarine environment may persist for long periods of time, creating large accumulations of estuarine sediments. However, estuaries are generally ephemeral features and, given their current rate of infilling and continued stability of sea level, they may be completely filled within several thousand years. Deltas will then become more prominent features of the coastal zone [Curray, 1977].

Lagoons

Lagoons are highly variable, shallow-marine environments separated from the open sea by barrier-island complexes composed largely of well-sorted sand. The barrier may be an island with dunes or no more than an offshore bar exposed only at low tide [Selley, 1970]. They form over half of the coastline south of New York, much of the Gulf Coast of the United States, and about one-third of the coast of Mexico [Phleger, 1969; Curray, 1977]. Lagoons are largely found along transgressive coastlines in association with estuaries. In fact it is sometimes difficult to differentiate between estuaries and lagoons. A typical estuary is a narrow, elongate embayment, while a typical lagoon is parallel to a sand barrier. Salinity does not vary much along a lagoon's length, so sediment transported over the bar during storms can be deposited uniformly. As lagoons are filled with sediment, they become tidal flats [Curray, 1977].

Lagoonal sediments are generally fine grained since they are deposited largely from suspension. However, sands are also deposited as washover fans and small-scale deltas. Carbonate muds can be deposited where sedimentation is especially low, and evaporites may form in hypersaline lagoons [Selley, 1970]. This variability is controlled by the climate in which lagoons are located. Characteristics of the lagoon are also strongly influenced by tidal range. Where tidal range is low, lagoons are marked by extreme salinities; they are brackish or hypersaline, because of limited communication with the open sea [Elliott, 1979]. Also, because tidal channels are absent or restricted, storm waves frequently flow over the barrier-forming washover fans. If these coalesce, they form a **back-barrier flat,** which projects into the lagoon [Elliott, 1979]. Where tidal range is high, salinity values are more like those of the adjacent ocean because of frequent interchange between the ocean and lagoon by way of tidal inlets. Flood tidal deltas form at these inlets, and tidal flats and salt marshes are widespread [Elliott, 1979].

Deltas

In spite of the Holocene transgression, a few large rivers have transported sediment in such abundance that their estuaries are filled and still there is more sediment than can be dispersed along adjacent coasts. Where this has occurred, sediment accumulates at the river mouth, forming a delta. Deltas occur off the Mississippi, Nile, Ganges-Brahmaputra, and Rhine rivers and the Rio Grande. The term *delta* was first applied by Herodotus in the fifth century B.C. to the triangular-shaped body of land between the branching channels of the Nile River [Moore, 1966].

Because deltas form where rivers transport more sediment to the sea than can be redistributed by longshore currents, they produce a regressive shoreline. Where longshore currents effectively redistribute this terrigenous material, bars and beaches parallel the coastline [Selley, 1970]. The river flow maintains enough momentum to carry the sediment load up to several kilometers into the ocean as a jet before the current dissipates. The jet loses momentum most rapidly near its margins due to shear stresses with the surrounding seawater. This leads to deposition of sediment at the margins, forming **submarine levees** (shallow embankments). Ridges may also form below the jet flow because of similar shear stresses developed along its lower surface.

Delta control is, therefore, largely a function of the rate of supply of sediments and wave energy or tidal range, or both, at the river mouth. Where sediment is deposited on the shelf, the offshore profile will tend to be flattened and, as a result, larger waves are required for effective erosion of the sediment. The distribution of deltas is therefore strongly related to the amount of energy at the continental margins. Deltas tend to be more common or well developed in semi-enclosed or enclosed seas where wave energy is limited; examples are the Danube, Mississippi, Po, and Rhine deltas [Elliott, 1979]. Deltas may not form off rivers where there are strong waves or relatively large tidal ranges, even where sediment input is high, such as in the Columbia River in the western United States [Gross, 1972]. A few rivers, however, supply so much sediment that a delta forms in spite of high wave energy and large tidal range. Examples of these are the Niger River in the African Bight and the Ganges-Brahmaputra rivers in the Bay of Bengal. Another factor that influences the character of deltas [Elliott, 1979] is the coarseness of sediment, which somewhat controls the location of sediment deposition within the delta complex. Coarse sediment tends to be deposited near the river mouth, while fine sediment is more widely dispersed. Rivers with brief, but high, rates of discharge will more readily transport coarse sediment into the delta than rivers with more even, seasonal flow. Furthermore, deposition will be enhanced if maximum sediment input coincides with times of reduced wave energy at river mouths. The delta shape is also strongly controlled by the ratio of sediment supply to energy. These shapes have been ranked by Curray [1977] as follows (highest to lowest rate of supply and energy conditions):

1. *Birdfoot.* The Mississippi delta is the only modern large example.

2. *Lobate.* In these, the distributary fingers of the birdfoot have been remolded into lobes.

3. *Cuspate.* V-shaped.

4. *Arcuate.* Rounded.

5. *Estuarine.* The shape of these is strongly controlled by the confining shape of the estuary in which they have formed.

The Mississippi delta exhibits a series of overlapping lobes that represent the modifications of a series of birdfoot deltas formed during the last 5000 years (Fig. 10–2). This delta is unique in having constructed lobes completely across a comparatively wide continental shelf, so that the sediment is now being directly supplied to the upper continental slope [Selley, 1970]. Massive amounts of sediment supplied to the deltaic region (annual rate of about 5×10^{11} kg/yr) have led to current subsidence rates of 1 to 4 cm/yr. The delta has subsided up to 150 m during the last 18,000 years alone; 10,000 m of sediment underlie the delta. Because deltas are so flat they are very sensitive to subsidence and to sea-level oscillations. A drop in sea level can lead to erosion of the deltaic sediments, while a rise in sea level results in transgression over the delta and progradation of the alluvial valley. When this process is completed, renewed delta growth may occur [Moore, 1966].

All deltas exhibit a system of branching **distributary channels** flanked by levees. The flatness of the deltaic plain leads to frequent breaching of the main channel levees and diversion of the distributaries. By this process an alluvial flood plain is built outward, lobate in plan and wedge-shaped in vertical cross section. The top of the delta consists of a radiating network of distributary channel sands flanked by levee silts [Selley, 1970]. These radiating series of distributary sands are referred to as **bar-finger sands.** Finer sediments and peats are deposited in the interdistributary areas, which are represented by a wide range of environments, including flood plains, swamps, and lagoons.

Coarse sediment deposition in deltas is concentrated at the subaerial **delta front** (Fig. 10–3). Beyond this, much finer sediments are deposited in the **prodelta** (Fig. 10–3), which is entirely below sea level [Moore, 1966]. Sediment is supplied to the prodelta area by jets of turbid river water. Where sediment accumulation seaward of the river is faster than subsidence or removal by longshore drift, progradation of the prodelta, delta front, and eventually the river mouth itself occurs [Wright, 1978]. Distributary channels are extended into the sea, initially by confinement within submarine levees and later in subaerial form.

As the deltaic system is prograded, it gives rise to the classic cross section of topset, foreset, and bottomset beds (Fig. 10–3). **Topset beds** are mixtures of coarse sands deposited in the delta front and deltaic plain areas and finer, organic-rich sediments of the interdistributary areas [Curray, 1977]. **Foreset beds** are fine, marine sediments of the prodelta facies. **Bottomset beds** are finer sediments deposited on the distal areas of the deltaic complex and grading into sediments of the adjacent continental shelf (Fig. 10–3).

Beaches

The most dynamic of all marine environments are located at the intersection of sea and land—beaches. Because beaches are constructed largely by wave processes, they are constantly changing at rates dependent upon individual waves, dif-

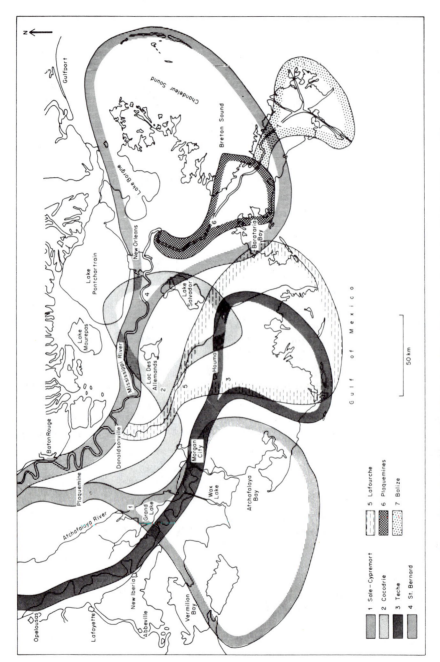

Figure 10-2 Subdeltas of the Mississippi River Delta complex during the last 5000 years. The site of major deltaic sedimentation has shifted several times. (From C. R. Kolb and J. R. Van Lopik, *Deltas in their Geologic Framework*, eds. Shirley and Ragsdale, p. 22, 1966)

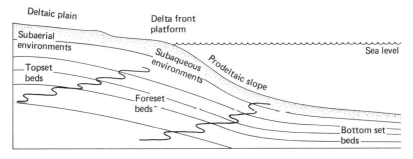

Figure 10-3 Gross features of delta morphology in greatly exaggerated cross section. (After R. K. Matthews, 1974)

ferent tides, different seasons, and different years. In the strict sense, a **beach** is an accumulation of unconsolidated sediment (sand, shingle, cobbles) from mean low tide to a physiographic change, such as a dune field. But a beach is considered by geologists to include the nearshore area below sea level (to a depth of 10 to 20 m), which is actively under the influence of surface waves [Komar, 1976].

Beaches and most other nearshore environments are dominated by wave action. As the waves shoal in this zone, they increase in height and steepness until they break along the beaches. The turbulence and currents generated by wave action violently stir sediments, causing their transfer along the shoreline by wave-related and tidal currents. Large volumes of suspended sediments are transported; thus the beach is shaped by supply and size of sedimentary material, height and period of waves, and by tidal range [Curray, 1964]. Most sediment is transported between the upper beach limits and depths of 15 m. Tides can play a major role in extending the depth range at which significant sediment transportation occurs, by constantly altering wave characters and the position of the breaker zone. The net effect of waves in general is to move sand toward the shore zone and to confine it there.

As sea level rose during the Holocene, beach sediments migrated shoreward with the transgression. Present-day distribution of beaches is, in part, related to the location of available sediment during this transgression. In areas like the central east coast of North America, where there is little sediment available, the modern beach sediments are derived from relict shelf deposits and have migrated across parts of the continental shelf. In some areas, complete migration was not possible, and ancient beaches became stranded on the continental shelf quite removed from the present nearshore area. They now form relict features.

As waves approach the nearshore area, they are transformed, depending upon the zone entered (Fig. 10-4). In the deeper water of the shelf, wave-induced orbital motions produce oscillatory flow, which may impinge on the sea floor—thus generating sediment motion [Elliott, 1979]. The depth at which waves begin to impinge on the sea floor is equal to one-half the deep-water wavelength, where the wavelength is the horizontal distance from wave crest to crest. The deep-water wavelength is given by the relationship $L = gT^2/2\pi$, where g is the acceleration of gravity and T is the wave period in sec [Inman, 1971].

When waves approach a shoreline, the crests are bent to conform with the

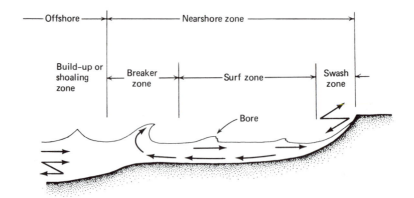

Figure 10-4 Terminology used to describe wave action in the nearshore regions. (From Paul D. Komar, *Beach Processes and Sedimentation* © 1976, pp. 12–13. Reprinted by permission of Prentice-Hall, Inc. Englewood Cliffs, New Jersey)

contours of the sea floor. This is called **wave refraction.** The effect of this process is to redistribute wave energy, focus energy on headlands, and reduce wave energy in coves or embayments. As a result, headlands erode and embayments become sites of deposition and of beaches; in time the coastline straightens. For straight coasts with parallel depth contours, refraction decreases the angle between the approaching wave and the coastline, causing a spreading of wave energy along the crests and a reduction in wave height. Periods of large ocean storm waves may be as great as 20 sec or more. Wave refraction, in these cases, will commence at depths as great as 300 m, much beyond the continental shelf. As the sea becomes more shallow, the waves become elliptical and set up a to-and-fro motion in the same direction as the wave movement. At depths where the waves cannot complete their orbital motions, they increase in height and steepness and eventually begin to break creating the **breaker zone** (Fig. 10–4). The **surf zone** is the zone in which a shallow, high-velocity translation wave, or **bore,** is propelled up the beach face [Elliott, 1979]. This uprush of water produces sediment transport and sorting. The landward flow may be balanced by a seaward return flow, although off certain beaches this return flow may be concentrated into rip currents. The **swash zone** represents the zone of maximum penetration of a shallow wave (Fig. 10–4). During the swash, all wave motion is lost and all forward movement is by inertia. Gravity and friction, along with percolation into the beach face, slow and then stop the forward movement. The most forward advance of the swash is marked by a line of sediment called the **swash mark.** Backwash is initially slow, but accelerates under the influence of gravity. Water is lost through percolation; the amount depends on the sediment type and affects the amount of backwash. Fine sediments allow slow percolation, while percolation is almost complete on gravel beaches. Percolation of the swash into coarse, permeable sediments enhances deposition and the formation of steep beach faces. When the beach is saturated, the backwash has a higher velocity and erosion can become important. Thus the slope of the beach face decreases with decrease in sediment size and increases in wave height. The degree of development of the wave zones differs among beaches. The surf zone is poorly developed in steep beaches because the waves break close to the shoreline, whereas broad surf zones occur on gently inclined beaches as waves break at greater distances from the shore.

Beaches develop a profile (Fig. 10–5) that reflects these changing conditions. Komar [1976] distinguishes four zones within the littoral or nearshore environments (Fig. 10–5). From landward to seaward these are backshore → foreshore → inshore → offshore. The **backshore** is that part of the beach that is covered by water only during storms, and thus is above normal high tide. The backshore area consists of one or more gently sloping **berms** deposited by the receding waves. The seaward limit of the berm is marked by an abrupt change in slope at the **berm crest,** or edge, which indicates the highest point of normal wave activity. Seaward of the berm crest is the beach face, which is the sloping section of the beach profile normally exposed to the action of the wave swash. Depending upon materials available and the processes acting upon it, this face may be inclined only 1° to 3° or as much as 30° [Davis, 1978]. Most beach faces are composed of fine, hard-packed sand, which forms gentle slopes. As grain size increases, the angle of the beach slope increases. Beaches composed of fine sand generally slope at about 3°, those with pebbles (4 to 64 mm) slope at about 15°, and those with cobble (64 to 256 mm) slope at 24°. The beach face is part of the **foreshore zone,** although the foreshore zone may include additional flat portions of the beach profile. The foreshore is the intertidal zone and is marked by swash-zone processes, although surf-zone processes may occur in certain cases. The **inshore zone** is the wider, subtidal part of the profile extending to just beyond the breaker zone (Fig. 10–5), or the limit of the bar and trough topography. The inner part of the inshore zone is dominated by wave-driven flow associated with shoaling waves, while the outer part of this zone is dominated by storm waves and tidal currents.

Not all beaches are composed of sand; they may consist of any coarse material available, from coarse silt to boulders, from shell fragments to carbonate skeletal debris. Quartz particles are most widespread, but carbonate particles become more prevalent on tropical beaches. There are several sources of beach sediments, including influx from rivers, erosion of cliffs and headlands, erosion of the sea floor,

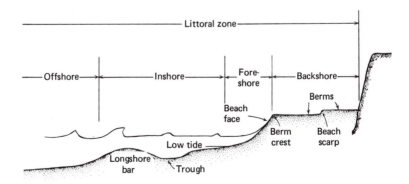

Figure 10–5 Terminology used to describe the beach profile. (From Paul D. Komar, *Beach Processes and Sedimentation* © 1976, pp. 12–13. Reprinted by permission of Prentice-Hall Inc., Englewood Cliffs, New Jersey

and biologically produced materials, such as mollusks and corals. Along the eastern United States, almost no river-transported sediment escapes from the estuaries. In this case most of the beach materials are derived from sands deposited offshore during lowstands of sea level and from the erosion of nearby cliffs. Under these conditions, the beaches are sandy where the waves erode sand and rocky where waves erode gravel or till. Sediments of the beach are usually well sorted. This particularly applies to the beach face. The sediments of other parts of the beach, like the berm, are much less well sorted. The primary sedimentary structures in beach sediments range from nearly horizontal plane laminae at the top of berms to seaward-dipping layers of the beach face to a diversity of cross stratification and channels [Friedman and Sanders, 1978].

Wave-generated nearshore currents

Three factors operate to transport sediment in the beach environment. As we have already seen, waves are transformed and generate sediment transport as they approach the shoreline. These waves, in turn, generate nearshore currents, which are also important in the transport of sediment. Thirdly, alternations in wave energies between fair weather and storm conditions cause alternations in sediment transport [Elliott, 1979]. Three main types of currents operate either singly or in combination; the most important are **longshore** currents, which move parallel to the shoreline. **Rip currents** generally move sediment offshore. **Wave-generated currents** move sediment in an onshore or offshore direction, depending on the specific conditions at any time.

The most critical parameters of wave-generated currents are wave height and angle of incidence in the breaker and surf zone. These currents are strongly dependent on wave steepness, as shown in Fig. 10-6, where the water flow induced by moderate and large waves is shown in profile. The onshore mass transport of water in waves must be balanced by the seaward return flow. Flatter waves of moderate height (Fig. 10-6(A)) tend to move sand shoreward because sand moves shoreward on the bottom as the return flow of sediment is in suspension at middle depths. Steeper waves (Fig. 10-6(B)) transport sediment on the bottom away from the shore, causing erosion of the beach area [Raudkivi, 1967].

Rip and longshore currents

When waves break so that an angle exists between the crest of the breaking wave and the shoreline, the momentum of the breaking wave has a component along the beach in the direction of wave propagation [Inman, 1971]. This creates **longshore currents,** which flow parallel to the beach inside the breaker zone (Fig. 10-7). In longshore currents, velocities are highest in the zone between the surf zone and the beach (that is, in the inshore and foreshore areas). The velocity of the current rapidly decreases to zero outside the breaker zone, showing that such currents are wave-induced rather than linked with ocean currents or tides (Fig. 10-7) [Komar, 1976]. Each oblique wave contributes a longshore component of movement to the surf zone as it breaks, leading to a buildup of water along the shoreline.

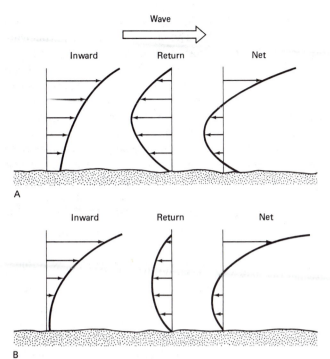

Figure 10-6   Vertical velocity profiles for water flow in large and moderate waves (lengths of arrows proportional to velocity of water). The net flow is determined by subtracting the inward from the return flow. Note larger return flow on sea floor in (B). (After A. J. Raudkivi, 1967)

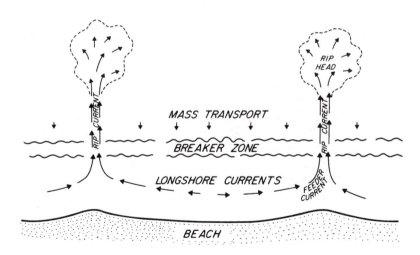

Figure 10-7   Schematic view of a nearshore, symmetrical circulation cell and its components resulting from breakers parallel to the shore. (From Shepard and Inman, 1950)

This buildup is balanced by a return flow out through the surf zone as relatively narrow rip currents (Fig. 10–7). A cell circulation system is thus formed in the nearshore areas by the combined components of mass transport, longshore currents, and rip currents [Shepard and Inman, 1951; Inman, 1971; Komar, 1976]. Even waves approaching the shoreline with their crests parallel to the shoreline can generate longshore currents. In this case, the shore currents increase in velocity from zero about midway between the two adjacent rip currents and reach a maximum just before turning seaward into the rip (Fig. 10–7) [Komar, 1976]. This type of circulation requires a slow, shoreward mass transport through the breaker zone in areas between the rip currents. Differences in the mean water level along a beach produce pressure fields, which require compensatory longshore and rip current flow. A decrease in mean water level is called a **wave setdown**; a rise in level is a **wave setup.** Longshore currents flow from regions of high water to regions of low water and thus flow away from zones of high waves [Inman, 1971]. Rip currents form where waves are lower.

On a smaller scale, alternate zones of high and low waves can occur along a coast due to interaction between the new waves entering the nearshore and trapped waves of oscillation within the nearshore zone, called **edgewaves** (Fig. 10–8). These waves can be either stationary or move progressively along a coastline. The edgewaves effectively create the changes in water levels along the shoreline, and hence these wavelengths control the spacing of the rip currents (Fig. 10–8).

The longshore cell circulation also appears to be linked to the development of beach cusps, which are rather uniformly spaced, crescentic seaward projections or mounds of sediment that trend at right angles to the shoreline [Komar, 1976]. Cusp spacing may be less than 2 m on shorelines with small waves to up to several hundred meters due to storm waves. The cusps are built of coarser sediment than the separating embayments. Maintenance of the coarser material in the cusps creates higher permeability, which leads to more rapid dissipation of swash energy. The less permeable embayment areas maintain higher swash energies, so the coarser material is more readily transported toward the cusps, keeping the intercusp areas clean. The formation of cusps appears to be associated with a wave-wave interaction similar to that associated with nearshore circulation cells [Inman, 1971].

Interaction of the currents with the surf zone results in sediment transport along the shoreline, as evidenced by sediment accumulation on the upstream side of artificial prominences called **groins,** designed to arrest this transport [Elliott, 1979]. Longshore sediment transport tends to straighten coastlines by eroding headlands, filling embayments, and creating spits and bars.

Additional longshore sediment transport is accomplished in the swash zone. In this zone, true longshore currents do not develop. Instead, movement of sediment is in a zig-zag, or sawtooth, pattern. The upslope component is in the direction of wave approach, whereas the downslope component is determined by gravity and thus is normal to the beach face. The returning swash interacts with the next incoming breaker to form a near-stationary wave marked by high turbulence and sediment suspension at the lower edge of the swash zone. This further assists in longshore sediment transport [Brenninkmeyer, 1976; Gorsline and Swift, 1977].

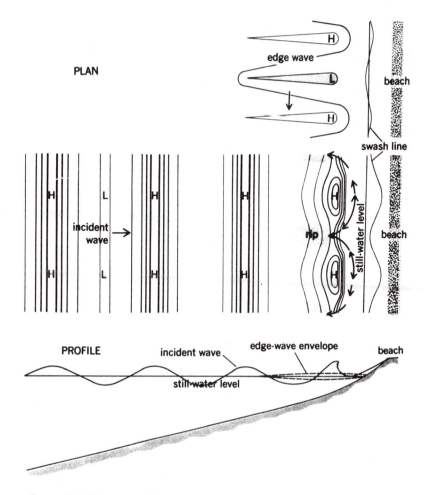

PLAN

edge wave

beach

swash line

incident
wave

rip

still-water level

beach

PROFILE   incident wave   edge-wave envelope   beach

still-water level

Figure 10-8 Formation of rip currents. The interaction of incident waves from deep water with edge waves traveling along the beach producing alternate zones of high (H) and low (L) breakers. Longshore currents flow away from zones of high waves where setup is maximum and converge on points of low waves, causing rip currents to flow seaward. (From D. L. Inman, *Encyclopedia of Science and Technology*, vol. 9, p. 29, 1971)

Compartmentalization of sand transport

Some coastlines exhibit broad-scale compartmentalization of sand, the best-known examples of which are in southern California (Fig. 10-9). Here a series of self-contained sediment compartments occurs along the coastline [Inman and Chamberlain, 1959; Inman and Frautschy, 1966]. Within each compartment, sediment is first supplied by river and cliff erosion. A typical cell begins with a rocky promontory or coastline with limited sand supply, which is eroded and transported by longshore currents toward the southeast. It is then dumped for removal from the

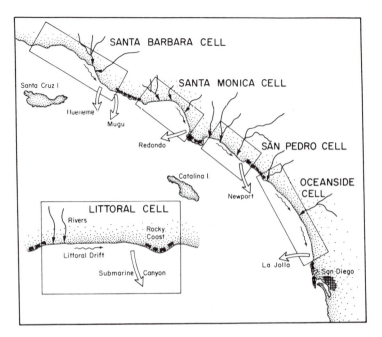

Figure 10-9 Coast of southern California showing cell divisions in which sand is introduced by rivers, transported southward as longshore drift, and trapped within the heads of submarine canyons which come close to the shoreline. Shallow sediments are then carried into the deeper ocean down the canyons. (After D. L. Inman and J. D. Frautschy, Coastal Engineers Santa Barbara Specialty Conference, pp. 511–536, 1966, courtesy the American Society of Civil Engineers)

beach system into the heads of submarine canyons that approach the beaches (Fig. 10–9). This system is important in some areas for transportation of sand to the deep sea; in some cases the sand terminates as dune fields. Beaches tend to be narrow at the upstream (northern) end of individual compartments and wide toward the downstream (southern) end of the compartment. Very important losses of sediment occur down the canyons [Shepard, 1973]. In fact, Curray [1977] believes that this is the single major mechanism for transport of sand beyond the shore zone. This mechanism accounts for much of the transport of sand to the deep sea during the Quaternary low stands of sea level.

Seasonal cycles

Beaches exhibit distinct seasonal cycles associated with changing fair and stormy weather. The largest and most rapid changes in beaches occur during winter storms or hurricanes. The effect on beaches is largely to move sand shoreward during fair weather, while eroding the beach during storms.

During storm conditions, longshore bars are displaced seaward. Inner sandbars may be eroded and displaced seaward by more than 15 m a day [Davis and Fox, 1972]. Strong longshore currents develop deep channels. Erosion is encouraged on the beach face as the water content rises, reducing grain-to-grain friction.

For most of the year, beaches are subjected only to moderate or low energy conditions. It is during these fair weather conditions that most of the sand is moved toward shore and the beach maintained. Much of this sand is from reservoirs of the submarine sandbanks built up in the offshore area during the winter [Davis, 1978]. This to-and-fro seasonal motion of sand is closely tied to wave height and steepness and is generally referred to as **wave climate.** These processes indicate that little sediment is permanently lost from most beaches during storms, as the changes that occur in the locus of sedimentation are temporary.

Barriers

Barriers are long, straight features parallel to the shore and separated from the mainland coasts by lagoons and bays. Individual barrier islands range from a few kilometers to more than 200 km long, and a few tens of kilometers in width. These barriers are widespread features along most of the lowland coasts of the world [Hoyt, 1967; Dickinson and others, 1972] including the east coast of the United States from Long Island to Florida and the Gulf Coast, and the coasts of the North Sea along the Netherlands, Germany, and Denmark. The longest barrier island in the world is Padre Island, Texas, which is about 200 km long and ranges from 1 to 8 km in width (Fig. 10–10) [Dickinson and others, 1972].

Barriers vary significantly in size. They may be only areas of beaches just above high water, which is called a **longshore bar,** or they may be major features up to 30 m in height with dunes and vegetation, which are called **barrier islands** with associated **barrier beaches.** These islands may consist of one or more ridges of dune sediments that mark successive shoreline positions during progradation. Such islands are broken at intervals by tidal inlets, kept open by tidal effects and storm waves. If connected to a headland, they are known as **barrier spits.** The lagoons vary considerably in width from a few kilometers up to several tens of kilometers. Barriers thus consist of three major clastic depositional environments: (1) the subtidal to subaerial barrier-beach complex; (2) the back-barrier region, or subtidal-intertidal lagoon; and (3) the subtidal-intertidal delta and inlet-channel complex [Reinson, 1979]. Facies of the barrier-beach and channel-delta environments are mainly sand and gravel, whereas lagoonal deposits can consist of both organic-rich mud and sand.

Other than the general controls over barrier evolution, their exact mode of formation is not well understood [Hoyt, 1967] and is controversial, probably because they are formed in several ways. Mechanisms proposed long ago include transportation of sand to the barrier from offshore areas [DeBeaumont, 1845; Johnson, 1919] and longshore transport of sand [Gilbert, 1885]. Other mechanisms include the buildup of submarine bars, spit progradation parallel to the coast and segmentation by inlets, and submergence of coastal beach ridges [Reinson, 1979]. It is clear, however, that the present-day barrier islands evolved during the late Holocene beginning about 5000 to 6000 years ago (Fig. 10–11) [Dolan and others, 1980]. As the sea rose during the Holocene and the shoreline moved across the continental shelf, large masses of sand were moved with the migrating shore zone as

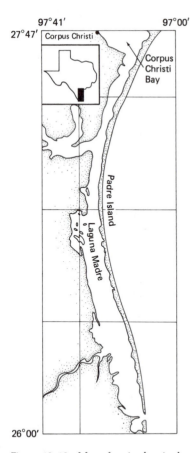

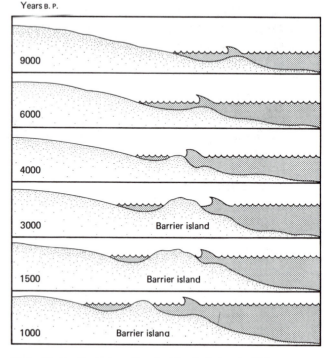

Years B. P.

9000

6000

4000

3000   Barrier island

1500   Barrier island

1000   Barrier island

26°00'

Figure 10-10 Map of major barrier bar (Padre Island) of the southern Texas coastline, with enclosed lagoon (Laguna Madre). (After Dickinson and others, 1972)

Figure 10-11 Evolution of barrier islands during the Holocene as sea level rose. About 3000 years ago large amounts of sediment may have been available to be reworked by shoreline processes to begin to form the modern barrier islands. (After R. Dolan and others, 1980)

beach deposits overlying a lagoonal carpet (Fig. 10–11). Once sea level began to stabilize about 4000 to 6000 years ago, the barrier islands began to evolve their present-day form. The bases of a large number of barriers throughout the world are at depths of 5 to 10 m, which is the depth at which the sea-level rise slowed drastically, about 6000 years ago. During the last 1000 years a continued slow rise of sea level has resulted in the further transgression of the barrier islands (Fig. 10–11), primarily by overwash and inlet formation, especially during severe storms [Dolan and others, 1980]. During transgression, individual spits and islands may come and go and lagoons may fluctuate in width, but once initiated, a barrier system will retreat as a steady-state phenomenon, as long as the variables controlling its behavior remain constant [Swift, 1975].

The evolution of barriers is, therefore, closely related to sand supply, sea-level history, including that resulting from subsidence or uplift, and the intensity of

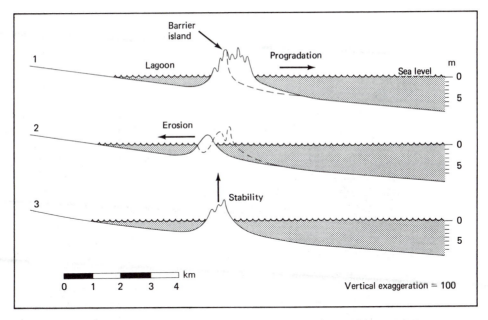

Figure 10-12 Schematic cross sections of barrier islands showing (a) progradation; (b) erosion; and (c) stabilization. These stages depend on rates of sediment supply, submergence, and on hydrodynamic factors. (After J. H. Hoyt, 1967)

waves and currents over the inner continental shelf. Under conditions of continued sediment supply, stable sea level, and low-to-moderate subsidence rates, barriers prograde seaward (Fig. 10–12). In contrast, a reduction in sediment supply, a rise in sea level, or a high rate of subsidence leads to the landward migration of barriers [Hoyt, 1967; Elliott, 1979] (Fig. 10–12).

CONTINENTAL SHELVES

A large number of interdependent factors influence the nature of sedimentary facies on present-day continental shelves [Johnson, 1978]. These include shelf width, rate and type of sediment supply, character of energy input, history of sea-level change, climatic regime both in the past and present, animal-sediment interactions, and chemical factors. Because of the large number of factors involved, it is not surprising that the character of shelves differ widely in both morphology and sedimentary facies.

Shelf Topography

The morphology of shelves includes negative relief features, such as submarine canyons, marginal basins, and linear depressions. Linear depressions include U-shaped channels, swales, shelf valleys, ice-flow gouges, and reef channels, all of which are open-ended at least on one side, implying that they are potential conduits for sediment transport [Gorsline and Swift, 1977]. Positive relief features of both structural and sedimentological origin include banks of approximately equal dimensions and linear ridges. Banks range from major, structurally controlled features to smaller features, such as reefs, diapirs, and coastal shoal complexes. Ridges include large features of structural origin to fields of small sand ridges common to the Atlantic shelves of the United States, Argentina, and Great Britain. Continental shelves also exhibit gradient changes, including terraces, shelf boundaries, and fault scarps [Gorsline and Swift, 1977]. Terraces are most pronounced on shelves of considerable relief, suitable for the incision of shoreline features. We have already seen that the configuration of the present surface of the continental shelf is largely a product of erosion and deposition during Quaternary sea-level fluctuations. Relict topography is very widespread at all latitudes, but it is most obvious on glaciated shelves of high latitudes, such as the Scotian and New England shelves. In these regions, dissected topography relief may exceed 200 m, as is the case in George's and Sable Island banks. The outer parts of these shelves may have several banks rising above the levels of the inner shelves. The banks represent morainic ridges formed by the deposition of sediments at the margin of the previous ice sheets and outwash plains built along the edges of the banks.

Shelves south of the direct effects of glaciation, but far enough north to receive meltwater drainage, exhibit relict river channels, as in the case of the Hudson Channel and the Block Channel [Swift, 1974]. The surface of shelves south of the ice advance tend to exhibit a subdued ridge and swale topography, which may represent relict beach-ridge topography preserved in Holocene transgressive sand sheets [Garrison and McMaster, 1966]. On the other hand, Uchupi [1968] interprets these linear features as large-scale longitudinal bedforms generated by storms [Swift, 1974].

On completely nonglaciated shelves, such as West Africa, a wide variety of geomorphic patterns resulted from the lowstand of sea level during the last glacial episode (Fig. 10–13) [McMaster and others, 1970]. At that time, rivers cut deeply into older shelf material and transported sediment to the shelf edge; deltas were built in the area of the shelf edge. Some streams cut deep valleys across the shelf, discharging sediment into submarine canyons. Barrier islands such as St. Ann's shoal (Fig. 10–13) were built close to the shelf edge by longshore sediment transport. The shelf morphology has been further modified by stillstands during the Holocene transgression represented by remnant barrier island-lagoon complexes and sea cliffs at levels of −90, −80, −55, −45, −35, and −25 m. Some of these were capped by accumulations of calcareous algae [McMaster and others, 1970].

Shelf Sediments

Sediments that veneer present day continental shelves consist of nearshore modern sediments in equilibrium with present-day environments and relict sediments that are not in equilibrium with the environment. About 70 percent of the sediment cover of the continental shelves has been classified as relict [Emery, 1968].

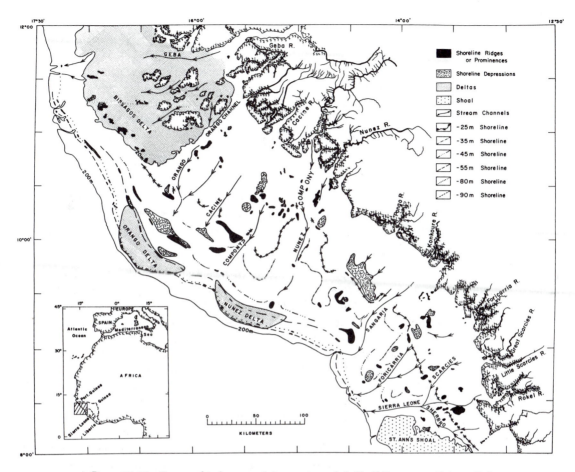

Figure 10-13 Geomorphic features of the continental shelf off Portuguese Guinea, Guinea, and Sierra Leone. (After R. L. McMaster, T. P. Lachance, and A. Ashraf, 1970)

The old-fashioned concept of decrease in grain size of modern sediments from nearshore across the continental shelf is now known to be largely restricted to certain nearshore sediment prisms (Fig. 10-14). Modern sands are generally deposited near the shore within about 6 km of the shoreline, where they are dispersed by longshore and other currents (Fig. 10-14) [Curray, 1977]. Beyond these are the relict sediments of the central and outer shelf (Fig. 10-14). These are differentiated from modern sands by their coarse character, iron staining, and dissolution pitting from subaerial weathering, and by their association with freshwater peat, oyster shells, and animal remains. Modern shelf muds occur only off rivers, in depressions, and in coastal areas and are probably dynamic accumulations. Although much may be carried to central and outer shelves, the general rule is that little such deposition occurs much beyond 30 km from the shorelines. In many areas, modern shelf sands are dominantly of biogenic origin. Thus shelf facies can be differentiated into a **shelf relict sand blanket** [Curray, 1964, 1965; Swift, 1969, 1970], which is a discontinuous veneer overlying Tertiary or older bedrock and modern sediment facies, a

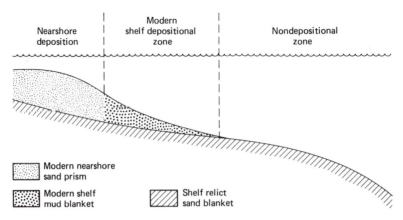

Figure 10-14 Schematic cross section of sediment facies of many modern continental shelves.

Legend:
- Modern nearshore sand prism
- Modern shelf mud blanket
- Shelf relict sand blanket

nearshore modern sand prism, consisting of beach sands and a seaward thinning and fining wedge of nearshore sand, and a **modern shelf mud blanket** further seaward (Fig. 10–14). The modern shelf mud blanket has resulted from bypassing of suspended fine material across the nearshore zone [Swift, 1974] and subsequent deposition on various parts of the shelf.

Tidal currents are also important in the redistribution of sediments in the nearshore area, especially in certain enclosed seas and blind gulfs, such as the seas of Northwest Europe, the Persian Gulf, the Gulf of Maine, and other similar areas. Tides result from changes in the gravitational attraction upon surface waters, mainly by the moon. Tidal fluctuations result from the interaction of a number of gravitational forces, both astronomical and terrestrial, and can occur twice daily **(semidiurnal tides),** once daily **(diurnal tides),** or as **mixed tides.** The strength of tidal currents are influenced by the amount of tidal fluctuation. Maximum current speeds induced by tidal fluctuations can reach up to 100 cm/sec [Johnson, 1978]. These currents can strongly affect sedimentation, forming linear sand ridges up to 30 m high and parallel with the tidal currents and sand waves (which are large ripple marks oriented perpendicular to the current direction [Off, 1963]).

East Coast North American sediments

The eastern coast of North America is made up of several environments, from shoreline to outer shelf, marked by specific morphologic, sedimentologic, and oceanographic parameters. These include the nearshore zone; the inner shelf, which exhibits large scale, active bed forms, and mobile coarse-to-fine sediments; the middle shelf zone, which is more stable than either the inner or outer shelf and thus veneered by relict sediments; and the outer shelf, which experiences frontal zone processes, such as boundary currents and internal waves making contact with the shelf edge. As a result, the sedimentary environment is perhaps more active here than commonly realized [Gorsline and Swift, 1977].

This zone pattern across the shelf is superimposed upon a much broader, latitudinally distributed pattern of **shelf provinces.** Cape Hatteras is a promontory that divides the shelf into a northern region dominated by detritus and a southern region dominated by calcium carbonate (Fig. 10–15). This difference gives rise to distinct color differences of shelf sediment: olive to the south due to the biogenic contribution, and brown and yellow to the north because of the greater glacial contributions. North of Cape Hatteras, shelf sediments are composed predominately of terrigenous sands that reflect glacial sources off New England and Canada (Fig. 10–15). Glacial till and gravelly outwash dominate the shelf only in the northern regions northeast of New York, the southern limit of glacial advance. The central Atlantic region is marked by fluvial sediment, in addition to reworked coastal deposits from the central Atlantic states. The inner shelf surfaces from Long Island to Cape Hatteras and the Georgia Bight also exhibit sand ridge topography as a result of intense current activity.

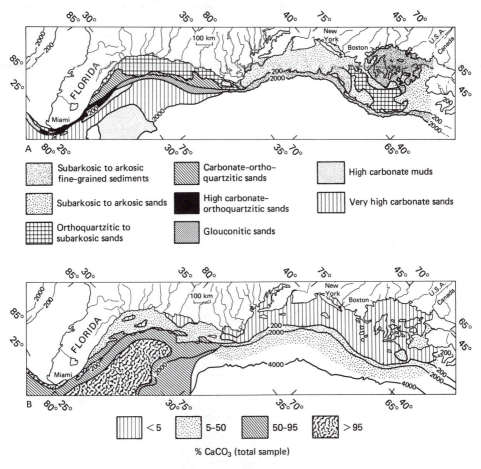

Figure 10-15 Sediment characteristics of the east coast of North America. Top figure shows sediment types; bottom shows calcium carbonate content. (From J. D. Milliman, O. H. Pilkey, and D. A. Ross, *Geol. Soc. Amer. Bull.*, vol. 83, pp. 1316, 1328, 1972, courtesy The Geological Society of America)

South of Cape Hatteras, skeletal sands and clean quartzitic sands are typical of the shelf sediments, with carbonates increasing south of Cape Canaveral and in the Bahamas (Fig. 10–15). Fine, silty sands form a nearshore belt and carbonates increase offshore (Fig. 10–15). Sediments are richer in quartz and poorer in feldspar because rivers eroded coastal plain areas, where chemical weathering is important. To the north of Cape Hatteras (Fig. 10–15), sediments are richer in feldspars because glacial erosion was dominant and chemical weathering less important.

Carbonate shelves

Although terrigenous sediments prevail over most continental shelves of the earth, there are some areas where carbonate sediments dominate. In areas where rates of terrigenous sediments are low, carbonate sediments accumulate largely from materials of biogenic origin. These typically accumulate on broad, shallow shelves where the hinterland is so low that little terrigenous detritus is available. They are most characteristic of shallow, tropical seas. Certain types of carbonate shelf sediments are restricted to depths shallower than 25 m, because at greater depths there is little or no development of the two most important types of shallow water carbonates: reef building corals and nonskeletal carbonates. Carbonate shelves, being largely of biogenic origin, differ markedly in many of their characteristics from terrigenous shelves. Unlike the clastic sediment wedges, the carbonate margins are created, influenced, and altered by a set of processes unique to themselves [Gorsline and Swift, 1977]. Production of carbonate material and its preservation is linked to climate, oceanographic factors, and especially the rate of terrigenous sedimentation [Milliman, 1974]. A dramatic southward increase in carbonate content of surface sediments of eastern North America (Fig. 10–15) reflects changing terrigenous input from the continent and an increase in water temperature and related increase in the diversity and productivity of organisms secreting carbonate. In the modern ocean, terrigenous sedimentation is more important in nearshore areas, so the carbonate content of shelf sediments increases offshore, reaching a maximum on the outer shelf. This increase in carbonate concentration does not reflect an increase in productivity of the carbonate-building organisms, which are often more productive at shallower depths, but instead reflects a reduction in the dilution of terrigenous detritus. Thus carbonate sediments predominate in areas where the terrigenous flux is small.

Most carbonate sediments are sand-sized, although mollusks and limestone fragments can contribute significant quantities of gravel in some areas [Milliman, 1974]. Nonskeletal grains form in shallow water and include pellet-aggregates and ooliths. **Ooliths** are small (about 0.5–1 mm), rounded, accretionary bodies of aragonite resulting from inorganic precipitation. The type of carbonate component depends both on the environment and on the age of deposition. A similarity of assemblages among tropical shelf carbonates of the world suggests similar environmental conditions for deposition. Reefs, so often thought of as the sole feature on carbonate shelves, often represent only a small part of the assemblages, where present. The inner-shelf areas contain lower carbonate content because of dilution by richer terrigenous sediments and also because substrates are softer—and therefore unsuitable—for development of attached, or epifaunal, assemblages. In contrast, the outer shelf often exhibits a harder substrate, which is required by coralline algae and epifaunal filter feeders.

The warmer carbonate shelf areas are marked by hermatypic corals and other warm-water organisms. Modern reefs contain only about 10 percent of corals in the original growth position. These, however, provide the reef framework. The remaining material consists of organisms inhabiting the intrareef frame and reef rubble and unconsolidated sediment [Sellwood, 1978]. Benthonic plants secreting carbonate add calcareous encrustations to the reef and provide the essential cement for the framework. Thus such reefs are more appropriately referred to as **coralgal reefs,** which—together with bryozoans and barnacles—form prominent facies along many outer shelves. Those corals that form in shallow subtropical-to-tropical seas are of the hermatypic types. These types are distinctive because they contain symbiotic zooxanthellae, which are largely dinoflagellates and which employ photosynthesis (thus limiting their growth to the photic zone). The main group of post-Paleozoic reef-building corals are the scleractinians which comprise about 700 modern species in the Indo-Pacific region and 50 species in the Caribbean. The principal environmental requirements for the growth of hermatypic corals as summarized by Ginsburg [1972] and Sellwood [1978] are as follows:

1. Shallow water (maximum depths of 100 m).

2. Warm water (total range, 18°–36°C).

3. Normal salinities (27–40‰).

4. Rather strong sunlight.

5. Adequate zooplankton productivity for food.

6. Stable substrate for attachment.

The number of genera and species increases with increasing temperature (Fig. 10–16) until optimum temperatures of about 23° to 27°C are reached. Coral reef growth tends to be retarded at higher temperatures. Optimal reef growth therefore occurs in the warm subtropics and tropics up to 20°N and 30°S of the equator. Coral reefs are best developed at the western margins of the oceans as compared with the eastern sides, which are under the influence of the colder limbs of the gyres and the upwelling of cold water (Fig. 10–16). Although the hermatypic corals are restricted to warm waters, many other components, such as coralline algae, barnacles, bryozoans, mollusca and sabellariid worms, can occur over a wide latitudinal range. In fact, the subtropical shelves of the southeastern United States are dominated by carbonate material other than corals. It is important to note that **ahermatypic corals,** those that lack symbiotic algae, can form reefs over much wider ranges and water depths than hermatypic corals because of their wide temperature tolerances (as low as −1°C). Like coralgal reefs, these reefs can dam sediment on the shelf and prevent its transportation to the continental slope. They are common on the continental shelf of Norway.

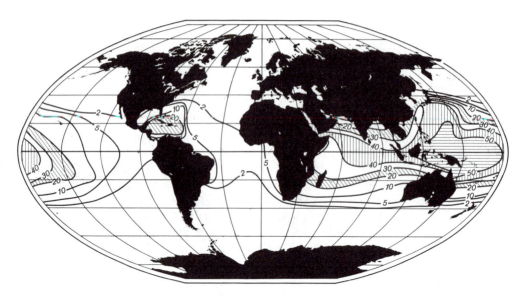

Figure 10-16 Number of genera of hermatypic corals in the world ocean based on data of Stoddart [1969]. The most prominent modern reefs occur in the western equatorial Atlantic and the Indo-Pacific. (From J. D. Milliman, *Part I Recent Sedimentary Carbonates, Marine Carbonates*, p. 155, 1974, Springer-Verlag New York, Inc.)

Although calcium carbonate solubility increases with decreasing water temperatures and hermatypic corals live only at temperatures higher than 18°C, shallow-water carbonate shelves do occur in colder areas, provided that the sediments are not extensively diluted by terrigenous sediments. Despite more terrigenous sediments at higher and middle latitude, important examples occur. The dominant epibenthonic filter feeders in such areas are often bryozoans. Bryozoan-rich sediments are typical of southern Australia, parts of Alaska, much of the Mediterranean, and several other regions [Milliman, 1974]. Barnacles may dominate in other areas, such as the shelves of eastern North America—including Grand Banks and George's Bank. It is not known why bryozoans dominate on some shelves and barnacles on others [Milliman, 1974].

In tropical to subtropical areas, carbonate shelves are of two major types [Sellwood, 1978].

1. **Protected shelf lagoons,** like those of the Great Barrier Reef and Bahama Islands, are enclosed within fringing barriers established by carbonate biogenic growth. The shelf-lagoon is generally a low energy environment, and thus muddy sediment is easily deposited.

2. **Open shelves,** such as Yucatan and western Florida, are similar to most other continental shelves in depth and structure and are exposed to greater energy because of the lack of a fringing barrier. Thus these shelves tend to exhibit coarse carbonate detritus since the finer material is transported to deeper depths.

Quaternary sea-level oscillations have played a major role in the sedimentation and character of carbonate shelves. This is particularly marked in open shelves where the Holocene sea-level rise created a major migration of sediment facies across the shelves as in their terrigenous-shelf counterparts. Modern shelf carbonates usually reflect a mixture of modern components and relict components of Pleistocene age [Milliman, 1974]. The age of these components is often difficult to determine because many of these organisms live over a wide depth range. Carbonate shelves are also largely protected from significant terrigenous sediment influx as this sediment tends to be trapped in adjacent estuaries, similar to the processes of modern terrigenous shelves.

Carbonate buildup is a general term used for significant accumulations of organic carbonate material. These accumulations are called *reefs* when they exhibit evidence of growth maintenance in the wave zone [Sellwood, 1978]. Layers and lenses of limestone are pieced together into a complex depositional and diagenetic marginal facies mosaic of reefs, sand, mud, and rock. Patch reefs, lagoon muds, and tidal and storm flats are developed on the landward side; seaward, a complex of material that includes cemented and uncemented rubble, mud, and sand, develops. Little sediment accumulates in the area downslope from the shelf margin because, as the buildup proceeds, the water becomes too deep, leading to a starved basin surrounded by fringing reefs.

Each carbonate buildup results from the growth of a particular suite of organisms. The modern coral reef is ecologically a steady-state system of extremely high organic productivity, intense carbonate metabolism, and complex food chains [Stoddart, 1969; Sellwood, 1978]. Reef-building organisms of the past were not always the same as those of the modern ocean. During the Phanerozoic, many groups of organisms that contributed to carbonate buildups evolved and became extinct. This created dynamic changes in the ecological interrelations of organisms through time. Hence modern analogies are not valid, especially those earlier than the Cenozoic. A second major difference between modern and past carbonate sedimentary environments concerns the width of shelves. Modern carbonate shelves are narrower than those built during certain parts of the Phanerozoic, such as the Cretaceous, when shallow, epeiric seaways covered vast areas of the continents depositing relatively thin (less than about 700 m) limestone sequences. The modern ocean contains no comparable environmental settings because of the relatively low sea levels that marked the late Cenozoic. A third factor of importance is that before about 100 Ma (middle Mesozoic) shallow-water carbonates were of greater importance in the global carbonate budget, because at that time two main groups of carbonate-secreting organisms that contribute to deep-sea sediments—the planktonic foraminifera and the calcareous nannoplankton—had not yet evolved. Since then, shallow-water carbonates have been less important. In the recent, shallow-water carbonates account for about 22 percent of the total. Carbonate buildups occur on both tectonically passive and active margins because they require only a suitable substrate and climate and isolation from terrigenous sediments for growth. Buildups occur in recently rifted margins such as the Red Sea and also at shelf margins during the mature stages of passive margin development. In these regions they exhibit upward growth or maintenance as the margin subsides or sea level rises and may eventually cap the continental terrace as the terrigenous influx is further diminished [Gorsline and Swift, 1977]. As these marginal masses form, they become buried and undergo a succession of biological, physical, and chemical processes. For instance, burrowing animals destroy much of the original reef structure. The resulting voids are filled in by the encrustations of carbonate-secreting organisms and by mud [Milliman, 1974]. Contemporaneous cementation consolidates the material, producing a rock quite different than the original reef structure. It is not critical that carbonate sediments be buried in order to be cemented [Ginsburg, 1957]. The submarine lithification of carbonate sediments remains a poorly understood process, but it is important because it consolidates carbonate sediments at early stages of their buildup, leads to further growth, and prevents their dispersal by submarine slumping.

CORAL REEFS

Coral reefs can be classified into three main types according to their morphology (Fig. 10–17). **Fringing reefs** lie adjacent to the land with essentially no lagoon. These reefs tend to occur in areas such as the Red Sea where little freshwater runoff, which would otherwise interfere with coral growth, occurs. **Barrier reefs** lie further offshore, with a lagoon of varying depths separating them from the land. The most spectacular example is the Great Barrier Reef of Australia, which is backed by shallow open sea several hundred kilometers in width. **Atolls** are subcircular reefs enclosing a lagoon; they are typically about 40 m deep with no island and only partially exposed, low-lying carbonate **cays** or **patch reefs**. Atolls, such as Bikini, are most common in the Pacific.

There has been much interest in the origin of coral atolls since 1842, when Charles Darwin formulated his theory about their development resulting from his observations during the voyage of H.M.S. *Beagle*. Darwin believed that the three main forms of coral reefs—fringing reefs, barrier reefs, and atolls—represent progressive stages of a single evolutionary process. Corals growing on the sides of a slowly subsiding volcanic or nonvolcanic island grow upward at a sufficiently rapid rate to remain as an atoll after the volcanic island has subsided below sea level (see Chapter 5). Corals do not grow very extensively in a lagoon, since nutrients are reduced. Thus lagoons tend to remain unfilled. Darwin's theory required two basic assumptions: that subsidence of oceanic volcanic islands was sufficiently widespread to account for the abundance of atolls, and that corals grow at rates fast enough to keep up with this subsidence. Much subsequent work has shown that both of these assumptions were well founded. Atoll drilling has indeed shown that the volcanic edifice does subside and that great thicknesses of reef materials have accumulated as subsidence occurred over millions of years. Drilling on Eniwetok Atoll revealed a coral accumulation of up to 1500 m; at Bikini Island, an accumulation of about 800 m was found. Furthermore, there is evidence that these materials were previously laid down close to, or above, sea level. Typical Caribbean reefs have accreted at rates of 0.5 to 1 m per thousand years on the forereef terrace, a

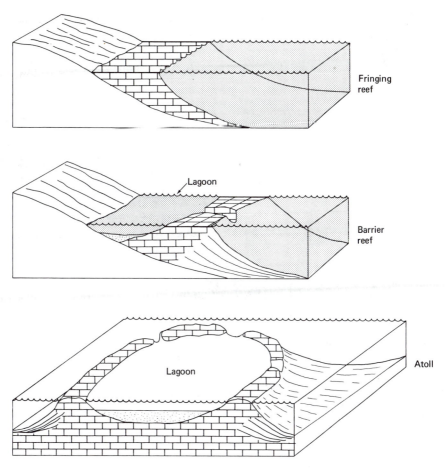

Figure 10-17 Schematic cross section of the three main types of modern reefs. (After R. C. Selley, 1970)

rate sufficient to keep up with any subsidence. The subsidence of ocean crust, associated with sea-floor spreading, is continuous from its origin at the mid-ocean ridges until its destruction in the subduction zones. At some point in time, and for a number of reasons, the subsidence rate of the volcanic island outstrips the upward rate of coral growth and the atoll sinks beneath the sea surface, forming a guyot (see Chapter 5).

Wells [1957] distinguished the following physiographic divisions of Indo-Pacific atolls.

SEAWARD SLOPE: Or **reef front,** a cliff consisting chiefly of reef-derived talus associated with a biota that is distinct but gradational, with the reef flat at shallower depths. The foot of the cliff is usually marked by talus.

REEF MARGIN: Contains an important algal ridge and coral growth.

REEF FLAT: The flat top of the reef formed because the organisms cannot survive prolonged subaerial exposure where wave action planes the surface. The reef flat is exposed to diurnal tidal cycles and has large fluctuations in pH, salinity, and oxygen content.

BACK REEF: Mostly a calm shallow area exhibiting a much more restricted biota due to higher environmental variability and reduced nutrients.

LAGOON: Lacks appreciable coral growth and is usually floored by finely divided carbonate mud in its deeper parts and sands in more turbulent areas.

Lagoonal sediments are also composed of fecal pellets, foraminiferal sands, coralgal sands of comminuted frame-building corals, and associated calcareous algae [Selley, 1970]. Atlantic reefs tend to lack the true inner lagoons of the Indo-Pacific.

Reef development is enhanced on the windward (or seaward) side of islands, where wave action is more intense and waters have more open-ocean qualities. On an ocean-wide basis, coral-reef development is limited or prevented at the eastern margin of ocean basins which are marked by colder surface currents and more intensified upwelling (see Chapter 8). Barrier reefs thus occur only at the western margin of the ocean basins.

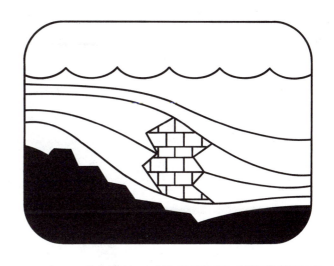

eleven
Continental Margin Types
and Divergent Margins

The researches of many commentators have already thrown much darkness on this subject, and it is probable that, if they continue, we shall soon know nothing at all about it.

Mark Twain

CONTINENTAL MARGIN TYPES

Introduction

The **continental** (or oceanic) **margins** mark the transition between thin, dense oceanic crust and the thicker, lighter, chemically different continental, or transitional, crust. Isostatic balance between the two crusts, modified by subsequent sedimentation or erosion, creates a major step in the seabed at the boundary, expressed by the continental slope [Laughton and Roberts, 1978]. Plate tectonics explains the evolution of the continental margins, which provide information about the formation of the continents because they represent the growing edge of the landmasses. The actual margins of continents are marked by submarine terraces and slopes at some distance from the seashore, evidence of the ephemeral nature of sea level.

Enormous thicknesses of marginal marine sediments comprise over half the total sediments of the ocean [Heezen and Tharp, 1965]. Much of the present tectonic activity takes place along continental margins, which enables interpretations of earlier phases of the earth's history observable in the folded, uplifted sequences of the inland mountain ranges. Ancient geosynclines, now transformed into mountain systems, often formed along the margins. Current hypotheses about continent and oceanic basin development require that major activity occur along the margins. The economic value of hydrocarbons in continental margins has encouraged the recent rapid increase in their study.

320

Continental margins encompass a wide transition zone that separates the oceanic and continental realms. This zone includes the continental shelf, slope, and rise and coastal plains. Continental shelves occupy about 7 percent of the sea floor (27×10^6 sq km). They are typically shallower than 130 m, but they can range to 550 m. Continental slopes occupy about 9 percent of the ocean floor (28×10^6 km²) and range from the shelf edge to depths of 4000 to 5000 m. The base of the slope is defined as the point where the sea-floor gradient drops below 1 in 40, producing a more gentle, seaward-sloping continental rise, extending from a depth of 4000 to 6000 m. Continental rises, at the base of the slopes, are immense accumulations of terrigenous sediment deposited by turbidity currents and other gravity flows and smaller quantities of pelagic sediments. Rises that occupy 19×10^6 km² of the ocean floor are modified by various types of bottom currents and occur only where the continental margins lie within crustal plates. Where deep-sea trenches mark zones of convergence at the edges of crustal plates, rises are absent. K. O. Emery [personal communication] has estimated the volume of post-Paleozoic sediments in various features, as shown in Table 11-1. Although the continental slopes and rises contain about 60 percent of the total sediment volume, they represent only about 10 percent of the earth's surface.

Classification of Margins

Eduard Suess [1885] recognized the fundamental differences between the continental margins that surround the Pacific and Atlantic oceans. He noted that the Pacific Ocean is bordered by folded mountain chains, island arcs, and volcanoes that parallel the coasts of the major continents; in contrast, the Atlantic margins exhibit a wide coastal plain, marking the landward continuation of the shelf. Using observations such as these, Suess classified continental margins into two types: *Pacific* and *Atlantic*. Early classifications of continental margins were based on topography, rather than structure. The knowledge of continental margins and their

TABLE 11-1 SEDIMENT VOLUMES STORED IN DIFFERENT GEOLOGICAL REGIONS[a]

	Volume ($\times 10^6$ km³)
Continents	45
Shelves	75
Slopes	200
Marginal basins	35
Rises	150
Deeper basins	25
Total	530

[a] After K. O. Emery [personal communication].

evolution advanced rapidly during the 1960s and 1970s as geophysical technology improved dramatically. Development of the continuous seismic-reflection profiler about 1960 (see Chapters 1 and 2) began to provide important detailed data about the internal structure of continental margins. This has led to further advances in their classification. Because of the enormous thicknesses of sediments on the continental margins, deep-sea drilling has provided information only about the upper part of the sequence in a few areas. Furthermore, the danger of drilling into hydrocarbon reservoirs without special blowout equipment has prevented drilling into many marginal sedimentary sequences. Nevertheless, deep-sea drilling has provided important insights into the nature and evolution of continental margins. Two principal types of continental margins, Atlantic and Pacific, are still recognized. The Atlantic, or passive margin, type is marked by stable continental blocks on the landward side that have been little deformed since Paleozoic time. They essentially lack earthquakes and widespread volcanism, unless very young, and are typically composed of a continental shelf, slope, and rise. Around the Indian Ocean, continental margins are mainly of the Atlantic (divergent) type, except in the northeast region, which is marked by the Java Trench and Indonesian Arc (Fig. 11–1). In contrast, **active margins** like those that border the Pacific Ocean (Fig. 11–1) are associated with a trench, volcanism, active mountains, and earthquakes that extend to a depth of as much as 700 km along the Benioff zone. The active margins are often associated with island arcs, marginal seas, and interarc basins [Laughton and Roberts, 1978].

The development of plate tectonics led to an understanding of the processes responsible for margin evolution and identification of their individual characters in more dynamic terms. A terminology and classification evolved based on the pro-

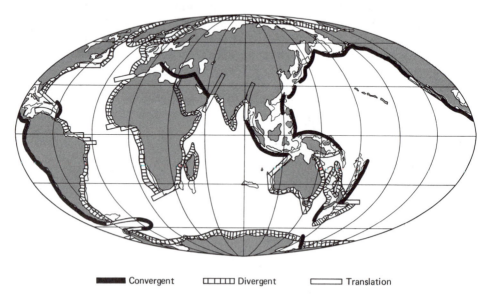

Figure 11–1 Distribution of convergent, divergent, and translation type margins around the world. (After K.O. Emery, 1980)

cesses of global tectonics, creation and destruction of oceanic crust. Destruction is related to subduction, the underlying cause for trench and mountain formation, seismicity and volcanism, crustal deformation and metamorphism, and opening of backarc basins. Several terms are applied to the same types of margins.

1. Divergent = passive = aseismic = Atlantic.

2. Convergent = active = seismic = Pacific.

3. Transform = translation. These can be active or passive and occur in both Pacific and Atlantic oceans.

Divergent, or passive, margins

Divergent margins mark the ocean-continental transition lying within a rigid lithospheric plate; that is, they are not plate boundaries. These margins develop when continents are rifted apart to form new oceans, so that the continent and adjacent ocean floor are part of the same plate. They form at divergent plate boundaries initially by rifting of the continental crust and, with time, move away from these boundaries, cooling and subsiding. Eventually they become sites of massive subsidence and thick accumulations of sediment. These boundaries are marked by extensional tectonics, a result of the "breakup" of the supercontinents of Gondwanaland and Laurasia during the Mesozoic and early Cenozoic (Fig. 11-2). The structure of these boundaries is relatively simple, although complications have resulted from thinning or stretching of the continental crust prior to fracture. Furthermore, changes in the position of rifting or the spreading axis have caused fragments of continental crust to be left behind as microcontinents surrounded and isolated by oceanic crust.

The morphology of passive margins is related to the trend of the initial rift, though older margins may have been substantially modified by subsidence, erosion, and deposition, leading to development of the classic shelf-slope-rise profile. This profile implies the dominance of subsidence tectonics (through sediment loading) over compressional tectonics, which dominates in the active margins. The tectonic character of passive margins allows the formation of the continental rise through sediment accumulation near the base of the continental slope. Continental rises form adjacent only to divergent margins.

Convergent, or active, margins

Active continental margins mark the boundaries between two converging plates, resulting in deformation or destruction of crust, often by subduction back into the interior of the earth. Unlike passive margins, the continent and the adjacent ocean floor belong to different plates. In most convergent margins, oceanic crust underthrusts continental crust (for instance, the west coast of South America); elsewhere, oceanic crust underthrusts modified oceanic crust (Lesser Antilles and South Sandwich Islands), and continental crust underthrusts other continental crust (Tibetan Plateau—two continents thick). These boundaries are also referred to as **seismic margins** because they are marked by shallow, intermediate,

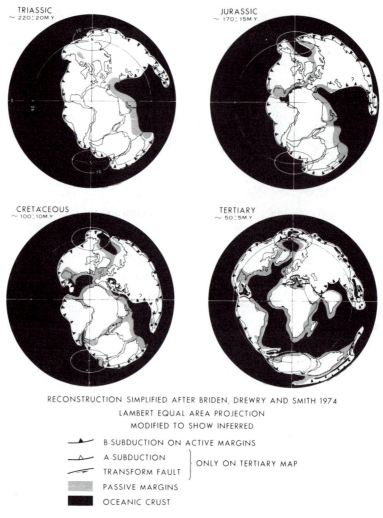

TRIASSIC
~ 220±20M.Y.

JURASSIC
~ 170±15M.Y.

CRETACEOUS
~ 100±10M.Y.

TERTIARY
~ 50±5M.Y.

RECONSTRUCTION SIMPLIFIED AFTER BRIDEN, DREWRY AND SMITH 1974
LAMBERT EQUAL AREA PROJECTION
MODIFIED TO SHOW INFERRED

B-SUBDUCTION ON ACTIVE MARGINS
A-SUBDUCTION   } ONLY ON TERTIARY MAP
TRANSFORM FAULT
PASSIVE MARGINS
OCEANIC CRUST

Figure 11-2 Stages in the breakup of Pangaea. (Reconstructions after Briden and others, 1974, simplified in Bally, 1979)

and deep earthquakes. They are associated with oceanic trenches and volcanic island arcs or mountain ranges, depending on whether the margin is an ocean-ocean contact or an ocean-continent contact. Most characteristic, however, is the formation of island arcs, with marginal basins developing on their concave sides. The transition from ocean to continent is much more complex and variable than in the Atlantic, and sedimentary accumulations do not appear to be as great. In the Pacific, the typical active margin is a continental shelf and slope bounded on the seaward side by a trough or trench. These topographic features act as effective sediment traps to block further seaward transport of sediment from the continents and explain the absence of continental rises around the Pacific. The structural, sedimentary, and volcanic histories of these subduction margins involve complex geologic relationships and are difficult to explore. In this book, active margins are used in their broadest sense to include the areas of backarc basins and intraoceanic arcs.

Transform, or translation, margins

Transform fault systems can intersect both passive and active margins (Fig. 11-1). Transform margins result from horizontal shear motion between plates and are marked by shallow focus earthquakes. Transform margins can become tectonically passive. During rifting, pieces of continental crust may move relative to adjacent oceanic crust, adding to difficulties in reassembling the original continental mass [Emery, 1977].

DIVERGENT, OR PASSIVE, MARGINS

Divergent margins bound the Arctic and Norwegian seas (Fig. 11-1), the North and South Atlantic, the Indian Ocean (with the exception of the Sunda Arc), and the Antarctic continent (except the Scotia Arc). Parts of the Mediterranean margin can also be considered to be of divergent type. Divergent margins have been extensively studied in the North Atlantic, although much is being learned elsewhere in the world, such as eastern South America.

Divergent margins are marked by rather smooth relief, due to the relative tectonic inactivity and large sediment accumulations. This sediment has accumulated in stable, but continually subsiding, tectonic environments with little deformation. Seismic refraction studies have revealed thicknesses of as much as 15 km but continental rise sediments are often so thick that most seismic systems are incapable of penetrating them. These great thicknesses of sediment cover the boundary between continental and oceanic crust, making its precise location and definition difficult. The contact has been obscured by extensive prograding of the continental slope and by deep burial of its base by terrigenous sediments. It may be further masked by topographic irregularities, such as submarine canyons and slumps on the continental slope, as well as by sediment dams along the slope, such as organic reefs, diapiric intrusions, and fault blocks [Emery and Uchupi, 1972].

Identification of the boundary between oceanic and continental basement is required for any precise predrift reconstruction; however, at present no satisfactory criteria exist for precise determination of continental boundaries [Rabinowitz, 1974]. Bullard and others [1965] selected isobaths for mathematically determining and evaluating the best geometric fits. Although these fits appear visually satisfactory, many gaps and overlaps are present [Rabinowitz, 1974]. Talwani and Eldholm [1973] demonstrated that the boundary between oceanic and continental basement is not necessarily associated with particular bathymetric contours, since subsidence of continental material along margins can complicate the picture.

The continental margin of the Northwest Atlantic Ocean, as an example, possesses several bordering and semiparallel geological and geophysical features (Fig. 11-3). Seismic refraction data off Nova Scotia have shown that a transition zone 50 km wide occurs near a distinct magnetic anomaly over the slope. This

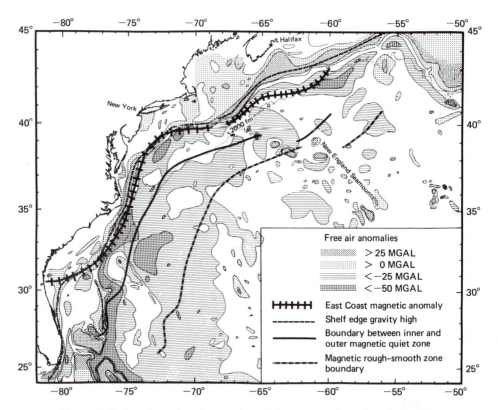

Figure 11-3 Location of major geophysical lineaments along the continental margins of eastern North America. (From P.D. Rabinowitz, doctoral dissertation, "The Continental Margin of the Northwest Atlantic Ocean," Columbia University, 1973)

anomaly is generally believed to be due to a magnetization contrast between oceanic and continental crust, a phenomenon known as the **edge effect.** The group of semiparallel lineaments (Fig. 11-3) recognized by Rabinowitz [1974] is as follows:

1. A subsurface ridge defined by seismic compressional velocities and located near the seaward edge of the continental shelf [Drake and others, 1959].

2. A continuous free-air gravity high anomaly located near the shelf break [Emery and others, 1970; Rabinowitz, 1974].

3. The nearly continuous high-amplitude magnetic anomaly, called the *East Coast magnetic anomaly* (ECMA) which is normally located over the slope, but in places dips as far seaward as the continental rise and as far landward as the coast line [Taylor and others, 1968].

4. A magnetic quiet zone situated seaward of the ECMA [Heirtzler and Hayes, 1967].

The magnetic quiet zone and the East Coast magnetic anomaly are two prominent characteristics of the magnetic field bordering this continental margin. The continuous free-air gravity high occurring near the shelf break is associated with a subsurface ridge called the **marginal basement high.** Local airy isostatic calculations show that this anomaly is not simply an edge effect resulting from thinning of continental crust; instead it results from an excess of mass below the sea floor. The gravity anomaly cannot be accounted for merely by calculating the thick accumulation of sediments and the subsurface ridge [Rabinowitz, 1974]. The East Coast magnetic anomaly is usually related to the marginal basement high. The magnetic anomaly labeled E (Fig. 11-3) occurs within the magnetic quiet zone, dividing it into inner and outer sectors. Rabinowitz [1974] considers the inner quiet zone, marked by a very subdued magnetic field, to be located over a subsided continental basement. In contrast, the outer quiet zone, marked by smaller-amplitude anomalies than those in the rough zones, is interpreted by Rabinowitz [1974] as being located over oceanic basement formed during the Newark interval (Triassic period) of dominantly normal geomagnetic polarity [Burek, 1970].

Evolution of Divergent Margins

Plate tectonics has provided a framework for interpreting the evolution of continental margins. It is generally assumed that **continental accretion,** or growth, has occurred throughout geologic time by sedimentation at both passive and active margins, giving rise to lateral extension of the continental crust. Whether or not the total volume of the ocean basins has been constant throughout geological history is unclear. In any case, some crustal thickening must have occurred to accommodate this lateral expansion. This requires changing geological and geophysical characteristics within these margins as they evolve tectonically [Drake and Burk, 1974].

Divergent continental margins undergo tectonic development of predominantly vertical type. This affects shelf and slope morphology of rifted margins. In addition, there is progressive widening of the area of transitional crust with maintenance of approximate isostatic equilibrium.

Most of the present-day divergent margins were formed within the last 200 m.y. (Fig. 11-2). About 200 m.y. ago, when all the present continents formed one supercontinent, Pangaea, there was only a small divergent margin extending from Northwest Africa to northern Australia (Fig. 11-2). The remainder of the margin of this large supercontinent was marked by active subduction of oceanic crust (Fig. 11-2). Since that time, Pangaea has fragmented and rifted margins have evolved. Sixty percent of the world's sediments are contained in the continental margins, and most are stored in divergent margins, monuments to 200 m.y. of slow subsidence and deposition in zones that originated in the interior of Pangaea [Fusod, 1977].

Evolution of continental margins can be classified into the following stages, shown in Fig. 11-4.

1 . *Rifting.* The rift-valley stage may not always occur, but it involves early graben formation prior to continental splitting (Fig. 11-4; Table 11-2). The present East African rift system may reflect this stage. Domal uplift caused by hot underlying upper mantle material may be associated with rifting, but

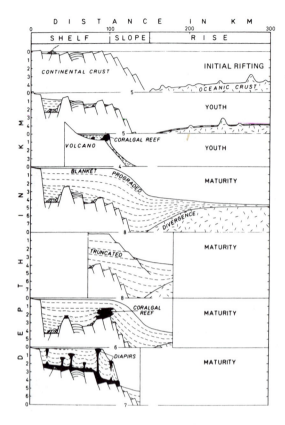

Figure 11-4 Evolutionary stages in the development of continental margins. Vertical exaggeration about 10×. (After K.O. Emery, 1980)

TABLE 11-2 STAGES OF CONTINENTAL MARGINS IDENTIFIABLE FROM CONTINUOUS SEISMIC REFLECTION PROFILES[a]

	Continental shelf	Continental slope	Continental rise
Initial	Pretectonic, tectonic volcanic, or glacial rocks, exposed or thinly covered.	Pretectonic, tectonic, or volcanic rocks exposed or thinly covered.	None.
Youth	Thick sediment fills basins and troughs.	Pretectonic, tectonic, or volcanic rocks exposed or thinly covered.	None or small.
Mature	Thick sediment cover: 1. broad blanket 2. coralgal reef dam 3. post-rift diapirs	Thick sediments: 1. prograded 2. truncated 3. post-rift diapirs	1. Thick for divergent margins. 2. Thin for translation margins.

[a]After Emery [1980].

328 Continental Margin Types and Divergent Margins

whether such doming is typical of passive margins or is mainly restricted to hot spot regions is unclear.

2. *Onset of spreading and youth.* This stage is characterized by separation of continental crust and accretion of oceanic crust in the gap between the continental blocks (Fig. 11–4; Table 11–2). It is a youthful stage, represented by the Red Sea, lasting about 50 m.y. after the onset of spreading. This stage is marked by rapid regional subsidence of the outer shelf and slope, while some graben subsidence may locally persist, with volcanic and thermal effects of the split dominating.

3. *Maturity.* Subdued regional subsidence begins after cessation of the initial thermal event (Fig. 11–4). Much of the present Atlantic-Ocean and Indian-Ocean margins are in this stage. Sedimentary thicknesses are vast and the sequences are strongly modified by slumping, canyon formation, and deep-current erosion and deposition.

Rifting

The ocean basins formed by the breakup of Pangaea provide partial records of their growth and paleoenvironments; however, records of their birth are well hidden beneath thick sedimentary deposits. Only along divergent margins can information be found on the birth of an ocean, and only divergent margins document the history of the transitional boundary between the continents and the oceans.

The shape of divergent margins is related to the trend of the initial rift, although subsequent modification takes place by subsidence, erosion, and deposition. The great East African rifts, the Rhine Valley graben, and the Rio Grande rift are believed to exemplify the basic rifting process [Bally, 1979], which sometimes leads to the formation of an ocean. Since the prerift sections in the ocean basins have rarely been penetrated, speculations on this evolutionary phase rely on the character of the modern rift-valley systems.

Rifting may be preceded by a period of regional domal uplift and possibly simultaneous volcanism. Uplifts are presumed to be thermal in origin. Subsequent erosion of the uplifted area and thermal metamorphism of the lower crust may cause crustal thinning. As the uplifted area expands, the crustal portion cracks and splits into subparallel and trilete (Y-shaped) fault patterns [Burke and Dewey, 1973]. The basic structural framework of the margin may be influenced by the preexisting structural fabric of the continent.

Intial rifting may not always occur in crust above sea level. In the Northeast Atlantic, rifting seems to have been submarine and not accompanied by widespread volcanism. This may have been because rifting took place on a Precambrian craton in the case of the Rockall Plateau and in an epicontinental basin in the case of Biscay. Figure 11–5 summarizes variation in basement structures resulting from differences in the early rifting phase in the development of oceans. This was marked by either volcanism in a narrow zone or a wider zone of dike intrusion.

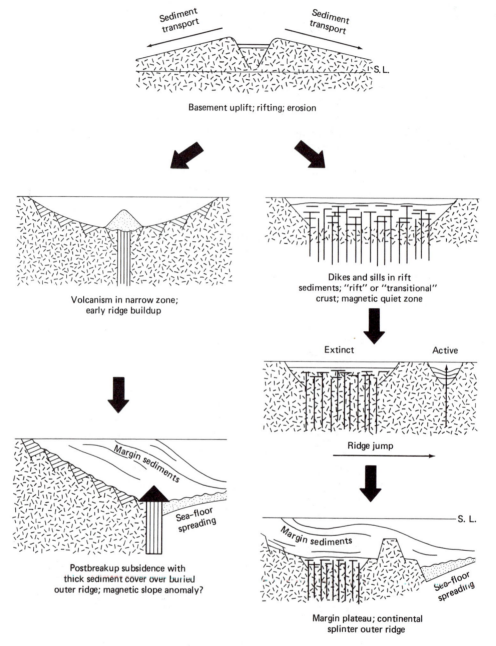

Figure 11-5 Evolution of passive margins. Style at left involves volcanism within a relatively narrow zone followed by early ridge buildup, postrift subsidence, and sea-floor spreading. Style at right involves more widespread initial intrusion of volcanic rocks, followed by a jump in the axis of volcanism and spreading.

Onset of Spreading

Following rifting, magma penetrates the continental crust, and oceanic crust begins to accrete at the edges of the separated blocks. Little is known about the character of changes from the rifting to the spreading phases. Continental reconstructions commonly exhibit overlaps and underlaps between different continents, suggesting width variations of the continent-ocean boundary zone. However, in many areas, oceanic magnetic anomalies and a characteristic isostatic gravity anomaly suggest that the continent-ocean boundary may lie within a narrow zone, linear over hundreds of kilometers. It is assumed that the continental crust is injected by dikes at this early stage and that sediments are interbedded with lavas.

The change from rifting to accretion marks a major change in the thermal regime of the margin. During rifting, the heat source remains beneath the rift axis, but the continental margins become successively remote from the heat source, allowing them to cool and subside (see Chapter 5). The initial line of spreading is not necessarily located within the zone of initial rifting (Fig. 11-5). For example, between Africa and North America, the initial spreading axis was located to one side [Rabinowitz, 1974]. Massive subsidence often follows the rifting phase of the passive margins, resulting from cooling or other processes unrelated to sediment loading. This is demonstrated by the existence of sediment-starved continental margins exhibiting considerable subsidence. Furthermore, subsidence curves derived from certain margins are similar to empirical cooling curves constructed for the oceanic crust.

Although occasional marine invasions occur during the earlier stages of rifting, real oceanic conditions do not begin until spreading commences. Even in the early stages of spreading, oceanic transgression may be prevented for a short period of time due to damming by the bounding continental blocks. A modern example is the Afar Depression of Ethiopia. Margins formed by recent rifting of the continent may have continental slopes steeper than normal because of the angle of the initial rift. The Gulf of California, for example, is a margin dominated by transform faulting, which began to open only about 5 m.y. ago. This margin exhibits some slopes steeper than 20°, which have not yet been modified by erosion and sedimentation.

Some recent models of divergent margin evolution postulate *thinning*, or attenuation, of the continental crust during the early stages of rifting. Other models suggest that attenuation continues over much longer periods of time and is linked to the overall subsidence history of the margin. Several different mechanisms for attenuation have been proposed, including: thermal expansion, uplift, and erosion; flow of the lower crust accompanied by subsidence; ductile thinning and metamorphism of the lower crust during any of these processes.

The time of breakup and formation of divergent margins may progress from one end of the margin to the other, as has occurred in the South Atlantic [Rabinowitz and LaBrecque, 1976]. Table 11-3 identifies the time of inception of several divergent margins [Bally, 1979].

TABLE 11–3   AGE OF INITIAL OPENING PHASE OF VARIOUS OCEAN BASINS[a]

1. East African rift system   (this may still be in the rifting stage).
2. Red Sea and Gulf of Aden   (about 10 Ma) [Laughton and others, 1970].
3. Gulf of California   (about 5 Ma) [Larson and others, 1968; Moore, 1973].
4. South Australian margin   (about 50 Ma) [Weissel and Hayes, 1972].
5. Labrador Sea and Europe; North America   (about 80 Ma) [Kristoffersen, 1977].
6. Norwegian Sea   (about 65 Ma) [Talwani and Eldholm, 1977].
7. South America/Africa   (about 130 Ma) [Larson and Ladd, 1973].
8. North America/Africa   (about 180 Ma) [Pitman and Talwani, 1972].

[a]After Bally [1979].

A proto-ocean: the Red Sea

The apparent fit of the Arabian and African coastlines of the Red Sea, combined with geophysical and geological evidence, has shown the Red Sea to be an incipient ocean resulting from sea-floor spreading. Therefore it is relevant to examine the geological character of the Red Sea, because of the possible insights it may provide about the early stages of sea-floor spreading and oceanic rifting.

The Red Sea consists of flat, shallow, coastal shelves (as deep as 370 m) from 30 to 120 km in width, which flank an axial trough with maximum depth of about 1000 to 2500 m along its length (Fig. 11–6). The trough, up to only 30 km in width, can be divided into a structurally simple marginal zone on either side of a deeper, structurally more complex axial zone [Ross and Schlee, 1973]. The main trough is similar to the rift valleys on mid-ocean ridges. It is a linear complex of valleys or depressions; in some areas, it is a flat-floored median valley (5–14 km wide), continuous over a few tens of kilometers. The valley exhibits symmetrical, high-amplitude magnetic anomalies, the most conspicuous anomaly being closely associated with the axial zone. Magnetic anomalies suggest that sea-floor spreading is occurring at the rate of 1 cm/yr. Positive Bouguer gravity anomalies of more than 100 mgal are associated with the main trough and result from the intrusion of dense basaltic rock. High heat-flow measurements in the main trough indicate the presence of molten material at shallow depths beneath the sea floor. Seismic profiles show little sediment cover in the axial zone, but as much as 500 m of sediment in other parts of the main trough (Fig. 11–6).

The Red Sea may have developed in two main stages. Initially, an early, or pre-Miocene, uplift and lateral extension resulted in crustal thinning, volcanism, and eventual formation of the main Red Sea basin. This was followed by sea-floor spreading during the Pliocene, which resulted in the axial zone of the Red Sea [Ross and Schlee, 1973]. By early or middle Miocene, the Red Sea probably had attained most of its present width, except for the amount of Pliocene and younger sea-floor spreading, which has given rise to the parallel magnetic-anomaly patterns. These extend only a short distance from the axial zone and suggest that continental crust of the separating continents remains on each side of the main trough. Thus magnetic, seismic-refraction, seismic-reflection, structural, and petrographic data show that oceanic crust is restricted to the main axial trough and therefore represents an incip-

I. Early or pre-Miocene

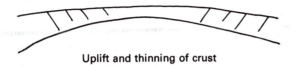

Uplift and thinning of crust

II. Middle Miocene

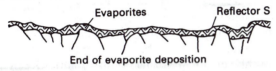

Sagging of flanks, rifting, volcanism, and evaporite deposition

III. Late Miocene

Evaporites   Reflector S

End of evaporite deposition

IV. Pliocene

Evaporites
Post-Miocene marine sediments   Reflector S

Erosion, uplift, salt deformation, and marine deposition

V. Late Pliocene-Holocene

Post-Miocene marine sediments

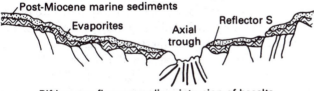

Evaporites   Axial   Reflector S
trough

Rifting, sea-floor spreading, intrusion of basalts,
and formation of axial trough

Figure 11-6 Evolution of the Red Sea since the middle Miocene (approximately 15 Ma). (From D. Ross and J. Schlee, *Geol. Soc. Amer. Bull.* vol. 84, pp. 3827–3848, 1973, courtesy The Geological Society of America)

ient phase of sea-floor spreading confined to a zone of up to only 30 km in width. It can be assumed that the Atlantic Ocean also began with a similar narrow trough.

Postrift Evolution

Subsequent evolution of divergent continental margins depends on age and the complex interaction between subsidence, sedimentation, climate, and ocean circulation. The lithology and volume of sediments that comprise the postrift sequence depend on oceanic paleoenvironment, climate, sea level, size, and geology of the continental hinterland. The ocean-basin margins disrupt the wind-driven latitudinal surface-water circulation to produce eastern and western boundary currents. These currents play a major role in the sedimentary history of the margins and have produced conspicuous unconformities in the record. The lithology and volume of sediments deposited on the margins are also strongly influenced by global changes in sea level. These and other factors that influence the sedimentary environment produce major regional differences in the sediment deposits. Overall continued subsidence provides a sequence of sediments, commencing with preuplift epicontinental sediments overlain successively by rift and block-faulting continental sediments, shore zone, and alternating shallow and deeper marine facies related to a long succession of marine regressions and transgressions. During the early spreading stage, sedimentation may be strongly influenced by barriers formed by fracture zones or volcanic ridges. Such barriers occurred in the early Cretaceous of the South Atlantic, for example, the Walvis ridge and Rio Grande rise.

Clastic material produced in the earliest rift stage has been observed only on land sections, such as the Triassic nonmarine red sandstones of eastern North America. The immediately overlying sediments, formed in the restricted environments typical of the early stages of ocean environment, are now well studied in the ocean basins. High evaporation rates in enclosed basins led to the deposition of thick evaporitic layers, such as in the narrow Mesozoic South Atlantic Ocean and the late Cenozoic Red Sea. The salt deposits are frequently accompanied by anoxic sediments, preserving high concentrations of organic carbon. A detailed discussion on the formation of these sediments is provided in Chapter 14. With subsequent oceanic growth and sedimentation, the early evaporitic sediments become deeply buried and may subsequently migrate upward through the overlying sediments to form the salt diapirs known to occur beneath the margins of the South Atlantic.

The history of subsidence largely controls the subsequent evolution of passive margins (Fig. 11-4). Observations on mature margins, such as eastern North America, show a subsidence of 3 km per 150 m.y. [Burke, 1979]. The rate of subsidence declines with time and is generally attributed to cooling of the margin as it moves away from the mid-ocean ridge. The principal effect of this subsidence is an increase in the volume of sediments that accumulates at the continental margin, which further increases the amount of subsidence (Fig. 11-4). Thus postrift sediments can become deeply buried as the margin progrades seaward. In the continental margins of North America, the deepest sediments are obscured by a nearly conformable series of marine Tertiary sediments capping Cretaceous and older accumulations.

Walcott [1972] quantitatively modeled the contribution of sediments on the slope and rise to margin subsidence. He used a simple flexural model, in which the lithosphere is represented by a thin elastic plate overlying a weak fluid layer and assumed homogeneous elastic properties for both oceanic and continental crust. This model shows that sedimentary sequences up to 18 km can be deposited before major flexure occurs. Such flexural models readily explain wide continental margins associated with deltas, but not narrower margins such as the North American East Coast [Bally, 1979]. Neither do the flexural models explain the predominance of shallow-water sediments in the East Coast marginal sequences.

A subtle interplay occurs between sea-level oscillations and subsidence of divergent continental margins. Subsidence results from both cooling and sediment loading, but sea-level oscillations can generate both relative upward and downward movement of the margins (as sea level falls and rises). The development of divergent continental margins resulting from transgressions and regressions should be predictable with knowledge about the history of sea-level change, subsidence rates from cooling (calculated from age), and subsidence rates from sedimentation [Burke, 1979].

As sedimentation proceeds, the original tectonic basins and troughs become fully or partly filled. At the same time, new ridges may form near the edge of the shelf, damming sediments and confining deposition to narrow troughs and trenches. Subsidence of the shelf allows the sedimentation to continue. These are important processes because they favor the continued construction of the continental shelf complex rather than the continental rises and abyssal plains at greater water depths. Sediment dams are primarily of tectonic origin in the convergent margins (Fig. 11–7). In divergent margins, however, ridges represent three principal features: the **marginal basement high** of probable tectonic origin, **coralgal reefs** (Fig. 11–8), and **diapirs** (Fig. 11–8). The origins of the marginal basement high are not known at this time because they are deeply buried and have not been drilled. This may represent the uplifted wall of the original rift-valley graben and thus may be of continental origin. The sediments of the shelf may eventually become thick enough to top and bypass the ridges (Fig. 11–4) [Emery, 1980].

Modern ridges composed of coralgal reefs are common in warm waters of the western tropics, especially in the Caribbean Sea. Where reef growth keeps pace with subsidence, as on fracture zones or marginal basement highs, an accumulation of thick sequences of shallow-water carbonates may result (such as the Blake Plateau)

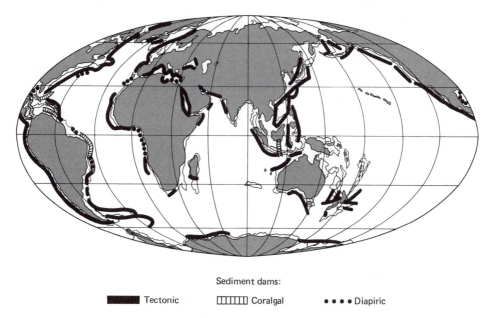

Sediment dams:

▬▬▬ Tectonic   ⊞⊞⊞⊞ Coralgal   • • • • Diapiric

Figure 11–7 Distribution of three types of sediment dams associated with particular continental margins. (After K.O. Emery, 1980)

Continental Margin Types and Divergent Margins   **335**

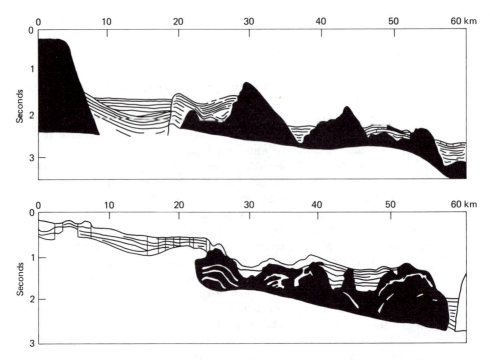

Figure 11-8 Continuous seismic reflection profiles exhibiting sediment dams. Top shows post-Miocene sediments dammed by fault blocks of volcanic rock (black) off the northwest coast of Mexico. Bottom shows Tertiary sediments dammed by diapiric masses of Jurassic salt (black) off the Gulf Coast of the United States. (After K.O. Emery, 1970)

without changing to deep-water facies. Buried coralgal reefs (Fig. 11-4) are known beneath the continental margins in the Gulf of Mexico and the western Atlantic. These reefs may be located on top of formed tectonic ridges, but seismic profiling cannot yet distinguish structures beneath the reefs [Emery, 1977]. The reefs in the Gulf of Mexico are important as sediment dams. These reefs occur beyond the edge of the carbonate platforms off the Florida and the Yucatan peninsulas; in both areas the continental slope is convex upward, steep at great depth, where the reef is best developed (Fig. 11-9), and gentler further up the slope, where the sediment may be retained by smaller reefs. Before the middle Cretaceous, the reef, largely algal in origin, almost surrounded the entire gulf, but enormous quantities of sediments from the continental interior deeply buried the northern, western, and southwestern sectors of the reef. The only exposed remnants of the reef border are the carbonate platforms, because these platforms produce less sediment than the interior of the continent. Eventually, the sediment overflowed the reef dams and built a continental rise west of the two carbonate platforms. About 0.5×10^6 km^3 of sediment were trapped by the reef dams before the middle Cretaceous around the perimeter of the gulf and subsequently along the carbonate platforms.

Diapiric dams (Fig. 11-8) consist of masses of low-density salt and or mud mobilized after being deeply buried by sediments. They rise through overlying

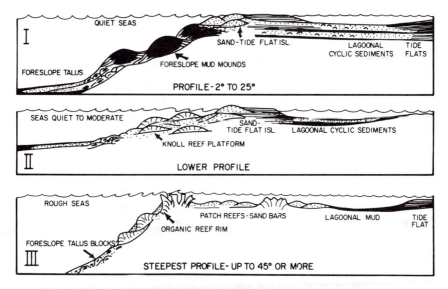

Figure 11-9 Three types of carbonate-shelf margins. (From J.L. Wilson, *AAPG Bulletin*, vol. 58, no. 5, p. 811, 1974, reproduced with permission of the American Association of Petroleum Geologists)

denser sediments, generally as bulbous units. Classic examples occur on the continental margins off Texas and northeastern Mexico and off Angola (Fig. 11-10(A)) [Baumgartner and van Andel, 1971]. A steep escarpment on the lower slope off Texas is a complex of salt intrusions and extrusions, which has trapped a thick sequence of sediments, subsequently deformed by continuing salt flowage. Many smaller salt domes also occur beneath the upper continental slope and shelf, and groups of these have also trapped and deformed sediments. Those diapirs that consist of mud are usually found off large river deltas, such as off the Mississippi and Niger rivers [Emery and others, 1970]. Various theories are invoked for initiation of salt domes, including a presalt irregular surface, variation of thickness and density of the overburden, faulting of the bedded salt, and externally applied compressive stresses [Humphris, 1979].

A substantial thickness variation in the overlying Tertiary and younger sediments is required to initiate growth of salt domes. Density logs from offshore wells indicate that a thickness difference of 1220 to 1525 m of Gulf Coast Tertiary or younger sediments is required before the sediment density surpasses that of salt and growth can begin. Where a salt feature begins to grow, the density contrast between the salt and the heavier sediments is sufficient to maintain growth [Humphris, 1979]. Structural growth is probably caused by lateral salt flowage, which results from stresses exerted by sediment-loading up-dip salt flowage as depicted by Humphris [1979] (Fig. 11-10(B)). Salt flowage resulting from the sediment loading on the shelf (Fig. 11-10(B)) may be a major mechanism for the initiation of salt domes on the continental slope.

The postrift sequence on divergent continental margins is a function of age and a poorly understood interplay between subsidence, sedimentation, ocean cir-

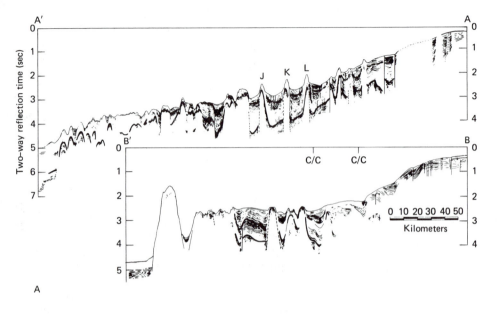

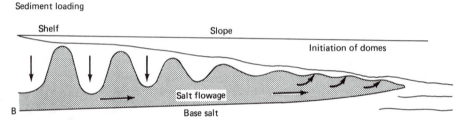

Figure 11-10 Salt dome growth. (A) Interpretations of seismic reflection records off the Angola margin, Northwest Africa, showing conspicuous diapirs. Inferred faults indicated by vertical straight lines. Vertical exaggeration is approximately 20×. (B) Diagrammatic representation of initiation of salt dome growth on continental slope as a result of sediment loading on shelf and upper slope. ((A) from T. R. Baumgartner and T. H. van Andel, *GSA Bulletin*, vol. 82, p. 799, 1971, courtesy The Geological Society of America; (B) from C. C. Humphris, Jr., *AAPG Bulletin*, vol. 63, no. 5, p. 789, 1979, reproduced with permission of the American Association of Petroleum Geologists)

culation, and climate. However, two basic types of margins are often contrasted: starved margins and mature margins.

Starved margins exhibit a thin, prograding cover (Figs. 11–11 and 11–12). They may be as young as the Neogene or as old as the Mesozoic. Typical examples are the west margins of Rockall Plateau and the Bay of Biscay and off western Australia (Fig. 11–11). Only starved margins have been drilled, due to the limited drilling capability of *Glomar Challenger* [Bally, 1979].

Mature margins are marked by a thick (about 10 km), prograding wedge of shelf sediments such as those on the U.S. Atlantic margin and its conjugate margin

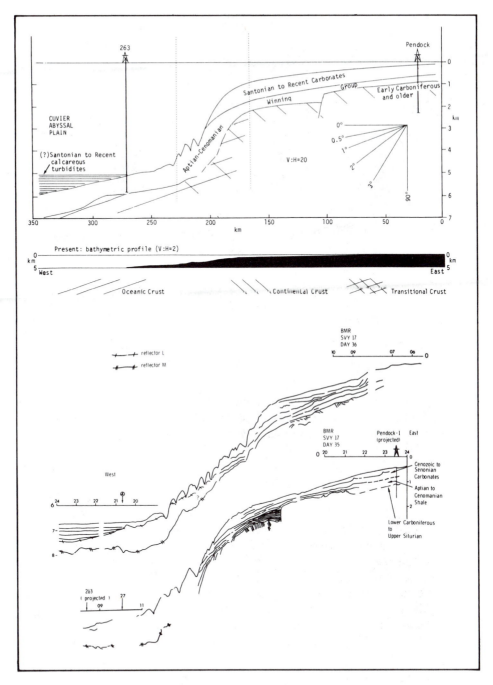

Figure 11–11 Top shows diagrammatic cross section of sediments draping across a starved divergent continental margin, northwestern Australia, and inferred distribution of continental, transitional, and oceanic crust. Drilled sites are indicated. Bottom shows seismic profiles across the same region as schematically illustrated in upper figure. (From J. J. Veevers, 1972)

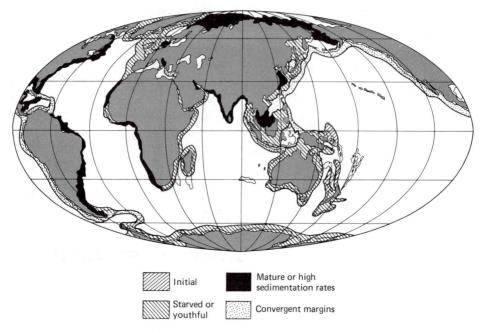

	Initial		Mature or high sedimentation rates
	Starved or youthful		Convergent margins

Figure 11-12 Distribution of four types of continental margins around the world as classified by Emery [1980]. (After K.O. Emery, 1980)

off Northwest Africa (Fig. 11–12). These margins contain some of the thickest and most continuous sedimentary sequences in all the oceans, although unconformities are still present, reflecting erosion or nondeposition due to bottom currents and sea-level change, or both. In maturity, the original topography of the shelves and most of the coralgal reefs are buried under a broad blanket of sediment. These thick sequences may cause mobilization as salt diapirs. At least some of the sediment that passes the shelves is deposited on the slopes, forming thick deposits. This causes a prograding of the slope during maturity. Other margins have been retrograding during the Cenozoic, so that shelf breaks are further landward of their Cretaceous positions [Schlee, 1977]. Instability of the slope is widespread during low sea level, causing slumping and the generation of subaqueous gravity flows, such as turbidity currents and debris flows. Submarine canyons serve to channel terrigenous debris to the continental rise and abyssal plains, especially during times of lowered sea level. Sediments that reach beyond the continental slope form continental rises either initially or by redeposition. Continuous seismic profiles exhibit numerous depositional irregularities, which include progradations, cut-and-filled channels, mass-movement scars, intraformational folds and faults, and unconformities.

Continental Rises

Continental rises are broad (100–1000 km wide), thick (up to 2 km), seaward-thinning, wedge-shaped aprons of sediment that lap against the bases of the continental slopes (Fig. 11–13; see Chapter 2). The upper surfaces of rises gently slope

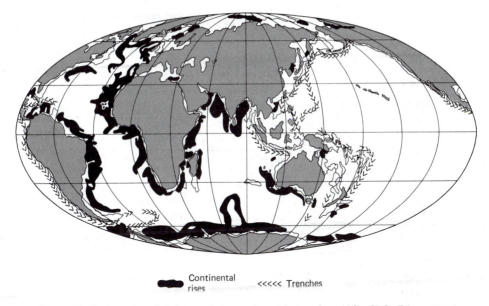

Continental
rises   <<<<< Trenches

Figure 11-13 Location of continental rises and oceanic trenches. (After K.O. Emery, 1980)

oceanward at the base of the continental slope and grade into the deep, flat abyssal plains. Compared with turbidite fans, where supply and emplacement of sediment are by turbidity currents, the supply of sediment to the continental rises is also by turbidity currents, but final redistribution and deposition may be by contour currents. Sedimentary processes of the continental rise are dominated by these factors. The sediments of continental rises consist of well-bedded muds and sands. The sands usually occur in deeper, seaward parts and near deep-sea channels that carry the turbidity currents.

Continental rises cover large areas of the ocean floor (Fig. 11-13) and comprise about 50×10^6 km^2, or 14 percent, of the area of the world ocean [Emery, 1977]. Those in the Atlantic cover 21×10^6 km^2, or 25 percent, of its total area. They are usually confined to divergent margins (Fig. 11-13), where the oceanic crust maintains a fixed relationship to the continental crust. They are virtually absent from convergent continental margins, where underthrusting tends to interfere with sediment buildup in the form of a rise, where tectonic dams may prevent seaward distribution of sediments or where the deep-sea trenches trap all sediment derived from the adjacent continents. In divergent margins, where dams tend to build up the continental shelves, the continental rises are reduced in size. Continental rises border about two-thirds of the eastern side of North America, extending from the southern United States to Greenland, being broken only by the southeastern Newfoundland outer ridge (Fig. 11-13) [Emery and Uchupi, 1972]. This has been called the *North American rise* by Emery [1969] and its formation dates from the late Cretaceous, well after the Triassic separation of the North Atlantic continents. Off New York, the continental rise is comprised of a progression of seaward-migrating units stacked against the base of the continental slope (Fig. 11-14) [Hollister and Heezen, 1972]. These have enabled the distinction of the

Continental Margin Types and Divergent Margins   **341**

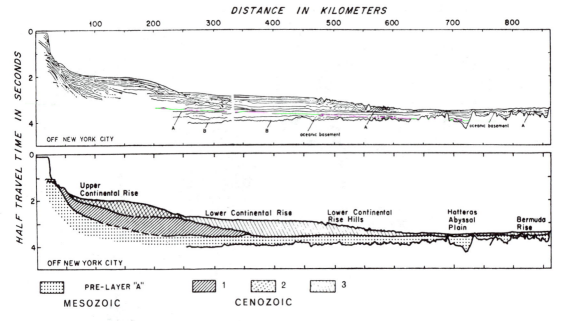

DISTANCE IN KILOMETERS

Figure 11-14 Continental margin off New York, showing migrating progression of continental rises stacked against the base of the continental slope. Top figure is seismic profile showing distinct reflection horizons marked A and B. Bottom figure is interpretation of sediment record in this profile. Numbers 1, 2, and 3 refer to successive episodes of sediment buildup within continental rise. (From C. Hollister and B. Heezen, *Studies in Physical Oceanography*, ed. A. Gordon, vol. 2, p. 59, 1972)

upper continental rise and **lower continental rise** (Fig. 11–14). The seaward part of the lower continental rise exhibits hilly topography, known as the **lower continental rise hills.** The Sigsbee rise in the Gulf of Mexico is much smaller (371,000 km²) and is absent or very narrow along the base of the Campeche escarpment.

The North American Continental Margin of the East Coast

The North American continental margin is the most extensively studied passive margin in the world; it is important because its structure and history have played a major role in the development of our understanding of passive margin evolution. This margin is paralleled on shore by the Appalachian Mountain system extending from the southern United States to the maritime provinces of Canada. The Atlantic margin is a Mesozoic-Cenozoic sedimentary wedge of up to 15 km, overlying a more complex structural foundation of igneous and metamorphic rocks, which are prerift continental rocks (Fig. 11–15). Drastic subsidence of the continental margin has occurred since the Jurassic, carrying shallow-water sediments to depths of many kilometers in a series of downfaulted basins beneath the continental shelf and slope, the Blake Plateau, and the Bahamas.

The older rocks (Triassic and Jurassic) remain largely unsampled; their characteristics are determined from seismic reflection and refraction profiles and

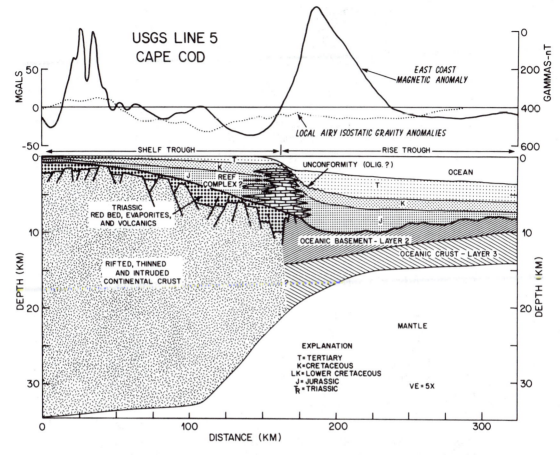

Figure 11-15 Interpretive stratigraphic cross section along section (CDP line 5) southeast of Cape Cod. Line runs across the northeastern part of the Long Island Platform (0 to 75 km distance) and across the southeastern end of the Georges Bank Basin (75 to 175 km distance). In this profile the top of the oceanic basement can be clearly traced to a depth of 10 km at the axis of the East Coast Magnetic Anomaly. (From J. A. Grow, R. E. Mattick, and J. S. Schlee, *Geological and Geophysical Investigations of Continental Margins*, AAPG Memoir 29, p. 77, 1979, reproduced with permission of the American Association of Petroleum Geologists)

extrapolation from deep wells. The oldest strata are believed to be Triassic continental redbeds that overlie downfaulted crustal blocks formed during the initial separation of North America and Africa (Fig. 11-15). Salt and evaporites may have accumulated in all the U.S. Atlantic margin basins during the early Jurassic Bahamas uplift. Late Jurassic and early Cretaceous rocks are largely thick, nonmarine sands and silty shales under the inner parts of the present shelves, but there are probably thick marine carbonates in the seaward parts of the basins (Fig. 11-15).

The general shape of the margins, as reflected by the 1000-m depth contour, bears little resemblance to the major basement structure and obscures the pattern of

Continental Margin Types and Divergent Margins **343**

basins and platforms along the margin [Klitgord and Behrendt, 1979]. The continental margin from Canada to the Gulf of Mexico is 10,000 km long and has an area of 1.07×10^6 km² [Emery and Uchupi, 1972]. Exceptionally shallow limits of the continental slope border the Straits of Florida and exceptionally deep ones border the Blake-Bahama basin.

The seaward margin of the Mesozoic shelf basins from South Carolina to Nova Scotia is marked by the ECMA [Klitgord and Behrendt, 1979] (Figs. 11–15 and 11–16). Seaward of the ECMA, the seismic-reflection and seismic-refraction data and long linear magnetic anomalies indicate typical oceanic crust (Fig. 11–15). In general, the seismic data indicate a greater depth to basement on the landward side of the ECMA than on the seaward side. A pattern of broad, intermediate-amplitude, nonlinear magnetic anomalies that characterize the sedimentary basins landward of the ECMA is probably caused by volcanic and metamorphic basement rocks. The geophysical data indicate an association of the ECMA with a marginal basement topographic high at depths of up to about 12 km. The character of this feature is unknown, with possibilities including carbonate banks or reefs, diapirs, volcanic ridges, or basement ridges.

Extensive block faulting has controlled the shape and location of the major marginal basins in the East Coast margin. Individual fault blocks are separated by former major transform faults (fracture zones) transverse to the continental margin and propagated in the oceanic crust (Fig. 11–16). These are now preserved in the offshore fracture-zone pattern (Fig. 11–16), the orientation of which has been determined from the offsets of sea-floor spreading anomalies. The fracture zones also cross the Jurassic quiet zone, with their trends clearly marked by basement relief (Fig. 11–16). The zones are aligned with displacements in the edge of the adjacent continental crust, which appear to partially control the geometry of four major structural basins that underlie the continental margin landward of the ECMA [Folger and others, 1979]. These postrift basins filled with Mesozoic sediments are, in order from north to south (Fig. 11–16), the George's Bank basin, the Baltimore Canyon Trough, the Carolina Trough, and the Blake Plateau basin. The boundaries between the basins are marked by disruption of gravity and magnetic anomaly trends paralleling the basin. The relationship between these initial transform faults and deep-sea fracture zones suggests that the original pattern of rifting and transform faults was propagated into the later, sea-floor spreading stage of continental separation.

The major structural elements of basins, platforms, and fracture zones for the East Coast continental margin are summarized in Fig. 11–16 [Folger and others, 1979]. The major platforms of shallow, pre-Jurassic continental crust along the margin border the syntectonic and postrift basins.

The continental margin of eastern North America has been compared to a reconstructed Paleozoic Appalachian geosyncline [Drake and others, 1959; Heezen, 1974]. The miogeosyncline is predominantly a shallow-water shelf complex of sandstones, shales, and limestones [Keen, 1968]. The flysch and graywacke sediments of the eugeosyncline are represented by continental-rise deposits, dominated by contourites and turbidites.

Differences in paleoclimatic and paleoceanographic history along the North

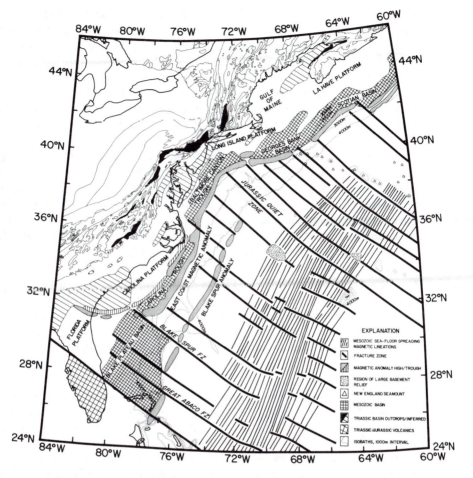

Figure 11-16 The major structural elements on the Atlantic continental margin, showing fracture zones and magnetic anomalies and the distribution of basins and platforms along the margin. (From K. D. Klitgord and J. C. Behrendt, *Geological and Geophysical Investigations of Continental Margins*, AAPG Memoir 29, p. 86, 1979, reproduced with permission of the American Association of Petroleum Geologists)

American continental margin have led to distinct differences in the structure and stratigraphy of the margins during the entire Cretaceous and Cenozoic. For example, the northern part of the margin has been strongly influenced by glacial sedimentary processes during the late Cenozoic, while the warm southern part of the margin has been influenced by coralgal reef formation during the entire history of this part of the margin. Keen [1968] divided the margin into three morphological regions: a southerly region from the Florida Panhandle to Cape Hatteras, a central region from Cape Hatteras to north of Long Island, and a northern region to the Grand Banks. The northern region, as far south as Cape Hatteras, is distinguished by its complexity, which is due to slope dissection by submarine canyons and deep, longitudinal troughs separated by shallow banks, such as the George's Bank. The continental margin in the central region is closest to the classical concept of a margin with shelf, slope, and rises, although it is cut by large numbers of submarine canyons. The central margin has been controlled by enormous terrigenous deposition from the North American ice sheets. The margin south of Cape Hatteras is complicated by the Blake Plateau. Slopes in the Gulf of Mexico are controlled by ancient reefs, by dams of diapiric salt intrusions, and by the huge Mississippi cone [Emery, 1977]. Thus a simple system of shelves, slopes, and rises does not exist in these regions.

The structure of the margin from Florida to Maine is dominated by basins and platforms (Fig. 11–16) [Folger and others, 1979]. The platforms are marked by thin sediment cover and numerous horsts and grabens, probably of Triassic age. The seaward edge of these platforms usually forms the landward edge of the major sediment basins over the margin. These continental platforms are probably the edge of continental crust not involved in the block faulting during the early stages of rifting. An integration of geophysical data is shown in a cross section across the southwestern part of the Long Island platform and the southwestern end of the George's Bank basin (Fig. 11–15) [Folger and others, 1979]. The age of the strata is based on correlations between reflectors and results from bore holes. In this part of the margin the shelf trough is wide, and a possible Mesozoic reef complex is associated with the inferred ocean-continental crust boundary. This area is unique because the top of oceanic basement can be traced to a depth of 10 km at the axis of the ECMA (Fig. 11–15). The shelf trough in this region is dominated by terrigenous sediments.

Further to the south, off the coast of New Jersey, the configuration of the basement beneath the Baltimore Canyon Trough is simpler than that beneath the George's Bank. The axis of the trough extends from the shelf off Long Island almost to the mouth of Chesapeake Bay. Average axial sediment thickness is 10 to 12 km. A carbonate bank, or reef, is inferred to have built up over basement. A basement ridge forming the seaward edge of the basins rises to an average depth of 8 km and in places to 6 km [Folger and others, 1979].

South of Cape Hatteras, the dominant topographic features of the continental margin are carbonate platforms between the Cape and the Puerto Rico Trench. The northern carbonate platform is the Blake Plateau, a surface 850 m deep, that projects from the Florida-Hatteras slope (Fig. 11–17). The southern carbonate province consists of the Bahama Banks, shallow topographic highs separated form the Florida Plateau, and the Greater Antilles by the Straits of Florida and Old Bahama Channel [Emery and Uchupi, 1972]. The eastern boundary of the Blake Plateau is an extremely steep **marginal escarpment** (the Blake escarpment) extending to depths of over 5000 m over a horizontal distance of less than 100 km (Fig. 11–17). The escarpment is the seaward edge of a linear Cretaceous reef, which formed a dam during the deposition of greater than 10 km of Cretaceous and older sediments of dominant carbonate facies at a time of rapid subsidence. The stratigraphic evolution of the Blake Plateau was controlled primarily by regional subsidence of the margin, acted upon by environment factors, such as the persistence of carbonate bank margins, which supported a carbonate platform; trapping of terrigenous

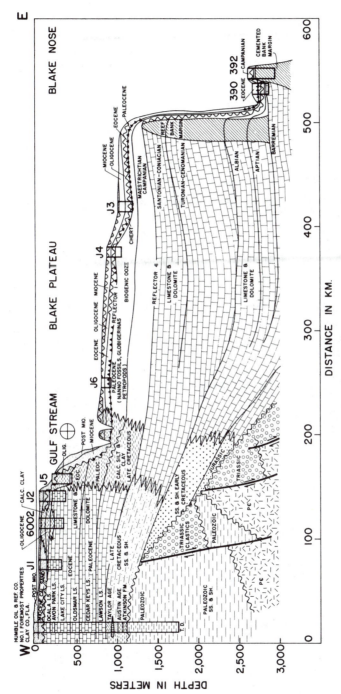

Figure 11-17 Interpretive stratigraphic cross section across the Blake Plateau based on data from drilled sections. (From R. E. Sheridan and P. Enos, *Deep Drilling Results in the Atlantic Ocean: Continental Margins and Paleoenvironment,* p. 119, 1979, copyrighted by the American Geophysical Union)

Continental Margin Types and Divergent Margins **347**

sediments in deltas and estuaries; erosion by the Gulf Stream; and eustatic changes in sea level, which shifted the locus of carbonate deposition [Sheridan and Enos, 1979]. Reef growth provided a dam for sediments until about the middle Cretaceous [Schlee and others, 1979]. After continental breakup, the main period of subsidence was during the Jurassic, when 6 to 8 km of sediments were deposited (Fig. 11-17). During the early Cretaceous, a thinner sequence of limestones was deposited, the reef bank margin ceased to grow in the middle Cretaceous (Fig. 11-17). The greater depth of the Blake Plateau compared with the rest of the Atlantic margin does not seem to have been created by major faulting, but from downward flexure resulting from the enormous sedimentary deposits. The adjacent oceanic crust was also dragged down by the sinking plateau. In the southern part of the plateau, the abyssal sea floor subsided much less than the plateau, and faulting partially decoupled the deep-sea basement from the plateau basement [Dillon and others, 1979].

The top of the Blake Plateau has been strongly eroded by the Gulf Stream (Fig. 11-17). This erosion began in earliest Cenozoic when the Gulf Stream began to flow through the Straits of Florida [Uchupi, 1967; Emery and Uchupi, 1972]. As a result of this activity, the Cenozoic sequence is very thin, with many unconformities. Carbonate ooze has been phosphatized, and the phosphorite nodules form lag deposits in certain areas. The phosphorite nodules are in turn covered by a continuous manganese pavement, possibly over an area of 5000 km^2 changing into manganese nodules to the south and phosphorite nodules to the west [Ramsay, 1977].

Seismic reflection profiles also indicate a thinning of the Cenozoic sequence across the Straits of Florida [Uchupi, 1970; Schlee, 1977], not unlike the Blake Plateau. The present outlines of the northern Straits of Florida have resulted from nondeposition beneath the Gulf Stream and carbonate accretion along the margins. It has been suggested by Chen [1965] that the Gulf Stream did not begin to flow through the Florida Straits until the early Tertiary (Eocene). Before this, the Gulf Stream flowed around the seaward edge of the Bahama platform and then over the central and northern part of the Blake Plateau [Schlee, 1977].

The Gulf of Mexico Margin

A number of basins are related to passive margins, yet are distinct as they do not face a spreading ridge. These basins include the Gulf of Mexico and the Gulf of St. Lawrence and share a similar evolutionary history with the Atlantic continental margins, beginning to form about the same time—during the late Paleozoic to early Mesozoic. The early phases of sedimentation were also marked by periods of evaporite formation (Fig. 11-18), indicating substantial isolation from the world ocean. Later history is marked by major subsidence and accumulations of thick clastic and carbonate sequences. Structural deformation styles are dominated by gravity tectonics, such as salt of shale diapirs (Fig. 11-18) and normal growth faults. These basins differ from the Atlantic passive margins because they were located within the Paleozoic foldbelts. For this reason Bally [1975; 1976] has suggested that they originated as backarc basins associated with continental collisions.

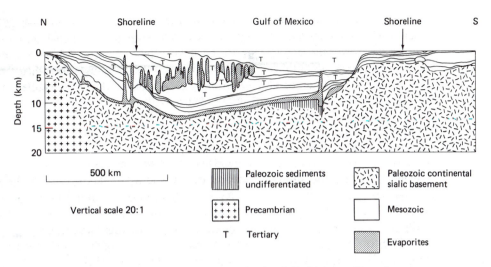

Figure 11-18 Geological cross section for the Gulf of Mexico. Vertical exaggeration is 20:1. (After Bally, 1979)

The western margin of the Gulf of Mexico is a unique passive margin. The portion of the province adjacent to eastern Mexico as far south as 19°N is marked by an outer slope exhibiting a series of gentle folds that parallel the shoreline. They are called the **Mexican ridges** and form long, linear, subparallel topographic features on the sea floor with average relief of 400 m. These folds have acted as sediment dams to the seaward flow of terrigenous sediment. As a result, the inner folds are buried, while the outer, exposed folds are blanketed by pelagic sediments.

Several hypotheses have been suggested for the origin of these folds [Ewing and others, 1970]: (1) decollement folding (deformed layer overlying undeformed surface), resulting from the gravity sliding of sedimentary rocks; (2) folding associated with compressional stresses developed by tectonism; (3) vertical movement of shale or salt masses related to static loading; and (4) folds controlled by faulting. Earlier work [Emery and others, 1970] suggested that the folds were caused by compression of a sequence of evaporites under a thin cover of sediments. However, Buffler and others [1979] have found that the top of a seismic unit containing inferred Jurassic salt continues relatively undeformed beneath the foldbelt, suggesting that the subparallel folds are not caused by salt tectonics. Instead, the folds appear to occur in competent beds above a possible decollement. The detachment from the underlying rocks is along a deformed zone occurring within a thick upper Cretaceous to lower Tertiary shale section. The uniform folding of most of the upper sedimentary section and numerous imbricate thrust faults associated with the folds suggests regional compression acting in an east-west direction [Buffler and others, 1979]. Involvement of beds as young as Pliocene and Pleistocene indicates that folding and thrusting is very young and may be continuing today. Seaward decreases in fold amplitude and sediment ponding in the synclines suggest that the zone of maximum deformation moved seaward through time. According to Buffler and others [1979], at least two separate mechanisms explain the tectonic style of the

Mexican ridge folds observed on seismic profiles: (1) massive sliding due to gravity, possibly triggered by regional uplift and supplemented by sediment loading at the head of the slide; and (2) compressional tectonism originating within the deeper crust beneath Mexico and transmitted into the fold area through deep thrust zones. In both cases, detachment and deformation take place above a decollement, or deformed zone, located within mobile substrata.

Summary of the Evolution of Atlantic Margins

By mapping the seismic stratigraphy of the North Atlantic margins and correlating reflectors with the stratigraphy of drilled sequences, it has been possible to develop a history for the margins from rifting to the present condition. The combined approach has been essential for development of this knowledge.

Both the North American and African margins are constructed of a series of offshore sedimentary basins that border older Paleozoic (222–600 Ma) foldbelts represented by the Appalachians and Mauritanides. The results of crustal separation have been similar on both the African and North American margins, and the evolution of both the margins has been similar up until the early Cenozoic, when paleoceanographic conditions began to diverge. Large differences in paleoclimatic conditions began to develop between North Africa and North America, especially in the Neogene, leading to differences in sediment sources and hence evolution of their respective margins. Figure 11–19 summarizes the relationships of continental crust, oceanic crust, and sedimentary character of the western North Atlantic from the margin to the mid-ocean ridge [Dewey and Bird, 1970].

Near the end of the Paleozoic era, the collision and welding of several continental plates formed a new megacontinent, Pangaea, on the present site of the North Atlantic Ocean. The continental welding was accompanied by right-lateral shear due to oblique contact of the Eurasian and North American plates, which produced a structural weakness along which later rifting took place [Ballard and Uchupi, 1975; Uchupi and others, 1976]. This rifting between North America, Africa, and Europe began during the late Triassic and early Jurassic (about 200–180 Ma). Some Permian tensional structures in North Africa suggest that the episode was initiated earlier on the African continent [Uchupi and others, 1976]. A long period of crustal stretching preceded final separation of the continents. The majority of this stretched continental crust was included on the North American side of the break, being represented by a number of Triassic rift basins on the continent [Schlee and others, 1979].

Following separation of the continents, the rifted margins began to subside rapidly, forming major basins along the present-day continental shelves. Sediments derived from the nearby uplifted blocks and continents rapidly accumulated in these basins, which were initially above sea level. These early stages of sedimentation during the Jurassic enjoyed high rates of deposition. Further subsidence below sea level allowed the seawater to invade basins of restricted circulation and of high evaporation causing the deposition of thick evaporitic deposits in some areas. Sediments also began to accumulate in lagoons, on alluvial plains, and in deltas. Off northernmost Africa, evaporites were deposited until the early Jurassic. By the

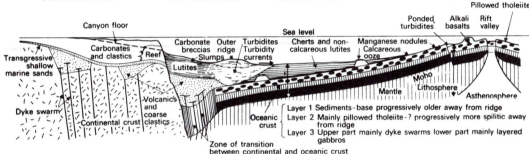

Figure 11-19 Geological cross section for the western North Atlantic Ocean showing relationships between continental crust, oceanic crust, and sediment facies. (From J. F. Dewey and J. M. Bird, *Jour. of Geophys. Res.*, vol. 75, p. 2632, 1970, copyrighted by the American Geophysical Union)

middle Jurassic, normal marine deposition had begun over most of the margins. From the middle to late Jurassic, a major reef formed over the Moroccan continental slope [Uchupi and others, 1976]. In the late Jurassic and early Cretaceous a vast shallow sea covered the margins. A series of reefs and carbonate platforms developed over almost the entire length of eastern North America from the Blake Plateau (Fig. 11-19) to Newfoundland [Bryant and others, 1968; Schlee and others, 1979]. These reef complexes built up along much of the continental margin over the oceanic basement high that had developed during the early stages of rifting [Folger and others, 1979]. By middle Cretaceous the reef system ceased to be an effective sediment dam, and the banks were overwhelmed by Cretaceous fluvial deltaic and shelf sediments, nearly filling the depression behind the reef complex. As the Cretaceous sediments began to prograde seaward over the former carbonate-shelf edge, extensive sedimentation began over the rise, developing its approximate present form (Fig. 11-19). At the beginning of the Cenozoic, the shelf edge break shifted 300 km to the west near the Florida Platform, apparently in response to sea-level lowering beginning in the early Cenozoic, and the Blake Plateau became a site for deeper-water sedimentation [Schlee and others, 1979]. The Gulf Stream shifted its position in the early Tertiary and began to flow through the Straits of Florida, causing erosion over much of the Blake Plateau.

During the late Cretaceous and Cenozoic, subsidence slowed. The most effective controls on sedimentation were regional warping and minor faulting, eustatic sea-level changes, and the activities of the Gulf Stream. Until the Miocene, the Gulf Stream was a strong influence along the eastern North American margin, resulting in carbonate-rich sediments. At that time, its influence began to wane in the northern part of the margin, causing an increase in the importance of clastic sediments. Furthermore, the late Cenozoic ice sheets of North America provided a major source of terrigenous sediments for the development of a prominent continental-rise wedge. Also, during lowstands of sea level during glacial episodes, clastic sediments bypassed the shelf and the shoreline moved to the present shelf break. As these sediments accumulated on the continental rise off North America, they were extensively reworked by contour currents.

Continental Margin Types and Divergent Margins   **351**

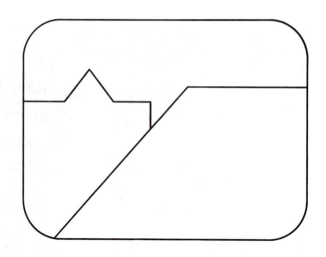

twelve
Convergent, or Active, Margins

Imagination is more important than knowledge.

Albert Einstein

CONVERGENT MARGINS

Introduction

Earth scientists are interested in convergent, or active, margins because the most active geodynamic processes are concentrated in these areas. However, only during the last few years has plate-tectonic theory made an integrated analysis of the structural evolution of convergent margins possible, and the full significance of their geodynamic character has only recently been understood. It is generally accepted that convergent margins, together with the mid-ocean ridges, are the two fundamental elements of global tectonics, with active margins being the inverse counterpart of sea-floor spreading. Active margins coincide with convergent plate boundaries and, in some areas, with transform-fault systems.

Definitions of convergent margins vary. Earlier definitions described them as narrow zones marked by active, deep-seated seismicity. However, the seismologic definition excludes backarc basins and intraoceanic areas, which are part of the convergent-margin system. Most workers now prefer to define convergent margins as the transition zone between the active subduction zone and the continental mainland [Bally, 1979]. A broader definition includes fossil subduction zones that now occur within the continents themselves, using the term **megasuture** to include all products of subduction-related processes during the Mesozoic and Cenozoic [Bally, 1979].

Since about 1966, when active margins became a vital part of plate tectonic theory and a basic convergent-margin model of active margins developed, most work has been concerned with the recognition of morphologic-structural variation around this model. However, deep-sea drilling in convergent margins has raised fundamental questions about the validity of existing models and has led to new approaches toward convergent-margin interpretation. Unfortunately, much of the deformation occurring in convergent margins is too deep and complex to be observed directly with existing technology. The few deep-drilled sequences that have been examined in combination with geophysical observations have led to further complications. To better understand crustal processes at convergent margins, more effort is required in the study of ancient subduction zones exposed on land. The interpretation of coastal orogenic belts in terms of convergent-margin models is a major theme linking the interests of the marine and continental geologist or geophysicist.

One of many diverse features of convergent margins is the arc region (Fig. 12–1). The simplest types of arcs are intraoceanic, arc-trench systems, such as the Mariana. In these, a volcanic arc separates the main ocean from an oceanic backarc basin. Other arc systems, like those adjacent to the North and South America cordilleras (sets of parallel mountain chains), are complex crustal accretions on older fold belts. Arcs such as the islands of Japan or New Zealand are a combination of continental fragments detached from large continents and zones of complex crustal accretion.

Active ocean margins can also include certain transform faults (Table 12–1). Several transform boundaries (long, strike-slip, or transcurrent faults) have been intensely studied on land, but few such studies have been carried out in the ocean. Some involve lateral motion of ocean crust against ocean crust, such as the Hunter

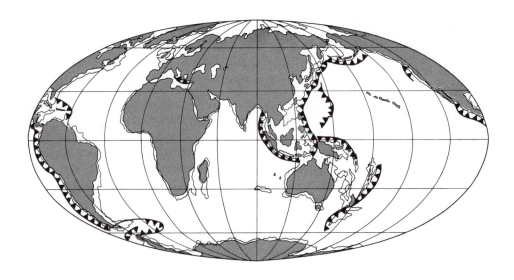

▲▲▲▲▲▲▲ Subduction zones

Figure 12–1 Distribution of subduction zones.

TABLE 12-1   CLASSIFICATION OF ACTIVE MARGINS[a]

	Examples
1. Convergent active margins	
Subduction of oceanic lithosphere	
Oceanic crust subducted below continent	Peru–Chile, Guatemala, Honduras, eastern Aleutian, Java–Sumatra
Ocean crust subducted below island arc	New Hebrides, Ryukyu, Japan, Kurile, Andaman, Bonin Mariana, Tonga, Caribbean, Scotia Arc
Oceanic plateau subducted below island arc	Ontong Java Plateau under Solomon Islands
Subduction of continental lithosphere	
Subduction of continental crust below island arc	Timor-Ceram
Subduction of continent below continent (continent collision)	
Without oceanic marginal basins	Gulf of Arabia
With formation of marginal basin	Eastern and western Mediterranean
2. Transform active margins	
Continent slipping past ocean	Queen Charlotte margin
Transform-rift system intersecting continent	Gulf of California
Ocean slipping past ocean	Hunter fracture zone
Continent slipping past continent	San Andreas fault

[a]After Bally [1979].

fracture zone at the south margin of the Fiji Plateau. Others, such as the Queen Charlotte fault zone, separate continental from oceanic crust (Table 12–1).

One of the best-known marine examples is associated with the **California borderland,** a complex of offshore and onshore basins off southern California associated with the San Andreas transform-fault system. This province is dominated by northwest-trending fault blocks, which are intercepted in the north by the east-west-trending Transverse Ranges and their offshore continuations.

Convergent margins are further subdivided into two main types depending on the nature of the converging plates: subduction and collision (Table 12–1). The Himalayas and the area north of Australia–New Guinea are typical collision zones, the result of continent-continent and continent-oceanic contact, respectively. The most widespread type of convergent margin, however, is the subduction type. Because continental collision is the inevitable consequence of oceanic convergence (see Chapter 5), an evolutionary sequence exists from subduction to collision-type margins.

Subduction Processes of Convergent Margins

Subduction is the tectonic process in which one segment of the lithosphere is inserted partially or totally below another, adjacent one. Under certain conditions, the sedimentary deposits overlying subducting oceanic crust may be subducted along with the underlying lithospheric rocks [Scholl and others, 1977]. This provides a mechanism for the downward displacement of deep-sea sediments of several kilometers beneath convergent margins. The tectonic removal into the subduction zone of the downgoing slab, including sediment and oceanic crust, is called **consumption.** Deep-drilled holes in the Japan, Mariana, and Middle America trenches show that the intense deformation associated with subduction may be confined to a very narrow zone. The narrowness of this zone may result from lubrication of the megathrusts by high pore-water pressure or some other mechanism that causes friction between plates to be reduced.

Offscraping involves the tectonic skimming of sedimentary and igneous rocks from the upper part of the oceanic lithosphere during the subduction process [Scholl, and others, 1977]. **Accretion** is the process in which material from the outer plate and trench is offscraped and added to the outer continental margin or arc by various mechanisms, such as imbricate thrusting or a combination of folding and thrusting [Kulm, and others, 1977]. Sometimes the outer margin is uplifted by varying degrees, with a topographic ridge often developing along the outermost margin, producing a basin between the ridge and the arc.

Basic Model of Convergent Margins

The history of the development of a basic model for convergent margins has been succinctly summarized in a report by the JOIDES Active Margin Panel [Fusod, 1977], upon which the following account is based.

The first model to account for active margin phenomena was the Tectogene [Vening Meinesz, 1952]; it involved a symmetrical downwarping of the ocean crust to explain the large negative anomalies associated with trenches in the western Pacific. Soon afterward, the model was changed to accommodate clear asymmetry of features associated with trenches and island arcs, and mantle convection was posited as a driving mechanism. Further additions include high seismicity below the trench and within the Benioff zones and calc-alkaline volcanism in response to thrusting of the ocean crust under the continents. Sediment and ocean crust on the downthrusting slab is accreted to the inner trench wall by thrust faulting the tectonic stacking, or it is subducted with the downthrusting oceanic crust. The discovery of sea-floor spreading required that large amounts of oceanic crust had to be disposed of, and subduction at convergent margins provided the mechanism. The model has also accounted for the development of coastal mountain ranges, a previously unexplained phenomenon. Thus the process of subduction of the oceanic crust is the underlying cause for trench and mountain formation, seismicity and volcanism, crustal deformation and metamorphism, and the development of back-arc basins. From front to rear, the principal units of a fully developed active arc system are as follows: trench, subduction complex, mid-slope basement high, forearc basin, frontal arc, volcanic chain, backarc basin, and a remnant arc (Figs. 12–2 and 12–3). Terminology may vary among different workers.

The subduction zone occupies the lower trench slope on the arc side of the trench axis. With continuing subduction, the deformed rocks begin to accrete to

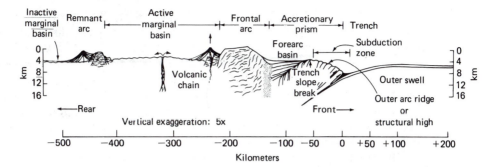

Figure 12-2 Cross section of typical island-arc system showing tectonic units and terminology. (After D. E. Karig and G. E. Sharman III, 1975, courtesy The Geological Society of America)

form an **accretionary prism,** or **subduction complex** (Fig. 12-3). This prism typically grows in width as successive increments are added by underthrusting at the trench [Dickinson and Seely, 1979], also growing upward to become a positive feature—an island chain, or uplands. In some cases, such as off eastern Japan, there is subsidence of the accretionary prism rather than uplift; thus the operating mechanisms must be quite different. The zone of active igneous activity is called the **volcanic arc,** or the **magmatic arc** if the cogenetic plutons lying beneath the volcanic arc are included [Dickinson and Seely, 1979]. A zone of uplift immediately in front of the volcanic arc is known as the **frontal arc** (Fig. 12-2). The **forearc** region includes all the features lying on the trench side of volcanoes, while the **backarc** region is behind the volcanoes (Fig. 12-3) [Dickinson and Seely, 1979]. The island-arc system is considered to have normal polarity if the trench is closer to the ocean basin than the volcanic chain. If the reverse is true, it has **reversed polarity.*** When this occurs, a trench has been abandoned and a new trench has formed on the opposite side of the system with opposite dip [Karig and Sharman, 1975]. This is typical of several western Pacific arcs.

Classification of Convergent Margins

Subduction-type margins are subdivided into **continental subduction-zone types,** lacking a backarc basin, and the **island-arc subduction-zone types,** including a backarc basin (Fig. 12-2; Table 2-1). There are many variations of these two types, which result in a wide range of arc-system patterns. Some examples have a well-developed accretionary prism and are called **accretionary trench systems** (Fig. 12-4). Others lack a prism, may have exposures of oceanic crust, and are called **nonaccretionary trench systems** (Fig. 12-4).

Classic island-arc subduction-zone types (Fig. 12-4), occurring in mid-oceanic settings and little influenced by terrigenous sediments, differ from the continental type (Fig. 12-4) by the lack of a succession of orogenic belts, which are believed to

* These terms are unrelated to those used in geomagnetism.

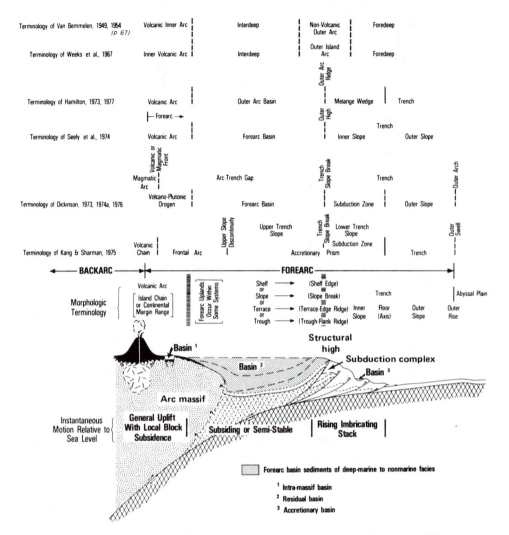

Figure 12-3 Generalized model of the forearc region showing terminology of different workers for the various structural features. (After Seely and Dickinson, 1977, reprinted with permission of The American Association of Petroleum Geologists)

be the product of former subduction zones along the edges of continents [Fusod, 1977]. Further subdivisions of active margins have been made on the basis of the tectonic relations between continental and oceanic crust (Table 12–1; Fig. 12–5).

Accretion of oceanic sediment and deeper crustal material onto the upper plate during subduction has been postulated from available data, but the structures that result from this process are still not well understood [Karig and Sharman, 1975]. Seismic methods and drilling have not yet adequately resolved the internal structure and the stratigraphy. Seismic reflection records reveal structure only in surface areas of forearc basins, ponds of sediment on the inner trench wall, and sediment overlying oceanic crust. The inferred deep-seabed thrust faults shown in

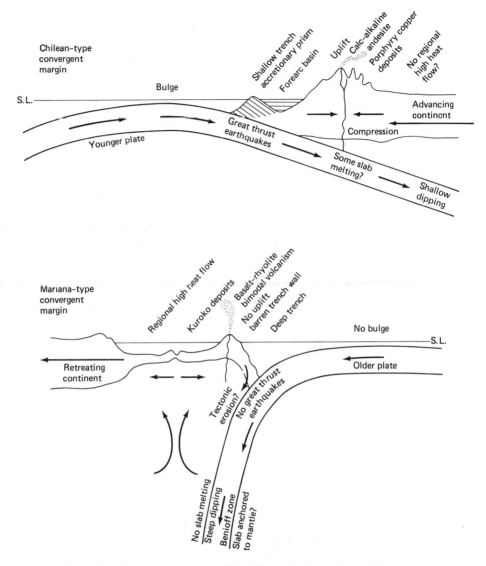

Figure 12-4 Diagrammatic sections showing two general modes of subduction and their possible tectonic implications and causes (not to scale). (After S. Uyeda, 1979)

Figs. 12–3 and 12–6 are now only beginning to be seen in high-quality multichannel seismic-reflection records. Furthermore, important information has been provided by direct observations in a few areas where the accretionary prism has been uplifted as islands. The best examples are Barbados, the Mentawai Islands in the Sunda Arc, and Middleton Island in the Aleutian Arc [Karig and Sharman, 1975]. All the Quaternary reefs or wave-cut terraces have been rapidly uplifted and all exhibit either highly deformed terrigenous sequences containing slices of basic or ultrabasic rocks or less disrupted terrigenous or carbonate sequences.

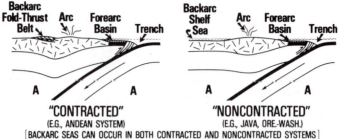

CONTINENTAL-MARGIN ARC-TRENCH SYSTEMS

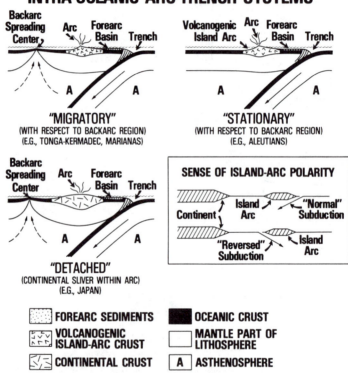

"CONTRACTED"
(E.G., ANDEAN SYSTEM)

"NONCONTRACTED"
(E.G., JAVA, ORE.-WASH.)

[BACKARC SEAS CAN OCCUR IN BOTH CONTRACTED AND NONCONTRACTED SYSTEMS]

INTRA-OCEANIC ARC-TRENCH SYSTEMS

"MIGRATORY"
(WITH RESPECT TO BACKARC REGION)
(E.G., TONGA-KERMADEC, MARIANAS)

"STATIONARY"
(WITH RESPECT TO BACKARC REGION)
(E.G., ALEUTIANS)

SENSE OF ISLAND-ARC POLARITY

"DETACHED"
(CONTINENTAL SLIVER WITHIN ARC)
(E.G., JAPAN)

FOREARC SEDIMENTS

VOLCANOGENIC ISLAND-ARC CRUST

CONTINENTAL CRUST

OCEANIC CRUST

MANTLE PART OF LITHOSPHERE

A ASTHENOSPHERE

Figure 12-5 Settings of arc-trench systems as defined by backarc features. (After Dickinson and Seely, 1977, reprinted with permission of The American Association of Petroleum Geologists)

The Forearc Region

The forearc region extends from the deep-sea trench to the volcanic arc and is at least 100 km wide. According to Seely and Dickinson [1977], this minimum width is controlled by the position of volcanoes, usually standing between 90 and 150 km above the Benioff zone. A basic model for accretion and deformation of material at

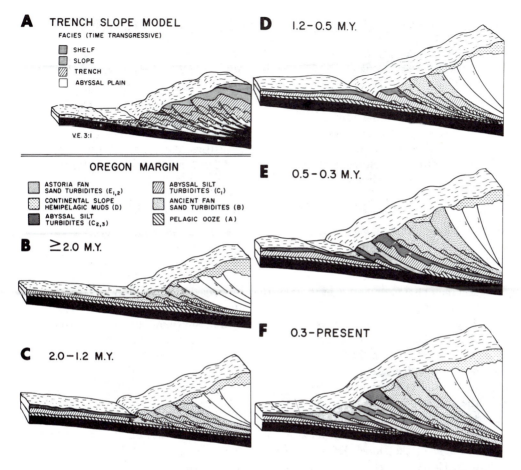

Figure 12-6 (A) Trench slope model of Seely and others [1974]. (B to F) Diagrammatic sections of the Oregon margin, showing structural and sedimentary evolution for the last 2 m.y. Note successive slices accreted to the prism. (From L. D. Kulm and G. A. Fowler, *The Geology of Continental Margins*, p. 278, 1974)

the inner walls of modern trenches was developed independently by Seely and others [1974] and by Karig [1974], and Karig and Sharman [1975] and has been called **underthrusting, offscraping,** or **plastering.** This model was developed using seismic data, revealing arcward-dipping reflectors within the accreted wedge. This wedge is highly deformed and irresolvable using acoustic methods. The accretionary prism is made up of arcward-dipping, wedge-shaped, thrust-bound packets characterized internally by complex, trenchward-verging folds. It also causes progressive landward tilting of accretionary basins, recorded by steeper landward dips of the older sediment units. With continued subduction, older, previously accreted material is uplifted and older thrusts are rotated arcward by the addition of successive wedges, or slices, of younger, newly accreted sediments at the base of the inner wall [Cowan and Silling, 1978].

Another major study was carried out by Seely and others [1974], presenting additional reflection profiles and emphasizing a hypothetical systematic repetition of imbricated trench, pelagic, and slope facies within the accretionary prism. This model (Fig. 12-6) has been further developed by Seely and Dickinson [1977], including a summary of nomenclature applied to forearc region features (Fig. 12-3). Deformation near the **trench-slope break,** or **structural high,** seems to mark the upper boundary of the zone of active deformation (Fig. 12-3). Surface deformation diminishes progressively upslope from the toe at a rate that seems to be related to the distance from the top of the underthrusting plate, its rigidity, and to the thickness of the sediment column [Seely and Dickinson, 1977]. The forearc basin between the trench-slope break and the frontal arc consists of relatively undeformed sediments lacking significant deformation.

The accretionary prism

The inner slope of the trench is a major feature of the forearc system, with a complex character and morphology that reflects the subduction and accretionary process. Between a zone of uplift, the **frontal arc** (Fig. 12-2), and the trench is the 50 to 400 km wide accretionary prism [Karig and Sharman, 1975] or arc-trench gap [Dickinson, 1974]. The accretionary prism may be a simple slope from the volcanic arc to the trench, as in the Mariana and Tonga arcs, or may be subdivided into an **outer-arc ridge** and an **outer-arc trough,** or **forearc basin** (Fig. 12-3). These two features meet at the trench-slope break (Figs. 12-2 and 12-3). The landward side of the accretionary prism, or the trench slope, is the upper slope discontinuity (Figs. 12-2 and 12-3). Deformation associated with subduction extends across the lower trench slope, from the trench axis to the trench-slope break; but it is most intense at the base of the slope. This region is classically considered to be a rising tectonic unit. However, drilling the accretionary prism to the east of Japan showed it to be an area of subsidence [von Huene and others, 1980]. The model of Karig [1973] and Seely and others [1974] indicates that the upper trench slope is a subsiding region made up of sediment, as represented by the forearc basin accumulation [Karig and Sharman, 1975]. This zone of subsidence is separated by the upper-slope discontinuity from the frontal arc (Fig. 12-2). The upper-slope discontinuity in young trenches is the upper part of the continental, or insular, slope that existed before subduction began. As the accretionary prism develops by sediment accretion, the distance between the upper-slope discontinuity and the trench axis increases. Substantial differences exist in the degree of development of the trench-slope break and forearc basins between different types of subduction zones. The greatest differences occur between subduction zones associated with continental blocks (for example, Peru–Chile, Alaska, and Sunda arcs), which exhibit prominent trench-slope breaks, and the Mariana and Tonga arcs, where the trench walls are almost devoid of sediments due to lack of a nearby source of terrigenous sediments.

Other than age of the accretionary prism, one of the major factors affecting the amount and style of accretion is the availability of sediment through time [Karig and Sharman, 1975]. Widening of the accretionary prisms of the Kermadec and

Caribbean arcs reflects high sedimentation rates from the adjacent continental areas rather than age or rate of subduction. Karig and Sharman [1975] believe the morphologic variations of the prism are more dependent upon relative rates of sedimentation from the various sources feeding the subduction zone than on the rate of subduction. The thickness of sediment varies with the age of the downgoing plate, with the productivity of the waters beneath which it has traveled and with available terrigenous sources. This ranges from as little as 200 m on the young crust of the Middle America Trench to greater than 3 km where Bengal fan sediments are subducted in the Sunda Trench [Karig and Sharman, 1975]. If, during the process of accretion, underthrusting is interrupted while high depositional rates continue, thick sediments can accumulate in the trench. After underthrusting resumes, these sediments are also incorporated into the accretionary prism.

Tectonic character has also been proposed to account for differences in the growth of accretionary prisms. Uyeda and Kanamori [1979] have suggested that the differences are due to the strength of coupling between different arc types. In those areas where the trench is adjacent to a continent, such as Chile, a strong mechanical coupling between the two plates causes extensive contortion and development of the prism (Fig. 12-4). In the island arcs (such as the Mariana), weak coupling enables subduction to occur without strong resistance from the upper plate, resulting in less contorted or less well-developed prisms (Fig. 12-4).

Two yet unanswered questions are how much sediment is offscraped and how much is subducted. If offscraping were a uniformly efficient process, there should be large amounts of sediment accreted in the active margins. During the last 100 million years, about 2.5×10^6 km^3 of oceanic sediment should have been carried into each 1000-km segment of the trench [Mitchell and Reading, 1978].

The compressional thrust model of Seely and others [1974] requires a substantial amount of uplift of the accretionary prism, resulting from underthrusting of the oceanic plate. In some areas the prism emerges above sea level to form an outer arc, such as Barbados in the Lesser Antilles, Mentawai Island off Sumatra, and Middleton Island off the Aleutian Arc. In other areas, such as the Oregon margin, uplift is also occurring but in a different form than in island arcs [Kulm and Fowler, 1974]. However, drilling in the area of the assumed accretionary prism to the east of Japan has shown clear subsidence rather than uplift. In the middle Cenozoic (about 22 Ma), a landmass called the *Oyashio ancient landmass* (to the east of Japan) began to subside below sea level and is now 2000 to 3000 m deeper than its previous level [von Huene and others, 1980]. This subsidence may be the result of tectonic erosion of rock near the Benioff zone. During subduction, these rocks have become impregnated with water, which assisted in the tectonic erosion of the base of the continental crust and part of the accretionary prism [von Huene and others, 1980]. This, in turn, caused subsidence of the accretionary prism areas, rather than uplift. Most material was subducted rather than accreted. It seems that this system has reversed since the middle Pliocene (about 2–3 Ma) however, and uplift has replaced the longer term subsidence [von Huene and others, 1980].

Off the Oregon margin, several lines of evidence indicate that Quaternary abyssal-plain and fan deposits of the Cascadia basin are being thrust beneath the earlier Cenozoic rocks that underlie the continental shelf. The earlier Cenozoic rocks have been uplifted more than 1 km and incorporated into the lower and middle continental slope, where they are either exposed or covered by the late Quaternary deposits [Kulm and Fowler, 1974]. The stratigraphic position of these deposits and their age relations suggest imbricate thrusting of thick slices of deep-sea terrigenous sediments (Fig. 12–6). The younger deposits may be lifting the older deposits higher onto the continental margin. These older deposits dip progressively steeper in the landward direction. Earlier on, the uplift occurred rapidly (1000 m/m.y.) in the lower slope (Fig. 12–6), but as accretion continued, the deposits were compacted and the rate of uplift of the older deposits on the outer shelf was reduced to only 100 m/m.y. [Kulm and Fowler, 1974]. During any particular time interval, the amount of uplift is greatest for the oldest deposits (Fig. 12–6).

The trench-slope break

A prominent trench-slope break, or structural high, marks the approximate inner edge of the zone of uplift and active deformation of the accretionary prism. Landward of the trench-slope break, deposition of undeformed sediment may occur within a forearc basin. The position of the trench-slope break seems to be largely related to the amount of compression and the amount of sediment being supplied to the forearc basin. Low rates of sediment supply produce narrow ridge types with the trench-slope break close to the frontal arc. The trench-slope break can migrate in either direction as sediment supplies vary, although migration is commonly seaward as the lateral extent of the forearc basin increases and the accretionary prism widens (Fig. 12–7) [Dickinson, 1973]. The contact between the sedimentary sequence of the forearc basin and the underlying subduction complex is a time-transgressive feature that records the lateral migration of the accretionary prism (Fig. 12–7) [Dickinson, 1977]. In arc systems where the present trench has been recently activated, such as the New Herbrides, the trench-slope break is absent or poorly developed [Karig, 1973]. In older arc systems, where the sediments on the subducting plate are thin, the trench-slope break appears as a bench. Where sediment supply is high, the trench-slope break is often a strongly developed ridge, at times reaching sea level; it is called the **tectonic, outer,** or **nonvolcanic arc** [Karig, 1973].

Forearc basins

Forearc basins typically consist of immature clastic sediments derived from rapid erosion of volcanic mountains and metamorphic rocks within the arc. They are deposited in water depths determined by the depth of basinal origin, rate of sedimentation, and rate of subsidence. Rapid rates of sedimentation tend to fill the basins and form a broad shelf. Lower rates of sediment input produce deeper unfilled basins [Dickinson, 1977]. Facies patterns involving turbidite and shelf sequences and fluvio-deltaic complexes are governed by water depth, sediment sources and rates, and subsidence of the basin. One of the best examples of a forearc basin is the "interdeep" of the Sunda Arc, lying west of the Burma-Sumatra-Java magmatic arc and east of a belt of flysch sediments and ultrabasic rocks, which form the outer arc of the Arakan and the Andaman-Nicobar-

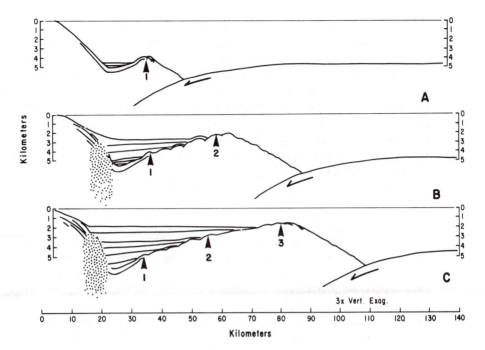

Figure 12-7 Inferred building out of accretionary prism showing migration of the morphology over the accreted material. The original trench slope break, shown by 1 in (A), becomes part of the subsiding basement of the upper slope area. Numbers mark arbitrary successive locations of the trench-slope break. The forearc basin shows steady growth. (From D. E. Karig and G. E. Sharman III, *Geol. Soc. Amer. Bull.*, vol. 86, p. 387, 1975, courtesy The Geological Society of America)

Mantawai line of islands [Mitchell and Reading, 1978]. Dickinson [1977] recognized several types of forearc basins, depending on the nature of the substratum, as follows:

1. Intramassif basins, lying within and resting upon continental basement terranes of the arc massif itself.

2. Residual basins, resting upon oceanic or transitional crust trapped between the arc and the site of initial subduction.

3. Accretionary basins, resting upon accreted elements of the growing subduction complex.

4. Composite basins, resting upon more than one of the foregoing basement types.

Intramassif basins are characterized by block faulting and marine or nonmarine arc- or backarc-derived sediments; accretionary basins are commonly characterized by compressional folds and listric (curvilinear) thrust faults and are filled with marine sediments derived from the uplifted subduction complex or arc terranes; residual basins are structurally and stratigraphically transitional between intramassif and accretionary basins, except that residual basins also include a basal sequence of abyssal-plain sediments [Dickinson, 1977]. The nature of the rocks beneath the sediments of the forearc basin remains uncertain, but beneath old forearc basins the rocks have been observed in uplifted, exposed sections [Dickinson, 1977]. Forearc basin sediments often occur in belts, flanked by terranes of blueschist facies rocks and melanges, representing subduction complexes on one side, and terranes of metavolcanics and batholiths, representing the magmatic arc on the other [Dickinson, 1977].

Forearc types

As subduction proceeds, forearc regions adopt a number of forms, which have been classified by Seely and Dickinson [1977] as **shelved, sloped, terraced,** or **ridged** (Fig. 12-8). Terraced forearcs exhibit a prominent terrace between the volcanic arc and the trench. These include ponded terraces and structural terraces (Fig. 12-8). Sloped forearcs exhibit a continuous, though regular, slope in the forearc region. Shelved forearc regions are similar to the ponded, terraced types, except that shelf sediments accumulate behind a dam. In ridged forearcs, the structural high lies on the seaward edge of an incompletely filled forearc basin or where terrestrial uplands form [Seely and Dickinson, 1977].

The Backarc Region

The backarc region includes the volcanic arc and backarc or marginal basins (Figs. 12-2 and 12-3). All active arc systems are associated with a volcanic arc, while backarc basins are not always present.

The volcanic arc

Volcanism pervades convergent margins. The first volcanoes appear abruptly inland from the trench and along the leading edge of a continent or island arc. This is called the **volcanic front** by Sugimura and others [1963]. The volume of recent eruptives is greatest at the front or immediately inland and decreases rapidly away from the trench. The active volcanic belt ranges from a single chain of volcanoes to a zone about 200 km wide (Fig. 12-9) although, at any single time, most volcanism takes place in a zone less than 50 km wide. Gutenberg and Richter [1954] observed the striking correlation between the principal structural arc with its active or inactive volcanoes and intermediate-to-deep earthquakes.

Well before the development of plate tectonic theory, Coats (in about 1960) recognized that arc magmas were being generated as a result of the thrust of oceanic crust beneath the Aleutian Arc. He suggested two mechanisms for magma generation: that the oceanic crust and its veneer of pelagic sediment could eventually melt with sufficient underthrusting, or that an aqueous fluid migrating upward from this underthrusted crust might induce melting in the overlying wedge of peridotite,

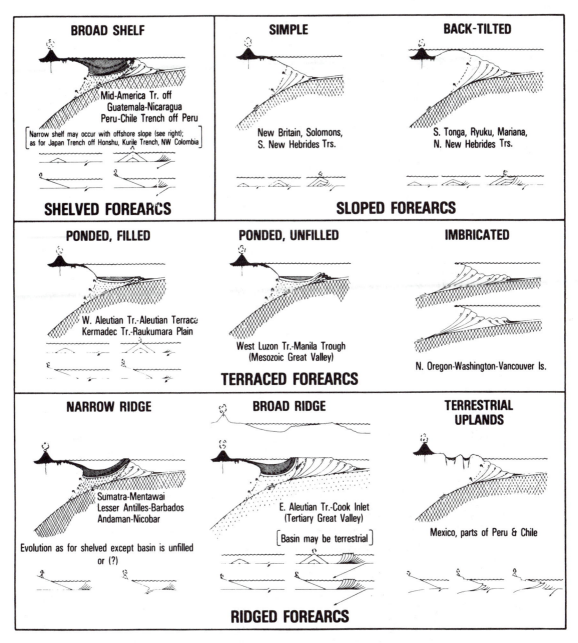

Figure 12–8 Classification of modern forearcs. Conceptual evolutionary cartoons are shown at the bottom of the diagrams. (From Seely and Dickinson, 1977, reprinted with permission of The American Association of Petroleum Geologists)

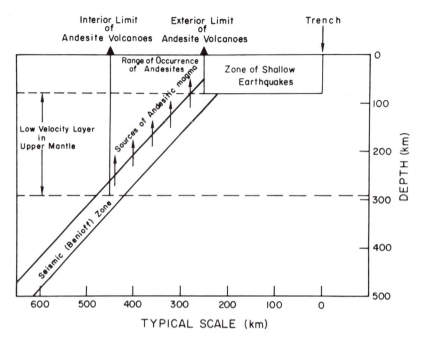

Interior Limit of Andesite Volcanoes
Exterior Limit of Andesite Volcanoes
Trench

Range of Occurrence of Andesites

Zone of Shallow Earthquakes

Sources of Andesitic magma

Low Velocity Layer in Upper Mantle

Seismic (Benioff) Zone

DEPTH (km)

0
100
200
300
400
500

600   500   400   300   200   100   0

TYPICAL SCALE (km)

Figure 12-9 Schematic diagram of model proposed by Hatherton and Dickinson [1969] for andesite production and distribution. (From T. Hatherton and W. R. Dickinson, *Jour. Geophys. Res.*, vol. 74, 5308, 1969, copyrighted by the American Geophysical Union)

because water greatly lowers the melting point of rocks. A controversy still exists over which of these mechanisms is more correct. Almost without exception, subduction is associated with earthquakes deeper than 100 km [Sykes, 1966] and with active volcanoes. Barazangi and Isacks [1976] discovered that, in the Andes, volcanoes occur only above those segments where the angle of the Benioff zone is steep enough to allow a wedge of the asthenosphere between the Benioff zone and the overlying lithosphere. On the other hand, volcanoes are rare or absent where the inclination of the Benioff zone is shallow and, as a result, remain within the lithosphere for large distances behind the Andes. These observations suggest that processes in the low-velocity zone between the lithosphere and the Benioff zone are suitable for the generation of magmas.

The mechanism for the generation of magmas under arcs is still not well understood. In the simple model, it is assumed that magmas are formed at some depth along the subducted slab and that the role of water in the subducted mafic oceanic crust must be important. Ringwood [1977] believes that the primary magmas produced near the Benioff zone at depths of 80 to 100 km are not andesitic but hydrous, tholeiitic basalts close to silica saturation. The tholeiitic magmas fractionate as they rise, principally by olivine separation, producing a range of basaltic andesites to andesitic magmas at shallow depths. Further shallow fractionation by separation of amphibole, pyroxene, and plagioclase produces dacites and rhyolites. Thus Ringwood [1977] believes that during subduction no material will be melted other than basalt or peridotite from the oceanic crust. Island-arc volcanism is but a link in the distillation of the mantle to produce continents [Marsh, 1979]. Basaltic oceanic crust (about 50 percent SiO_2) is differentiated from peridotite (about 47 percent SiO_2), which in turn yields the island-arc andesitic suite (about 55 percent SiO_2).

Once formed, the magma collects as a ribbonlike body near the upper plate edge and sends up magma in fingers that produce regularly spaced (often about 70 km) volcanic centers at the surface [Marsh, 1979]. It is unclear how the magma moves to the surface. It may travel as a propagating, magma-filled, elastic crack or as a hot, viscous glob slowly forcing its way along. In order to reach the surface, the magma must remain less than 50 percent crystallized. A few million years after establishment of the primary front, a weak secondary front of volcanoes may appear about 50 km behind the front. The development of this secondary belt of volcanoes may result from the downward motion of the initial melting ribbon as subduction proceeds. The geometry of the volcanism in relation to the Benioff zone implies that melting cannot be sustained for more than 100 km down the zone.

It is common for the axis of volcanism to shift laterally in rather rapid steps, although no systematic patterns are apparent in the direction of migration. In the Lesser Antilles Arc, the younger volcanic chain has migrated further from the trenches than the older arc; in Indonesia there has been a progressive outward migration with time; while in the Andes, migrations have occurred in both directions [Williams and McBirney, 1979].

It is a common belief that andesites are volumetrically dominant in island arcs. This was based on early twentieth-century observations, such as those by Marshall [1912], who distinguished an **andesite line** between the Pacific intraoceanic province dominated by basalts and circumoceanic provinces supposedly dominated by andesites. Although certain island arcs, such as Japan and the Lesser Antilles, are dominated by andesite, many are dominated by rocks closer to basalt, especially those island arcs built upon oceanic crust, such as the Tonga and Mariana arcs.

The andesite suite includes rocks with a wide silica compositional range. At one extreme are the aluminous basalts, containing about 50 percent silica; at the other are the dacites, with about 65 percent silica. The relative volumes of major rock types differ widely, both spatially and temporally. In certain arcs, although andesitic cones are conspicuous, their total volume is small when compared even with individual flows of plateau basalts [Williams and McBirney, 1979]. Island-arc basaltic lavas differ from other basalt lavas in containing less TiO_2 (less than 1 percent), low abundances of MgO (less than about 6 percent), high abundance of Al_2O_3 (greater than about 16 percent), and several other factors. Island-arc volcanoes have been found to exhibit wider ranges in variation in K_2O than in any other oxide. During the 1950s, Japanese petrologists, notably H. Kuno, were developing the idea that composition of volcanic rocks in active margins might be related to focal depths of intermediate and deep earthquakes beneath the volcanoes. They found a general increase in alkali content with increasing depth to the Benioff zone. Later Dickinson and Hatherton [1967] demonstrated a systematic correlation of potash and depth to the underlying Benioff zone. Volcanic centers over deeper parts of the Benioff zone contain lavas richer in K_2O. Thus the K_2O content in-

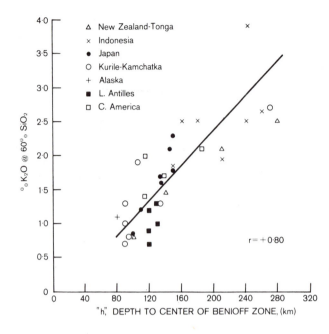

Figure 12-10 Relationship of K₂0 content of andesite to depth of earthquakes (within Benioff zone) beneath the volcanoes. (From T. Hatherton and W. R. Dickinson, p. 5302, *Jour. Geophys. Res.*, vol. 74, p. 5302, 1969, copyrighted by the American Geophysical Union)

creases in a backarc direction (Fig. 12–10). This correlation probably arises from the effect of pressure on the equilibrium constant describing the distribution of potash within magmas. This empirical relation serves as a guide to reconstruct paleoseismic subduction zones [Bally, 1979], although it is still unknown if the K_2O-depth relationship has remained constant through geologic history.

Volcanic rocks occurring with increasing distance behind the volcanic arc become less siliceous and more alkaline. For instance, the andesitic and tholeiitic basaltic rocks of the Japan volcanic arc give way to rocks richer in alkalies and poorer in silica in the Japan Sea, and in time to feldspathoidal rocks, such as leucite and basanites, in Korea and Manchuria [Williams and McBirney, 1979]. These trends result from decreasing SiO_2 and increasing K_2O content.

Backarc, or marginal, basins

One of the more striking geologic characteristics of the western Pacific Ocean is a large and complex group of backarc, or marginal, basins set between the intraoceanic arcs and the continents. In fact, most of the area of active margins is encompassed by marginal basins. These include areas such as the Coral Sea, New Hebrides basin, Philippine Sea, and Japan Sea. Except for the Scotia and Antilles arcs of the Atlantic, these marginal basins are peculiar to the western Pacific Ocean. Not all trench-arc systems have backarc basins, however. The Peruvian and Chilean arcs do not exhibit backarc basins, and in some areas, such as the Japan Sea, the process of backarc basin formation appears to be almost inactive. Clearly, subduction of an oceanic plate is not sufficient by itself to create backarc basins, although it seems to be a necessary factor [Uyeda and Kanamori, 1979]. Most of the basins shown in Fig. 12-11 are of Cenozoic age, having begun to form in late Paleogene time and subsided during the Neogene [Fusod, 1977].

Alfred Wegener [1924] identified the geometric relations between marginal basins and arcs and postulated that island arcs, particularly those of Southeast

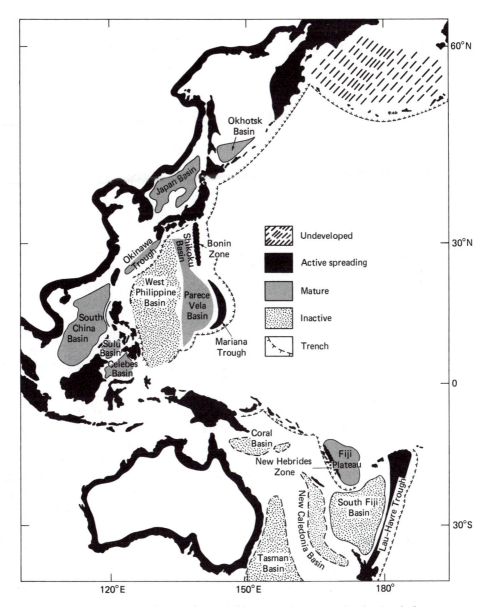

Figure 12-11 Distribution of marginal basins in the western Pacific classified according to their evolutionary stage. (After D. E. Karig, 1971)

Asia, are marginal chains, which detached from the continental masses when the latter drifted westward and remained attached to the old sea floor. The marginal basin was younger ocean floor produced from continental drift. Thus, Wegener inferred that marginal basins were of extensional origin with migration of an active arc system away from the continent. Since Wegener's work, other theories have been examined. Menard [1967] demonstrated large variability in thickness of the crust beneath marginal basins; although the crusts usually have thicknesses approaching those of the continental crust, the seismic velocities are closer to those of oceanic crust. These basins have been interpreted as either continental crust that has subsided to marginal-basin depths (a popular theory among Russian workers) or as oceanic crust that has become more continental as a result of large accumulations of sediment [Mitchell and Reading, 1978]. Other theories include the entrapment of a marginal part of preexisting ocean by the formation of an island arc, opening related to leaky transform fault, or opening related to the subduction of a ridge.

Most marginal basins are now known not to represent old ocean floor trapped behind an island arc. One clear exception is the Bering Sea [Cooper and others, 1976]. We have since returned to the thoughts of Wegener, believing that most basins have formed by spreading behind the volcanic arc with apparent simple migration of the arc away from the continental margin [Karig, 1970; Packham and Falvey, 1971]. This is possibly driven by the heating of the basin floor from below. A number of geophysical and geological observations have shown that extensional opening of backarc basins has been substantial in recent geologic time. Several models have been proposed and summarized by Bibee and others [1980]: One possibility is a sea-floor spreading process similar to mid-ocean ridges with discrete ridge transform sections, as proposed by Karig [1970], Sclater and others [1972], and Weissel [1977] for the Lau basin and Packham and Falvey [1971] for the Mariana Trough. The other extreme is a diffuse spreading model proposed by Sclater and others [1972], in which volcanism occurs contemporaneously throughout the basin. An intermediate model, which restricts volcanism to the central part of the basin but is less coherent than a ridge-transform model, has been proposed by Karig and others [1978].

Ocean-floor depth in most active marginal basins is similar to that of spreading ocean-ridge systems. Furthermore, rocks dredged in marginal basins do not differ in the geochemistry of either major or minor elements from those on spreading ocean ridges. Direct evidence from drilling and indirect evidence based on the amount of sediment cover, geologic trends, paleomagnetic studies, and the fit of predrift continental margins demonstrates that marginal basin crust is young and could not have formed at the mid-ocean ridge system.

If sea-floor spreading is occurring in marginal basins due to the separation of rigid lithosphere plates, there should be symmetric magnetic anomaly patterns similar to those formed at mid-ocean ridges. Magnetic lineations have been mapped in several marginal basins but have been of limited value in unraveling their history. This limited success has resulted from the relatively small size of the basins, which provide only a few anomalies, and lower, less-consistent magnetic lineations compared with mid-ocean ridge anomalies. Magnetic lineations have been mapped in the Japan Sea, Coral Sea, and the South China Sea, but correlations with the geomagnetic reversal time scale have proven difficult.

Seismic-refraction measurements of marginal-basin crust and petrology of dredged and drilled rocks indicate that the crust is oceanic in character. On the basis of sediment thickness, depth and character of acoustic basement, and mean heat flow, Karig [1971] separated basins where active crustal extension is occurring, such as the Lau-Havre basin, Mariana Trough, Scotia Sea, and the Andaman Sea, from those where extension has ceased, such as South Fiji, Shikoku, West Philippine, Coral Sea, Tasman Sea, and Japan Sea [Watts and others, 1977]. Because of the difficulty of correlating magnetic anomalies in the marginal basins, details of crustal history have not been determined, although geological evidence from adjacent islands and deep-sea drilling has provided general ages [Watts and others, 1977]. Those basins with well-defined and correlated magnetic lineations include the West Philippine basin, the Shikoku basin, the South Fiji basin, and the Bering basin [Watts and others, 1977].

Marginal basins have been found to be areas of unusually high heat flow compared with the broader ocean basins [Menard, 1967]. Packham and Falvey [1971] suggested that the high heat flow of marginal seas indicated regions of high-energy tectonic processes. This high heat flow, however, may not reflect higher rates of energy being supplied to the marginal basins, but may instead reflect other factors. Molnar and Atwater [1978] suggest that higher heat flow in marginal basins is not unusually high for the ages of their ocean floors. On the mid-ocean ridge, the character of heat flow differs because hydrothermal circulation is inferred to be more of an influence, and measurements tend to underestimate the total heat flow. Only in regions containing thick sediments, such as marginal basins, is most of the heat carried to the surface by conduction. Heat-flow measurements in these regions are probably more representative of the total heat flow from the sea floor of a given age than measurements of the same age from the mid-ocean ridges.

Little is known about the mechanism causing extension in marginal basins. Compressional tectonics would be the expected mechanism, but marginal basins are clearly under an extensional stress regime. Various models have been proposed. In a mechanism proposed by Sleep and Toksoz [1971] (Fig. 12–12), the subducting lithosphere drags part of the low-viscosity asthenosphere until the flow is deflected by an increase in viscosity, density, or both. This generates a flow pattern that brings hot asthenospheric material to the base of the lithosphere under the marginal basin. A combination of the upwelling of material, heating of the lithosphere, and flow-induced tension initiates rifting and causes spreading in the marginal basins. This theory has been extended by Bibee and others [1980] to explain the complex magnetic anomaly patterns in the Mariana basin (Fig. 12–13). This model requires spreading from a ridge within the marginal basin at any one time, but temporal variation in ridge position (or ridge-jumping), produces the complex magnetic anomaly patterns. Figure 12–13 shows a possible sequence of events in the opening of a barkarc basin, as proposed by Bibee and others [1980]. Initially, tensile stresses in the island arc cause rifting to occur and the establishment of a spreading center within the basin. As symmetric spreading continues, the spreading axis moves away

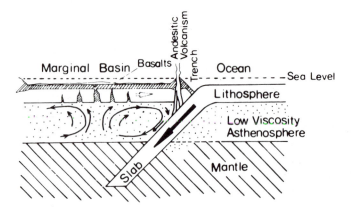

Figure 12–12 One possible model to explain the opening of backarc basins. This is called the secondary flow model. (From N. Sleep and M. N. Toksöz, reprinted by permission from *Nature*, vol. 233, pp. 548–550, 1971, copyright © 1971, Macmillan Journals Ltd.)

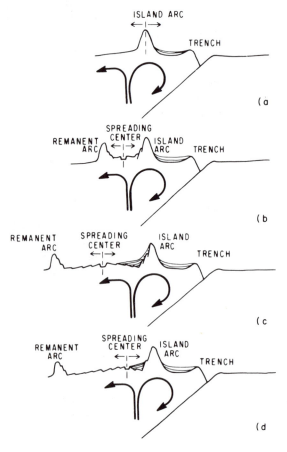

Figure 12–13 Development of backarc basin and jump in its axis of spreading. A possible model to explain the offset of the spreading axis within the back-arc basin toward the active island arc. In (a) a driving mechanism of some type sets up a tensional stress field which causes the active arc (a zone of weakness) to split. In (b) a basin has formed with a spreading center located symmetrically between the two arcs. In (c) the symmetric accretion of lithosphere results in the motion of the spreading center away from the trench and tensional field. Finally, in (d) the change in stress field near the spreading center causes the axis of spreading to jump back toward the active arc resulting in its asymmetrical position within the basin. (From L. D. Bibee, G. G. Shor, Jr., and R. S. Lu, *Marine Geology*, vol. 35, p. 95, 1980)

from the trench as new lithosphere is accreted, maintaining its integrity because it is a zone of weakness. Eventually the spreading center will move so far from the trench that the stress pattern will change considerably. At this point, the spreading center will jump back toward the active arc, where the driving forces are localized. This may be a small jump within the basin resulting in the asymmetrical location of the spreading center in the Mariana Trough or it may be a jump all the way back into the arc itself, resulting in a new remanent arc and a new basin. Such a mechanism may explain the formation of the series of interarc basins observed in the Philippine Sea. The result in terms of a magnetic signature would be anomalies with local, but no large-scale, correlation [Bibee and others, 1980].

We have already seen that not all subduction boundaries exhibit backarc basins and not all subduction boundaries that have such basins are actively opening at present. The extreme cases of each mode are represented by the Chilean Arc, which lacks backarc basins, and the Mariana Arc, which exhibits these basins (Fig. 12–4). Uyeda and his colleagues consider that these modes represent differences in the state of tectonic stress in the backarc region caused by variations in the character of subduction. Uyeda and Kanamori [1979] discovered a fundamental difference between the interplate thrust-type earthquakes (from focal mechanism solutions) in the two types of subduction zones. Although the numbers of earthquakes are similar between the two regions, the energy released by earthquakes is almost two orders of magnitude different between the two modes of subduction. More than 90 percent of the global seismic energy is released in subduction zones lacking active back-arc basins. Uyeda and Kanamori [1979] also found that earthquakes of very large magnitude occur in the subduction zones lacking active backarc basins. These fundamental differences in stress regions seem to result from the extent of mechanical coupling between the two interreacting plates. In the Chilean-type boundary, the mechanical coupling between the upper and lower plates is strong, and the subducting slab forces itself into the mantle, overcoming the strong resistive interactions with the upper plate. This results in large-magnitude earthquakes. In contrast, Mariana-type boundaries have coupling that is either much weaker or effectively nonexistent, and the subducting plate falls into the mantle rather freely. This allows tension in the area behind the arc and decoupling of the arc, which produces oceanward retreat of the trench and the development of backarc basins. No decoupling can occur in the Chilean margins because the trench remains fixed with respect to the mantle.

Toksoz and Bird [1977] have classified marginal basins according to their evolutionary development, based on the amount of heat flow, seismic-wave attenuation, and general spreading characteristics. Four stages of evolution of marginal basins have been distinguished by Toksoz and Bird [1977]: undeveloped, active spreading, mature, and inactive. Undeveloped basins exhibit typical basin morphology, but without high heat flow. Subduction has commenced and induced convection. However, insufficient time has elapsed to heat, widen, and rift the crust beneath the basin. Examples may be the Aleutian and Caribbean subduction zones. Active basins display clear evidence of active spreading centers with an axial topographic high, generally parallel to the axis of the island arc; high heat flow; and partly correlatable magnetic anomalies. Examples include the Lau-Havre and the

Mariana basins. Active basins may evolve into mature basins as more sea floor is created by spreading and as convection heats a broad region under the basin floor. Mature basins exhibit a broad heat-flow anomaly and evidence of at least a past spreading phase with widening resulting from spreading. The asthenosphere has been cooling because of the convection and rise of hot material toward the surface. Spreading may have slowed or it may have become less organized. Mature basins are the same as those classified by Karig [1971] as "inactive with high heat flow." Examples include the Sea of Japan, Okhotsk basin, and the North Fiji basin. Inactive basins are in the final stage when subduction has ceased and the crust loses its thermal energy and heat flow values return to normal: The basins have reached old age. Examples include the South Fiji basin and the West Philippine basin. According to Toksoz and Bird [1977], spreading of the basin floor moves the island arc and trench oceanward, decreasing the dip of the downgoing slab. The behavior of the induced convection at this stage depends on whether it can draw upon a large mass of fresh, hot asthenosphere from a wide region. If this occurs, the hot asthenosphere will also move and begin warming a new, future basin region. If the flow is trapped, its gradual cooling will produce increasing resistance to subduction and ultimately create a new subduction zone seaward of the older one.

Considerable variation exists in the character and thickness of sediments in marginal basins. The quantity of sediment within marginal basins is primarily a function of available source areas and, secondarily, of age. Karig and Moore [1975] developed an evolutionary model for sedimentation for marginal basins largely isolated from any important terrigenous input. Four sediment types dominate, including volcanically derived material from the volcanic arc, volumetrically the most important; montmorillonite clays derived from the volcanic arc; biogenic ooze; and windblown dust from the continents. An apron of volcanically derived sediments commonly develops adjacent to the volcanic arc. This apron grades outward into deep-sea clays or biogenic oozes where turbidite sequences have not accumulated.

Active Arc of New Zealand

The New Zealand region has been at the boundary of the Indian and Pacific plates for the last 40 m.y., and its structure contains most of the elements of an active margin. Several margins have been uplifted above sea level, making them more accessible for study than many other trench-arc systems [Lewis, 1980; Cole and Lewis, 1981]. The boundary between the Indian and Pacific plates extends from the Tonga-Kermadec Trench, through New Zealand to the Puysegur Trench (Fig. 12-14). The character of this boundary changes significantly from north to south. The Hikurangi margin-trough system on the northeastern side of New Zealand (Fig. 12-14) is mainly an extension of the Tonga-Kermadec arc-trench subduction system into a continental environment. The Taupo-Hikurangi system is part of a complex interaction between an obliquely compressive plate boundary and a small piece of continental crust. This system has been interpreted by Lewis [1980] and Cole and Lewis [1981] within an active-margin framework from which the following account is based.

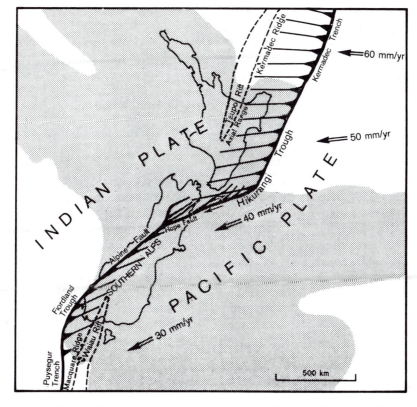

Figure 12–14 Major elements of the Indian-Pacific Plate boundary in the New Zealand region. Stippling represents continental crust. Arrows show relative motion of Pacific Plate with respect to the Indian Plate. Lines represent direction of motion of underthrusting plate. (From K. Lewis (1971) and J. W. Cole and K. Lewis, 1981, modified from Walcott, 1978)

To the north of New Zealand, along the Tonga-Kermadec Trench, ocean crust of the Pacific plate is subducted beneath ocean crust of the Indian plate. To the south, in the Puysegur Trench, there is a mirror image with oceanic Indian plate being thrust beneath the oceanic Pacific plate. Between these two subduction zones, motion becomes increasingly oblique, with the whole system truncated at a strike-slip, transform boundary that extends as a complex system of transcurrent faults extending through New Zealand, including the Alpine fault (Fig. 12–14). Shearing is due to more rapid northward motion of the Indian plate from the spreading Pacific-Antarctic ridge than the Pacific plate [Molnar and others, 1975]. Total shear between the plates has been about 1000 km, although the character of the shear has been complex and changeable. During the last 10 m.y., most of the plate boundary has experienced oblique compression. This compression has intensified during the last few million years, when most of the uplift of the southern Alps has occurred (**the Kaikoura orogeny**). Subduction is associated with the andesite-dacite volcanism in the Taupo graben 250 km to the west and with a Benioff zone of seismicity that begins to dip steeply about 200 km to the west of the trough.

The North Island of New Zealand contains most elements of a forearc region (Figs. 12–15 and 12–16). Landward of a relatively shallow trench is a 150-km wide accretionary prism, trench-slope break (mid-slope high), a forearc basin, a frontal arc (the main graywacke ranges), a volcanic arc, and a narrow, backarc basin. Subduction, combined with rapid detrital sedimentation, has produced a 150-km, imbricate-thrust controlled, accretionary wedge of seaward-faulted, anticlinal ridges and landward-tilting basins. The basins are generally 5 to 30 km wide and 10 to 60 km long and contain sediment sequences up to 2000 m thick.

Accretionary prism

The continental slope of eastern North Island is marked by a series of synclinal sedimentary basins and anticlinal ridges aligned approximately parallel to the slope (Figs. 12–16 and 12–17). These basins are typically 15 km wide and 30 km long. The folds have been estimated to be growing at a rate of 3 m per 1000 years between ridge crest and trough [Lewis, 1971]. Paleodepth data indicate that the upper continental slope has been uplifted from 700 to 1600 m during the last few million years. Most of the folds are steeply dipping or faulted on their seaward side, and low-angle underthrusting is evident near the trench. These are believed to represent thrust faults resulting from sedimentary offscraping upon subduction. The addition of new wedges at the toe of the slope uplifts the accretionary prism. The sub-

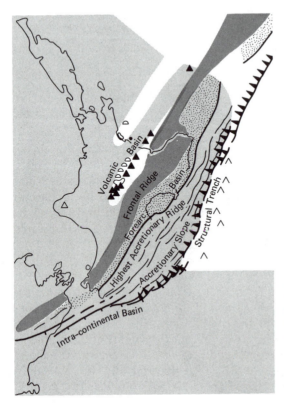

Figure 12–15 Major structural elements of the Hikurangi Margin of North Island, New Zealand, showing onshore propagation of some elements from the Kermadec system to the north and termination of the oblique-subduction system at the transform boundary to the south. (From K. Lewis, 1971, and J. W. Cole and K. Lewis, 1981, *Tectonophysics*, vol. 72, pp. 1–22)

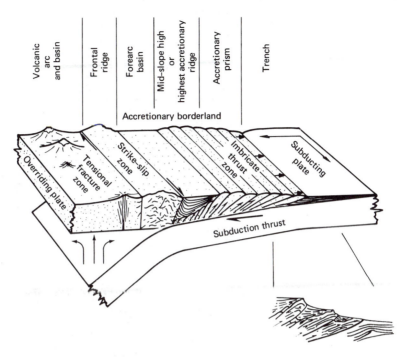

Figure 12-16 Schematic model showing major elements of the obliquely subducting margin of eastern North Island, New Zealand. (After J. W. Cole and K. Lewis, 1981)

parallel anticlines act as sediment dams, causing an accumulation of thick sediment sequences of turbidites, hemipelagic muds, and volcanic ash layers. In some places the basins have become totally filled, and sediment is draped over the growing seaward-faulted anticlines (Fig. 12–17).

Trench-slope break and forearc basin

Eastern North Island is unusual because most of the highest accretionary ridge and forearc basin are on land, forming a line of coastal hills. In some areas, parts of the trench-slope break are exposed as islands of small or medium size such as Barbados and the Car Nicobar Islands. In most areas the trench-slope break is at bathyal depths. Few other places besides New Zealand have elements of the accretionary wedge so well exposed. The trench-slope break is bounded on the landward side by a wide, deep, forearc basin.

Frontal arc

This feature is expressed as a series of high ranges extending for much of the length of New Zealand. Sediments of the frontal arc are usually graywacke of turbidite origin and late Paleozoic and Mesozoic age. Much of the frontal ridge has been rapidly uplifted (4–7 mm/yr) [Wellman, 1970] during the last few million years.

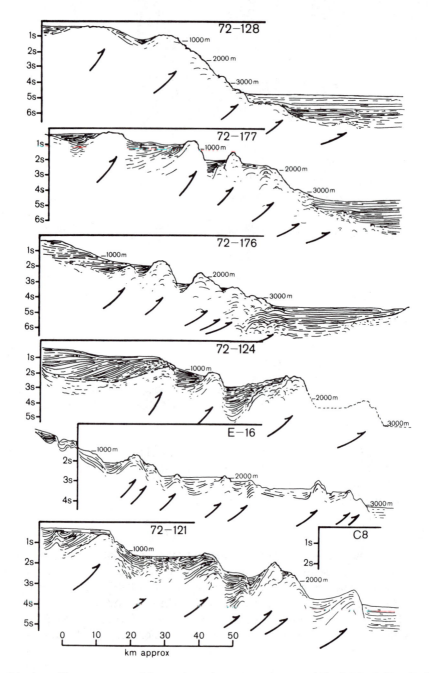

Figure 12-17 Seismic profiles across part of the continental margin to the east of North Island, New Zealand. These are within the accretionary borderland (Fig. 12-16) and clearly exhibit successive steps within the accretionary prism. Top profile is most northern and is located to the east of northern Hawkes Bay; bottom profile is the most southern and is to the southeast of Wellington. Hikurangi Trough sediments to right in each profile are Plio-Pleistocene in age and almost certainly turbidites. Slope strata at left are mainly late Mesozoic and early Tertiary in age. Arrows represent probable position of reverse faults with major strike-slip component of motion. At left, two-way travel time in seconds. On slope, depth to seabed in meters. Horizontal scale is approximate only. A line sloping at 45° on the profile represents a surface with a dip of about 6° in the line of the profile. (From Cole and Lewis, 1981, *Tectonophysics*, vol. 72, pp. 1–22)

Volcanic arc

Behind the frontal arc exists a line of andesite-dacite volcanoes extending from the Tongariro volcanic center in the center of the North Island to White Island offshore to the northeast. Most of the andesites are olivine bearing and have been erupting for the last 50,000 years. The volcanic arc occupies the eastern side of a backarc basin 40 km wide.

Backarc basin

This is an intensely faulted graben filled with more than 2 km of predominantly rhyolitic pyroclastic sediments. The basin began opening about 2 m.y. ago.

This active margin has been evolving to its present form only during the last few million years, before which there existed an older (middle Cenozoic) subduction system complete with volcanic arc to the west and extending northward through the Northland Peninsula (Fig. 12-14). Although the accretionary prism is well developed in the modern system, the New Zealand region is not typical of many convergent margins. This is because it is transitional between a convergent and a transform margin. Also, the backarc basin is anomalously narrow and volcanically is highly influenced by its setting within continental crust.

Inception of Convergent Margins

Arc trenches are formed by contractural activation of previously rifted, divergent continental margins (Fig. 12-18), by breakage across an intact oceanic plate, by reversal of polarity of an isolated interoceanic arc, or by reversal of polarity following accretion of an intraoceanic arc to a previously divergent continental margin by crustal collision (Fig. 12-18) [Seely and Dickinson, 1977; Dickinson and Seely, 1979]. Little is known about the conditions that trigger the subduction process. However, under certain circumstances the oceanic plate seems to lose its support; it collapses and forms a new subduction zone. Molnar and Atwater [1978] have pointed out that the style of tectonics of subduction zones generally correlates with the age of the ocean floor being subducted. Subduction of ocean floor older than 50 Ma, and especially of those older than 100 Ma, is commonly asociated with interarc spreading. Subduction of ocean floor younger than 50 Ma is often associated with Cordilleran tectonics—broad parallel zones of deformation and high mountains. Young oceanic lithosphere is more buoyant than old oceanic lithosphere, so that it sinks less readily into the asthenosphere [Molnar and Atwater, 1978]. Heavy, old oceanic lithosphere may be pulled down more rapidly than the plates can converge toward each other. Consequently, the oceanic plate loses its support and begins to sink before it reaches the existing trench. Such a collapse of the oceanic plate may lead to the formation of backarc basins. Alternatively, a new subduction zone may begin to form in the backarc region, beginning at the volcanic arc where a preexisting zone of weakness exists. Subduction of young oceanic lithosphere may require an extra force, which may cause the high mountains and broad zones of deformation associated with Andean-type margins. Conversely, if ancient Andean margins can be clearly recognized in the geologic record, they may

Plate Rupture

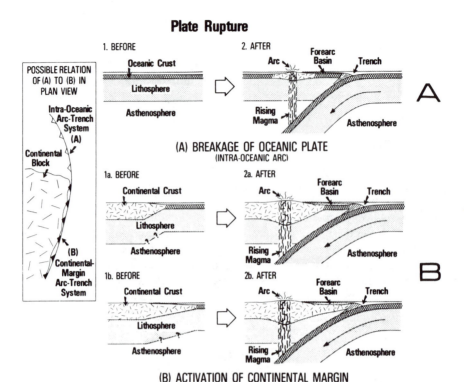

Figure 12–18 Two models for the inception of a convergent margin. (After Dickinson and Seely, 1977, reprinted with permission of The American Association of Petroleum Geologists)

indicate subduction of relatively young oceanic lithosphere [Molnar and Atwater, 1978].

Episodicity of Arc Volcanism and Tectonism

Tectonism and volcanism have been episodic throughout geologic time [Stille, 1924; Umbgrove, 1947] and not constant, as often assumed during early phases in the development of plate tectonics. The processes responsible for island-arc formation appear to be episodic, marked by pulses of volcanic activity [Karig, 1973]. The generation of calc-alkalic magmas is believed to result from melting near a descending crustal slab and the mantle above this slab in the subduction zone. The record of volcanic activity in the island-arc regions may provide vital information on whether subduction, and hence spreading rates, are nearly constant or if they exhibit fluctuations in activity. Such information relates directly to problems concerning the mechanics of plate tectonism. Karig [1975], for instance, determined that the marginal basins of the Philippine Sea developed most rapidly during times of maximum volcanism along the associated arc system. Extension of these marginal basins has not been constant during the Cenozoic, but seems to have occurred during only one-third of the time during the last 50 m.y. (since middle Eocene). Arc volcanism along the eastern periphery of the Philippine plate has also been of uneven intensity during the Cenozoic, with the pulses of arc volcanism of much longer duration than the extensional phases. The significance of changing rates of volcanic extrusion along convergent plate margins is not well understood, but it is assumed that intensity of arc volcanism is crudely proportional to the rate of subduction [Karig, 1973].

Unfortunately, there are few quantitative data on the rates of eruption of igneous rocks over prolonged periods in different structural settings. Stratigraphically controlled volumetric data for volcanic rocks have been compiled for part of the Cascade Range in Oregon [McBirney and others, 1974] and for part of the Japanese Arc [Sugimura and others, 1963]. Interpreting the history of volcanism from land geology by itself is difficult because gaps in the record due to erosional unconformities or superposition of later volcanics are not always distinguishable from periods of volcanic quiescence. Therefore direct comparision between subaerial and submarine deposited volcanic material is valuable. Initial studies on the longer term stratigraphic record of explosive volcanism reflected in volcanic ash layers in deep-sea cores have been carried out by Kennett and Thunell [1975; 1977A,B] and correlated with the subaerial record of volcanism [Kennett and others, 1977]. These limited measurements indicate that volcanism is not uniform, but has occurred in distinct episodes that seem to be synchronous over broad regions and are perhaps even worldwide. The cause of this synchroneity is not yet known, but it may be due to discontinuous plate motions manifested as changing rates of explosive volcanism.

The record of subaerial volcanism [McBirney and others, 1974] shows that most recently active basaltic and andesitic volcanoes around the Pacific margins have grown to their present size during the Quaternary. Similar episodes of Tertiary volcanism were also concentrated in brief pulses with duration of only 1 to 2 m.y.

Examination of layer sequences of volcanic ash and their alteration products in deep-sea sediments through the Tertiary and much of the late Mesozoic has been made possible by the DSDP. These studies are among the best for determining the long-term eruptive histories of regions in different tectonic settings. Studies of offshore tephra-layer stratigraphy in deep-sea drilled sections [Kennett and Thunell, 1975; 1977A, B] show strong evidence for widespread synchronism or near synchronism of volcanic acivity over large regions around the Pacific and perhaps other oceanic areas [Kennett and others, 1977; Vogt, 1972; 1979]. Large segments of the circum-Pacific were affected by increased volcanic activity during particular intervals within the Cenozoic. Within limits, their data provide a reliable record of extrusive and explosive volcanism (Fig. 12–19). Although terrestrial and marine records for individual regions reveal important differences in the episodicity of volcanism, correlation is found between activity in the Southwest Pacific, Central America, and the Cascade Range of western North America. The two most important pulses (Fig. 12–19) of Neogene volcanism occurred during the Quaternary, called the **Cascadian episode** [Kennett and others, 1977], and within the middle Miocene (16–14 Ma), called the **Columbian episode** [Watkins and Baksi, 1974].

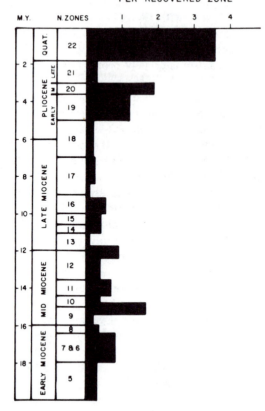

GLOBAL COMPILATION
AVERAGE NUMBER OF ASH LAYERS
PER RECOVERED ZONE

Figure 12-19 Global compilation showing the average number of volcanic ash layers per recovered N zone (planktonic foraminiferal zone) for suite of Neogene DSDP sequences up to Leg 34. These layers contain volcanic ash other than those that have clearly undergone plate motions (normal to potential volcanic ash sources). (After Kennett and others, 1977)

Less-important episodes occurred during the latest Miocene to early Pliocene (6–3 Ma), called the **Fijian episode,** and during the late Miocene (11–8 Ma), called the **Andean episode** [Kennett and others, 1977]. Dating of terrestrial sequences indicates that these episodes of intense volcanism took place over short intervals of time separated by longer, more-quiescent periods.

Thus strong evidence is emerging for general episodicity of volcanism at plate boundaries over large areas, that volcanic episodes may act in unison over wide areas, and that geochemical similarities or historical characters of volcanism may show similarities over large areas. The major increase in explosive volcanism during the latest Cenozoic may also be related to increased tectonism of landmasses. It seems that many of the world's major mountain systems have been subjected to substantial uplift during the last 2 Ma. These include the Alps and the Himalayas in Eurasia, and the Sierra Nevada and Andes in the Americas. Such tectonism is probably linked with increases in volcanism recorded on land [Kennett and Thunell, 1975; McBirney and others, 1974].

The causes of such episodic volcanism and tectonism of the island-arc and continental-margin areas are not known, but are probably due to variations in the rate of plate convergence [Brookfield, 1971; Chappell, 1975; Kennett and Thunell, 1975; Scheidegger and Kulm, 1975]. Synchronous episodic activity around much of the Pacific basin [Kennett and others, 1977] suggests large-scale tectonic controls. The problem of episodic volcanism during times of near-uniform sea-floor spreading is a fundamental and unsolved problem confronting the concepts of global tectonics [McBirney, 1971; Scheidegger and Kulm, 1975; Hogan and others, 1978]. The broad scale of episodic events suggests global rather than regional or local controls. Although the rate of volcanism should show some relation to rates of subduction and sea-floor spreading, the magnitude of changes in volcanism with time cannot simply be due to comparable changes in sea-floor spreading rates; no evidence has been found for such drastic changes. For this reason Kennett and Thunell [1975] suggested that explosive volcanicity may represent a sensitive amplifier of changes in the rates of sea-floor spreading and subduction.

If episodicity of volcanism is related to changes in subduction rate, evidence should exist for some change in the rate of sea-floor spreading. During times of more rapid spreading, increased shear and melting along lithospheric boundaries should increase explosive volcanism, whereas during times of less rapid spreading, volcanic activity may be less intensive. The most accurate record of horizontal plate motion is found at the oceans' most rapidly spreading ridges, where the largest amounts of deep-sea floor are formed during a given polarity interval and where there is the least amount of topographically induced noise in the geomagnetic signal [Rea and Scheidegger, 1979]. The East Pacific rise along the western edge of the Nazca plate is the most rapidly spreading portion of the oceanic ridge system, with an average half-rate of spreading of about 8 cm/yr. Rea and Scheidegger [1979] carried out a detailed examination of spreading rates over several locations along the Nazca-Pacific plate boundary and presented evidence for small changes in the rates of spreading during the last 4 m.y.

The problem of episodic volcanism has taken on broader implications since the discovery [Vogt, 1972; Kennett and Thunell, 1975] of an apparent synchronism of Cenozoic volcanic activity in the circum-Pacific region with that of the Hawaiian-Emperor volcanic chain [Vogt, 1972, 1975; Shaw, 1973; Jackson and others, 1975; Rea and Scheidegger, 1979] where increase in volcanic activity is well documented for the middle Miocene and latest Cenozoic (Fig. 12-20). This last episode began about 5 Ma and has increased toward the present day. Plots of extrusion rates through time derived from the topographic expression of the ridge by Vogt [1972], Jackson and others [1975], Rea and Scheidegger [1978], and Vogt [1979] are shown in Fig. 12-20. The curves correlate reasonably well with radiometric and ash-frequency data of Kennett and others [1977]. Limited dating from other hot spots, including those in the Atlantic Ocean, suggest similar patterns [Vogt, 1972; 1975]. Rea and Scheidegger [1979] present temporal relations among a variety of volcanic expressions, including Hawaiian extrusion, silica variation, and ash accumulation in deep-sea sedimentary sequences, circum-Pacific volcanic episodes, and sea-floor spreading rates. This also shows similarities to the tempo of volcanic activity in regions of distinctly different tectonic regimes, including the

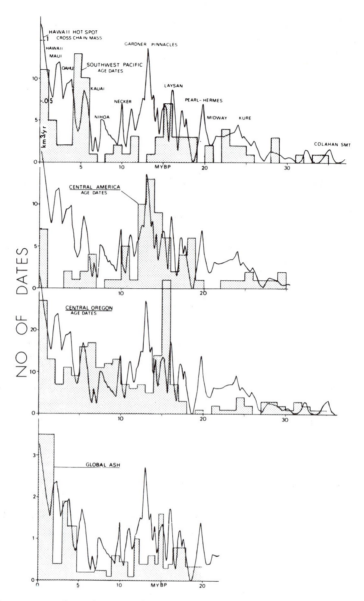

Figure 12-20 Cross-chain mass (single line—expressed as discharge rate in km³/yr) along the Hawaiian volcanic chain [Vogt, 1972, 1975] compared to three regional indices of circum-Pacific volcanism and global ash frequency [Kennett and others, 1977], all as a function of age. Note the oscillations rather than constancy of all parameters measured. Age scale along Hawaiian chain was obtained by smooth fit to published radiometric ages. The central Oregon data are probably biased toward young rocks; tephra in the 30 to 20 Ma range is underrepresented. (From P. R. Vogt, *Geology* vol. 7, p. 94, 1979, courtesy The Geological Society of America)

mid-ocean ridge crest, mid-plate volcanism, subduction zones, and continental margins. All measured parameters show increases in volcanicity during the latest Cenozoic (about 2 Ma to the present). As discussed in Chapter 9, possible correlations also exist between changes in oceanic-ridge volcanism and sea-level change [Vail and others, 1977; Kennett and others, 1977].

In summary, similarities in volcanic tempo in regions of different tectonic regimes (Fig. 12–20) indicate that episodicity is not always due to local factors, such as regional differences in subduction rates. Volcanic activity associated with convergent plate boundaries and hot spots, such as Iceland and Hawaii [Vogt, 1972], and of the mid-ocean ridges are, at times, possibly synchronous. Variations in the dynamics of lithospheric plate motion on a worldwide basis may be ultimately recorded in volcanic-ash sequences [Kennett and Thunell, 1975] and perhaps the record of marine regressions and transgressions over the continental margins.

Basic Problems of Convergent Margin Processes

Because even the most powerful, multichannel seismic-reflection techniques have barely been able to discriminate deep structure in subduction zones and because little deep drilling has been carried out in the active margins, major questions remain to be answered about processes that mold the convergent margins. Although investigations on land and at sea are bound together by a unifying theory, an enormous gap still exists between the simple subduction model for convergent margins and geologic relationships on land. As the ability to discern structure at depth in the active margin zones increases, major problems will develop and the picture will become more complex. Among the numerous questions that remain unanswered [Fusod, 1977] are the following:

1. How does simple tectonic stacking produce the complex structures observed in coastal mountain belts?

2. What controls locations of island-arc subduction zones?

3. Why does interarc spreading occur where it does instead of along lines of weakness associated with the subduction zone closer to the trench? Why are interarc basins small instead of large?

4. What proportion of sediment is accreted and what proportion is subducted in different kinds of convergent margins? How does this affect the structure of the accretionary prism?

5. How valid are present interpretations that flysch and graywacke sequences are ancient accretionary prisms and that ophiolitic complexes are the roots of ancient subduction zones? Where are present-day geosynclines located? Where, within modern convergent margins, are the tectonic settings inferred from land-based outcrops?

Continental Borderland Province: California

The morphology of the ocean off southern California reflects the distinctive tectonic character of the region. In this area, the true continental shelf is only a few kilometers wide and is adjacent to a long (1000 km) and wide (200 km) sequence of basins, banks, and islands called the **continental borderland** [Emery, 1960]. Seaward of the borderland province is the continental slope. The continental borderland is located on continental crust and contains about 20 basins arranged in 3 irregular rows to form a checkerboard pattern [Gorsline, 1978] (Fig. 12–21). The basins range in axial length from 50 to about 200 km and in width from 20 km to 100 km. All but one of the basins (San Diego Trough) are closed depressions of 1000 m to greater than 2000 m depth and have sill depths that range in depth from 200 m to more than 2000 m [Gorsline, 1978].

The block-and-basin topography is believed to originate within the tectonic regime of a broad and soft transform fault zone associated with the San Andreas fault [Crowell, 1974]. The pattern of escarpments, the shapes of banks, ridges, and islands, and the configuration of basins suggests a pull-apart origin of some of the basins [Crowell, 1974]. The basins are believed to have opened along a system of anastamosing, subparallel, transcurrent faults, all with the same direction of offset as the San Andreas fault. Because the tectonism is of late Cenozoic age, there has been insufficient time for basinal infilling and topographic smoothing, and hence the topography is relatively rugged compared with most continental platforms. Initial formation of the borderland began during the late Miocene (10 to 5 Ma); since that time, periodic tectonism has continued to alter the sedimentary deposits. The

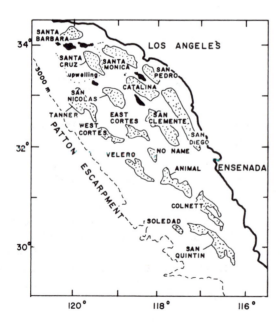

Figure 12-21 The California Continental Borderland showing the outlines of the major basins. Upwelling is centered in the area south of the Northern Channel Islands and extends south to 33° N latitude. (From D. S. Gorsline, *Journal of Sedimentary Petrology*, vol. 48, p. 1056, 1978, reprinted with permission of the Society of Economic Paleontologists and Mineralogists)

contemporary basins contain conformable, continuous, and relatively undisturbed deposits dating from the late Pliocene (about 3 Ma) and overlying, increasingly deformed older late Cenozoic sediment prisms [Gorsline, 1978].

The borderland represents an unusual modern sedimentary environment because it consists of a set of margin basins arranged at increasing distances from the source of sediments. The offshore basins are relatively starved of clastic material and contain only pelagic and hemipelagic sediment. Seismic profiles show that the thickest sediment deposits are in the nearshore basins. Sediment spills from these basins into adjacent basins farther seaward occur only when sediments in the inner basins have filled to sill depths [Emery, 1960]. A succession of older (Miocene-Pliocene) filled basins occurs on land as an inland extension of the borderland province. Several are important for oil production, including the Los Angeles and Ventura basins.

COLLISION PROCESSES ON CONVERGENT MARGINS

Collision margins are the most complicated and least understood element of plate tectonics. These areas have the most complex topography and structure which, according to Peter Molnar, reflect complicated tectonic weather within a regime of relatively calm tectonic climate. Continental collision is an inevitable consequence of plate motion and is thus a *general, not rare,* phenomenon [Dewey, 1980]. In 1965, J. Tuzo Wilson recognized that an evolutionary cycle acts upon the ocean basins, opening them during one phase and later closing them, creating mountain belts such as the Appalachians (see Chapter 6). These mountain belts mark the locations where oceans have closed, effectively eliminating much of their history from the geological record. Evidence from northwestern Canada indicates that ocean opening and closing processes 2 billion years ago were similar to those of the present time [Burke, 1979]. Thus plate tectonic processes have been operating on the earth for at least this length of time, and it is unnecessary to invoke the operation of nonuniformitarianism mountain-building processes to explain the ancient fold belts. Collision margins are marked by **obduction** (plate accretion resulting from thrusting), by extensive ophiolite emplacement (see Chapter 7), and sometimes by arc reversals.

There has been a considerable evolution of thought concerning mountain building. In 1857, James Hall observed that the Appalachians were made of intensely folded sediments of much greater thickness than sediment of the same age in undisturbed regions. This led to the concept that no mountain range could form without first developing thick sediment deposits, a central tenet of the **geosynclinal theory.** Later, workers in the European Alps discovered deep-sea sediments in folded belts, and the association of radiolarian-rich sediments with ophiolites led to the concept of a deep basin in the region of the Alpine geosyncline flanked by an African and a European shelf and slope. Much later, Drake and others [1959] related geosynclinal regions to continental margins.

Collision margins, expressed as uplifted mountain belts, are primarily within the realm of the field geologist on land. Nevertheless, they are of interest to the marine geologist because any record of ancient, destroyed oceans is preserved in these regions. They are especially important in the reconstruction of Paleozoic and Precambrian oceans. There are two principal types of collision margins: those resulting from collision of a continent with an island arc (Fig. 12–22) and those resulting from continent-to-continent collision (Fig. 12–23). There is general agreement, for instance, that the Banda Arc of eastern Indonesia represents the collision between an Asian arc and the margin of northern Australia. This type of collision forms a series of parallel mountain chains of the cordilleran type. Other folded belts, such as the Alps and the Himalayas, have resulted from the collision of two continents. The sequence of events involved in the approach and collision of an Atlantic-type continental margin (such as off North Australia) with a subduction margin of another continent is shown in Fig. 12–23 [Dewey and Bird, 1970]. A subducting (and thus trenched) margin may be associated with an existing or developing cordillera or with a margin resulting from an island-arc-to-continent collision [Dewey and Bird, 1970]. The pattern is even more complex in the Mediterranean region with the consumption of the northward moving African plate down into the Ionian Trench south of the Aegean Arc. Given the present plate motions, a collision between Greece and Turkey seems inevitable [Dewey and Bird, 1970]. The Alpine fold belt, between the Ionian Trench and the European platform, is a complex of sutures representing the sites of old trenches and constructed of flysch, blueschist facies rocks, and ophiolites. Crystalline massifs between these sutures are probably

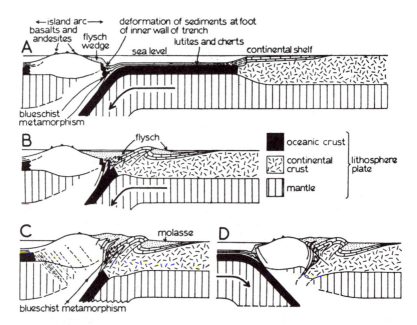

Figure 12-22 Schematic sequence (A to D) of cross sections showing the collision of a continental margin of Atlantic type with an island arc, followed by change in the direction of plate descent. (From J. F. Dewey and J. M. Bird, *Jour. Geophys. Res.* vol. 75, p. 2641, 1970, copyrighted by the American Geophysical Union)

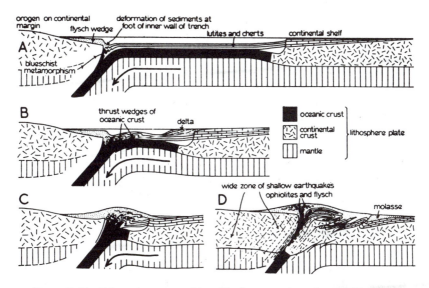

Figure 12-23 Schematic sequence (A to D) of cross sections showing the collision of two continents. (From J. F. Dewey and J. M. Bird, *Jour. of Geophys. Res.* vol. 75, p. 2642, 1970, copyrighted by the American Geophysical Union)

previous microcontinents or island arcs accumulated by a series of collisions at the northern margin of the Tethys Sea [Dewey and Bird, 1970].

After collision of a previous Atlantic-type margin with a trench, the continent is drawn against, and partly beneath, the accretionary prism. As the suture belt closes tightly, rocks of the previously passive continental margin may become involved in thrusting associated with the subduction zone (Fig. 12-23). Partial subduction continues until the buoyancy of the subducted continental edge arrests plate descent, after which additional major thrusting develops (Fig. 12-13) [Dickinson, 1977]. In the Himalayas, the thrust complex, with width of more than 100 km, is composed of peninsular Indian crustal elements including basement rocks.

Although it is well known that continental crust is too light to be subducted, geologic evidence exists for large-scale crustal shortening of up to 1000 km by thrust faulting in mountain belts such as the Himalayas. Thus some underthrusting of continental crust may occur [Molnar and Atwater, 1978], although the amount involved is difficult to determine. Two processes probably contribute to the underthrusting of continental crust. First, the lower part of the lithosphere is cold and dense enough to include sinking if partial decoupling could occur with the overlying lithosphere. Secondly, the gravitational force acting on the subducted oceanic crust may continue to act upon the continental crust. Eventually, however, the buoyancy of the underthrust continental rocks will prevent further destruction, and the descending plate may break off and sink into the asthenosphere [McKenzie, 1969]. Elimination of the trench will be followed by intense deformation over a wide area. The structures resulting from a collision are highly variable and may depend on the character and thickness of the sedimentary sequence deposited before the collision. This variation has been extensively described by numerous workers but is beyond the scope of this book. A variety of high-strain zones associated with wide zones of basement reactivation makes it difficult to recognize the line along which continental masses originally divided particularly at deep structural levels in eroded older orogenic systems. This creates difficulty in reassembling continental fragments to former positions.

Mountains are formed by several processes that do not often occur singly, but in combination [Dewey and Bird, 1970]. For instance, the evolution of the Appalachian Mountain belt involved formation of both Ordovician island arcs and Cordilleran mountains and Devonian continental collision [Bird and Dewey, 1970]. Thus the mountain belts presently lying within continents are probably a complex of cordilleran belts, microcontinents, and volcanic arcs of different ages, consolidated by complete closure of a former ocean.

Mediterranean Tectonic Evolution

The Mediterranean Sea is a product of a near collision between Africa and Eurasia. This convergence is reflected by modern tectonism, volcanism, and earthquakes in both the western and eastern parts of the Mediterranean Sea. This sea is the classic example of the relatively small, complex, and deep marine basin largely surrounded by continents that have been called **mediterraneans** by Shepard [1948]. Other examples are the Gulf of Mexico, Caribbean, and the China Sea, although their tectonic settings are each very distinct. Water depths of these enclosed seas are somewhat shallower than those of the large oceans, although they can extend to depths as great as 4000 m. The Mediterranean Sea is one of the most geologically complex of the world's convergent margin areas. This results primarily from the fact that it lies as a trapped basin within a collision boundary caused by the northward movement of Africa toward Eurasia. The tectonic evolution of the Mediterranean is a very controversial issue, resulting from this complexity and the interplay between a large number of European workers of different nationalities and geological backgrounds! Also, because the tectonic evolution of the Mediterranean Sea is linked to a continent-continent collision, much of the geological evidence lies within the folded and thrusted mountain belts surrounding the region.

The Mediterranean Sea occupies a broad furrow extending from the Strait of Gibraltar eastward to the coast of the Levant, a distance of 4000 km. A basic threefold physiographic subdivision can be recognized for the basin.

1. The Balearic Basin, marked by a flat, abyssal plain.

2. The Tyrrhenian and Aegean seas, exhibiting numerous seamounts and active volcanoes.

3. The eastern Mediterranean province, dominated by an arcuate submarine ridge (the Mediterranean ridge).

These basins are similar because all contain a Pliocene-Quaternary clastic sequence, a late Miocene evaporite sequence, and a deep-water pre-late Miocene sequence beneath the evaporites. The basins, however, differ markedly in age and structural

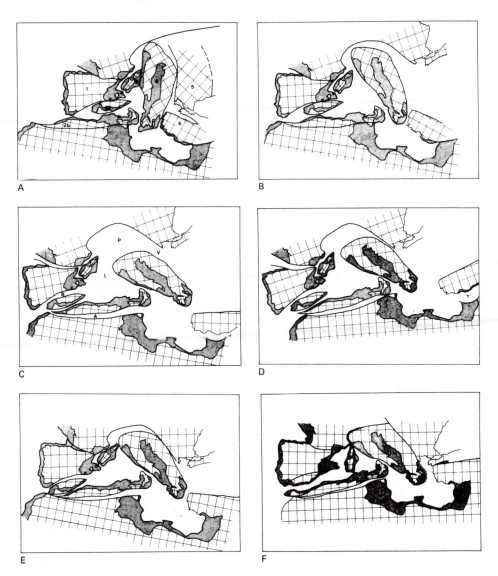

Figure 12-24 One model of the tectonic evolution of the Mediterranean [Hsu, 1977]. (a) Reconstruction of microcontinents in the late Triassic: (1) Iberia, (2a) Alboran (Rif-Betic), (2b) Alboran (Kabylia), (3) Corso-Sardinia, (4) Italo-Dinaridian, (5) Bulgarian, (6) Anatolia. (b) Alpine Tethys in the late Jurassic. (c) Alpine Tethys in the early Cretaceous—(A) Atlas geosyncline, (B) Betic geosyncline, (L) Liguria eugeosyncline, (P) Pennine eugeosyncline, (V) Vardar eugeosyncline. The Alpine geosyncline has reached its widest extent. (d) Alpine Tethys in the late Cretaceous. Note the elimination of the Vadar and the narrowing of the Pennine by plate consumption during the Cretaceous. The Tauride eugeosyncline (T) was fully developed, and sea-floor spreading created the Troodos Massif of Cyprus. (e) Alpine Tethys in the late Eocene. The collison of the Italo-Dinaridian block and Eurasia led to the late Eocene deforma- tion of the western Alps, when the subduction zone flipped from north to southdipping. Movement of the Alboran block led to deformation of the Betic, Rif, and Tellian Atlas. (f) Mediterranean in the middle Miocene after the rotation of Corsica and Sardinia. The gaps be- tween Eurasia and microcontinents were all but eliminated by widespread early and middle Miocene orogenies. The shape of the Mediterranean began to assume a geometry similar to that of the present sea. Only the movement of the Alboran block during the Tortonian was still to take place, to enlarge the Tyrrhenian Basin. (From K. J. Hsu, *The Ocean Basins and Margins* vol. 4A, pp. 54–55, 1977)

development. The Balearic, Tyrrhenian, and Aegean basins are relatively young (Oligocene to Recent) marginal types, while the eastern Mediterranean is a remnant of an early Mesozoic ocean and its southern continental margins.

Argand [1924] was the first to explain the complexity of Mediterranean tectonics by continental drift and rotation of microcontinents. This collision led to the closure of the Tethys Sea. The Mediterranean sea floor and adjacent mountain belts represent the westward extension of the much larger Tethyan tectonic belt that stretches eastward through the Middle East and Asia. Neumayr [1883] first proposed the idea of an ancient Mediterranean basin extending from India to Central America based on Jurassic faunal biogeography—the Tethys of Suess [1885]. Du Toit [1937] proposed that the Tethys separated Laurasia from Gondwanaland from the mid-Paleozoic onward.

The present Mediterranean Sea effectively conceals the structural relationships between the tectonic belts of North Africa and those of southern Europe. Recent geophysical studies, deep-sea drilling, and reinterpretations based on plate tectonics demonstrate a constantly changing mosaic of small and large plates, which produce ridges, trenches, backarc basins, and island arcs. Modern interpretations favor the concept that the present Mediterranean is a composite tectonic unit combining a relict Tethys with later basinal development associated with a **neo-Tethys** [Hsu, 1977]. This interpretation is based on great tectonic contrasts between the western basins, created by extensional rifting, and the eastern Mediterranean basin, which has long been influenced by compressional forces [Hsu and Ryan, 1973; Hsu, 1977]. The sedimentary record in the eastern Mediterranean indicates that this sea was already in existence during the Mesozoic as part of the Tethys and has been steadily restricted ever since. The most recent compressional phases in the eastern Mediterranean have resulted from movement of the African, Arabian, and Eurasian plates, causing motion in the intervening microplates. The remainder of the Mediterranean, the western Mediterranean and the Aegean Sea, are backarc basins that developed during the middle Tertiary behind the subduction zones now represented by the Apennines-Atlas and the Hellenides mountains.

An analysis of the tectonic evolution of the Mediterranean carried out by Hsu [1977] is summarized as follows and is shown in Fig. 12–24. It should be emphasized that this is only one view, although recent, of this extremely structurally complicated region.

During the early Jurassic, Pangaea was split into a northern and a southern continent. At that time three microcontinents developed between Europe and Africa: (1) the Alboroin; (2) Italo-Dinaridian; and (3) the Bulgarian (Fig. 12–24 (A), (B)). The creation of Jurassic and Cretaceous basins in the west was accompanied by the destruction of the Triassic Tethys in the east along a subduction plate margin. As Africa moved eastward during the Jurassic and early Cretaceous, microcontinents became detached (Fig. 12–24(C)) and moved toward Eurasia ahead of the main mass of Africa, which followed later (Fig. 12–24(E), (F)). This trapped the ancestral eastern Mediterranean (Fig. 12–24(D), (E)). Continued northward movement of Africa has continued to compress the eastern Mediterranean Sea, which ultimately faces destruction.

Oceanic Sediments
and Microfossils

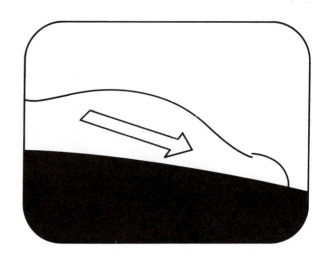

thirteen
Terrigenous
Deep-Sea Sediments

There is no sound, no echo of sound, in the deserts of the deep,
Or the great grey level plains of ooze where the shell-burred cables creep.
Here in the womb of the world—here on the tie-ribs of earth
Words, and the words of men, flicker and flutter and beat.

Rudyard Kipling

CLASSIFICATION OF DEEP-SEA SEDIMENTS

Introduction

Deep-sea sediments are those deposited at depths greater than about 500 m. These sediments are dominated by biogenic (fossil) components and pelagic clays, although terrigenous sediments are widespread in some deep-sea basins. Emiliani and Milliman [1966], Lisitzin [1972], and Berger [1976] present useful summaries of deep-sea sediments.

When deep-sea sediments were first comprehensively explored during the *Challenger* expedition, they were classified in a combined descriptive-genetic fashion as follows:

1. *Pelagic deposits.* Red clay, radiolarian ooze, diatom ooze, globigerina ooze.

2. *Terrigenous deposits.* Blue mud, red mud, green mud, volcanic mud, coral mud.

Red clay, siliceous ooze, and carbonate ooze were the most widespread deposits mapped during the *Challenger* expedition. The red clay is dark brown to red due to oxidation of iron minerals and consists of a variety of clay minerals. Minor components include volcanic ash, cosmic spherules, fish teeth and bones, and sometimes traces of carbonate and silica.

397

Siliceous ooze is radiolarian ooze (greater than or equal to 30 percent radiolarian tests) or diatom ooze (greater than or equal to 30 percent diatom frustules). Carbonate ooze is nannofossil ooze (greater than or equal to 30 percent nannofossils), foraminiferal ooze (greater than or equal to 30 percent foraminiferal tests), or pteropod ooze (greater than or equal 30 percent pteropod shells).

Research has shown that calcareous sediments are draped over oceanic rises and platforms, while red clays are distributed throughout the deep basins. Siliceous deposits are characteristic of areas of high biological fertility, especially the ocean margins, the equatorial divergence, and south of the Antarctic convergence. Terrigenous deposits are transported by several processes to the ocean floor, some at great distances from the source regions.

The distribution patterns of marine sediments are shown in Fig. 13-1. This shows the dominance of topographically controlled calcareous ooze and clays throughout the basins. Siliceous oozes are concentrated at high latitudes, in the equatorial Pacific and Indian oceans, and beneath some coastal upwellings, such as off western South America. Glacial-marine sediments are dominant at high latitudes.

There are four primary mechanisms for deep-sea sedimentation [Emiliani and Milliman, 1966]: settling from the water column; bottom transportation by gravity flows, including turbidity currents, debris flows, grain flows, and slumping; transportation by geostrophic bottom currents, including contour currents; or chemical and biochemical precipitation on the ocean floor.

A classification scheme needs to be flexible enough to incorporate the conspicuous gradations that occur between marine sediments. Terrigenous sediments are those derived from land sources, including nearshore deposits of great variety,

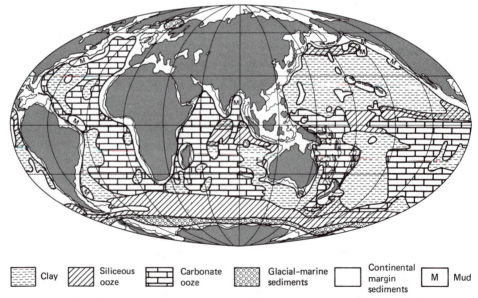

Figure 13-1 General patterns of Recent sediments on the deep-ocean floor. (After W. H. Berger, 1974)

turbidites, channelized deposits, such as deep-sea fans, and eolian and glacial marine deposits. Biogenic sediments are of biological origin. Nearshore types include the calcareous sands and limestones of corals, bryozoans, or mollusks, while deep-sea types include carbonate and siliceous sediments and organic-rich sediments. Pelagic sediments settle through the water column and include biogenic material, terrigenous clays and silts, pyroclastic materials blown through the air to the oceans, ice-rafted debris, and extraterrestrial material. Hemipelagic sediments are combinations of terrigenous and biogenic components. Volcanogenic deposits include wind-blown volcanic ash, submarine pyroclastic flows, hyaloclastites formed by the fracturing of volcanic rocks erupted beneath the sea, and reworked volcanic debris. A variety of materials are reworked and redeposited on the ocean floor, including terrigenous materials (mostly from rivers) transported downslope by gravity flows, volcanic materials produced by submarine volcanism and reworked, and deep-sea deposits reworked by bottom currents.

Oceanic sediments have been classified in various ways. Most sediment is classified either by the size of particles, the description of components, the process of formation, or some combination of these three. But the most successful classifications combine description and process of formation. The most widely used classifications are based on the more objective descriptive characteristics of the components [Lisitzin, 1972]. These schemes must be useful in discriminating mixed sediment types. Although descriptive classifications are not established to identify processes of sedimentation, the descriptive and the genetic classifications are not mutually exclusive, as much overlap naturally exists between sediment character and mode of formation.

The ultimate aim of classification is to understand processes and relationships. For this reason, one goal in sedimentology is to develop a system of classification reflecting both the origin and the history of the sediments. Unfortunately, classifications based on the genesis of sediments tend to be controversial because many sedimentary processes are still poorly understood. It may be difficult to select one process of formation from several alternatives. For this reason, the descriptive classifications are still the most useful and thus the most widely used.

Methods of Classification

Size classification

Classifications based on grain size have been widely used in marine geological research. They represent the systematic division of continuous ranges of sizes into classes or grades and may be arithmetic or geometric. Although an arithmetic progression is simpler, it does not reflect relative differences. There are great differences between fragments less than 1 mm in diameter, whereas the difference between fragments 11.5 mm and 11.6 mm is minimal. Therefore geometric grade scales are preferred. The most commonly used scale is that of Krumbein [1936], who based his classification on the systems of Wentworth [1922] and Udden [1914]. Krumbein's scale commences at 1 mm and increases or decreases by powers of 2; each grade limit is twice as large as the next smaller grade limit. Particle sizes are normally expressed as *phi* (φ) *units*. The value of φ is equal to the negative logarithm to the base 2 of the particle diameter in millimeters and is a more convenient expression of size. The subdivisions are as follows.

	φ units
Granule	-1
Very coarse sand	0
Coarse sand	+1
Medium sand	+2
Fine sand	+3
Very fine sand	+4
Silt	+8
Clay or mud	

The broad subdivisions commonly used for the finer sediments are: sand, 2000 to 62 μm; silt, 62 to 4 μm; and mud, less than 4 μm.

Compositional classification

Various classifications are based primarily on sediment composition, although these classifications also often reflect the origin of sediments. Lisitzin [1972] has grouped deep-sea sediments into terrigenous (less than 30 percent calcium carbonate and silica), biogenic (greater than 30 percent biogenic calcium carbonate and silica), chemogenic (chemical precipitates from seawater), volcanogenic (composed mainly of pyroclastic material, and polygenic (red clay)). Berger [1976] presented a classification (Table 13-1) of pelagic and hemipelagic origins, which is one of the most useful available for those sediments that dominate ocean basins; however, it does not include a variety of other kinds of sediments.

Important classification systems have been used extensively by scientists participating in the DSDP. The bulk of deep-sea sediments have now been described using these classification schemes and published in the initial reports of the DSDP. During the first three phases of the DSDP, from 1968 to 1976, the description and classification of sediment cores obtained at 392 sites evolved from a poorly defined, qualitative, and generalized nomenclature to a precise system of sediment names and classes based on standardized descriptive parameters [van Andel, 1981].

Three main classifications developed successively, of which the last one—summarized by van Andel [1981]—was officially adopted by the DSDP and used beginning with Leg 38 in 1974. Procedures used in the early classifications scheme are summarized in Fig. 13-2 [Weser, 1973]. This classification system is largely descriptive, although genetic implications are included when they do not conflict with the descriptive aspects or when they are not controversial, such as with the recognition of particular products of the various current transport mechanisms. No attempt is made to distinguish between pelagic and terrigenous sediments, allowing the relative importance of all major and minor constituents to be recognized. For this purpose smear slide analyses of sediments have been routinely performed, using the

TABLE 13-1 CLASSIFICATION OF DEEP-SEA SEDIMENTS[a]

I. (Eu-) pelagic deposits (oozes and clays).
 < 25% of fraction > 5 μm is of terrigenic, volcanogenic, or neritic origin.
 Median grain size < 5 μm (except authigenic minerals and pelagic fossils):
 A. Pelagic clays. $CaCO_3$ and siliceous fossils < 30%.
   1. $CaCO_3$: 1–10%; (slightly) calcareous clay.
   2. $CaCO_3$: 10–30%; very calcareous (or marl) clay.
   3. Siliceous fossils: 1–10%; (slightly) siliceous clay.
   4. Siliceous fossils: 10–30%; very siliceous clay.
 B. Oozes. $CaCO_3$ or siliceous fossils > 30%.
   1. $CaCO_3$: > 30%; < ⅔ $CaCO_3$: marl ooze; > ⅔ $CaCO_3$: chalk ooze.
   2. $CaCO_3$: < 30%; > 30% siliceous fossils: diatom or radiolarian ooze.

II. Hemipelagic deposits (muds).
 > 25% of fraction > 5 μm is of terrigenic, volcanogenic, or neritic origin.
 Median grain size > 5 μm (except authigenic minerals and pelagic fossils):
 A. Calcareous muds. $CaCO_3$ > 30%.
   1. < ⅔ $CaCO_3$: marl mud; > ⅔ $CaCO_3$: chalk mud.
   2. Skeletal $CaCO_3$: > 30%: foram; nanno; coquina.
 B. Terrigenous mud. $CaCO_3$ < 30%. Quartz, feldspar, mica dominant.
   Prefixes: quartzose, arkosic, micaceous.
 C. Volcanogenic muds. $CaCO_3$ < 30%. Ash, palagonite, and so on dominant.

III. Pelagic or hemipelagic deposits.
   1. Dolomite-sapropelite cycles.
   2. Black (carbonaceous) clay and mud: sapropelites.
   3. Silicified claystones and mudstones: chert.
   4. Limestone.

[a]From Berger [1974], based on the scheme of Olausson, modified by Berger and von Rad [1972].

petrographic microscope. A thin smear of sediment on a glass slide enables all components other than coarse sands to be quantitatively analyzed, using normal and polarized light.

A summary of the rules used in this classification are as follows [Weser, 1973; Hayes, Frakes and others, 1975]:

I. *Ranking of Components.*

A. Major Constituents.

  1. Sediment assumes the name of those constituents present in major amounts (major defined as greater than 25 percent).

  2. Where more than one major constituent is present, the one in greatest abundance is listed farthest to the right, with the remaining major constituents listed progressively farther to the left.

  3. Class limits, when two or more major constituents are present in a sediment, are based on 25 percent intervals. For example:

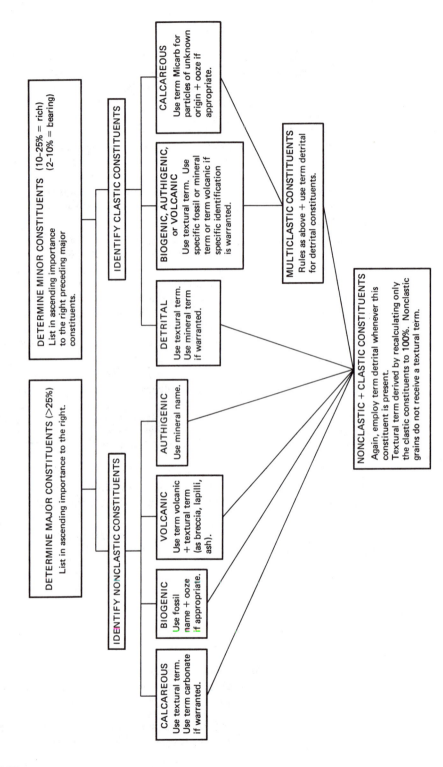

Figure 13-2 Approach to the classification of deep-sea sediments used by shipboard sedimentologists of the DSDP through Leg 37 and published in *Initial Reports*. Steps in the process begin at top.

% Zeolites	% Nannos	
0–25	75–100	= Nanno ooze
25–50	50–75	= Zeolitic nanno ooze
50–75	25–50	= Nanno zeolitite
75–100	0–25	= Zeolitite

B. Minor Constituents.

1. Constituents present in amounts 10 to 25 percent prefixed to the sediment name by using the term *rich*.
Example: 50 percent nannofossils, 30 percent radiolarians, 20 percent zeolites is called a zeolite-rich rad nanno ooze.

2. Constituents present in amounts 2 to 10 percent prefixed to the sediment name by using the term *bearing*.
Example: 50 percent nannofossils, 40 percent radiolarians, 10 percent zeolites is called a zeolite-bearing rad nanno ooze.

II. *Rules for Biogenic Constituents.*

A. Nannofossil is applied only to the calcareous tests of coccolithophorids and discoasters.

B. Abbreviations and contractions like *nanno* for nannofosssil, *foram* for foraminifera, *rad* for radiolarian, and *spicule* for sponge spicule may be used in the sediment name.

C. The term *ooze* follows a microfossil taxonomic group whenever one of these groups is the dominant sediment constituent.

D. Chalk is used to describe a semilithified carbonate ooze and limestone carbonate ooze.

E. Semilithified diatom and rad oozes are called *diatomite* or *radiolarite,* respectively. Lithified siliceous oozes are called *chert.*

III. *Rules for Clastic Sediments.*

A. Clastic components, whether detrital, volcanic, biogenic, or authigenic, are given a textural designation. When detrital grains (those derived from erosion of preexisting rocks) are the sole clastic constituent of a sediment, a simple textural term suffices for its name. The appropriate term is derived from Shepard's [1961] triangle diagram (Fig. 13–3(A)).

B. When biogenic or authigenic components occur with detrital grains, the biogenic or authigenic material is not given a textural designation. The detrital material is classified texturally by recalculating its size components to 100 percent. Since nondetrital fractions are present, the detrital fraction now requires a compositional term, *detrital*. For example:

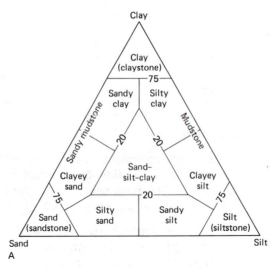

Pelagic clay <30% siliceous fossils	Auth. comp common	Uncommon sediment types				Auth. comp rare
>30% siliceous fossils Pelagic siliceous sediments < 30% CaCO₃	<30% Silt and clay >	>30% Silt and clay <	Transitional siliceous sediments < 30% CaCO₃	>10% Diatoms <	<10% Diatoms >	Terrigenous and volcanic detrital sediments
>30% CaCO₃ Pelagic calcareous sediments	<30% Silt and clay >	>30% Silt and clay <	>30% CaCO₃ Transitional calcareous sediments	<30% CaCO₃ >	>30% CaCO₃ <	

B

Figure 13–3   (A) Categories of clastic sediments according to relative proportions of sand, silt, and clay. (After Shepard, 1961) (B) Summary diagram of JOIDES sediment classification employed after Leg 38 of the DSDP. (After T. H. van Andel, 1981)

Foraminifera	= 40 percent	
Detrital fraction-clay size	= 40 percent	(recalculate to 67 percent)
silt size	= 20 percent	(recalculate to 33 percent)

The proper sediment name is *foram detrital silty clay*, which identifies all the important constituents and ranks their importance.

The official JOIDES classification adopted in 1974 was designed to simplify sediment names, to encourage more uniform usage, and to reduce the number of quantitative boundaries, therefore decreasing errors of estimates of the various sediment components. This is summarized in Fig. 13–3(B).

TERRIGENOUS DEEP-SEA SEDIMENTS

Terrigenous sediments are derived from the land. The material is transported throughout the oceans by a variety of processes, including various gravity flows (such as turbidity currents), the wind, and an important ice-rafted component from certain high-latitude regions. The sediment **provenance** is its source region. At all stages of the sediment's origin, transportation, deposition, or redeposition, changes may occur in the physical, chemical, and mineralogical character of terrigenous sediments. These characteristics provide important information about past environmental conditions.

Nearly all particles are introduced into the oceans at the boundaries. The largest input is from rivers, which not only supply terrigenous material but add nutrients necessary for biological productivity. Along the east coast of North America at the present, most of this material is trapped in estuaries, and the amount of sediment moving across the continental shelf to the deep sea and the process by which it is moved are uncertain. During glacial episodes, when sea level was lower, rivers crossed the continental shelf and disgorged their loads directly into the deep sea, mostly through submarine canyons. Terrigenous material is otherwise carried to the open ocean in icebergs in the polar regions and as windblown dust.

Once in the ocean, terrigenous sediments are transported to deeper waters, after initial storage in many areas on the continental shelves and in lagoons and estuaries. Mechanisms capable of transporting coarse sediments into deeper waters include slumps, slides, and mass flows. Sediment flows pulled downslope by gravity have been called **sediment gravity flows** by Middleton and Hampton [1976]. They distinguish four main types, of which the most widely known is *turbidity current flow*. Sediment gravity flows are distinguished from gravity sliding or slumping largely by the amount of internal deformation. Deformation is well developed in flows, intermediate in slumps, and minor in slides.

Gravity Transport to the Deep Sea

Slides and slumps

There is evidence that submarine slumping, sliding, and related forms of gravity-induced slope instability are major methods of sediment transport down continental slopes, and so are important in the shaping of continental margins. Emery and others [1970] estimated that such deformations have affected at least 50 percent of the sediment underlying the continental rise off eastern North America. Seismic profiles of the continental slope show areas of hummocky to chaotic surface geometry, abrupt slopes, deformed and discontinuous subbottom reflectors, and rotated sedimentary units (Fig. 13–4). The downslope movement of sediment

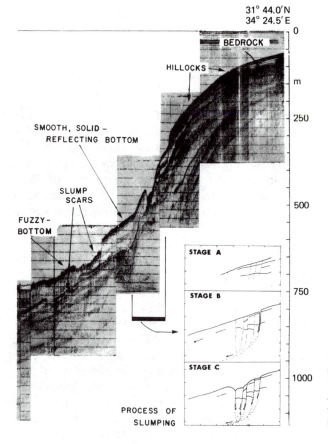

31° 44.0'N
34° 24.5'E

BEDROCK

HILLOCKS

SMOOTH, SOLID –
REFLECTING BOTTOM

SLUMP
SCARS

FUZZY-
BOTTOM

STAGE A

STAGE B

STAGE C

PROCESS OF
SLUMPING

m
250
500
750
1000

Figure 13-4 Seismic profile off southern Israel showing slump features. (From G. Almagor and Z. Garfunkel, *AAPG Bull.* vol. 63, p. 329. reproduced with permission of American Association of Petroleum Geologists)

by gravitational processes is enhanced by increased sediment supply and rapid deposition during glacial periods. The steepness of the bottom and the rate of sedimentation control the amount of material which can be deposited. Only a few meters of rapid sediment deposition are stable on the steep, lower continental slope. Especially large slumps occur at the increase in gradient from the upper continental slope to lower continental slope.

A slump is a downslope translocation of a sediment parcel along discrete shear planes, which form a curved surface of displacement. Strata in a slump will normally dip back toward the slope as a result of the rotation of the unit. Typically the lower edge of the slump becomes fluidized, destroying stratification and becoming potential debris flows. Large-scale slumping (involving blocks hundreds of meters thick) is most apparent at the transition between the gentler, upper continental slope and the steeper, lower continental slope [MacIlvaine and Ross, 1979].

In **slides,** large blocks of material move on only a few, well-defined slippage planes. This process is less exaggerated than in slumps, leading to less breakage and internal deformation. Slides and slumps are particularly important because they may be the initial stages of larger scale downslope sediment movement, or sediment gravity slides. Translations almost certainly occur between different kinds of gravity flows.

One of the largest known slump deposits is off the Grand Banks, in the Northwest Atlantic; it is almost 400 m thick and 50 km long. The lateral extent of large submarine slumps is generally not determined, but the few surveys made in areas marked by slumps indicate that they range from 20 to 170 km in length, thus similar to large, gravity-slide deposits on land. Off Israel, slump structures extend from the shelf break to the base of the continental slope (Fig. 13-4). In this region, movement was contemporaneous with sedimentation, so the surface failures appear as normal growth faults in the upper parts of the sediment sequence. Many of these breaks are still active, producing a terraced topography on the outer continental shelf and slope. Although little downslope sediment translation is occurring, the long-term effect could be significant.

The causes of slope instability are seldom evident on seismic-reflection records. The greater steepness (average 4°) of the continental slope is important. Slumping is most common on slopes exhibiting gradients from 3 to 9° (average 5.5°). Slides may be more important on less steep profiles. High rates of sedimentation are also necessary to induce slumping. Sedimentation creates oversteep slopes and differential compaction. Where sedimentation rates are particularly high, such as in the Mississippi River delta region, slumping can be widespread even on slopes of low inclinations (0.2 to 1.5°). Widespread slumping in the Mississippi River delta and its continental shelf may be caused by high sedimentation rates, differential loading as coarser-grained sediments are deposited near the mouth and finer materials in offshore areas, high water content and underconsolidation with the formation of large excess pore fluid pressures, and rapid biochemical degradation of organic material leading to the formation of gas [Emery and others, 1970]. All these factors combine to create instability. In many cases slumping and sliding is triggered by earthquakes.

Instability on the slope can also be created by structural changes in deeper sediments on the continental slope. Off Israel, flowage of late Miocene evaporitic rocks at some depth has been important in the formation of slump structures [Almagor and Garfunkel, 1979]. The evaporite undulations flex and break overlying sediments, causing slumps. The evaporites flow due to the mass of overlying sediment (greater than 6–8 km). In areas of thinner overburden, mobility is caused by the buildup of pore pressure in porous sediments interbedded with impervious evaporites. Instability results because pore pressure cannot be evenly distributed.

Other gravity-induced processes, which may be important but about which little are known, are **subaqueous rock falls, slides,** and **avalanches.** Recent observations from submersibles have shown that rock falls may be important in areas of very steep topography, like the walls of submarine canyons. **Talus** or **scree deposits** at the bases of these slopes are marked by a wide range of lithology and age.

Sediment gravity flows

In sediment gravity flows, the sediment is transported under the influence of gravity and the sediment motion moves the interstitial fluid [Middleton and Hampton, 1976]. This differentiates them from **fluid gravity flows,** such as rivers, where the fluid is moved downstream under the influence of gravity and transports the sediment, or from wind, which blows particles along. Mechanisms, such as **suspension** (by turbulence), **saltation** (by hydraulic lift forces and drag), and **traction** (by dragging or rolling particles on the bottom), are characteristic of sediment gravity flows and of some types of fluid gravity flows. However, in sediment gravity flows additional processes, which appear to be minor in fluid flows, are important. These include upward intergranular flow, direct interaction between grains, and the support of grains by a cohesive fluid.

Sediment gravity flows are of two main types: highly concentrated flows of sediment supported by a range of different mechanisms, including turbulence; or turbidity currents, which are flows of sediments of relatively low concentration, supported by turbulence. Densities of the highly concentrated flows are only slightly less than those of unconsolidated sediment (perhaps 1.5 to 2.4 g/cm³), whereas the densities of the currents with low concentration (turbidity currents) often range from 1.03 to 1.3 g/cm³. The classification of sediment gravity flows by Middleton and Hampton [1976] is based on the character of the dominant sediment support mechanism. They recognized four main categories (Fig. 13–5):

1. *Fluidized sediment flows,* in which the sediment is supported by the upward flow of fluid escaping from between the grains as the grains settle by gravity.

2. *Grain flows,* in which the sediment is supported by direct grain to grain interactions.

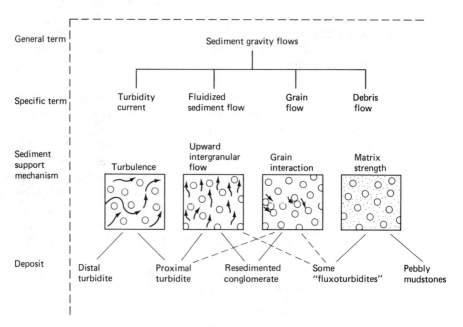

Figure 13–5 Classification of subaqueous sediment gravity flows. (From G. V. Middleton and M. A. Hampton, *Marine Sediment Transport and Environmental Management,* eds. D. J. Stanley and D. J. P. Swift, John Wiley & Sons, Inc., 1976)

3. *Debris flows,* in which the larger grains are supported by a matrix consisting of a mixture of interstitial fluid and fine sediment possessing a finite yield strength.

4. *Turbidity currents,* in which the sediment is supported mainly by the upward component of fluid turbulence.

Transitions are likely among the first three transportation modes, at different stages of velocities and turbulences of the flow as they mark the evolution of a flow from inception to final sediment deposition. Though the characteristics of the different types of flows are speculative, the full spectrum of transport mechanisms has probably now been recognized. What remains is to define clearly and recognize the various deposits that result from the different types of flow. As only small amounts of water need to be added to reduce the strength of unconsolidated sediment to produce fluid flow, slumping and sliding may commonly become flows of various kinds.

Fluidized sediment flows Fluidized sediment flow is liquefied, cohesionless particle movement. Upward flow of fluid through soft sediment expands the material enough to cause it to behave as a viscous fluid (Fig. 13-5). This may occur in loosely packed sand as pore pressures rise above normal hydrostatic pressure. Rather than remaining supported by grain-to-grain contact, the grains become supported by the pore fluid, causing the sand to move as a traction carpet downslope, even on gentle slopes. Deposition occurs as pore pressures drop.

Grain flows Grain flows result from grain-to-grain contact rather than from turbulence within the fluid (Fig. 13-5). The dispersive pressure developed is proportional to the shear stress transmitted between grains, and it must override the tendency for the grains to settle [Middleton and Hampton, 1976]. Grains remain in suspension by bouncing upward off one another, and dispersive pressures are generated by gravity. Deposition from grain flows is by mass emplacement, which is the sudden cessation of the flow due to the simultaneous deposition of a layer several grains thick. Traction transportation involves individual grain settling. Sands flowing down the upper reaches of submarine canyons may offer examples of fluidized sediment flow, or grain flow. Coarse, well-sorted gravel in the channels of such canyons may have been transported by this mechanism [Shepard and Dill, 1966].

Debris flows Debris flows are gravity-induced, downslope movements of mixtures of coarse and fine debris and water, resembling the flow of wet concrete [Hampton, 1970; 1972A, B]. Grains are supported by strength and buoyancy (Fig. 13-5). The clay minerals and the water combine as a muddy fluid, possessing a finite cohesion (strength) that supports the flow. Support of grains by the strength of the surrounding fluid distinguishes true debris flows from grain flow and turbidity currents.

Deposition from debris flows occurs by rapid mass emplacement, as the driving force of gravity decreases below the strength of the debris. Textures of bouldery debris-flow deposits may resemble glacial tillite deposits, causing misinterpretation of some ancient deposits. Bouldery debris-flow deposits are typically massive with random, large boulders in a fine-grained matrix. Sharp angular contacts and muddy clasts of various colors may be important. Debris flows can be quite mobile, even over slopes as low as 0.1°. They may be a common phenomenon in the ocean and are certainly widespread where they occur. Sediment is known to have been transported for distances as great as 700 km on the continental rise off the Spanish Sahara, forming deposits covering an area of up to 30,000 km². Bottom photographs on the Gillis Seamount in the North Pacific reveal downslope-trending, anastomosing streams of sand and gravel of a meter or more in width. It is not clear if these represent debris flows or grain flow. Inferred steady downslope sand transport in such flows may explain the small amount of sediment accumulation on the upper flanks of seamounts revealed by seismic profiling.

Turbidity Currents   Turbidity currents are short-lived, powerful, gravity-driven currents consisting of dilute mixtures of sediment and water of density greater than the surrounding water, the motion of which is maintained by internal turbulence (Fig. 13–5) [Kuenen and Migliorini, 1950]. They can contain vast amounts of material and are of major importance in the transportation of terrigenous sediments from shallow water to the deep-ocean basins, where in many places they have built up the relatively flat abyssal plains. The deposition of sediment from turbidity currents forms **turbidites,** which are marked by graded bedding, moderate sorting, and well-developed primary structures. The mechanics of the flow of turbidity currents and the characteristics of turbidites have been well studied, but the origin of these currents is still not clearly understood. Turbidites and submarine slides and slumps are associated, and it is possible that turbidity currents are generated by slides or slumps. The bulk of sediment of submarine slides ranges from about 1.5 to 2.4 g/cm³, which is much higher than inferred for turbidity currents (1.03 to 1.3 g/cm³). A major problem concerning the generation of turbidity currents is how marine sediments become diluted to the low densities necessary to generate a turbidity current. An intermediate step is called for in the transport of sediment between the slumping or sliding and the turbidity current. This step is probably debris flow.

In a turbidity current, the head pressure is greater than that of the surrounding seawater, and it seems that this excess pressure drives the current (Fig. 13–6). Head pressure is maintained by the tailing body, which—because it is in effect moving faster than the head—forces fluid into it [Middleton and Hampton, 1976]. The additional fluid that is transferred to the head from the body is apparently raised within the head and then returned to the body by means of waves breaking along the interface directly behind it (Fig. 13–6). The current must maintain a certain velocity to maintain momentum, and thus continually compensates for the drag that is generated within the flow. It seems that minimum speeds of turbidity currents are about 13 cm/sec, while maximum speeds can attain 870 cm/sec. In general, turbidity currents can travel over 90 km/hr, carrying up to 3 kg/m³ of material and spreading the material over a distance of 1000 km from the source. At

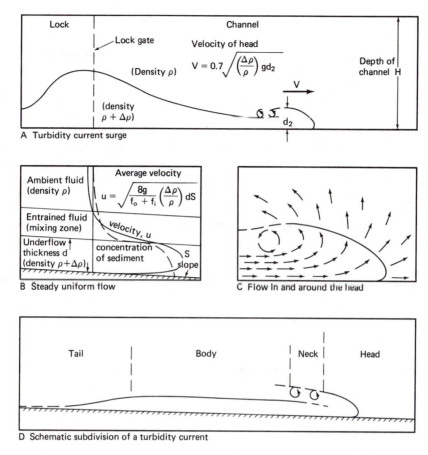

Figure 13–6 Hydraulics of turbidity currents based on laboratory experiments in sediment flumes. (A) Turbidity current surge, as observed in a horizontal channel after releasing suspension from a lock at one end. The velocity of the head, V, is related to the thickness of the head, d_2, the density difference between the turbidity current and the water above, $\triangle\rho$, the density of the water, ρ, and the acceleration due to gravity, g. (B) Steady, uniform flow of a turbidity current down a slope, S. The average velocity of flow, u; is related to the thickness of the flow, d, the density difference, and the frictional resistance at the bottom (f_o) and upper interface (f_i). (C) Flow pattern within and around the head of a turbidity current. (D) Schematic division of a turbidity current into head, neck, body, and tail. (From G. V. Middleton and M. A. Hampton, *Marine Sediment Transport and Environmental Management*, ed. D. J. Stanley and D. J. P. Swift, John Wiley & Sons, Inc., 1976)

various stages in its evolution, a turbidity current can be neutral, eroding, depositing-reworking, or depositing-deforming in character.

Important evidence on the force and speed of turbidity currents comes from the study of submarine cable breaks caused by turbidity current flows. The most famous is that associated with the earthquake at Grand Banks on November 19, 1929, in which several submarine cables were broken. The development of this flow was studied by B. Heezen and M. Ewing [1952]. During the earthquake, several telegraph cables were broken in an apparently systematic, progressive manner over a 13-hour period. The current traveled down and across the continental slope, continental rise, and ocean-basin floor, continuing far out into the abyssal plain and reaching a limit of well over 720 km from its source area on the continental slope. The sequence is believed to be: earthquake; slumping and sliding, causing the instantaneous breakage of about a dozen cables; and the generation of a turbidity current, which flowed down the slope along several channels and later converged into a broad front that progressively cut the remaining cables. Maximum speeds reached during this flow have been debated, but on the continental slope, where the bottom gradient is 1:10 to 1:30, the velocity of the current reached approximately 40 to 55 km/hr. After the cable break, B. Heezen discovered a turbidite layer of about 1 m overlying the normal pelagic sequence over an area of at least 100,000 km^2.

Turbidity currents may originate in several ways. They can be triggered by heavily charged rivers, by an earthquake setting off a slide, or by oversteepening of a depositional slope through sediment buildup. The multiple lowerings of sea level during the Quaternary greatly stimulated the action of turbidity currents as large amounts of materials stored on the continental shelf and in nearshore areas were transferred down the continental slope. Furthermore, rivers discharged their sediment loads at the shelf edge rather than over the broader continental shelves and large erosion valleys formed.

Turbidity currents are commonly associated with systems of submarine canyons along the continental margins. Most canyon systems lead from river systems and carry predominantly river-borne sediments.

Sediment deposits associated with turbidity currents are of three types: (1) deep-sea channels; (2) deep-sea fans; and (3) abyssal-plain deposits. Complete gradation occurs between these deposits; indeed some of the strongest evidence for turbidity currents comes from the topography of the ocean floor. At the mouths of submarine canyons, extensive overlapping fans, exhibiting gentle slopes, have been built. These are incised radially by one or more **leveed channels.** The channels contain sediments of shallow-water origin transported by various gravity flows, but inferred to be largely of turbidity current origin. At their outer limits the submarine fans gradually merge and coalesce in adjacent abyssal plains.

Turbidites   Almost all sediment that has been supplied toward the building of the abyssal plains has been derived from turbidity currents. Because the currents are episodic, the resulting **turbidites** normally consist of sands, interbedded with finer pelagic sediments. Reflection profiles show that single coarse beds in abyssal plains extend over several hundred kilometers, reflecting one depositional event. The coarse layers are not encountered on isolated hills or ridges, or even on those ridges rising no more than a hundred meters in the center of abyssal plains.

Turbidites have been differentiated into four main facies, each representing deposition in a different environment [Blatt and others, 1972]:

1. *Channel deposits* of sands and pebbles, which may be grain flow deposits rather than turbidites.

2. *Proximal turbidites,* which are relatively close to the sediment source. They are marked by massiveness, poorly developed tractional structures, relatively weak grading, and little interbedded pelagic clay and terrigenous mud.

3. *Turbidites* of the classic type, which have distinct graded bedding, oriented erosion and fill markings at the base of the sand layer known as *sole marks,* interbedded pelagic clays, and a characteristic succession of sedimentary structures now known as the *Bouma sequence* (Fig. 13-7).

4. *Distal turbidites,* which are most distant from the source region and consist of thin, fine-grained layers, lacking massive intervals and laminations but with well-developed cross lamination.

In general, criteria for the recognition of turbidites remain controversial but include the presence of the Bouma sequence, evidence of rapid deposition, and displaced faunas. The Bouma sequence is a fixed characteristic succession of five layers, making up a complete sequence of a turbidite (Figs. 13-7 and 13-8) [Bouma, 1962]. One or more layers may be missing for various reasons. The first four (A to D) are deposited from turbidity currents and decrease in grain size upward through a sequence interpreted as a result of deposition from waning current velocity. The uppermost division (E) consists of very fine sediments, resulting from either pelagic sedimentation or deposition from lower density suspension, such as deep-water muddy flows. This part of the sequence is believed to accumulate between phases of turbidity current action. A complete Bouma sequence is not often found. In the Gulf of Mexico, for instance, Divisions A and B are most often missing, reflecting the finer sediment deposition in this area. A lack of coarse-grained detritus means that normal gradations between proximal and distal turbidites with increasing distance from the source are not preserved in the record in the Gulf of Mexico. Differences between proximal and distal turbidites are due to differences in grain segregation within the turbidity currents. Lateral grading does not develop during the early stages of flow; hence proximal turbidites tend to be poorly graded and of

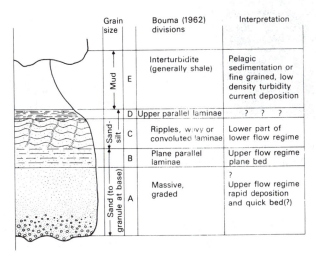

Figure 13-7 An idealized stratigraphic sequence of a turbidite bed, often called a Bouma sequence after Arnold Bouma who first associated it with deposition from turbidity currents. Interpretation of flow regime is given at right. (From G. V. Middleton and M. A. Hampton, *Marine Sediment Transport and Environmental Management,* eds. D. J. Stanley and D. J. P. Swift, 1976, John Wiley & Sons, Inc.)

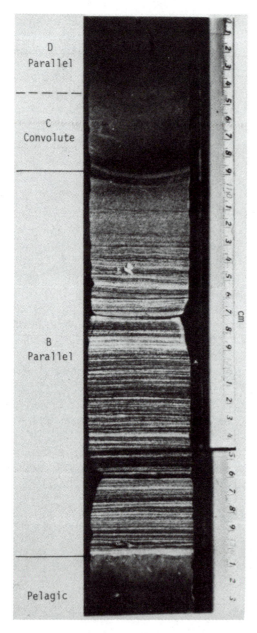

Figure 13–8 Photograph of turbidite in DSDP cores from the Southeast Pacific showing intervals of the Bouma sequence. (From B. Tucholke and others, 1976)

variable thickness and are frequently formed in currents that have eroded the underlying sediments. Distal turbidites are fine grained, containing little or no sand, and are more consistently graded; the currents did not erode the underlying sediments.

Shallow-water benthonic microfossils are transported to the deep ocean in turbidity currents; indeed, the presence of displaced species has always been strong evidence in distinguishing turbidity currents and tracing displaced marine

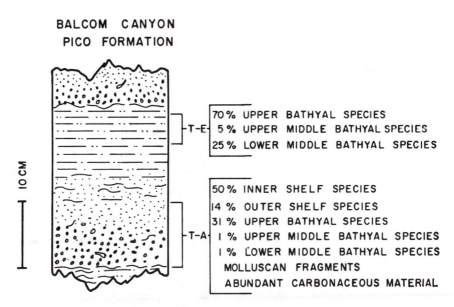

BALCOM CANYON
PICO FORMATION

T-E
- 70% UPPER BATHYAL SPECIES
- 5% UPPER MIDDLE BATHYAL SPECIES
- 25% LOWER MIDDLE BATHYAL SPECIES

T-A
- 50% INNER SHELF SPECIES
- 14% OUTER SHELF SPECIES
- 31% UPPER BATHYAL SPECIES
- 1% UPPER MIDDLE BATHYAL SPECIES
- 1% LOWER MIDDLE BATHYAL SPECIES
- MOLLUSCAN FRAGMENTS
- ABUNDANT CARBONACEOUS MATERIAL

10 CM

Figure 13-9 Sources of benthonic foraminifera within a turbidite bed in Pliocene rocks of southern California. Middle bathyal species are inferred to represent those that were living in place; others were transported in by the turbidity currents. (After J. C. Ingle, Jr., *Bulletins of Amer. Paleontology,* vol. 52, 1977)

sediments. Turbidites characteristically contain a high percentage of species displaced from shallow water. The percentage of displaced species has been found to increase progressively downslope as faunal elements from successively deeper environments are added to the turbidity current [Ingle, 1967]. For instance, in a Pliocene turbidite (Fig. 13-9), the coarser, lowermost part is dominated by species from the continental shelf, while upper units contain mainly deeper water species. The difference in species assemblages through the turbidite suggests an inverse time of arrival at the site of deposition; the shallowest elements arrive first with the coarsest sediment, while upper bathyal assemblages arrive last with sediments of lower settling velocity.

Submarine Canyons

Submarine canyons serve as major conduits of terrigenous sediment from the continents to the deep basins. These are among the most easily recognizable and studied features of the continental margins (Fig. 13-10) with steep walls, sinuous valleys with V-shaped cross sections, steps, and considerable irregularity of the floor. Most canyons commence on the continental shelf, commonly at the mouths of large rivers, and branch out normal to the coastline (Fig. 13-10). Others are deflected along structural boundaries. Most canyons are cut to the base of the continental slope and extend further as channels into fan systems. Some submarine canyons are dynamically active features eroding at the present time. These are generally associated with active margins. Most canyons found on passive continen-

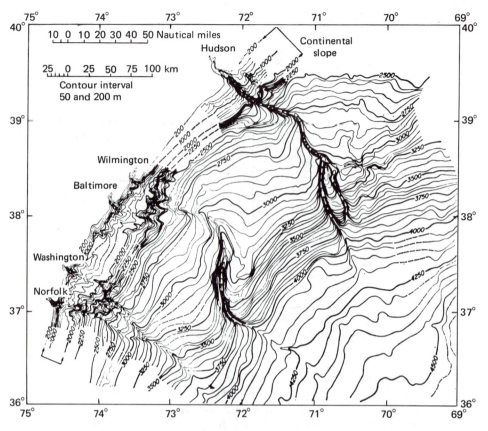

Figure 13-10 Bathymetry of shelf and slope off the eastern United States showing details of submarine canyons and their extensions across part of the continental rise. (Reprinted with permission from *Deep Sea Research*, vol. 14, R. M. Pratt, "The Seaward Extension of Submarine Canyons off the Northeast Coast of the United States," copyright 1967, Pergamon Press, Ltd.)

tal margins are relatively inactive in the present time. The erosion has often exposed outcrops of rock sequences of the continental margins. Canyons vary considerably in length, from where they begin to where they merge with the continental rise or fans. Shepard and Dill [1966] have estimated an average length of 50 km, with the longest canyons, of up to 370 km, in the southern Bering Sea. Submarine canyons are often present off large rivers, including the Congo, Columbia, Hudson, and Rhone. Other deep trough-shaped valleys occur off the deltas of the Ganges, Indus, and Mississippi [Shepard and Emery, 1973]. The Congo Submarine Canyon is unique because it penetrates as a deep estuary into the continent at the river's mouth. This canyon is a critical part of the constant flushing of the large amounts of sediment introduced by the river and also accounts for the absence of a broad delta at the mouth of the Congo River. The average slope of canyon floors is quite steep: 58 m/km [Shepard and Dill, 1966]. However, this estimate is weighted by a predominance of short canyons, characteristically exhibiting steep, sloping floors.

Longer canyons typically have lower gradients, between 8 and 13 m/km [Shepard and Dill, 1966].

Sedimentation in canyons

Research submersibles have allowed direct observations of sediments and minor topographic features in submarine canyons. These observations have shown that canyon floors commonly are flat and have transverse ripple marks at all depths and scour marks around boulders, reflecting bottom current activity. Numerous boulders lying on flat canyon floors have been derived from mass movements of bedrock down the steep canyon sides. Some canyons, such as the Scripps Canyon in southern California, act as chutes for sand from nearby beaches. The head of the Scripps Canyon comes to within only 60 m of the low-tide shoreline and leads into a system of narrow, steep gorges acting as conduits for the seaward transport of beach sand (see Chapter 10). Sediment cores taken from canyon floors are commonly graded, laminated, or cross laminated, reflecting a dynamic and episodic sedimentary regime. The movement of sediment down canyons is probably by a combination of slow creep [Emery and Uchupi, 1972] and turbidity currents.

Turbidity currents deeply incise and erode the canyons, sweeping loose sediments downslope. Underwater photographs and observations from submersibles have revealed polished canyon walls and the talus slopes of collapsed walls of canyons undercut by eroding gravity currents. Evidence suggests that turbidity-current frequency accelerated downcutting of canyons during the low stands of sea level associated with Quaternary glacial periods. During these times, submarine canyons, especially those on the Atlantic margins, were more active conduits for sand and gravel funneled between the continental shelf to the rises and abyssal plains. The postglacial rise of sea level has drowned the heads of the canyons and the wide shelf has separated them further from sediment supplies traveling along the shore. Numerous estuaries and lagoons along the Atlantic coast have also served to trap sediments before they can reach the canyons [Emery and Uchupi, 1972] (see Chapter 10). In contrast, the canyons along the Pacific margin of North America have been more active during postglacial times. These canyons have a large, steady supply of sediment from streams, almost filling their estuaries. These streams commonly empty only a few hundred meters from the heads of canyons or contribute their sediment to beaches, where longshore movement carries the sediment to the canyon heads [Emery and Uchupi, 1972]. In the Astoria Canyon, seaward of the Columbia River, the last major turbidity current events in the canyon-fan system derived much of their sediment from volcanic debris produced during cataclysmic eruptions of Mt. Mazama about 6600 years ago [Nelson and others, 1968].

Turbidity currents are deflected by the Coriolis effect. This was discovered by Menard [1955], based on his observation that deep-sea channels cut by turbidity currents tend to bend toward their left (at least initially) in the Northern Hemisphere. He suggested that the currents deviate, incline upward toward the right, and develop higher levees on this side through deposition (Fig. 13–10). As a result, large turbidity currents tend to spill over the lower levees on the left and develop channels that hook toward the left. This theory is supported by Pratt's

[1968] observations that the right-hand (southern) banks of the Hudson and Wilmington canyons are higher than the left-hand banks (Fig. 13–10).

Shepard and his co-workers used remote current-recording meters to demonstrate that tidal forces operating within canyons produce oscillating flows with periods typical of tides in deeper parts of the canyons and progressively shorter periods in the shallow parts of the canyon. At depths of several hundred meters, they have periodicities from 20 min to several hours, probably due to internal waves, while in progressively deeper waters the periodicity approaches that of the semidiurnal tide. Current velocities of up to 30 cm/sec have been measured and are capable of transporting most types of sediment. The data show that there is a net down-canyon flow of water and sediment [Laughton and Roberts, 1978]. At the base of the continental slope, canyons empty onto cones or fans, where the slope decrease causes sediment settling. Many large submarine canyons reaching the base of the slope continue on the rise as deep-sea channels, a few tens of meters deep and a few hundred meters across. Some of these channels exhibit broad bends and distributaries and some are bordered by levees [Emery and Uchupi, 1972]. Many canyons adjacent to the continental slope join as tributaries at the continental rise. For instance the Hatteras and Pamlico, along with several unnamed canyons, join on the rise off Cape Hatteras and continue as a single deep-sea channel to the edge of the Hatteras abyssal plain at a depth of 5300 m.

A number of mechanisms have been proposed for the origin and maintenance of submarine canyons. Although most workers now believe that most have been cut by the action of downslope sediment movement, major controversies between opposing theories prevailed for many years [Whitaker, 1974]. By the 1930s two major theories had become established. Shepard favored deep subaerial erosion when the continental margins were temporarily uplifted, followed by rapid drowning, then filling, and reopening by landslides. The other theory, developed by Daly, stated that canyons were cut by density currents during lowstands of sea levels when abundant sediment was available. Kuenen [1937] further developed this theory and became convinced that turbidity currents were capable of eroding the canyon systems. Subsequent work has shown that both theories apply to many canyons. The heads of many canyons approaching the present coastline were probably partly cut during lowered sea level; some are extensions of present-day rivers, which cut across the continental shelf at times of low sea level. These canyon heads have since been drowned and kept open or actively eroded by turbidity currents, gravity flows, slumping, or other submarine processes.

Erosion of submarine canyons requires an abundant supply of continental sediments to the canyon head. Close associations between large rivers and continuing canyons have shown that point sources are critical. Little is known about the age of submarine canyons, but the initial cutting of many probably occurred during the late Tertiary in response to tectonic and eustatic events. A highly active period of canyon cutting seems to have been during the last 2 to 3 m.y., when large ice sheets oscillated in the Northern Hemisphere. Associated drops in sea level exposed the continental shelves, and rivers cut across the shelves and emptied their loads onto the upper continental slope. Spectacular canyon cutting also occurred during times of low sea level in the earlier Cenozoic [Seibold and Hinz, 1974].

Particularly spectacular canyons are located in the southeastern Bering Sea, including the Bering Canyon, the Zhemchug Canyon, perhaps the world's largest with the volume of 8500 km^3, and the Pribilof Canyon [Scholl and others, 1968]. The Bering Canyon was periodically cut and filled by axial sedimentation during the late Tertiary and Quaternary. Excavation of the Zhemchug and Pribilof canyons, however, took place entirely during the Quaternary low sea-level stands, when major Alaskan rivers emptied directly into their heads, carrying enormous amounts of glacially derived sediment [Scholl and others, 1968].

On the east coast of North America, the number of canyons and the proximity of the extent of the ice cap during the Quaternary ice ages [Emery and Uchupi, 1972] are closely related. At least 190 canyons occur between Labrador and Cape Hatteras, but the number drops considerably further south. North of Cape Hatteras, active canyon cutting occurred when an abundant supply of sediment was brought to the shelf edge by the ice sheet. In general, canyons are largest in areas where the ice sheet reached the edge of the shelf [Emery and Uchupi, 1972]. Although active erosion has largely ceased since the ice sheets disappeared, a lack of infilling suggests that the canyons are not receiving much sediment and may still actively channel offshore any material that is received.

Depositional Sites

Turbidity fans

When turbidity currents reach the foot of the continental slope, they deposit their loads in a number of environments, including deep-sea fans, continental rises, abyssal plains, and trenches. These are all natural traps because of the decrease in gradient. The most conspicuous and rapidly formed features occur seaward of large rivers and submarine canyons. Most sediment contributed to fans and continental rises probably bypasses the slope and is transported down the axes of canyons. From the mouth of the canyon, they enter a fan valley or fan-valley system for distribution to the area of active deposition of the fan or rise. Those features at the mouths of canyons off the Pacific coast of North America were called **deep-sea fans** by Menard [1955]. Similar features off large river deltas in the Atlantic were called **abyssal cones** by Ewing and others [1958]. Collectively, all types of fans and cones formed in the deep sea by sedimentation of clastics have been called **turbidite fans** by Rupke [1978]. A complex sequence of turbidite basinal depositional history results from the interplay of sediment source and basin geometry. Basin infilling and resulting sediment overflow can occur between previously isolated basins.

The processes controlling fan morphology are not well understood [Normark, 1970]. Studies based on fan morphology and distribution of near-surface sediments are inadequate to determine growth processes of deep-sea fans. Major criteria for constructing the growth patterns of turbidite fans were established by Normark [1970] and include understanding the origin of terraces, levees, and slump features of the fan valleys; the character of the termination of fan valleys; and the origin of fine-scale structures and morphology of the fan surface outside the fan valley. From this Normark recognized that the fan surface consists of several distinct

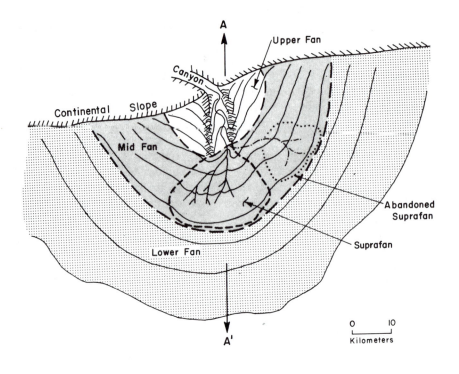

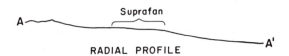

Figure 13-11 Schematic representation of model for submarine fan growth emphasizing active and abandoned depositional lobes, termed suprafans. (From W. R. Normark, *AAPG Bull.*, vol. 54, p. 2189, 1970, reproduced by permission of the American Association of Petroleum Geologists)

depositional environments. In transverse profile, a fan surface includes: (1) channels; (2) levees; and (3) interchannel areas (Fig. 13–11) [Rupke, 1978]. In radial profile it is possible to differentiate upper-, middle-, and lower-fan environments (Fig. 13-11). This tripartite division is not always evident, especially in most abyssal cones [Rupke, 1978]. Normark emphasized relict and buried channels as representing former positions of the fan valley and its distributaries. Much of the interchannel surface of modern fans is covered by a thin blanket of hemipelagic mud, while the channel floors are fine-grained turbidites. The structure of modern fans was determined during the high rates of clastic sedimentation during the Quaternary low stands of sea level. Rates of sedimentation since the Pleistocene have decreased from about 100 cm per 1000 years to as low as 5 cm per 1000 years [Damuth and Kumar, 1975; Nelson, 1976; Rupke, 1978].

Rupke [1978] has classified turbidite fans according to sediment source as follows:

1. *Deep-sea fans.* Off the mouths of submarine canyons.

2. *Abyssal cones.* Major river deltas.

3. *Short-headed, delta-front fans.* Off short-headed deltas in lakes fed primarily by river-generated turbidity currents.

4. *Continental rise fans.* Formed under the influence of redepositing contour currents, such as existing off the east coast of North America.

5. *Mixed-type fans.*

Deep-sea fans

Because the source of sediments for deep-sea fans is a canyon at a fixed point, deep-sea fans have a distinct apex (Fig. 13–11). This shape becomes more complex with the coalescing of adjacent fans. The radii of fans vary from tens of kilometers for those in the California borderland to about 300 km for the Monterey fan [Nelson and Kulm, 1973; Rupke, 1978]. Sediment thickness is rarely greater than 1 km. The growth pattern of a fan relates events in and around fan valleys to the structure and morphology of the open fan [Normark, 1970]. The growth pattern cannot be determined without knowledge of the origin and recent history of the fan-valley system. Is the present fan-valley system in equilibrium with the existing fan morphology? If not, the growth of a fan can be reconstructed only after determining the changes that have occurred since growth ceased [Normark, 1970]. Many fans, such as the Astoria fan, are not in equilibrium with present-day conditions because they developed during lowered sea level [Nelson, 1976].

From studies of the San Lucas and Astoria fans, Normark [1970] predicted that fan growth involves the formation of a leveed fan valley or upper fan, extending from the mouth of the submarine canyon (Fig. 13–11). The valley is marked by natural levees and the valley floor may be built above the general level of the fan surface. The levees decrease in height and disappear down the fan. Rapid radial deposition at the end of the leveed valley accelerated growth in this area, forming a depositional bulge, or lobe, on the fan surface called a **suprafan** [Normark, 1970]. The suprafan (Fig. 13–11) is a small delta, or fanlike deposit, probably formed as turbidity currents spread out upon leaving the confinement of the leveed valley. In radial profile the suprafan appears as a low, convex-upward segment of an otherwise concave-upward profile. Channels are numerous on the surface of the suprafan [Normark, 1970], forming a rapidly changing distributory system (Fig. 13–11). Migration of the fan lobes produces a fan complex constructed of several individual fans. The suprafan grades downward into a nearly flat-lying lower fan lacking even small channels (Fig. 13–11).

Abyssal cones

Abyssal cones develop off the deltas of major rivers marked by extensive drainage basins and large sediment load. Large deltas that prograde the continental margin usually occur on divergent margins. In the Atlantic seven cones are important: St. Lawrence, Hudson, Mississippi, Amazon, Orange, Congo, and Niger

[Emery, 1977]. Three (Mississippi, Amazon, and Niger) consist of a delta deep-sea cone couplet. The other four consist only of deep-sea cones that have little physiographic connection with their present rivers, because of little present-day activity (St. Lawrence, Hudson, and Orange) or because estuaries, deeply incised during times of low sea level, are still being filled in (St. Lawrence, Amazon, and Congo) [Emery, 1977]. Three of the great rivers of the world (Amazon, Ganges-Brahmaputra, and Mississippi) have built the largest cones. Abyssal cones widen seaward and generally grade into large abyssal plains. They can range from 300 km (Rhone cone) to almost 3000 km in length (Bengal cone) and up to 1000 km in width (Bengal cone). The inner part of the cone can range in thickness up to 12 km, as is the case of the Bengal cone [Curray and Moore, 1971]. The Bengal cone is formed by sediment redeposition from the Ganges-Brahmaputra rivers and is influenced by tectonic deposition begun after the India plate collided with Asia in the Eocene, ultimately forming the Himalayas by the middle Miocene. Extensive Quaternary uplift of nearly 2000 m has occurred, counteracted by tremendous erosional activity, which supplies immense amounts of sediment to the Bay of Bengal via the river system. The present rate of sediment influx suggests a denudation rate of more than 70 cm per 1000 years for the Himalayan chain.

The largest submarine cone off North America is the Mississippi cone, seaward of the topset surface of the Mississippi delta (see Chapter 10). This is a broad, concave feature with an area of about 200,000 km², about the size of New England [Moore and others, 1978]. The cone consists of the three major parts typical of turbidite fans. The inner part of the cone contains a partly leveed channel cut into older cone sediments. The channel is filled with the late Pleistocene, fine clastic sediments. The middle part of the cone is constructed of a massive complex of fan channels that has constructed a suprafan up to 500 m above the surrounding fan surface. The lower part of the fan is marked by smooth, gentle slopes that contain depositional distributary channels.

The cone's depositional pattern has changed with sea-level fluctuation. During high sea-level stands, the Mississippi River built up successive deltas, lobate or birdfoot on the shelf, at its numerous mouths. During the latest Pleistocene (18,000 years B.P.), when sea level was lowered by more than 100 m, the ancestral Mississippi River cut canyons into the continental shelf and disgorged its sediment load directly onto the slope for distribution to the cone.

Influence of contour currents on rises

In the 1950s and 1960s the demonstration of large and active sediment-drift deposits on the rise by B. Heezen and his colleagues provided compelling evidence that at the depositional surface, the sediment of the continental rise is very mobile and may move thousands of kilometers parallel to the rise before being finally deposited. Heezen and others [1966] proposed that the continental rise of eastern North America has been built and shaped primarily by the combined effects of turbidity currents and contour currents. The contour currents (see Chapter 8) produce a suite of bedforms ranging from small current ripples to large sediment waves or dunes with amplitudes of tens of meters and wavelengths of thousands of meters and can redeposit sediments as constructional ridges parallel to the current. These ridges may be as large as the Blake-Bahama outer ridge and the Greater Antilles outer ridge. Near-bottom velocities of up to 18 cm/sec have been observed in the southerly flowing Western Boundary Undercurrent off eastern North America [Heezen, 1974]. Current velocities this high can transport all sediment sizes found on the rise. However, contour currents are permanent flows with current velocities closer to the minimum values required for transportation of rise sediments. The thickest sediments in the ocean occur near the axes of deep boundary currents, and these deposits become thinner with increasing distance from the current axes [Heezen, 1974]. The axis of the highest velocity and greatest volume transport is deflected to the right side of the current, due to the Coriolis force; sediment transport is greatest on the continental rise and decreases with increasing depth from the current axis. As long as the current velocities remain lower than required for erosion, the quantity of sediment that is deposited will be roughly proportional to the volume transport of the current. Off the east coast of North America, the effects of contour currents are most visible in middle Miocene and younger sediments deposited along the continental rise [Tucholke and Mountain, 1979].

Sediments deposited by contour currents have been called **contourites** and exhibit a suite of features different from turbidites. They have thin beds with sharp contacts between beds; they are persistently laminated and well sorted and graded. Contourites commonly occur in sediment sequences of the continental rise but are not usually found in the abyssal plains dominated by turbidites.

Lateral transport and erosion by contour currents controls sedimentation on the continental rise of eastern North America. However, great variation exists in the importance of contour currents in sedimentary processes, and the processes operating on the continental rise of eastern North America are not necessarily typical of rises elsewhere in the oceans. For example, contour current deposition is apparently not an important process on much of the West African margin, where the contour currents themselves are not as well developed as in the western Atlantic. Young and Hollister [1974] and Embley [1975] found no evidence for contourites in late Quaternary sediments on the Northwest African continental margin. However, typical contourites have been described from the Antarctic continental rise [Piper and Brisco, 1975; Tucholke and others, 1976]. A more detailed discussion about bottom currents and their geological effects is given in Chapter 15.

Abyssal plains

The abyssal plains are formed by the accumulation of turbidites, which have buildup deposits over 1000 m thick. The formation of abyssal plains is largely a function of sediment availability and topography adjacent to sediment-source areas. The greatest development of turbidite sequences occurs off major drainage basins. In general, abyssal plains have formed off divergent margins, as typically represented in the Atlantic Ocean. However, many deep-sea trenches associated with active margins, such as those around the periphery of the Pacific Ocean, are also floored by narrow abyssal plains. In the Pacific, the abyssal plains are restricted to broad fringes along the North American continent about 2000 km across and to narrower archipelagic aprons around volcanic islands. Elsewhere the plains are missing. This is because, in most places, trenches, basins, and ridges intervene between the continental supply centers and the abyssal floor, preventing the wide dispersal of turbidity currents.

In the Atlantic vast amounts of materials were eroded from adjacent land-masses to form the Sohm, Hatteras, and Mares abyssal plains in the Northwest Atlantic. Likewise, the area between the east flank of the mid-Atlantic ridge and the continental slopes of Africa and Europe has been topped up to form the Biscay, Iberian, Tagus, Horseshoe, Maderia, and Cape Verde abyssal plains [Horn and others, 1971A; 1972]. Turbidites of the North Atlantic are slightly coarser than those in the North Pacific, due to the availability of larger amounts of coarse glacial debris and less-obstructed transfer of terrigenous sediment.

The Hatteras abyssal plain is typical and lies at the base of the continental rise off the east coast of the United States (Fig. 13–12). It is 1000 km long and 150 to 300 km wide and the bottom gradient is toward the south. Textural patterns suggest that the Hatteras and Hudson canyons are the principal sources of most turbidity currents [Horn and others, 1971]. The sediment parameters indicate a southward direction of transport off the eastern end of Hatteras Canyon. It seems that during an earlier stage of the development of this plain, turbidity currents flowed north and filled a depression to the north of Hudson Canyon. After this was filled, the flow was diverted to the south (Fig. 13–12).

The deep-sea sands of the Hatteras abyssal plain are medium grained near the mouths of contributing canyons and become finer seaward. As distances from the canyons increase, fine-grained silt begins to dominate the basal portions of turbidites. Seismic profile records across the plain reveal a thicker turbidite fill in the north. At the extreme southern end, a few cores contain turbidites of very fine-grained sediment. These reflect either maximum distances of turbidity-current transportation over the Hatteras abyssal plain or overflow from smaller plains, which are rich in carbonate sediment, to the south.

Spectacular increases in the rate of formation of abyssal plains, especially those in the North Atlantic and North Pacific, have occurred during the late Cenozoic associated with the inception of widespread continental glaciations and frequent lowerings of sea level (see Chapter 9). Turbidite deposition dramatically increased. Turbidity currents were frequent, occurring every few years. During the Holocene high sea level, their frequency was drastically reduced to perhaps as few as one every 1000 years. There is relatively little new sediment being added to the abyssal plains today, although much evidence exists for extensive reworking and erosion by bottom currents.

The tectonic control of trench filling is also important. For example, although turbidites of terrigenous detritus have been reaching the abyssal floor of the Gulf of Alaska since at least the early Eocene, far less would have gone to this area during the late Cenozoic had there not been orogenic uplift over the Pacific margin to form partially glaciated mountains in the Neogene [Stewart, 1976]. Trench filling during the Quaternary to form turbidite wedges in the Washington, Oregon, and Aleutian trenches would probably also have been less rapid if adjacent coastal mountains

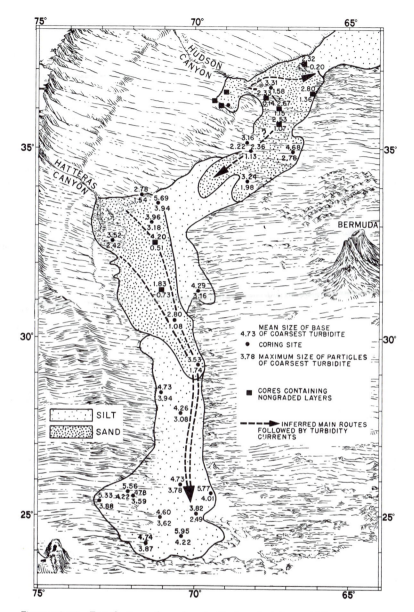

Figure 13-12 Distribution of sand and silt deposited by turbidity currents in abyssal plains fed by submarine canyons off part of the eastern United States. Numbers represent mean and maximum grain size (in phi units) of samples from the base of the coarsest turbidite in each piston core. (From D. R. Horn and others, *Marine Geology*, vol. 11, p. 291, 1971)

capable of nourishing extensive ice fields had not formed. Likewise, turbidite deposition in the Coral Sea basin of the Southwest Pacific was closely tied to the tectonic uplift of New Guinea–Papua during the Miocene, which provided the source of terrigenous sediments.

Hemipelagic Sediments

The hemipelagic sediments are important and widespread sediments draped over the middle and upper continental slopes throughout the world (middle and upper bathyal zones). They generally lack the coarse sediments typical of the continental shelf, yet are deposited close enough to land to be dominated by terrigenous muds and silts. Biogenic components, especially foraminifera, are important. At greater depths on the continental slope, the hemipelagic sediments often grade into biogenic oozes. The dominance of the terrigenous component in hemipelagic sediments makes them much darker than pelagic oozes. Clay types vary according to the source region.

In the convergent margin areas, hemipelagic sedimentation sequences are often eventually incorporated within the accretionary prism and uplifted as thick marine sections on land. In New Zealand, for instance, such uplifted sequences of dark, hemipelagic mudstones and siltstones (called *papa* by the Maoris), have provided the basis for classic studies of Cenozoic biostratigraphy and paleoclimatology. These sequences can be represented by several hundred meters of very uniform, largely unbedded (massive) mudstone containing scattered mollusk fossils and occasional tephra layers. They have attracted little attention from sedimentologists because of their uniformity but are of great interest to paleontologists.

Deep-Sea Clays

The most widely distributed terrigenous minerals in pelagic sediments of the oceans are clays. Clays are a diverse mineral group of hydrous silicates of aluminum. Some are formed by weathering or as alteration products of primary silicate minerals while others, including the micas, are primary minerals. Although clay minerals are by far the dominant components of sedimentary rocks on land or in nearshore deposits, in the ocean they are normally masked by other sediment types, except in some deep basins. All clay minerals are very fine grained, as their crystal structure is such that they occur as flakes of only a few microns or less in diameter. Four types of clays dominate ocean sediments: chlorite, illite, kaolinite, and montmorillonite (smectite). The relative proportions of these major clay minerals in a sediment will vary according to the prevailing climatic regime in the source region and to the mixing processes that occur in the ocean.

Brown or red clay

Brown or red clay is an extremely fine-grained pelagic deposit of bright reddish brown to chocolate color, which is formed by the slow accumulation of materials in the deeper parts of the ocean (normally greater than 4000 m), away from turbidite deposits. The color is due to amorphous or poorly crystalline coatings of iron oxides on the sediment particles. These deposits are very widespread in the oceans and were first mapped and described as red clays during the *Challenger* expedition. Although they are brown, the name applied by the *Challenger* scientists is still widely used. Brown clays consist mostly of windblown clay minerals of various kinds, as well as fine-grained minerals, such as quartz, feldspar and pyroxenes, meteoric and volcanic dust, fish teeth and bones, whale earbones, and manganese micronodules. The calcium carbonate content is normally extremely low and by definition is less than 30 percent. These deposits are the very fine-grained components of pelagic sediments after the removal by dissolution of all but the most solution-resistant material, such as shark teeth. Composition varies with the different climatic zones of adjacent continents. Sedimentation rates are normally less than about 1 mm per 1000 years. Because sedimentation rates are so low, even thin sequences of brown clay represent large intervals of time. The amount of geological information contained in red clays is limited but useful. Chronologies have been developed by using fish-teeth biostratigraphy and magnetostratigraphy. Quartz is often present and its abundance provides information on paleowind patterns and intensity. Clay mineralogy also indicates the climatic regime of the source region, suggesting that sources of clays were dominantly volcanic in the early Cenozoic and terrigenous in the late Cenozoic.

Origin and sources of clays

Sources of ocean clays are intimately linked to the actual formation of these minerals. There have been two principal schools of thought regarding the origin of clay minerals in ocean sediments. The first speculates that clay minerals form in the oceans through the interaction between silica and aluminum compounds. Hence clay mineralogy is closely related to the environmental conditions at the site of deposition. The second philosophy is that clay minerals are formed by the weathering of rocks on land. They are then transported to the oceans with little or no subsequent alteration. Clay origin is important geochemically, because if clays are formed in the oceans, they would form very important sinks for certain ions, such as K^+, Mg^{2+}, Ca^{2+}, and Na^+. Three general areas have been investigated in the last two decades to determine where clays are formed.

1. The character of clays in rivers and their relationships to the source areas of river sediments.

2. Whether clay minerals exist throughout ocean sediments and whether there is a relation with climatic zones.

3. Whether clays are newly formed or if their ages reflect the old ages of most of the potential source rocks on the continents, using radiometric dating.

The distribution of clay minerals carried in the suspended loads of the Amazon River system as studied by Gibbs [1967] has been important in helping to resolve the question. Mineral composition was related to a number of environmental factors, including the type of source rock, relief, temperature, precipitation, and the nature of vegetation. It was discovered that the distribution of illite and chlorite clays is closely dependent on parent-rock composition, and these minerals are more important (about 75 percent of suspended load) in the upper reaches of rivers of high relief, closer to the source regions. Kaolinite and montmorillonite (smectite) are more important in the lower basins because of greater weathering. These are the newly altered (crystallized) minerals and make up as much as 60 percent of the suspended load in the warm, humid lower reaches of the river. However, the Amazon River drains a cooler, less humid region of extensive mountainous terrain, so an average of 80 percent of the total particulate flux is relatively newly eroded material showing little alteration. Thus the clay types clearly reflect the major soil types in the source areas which, in turn, are a response to different climatic regimes.

Because the type of clay in soil so strongly reflects climatic control, it is reasonable to expect the distribution of clay minerals in modern depositional basins to be related to climatic patterns on adjacent landmasses. This assumption has been tested by several investigators and found to be generally valid. For instance, in the Gulf of Mexico, the Apalachicola River—which drains the subtropical region of southeastern United States—carried 60 to 80 percent kaolinite. In contrast, the Mississippi River—which receives most of its suspended sediment from the temperate northern half of the United States—contains only 10 to 20 percent kaolinite.

Although much clay is transported to the oceans by rivers, most of the clays occurring in the areas remote from land were transported by the wind. The composition of the windblown clays, like those transported by rivers, is closely related to the climatic regime in the source area.

Oceanic clay-mineral suites are often a continuation of the terrestrial clay-mineral zones [Biscaye, 1965; Griffin and others, 1968; Windom, 1976]. The relationship is obvious and supports the idea that most clays are carried to the sedimentary basins and deposited without being changed. Oceanic sediments contain clays that reflect the weathering process that produced them: Chlorites and illites are in colder areas and in the oceans near temperate zone continents; kaolinites are in tropical zones (Figs. 13–13 and 13–14). Montmorillonite (smectite) distributions are less distinct and are more closely associated with areas of volcanic ash, which are commonly transformed into montmorillonite in marine environments. The zonation, however, is not always simple; any clay mineral may occur in the soils and sediments of any climatic zone, but the fact that certain clay minerals tend to form in particular climatic regimes leads to concentrations of certain clay minerals. Radiometric ages of clay-rich sediments also support the terrigenous origin of most clays without alteration. If clay minerals come from continental weathering sources without major chemical changes in seawater, their radiometric ages should be close to those of the source rocks. Tests show that the ages of the clay deposits are comparable to those of the source rocks. For instance, the ages of clays in the western North Atlantic are very old, such as those of the source rocks of the Canadian Shield region, while clays in the Argentine Basin reflect the weathering of younger rocks in southern South America.

Clay-mineral distributions, as shown by recent studies of clay distributions in the Indian Ocean, are also related to the bottom currents that transport and mix the clays from different climatic regimes. Furthermore, montmorillonite has been shown to form from the submarine alteration of basalts with little or no subsequent

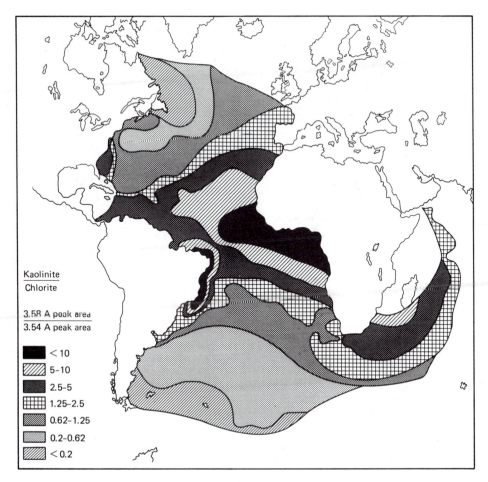

Kaolinite
Chlorite

3.58 A peak area
3.54 A peak area

- ■ < 10
- ▨ 5–10
- ▨ 2.5–5
- ▦ 1.25–2.5
- ▨ 0.62–1.25
- ▨ 0.2–0.62
- ▨ < 0.2

Figure 13–13 Kaolinite-to-chlorite ratio distribution in the greater than 2μm fraction of surface sediments of the Atlantic Ocean. High kaolinite-to-chlorite values are found in the tropics and reflect the intense weathering on the adjacent continents. (From P. E. Biscaye, *Geol. Soc. Amer. Bull.*, vol. 76, p. 810, 1965, courtesy The Geological Society of America)

transportation. This is evidenced by the abundance of this clay in those areas showing evidence of volcanic activity.

Distribution of clays

Chlorite Chlorite is an unstable clay mineral found where erosion is mechanical rather than chemical. The acidic, warm, humid conditions of the tropics enhance the weathering of chlorite to kaolinite. Latitudinal dependence is strong, and high concentrations of chlorite are found at high latitudes only when chemical weathering processes are less important than mechanical. Thus chlorite is more

Terrigenous Deep-Sea Sediments   **429**

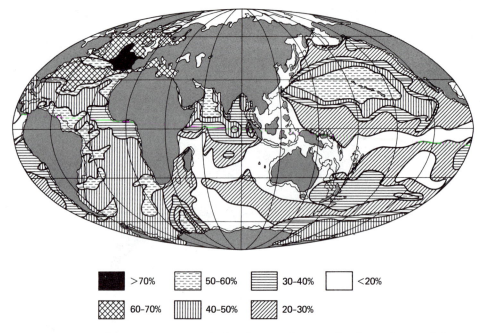

Legend							
■	>70%	▦	50-60%	▤	30-40%	□	<20%
▨	60-70%	▥	40-50%	▧	20-30%		

Figure 13-14 Illite concentrations in the less than 2μm fraction of surface sediments of the world ocean. Note maximum concentrations in certain middle latitude areas. (After H. L. Windom, 1976)

abundant in polar than in tropical areas. It is transported to the ocean by ice or water and is subjected to little or no subsequent chemical weathering. Chlorite ranges from 10 to 18 percent (average 13 percent) of the clay-mineral assemblage in ocean basins. Chlorite can be considered a primary clay mineral formed by metamorphic processes and not a secondary mineral product of chemical weathering or alteration during soil formation.

Illite   Illite, which is a general term for the mica group of minerals, is the most pervasive clay mineral, often composing 50 percent of the total sediment. In the Southern Hemisphere, illite makes up 20 to 50 percent of the sediment; in the Northern Hemisphere it generally forms more than 50 percent, reflecting the continental input (Fig. 13-14). Illites are mostly mechanically ground primary minerals, rather than recrystallization products formed during weathering. Illite concentrations delineate the extent and the amount of contribution of river-borne solids. This can be seen by the outlines of the areas receiving materials from the South American rivers (Fig. 13-14). These rivers are marked by high concentrations of illite, resulting from erosion of Andean rocks with little subsequent weathering. The highest illite concentrations occur in the North Atlantic, reflecting the abundance of this mineral in soils and sediments of eastern North Amercia.

The most characteristic feature of illite distribution in the Pacific Ocean is a wide band of high concentration between 20 and 40°N, associated with the high quartz abundance resulting from transportation by the jet stream (Fig. 13-14).

Kaolinite   The distribution of kaolinite, like chlorite, is latitudinally dependent, being 10 times more abundant in tropical than in polar regions. The distribution and amounts of this mineral reflect the intensity of the soil-forming processes in the source areas. In polar regions, where soil formation is slow and retarded by lack of chemical weathering, the soils and sediments show only small amounts of kaolinite. The high kaolinite abundances in recent marine sediments are confined to equatorial sediments, and kaolinite may be considered the low-latitude clay mineral. The reciprocal relation existing between kaolinite and chlorite distributions is shown in Fig. 13-13. Kaolinite abundances range from 8 to 20 percent (average 13 percent) of the total clay suite in the ocean basins. Gibbsite is another clay mineral often associated with, but less abundant than, kaolinite. Maximum concentrations of kaolinite are found off equatorial West Africa (Fig. 13-13). There are well-defined concentration gradients, decreasing with increasing latitudes, with high concentrations between 25°N and 25°S.

The eastern part of the Pacific exhibits low concentrations of kaolinite, probably due to the lack of a source area on the west coast of the Americas. The mountain ranges that extend the entire length of the American continent do not provide appropriate conditions for the formation of these lateritic soils. Therefore any runoff, which is quite small, will be characterized by products of high-elevation weathering processes, products normally very low in kaolinite. High concentrations of kaolinite off the coast of western Australia are derived from extensive kaolinite-rich lateritic deposits of the Australian desert.

Montmorillonite (smectite)   Montmorillonite is an alteration product of volcanic material on land and in the oceans. It is common in areas of low sedimentation, close to sources of volcanic materials. The source can be either windblown volcanic ash or basaltic glass on the ocean floor. Layers of volcanic ash in deep-sea sequences are often altered to montmorillonite in as little as 20 m.y. It is more abundant in the Pacific and Indian oceans than in the Atlantic, which is less volcanically active. It is most abundant in the South Pacific, where it forms up to 50 percent of the clay-mineral suite.

The oceanic clay minerals form two representative associations: (1) chlorite-illite suite in cold and temperate zones; and (2) kaolinite-gibbsite-montmorillonite suite in equatorial and warm subtropical areas. All oceanic clays but montmorillonite are detrital, being carried in directly from continental sources and not precipitated out of the oceans. Chamley [1979] has shown relationships between changes in clay-mineral suites and glacial-interglacial oscillations during the Quaternary. Furthermore, long-term changes in clay-mineral suites in the equatorial Pacific have been shown and are inferred to be associated with climatic changes during the late Cenozoic. An increase in illite and chlorite during the last 3 m.y. reflects reduced weathering in source areas due to increasingly cooler climates and more glaciation in the Northern Hemisphere and increased elevation of the land-masses supplying the sediment adjacent to the Pacific Ocean. But studies have been few despite the clear climate-clay relationship. Perhaps in the future there will be more research carried out using clay-mineralogy for paleoclimatic interpretations.

Windblown Sediments

Sediments transported by the wind are called **eolian** sediments after Aeolus, the Greek god of the winds who, in Homer's *Odyssey,* lived on the island of Aeolia. To help Ulysses' return to Ithaca, Aeolus presented all the adverse winds securely tied up in a bag, but Ulysses' curious companions opened them and his ship was driven off course.

Windblown material is of both nonvolcanic and volcanic origin. In this section we shall be concerned with the windblown nonvolcanic component. Wind has long been recognized as an important mechanism for sediment transport to the ocean floor. Charles Darwin [1846] was probably the first to recognize this; he observed dust storms emerging from the Saharan region, as other seafarers had observed for centuries. Quartz was recognized as a significant component of deep-sea sediments by Murray and Renard [1884], who drew attention to the importance of wind transport of this material. Eolian sediments are important regionally, particularly in areas adjacent to arid regions, such as North Africa, Australia, Arabia, and western North America. Except in areas protected from other terrigenous sediment, such as parts of the mid-Atlantic ridge, eolian sediments do not make significant contributions to deep-sea sediment. Other processes easily mask them. Nevertheless, the very fine-grained component of deep-sea sediment is largely of windblown origin.

Eolian dust is an insignificant source for the dissolved matter in the oceans. Trace elements may be stripped by desorption processes from eolian dust on entry into the ocean, but this is much less important than in other terrigenous sediments. Eolian sediments do transport soluble constituents to the ocean, but most materials are insoluble quartz and biogenic silica. Excess quantities of trace metal constituents in certain marine sediments must be derived from processes other than eolian transportation.

Dust transportation

Regions most favorable for creation of eolian sediment are marked by dry air and persistent trade winds. In humid zones eolian particles are washed from the atmosphere by rain and deposited locally. Principal conditions necessary for long-distance eolian transport of desert material include dry surface conditions, strong surface winds, small particles, appropriate trajectory into the troposphere, vigorous vertical mixing in the troposphere to maintain air-suspension of particles, and onset of rain in the area of deposition. High surface temperatures in the desert create much mixing in the overlying troposphere. Because of a temperature inversion that divides the troposphere from the stratosphere, particle transportation is controlled by prevailing wind directions and velocities in the troposphere. Material that does enter the stratosphere can be globally transported by winds of up to 500 km/hr. Residence time in the jet streams is about 2 weeks for particle sizes from 2 to 10 μm.

Dust storms over the North Atlantic from the Sahara area are common. The Sahara is a major source of eolian sediment, being one of the world's largest desert regions with an area of 8.7 million km^2. Estimates of the amount of windblown material leaving the Sahara associated with dust storms range from 60 to 200 × 10⁶ tons per year. Most of this material becomes a significant fraction of north equatorial Atlantic deep-sea sediments [Folger, 1970]. This material is transported great distances across the equatorial Atlantic (Fig. 13–15); high-haze frequencies in the region are associated with this tropospheric dust transport.

One of the earliest studies to offer definite proof of the long-range transport of dust was by Parkin and others [1967]. They collected surprisingly large quantities of red-brown (summer) and gray (winter) dust over Barbados. The particle size and mineralogy of the dust are similar to that in underlying deep-sea sediments. This dust is carried about 6000 km by the northeast trade winds from arid areas of Africa and Europe. In the Barbados region, dust particles are normally fine grained (40 percent are less than 2 μm) as compared with coarser dust collected in dust storms closer to Africa (25 percent are larger than 10 μm). These observations indicate that dust storms are capable of transporting rather large particles into the atmosphere, accounting for the occurrence of relatively coarse material in pelagic sediments. Vigorous vertical mixing in the lower atmosphere is required to keep coarse particles suspended, because these normally settle rapidly. Dust storms in Australia are another vast source area for eolian sediments and are also capable of transporting large particles. A large storm, in 1928, produced an average surface distribution of about 3 to 30 g/m² over New Zealand about 2000 km away from Australia [Glasby, 1971]. This material was also deposited over much of the Tasman Sea, which suggests that the fine-grained material in this region has been deposited by this process.

Eolian sediment components

Dust collected from the atmosphere is very fine grained and is generally reddish brown in color; quartz is the major component. Eolian sediment is associated with other minerals, such as calcite and biogenic components and especially *opal phytoliths*, freshwater diatoms, and even fungus spores. Phytoliths are accumulations of amorphous opal ($SiO_2 \cdot nH_2O$) in plants. They occur in greatest abundance in the leaves of members of the grass, sedge, and rush families and, as they resist dissolution, they persist in soils, dust, and lakes. Baker [1959] reported the first fossil phytoliths, and they have since been found in paleosoils and in land and deep-sea sediments. They are small (10–200 μm) but so abundant that they can make up 4 percent of the top soil by weight and up to 5 percent of windblown dust. The different shapes of phytolith assemblages—rods, dumbbells, hats, and trapezoids—encourages classification according to their morphology. Relationships established between phytolith assemblages and vegetation type are potential diagnostic tools for paleoclimatologists.

Phytoliths and freshwater diatoms are most abundant in equatorial regions, and fungus spores are dominant in mid-latitude regions. In mid-latitudes, westerly winds transport considerable terrigenous material, including fungus spores, to the western Atlantic but little to the eastern Atlantic. Phytoliths and freshwater diatoms have been described in deep-sea cores in the equatorial regions where dust storms are frequent, and many of the components are present in similar proportions

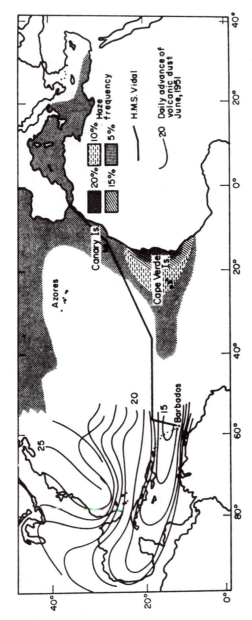

Figure 13-15 Average haze frequency for summer months in the North Atlantic indicative of tropospheric dust and contours showing the daily advance of a volcanic dust cloud between June 15 to June 26, 1951, that originated in the Cape Verde Islands. (Reprinted with permission from *Deep Sea Research*, vol. 17, D. W. Folger, "Wind Transport of Land-Derived Mineral, Biogenic and Industrial Matter over the North Atlantic," Copyright 1970, Pergamon Press, Ltd.)

in the dust and in the deep-sea sediments. Therefore phytoliths and freshwater diatoms may indicate the amount of eolian material contributed by the trade winds to the equatorial Atlantic.

Despite the paleoclimatic potential of phytoliths, the component that has been most useful in tracing wind activity is quartz. Quartz originates only on land and is considerably more resistant to dissolution than phytoliths or marine-biogenic silica. It therefore is preserved in sediments well after most of the biogenic (SiO_2 and $CaCO_3$) components have been dissolved. The importance of quartz as an eolian component in deep-sea sediments was recognized by Radczewski [1937] in the Atlantic and by Rex and Goldberg [1963] and Biscaye [1965] in the Pacific; the latter found a remarkable correlation between the latitudinal location of arid areas in the Northern and Southern hemispheres and the percentage of quartz in deep-sea sediments at the same latitudes. The Pacific Ocean is a particularly good area to study windblown quartz distributions, because much of the ocean floor is protected from the effects of turbidity currents by trenches or ridges. More detailed work has shown a high correlation between quartz concentrations across the Pacific and the high-altitude jet stream [Biscaye, 1965]. A narrowly defined mid-latitude band between 25° and 50°N is separated from a secondary lobe of high values at about 10°N associated with the northeast trade winds and extending off the arid regions of Southwest North America (Fig. 13–16) [Rex and Goldberg, 1958].

Paleoclimatic value

The relationship between eolian sediments, present-day wind patterns, and arid-source regions provides a key for monitoring past paleoclimatic conditions. It

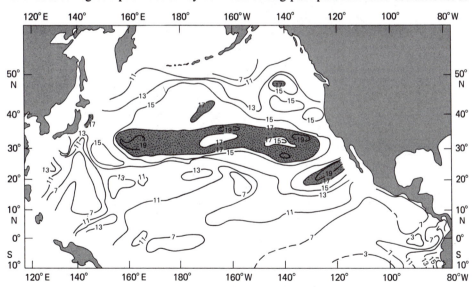

Figure 13-16 Distribution of quartz concentrations in the North Pacific. Contours are concentrations in weight percent on a carbonate-free and opal-free basis. Note the narrowly defined mid-latitude band of high values which is distinctly separated from the lobe of high values associated with the N.E. trade winds. (Courtesy T. C. Moore, Jr. and G. R. Heath, 1978)

can supply information on the distribution of arid areas (some eolian particles such as phytoliths are sensitive indicators of certain climates), the position of trade-wind belts transporting the dust, and the strength of the wind. Areas of upwelling, which are in large part dependent on wind strength, are the principal zones of biological productivity in the oceans. Intensity of upwelling is controlled by changes in wind strength and direction. Thus estimates of past wind strength and direction can provide constraints upon inferences about paleoproductivity. Such information is necessary to distinguish changes in the accumulation of biogenic debris, which are due to variations in upwelling, from fluctuations due to changes in the supply of nutrients, such as silica, to the oceans.

Past trade-wind positions have been studied using changes in patterns of quartz abundance and grain-size distributions in the Atlantic and Pacific. During glacial episodes, the bands of windblown sediments shift position several degrees to the south, reflecting the changed position of the trade-wind belts. Also, particle sizes increased during the glacial episodes as winds were more vigorous. The abundance of eolian sediments also shows changes in response to changing aridity and vegetation cover in the source regions. For instance, in the equatorial Atlantic, the concentration of phytoliths and freshwater diatoms is consistently lower prior to 1 Ma because conditions were warmer and the Saharan area was probably less arid [Parmenter and Folger, 1974]. Since that time, the abundance of eolian biogenics have fluctuated as paleotemperatures exhibited greater variation during the late Quaternary. During glacial episodes, conditions were much more arid, so eolian biogenic material was more readily transported to the oceans. During interglacials, freshwater diatoms in lake sediments were more often covered by water; phytoliths were trapped in vegetal mats; and with less soil erosion, total atmospheric dust-loads diminished. These patterns were reversed during the glacial epsiodes.

Volcanic Marine Sediments

Marine sediments of volcanic origin, or **volcanogenic** sediments, are either the primary or secondary result of volcanic activity. These include the following:

1. Sediments formed as a result of aerial volcanic explosions, subsequently transported to the ocean floor and called marine **pyroclastic** sediments.

2. Sediments that are reworked fragments of volcanic rocks, called marine **epiclastic** sediments. These can be derived from either preexisting pyroclastic deposits or as erosion products of submarine volcanic rocks.

3. Sediments formed in place on the ocean floor either as a result of submarine eruptions or from precipitation from hydrothermal activity. These are marine **authigenic** sediments or rocks. Because they are nonterrigenous, they are discussed in Chapter 14.

Volcanogenic sediments are important contributors to marine sediment, particularly near island arcs where sedimentary wedges consisting largely of volcanic components can be several thousand meters thick [Lisitzin, 1972]. Air-fall deposits, which are the most important, are commonly referred to as **tephra**. Tephra includes air-fall material, as well as a wide range of flow pyroclastic materials like ignimbrites (subaerial acid-volcanic sheets). Tephra is a Greek word meaning *ash* and was introduced into modern usage by Thorarinsson [1944]. In 300 B.C., Aristotle used the term for volcanic ash produced by an active Italian volcano. In deep-sea sequences we are concerned largely with air-fall volcanic material, most of which (about 90 percent) is *volcanic glass*, or *ash*. Tephrochronology is concerned with sequence, correlation, and age of these tephra deposits (see Chapter 2). Very fine ash can be transported globally, and coarse ash can be carried over several thousand kilometers. Primary materials include feldspar, pyroxene, and other minerals, while secondary minerals include clays derived from volcanic materials, such as smectite and palygorskite, and other minerals like zeolites. Pumice is a frothy (highly vesicular) primary volcanic rock, and although minor volumetrically, is important because its buoyancy gives it very wide dispersion in the oceans. Pumice also may float attached organisms vast distances, with important biogeographic consequences. Submarine volcanoes are nonexplosive, so although they may produce extensive pyroclastic flows, they do not produce much material that would be classified as sediment. Some fragmentation occurs, which can form secondary epiclastic sediments.

In areas distant from volcanic centers, the importance of volcanogenic materials has long been suspected but is masked by other sediment types and diagenetic alteration. Despite difficulties, attempts have been made to determine the percentage of volcanic materials blown to various oceanic areas. Although volcanic and other windblown sediment were not differentiated, the North and South Pacific and the central Atlantic oceans may now receive as much as 25 to 75 percent of their detrital phases from atmospheric dust fallout [Lisitzin, 1972]. A large proportion of this material is of volcanic origin. The importance of pyroclastic material can be gauged from the number of volcanoes, 450 of which have been active since A.D. 1500. These volcanoes have ejected some 330 km^3 of pyroclastic material and 50 km^3 of lava during this time.

Most Neogene volcanic activity occurred around the Pacific Ocean, in Indonesia, and the eastern Caribbean; hence, these areas are under the greatest influence of volcanic sedimentation. Material is derived from three types of volcanism: (1) island-arc and ocean-margin volcanism; (2) mid-ocean ridge and oceanic intraplate volcanism; and (3) continental volcanism.

Volcanism associated with convergent margins is marked by a large number of foci, the largest scale eruptions, and high explosive capacities. Lavas are highly viscous, so eruptions are highly explosive. Pyroclastic material makes up to 90 percent of the solid products (even 99 percent, as in the Indonesian volcanoes). Island-arc volcanism dominates the production of marine volcanic components. Mid-ocean ridge and oceanic intraplate volcanism contributes much less tephra to the oceans. Eruptions are much less explosive due to low lava viscosity, so pyroclastic material is concentrated close to volcanic centers. It has been calculated that, of the 330 km^3 of total pyroclastic material produced since A.D. 1500, 310 km^3 was produced in the island-arc and continental-margin areas. Only 19 km^3 were produced in mid-ocean sites, of which 10 km^3 were produced from Iceland alone [Lisitzin, 1972].

Most volcanoes erupt tephra to a height of less than 6 km, thus distributing the ash only in a local area. Occasionally ash is ejected into the stratosphere to heights greater than 10 to 15 km, resulting in tephra transport 3000 to 6000 km from volcanic centers. Very fine material (0.3–1 μm) may be distributed over vast areas meridionally. These could be called **global ash falls.**

Physical and mineralogical characteristics of tephra

Tephra varies widely in size, shape, and composition. It may consist of dense or porous particles (glass or crystals) crystallized from the magma, solidified volcanic rocks of previous eruptions, and accidental fragments of country rock. Tephra is often highly distinctive, reflecting a single particular eruptive episode. Much work has gone into establishing criteria to differentiate between tephra layers and to identify the character of the eruption that produced the tephra and the tephra's subsequent mode of transportation. Parameters that have been employed include physical, mineralogical, and geochemical characteristics. Physical characteristics of the ash layers may include color, thickness, and bedding characteristics; individual shards may be characterized by size, shape, and internal structure. Mineralogical characteristics include mineral composition, type and character of phenocrysts (heavy minerals), and color of minerals, while geochemical characteristics are in part reflected by the refractive index of the glass. Composition is best determined using the electron microprobe.

Steen-McIntyre [1977] grouped the petrographic characteristics employed by tephrochronologists to categorize tephra as either essential or supplemental. **Essential petrographic characteristics** reflect the physical and chemical nature of the parent magma at the time of eruption. Criteria include crystallographic properties, morphology, internal structures, and chemical composition. **Supplemental petrographic characteristics** are strongly influenced, for instance, by reworking or by winnowing and differential settling of the components during transportation. Criteria used include size of particles, relative abundance of phenocrysts, and the degree of alteration of glass and minerals. Thus essential characteristics assist in correlation, while supplemental characteristics provide information on the explosivity of eruptions, wind patterns, and other transportational and depositional histories.

The nature of particle-size distribution is of importance in the classification of volcanogenic sediments. Tephra may be

Bombs	—	greater than 64 mm in diameter
Lapilli	—	2 to 6 mm
Ash	—	less than 2 mm

Various studies have examined size changes relative to distance from the eruptive center. One of the earliest studies, on Crater Lake, Hekla, and other eruptions (Fig. 13-17), showed that median grain size generally decreases with increasing distance from the source [Fisher, 1964]. Sorting of fragments increases with increasing distance from the source because large fragments drop out close to the source, thereby increasing the total range of sizes closer to the source. A number of

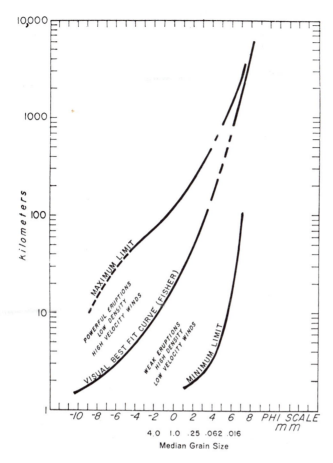

Figure 13–17 Empirically derived relationship between volcanic particle size and distance from volcanic eruptions with superimposed fields described in terms of eruption character, particle density, and wind strengths. (After Fisher, 1964, simplified by D. M. Shaw, N. D. Watkins, and T. C. Huang, *Jour. Geophys. Res.*, vol. 79, p. 3088, 1974, copyrighted by the American Geophysical Union)

variables control grain-size characteristics of a tephra layer, including viscosity and gas content of the magma and explosivity of eruptions (Fig. 13–18). These, in turn, control the size, shape, and density of fragments produced, as well as their angles and heights of projection. Other important variables include wind velocity, turbulence, air density, pressure, and humidity.

A wide range of mineralogical criteria have been used to mark different tephra layers. These include characteristics of glass, pumice, scoria, phenocrysts, zoned feldspars, quartz, pyroxene, hornblende, and heavy, opaque minerals, such as magnetite, as well as mineral color. Minerals associated with volcanic ash are also well sorted by the wind.

Source regions

Lisitzin [1972] distinguished three major types of ashfalls.

1. *Local.*

2. *Tropospheric.*

3. *Global.*

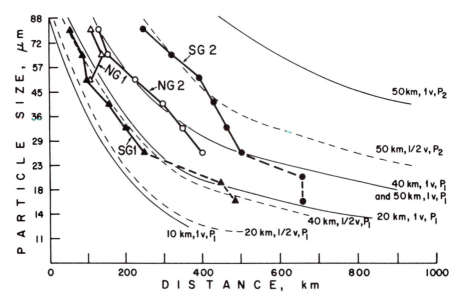

Figure 13-18 Particle size changes in volcanic ash of four separate volcanic eruptions preserved in piston cores at increasing distance from the Azores Islands (the source region). Graph shows the downwind accumulation—the rate maxima of eight selected size ranges. Thick lines with data points are the observed positions of the maxima for the four eruptions, NG1, SG1, NG2, and SG2. The theoretical lines (fine lines) are the computed maxima positions for four eruptive cloud heights: 10, 20, 40, and 50 km, assuming the average wind speeds are equal to (1 v) or half of (½ v—dashed lines) the present North Atlantic speeds. The theoretical curves are labeled P_1 and P_2 to indicate which of the computed maxima has been used. (From T. C. Huang, N. D. Watkins, and L. Wilson, *Geol. Soc. Amer. Bull.*, Part I, vol. 90, p. 132, 1979, courtesy The Geological Society of America)

The categories are largely determined by the explosivity of the eruptions, volume of material produced, and duration of the eruption. Wind strength is secondary in determining which type of ashfall deposit will result from a volcanic event.

Local ashfalls are deposited within a few hundred kilometers of the source. Material derived from these eruptions is generally diverse and poorly sorted. Heavy mineral concentrations decrease away from the source, and size-sorting by wind is typical (Figs. 13-17 and 13-18).

Tropospheric ashfalls are deposited a few hundred to several thousand kilometers from the source region. Ash may remain in the atmosphere from several days to 1 month. Sufficient volcanic explosivity is required to eject particles to heights of 5 to 12 km. These eruptions are geologically common phenomena, occurring about 5×10^5 times every million years. Clear size-sorting of particles is closely associated with explosivity and distance from the source (Fig. 13-18). Deposits at distances greater than 1000 km from the source are thin, of relatively uniform thickness, and composed of particles dominated by glass of size 20 to 1 μm, with an admixture of crystals. Deposits closer than 1000 km contain particles dominated by sizes in the range 20 to 100 μm (Fig. 13-18) [Huang and others, 1979].

Global ashfalls are formed from the settling of extremely fine-grained ash that has encircled the earth with an extremely long residence time (up to several years). Distribution is thus vast and meridional. These materials are deposited irrespective of their source. Volcanism that produces global ashfalls is important in distributing fine pelagic sediments to the deep sea. The median diameter is about 0.3 to 1 μm, and material coarser than 3 to 5 μm is rare. The fineness of the material makes it extremely difficult to distinguish from other sedimentary components.

After eruption there are various modes of transportation that can control the final distribution of tephra to marine sequences. These include wind, oceanic currents, ice-rafting, and redeposition by turbidity and other bottom currents.

Distribution of marine tephra deposits

Volcanogenic sediments are transported to the ocean by a number of processes: wind; subaerial and submarine pyroclastic flows and falls; subsequent subaerial or submarine erosion; streams; wind and ocean currents, including bottom and turbidity currents; and ice from polar regions including sea ice and icebergs.

Contributions from basaltic volcanoes are normally only locally distributed, so they are of limited use in tephrochronological investigations, while highly silicic volcanism is prone to pervasive distribution. The amount of volcanic material erupted from seamounts is unknown, but is governed by explosivity of the magma and the depth of eruption below sea level.

Global wind patterns are critical in both the direction and extent to which tephra is deposited from volcanic eruptions (Fig. 13–19). The major wind-distributed belts of tephra are apparent; there is a westerly belt in the zone of the westward-flowing tropospheric winds between 20°N and 20°S and easterly distributed belts at latitudes higher than 20°N and 20°S in the zones of prevailing eastward-flowing winds (westerlies). Prevailing winds in these belts concentrate most ashes in preferred directions (Fig. 13–19), although distributions opposite these patterns are important because of opposite wind directions at different levels in the atmosphere. Ash distribution patterns provide useful criteria for determining past changes in the latitudinal positions of boundaries between the major wind-belt systems.

In general, ash is deposited from an eruption column on all sides of a vent, but it is characteristically distributed on the leeward side if steady winds are blowing during the eruption (Fig. 13–20). Thus tephra accumulations tend to be elongated away from the source with the wind direction. If wind directions above the vent are similar at all elevations, the pattern of ash accumulation will be very elongated. If the pattern of wind direction varies with altitude, ash distribution may be irregular. If winds are light, no preferred orientation will occur and there will be a circular fallout area.

Volcanic ash is deep-sea sediments may be either in discrete layers or dispersed throughout other sediments. In contrast to dispersed ash, discrete ash layers clearly mark maxima in explosive volcanism in the source regions. In marine sequences east of New Zealand, distinct ash layers formed only when eruptions deposited material

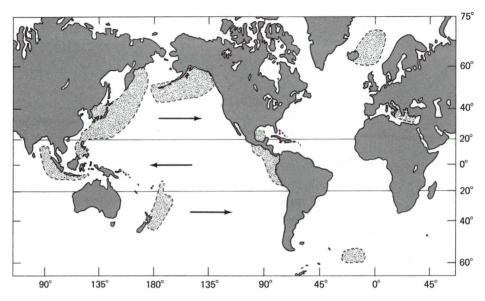

Figure 13-19 Generalized figure showing important areas of present-day abundant coarse volcanic glass distributions over the ocean floor. Distribution patterns are controlled by source regions and prevailing wind directions in the lower troposphere. Major changes in vertical wind patterns within the troposphere and between the troposphere and stratosphere have led to more complex patterns of distribution in certain areas not shown in this figure. (Courtesy D. Ninkovitch)

at a rate greater than 100 mg per 1000 years per square centimeter in the 88 to 11 μm fraction [Watkins and Huang, 1977]. Small volumes of material arriving on the sea floor that could potentially form thin ash layers are mixed with other sediment by bioturbation. Therefore any marine tephra sequence is certainly an incomplete record of explosive volcanicity in the source regions. Minor eruptions, while possibly of significance to the depositional history within about 100 km of the source, are generally not recorded in the marine realm, except as fine-grained disseminated material.

Glacial Sediments in the Ocean

The earth is a conspicuously glaciated planet. It currently has ice sheets in both polar regions: the immense ice sheets on Antarctica up to 4300 m thick and a much smaller ice sheet on Greenland. During the glacial episodes of the Quaternary, the ice sheets expanded over the Northern Hemisphere and played a major role on sediment processes in the oceans. The present interglacial (Holocene) is not typical of the latest Cenozoic, because glacial sediments have usually been much more widespread. Nevertheless, even during interglacials, given a permanent Antarctic ice sheet, glacial sediments dominate large areas of the ocean floor and are the dominant source material from one of six continents (Antarctica). During glacial episodes glacial sediments were the dominant sediment materials produced by two of the continents (Antarctica and North America) and one subcontinent

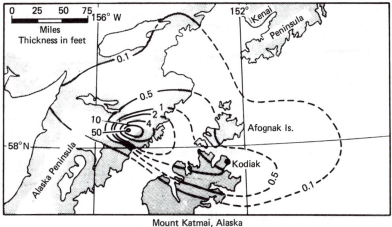

Mount Katmai, Alaska

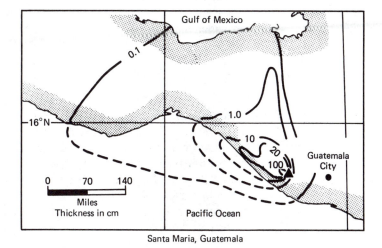

Santa Maria, Guatemala

Figure 13-20 Isopachs (in cm) of volcanic ash accumulations resulting from eruptions of Mount Katmai, Alaska, and Santa Maria, Guatemala. (From G. P. Eaton, *Jour Geol.*, pp. 14–15, 1964, reproduced by permission of The University of Chicago Press)

(Europe). One of the major processes of transporting terrigenous sediments to the oceans is from melting ice. Ice-rafting of sediment is an important depositional process around Antarctica, creating **glacial-marine sediments.**

Ice and icebergs

In Antarctica, the leading role of sediment transportation in ice is by icebergs. Sea ice is of almost no importance, since the wide development of ice shelves essentially precludes the capture of sediments in sea ice. Icebergs are generated largely by the immense ice shelves that encompass almost half of the Antarctic coastline. Annual ice discharge is about 1450 km³, most of which (about 60 percent) flows into and forms the ice shelves. Icebergs contain about 1.6 percent sedimentary material by volume, giving a total annual maximum discharge of between 35 to 50 × 10⁹ tons [Lisitzin, 1972]. The total volume of sediments contained in icebergs has changed as the evolution of glaciation and glacial erosion proceeded on the continent. For instance, it has been suggested that Antarctic glaciers today seem to play a protective role, rather than acting as agents of erosion. Earlier periods of more widespread erosional efficiency were gradually phased out when easily erodable material was no longer available, or when the topography of the continent became more subdued. Thus Quaternary glacial-sediment concentrations in the seas surrounding Antarctica do not reflect its long-term glacial conditions. The highest iceberg densities and the largest icebergs occur off the large ice shelves, especially those at the inner margins of the Ross and Weddell seas. The gyre circulation in both of these seas transports the icebergs northward into the Circumpolar Current. Iceberg density decreases rapidly northward and is negligible north of the Antarctic convergence, although ice-rafted clasts have been discovered on the ocean floor far to the north of present-day iceberg limits, indicating that the range of icebergs has, at times in the past, been wider than at present.

Glacial-marine sediments—Antarctica

In 1910, G. T. Philippi proposed the title *glazial marine ablagerungen* (glacial-marine sediments) for Antarctic marine sediments that contain an abundant silt fraction composed of rock flour, coarser, poorly sorted debris, little calcite, and little biogenic material. A random distribution of sand grains and pebbles in a clay matrix is indicative of deposition from icebergs. Glacial-marine sediments form a wide halo around Antarctica (Fig. 13–21) and are distributed in the Arctic and in smaller areas in the North Atlantic and North Pacific. The circum-Antarctic belt is about 300 to 1000 km wide and consists of sediments ranging from thin clay muds to boulder deposits. The northern limit of glacial-marine sediments is controlled approximately by the 0°C surface-water isotherm, which controls the rate of melting of icebergs. Glacial-marine sediments have been defined by Goodell and others [1973] as those with more than 30 percent sand and larger grain sizes and with a silt to clay ratio exceeding 1.0 and often exceeding 2.0. Such a classification is considered too restrictive by some as it excludes sediments of apparent glacial origin lying within the belt of glacial-marine sediments and exhibiting no evidence of postdepositional reworking. Goodell distinguishes four zones of glacial-marine sediments surrounding Antarctica, each distinguished on textural and genetic criteria and each merging latitudinally with the other zones. The southerly zone consists of undifferentiated coastal deposits of submarine glacial till, gravels, sands, and biogenic deposits. The tills are unchanged by marine processes and occupy the inner third of the Antarctic continental shelf. They have a coarse fraction with angular, faceted, and striated pebbles and granules. This zone merges northward into sand-silts, which are similar to sediments to the south but transitional between the coarser, more readily identifiable glacial-marine deposits of the outer continental shelf and slope and the pelagic clays of the abyssal floor. They reflect the

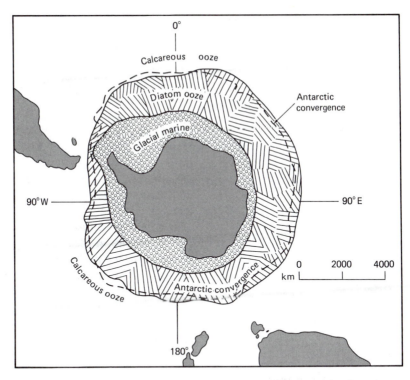

Figure 13-21 General circum-Antartic patterns of surface sediment distributions in the Southern Ocean. (After J. Hays, 1967)

decrease of coarse debris away from the continent. This zone merges northward into clay-silts and pelagic clays, which also contain occasional ice-rafted detritus. The silty clays, in turn, merge northward into biogenic deposits of either siliceous or calcareous oozes. Thus the rate of ice-rafted sedimentation and the coarseness of the sediments is closely related to the distance from Antarctica. Rates also depend on preferred paths of icebergs controlled by general circulation, storm tracks, and pack-ice distribution. Various schemes have been proposed to classify glacial-marine sediments. Such classifications are largely concerned with differentiating the relative role and proportions of direct ice-rafting, reworking, and normal sedimentation processes. The following classification is after Anderson and others [1977] and Kurtz and Anderson [1979].

ORTHOTILLS: Sediments deposited from grounded ice.
Texture—unsorted; boulder to clay-sized material; gravelly sandy muds.
Structure—massive, no clast fabric.
Fossils—not normally present, except reworked forms at times.
Distribution—restricted to continental shelf; may attain considerable thickness (to tens of meters).

PARATILLS: Sediments deposited from floating ice which are subsequently affected by marine agents. Two types exist:

Residual paratills—paratills winnowed by marine currents, or by bottom currents to form a residual lag deposit.

Texture—coarser than orthotills; generally lacking in fine-grained material; gravelly sands.

Structure—crude to well-developed stratification.

Distribution—occur on the continental margin with decreasing importance far from the ice front.

Compound paratills—paratills resulting from the combination of ice-rafting and normal hemipelagic sedimentation.

Texture—finer grained than orthotills; can be bimodal; pebbly or sandy muds.

Structure—crude to well-developed stratification.

Distribution—can occur anywhere within the zone of ice rafting.

Four main factors control the rate of ice-rafted sedimentation.

1. Rate of continental erosion.

2. Thermal structure of glaciers or ice shelves, and hence both their erosional and melting behavior.

3. Size of ice shelves.

4. Sea temperatures.

The importance and interaction of these processes is still problematic. Various models have been proposed [Anderson, 1972]. A **dry base ice shelf** should result in basal freezing and slower melting of icebergs (Fig. 13–22). Sediment would be deposited at greater distances from the source. **Wet-base ice shelves** would melt rapidly in the nearshore environment, losing most of their sedimentary load far back beneath the ice shelf where basal melting occurs (Fig. 13–23). The scarcity of ice-rafted detritus in the surface muds of the abyssal floor adjacent to the Weddell Sea suggests that glacial marine sedimentation is at a minimum in this area and that ice shelves in the eastern Weddell Sea are wet-base shelves. Figures 13–22 and 13–23 illustrate the changes in glacial-marine sedimentation in the ice-shelf environment occurring during a cycle of an advancing dry-base stage (possibly representing an early glacial stage) to a receded wet-base stage (possibly representing a late interglacial stage).

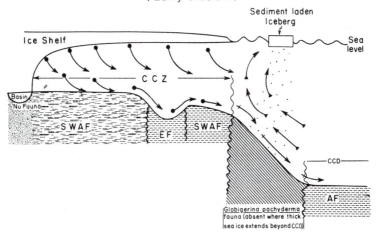

ADVANCING DRY BASE
(Early Glacial?)

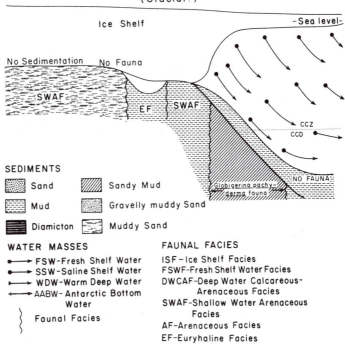

ADVANCED DRY BASE
(Glacial?)

SEDIMENTS

- Sand
- Mud
- Diamicton
- Sandy Mud
- Gravelly muddy Sand
- Muddy Sand

WATER MASSES
- FSW—Fresh Shelf Water
- SSW—Saline Shelf Water
- WDW—Warm Deep Water
- AABW—Antarctic Bottom Water
- Faunal Facies

FAUNAL FACIES
ISF—Ice Shelf Facies
FSWF—Fresh Shelf Water Facies
DWCAF—Deep Water Calcareous-Arenaceous Facies
SWAF—Shallow Water Arenaceous Facies
AF—Arenaceous Facies
EF—Euryhaline Facies

Figure 13-22 Schematic sections off Antarctica showing distribution of sediment facies and benthonic foraminiferal faunas as an ice shelf undergoes (a) advancing dry-base conditions and (b) advanced dry-base conditions. The CCZ is the region in which calcium carbonate is believed to be neither precipitated nor preserved. CCD is calcite compensation depth. (From J. B. Anderson, doctoral dissertation, Florida State University, 1972)

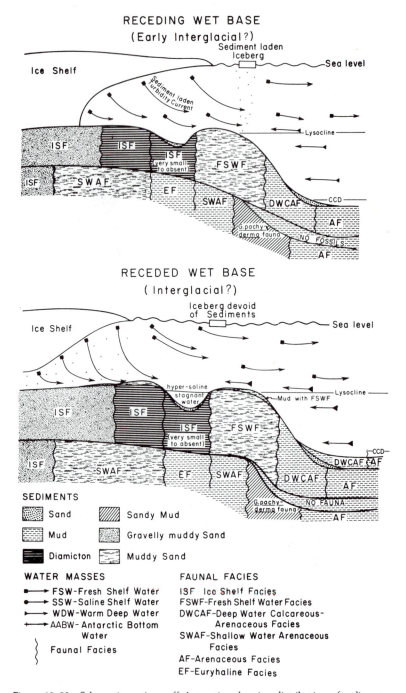

Figure 13-23 Schematic sections off Antarctica showing distribution of sediment facies and benthonic foraminiferal faunas as an ice shelf undergoes (a) receding wet-base conditions and (b) receded wet-base conditions. (From J. B. Anderson, doctoral dissertation, Florida State University, 1972)

Sedimentation from a freezing ice shelf is probably minimal because incorporated debris is not released in the cold, nearshore regions. Muds and muddy sands would dominate the continental shelf areas adjacent to freezing ice shelves. Basal melting of icebergs would begin at the outer limit of thick sea ice, with deposition of sandy muds occurring in this distal region. In the Weddell Sea, this region coincides with the continental slope. During the wet-base stage, basal melting of an ice shelf would result in rapid sedimentation of gravelly muddy sands near the ice shelf. In the geological record, the initial transition from dry- (cold) to wet- (warm) base conditions would result in pronounced calving of sediment-laden icebergs and reduced extent of thick sea ice by more rapid melting. This would appear as an increase in ice-rafted debris near the continent. It is not known what effect the nearshore debris would have on deep-sea sediments, but the northward extent of ice-rafting would be greater during dry-base conditions.

Glacial-marine sediments—Arctic

Because the Northern Hemisphere has lacked extensive ice sheets during the Holocene (last 10,000 years) and because the North Polar region is oceanic and surrounded by continents, the nature and distribution of glacial-marine sedimentation in the Arctic is somewhat different from the Antarctic. The sedimentation process in the Arctic Ocean is distinctive because transport of terrigenous and shallow-water sediment by surface currents and winds is inhibited by sea ice for most of the year. In the summer, when the sea ice breaks up, debris-laden ice islands, icebergs, and ice flows drift across the Arctic as dictated by prevailing currents and winds, distributing sediment to the sea floor as they melt. The near-absence of ice shelves surrounding the landmasses allows ice flows ready access to shallow-water sediments in the nearshore environment. Ice flows that scrape or freeze to the bottom pick up sediment, which then makes its way upward as surface ice melts and new ice accretes on the bottom of the flow.

Sediment frozen to pack ice near river mouths and coastlines also adds to Arctic sediment. River pack ice is broken up by spring floods before the nearshore pack ice melts; the rafted ice then carries large loads of sediment downstream. The sediment load is transported until the flows melt or overturn. Arctic sediments tend to be much finer (clays and silts) than in the Antarctic, reflecting the higher proportions of mud carried by ice. The Arctic sediment sources are largely permafrost areas, which produce fine-grained soils. Many major rivers flowing into the Arctic have very low gradients as they cross the normally wide coastal plain; the coarsest material transported is deposited on the shelf as the rivers rapidly lose competency. The deltas and nearshore areas trap most of the sandy and pebbly sediments.

Quaternary production of glacial-marine sediments

The distribution of Quaternary glacial-marine sediments shows when, where, and at what rates sediment dropped from melting ice into polar seas, how paleocirculation changed, and the character and size of ice sheets and shelves. Ruddiman [1977A] calculated the deposition rate of glacial-marine sediments in different oceans during the Quaternary to discern the changing role of deposition as ice-cap distributions have changed. His work showed that the subpolar North Atlantic (62 percent) and the circum-Antarctic (12 percent) store most (74 percent) of the world ice-rafted deposition. The remainder occurs in the Arctic Ocean (6 percent), the North Pacific (8 percent), and the Norwegian Sea (12 percent). The North Atlantic Ocean, at latitudes south of Iceland, receives roughly 60 percent of the world's ice-rafted sand. The low input from the Southern Hemisphere is surprising. It cannot reflect ice passage northward beyond the Antarctic seas and deposition at lower latitudes because the calculations cover both glacial and interglacial depositional maxima. It seems there is very little ice-rafted deposition of terrigenous debris for the Southern Ocean as a whole: Despite an area six times greater than the subpolar North Atlantic, the total input of continental detritus is five times less. Depositional rates are low, in part because Antarctica is a polar desert with generally low precipitation and the glaciers are probably mostly dry-base ice, carrying only small sediment loads. Also, ice cover on Antarctica during the late Quaternary largely prohibited the various kinds of subaerial weathering that provides continental detritus to fluctuating ice sheets. Sources of icebergs in the North Pacific were also minor during the Quaternary, resulting in low depositional rates of glacial-marine sediments.

Sediments of Extraterrestrial Sources

An insignificant fraction of deep-sea sediments are made up of micrometeorites, which continually shower the earth's atmosphere. They are thus separate from terrigenous sediments. These are generally conspicuous only in deep-sea sediments of very low rates of sedimentation, particularly the brown clays. Micrometeorites are represented by small, glassy bodies called **microtektites.** They were first discovered by Billy Glass in 1967 in Indian Ocean sediments adjacent to the previously well-known *tektite* field of western Australia. Microtektites are generally 1 mm to about 30 μm in diameter, of varying shapes (Fig. 13–24), such as ovoid, teardrop, or dumbbell shaped, and of various colors, but commonly yellow brown to brown. They exhibit a wide variety of surface features, from glassy smooth to rough, with irregular pitting. Most are smooth with shallow pits. Accumulation rates of micrometeorites have been estimated at about 0.00002 mm per 1000 years, though extraterrestrial material occasionally occurs in much greater abundance in well-defined levels in deep-sea sediment sequences. More material is particularly important, as it possibly indicates catastrophies.

Tektites are small, generally 2 to 4 cm in diameter, glass bodies ranging in color from black to translucent green; they are found in several localities, referred to as **strewn fields,** on the earth's surface. Tektites are similar to obsidian but can be distinguished from these terrestrially formed igneous glasses by their distinct geochemistry and a general lack of crystalline implosions. There are four known tektite strewn fields (Fig. 13–25): the Australasian (Australia, Indonesia, and northern Philippines), 0.69 Ma; the Ivory Coast of Africa, 1.1 Ma; the Czechoslovakian, 14.7 Ma; and the North American (Texas - Georgia), 35 Ma.

Microtektites have been found in deep-sea sediments adjacent to three of the

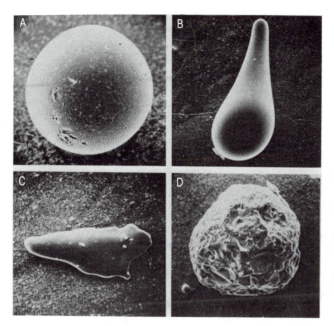

Figure 13-24 Scanning electron micrographs of microtektites from the Australasian strewnfield. (A) Spherule is 470 μm in diameter. (B) Teardrop is 880 μm long. (C) Irregular specimen is 535 μm long. (D) Form is 250 μm long with irregular surface. (From B. P. Glass, *Antarctic Oceanology II The Australian–New Zealand Sector*, p. 336, 1972, copyrighted by the American Geophysical Union)

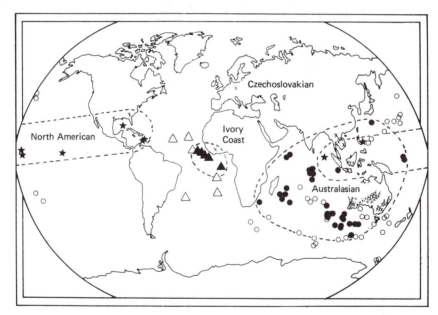

Figure 13-25 The Australasian, Ivory Coast, and North American microtektite strewnfields. The location of cores containing Australasian, Ivory Coast, and North American microtektites are indicated by solid circles, solid triangles, and stars, respectively. Cores in which Australasian and Ivory Coast microtektites were not found are indicated by open circles and open triangles, respectively. Tektite locations on land are indicated by X's. Boundaries of strewnfields (indicated by dashed lines) are drawn to include all known tektite and microtektite occurrences for each strewnfield. (From Glass, Swinki, and Zwart, 1979)

Terrigenous Deep-Sea Sediments **451**

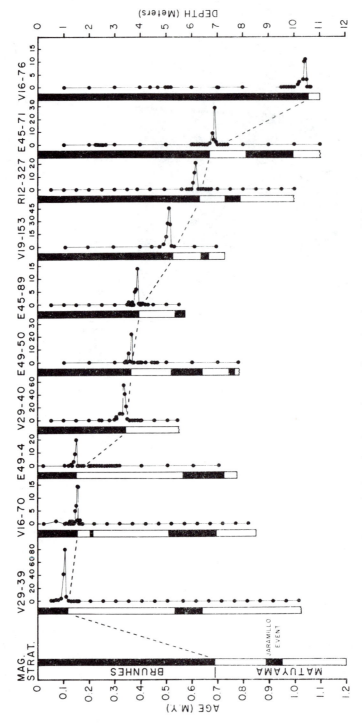

Figure 13-26 Absolute abundance of microtektites (number per 8 cm³ sample) from Australasian strewnfield in deep-sea cores correlated with paleomagnetic stratigraphy. Dashed line shows correlation of Brunhes/Matuyama boundary. Note that peak in microtektite abundance is always within 30 cm of Brunhes/Matuyama boundary. (From Glass, Swinki, and Zwart, 1979)

known tektite strewn fields: Australasian, Ivory Coast, and North American (Fig. 13–25). Identification of microtektites and association with a known strewn field is based on geographic location, age, general appearance, petrography, physical properties, and chemical composition. In eastern Indian Ocean cores, they are generally concentrated in a layer at least 20 to 40 cm thick (Fig. 13–26). The thickness of the layer reflects the degree of reworking by bottom animals since deposition. In fact, the stratigraphic distribution, of microtektites have served as valuable markers for estimating the extent of reworking by bioturbation (see Chapter 14). Microtektite layers record geologically instant events and thus form chronostratigraphic layers in deep-sea sediments (Fig. 13–26).

The recognition of microtektites in deep-sea sediments has greatly extended the geographic limits and revised upward the estimates of total masses of tektites in the different strewn fields. Microtektites also have been useful in geochemical analysis, showing greater range of chemical composition than the larger tektites. Their stratigraphy is better defined in deep-sea sequences, providing more precise ages of the strewn fields. Microtektites are dated directly using the fission-track method and indirectly using magnetostratigraphy and biostratigraphy. The Australasian and Ivory Coast microtektite layers coincide with the Brunhes-Matuyama Paleomagnetic reversal boundary (0.69 Ma; Fig. 13–26) and the Jaramillo Paleomagnetic event (1.1 Ma), respectively, and have assisted in confirming the ages of these events through fission-track dating. It is not known if the magnetic reversals and the tektite origins are connected, although speculations have been presented by Glass and Heezen. Most investigators believe that microtektites have resulted from high-velocity meteorite impacts on the earth's surface. Other theories credit their origin to disintegration in the upper atmosphere of a large meteoritic body. Or a meteorite may have exploded in the outer atmosphere, so that the glassy debris was first cooled and then reheated in its passage through the lower atmosphere, producing aerodynamically eroded tektites like those found in Australia.

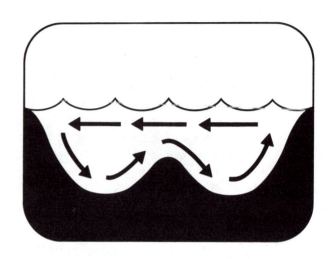

fourteen
Biogenic
and Authigenic
Oceanic Sediments

Into a melody of deepblue;
Shells drifting downward
To sink and settle on the seafloor.
An earth song of creation
Kneads them into stone,
Draws aside the waves
And lifts them into mountains.

Nancy Penrose

BIOGENIC SEDIMENTS

Biogenic sediments are those which are formed primarily from the remains of marine organisms. Although marine plants and animals are incredibly diverse, only a small number of groups produce hard parts capable of contributing to sediments and, of these, even fewer produce specimens in large enough quantities to form major sediment types.

On the continental shelf, calcareous sediments are formed by the accumulations of the skeletons of three or four groups of megafossils: corals and calcareous algae form *coralgal reef limestones,* mollusks form *coquina limestones,* and bryozoa form *bryozoan limestones.* Microfossils do not generally occur in large enough numbers on the shelf to form sediments. They are simply too small to overcome the large dilution factor of terrigenous sediments or the megafossils that often have large volumes per specimen. Though some groups, including Foraminifera, can add substantially to the volume of shallow-water sediments, no major siliceous biogenic deposits are known from the continental shelves or inshore environments.

By far the most widespread and voluminous oceanic biogenic surface sediments are the **biogenic oozes.** Oozes are pelagic sediments of at least 30 percent skeletal remains of pelagic organisms, the remainder being clay minerals. *Carbonate* or *calcareous* ooze is made up of microfossils with tests of calcium carbonate; *siliceous ooze* is siliceous microfossil material. Oozes are further subdivided by their characteristic organism. The term *ooze* was first coined during the *Challenger* expedition to describe this sediment because of its soft, soupy character.

455

Four major processes control the final character of biogenic oozes: *supply* of biogenic material, *dissolution* of biogenic material, *dilution* of the sediment by non-biogenic components, and *diagenetic alteration* of the ooze.

The supply of biogenic material (carbonate and silica) is controlled by the relative fertility of the organisms producing the material which, in turn, is related to the availability of nutrients needed for growth. Primary production depends upon both sunlight and nutrients. In the photic zone, large areas of the ocean are marked by low levels of nutrients, because most supplies have already been used by the existing phytoplankton. Nutrients are transported to intermediate water depths as dead organisms or the products of organisms. These nutrients are not thoroughly used at greater depths because of the lack of sunlight. As a result, nutrients build up. The permanent thermocline exists between nutrient-rich intermediate depths and nutrient-poor surface waters (see Chapter 8). This thermocline prevents the efficient replenishment of nutrients back to the photic zone. Areas in the oceans where the thermocline is shallow or indistinct are richer in nutrients at or near the surface. These include zones of coastal upwelling and oceanic divergence where mixing takes place. These areas are often marked by the highest rates of formation of biogenic oozes. The central gyre regions of the oceans are the least mixed and possess a more permanent thermocline. These are the areas of lowest biological productivity and biogenic sediment formation.

Although primary productivity is important for the formation and character of oozes, other factors are also important. Extensive dilution of the biogenic component, even when high, takes place on the continental margins because of large terrigenous sediment input. Continental-margin (slope and rise) deposits are often mixtures of terrigenous clays and biogenic material—the **hemipelagic sediments.** In these sediments, the biogenic component is masked; a complete gradation between hemipelagic sediments and the biogenic oozes begins only at greater depths on the continental slope or where terrigenous sediments begin to decrease in importance.

The most important controlling factor of the distribution and character of oozes is its preservation. A vast majority of the biogenic material produced in the surface waters of the ocean is dissolved before it even forms sediment. Most of the dissolution of silica takes place in highly silica-undersaturated surface waters (Fig. 14–1). The solubility of silica decreases with increasing pressure and decreasing temperature. Therefore, although silica is undersaturated everywhere in the oceans, it is less undersaturated in deep water. In contrast, carbonate dissolution increases with depth (Fig. 14–1) as bottom waters become more undersaturated in calcium carbonate. Therefore carbonate biogenic distribution is largely a preservation signal. Carbonate and silica biogenic patterns reflect fundamentally different processes of formation.

Siliceous biogenic sediments reflect the biological fertility of surface waters. Because surface waters are undersaturated in silica, increased nutrient and silica supply by upwelling and mixing increases the productivity of siliceous organisms which, in turn, helps counteract the tendency to dissolve and recycle siliceous organisms immediately after death. Increased preservation leads to an accumulation of the siliceous biogenic component in the underlying sediments.

Carbonate biogenic sediments reflect the preservation of calcium carbonate at

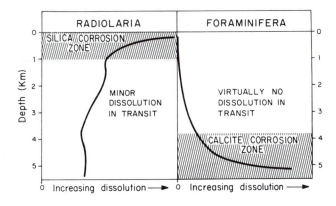

Figure 14-1 Comparison of dissolution profiles of radiolaria and planktonic foraminifera based on field experiments. Most dissolution of radiolaria (and diatoms) occurs in shallow waters. In contrast, most dissolution of carbonate microfossils occurs on the sea floor at depths greater than 3.5 km. (From W. H. Berger, 1976)

depth. Because most surface waters are saturated with calcium carbonate, there is no tendency for dissolution in upper waters immediately after death of the organisms. However, the subsurface waters become increasingly undersaturated in calcium carbonate. Increased calcium carbonate solution at depth is the primary factor controlling carbonate distribution and is largely independent of the amount of biologic productivity occurring in surface waters.

Major differences between the oceans occur in the proportions of siliceous and carbonate-rich sediments. The Atlantic has widespread carbonate oozes, while the Pacific has widespread siliceous oozes; the Indian Ocean has a combination of both (Table 14-1). The character of each ocean results from differences in bottom-water circulation, which controls both dissolution patterns and, through upwelling, the patterns of fertility throughout the oceans. The deep-sea circulation patterns have led to a fractionation process of silica and carbonate, which occurs between the ocean basins. This is called **basin-to-basin fractionation** and was largely developed by Wolfgang Berger, whose studies over the last decade have provided much understanding of the formation of deep-sea biogenic oozes.

The North Atlantic is marked by an *antiestuarine,* or lagoonal-type, *circulation,* while the North Pacific has an *estuarine circulation* [Berger, 1970A; 1974]. In

TABLE 14-1 PERCENTAGE OF PELAGIC SEDIMENT COVERAGE IN THE WORLD OCEANS[a]

	Atlantic	Pacific	Indian	Total
Foraminiferal ooze	65	36	54	47
Pteropod ooze	2	0.1	—	0.5
Diatom ooze	7	10	20	12
Radiolarian ooze	—	5	0.5	3
Brown clay	26	49	25	38
Relative size of ocean (%)	23	53	24	100

[a]From Berger [1976]. Data from Sverdrup and others [1942]. Pacific pteropod ooze area from Bezrukov [1970]. Area of deep-sea floor = 268.1 × 10^6 km^2.

Biogenic and Authigenic Oceanic Sediments   **457**

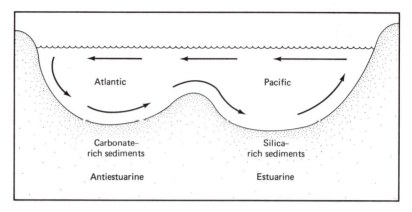

Figure 14-2 Concept of antiestuarine-estuarine circulation system between the Atlantic and Pacific oceans. The Atlantic forms young oxygenated bottom waters at high latitudes allowing carbonate to accumulate over most parts of the ocean basin. The Pacific receives older deep water at depth and gives up surface waters (like an estuary). Carbonate-poor sediments accumulate over much of the North Pacific since older deep waters are rich in CO_2 and thus are more acidic. Silica-rich sediments accumulate in areas of upwelling.

antiestuarine circulation, deep water flows outward and is exchanged with shallow water flowing into the system (Fig. 14-2). A system which exchanges deep water for shallow water becomes depleted in nutrients, while basins that exchange shallow water for deep water become enhanced in nutrients by increased upwelling at the surface. Typical antiestuarine systems are driven by excess evaporation, so that waters are highly saline. Bottom waters in these basins are young and well oxygenated, so they tend to be close to calcium carbonate saturation. The North Atlantic is such a basin. In contrast, in estuarine circulation shallow water is exchanged for deep water, increasing the upwelling of nutrient-rich waters. This process favors higher fertility, which leads to greater siliceous biogenic production and preservation. On the other hand, the bottom waters are further from the areas where they were formed and, because of their greater age, are enriched in CO_2 (through metabolism of deep-sea organisms). These waters tend to be undersaturated with calcite throughout the water column, leading to increased calcium carbonate dissolution. Thus estuarine basins, such as the North Pacific, exhibit both production and preservation characteristics that favor deposition of siliceous biogenic sediments. Over extended intervals of geologic time, links between different ocean basins change, leading to major circulation differences of both deep and shallow waters. As a result, past basin-to-basin fractionation has differed from the present. The distribution and nature of biogenic oozes helps provide clues to such changes in the past.

Carbonate Oozes

Carbonate oozes cover about 50 percent, or 140×10^6 km^2, of the ocean floor and account for about 67 percent of calcium carbonate in surface sediments of the ocean. Shallow-water carbonates occurring in modern reefs and on continental shelves account for about 9 percent of carbonate deposition, while those on the continental slope account for 24 percent. Distribution patterns of carbonate in the ocean have been described since Murray and Renard discovered in 1873 that carbonate was largely absent below depths of about 4500 m and that this was caused by selective dissolution of the carbonate. Carbonate is thus draped over topographic highs in the ocean basins; if all the water were removed, it would be easy to observe the off-white carbonate oozes covering elevated regions like snow on mountains. For this reason, some authors have referred to the level at which carbonate disappears from surface sediments at depth as the **snow line.** The snow line is normally called the **calcite compensation depth** (CCD) and is that level in the ocean where the rate of dissolution of carbonate balances the rate of accumulation. In most regions the boundary is marked by a transition from carbonate ooze to brown clay, although there may be a transition into siliceous ooze in some areas. As with snow lines on land, the slopes of the CCD are different between ocean basins; the reasons for this are described later.

Carbonate sediments in the deep sea are the primary record of the geochemical cycles of calcium and carbon dioxide on the earth's surface (atmosphere, hydrosphere, and lithosphere). The oceans act as a primary reservoir for calcium and carbonate. Exchange of CO_2 across the air-sea interface, river input of bicarbonate and calcium, and removal of calcium carbonate by precipitation are important elements within these cycles. Ultimately, the amount of calcium carbonate deposited is limited by the input of calcium carbonate into the oceans (Fig. 14–3). Thus the average rate of sedimentation of calcium carbonate (1 cm/yr) is a reflection of the rate of weathering of continental rock and of supply from hydrothermal vents on the mid-ocean ridges.

If it is assumed that the oceans are in a steady state, the amount of calcium carbonate deposited annually should equal the amount of new material brought into the oceans. This transportation to the oceans is almost entirely by river water

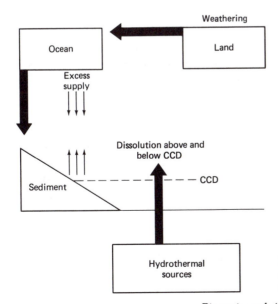

Figure 14–3 Major components of the calcium carbonate system in the ocean. (After A.T.S. Ramsay, 1974)

inflow and the hydrothermal activity of the mid-ocean ridges (see Chapter 7), which is 0.11 g/cm^2 per 1000 years [Edmond and others, 1979]. The rate of deposition of calcium carbonate by organisms is much greater (1.3 g/cm^2 per 1000 years) [Berger, 1976] than the river input. Therefore 90 percent of the deposited calcium carbonate must dissolve. This dissolution is manifested by the CCD (Fig. 14–3). If, for any reason, increased carbonate deposition occurs in shallow waters or if change occurs in the amount of carbonate being brought into the oceans by rivers or hydrothermal activity, the calcium-carbonate budget will change, causing a shift in the depth of the CCD. Large shifts of up to several thousand meters of the CCD have occurred during the late Phanerozoic.

There are two major and one minor types of carbonate ooze. These largely reflect the amount of dissolution of the original biogenic material.

Foraminiferal ooze is dominated by tests of planktonic foraminifera. These were called *Globigerina ooze* by Murray and Renard in 1873; this is a misnomer because these oozes consist of several other foraminiferal genera besides *Globigerina*. Foraminifera are largely of sand-size fraction (greater than 61 μm), although many foraminiferal oozes exhibit bimodal size distributions because of the presence of a larger foraminiferal-size component and a much finer component of calcareous nannofossils (less than about 30 μm).

Nannofossil ooze, or coccolith ooze, is dominated by calcareous nannofossils of the recent algae family Coccolithophoridae and the extinct group called discoasters. Because calcareous nannofossils tend to be slightly more resistant to dissolution than planktonic foraminifera, nanno oozes are often found at slightly greater depths, immediately above the CCD. These are fine-grained oozes, although there is often a coarse-grained planktonic foraminiferal component.

Pteropod ooze is dominated by pteropods and heteropods, which are planktonic mollusks, slightly larger than most planktonic foraminifera and which construct their shells of aragonite. Tests constructed of aragonite are easily destroyed by dissolution. Pteropod oozes are restricted to shallow tropical areas at depths less than 3000 m in the Atlantic and even shallower in the Pacific. The calcite supersaturated shallow waters of the tropics are favorable for pteropod preservation. On the Bermuda platform, pteropod ooze is best developed at depths close to 2000 m. These oozes are particularly rich in carbonate (greater than 95 percent) because much of the original carbonate fraction remains preserved.

Distribution of carbonate oozes

The distribution of carbonate oozes in the ocean basins is not uniform. Many secondary factors control the apportionment of the carbonate sediments and their preservation within ocean basins. A number of workers have mapped the distribution of the percentage of calcium carbonate in surface sediments of the oceans (Fig. 14–4).

Atlantic Ocean   The simple bathymetric control on the distribution of calcium carbonate in the Atlantic is evident in Fig. 14–4. Carbonate percentages are highest on the mid-Atlantic ridge and associated rises and plateaus, such as the Walvis

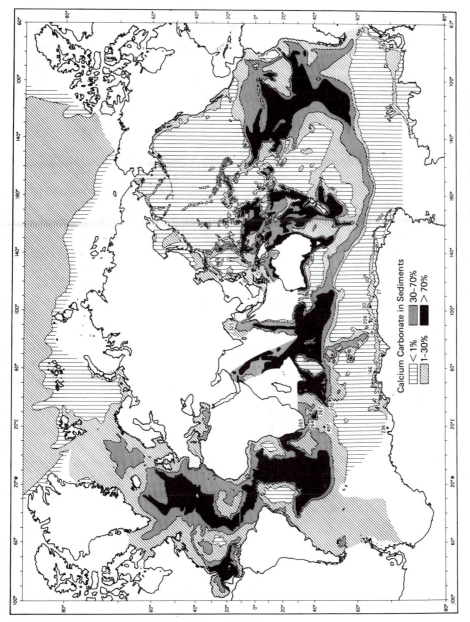

Figure 14-4 Calcium carbonate distribution in surface sediments of the world ocean (percent of dry sediment). (From A. P. Lisitzin, *The Micropaleontology of Oceans*, p. 204, 1971, Cambridge University Press)

Calcium Carbonate in Sediments

< 1%
1-30%
30-70%
> 70%

ridge, the Rio Grande rise, and the Falkland Plateau. Nevertheless diverse sources perturbate this signal, the most important being dilution by terrigenous sediments and bottom-water distribution patterns. Dilution by shallow-water detrital material decreases at more remote sites. Further dilution takes place at greater depths in some areas by detrital, sediment-laden bottom currents. The Atlantic is the most carbonate-rich ocean (Table 14–1). A lack of marginal trenches provides ready access to the deep basins for continental material. The North Atlantic is more carbonate-rich than the South Atlantic and the Southeast Atlantic is more carbonate-rich than the Southwest Atlantic (Fig. 14–4). Dilution is by terrigenous debris, which is derived from several areas. Surface sediments of the South Atlantic south of about 45°S are low in carbonate, which is typical for all sectors of the Antarctic Ocean.

Pacific Ocean   The Pacific Ocean is not a carbonate-rich ocean (Fig. 14–4; Table 14–1). Areas of high percentages of carbonate include the East Pacific rise and other elevated areas, such as the Chile, Carnegie, and Cocos ridges in the Southeast and equatorial Pacific. High percentages of carbonates in the Southwest Pacific are associated with the Lord Howe rise and the New Zealand platform. In the North Pacific, the only modern carbonate-rich province is in the far western sector, to the east of New Guinea (Fig. 14–4) and in the Philippine Sea. Vast areas of the Pacific Ocean are marked by very low concentrations of calcium carbonate. Instead the basins are dominated by extensive brown clay and siliceous ooze deposits. The South Pacific, north of 50°S, is marked by much more extensive areas of carbonate sediment than the North Pacific.

Indian Ocean   The Indian Ocean is intermediate between the carbonate-rich Atlantic Ocean and the carbonate-poor Pacific Ocean (Fig. 14–4; Table 14–1). Carbonate is abundant (greater than 75 percent) on topographic highs but occurs in low concentrations (less than 10 percent) in the basins. Dilution by terrigenous debris is conspicuous in the Bay of Bengal and the Arabian Sea adjacent to the Indian subcontinent and on the Mozambique basin off Southeast Africa. Kolla and others [1976] define a **carbonate critical depth** (CCRD) as the level below which carbonate is less than 10 percent. This level is, by definition, systematically shallower than the CCD, where carbonate is essentially absent. They prefer to use the CCRD because it is easier to identify analytically. The CCRD is deepest in the equatorial region, becoming shallower toward the south and reaching its shallowest depths (3900 m) between 50 and 60°S.

Production of pelagic carbonate

A tenfold variation in supply rates to the sea floor is apparent throughout the ocean basins. Carbonate production is greatest in surface waters having high biological productivity. These areas of upwelling include the polar regions (Fig. 14–5), the equatorial divergence, certain coastal areas, such as off West Africa, and areas peripheral to the major gyres of oceanic circulation. Lowest production occurs in central gyres like the Sargasso Sea (Fig. 14–5). Within each oceanic realm,

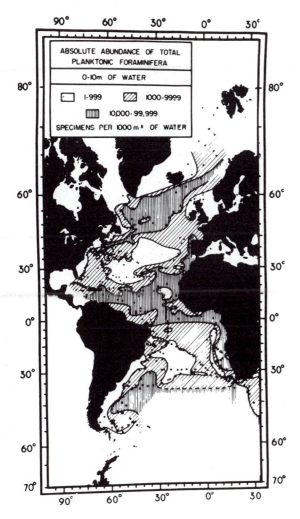

Figure 14–5 Absolute abundance of planktonic foraminifera (specimens per 1000 m³ of water) in the upper 10 m of the water column of the Atlantic Ocean. (After Bé and Tolderlund, 1971)

concentrations of planktonic foraminifera are at least ten times greater in the fertile, high-latitude coastal and equatorial regions than in the gyres. Seasonal variations in productivity are very conspicuous, so the sediment record is an averaging of long-period seasonal variation and patchiness in the living plankton. Calcareous nannofossils are more important in areas of low productivity than foraminifera, and foraminiferal oozes are more important where productivity is high. Although coccoliths reproduce much more rapidly than foraminifera, they are much smaller so sediment accumulates more slowly, especially where most of the foraminifera are dissolved.

Reported concentrations of planktonic foraminifera vary from 0 to 100×10^6 specimens per 1000 m³ [Bé and Tolderlund, 1971]. Highest reported concentrations in the North Pacific are about 1000 times as great as the lowest ones in the Sargasso Sea. Total abundances in the Atlantic are about half those for the Indo-Pacific, yet

Biogenic and Authigenic Oceanic Sediments   **463**

carbonate accumulation rates are less than half in the Pacific, demonstrating the importance of preservation in controlling carbonate distribution. Rates of productivity and accumulation are based on about ten planktonic foraminifera per cubic meter of seawater, which is an approximate world average [Berger, 1976]. If a reproducing assemblage is 2000/m^2 in the upper 200 m of the water column, then there are 200,000 specimens/m^2/yr or about 20 tests/cm^2 of ocean bottom per year. One test/cm^2/yr would provide a sedimentation rate of about 1 mm of foraminiferal ooze per 1000 years. Twenty tests per year should therefore produce a sedimentation rate of about 2 cm per 1000 years assuming no dissolution of carbonate. An average rate of calcium carbonate supply to sediments is 2g/cm^2 per 1000 years.

Carbonate oozes generally are deposited at rates of about 1 to 3 cm per 1000 years. This is the traditional method of expressing rates, but they are sometimes better expressed in g/cm^2/yr.

Production of coccoliths in the photic layer is variable according to season, local nutrient supply, and other hydrographic factors. In equatorial central Pacific waters, where the seasonal fluctuation is small, annual production is estimated as 7 to 22×10^{12} m^2 of coccoliths (typically 8 g/m^2 per year) [Honjo, 1976]. Growth rate of phytoplankton is estimated to be about 0.2 to 0.3 doublings per day in the surface waters of the central gyre of the North Pacific. Conservatively, this growth rate implies turnover rates of coccolith assemblages every 4 to 10 days in temperate to tropical water.

Large areas of the Pacific and Indian oceans have sedimentation rates less than 1 cm per 1000 years. The deepest parts of the ocean basins, isolated from the influence of turbidity currents, often have the lowest sedimentation rate and red-clay deposits. The carbonate or siliceous-rich facies over much of the shallower Atlantic, on the mid-ocean ridges, and in high biogenically productive areas in the high latitudes reflect intermediate sedimentation rates (1 to 3 cm per 1000 years). Highest sedimentation rates occur on the continental margins and inland seas because of continental input, in the abyssal plains due to turbidity currents, and in the North Atlantic as a result of glacial erosion and transport from North American and European glacial ice caps. The actual sedimentation rates vary considerably, for example, 10 cm per 1000 years in the turbidite sequences of the Aleutian abyssal plain and the pelagic sequences of the Gulf of Mexico to well over 20 to 30 cm per 1000 years in the continental margin and borderland provinces. Sedimentation rates of sequences are a principal factor is determining the amount of historical resolution obtainable from a sequence.

Dissolution of calcium carbonate

The major factor modifying the productivity pattern of deposition is the dissolution of carbonate microfossils. Larger specimens of planktonic foraminifera (greater than 100 μm) undergo little dissolution before deposition on the sea floor at typical oceanic depths (4000–4500 m). Thus most modification of the original plankton assemblage in the coarser carbonate particles occurs at the sediment-seawater interface. This also seems to be the case with the smaller carbonate microfossils, which are transported rapidly to the sea floor within fecal pellets. Dissolution on the sea floor is slowed as new sediment is accumulated, producing a diffusion barrier of increasing thickness separating the carbonate from the dissolving solution.

Early predictions based on laboratory experiments and thermodynamic theory indicate that the ocean should be undersaturated with calcium carbonate at all depths below the upper few hundred meters (Fig. 14-6). These predictions were later verified by field experiments in the central Pacific region, in which calcite spheres and foraminiferal assemblages were suspended on moorings and the weight loss resulting from dissolution of material was measured [Peterson, 1966]. These experiments showed that the transition from calcite supersaturation to undersaturation occurs within several hundred meters of the surface; seawater is undersaturated at all depths beneath this (Fig. 14-7). However, a sharp increase in the rate of

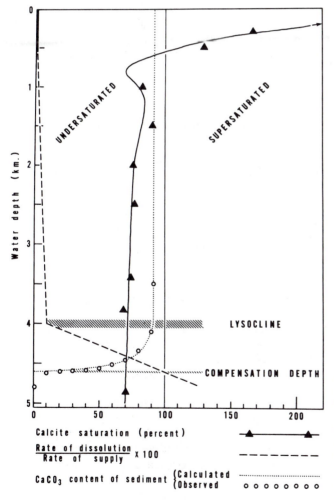

Figure 14-6 Parameters influencing the distribution of calcium carbonate with increasing water depth in equatorial Pacific sediment. (From Tj. H. van Andel, G. R. Heath, and T. C. Moore, Jr., *Cenozoic History and Paleoceanography of the Central Equatorial Pacific Ocean*, GSA Memoir 143, p. 40, 1975, courtesy The Geological Society of America)

Calcite saturation (percent)

$\frac{\text{Rate of dissolution}}{\text{Rate of supply}} \times 100$

CaCO$_3$ content of sediment $\begin{cases} \text{Calculated} \\ \text{Observed} \end{cases}$

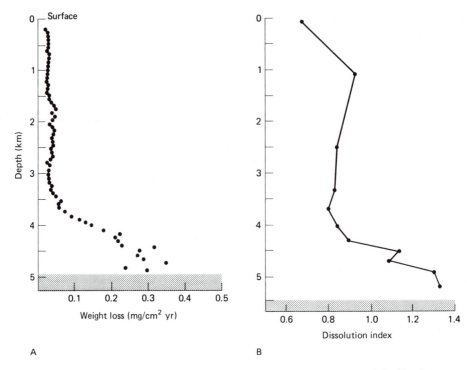

Figure 14-7 Dissolution curves of calcium carbonate in the oceans. (A) Profile of dissolution rate of calcite on a moored buoy in the central Pacific demonstrating the existence of a level of rapid increase of dissolution near 3700 m, as shown by Peterson [1966]. (B) Profile of preservation aspect ("dissolution index") of foraminifera in the central Atlantic showing an abrupt transition from well-preserved to poorly preserved foraminiferal assemblages at nearly 4500 m (foraminiferal lysocline). (After W. H. Berger, 1975)

dissolution occurs in the water column at a depth of about 3700 m (Fig. 14–7), which corresponds closely to depths (3700–4000 m) at which calcite is rapidly removed from the sediments through dissolution. Thus although waters are undersaturated below the ocean surface, effective dissolution of foraminifera begins at much greater depths (about 3700–4500 m) in this and other areas [Berger, 1975; van Andel and others, 1975].

Dissolution results from corrosiveness of the water, which increases with decrease in carbonate-ion content, low temperature, high hydrostatic pressure, increased water flow through the sediment, and high partial pressure of CO_2. Carbon dioxide in the water makes carbonic acid, which dissolves the calcium carbonate:

$$CO_2 + H_2O + CaCO_3 \rightleftharpoons Ca^{2+} + 2\ HCO_3^-$$

Carbon dioxide is produced by the respiration of benthonic organisms. Higher partial pressures develop as the water remains close to the sources of respiration products and to areas where organic materials have been oxidized, close to the ocean floor.

466 Biogenic and Authigenic Oceanic Sediments

At least three regionally varying levels related to calcium-carbonate preservation in deep-sea sediments have been recognized. From shallowest to deepest they are as follows:

1. *Lysocline.* The depth which separates well-preserved from poorly preserved assemblages of planktonic foraminifera (foraminiferal lysocline), pteropods (pteropod lysocline), and coccoliths (coccolith lysocline).

2. *Carbonate critical depth (CCRD).* The level below which calcium carbonate forms less than 10 percent of the sediment.

3. *Calcite compensation depth (CCD):* The depth above which carbonate-rich sediments accumulate and below which carbonate-free sediments accumulate (Fig. 14–6).

The two levels most widely recognized by marine sedimentologists are the CCD and the lysocline (Fig. 14–6). Because the CCD is a simple measure of the percentage of calcium carbonate in a sediment, this level will be discussed first even though it represents the final stage of carbonate removal from sediments. The concept of the lysocline is based on the observation of the preservational character of microfossil assemblages.

Calcite compensation depth

The CCD has a mean depth of about 4.5 km, located at about mid-distance between the crests of the mid-ocean ridges and the deepest parts of the ocean basins (other than the trenches). The CCD lies in the region of about 50 percent saturation of CO_3^{2-}, but the meaning of this is not understood. Large differences in the CCD level occur between the Atlantic and Pacific oceans (Fig. 14–8) [Berger and Winterer, 1974]. In the Pacific, the CCD is typically at shallow depths of about 4200 to 4500 m (Fig. 14–8), whereas in most of the North Atlantic and part of the South Atlantic, it is at or deeper than about 5000 m (Fig. 14–8). The shoaling of the CCD in the Pacific results from the greater age and CO_2 content, thus accounting for the increased corrosiveness of bottom waters. The CCD is significantly depressed to about 5000 m in the equatorial Pacific zone of high productivity (Fig. 14–8), except in the eastern part. The depth increase is a result of increased biological productivity and supply of calcareous biogenic material to the ocean floor. There is a rise in the CCD toward the continental areas (Fig. 14–8) as organic productivity—in a form that tends to increase CO_2 in bottom waters—increases, producing carbonic acid and increasing dissolution. Thus in areas of upwelling and increased productivity distant from continents, the CCD is depressed, while the CCD becomes shallower in areas close to continents where productivity is increased.

The depth of the CCD and the sharpness of the transition between calcareous and noncalcareous deposits is a function of three main variables: the depth of the abyssal thermocline; the rate of increase of dissolution with depth; and the rate of supply of carbonate and noncarbonate material to the sediment. The influence of

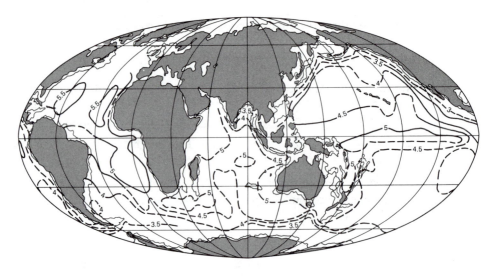

Topography of calcite compensation depth

Figure 14-8 Topography of calcite compensation depth (CCD) in the world ocean in km. (After W. H. Berger and E. L. Winterer, 1974)

bottom-water circulation is particularly important and is well illustrated by the calcium carbonate distribution in the South Atlantic [Turekian, 1964; 1965]. In this area, the CCD varies from basin to basin [Ellis and Moore, 1973]. In the Brazil and northern Argentine basins, it lies at 4800 m, in the Cape basin at 5100 m, and in the Angola basin at or below 5400 m (Fig. 14-9). The different depths of the CCD are controlled by the distribution of the cold Antarctic bottom water, which is rich in dissolved carbon dioxide, has low concentrations of the carbonate ion, and thus tends to dissolve calcium carbonate. Antarctic bottom water flows northward through the Argentine Basin, passing through breaches in the Rio Grande rise into

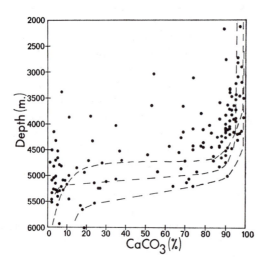

Figure 14-9 Plot of calcium carbonate content in bottom sediments versus depth for the South Atlantic. Dashed lines represent CCD's for the Brazil (top), Cape (center), and Angola (bottom) basins. (Courtesy Ellis and Moore, 1973)

468 Biogenic and Authigenic Oceanic Sediments

the Brazil basin. East of the mid-Atlantic ridge, it flows into the Cape basin; it is barred from entering the Angola basin from the south by the Walvis ridge. From the Brazil basin, the water passes eastward through the mid-Atlantic ridge via the Romanche fracture zone, causing an increase in dissolution in that area. However, a sill in the Romanche fracture zone at 4300 m prevents the deepest and coldest water from passing through, and the bottom water entering the Angola basin is 0.5°C warmer than that of the Brazil basin. True AABW thus does not enter the Angola basin. This restricted flow pattern accounts for the presence of calcareous oozes in the deep Angola basin and for their absence in the deepest parts of other basins.

Calcite and aragonite lysoclines

A lysocline is a level separating well-preserved from poorly preserved microfossil assemblages (Fig. 14-10) or the depth at which a noticeable decrease in the percentage of carbonate sediments occurs. Because different calcareous microfossil

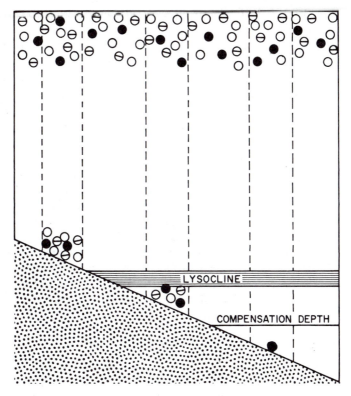

Figure 14-10 Schematic diagram shows increasingly selective dissolution of planktonic foraminiferal species with increasing water depth. Most species remain in sediments above lysocline, below which there is rapid decrease in diversity of species and morphotypes and concentration of robust forms. Below the CCD virtually no species remain. (With permission from A.W.H. Bé, *Oceanic Micropalaeontology*, vol. 1, p. 49, 1977. Copyright by Academic Press, Inc. (London) Ltd.)

groups, as a whole, dissolve at different rates, three types of lysoclines have been distinguished: pteropod, foraminiferal, and coccolith. Since aragonite, the shell material of pteropod, dissolves more easily, the pteropod, or aragonite, lysocline is usually several hundred to several thousand meters shallower than the foraminiferal lysocline; the coccolith lysocline, because of greater dissolution resistance of the coccoliths, is several hundred meters deeper.

Of the three, the foraminiferal lysocline is the most easily mapped surface. Dissolution experiments, using planktonic foraminifera from a well-preserved tropical sediment, suggest that the foraminiferal lysocline corresponds to the level at which about 80 percent by weight of foraminifera is lost by dissolution. Lysocline surfaces slope between and within the oceans. The most conspicuous characteristics of the lysoclines are that they lie at the deepest levels in the North Atlantic and the shallowest levels in the North Pacific. They also slope toward intermediate depths in the Antarctic. Most workers believe that the depth variation in the lysoclines in the oceans is due primarily to a difference in carbonate-ion concentrations in the bottom waters from various regions [Pytkowicz, 1970; Morse and Berner, 1972; Morse, 1974; Broecker and Takahashi, 1978]. Broecker and Takahashi [1978] have further demonstrated that to within 200 m, the depth of the calcite (foraminiferal) lysocline can be explained in terms of the carbonate-ion concentrations in deep waters (Fig. 14–11). The carbonate-ion concentrations differ amongst different deep-water masses, and hence the depth of the lysocline will vary according to the mix among three water types (North Atlantic deep water, AABW, and North Pacific deep water). The lysocline has its greatest depth where the high carbonate-ion content NADW extends to the bottom. It has an intermediate depth where the waters of Antarctic origin bathe the sediment. The shallowest lysocline occurs in the North Pacific, where low carbonate-ion (nutrient-rich) deep water overlies the AABW [Broecker and Takahashi, 1978].

Little agreement exists about any secondary regulatory mechanisms for carbonate dissolution and hence the depth of the lysocline. Suggested mechanisms include the rate of rain of calcareous and organic matter [Berger, 1970B; Heath and Culberson, 1970], the existence of organic or bacterial coatings on calcareous grains [Chave and Suess, 1970], concentrations of kinetic inhibitors, such as phosphate ions, in deep-ocean water [Morse, 1974], and the turbulence or velocities of bottom waters [Edmond, 1974]. Such influences may locally change the depth of the lysocline, but Broecker and Takahashi [1978] believe that the primary control is the carbonate-ion concentration of the intermediate- and deep-water masses.

Differential dissolution of calcareous microfossils

Dissolution enriches the foraminiferal assemblages with resistant forms (Fig. 14-10). This also applies to other groups, including coccoliths, and siliceous groups, including radiolaria and diatoms. The degree of enrichment of resistant forms is a measure of the amount of dissolution that has taken place. Both coccoliths and benthonic foraminifera are generally more resistant to dissolution than the most resistant planktonic foraminifera. As dissolution of microfossil

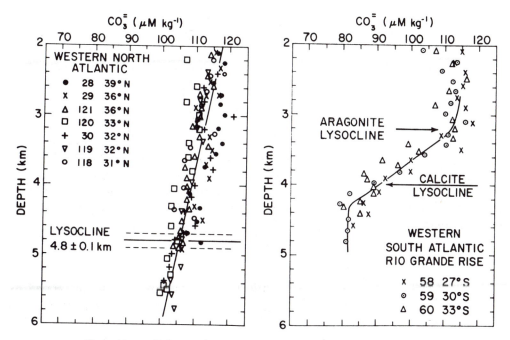

Figure 14-11 Carbonate ion versus water depth in the Atlantic Ocean. Left, western basin of North Atlantic (31 to 39° N). Depth of calcite lysocline from Kipp [1976] based on changes in foraminiferal assemblages in surface sediment samples. The carbonate ion concentration at the lysocline is 106 ± 2μ Mkg^{-1}. Right, South Atlantic near Rio Grande rise. (Reprinted with permission from *Deep Sea Research*, vol. 25, W. S. Broecker and T. Takahashi, "The Relationship between Lysocline Depth and Insitu Carbonate Ion Concentration," copyright 1978, Pergamon Press, Ltd.)

assemblages increases, the interpretation in terms of surface-water ecology becomes increasingly more difficult.

Foraminifera Most foraminiferal shells reaching the ocean floor are soon dissolved. This process greatly influences the nature of foraminiferal assemblages preserved in deep-sea sediments. The assemblages, as a result of dissolution, winnowing, and mixing, can be drastically different from the original plankton assemblage in overlying waters (Fig. 14-10). Examination of the quality of preservation of planktonic foraminiferal shells in deep tows shows that most dissolution takes place on the sea floor rather than when settling through the water column.

Present-day species have been grouped in differential dissolution rankings. One example is shown in Fig. 14-10. All assemblages from different latitudes contain elements with differing dissolution susceptibilities. Differential dissolution is controlled by test thickness, size, shape, and other factors. Shell thickness seems to be most important factor. Preservational aspects are quantified as a *solution index*, which can be based on any number of factors, such as altered faunal composition, fragmentation of specimens, or benthonic to planktonic ratios.

In general, species that live high in the water column are often more fragile than those living at greater depths, so they are more susceptible to dissolution. Adult populations of several planktonic foraminiferal species, which live at relatively deep depths, secrete additional calcite in the form of keels and crusts. These features increase their resistance to dissolution.

Calcareous nannofossils   The majority of coccoliths are dissolution resistant in both the plankton and in a well-preserved sediment assemblage, whereas well-preserved planktonic foraminiferal assemblages contain a high percentage of delicate forms. Thus dissolution patterns of the two groups are dissimilar. Coccolith assemblages begin to show evidence of dissolution above 3000 m, and there are a series of changes in composition of the assemblages between 3000 and 5000 m leading to decreasingly diverse coccolith floras and finally to the disappearance of coccoliths at 5300 m.

Chalk and limestone formation

Chalks and limestones are lithified carbonate oozes. They are formed by dissolution, reprecipitation, and recrystallization, in addition to the gravitational compaction of biogenic carbonate. Porosity is reduced from about 70 percent in oozes to about 10 percent in cemented limestones. Volume decreases by about one-third. Existing information indicates that carbonate ooze is transformed into chalk under a few hundred meters burial, while limestones are produced by further cementation under about 1 km burial. Calcareous rocks have been drilled in all ocean basins and are 20 to 120 m.y. old.

The broad picture of progressive carbonate diagenesis, from ooze to chalk to limestone, has become well known through use of the scanning electron microscope. A gradual breakup of foraminifera occurs with depth, not as a result of crushing but through dissolution. Calcareous nannofossils are highly susceptible to selective dissolution and test overgrowing. Sediment volume decreases as a result of dewatering in early stages, changes in packing, loss in porosity due to the dissolution of foraminiferal tests, and selective dissolution of nannofossils and reprecipitation of calcite [van der Lingen, 1977]. Discoasters begin to develop secondary calcite overgrowths shortly after deposition. Calcite precipitation on other microfossils begins only after compaction and dewatering have advanced to the stage that a grain-supporting framework is established. This stage is reached in the lower part of the stiff-ooze interval, where the sediment becomes crumbly. Calcite begins to precipitate along the edges and in the central areas of coccolith placoliths and eventually between the proximal and distal shields. Free-growing euhedral calcite crystals begin to grow inside foraminiferal chambers. Pore space fills and particles are welded together. In advanced stages of diagenesis, nearly all microfossils are covered with subhedral to euhedral calcite overgrowths; the central areas of coccoliths are filled to overflowing with granular calcite, and remaining foraminiferal chambers are filled with secondary calcite. Also, a large fraction of pore space is filled with granular calcite cement [Schlanger and others, 1973; Davies and Supko, 1973; van der Lingen, 1977].

We have seen that there is an overall change from ooze to chalk to limestone with increasing depth and time, but there are numerous smaller scale reversals in lithification. To explain these, Schlanger and Douglas [1974] introduced the concept of *diagenetic potential*. According to this concept, different sediments take varying lengths of time to reach equal stages in lithification. The diagenetic potential measures how much diagenesis a sediment can be expected to undergo in the normal course of geologic history. This is governed by original depositional sediment character, such as proportions of small-to-large microfossils, proportions of coccoliths and foraminifera, amount of dissolution the assemblages have been exposed to below the lysocline or CCD, sedimentation rates, biogenic productivity, and numerous other subtle factors. Differences in diagenetic potential result from changing paleoceanographic conditions. Seismic reflectors in deep-sea sequences are related to diagenetically formed horizons capable of acoustic reflection. Seismic reflectors, in this context, record paleoceanographic events; because these can affect wide areas of the oceans, seismic reflectors should be widely correlatable. This also indicates that acoustic reflectors are correlatable and they should record the stratigraphic distribution of the diagenetic potential that the reflecting horizon possessed upon burial.

In addition to changing microfossil assemblages, diagenesis plays a major role in changing the isotopic composition of biogenic calcium carbonate. The isotopic composition of microfossils must remain unaltered if it is to have any paleoclimatic significance [Savin, 1977]. Encrustation of coccoliths or foraminifera by secondarily precipitated calcite can bias paleotemperatures to colder values [Douglas and Savin, 1978]. Analysis of the oxygen isotopic record must take into account possible diagenetic alteration [Keigwin, 1979]. Keigwin [1979] found that foraminiferal recrystallization was well advanced in sediments as young as the late Miocene (about 6 Ma); this clearly has affected the oxygen-isotope record in the drilled sequences examined (Fig. 14–12).

Siliceous Oozes

Silicon is one of the most abundant elements on the earth. It occurs most often in nature combined with oxygen; in this form it is referred to as *silica*. The silica-rich terrestrial rocks are the best sources of silica, and weathering and erosion of these materials supplies much silica to rivers. Within the oceans, this silica is primarily deposited as nearshore sediment, but some does reach deep-sea areas. Some of the dissolved silica is trapped in pore waters during sedimentation, but most silica removed from seawater is used by living organisms. The biological removal of silica is the first of two dominant factors in the cycle of marine silica. Several groups of organisms construct their skeletons of opaline biogenic (amorphous, hydrated) silica. These groups include the diatoms, radiolarians, siliceous sponges, and the silicoflagellates. Dissolution is the second dominant influence in the cycle of marine silica. After death, the skeletal material of the planktonic siliceous organisms settles through the water column, where most of the silica redissolves. A small proportion of the siliceous material accumulates. When this is

Biogenic and Authigenic Oceanic Sediments **473**

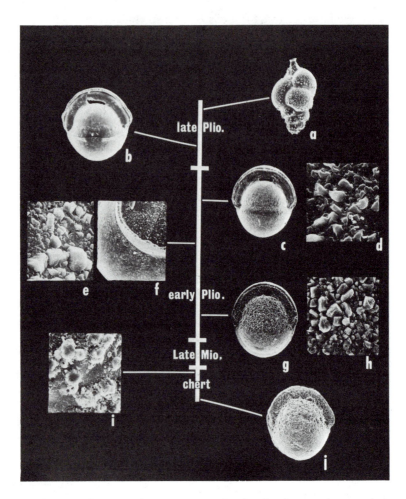

Figure 14–12 Scanning electron micrographs showing increase in secondary calcification (overgrowths) of benthonic foraminifera with increasing depth and age in DSDP Site 157. Two specimens at top of late Pliocene age lack overgrowths. Early Pliocene to late Miocene forms exhibit increased secondary calcification as diagenesis proceeds. (From L. D. Keigwin, Jr., *Earth and Planetary Science Letters*, vol. 45, pp. 361–382, 1979)

in concentrated form (greater than 30 percent of the sediment), they are called *siliceous oozes;* they are called *diatom oozes* when dominated (greater than 30 percent) by diatoms and *radiolarian oozes* when dominated (greater than 30 percent) by radiolarians. Diatom oozes are typical of high-latitude areas and some continental margin regions, while radiolarian oozes are typical of the equatorial areas of divergence. Siliceous sedimentation patterns closely match those areas with high biological fertility and high concentrations of phosphate due to upwelling and surface mixing. Conditions that favor the production of protoplasm also favor the biogenic fixation of opaline silica. The correlation between siliceous sediments and biological productivity is also enhanced because areas of high biological productivity and silica fixation also supply much organic matter to the sediments, which results in pore-water chemistry that favors the preservation of biogenic silica. The conditions that favor preservation of silica and calcium carbonate are different, resulting in rich carbonate oozes that are relatively poor in biogenic silica, and vice versa. The deposition of biogenic silica is an oceanographic-controlled, rather than a geochemically controlled, phenomenon. Regardless of where dissolved silica is transported into the oceans, it is deposited as biogenic silica only beneath nutrient-rich surface waters where biological productivity is high.

The river influx of silica has been calculated to be about 4.3×10^{14} g/yr [Heath, 1974]. It is believed that the modern and the late Quaternary oceans are in steady state, with silica deposition rates balancing the amount of new material entering the oceans. The modern steady-state system is characterized by silica undersaturation at all oceanic depths because siliceous organisms, which are oblivious to the maintenance of geochemical balances, annually produce biogenic silica greatly exceeding the annual river input. This has lead to an undersaturation of amorphous silica on an ocean-wide basis. Despite this undersaturation of silica in the oceans, organisms continue to secrete skeletal material. But in all areas, except for perhaps the North Atlantic where dissolved silica concentrations may approach zero, the availability of silica is not a limiting factor on the growth of siliceous organisms. The primary limiting factor in biological productivity is the availability of nutrients, which are supplied through upwelling in areas of high productivity. Each year these processes turn over more silica than is supplied to, or removed from, the oceanic reservoir. The undersaturation in the ocean causes the destruction of most siliceous skeletons immediately upon death of the organism. Usually only a small proportion (1 to 10 percent) are deposited as sediment on the ocean floor. Those skeletons that are deposited are thicker or, for other reasons, more resistant to dissolution. The concentration of dissolved silica in the oceans, hence the dissolution rate, is in direct proportion to the excess of biogenic production in surface waters due to the influx of new silica into the ocean. This is like the calcium-carbonate cycle, with one important difference: The solubility of calcium carbonate increases with pressure (and therefore depth), whereas shallow waters are preferentially undersaturated in silica (Fig. 14–1). The tendency for silica-rich sediments to accumulate in the deeper parts of the basins and carbonates to dissolve minimizes the importance of carbonate as a dilutant in siliceous oozes. It takes from 200 to 300 years for dissolved silica to be biologically utilized and up to 18,000 years for silica to be incorporated into sediment. So, from a geological viewpoint, rates of uptake and deposition reflect changes in supply rate almost instantaneously.

In general, the oldest deep waters (those furthest from their last contact with the photic zone) contain the highest silica concentrations. In the Pacific Ocean, where the bottom waters come from the south, there is a northward increase in silica. This buildup of silica results from increasing dissolution of biogenic material from sediments and water sources and makes the Pacific a siliceous ocean. Because the Atlantic exchanges deep water for silica-poor surface waters, it is a silica-poor ocean. The Indian Ocean contains bottom water of intermediate age between the Atlantic and North Pacific, and hence is moderately siliceous with only small areas of radiolarian oozes occurring near the equator.

Distribution of siliceous oozes

As silica removal in the oceans is through biological processes, where does biogenic-silica accumulation take place? The answer to this question is controversial. Some believe that most biogenic silica is deposited in nearshore sediments, like in estuarine mixing areas [Heath, 1974]. Others believe that about 75 percent of the oceans' silica supply is deposited as deep-sea sediments in the Antarctic [deMaster, 1979]. Regardless of the answer, there is a clear correlation between siliceous planktonic productivity (Fig. 14-13) and silica accumulation in bottom sediments (Fig. 14-14). Besides the upwelling areas south of the Antarctic convergence around Antarctica (Figs. 14-13 and 14-14), the principal sites of deposition are in high latitudes of the Northern Hemisphere (Bering Sea, North Pacific, and the Sea of Okhotsk; Fig. 14-14) and in certain areas of coastal upwelling of the eastern boundary currents [Lisitzin, 1972; Calvert, 1974]. Concentrations are much lower in the high-latitude Northern Hemisphere areas than in Antarctica, partly because of greater dilution by terrigenous debris. In regions adjacent to the western parts of continents, coastal upwelling is produced by the transport of surface water away from the coastline. Upwelling in these regions is seasonal and the upwelled water originates from a depth of a few hundred meters, for example, the California, Peru, and Canary currents. These areas sustain high rates of silica accumulation, the highest rates having been recorded in the Gulf of California, where up to

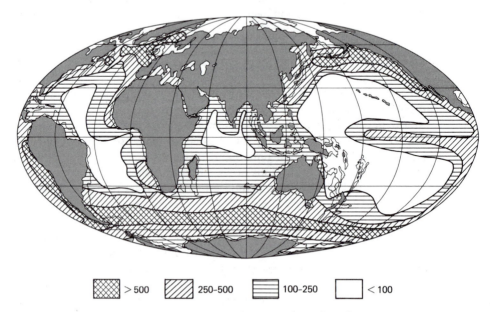

$\boxtimes$ > 500 $\diagup\!\!\!\diagup$ 250-500 $\equiv$ 100-250 $\square$ < 100

Figure 14-13 Rate of extraction of dissolved silicon (g SiO_2^{-2}/yr^{-1}) by phytoplankton in near surface waters of the world ocean. (After Calvert, 1974)

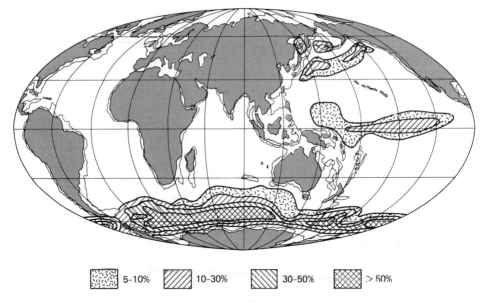

Figure 14-14 Distribution and percent concentration of biogenic silica in surface sediments of the world ocean on a CaCo₃ free basis. (After Calvert, 1974)

8×10^{-2} g/cm²/yr have been deposited. This is comparable to the highest rates in the deep sea south of the Antarctic Convergence, which are 2×10^{-2} g/cm²/yr. Despite the high rates of silica deposition in marginal basins, such as the Gulf of California, their total area is quite small so they account for less than 10 percent of total silica removal [deMaster, 1979].

The important siliceous sediment belt surrounds Antarctica (Fig. 14–14) and is 900 to 2000 km wide. The northern boundary of this belt coincides with the Antarctic convergence; to the south, siliceous sediments grade into glacial-marine deposits. More than 75 percent of all oceanic silica accumulates in this region [deMaster, 1979], and the biogenic siliceous component, dominated by diatoms, can form 70 percent of the sediment by weight. The upwelling is created by strong winds blowing from and around Antarctica, which displace surface waters northward. These are replaced by upwelling, nutrient-rich intermediate waters (see Chapter 8).

In equatorial regions, divergence of surface waters, caused by an asymmetric pattern of surface winds about the geographic equator, leads to extensive upwelling, causing high biological productivity (Fig. 14–13) and results in the deposition of carbonate and siliceous material. The siliceous material is dominated by radiolarians and the rates of siliceous deposition are much less (0.089 g/cm² per 1000 years) than in high-latitude regions. On the deep-ocean floor west of the crest of the East Pacific rise, the biogenic deposits grade from carbonate-rich oozes near the equator to dominantly siliceous oozes immediately to the north and south. The carbonate sediments accumulate at rates of 10 to 20 cm per 1000 years depending on the water depth, whereas siliceous sediments accumulate at 4 to 5 cm per 1000 years. The decreasing gradient of silica concentration east of about 110°W (Fig.

14–14) is caused by dilution by terrigenous material derived from Central and South America. The decreasing gradient toward the west (Fig. 14–14) corresponds to a general east-to-west decrease in biogenic productivity as upwelling becomes less intensive. North and south of the biogenic-rich equatorial belt, the siliceous oozes grade into brown clays. These clays occur beneath relatively infertile central-water masses and at greater water depths, where dissolution of siliceous material and carbonate material becomes important. Nevertheless, the transition from siliceous ooze to brown clay is due mainly to the decline of productivity away from the equator. Sedimentation rates are less than 2 mm per 1000 years in these areas.

The marine biogenic silica cycle

Figure 14–15 illustrates a number of sources, sinks, and transformation processes that can influence the marine geochemistry of silica, including river influx, submarine volcanism (from hydrothermal solutions), submarine weathering (*deuteric alteration*), which includes dissolution of siliceous tests prior to burial, and upward migration of silica from interstitial waters [Heath, 1974]. Of the various primary sources (Fig. 14–15; Table 14–2) river influx (4.27 × 10^{14} g per year) is the most important. Submarine volcanic material reacts with seawater to form silica as a leaching product, but the amount of material produced by hydrothermal activity seems negligible compared with river water (only about 20 percent according to Corliss and others [1979]). Low-temperature reactions on the sea floor that tend to release silica can be grouped into three main classes (Fig. 14–15): dissolution of opaline silica tests, low-temperature alteration of oceanic basalt, and low-temperature alteration of detrital silica particles. The low-temperature alteration of basalts may yield about 20 percent of the amount of silica annually supplied by rivers (Table 14–2). Dissolution of opaline tests at the sea floor also must be important but cannot be distinguished from dissolution in the

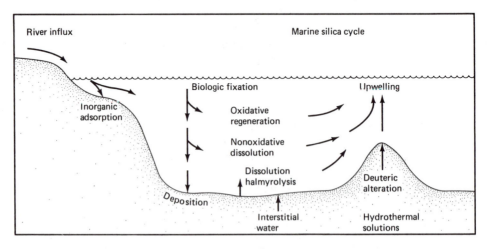

Figure 14–15 Components of the cycle of dissolved silica in the oceans. (After G. R. Heath, 1974)

TABLE 14-2 GEOCHEMICAL BALANCE OF DISSOLVED SILICA IN THE MODERN OCEAN SHOWN IN UNITS OF 10^{14} g SiO_2/YR[a]

Input		Removal	
Process	Amount	Process	Amount
Dissolved in rivers	4.3	Burial of opaline tests	10.4
Submarine weathering	0.9		
Diffusion out of the sea floor	5.7	Inorganic adsorption at river mouths	0.4
	——		——
	10.9		10.8

[a]From deMaster [1979], after Heath [1974].

water column. Silica derived by low-temperature alteration and released from interstitial water in deep-sea sediments is another major source, and the amount is probably greater than annual input from rivers [deMaster, 1979] (Table 14-2).

The major processes of silica transformation in the ocean (Fig. 14-15) include the following:

1. *Biogenic fixation by organisms.* This has been calculated to be 40 to 75 times the annual river influx. The rate of transformation of silica in the ocean by this process greatly exceeds the rate of influx or removal.

2. *Oxidative regeneration.* The rapid postmortum dissolution of fragile siliceous tests once their protective organic covering disappears. This takes place in the upper 1000m of the water column and is associated with increased dissolved phosphate due to the oxidation of organic matter.

3. *Nonoxidative dissolution.* Acts upon opaline tests either falling through the water column or on the sea floor. Because deep waters are everywhere undersaturated with opaline silica, the dissolution rate must be limited by chemical equilibrium. Nonoxidative dissolution is slow, so most of this dissolution takes place on the sea floor. As bottom waters become older, they accumulate dissolved silica through dissolution.

Silica sinks in the ocean (Fig. 14-15) include regions of accumulation of biogenic siliceous material, siliceous material adsorbed from dissolved silica related to submarine volcanism, and inorganic adsorption of silica on particulate matter as it enters the ocean. The most important sink is the deposition of biogenic material. Most silica removal is biogenic and occurs in the deep-sea sediments around Antarctica [deMaster, 1979]. Because biological processes are the dominant influence on the marine-silica cycle, supersaturation and consequent inorganic precipitation of amorphous silica appears improbable since the Cambrian, when siliceous organisms became common in the fossil record and virtually impossible since the late Mesozoic, when diatoms became abundant in the fossil record.

The major elements in the marine-silica budget are summarized in Table 14–2. An annual balance sheet is shown [deMaster, 1979], in which the various sources (rivers, sea-floor diffusion, and submarine weathering) balance the principal processes of removal (deposition of opaline tests and inorganic adsorption at river mouths).

Dissolution of biogenic silica

Only a small fraction (less than 4 percent) of the silica incorporated into opaline skeletons in the near-surface waters of the oceans ever reaches the sea floor; the bulk returns to the water column as dissolved H_4SiO_4, primarily in the upper 1000 m (Fig. 14–1). Those siliceous microfossils that do reach the sea floor, even in the form of isolated individuals or numerous specimens aggregated in fecal pellets, are subject to further dissolution both before and after burial, because the bottom waters and interstitial waters are also undersaturated with amorphous silica. Calculations for the central equatorial Pacific suggest that 90 to 99 percent of the biogenic opal produced in surface waters dissolved before reaching the sediment-water interface; an additional amount dissolves within the sediment and diffuses into bottom water, leaving about 2 percent of the original amount of opal produced by organisms in the sedimentary record.

The most fragile, thinly silicified skeletons tend to disappear entirely during settling through the water column, while the more robust, thickened forms survive longer during settling or while on the sea floor. The broader taxonomic groups can be ranked in increasing order of resistance to dissolution as silicoflagellates, diatoms, delicate radiolarians, robust radiolarians, and sponge spicules. So many assemblages dissolve that bottom sediment faunas are poor relicts of the original surface-water assemblage. Diatom frustules produced in the North Atlantic are possibly weakly silicified compared with those produced elsewhere. North Atlantic waters contain very low concentrations of silica, even approaching zero in the photic zone where organisms have extracted almost all the available silica. Atlantic diatoms dissolve readily, and this may either be a response to weaker silicification or to a decrease in robust species. Biogenic opal is better preserved in sediments underlying surface waters of high productivity.

Chert formation

Deep-sea cherts usually occur in very hard, compact beds or nodules of cryptocrystalline and microcrystalline cristobalite and quartz (both are SiO_2 polymorphs), derived from the reprecipitation of amorphous silica. The rocks are very well cemented and porosity is extremely low. One of the major early discoveries of the DSDP was the widespread occurrence of chert in deep-sea sediment sequences. Because of their hardness, cherts created major problems during the early stages of the drilling program and prevented deep drilling at many sites. Fortunately, development of more efficient drill bits and reentry capability alleviated this problem. Cherts have been sampled in all ocean basins, generally several hundred meters below the sea floor. Although widespread, they form only a small volume of oceanic sediments. Sediment associations show that only a small fraction of siliceous sediments on the sea floor have been converted to chert; the rest remain relatively unconsolidated [Hsu and Jenkyns, 1974]. Chert is associated with a wide variety of sediment types and has been reported from rocks of all ages back to the Jurassic, although it becomes particularly important in sediments older than the Oligocene. Over one-third of Eocene deep-drilled sequences contain at least some chert. A sharp decrease in the amount of siliceous sediments and associated cherts at the end of the Eocene suggests that a major change occurred in the chemical balance of the oceans.

Silica diagenesis and the formation of chert has received much attention. Several problems have been examined, including the stages through which silica diagenesis proceeds, the mechanism by which chert forms, and the origin of silica in the oceans that eventually forms chert. We shall examine each of these questions.

One of the first signs of silica diagenesis is the welding of radiolarian tests. In general the formation of cherts proceeds from solution and reprecipitation of opal, forming a spongy matrix of disordered cristobalite with unfilled opaline skeletons. This eventually is transformed into a rock composed entirely of quartz [Heath and Moberly, 1971]. Heath and Moberly [1971] have recognized four stages in the formation of chert within carbonate sediments: infilling of empty foraminiferal chambers with chalcedony or chalcedonic quartz, replacement of the groundmass of the carbonate rock with fine grained cristobalite, replacement of foraminiferal tests themselves by chalcedony and chalcedonic quartz, and final infilling with silica of all the voids in the rock and the inversion of the cristobalite matrix to quartz.

The exact process of chert formation is yet to be determined in detail, although two theories have received considerable attention. The first is the **maturation theory** of Wise and Weaver [1974] and von Rad and Rosch [1974], which states that given time, silica will eventually proceed through the various stages of diagenesis such as those described above. The second theory relates chert formation to specific sedimentological and mineralogical associations [Lancelot, 1973] and is called the **mineral association theory.** In the maturation theory, chert can form at various rates. For instance Wise and Weaver [1974] have observed the transformation of siliceous ooze into true chert in a particularly young core of Pliocene age (3–4 Ma) in the Southern Ocean. The rate at which chert forms is related to the diagenetic potential, which is not well understood. In contrast, Lancelot [1973] placed much more emphasis on the role of clay minerals and sediment permeability. He considered that different types of chert occur in different host sediments: quartz and chalcedony cherts in carbonate sediments and disordered cristobalite cherts in clays. As a result of these observations, Lancelot suggested that because clay sediments contain a high metallic-cation content and low porosity, cristobalite forms because its disordered structure is open enough to accommodate large cations. As permeability increases in carbonate sediments, chalcedony and quartz can precipitate because of a related sharp increase in the ratio of silica to metallic cations. Wise and Weaver [1974] contest this theory on the grounds that disordered cristobalite can form regardless of the lithology, as long as there is an adequate source supply of biogenic silica. Both Calvert [1974] and Heath [1974] believe that there must be a substantial flux of silica from bottom sediments into the overlying seawater [Davies and Supko, 1973]. This leads some workers to believe that cherts form as a result of retardation of the flux by rapid burial of siliceous sediment by turbidites of ashfalls [Hsu and Jenkins, 1974].

A major problem concerning the origin of chert has been the identification of the silica source. At least three sources have been proposed: Volcanogenic theories have been postulated for the origin of some Cenozoic cherts [Calvert, 1971; Heath and Moberly, 1971; Gibson and Towe, 1971; Mattson and Pessagno, 1971]. A second theory is that the diagenetic transformation of smectite to illite liberates silica, which can be utilized for silicification [Gibson and Towe, 1971]. The most widely accepted theory of chert origin, supported by many DSDP studies, is through the diagenesis of biogenic silica [Calvert, 1971; Lancelot, 1973, Heath, 1974].

Gibson and Towe [1971] and Mattson and Pessagno [1971] suggested at least a partial volcanic origin for Eocene cherts in the North Atlantic and Caribbean regions. This conclusion was partly based on association of the cherts with smectites and zeolites and primary volcanic debris. These conclusions were debated by Wise and Weaver [1974], who contended that the vast majority of oceanic chert formed since the middle Paleozoic is of biogenic origin. Thus cristobalite, tridymite, and quartz are created in sediments rich in excess silica derived from the dissolution of siliceous microfossils. They believe that the occurrence of a few traces of siliceous microfossils in a chert sequence may well indicate that many more such fossils were present in the original sediment than are presently preserved, the majority having been dissolved during diagenesis. Garrels and MacKenzie [1971] pointed out that in the formation of chert, the original siliceous skeletal fragments can be so thoroughly altered by dissolution and redeposition that they are no longer recognizable. A siliceous biogenic sediment can be transformed to a rock containing no visual evidence of the original biogenic nature of the deposit. In Eocene cherts, radiolarian abundances rapidly decrease near levels of chert formation. It is likely that cherts form in a number of ways and from a number of sources, including dissolution and reprecipitation of biogenic silica, replacement of chalks, alteration of volcanic debris, and precipitation from hydrothermal solutions [Calvert, 1974; Davies and Supko, 1973]. Nevertheless, transformation from biogenic materials must be the most important because of the close association of cherts with opaline sediments. An absence of recognizable volcanic debris in most cherts and knowledge of the geochemistry of silica in the sedimentary cycle generally oppose the necessity for an exotic source of mobile silica [Heath, 1974; Wise and Weaver, 1974].

Because of the close association of siliceous sediments with cherts, some theories necessitate major paleoceanographic events to cause chert formation [Heath and Moberly, 1971; Berggren and Hollister, 1974; Ramsay, 1971; Herman, 1972]. Such speculations are not unfounded. Biogenic siliceous sedimentation was particularly important at certain times in the Eocene—cherts are widespread in the middle Eocene forming seismic Horizon A (see Chapter 2).

Berger [1974] speculated that during the middle Eocene and late Cretaceous when sea levels were high, the flooded shelves would have tied up much carbonate but released the available silica to the deep ocean. High sea levels also would have reduced the amount of terrigenous sediments entering the deeper ocean, thus decreasing dilution of oceanic siliceous sediments. During these times oceanic sediments would become silica-rich, providing opportunity for chert formation. Conversely, during times of marine regression and narrow shelves, such as the Oligocene, the ocean basins would have become carbonate-rich and silica-poor. Silica deposition might, instead, become more important at continental margins and shelf edges due to increased upwelling by intensified seasonal winds. If the long-term regression-transgression cycles are at least partly a response to glacioeustasy, chert formation and warm climates would be associated. McGowran [1978] also believes that the underlying cause of chert formation is paleoceanographic change but, conversely, he believes that cherts are associated with cool, rather than warm, climates.

Sinking of Pelagic Sediments

Fine-grained, windblown sediment and biogenically produced particles in the upper part of the ocean must sink through several thousand meters of water to form deep-sea pelagic sediments. For particles with diameters greater than 30 μm, sinking to such depths takes just a few days to a few weeks. If biogenic material is as highly susceptible to dissolution as are a large proportion of diatoms, sinking rates of only a few weeks are long enough for the skeletons to be dissolved in the upper part of the water column. Most planktonic foraminifera, which are greater than 50 to 60μm in diameter, sink at about 2 cm/sec in laboratory tanks, which means they would reach the deepest parts of the ocean basins in only a few days. Even allowing for greater turbulence, the planktonic foraminifera and particles of similar size, such as many radiolaria, sink rapidly to the ocean floor.

However, many components in pelagic sediments are much smaller than foraminifera and radiolaria. These include the calcareous nannofossils (average size 10 μm), diatoms (average size about 50 μm), and windblown sediments and volcanic material (less than 5–10 μm). Such small particles sink very slowly. From sinking rate experiments in the laboratory, the average coccolith should take about 100 years to sink to the deep-ocean floor. In fact, typical ocean turbulence should theoretically prevent individual particles of this size from sinking much at all. However, the fact that fine-grained sediment has formed large accumulations on the ocean floor clearly demonstrates that these particles do sink by some process. A process for sinking was first suggested by Lohmann [1902]; sinking may be accelerated by fine particles combining into small bundles derived from small zooplankton, called **fecal pellets.** Fecal pellets are small (50–250 μm) aggregations of fecal material covered by a pellicle (Fig. 14–16), which protects the enclosed material. They contain large numbers of empty phytoplankton skeletons, which evidently survive ingestion by the zooplankton [Schrader, 1971]. A single pellet may contain 10^5 coccoliths, or approximately 1 microgram of calcium carbonate. Sinking rates of fecal pellets range from 40 to 400 m/day, or from 1 to 3 orders of magnitude greater than for most phytoplankton cells. Thus, in less than a couple of months, pellets can reach depths of 5000 m. Descent within the pellet almost assures the preservation of material through the water column as long as the pellicle re-

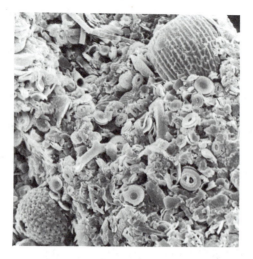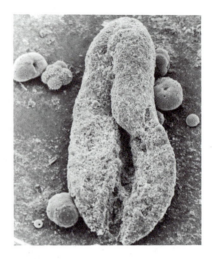

Figure 14-16 Fecal pellet of pelagic copepod. Pellet is about 180 μm in length. Enlargement is same pellet looking into area of crack in surface, showing packing of coccoliths. Diameter of a typical coccolith disc is 3 μm. (Courtesy S. Honjo, Woods Hole Oceanographic Institution)

mains intact. In tropical water, the pellicle is more readily biodegraded and material can be shed into the water column, where the previously enclosed phytoplankton are easily dissolved (Fig. 14-17) [Honjo, 1977].

Accelerated sinking of plankton in fecal pellets (Fig. 14-17) has two important geological consequences.

1. Surface-water plankton assemblages can be replicated in the underlying surface sediments without a significant time lag between production and deposition. Lateral drift by ocean currents is minimized. Thus biogeographic patterns exhibited by plankton are not disturbed by lateral oceanographic drift, which would occur if the fecal-pellet mechanism were not so efficient. Honjo [1977] has estimated that 92 percent of coccoliths produced in the North Pacific sink to depths of several thousand meters within fecal pellets, after which the majority dissolve on the ocean floor.

2. Accelerated sinking prevents the dissolution of high proportions of microplankton in their descent through the water column. The surface area of a typical coccolith exposed to seawater is approximately 80 μm². The dissolution rate of calcite, measured by Peterson [1966] at a station in the central Pacific, was 0.3 mm/cm²/year. At this rate coccoliths would dissolve within 1 year while sinking less than 100 m below the calcite saturation depth. The presence of vast areas of coccolith ooze at depths in the deep ocean demonstrates the effectiveness of accelerated sinking. This process is probably even more important for preserving siliceous microfossils, because of the undersaturation of surface waters and the susceptibility to dissolution immediately after death of the individual. Sedimentation in pellets may thus

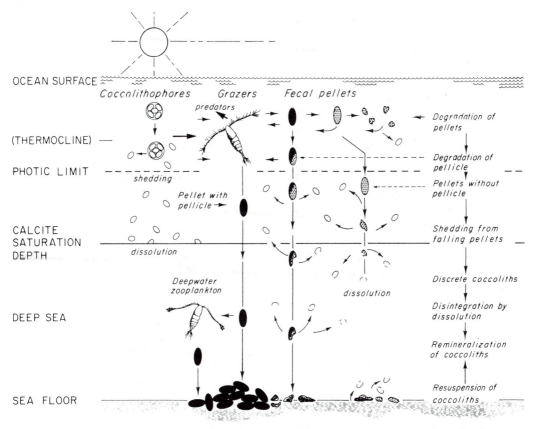

OCEAN SURFACE

Coccolithophores   *Grazers*   *Fecal pellets*

predators

(THERMOCLINE) —

Degradation of pellets

PHOTIC LIMIT

shedding

Degradation of pellicle

Pellets without pellicle

Pellet with pellicle →

CALCITE
SATURATION
DEPTH

dissolution

Shedding from falling pellets

Deepwater zooplankton

Discrete coccoliths

dissolution

DEEP SEA

Disintegration by dissolution

Remineralization of coccoliths

SEA FLOOR

Resuspension of coccoliths

Sinking rates of coccoliths; in a pellet – 160m day; a discrete coccolith – 0.15m day

Figure 14-17 A model of the relationship between the production, transportation, dissolution, and deposition of coccoliths in the open, deep ocean. Scales are not in proportion. (From S. Honjo, *Marine Micropaleontology*, vol. 1, p. 76, 1976)

supply an excess of silica to the ocean floor, helping to drive surface waters into undersaturation. Successful transit undamaged through the water column, though, does not normally mean preservation in bottom sediments, because once on the ocean floor, the pellets are biodegradable and the microfossil skeletons may be dissolved by undersaturated bottom waters if present (Fig. 14–17).

For fine-grained, nonbiogenic materials, such as clay minerals and quartz, which are resistant to dissolution, accelerated sinking is a much less important geological process. Nevertheless, these particles are probably also included in fecal pellets of nondiscriminating zooplankton species and assisted in their passage to the ocean floor.

Organic-Rich Sediments of Anoxic Waters

Anoxic, or anaerobic, sediments are deposited on the sea floor, where water is depleted of oxygen. Sediments become enriched by organic matter, which remains unoxidized due to lack of ventilation. Sediments resulting from these reducing conditions at the sea-bottom environment are invariably black because of rich organic content and pyrite. Anoxic sediments are not extensive in the modern ocean, but they have been very widespread in the past, especially in young, widening ocean basins, such as the early, highly restricted Atlantic Ocean. In the modern ocean, there are anoxic sediments in the Black Sea, the semiisolated basins off southern California, and a single, brine-filled small basin on the continental slope off Louisiana in the Gulf of Mexico. The eastern Mediterranean basin has also experienced intervals of bottom-water anoxia during the Quaternary, although the modern environment is oxygenating.

Anoxic conditions develop from a number of factors, which effectively deplete the oxygen concentration at the depth of sediment deposition. This occurs in semiisolated basins with restricted bottom-water circulation or in areas beneath the oceanic mid-water, oxygen-minimum zone under highly fertile, productive surface-water masses along certain continental margins. The sediments produced in these distinctly different environments are difficult to distinguish because organic-rich, laminated sequences are the norm. Organic-rich sediments formed in anaerobic conditions are called **sapropels** and include some black, bituminous shales and mudstones. At times large accumulations of organic carbon and sulfur in the form of pyrite have been laid down. These are of major economic importance. A lack of associated sediment disturbance suggests the absence of burrowing benthonic organisms. The mode of formation of sapropels, which cover large areas of the ocean floor, are diverse and poorly known. Thorough reviews on this subject are by Ryan and Cita [1977] and Thiede and van Andel [1977]. Various mechanisms have been proposed to account for the origin of sapropels of varying ages.

A vertical expansion of the oxygen-minimum layer on an ocean-wide basis at times of either increased surface-water productivity (such as during upwelling) or at times of reduced thermohaline circulation has been proposed. This process requires oxygen depletion only over a limited depth range, not necessarily including the deepest parts of a basin (Fig. 14–18). The presence of an oxygen-minimum layer best explains the distribution of ancient, organic-rich sediments found on topographic highs.

Stagnation through enclosure or semienclosure in basins and the presence of barriers to reduce circulation are other major factors in the formation of organic-rich sediments (Fig. 14–18). Stagnation of the Black Sea in the late Quaternary and of the North Atlantic basin during the Cretaceous probably resulted from ineffective abyssal circulation and poor vertical circulation. Anoxic conditions may, in these cases, result from insufficient amounts of oxygen to resupply the bottom waters. Even under open-ocean conditions, where bottom waters are continually resupplied, the oxygen tends to be depleted as the water becomes older and has traveled greater distances from the source regions at high latitudes. When first formed around the Antarctic, present-day bottom waters contain 8 ml/l of dis-

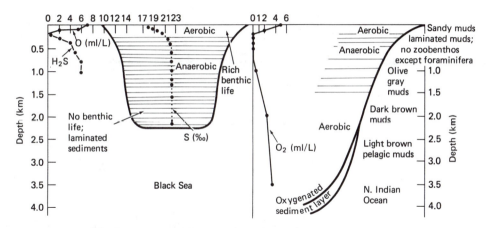

Figure 14-18 Schematic representation of aerobic-anaerobic water masses and their impact upon the distribution of sediments in the Black Sea and in the northern Indian Ocean. (After J. Thiede and Tj. H. van Andel, 1977)

solved oxygen. However, by the time these waters have traveled to the North Pacific, this has been reduced by respiration of abyssal organisms to only 2 m*l/l*.

Another mechanism requires stable water-mass stratification for establishment of deep-water anoxia and for the preservation of unusual amounts of organic matter in deep-sea sediments. Vertical water-mass stability results from increased stratification of salinity and temperature. Salinity and temperature depend on such factors as average air temperature, the balance of evaporation versus precipitation over the ocean and adjacent land areas, and the amount of restriction of an ocean basin. The vertical mixing of seawater weakens due to strong density stratification. These result from formation of low-salinity surface layers, like those which occurred in the eastern Mediterranean during sapropel formation.

Examples from modern and late Quaternary times

Black Sea sediments   Deep-water sediments of the Black Sea of late-Quaternary age record sapropel deposition [Degens and Ross, 1972]. Present-day deposition is largely aerobic, with the deposition of coccolith oozes. Stagnation in the Black Sea during the late Quaternary resulted from the basin's inability to overturn its water masses and ventilate the deeper regions with the oxygen vital for the consumption and decay of organic matter. This led to organic-carbon content of nearly 20 percent of the bulk sediment in a Holocene sapropel about 7000 years old. The late Quaternary sedimentary sequence in the Black Sea is intimately linked with the history of sea-level change (Fig. 14–19). The sediment sequence is controlled by changing history and balances between freshwater, mixed freshwater, marine, and true marine conditions. About 23,000 years ago, when sea level was about 40 m below that of present day, the Black Sea was completely isolated from the Mediterranean and acquired an entirely freshwater character. Aerobic conditions prevailed throughout the entire water column (Fig. 14–19); this phase lasted for 12,000 to 13,000 years [Degens and Ross, 1972].

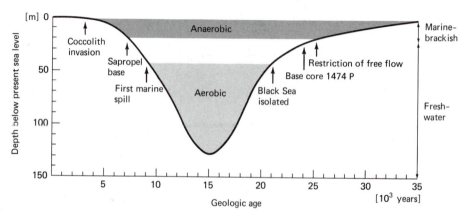

Figure 14-19 Schematic diagram of sea-level history in the Black Sea during the last 35,000 years and major paleoenvironmental events related to changes in sea level. (From E. T. Degens and D. A. Ross, *Chemical Geology*, vol. 10, p. 13, 1972)

From about 9000 to 7000 years B.P., seawater occasionally spilled over the Bosporus Sill into the Black Sea as a result of sea-level rise. These marine spills formed layers of salty water at the bottom of the Black Sea but did not have much effect on the salinity of the surface waters. Vertical circulation was decreased and anaerobic conditions began to develop at the sediment-water interface, promoting the preservation of organic debris. This interval marked the beginning of a shift from the fresh to the marine realm and from well-aerated to stagnant conditions. The sill depth at the Bosporus Straits was still too shallow to prevent two-way circulation, but the outflow of freshwater largely inhibited the penetration of oxygen-rich saltwater in required amounts. About 7000 years B.P. a hydrogen-sulfide zone became so well established in the deep basin, perhaps several hundred meters above the sediment-water interface, that incoming oxygenated water was no longer able to break up the stagnant water mass (Fig. 14–19). By about 3000 years B.P. true marine conditions finally prevailed. Two-way circulation became established between the Mediterranean and the Black Sea, and deep-sea marine faunas returned (Fig. 14–19).

California borderland basins Isolation of the California borderland basins (Fig. 12–21) from the normal vertical water-mass gradients of the open ocean have led to sluggish circulation as a result of the breakdown of these gradients. Deep circulation enters the outer central borderland and moves north over progressively shallower sills. Basin water below sill depth is uniform in density [Emery, 1960] and the succeeding northern basins are filled by progressively shallower levels of the entering deep-water flow. This flow at the present is from intermediate waters of the North Pacific, with very low oxygen content. Therefore the basin floors are all covered by waters with initially low oxygen content, which is further decreased by oxidation of the organic matter contributed from upwelling [Gorsline, 1978]. Partly anoxic conditions in the Santa Barbara basin have effectively excluded bottom dwellers and allowed undisturbed accumulations of organic-rich sediments. These are marine varves (or laminae) and are annual sediment cycles of particular importance because they enclose siliceous microfossil sequences for the last 10,000 years. These have been studied at high resolution and provide detailed paleoclimatic and paleoceanographic records.

Brine basin, Gulf of Mexico   Anoxic sediments are also known to form in basins of brine concentrations. The best known modern example is the Orca Basin at a depth of 1800 m in the continental slope of the northern Gulf of Mexico. Salt leakage from a salt diapir during the late Quaternary created a concentration of brine, filling the lower 200 m of this enclosed basin. The high density of this bottom water prevents vertical overturn of the water and causes stagnation. This prevents oxidation of the sediments and allows the accumulation of organic material. Virtually no benthonic life is known to exist at the levels of brine in this basin. Sediments are black, organic-rich, and commonly laminated.

Upper continental slope   Sediment under anaerobic conditions can also occur in the open ocean on the upper part of the continental slope (Fig. 14–18), where an oxygen-minimum layer under highly productive surface waters impinges on the ocean floor. Oxygen levels are reduced, due to oxidation of large quantities of organic matter settling through the water column. This occurs in the Northwest Indian Ocean (Fig. 14–18), where there is virtually no oxygen in water depths between 150 and 800 m. Organic-rich sediments are deposited in this zone. Oxygen levels increase again below the oxygen-minimum zone with related changes occurring in sediments (Fig. 14–18).

Fossil examples

Eastern Mediterranean sapropels   Sapropel deposition has been periodic in the eastern Mediterranean throughout the late Cenozoic. Sapropel layers are prominent features in sediments from the Nile cone, the Straits of Sicily, and the Adriatic Sea. They are presently not being formed. The late-Cenozoic sapropels contain organic carbon ranging from about 1 to 18 percent (average 4 percent). Several theories have been proposed to account for the deposition of sapropel layers in the eastern basin. However, any attempt to understand the mechanisms for formation of sapropels in the eastern Mediterranean must consider the following: the total lack of benthonic faunas and bioturbation; the presence of an abnormal planktonic foraminiferal assemblage within the sapropel layers; that the organic material is of marine origin; that deposition of sapropel layers; occurs during both warming and cooling trends within interglacial episodes; and the remarkable uniformity of some individual sapropels over large areas (more than 1000 km) and over a wide depth range, from the deepest basin to as shallow as 300 m on the continental slope [Thunell and others, 1977].

The absence of benthonic microfossils and the high organic content of the sapropels suggests a temporary depletion of oxygen in the bottom waters. A planktonic foraminiferal assemblage containing abundant *Neogloboquadrina dutertrei* occurs in the sapropel layers. The increase in this species, which is associated with relatively low salinities, suggests that large amounts of freshwater were introduced to the sea surface during this type of sapropel formation. Oxygen isotopic changes also mark particularly light isotopic episodes associated with the sapropels, which support the existence of a low-salinity surface layer. Stagnation was brief, lasting only a few thousand years [Thunell and others, 1977].

Most theories include the production of a widespread and persistent low-salinity surface layer that would interrupt normal thermohaline vertical circulation and produce anoxic conditions in the deep basins. Currently, Olausson's [1961] model, which is supported by Ryan [1972], is the most widely accepted. This theory relates sapropel deposition to deglaciation and high sea-level stands. As sea level rose above the sill depth at the Bosporus Straits (40 m), freshwater from the Black Sea (which was full of water from the melting ice sheet) flowed into the Aegean Sea in the eastern Mediterranean. The combination of reduced surface salinities and increased temperatures was sufficient to decrease vertical circulation and to produce stagnation in deep basins, allowing the accumulation of sapropel layers.

The early Atlantic Ocean During the upper Jurassic to Cretaceous, very extensive, thick sequences of black shales (organic-rich shales) were formed under anoxic conditions during the early stages of the opening of the Atlantic Ocean, when paleocirculation in the more-enclosed basins was restricted [McCoy and Zimmerman, 1977; Thiede and van Andel, 1977; Arthur and Natland, 1979]. Organic-rich sediments were deposited in the North Atlantic at discrete times from about early Cretaceous (Hauterivian) through middle Cretaceous (Cenomanian) and locally during the late Cretaceous (Turonian). In the South Atlantic, they are widespread in layers deposited during the late Jurassic (Oxfordian) through middle Cretaceous (Cenomanian) and locally even during the late Cretaceous (Coniacian). The shales are made up of terrigenous clays rich in organic material, including plant fragments. The organic matter is often concentrated a few tens of meters thick, interbedded with calcareous sediments containing distinctive benthonic foraminiferal assemblages. The original depths of deposition of the early Cretaceous anoxic sediments range from 2500 to 3000 m, while those of the late Cretaceous are at about 2500 m (Fig. 14–20). The anoxic environments may have developed at oceanic intermediate waters, in an oxygen-minimum zone. The full depth of the oxygen-minimum zone during the Cretaceous still needs to be determined by additional drilling over wider paleodepth ranges. But while the late-Cretaceous anoxic sediments were being deposited, contemporaneous oxygenated sediments may have been deposited at deeper depths in the same basins (Fig. 14–20). This could be a very important, albeit debatable, factor in the genesis of the black shales. The nature of the formation of such widespread anoxic sediments still remains a problem and several theories have been proposed. Different workers place different stress upon a number of factors, including the age, source, and mode of formation of bottom water, the upwelling and fertility of overlying waters, and the amount of degradable organic material entering the basins. Some believe that these sequences were not formed like other sapropels and that their high organic content resulted from high input of organic-rich sediments from continental regions [Arthur and Natland, 1979]. The widespread occurrence of black shales during the Cretaceous may have been due to high rainfall and the development of large deltas, but even

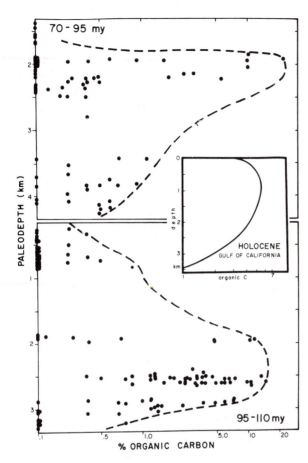

Figure 14-20 Plots of variation of the organic carbon content of sediments against paleodepth (km) for two Cretaceous time intervals (70–95 Ma and 95–110 Ma) in the South Atlantic. Insert shows plot of organic carbon content versus water depth of modern sediment deposition in the Gulf of California. Note highest organic carbon content at mid-water depths, which is associated with the oxygen minimum layer in this region. (After J. Thiede and Tj. H. van Andel, 1977)

large amounts of organic material will be oxidized if anoxic conditions are absent. Perhaps young ocean basins progress through an early stage of distinct salinity stratification, allowing preservation of organic carbon in sediments regardless of the supply rate from rivers or surface productivity. The effects of salinity stratification diminish as the basin widens and the barriers preventing effective circulation disappear.

The major question remaining concerns the factors that gave rise to strong density stratification in the water column. It is possible that during the Cretaceous, episodes of increased river input and rainfall not only supplied large amounts of organic debris but also created an intensified density stratification by producing low-salinity surface waters. Replacement waters from the Pacific Ocean through the Caribbean may have been depleted of oxygen by highly productive organisms in the equatorial belt, or evaporation may have exceeded precipitation and a stable density stratification developed by the sinking of dense, saline surface or shelf waters. Anoxic conditions would result from long residence times of bottom water and low initial oxygen concentrations in the highly saline waters. Intermediate conditions may have developed in sluggishly circulating but more open basins. Dense saline

Biogenic and Authigenic Oceanic Sediments **491**

seawater produced by evaporation on low-latitude marginal coastal platforms may have spilled into basins, causing stable stratification, and producing anoxia. These spills, which may have been climatically modulated, could have caused intermittent anoxic conditions in larger ocean basins.

AUTHIGENIC SEDIMENTS

Authigenic minerals are formed in situ in sediments and on the ocean floor. They record the physiochemical and biological reactions that operated during deposition and alteration of ocean-floor sediments. Most form from the slow precipitation of minerals from seawater. The origins of the ions are diverse, including hydrothermal activity and biogenically produced materials. Compositional changes between the layers that make up manganese nodules can provide information of geochemical changes in the environment during formation. Once detailed relationships are established between authigenic mineral formation and present-day environments, it should be possible to reconstruct paleoenvironments using authigenic minerals.

Five groups persist in deep-sea sediments: metal-rich sediments and iron oxides, manganese nodules, phosphorites, zeolites, and barite. An excellent review of authigenic minerals in deep-sea sediments is given by Cronan [1975].

Metal-Rich Sediments and Iron Oxides

Transition metal-enriched sedimentary deposits are closely associated with actively spreading mid-ocean ridges. These sediments are enriched in iron, manganese, copper, chromium, lead, and other metals in areas of high heat flow near the ridge crest, and their genesis is closely linked with the generation of new ocean floor. Iron concentrations may be higher than 20 percent on a carbonate-free basis. Sediments immediately overlying oceanic basement on the flanks of the ridges are also enriched in metals (Fig. 14–21). These are called **iron-rich basal sediments** and are the Tertiary equivalents of those currently forming on the crests of the mid-ocean ridges. These basal sediments, originally formed at the ridge crests, moved to deeper locations on the flanks of the ridges as a result of sea-floor spreading. Metalliferous sediments also are known to form in rift zones—the earliest stages of continental drift and sea-floor spreading. The best example, the Red Sea, exemplifies this, although the metalliferous sediments in this region tend to be masked by terrigenous debris input from nearby continents (see Chapter 11). Deposition of sediment on mid-ocean ridges is otherwise almost exclusively of pelagic origin, because the ridges are protected from terrigenous components.

Three distinct types of metal-rich deposits have been identified in present-day ridge-crest environments [Edmond and others, 1979]. There are iron manganese–rich [Bostrom and Peterson, 1965]; manganese-rich (essentially pure MnO_2) [Scott and others, 1974; Moore and Vogt, 1976]; and iron sulphide-rich but manganese-depleted [Edmond and others, 1979]. The iron-manganese association seems to be the most widespread and forms the iron-rich basal sediments in the

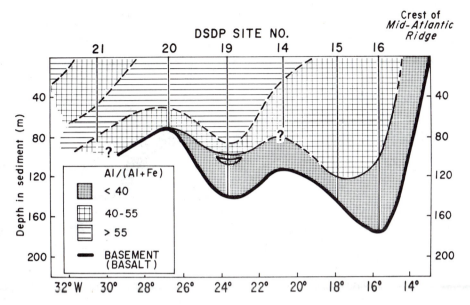

Figure 14-21 Schematic section across the western flank of the mid-Atlantic ridge in the South Atlantic showing iron enrichment of basal sediments on the ridge. (After K. Boström, reprinted by permission from *Nature*, vol. 227, p. 1041, 1970, copyright © 1970 Macmillan Journals, Ltd.)

oceanic sediment column. All three have been ascribed to hydrothermal activity, either surficial or deep-seated associated with ocean-ridge volcanic activity (Fig. 14–22).

Edmond and others [1979] have demonstrated that the suites of metal-rich deposits formed on new oceanic crust are different manifestations of a single phenomenon—progressive mixing of a primary, high-temperature, acid, reducing solution during its ascent through the hot volcanic rock beneath the ridges with seawater percolating through the rock column. Local variations in this flow regime determine the temperature and chemistry of the waters, which finally upwell onto the ocean floor.

Sulphides are deposited as a result of minimum dilution of the primary solution, leading to debouching of high-temperature, acid, sulphide, and metal-rich solutions which mix with the cold, alkaline-oxidizing ambient water (Fig. 14–22). If flow rates are sufficiently high to dominate the local bottom-water regime, massive sulphides are formed. Conversely, where the hydrothermal solutions are strongly diluted by actively circulating "ground waters," the deposition of the sulphides occurs within the deep conduits rather than at the ocean surface. Instead, manganese crusts are deposited at the ocean floor from the cooler, oxidizing solutions containing only a small percentage of the original fluid. An intermediate range of mixing of hydrothermal solutions and ground waters leads to the precipitation of iron manganese typical of the iron-rich basal sediments [Edmond and others, 1979].

Iron- (and manganese-) rich basal sediments are strongly associated with fast-spreading ridges, iron sulphide-rich deposits with ridges of intermediate spreading-

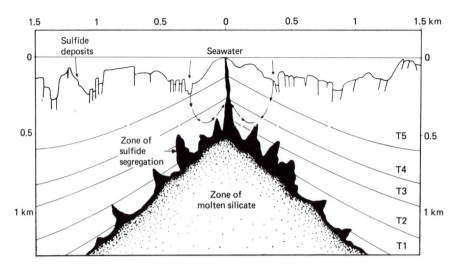

Figure 14-22 Schematic representation of the axial zone of a rapidly spreading ridge system—areas of sulfide deposition. The hypothetical zone of molten material has the composition of a basaltic melt capped by a zone of crystal-liquid segregation of sulfide and high-temperature solutions. Block-faulted areas are preferential sites for seawater fluid circulation system. Superimposed temperature curves (T) are the calculations of Sleep [1975] for a ridge spreading at a half-rate of 5 cm/year. Because the rate of spreading near 21°N is around 3 cm/year, the isotherms will be steeper and the top of the molten zone 1 to 2 km deeper in the crust than shown. Temperatures T1, 1185°C; T2, 1000°C; T3, 800°C; T4, 600°C; and T5, 300°C. Vertical exaggeration of the bottom profile is 2×. The zone of molten silicate representing a magma chamber has no vertical exaggeration. (From B. Hekinian and others, *Science*, vol. 207, 1980, p. 1443, copyright © 1980 by the American Association for the Advancement of Science)

rate, and manganese-rich deposits with slow-spreading ridges. The paucity of iron manganese–rich sediments on very slow spreading ridges may be related to their more intense tectonism, which may increase the permeability of the crust and hence allow more ground-water circulation and lower the temperatures of solutions upwelling onto the ocean floor. At centers of intermediate rates of spreading, such as the Galapagos spreading ridge, permeability is relatively low because of a combination of less tectonic fracturing and sufficiently slow rates of volcanic-rock accretion. In these areas hot waters reach the ocean floor to deposit sulphides (for example, pyrite and marcasite) [Edmond and others, 1979; Corliss and others, 1979]. Massive sulphide deposits found on the East Pacific rise near 21°N form roughly conical and columnar structures of variable size (3–10 m in height) aligned approximately parallel to the axis of the accreting plate boundary (see Chapter 7).

The metal-rich hydrothermal solutions can be transported over considerable distances. The iron- and manganese-rich sediments in the Bauer basin are thought to have formed on the East Pacific rise 1000 km away and to have been transported by bottom currents. The accumulation rate of hydrothermal sediment at any site depends, therefore, not only on the activity at the ridge crest but also on the intensity and direction of bottom-water transport away from the ridge crest area.

The formation of metal-rich sediments in the Red Sea basin is accomplished by a slightly different process from those associated with the mid-ocean ridges, but it is also the result of hydrothermal activity associated with the formation of new crust. A major diffference is that the adjacent continents are nearby sources of heavy metals. In the Red Sea, certain basins contain hot (50–60°C) and highly saline brines at water depths of about 2000 m. The salt for the brines is derived from leakage of underlying evaporitic deposits. The density of the brines prevents vertical circulation in the water column. This lack of circulation, together with the oxidation of large amounts of organic material, creates anoxic conditions and traps metal-rich solutions in the basins. The reducing conditions have allowed very high concentrations of metal ions to accumulate in the water, a thousand times more than in overlying surface waters. As the brine and overlying aerobic water meet, ferric hydroxide is precipitated, which then scavenges copper, zinc, cobalt, manganese, lead, and other metals from the water. In the anoxic conditions at greater depths, most of the metals react with hydrogen sulphide, producing metallic sulphides of bright colors in the sediments, containing concentrations of copper, zinc, silver, lead, iron, and manganese [Emery and Skinner, 1977].

Iron-rich basal sediments away from the ridge crests are considered ancient analogs of similar sediments currently forming on the ridge crests. This process has occurred at varying intensity since the early Cenozoic. The basal sediments consistently contain metal concentrations higher than those in the sediments above them or in nonridge crest sediments of equivalent age. Manganese and iron are the elements most constantly enriched in ridge crest sediments deposited during the past 50 m.y. The accumulation rates of Fe in basal sediments indicate that there have been changes in the rate of hydrothermal activity. At times of intense hydrothermal activity, the average rate of Fe accumulation at nonridge crest sites was also higher than average. The distribution of elements associated with hydrothermal sediment, like Mn and Zn, also reflects variations in the intensity of hydrothermal activity at the time. The times of highest Fe accumulation along the East Pacific rise closely correspond with intervals of major change in the spreading rate or direction (or both) along the ridge crest [Leinen and Stakes, 1979].

Manganese Nodules

Perhaps the best known of the few household words connected with marine geology is **manganese nodule.** For decades they were considered only a scientific curiosity, but are now, because of their great economic value, the focus of much attention. They contain billions of tons of metals and besides manganese, are rich in nickel, copper, cobalt, iron, and traces of two dozen other metals. These deposits, more than any other, have stimulated political action regarding the international law of the open ocean and its sea floor. Manganese nodules are economically attractive due to their high concentrations of copper and nickel. Nodules in a broad belt south of Hawaii contain more than double the average concentration of copper and nickel deposits mined on land. Nickel is particularly important because few deposits are available and because it occurs in copper-rich nodules. Cobalt is also concentrated in many nodules, but its distribution is more uneven and tends to occur with concentrations of iron and lead. A correlation exists between sediment type and metal enrichment. For example, copper and nickel are most common in areas of siliceous oozes, while iron is largely of hydrothermal origin. Sophisticated technology is being developed for the extensive mining of manganese nodules from the deep-sea floor.

Manganese deposits are very widespread on the sea floor, occurring as nodules, crusts, or thin coatings on rocks. Nodules, the commercially significant type of deposit, are black or brown agglomerations of manganese oxide and iron oxide minerals, commonly from 1 to 10 cm in diameter (range 20 μm to 15 cm). The ferromanganese oxides occur as a fine-grained silicate or iron oxide–rich groundmass associated with detrital mineral grains and biogenic components. Nodules are layered concentrically, like an onion, around a central nucleus of variable composition (Fig. 14–23).

Deep-sea manganese nodules were first recovered from the ocean floor in February, 1873, 160 miles to the southwest of the Island of Ferro in the Canary Islands during the *Challenger* expedition. They have since been mapped, mostly by the use of bottom photographs, in variable detail in all oceans except the Arctic. Their distribution is ocean-wide except in areas of high sedimentation, such as where turbidites or hemipelagic sediments are deposited. They are most commonly

Figure 14-23 Cross section of manganese nodule from the Pacific. In this specimen two nodules coalesced as they grew. The nodule is about 6.5 cm in length. (From Sorem and Foster, 1972)

found in areas where rates are less than about 5 mm per 1000 years and in well-oxygenated bottom conditions.

The concentric layers of manganese nodules represent compositional and mineralogical boundaries and range in size from easily visible rings to microscopic structures (Fig. 14-23). These variations may reflect changes in the environment of formation and as such represent paleogeochemical indicators related to compositional changes of seawater during nodule growth. Given the compositional changes recorded by individual nodules, it is not surprising that there are systematic differences between the average composition of manganese nodules from the different oceans. Biogenic sediment grains often are incorporated in the nodule matrix during periods of low metal-oxide accretion or high pelagic sedimentation. The vertical distribution of nodules in sedimentary sequences is not well known, though it does seem that disproportionally high nodule abundances occur on the surface of the ocean floor [Horn, 1972]. It seems that for every nodule located on the surface of the ocean floor, only one will be found in the upper 4 m of the sediment column. Several important questions about manganese nodules still need to be answered satisfactorily. Most importantly, what is the mechansim by which nodules are formed? Also, how do they retain their positions on the sea floor? What environmental parameters determine their compositions? What is the source of the numerous components? What is the stratigraphic record of nodules through the Mesozoic and Cenozoic ocean sediments?

Geochemistry of manganese nodules

Deep-sea nodules generally accumulate slowly, causing a prolonged contact between the ferromanganese oxides and seawater. This allows increased scavenging of minor elements. Seawater is supersaturated in manganese (Mn^{+2}), which is the principal source for the precipitation of MnO_2:

$$Mn^{+2} + \tfrac{1}{2} O_2 + 2 OH^- \rightleftharpoons MnO_2 + H_2O$$

The MnO_2 will not precipitate in free solution. It requires the catalytic effect of a precipitation surface, which requires a nucleus, nodule, or rock surface to begin. The two principal minerals in nodules are birnessite and todorokite, which differ primarily in the degree of oxidation and hydration, controlled by the Eh of the environment of deposition. Both nickel and copper may be able to substitute for manganese in todorokite, which accounts for their abundance in todorokite-rich nodules. The concentrations of five metals in nodules are shown in Table 14-3.

Nodules differ in morphology and chemical composition in the upper (seawater) and lower (sediment) sectors. In the upper sector the chemical reactions occur between seawater and the nodule, while in the lower sector, reactions occur between seawater, the nodule, and the sediment. Manganese and copper are more greatly enriched in the lower sector, with greater iron and cobalt enrichment in the upper sector [Cronan, 1977]. Manganese coatings on the surfaces of chert layers protruding from rock outcrops not in contact with bottom sediment exhibit similar chemical compositions to upper sectors of nodules. Nodule composition varies throughout the world ocean (Fig. 14-24), resulting from variations in mineralogy,

TABLE 14-3 AVERAGE ABUNDANCE OF Mn, Fe, Co, AND Cu IN
FERROMANGANESE OXIDE CONCRETIONS FROM
THE ATLANTIC, PACIFIC, AND INDIAN OCEANS, IN WEIGHT
PERCENT AND OCEANIC AVERAGE[a]

| | | ATLANTIC | | | PACIFIC |
	Average	Maximum	Minimum	Average	Maximum
Mn	16.18	37.69	1.32	19.75	34.60
Fe	21.2	41.79	4.76	14.29	32.73
Ni	0.297	1.41	0.019	0.722	2.37
Co	0.309	1.01	0.017	0.381	2.58
Cu	0.109	0.884	0.022	0.366	1.97

| | | INDIAN | | | OCEANIC |
	Minimum	Average	Maximum	Minimum	Average
Mn	9.87	18.03	29.16	11.67	17.99
Fe	6.47	16.25	26.46	6.71	17.25
Ni	0.161	0.510	2.01	0.167	0.509
Co	0.052	0.279	1.04	0.068	0.323
Cu	0.034	0.223	1.38	0.029	0.233

[a]After Cronan [1977].

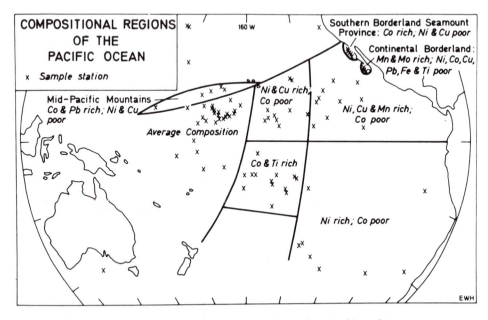

Figure 14-24 Some general differences in geochemical composition of manganese
nodules throughout the Pacific Ocean. (From D. S. Cronan, "Deep-Sea Nodules:
Distribution and Geochemistry," in *Marine Manganese Deposits*, ed. G. P. Glasby,
pp. 11–44, 1977, Elsevier Scientific Publishing Co.)

source of elements, and the depositional environment. Element sources can change in response to variations in biological input, changes in the intensity, location and type of volcanism, and other factors.

Rates of formation

Manganese nodules grow slowly. Radiometric dating techniques indicate growth rates of between about 1 to 4 mm per 10^6 years or from 0.2 to 1.0 mg/cm^2 per 1000 years. By comparison, sediment accumulates at a rate exceeding 1 m per 10^6 years even in areas of slow sedimentation far from sources of continental debris and outside areas of high biogenic productivity. Growth may be discontinuous as times of burial and exposure on the ocean floor alternate and as the availability of the metals that form them oscillate. Krishnaswami [1976] found that authigenic precipitation of Mn is nearly uniform on the Pacific Ocean floor and that variations in absolute concentrations in sediments probably reflect local differences in rates of sedimentation. Nodules are commonly found on the surface or partially buried. The very slow rates of accumulation confirm that manganese nodules exist not because of some unusual ability to attract available manganese, but because some mechanism prevents them from becoming buried. Nodules commonly remain at the surface and accumulate manganese at about the same rate as the surrounding sediment. If they become buried, they cease growing. The presence of subsurface nodules in some areas demonstrates that they do eventually become buried. How do they stay at the surface for so long? There is little agreement but possibilities include: the upward migration of nodules at the same rate as the surface, discontinuous growth rates with times of rapid growth during erosion of surrounding sediments interrupted by slow nodule growth during times of sedimentation, and nodule growth as the sediment surface moves downward.

Buried nodules remain unaltered at depth. They are, therefore, not formed by the upward diffusion of elements from previously buried nodules. Rolling by bottom currents or nudging by foraging benthonic organisms may keep them at the surface. Because sediment collects on the tops of some nodules, the top is identifiable. It has also been shown that many nodules have turned over because their tops are clear. The second mechanism is a strong possibility, because a distinct correlation between nodules and benthonic organisms exists in bottom photographs. The correlation is weaker between nodule occurrence and high-velocity bottom currents. Many areas of high abundance of nodules, such as the equatorial Pacific, show no evidence of strong bottom-current activity; therefore nodules in these areas cannot have been formed by such a mechanism. On the other hand, there is good evidence of rich nodule fields in certain areas of the ocean floor, such as in the Southern Ocean, which are swept by high-velocity currents. In these regions, bottom-sediment erosion and nondeposition provide time for manganese nodules to form. It is possible that as the bottom currents sweep the sediment away, the nodules shift and rotate, exposing all surfaces so that the onionlike layers are formed. Initially, the nodules are spaced more or less evenly without coalescence, probably through the action of burrowing organisms between the nodules. As nodule concentrations increase, they begin to limit the territory and food supply of burrowing organisms. Some areas show manganese nodules covering almost 100 percent of the sediment surface. These areas, called **manganese nodule pavements,** separate the burrowing organisms from the food supply and form an armored surface, which prevents erosion. In most of these pavements, individual nodules remain separate as cobbled surfaces, probably because of continued gentle shifting of the individual nodules. Some nodules do join to form a solid surface.

Distribution

Large concentrations of nodules occur in the North and South Pacific (Fig. 14–25), associated with brown clays and slowly depositing siliceous oozes [Cronan, 1976; Glasby, 1973; Menard, 1976]. In the North Pacific, the greatest concentrations occur near the southern edge of the vast region of brown-clay deposition dominating much of the North Pacific basin. Extensive nodule provinces occur in areas of lowest sedimentation. The outer limits of the regions of high nodule concentration coincide with increased sedimentation rates from terrigenous or biogenic contributors. In the Pacific Ocean, where the mid-ocean ridges exert less control on the bottom-water circulation, the nodules are most abundant in east to west provinces, lying north and south of the biogenic oozes of the equatorial Pacific (Fig. 14–25). The Southern Ocean associated with the Circum-Antarctic Current is marked by extensive nodule areas concentrated by bottom-current activity. In the Atlantic (Fig. 14–25) and Indian oceans, where there are higher biogenic sedimentation rates (especially in the Atlantic) and turbidite deposition, areas of low sedimenta-

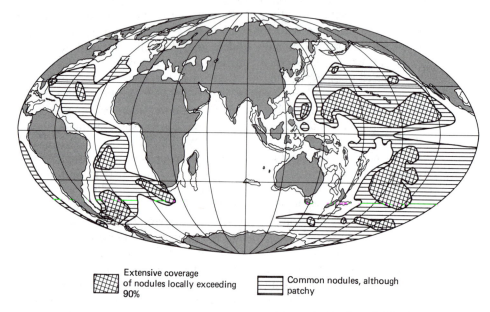

Extensive coverage of nodules locally exceeding 90%

Common nodules, although patchy

Figure 14–25 General distribution of manganese nodules in the Pacific and Atlantic oceans. (After D. S. Cronan, "Deep-Sea Nodules: Distribution and Geochemistry," in *Marine Manganese Deposits*, ed. G. P. Glasby, pp. 11–44, 1977, Elsevier Scientific Publishing Co.)

tion rates (where brown clays are deposited) occur only between the abyssal plains and the mid-ocean ridges. The highest concentrations of nodules outside the influence of the Circum-Antarctic Current occur in these areas (Fig. 14-25).

Origin of nodules

Nodule origin is under considerable discussion. The *Challenger* expedition produced four major hypotheses of nodule formation: hydrogenous (precipitation from seawater); hydrothermal (precipitation from hydrothermal solutions from submarine volcanoes); halmyrolitic (weathering of submarine volcanic rocks and debris); and diagenetic (remobilization of Mn in the sediment column and reprecipitation at the sediment-water interface). The first two hypotheses have received the most attention. Precipitation from seawater would be slow and steady, while precipitation from hydrothermal sources would be more rapid at times of submarine volcanism. Most evidence indicates slow rates of formation. Manganese, from any source, may be precipitated as nodules. Many nodule fields are distant from any local volcanic sources, but such sources are probably important where present.

Thus in summary, most nodules grow while being rolled over the sediment surface or being nudged by benthonic organisms. Continuous nodule pavements require the erosion of the sedimentary surface and settling of the nodules. Many fields in the oceans are relict lag-nodule deposits. Nodules tend to form in environments where the sedimentation rate is 4 to 8 mm per 1000 years and which are marginal to the equatorial biogenic province and the flanks of the mid-ocean ridges. Sedimentation rates in the brown-clay provinces are so slow that any nodules that do begin to form can often continue to grow for long periods of time.

Distribution of nodules through time

Nodules occur in sediments of all epochs from Eocene to the present. By far the greatest numbers of nodules occur at the sediment-water interface, though this does not mean that they have been formed mostly in recent geological times. Many of the nodules are the result of concentration from erosion.

Subsurface nodules are abundant in oceanic sequences, and some deeply buried horizons in the eastern equatorial Pacific contain nodules in higher numbers than occur on an average in the top few meters of the sediment column. This suggests that a larger proportion of the sea floor was covered with nodules during the Tertiary than the Quaternary. This may have been caused by slower rates of sedimentation due to slower biogenic sedimentation rates and lower terrigenous input or more concentrated source material. The latter seems unlikely because higher terrigenous input and increased global explosive volcanism during the late Cenozoic should have increased sources of ions for nodule formation. Thus the concentration of nodules on the modern sea floor is a secondary artifact not reflecting increased rates of nodule formation. The chemistry of the nodules has also changed through time. Nickel and probably copper are more concentrated in nodules of mid-Tertiary age (Miocene-Oligocene) than in younger or older sequences.

Phosphorites

Phosphorites are sedimentary rocks composed mainly of phosphate (P_2O_5) minerals, principally microcrystalline carbonate fluorapatite. The first reported phosphorites recovered from the ocean floor were dredged from the South African continental margin in 1873 during the *Challenger* expedition. They occur in two broad tectonic settings: either in thin sequences (usually less than 30 cm thick) on continental shelves and upper continental slopes (200 to 500 m), plateaus, and shallow-water rises, often associated with shallow-water limestones and calcareous muds, or in thick sequences (greater than several hundred meters) in geosynclines associated with organic-rich shales, cherts, and dolomites with subordinate limestone.

Phosphates are most abundant on the continental shelf and upper continental slope, sometimes occurring in nodular form as **phosphorite nodules** more than 25 cm across, but more often as large aggregate slabs and pebbles. In this facies, the phosphorites are represented in conglomeratic or nonconglomeratic form. Conglomeratic types are composed of phosphatized limestone pebbles and megafossils set in a matrix of glauconite and other minerals. Phosphate content in these rocks averages 18 percent. The nonconglomeratic types are phosphatized limestones, containing up to 15 percent phosphate. Most phosphoritic deposits on continental margins appear to be *relict* occurrences; in the modern ocean they seem to be forming beneath highly biologically productive coastal zones off Southwest Africa and Peru, where young, laminated, unconsolidated phosphorites are mixed with relict phosphoritic nodules. Abundant phosphorus in seawater is essential for phosphorite formation. Thus phosphorites are often associated with the areas of upwelling or historical upwelling of nutrient-rich waters. Off South Africa, upwelling of phosphate-rich waters is associated with the Benguela Current system and responsible for the formation of phosphorites along the west coast [Siesser, 1978]. The phosphorites deposited on the Agulhas Bank, south of South Africa, are one of the world's largest sea-floor sediment concentrations and have been linked with ancient, major upwelling.

Phosphorite may be formed by the replacement of carbonate by phosphate in anoxic biogenic sediments, the phosphate being derived from the production of plankton in surface waters. Most believe that the high organic productivity necessary for the formation of phosphorite deposits occurs in the plankton, resulting in very high sedimentation rates. This leaves a high proportion of the organic debris unoxidized in the still semiconsolidated sediments which is later transformed into phosphorites. The progressive replacement of carbonate by phosphorite with increasing depth has been observed in surface sediments.

The second model suggests formation of phosphorites directly from nutrient-rich intermediate waters, bypassing the intermediate process of high primary-plankton productivity in overlying surface waters. On the Agulhas Bank, nutrient-rich waters upwelled to the shelf edge (about 150 m) by dynamic upwelling. Phosphate-rich waters in this area extend from the thermocline down to the sea floor. Marine phosphorite deposits are widespread throughout the world in late Miocene sequences, suggesting significant upwelling at that time over shallow-water continental platforms. Other environmental changes besides upwelling must have occurred to create such widespread deposits, but these are unknown at this time.

Zeolites

Zeolites are a common group of white or colorless hydrous aluminosilicate minerals similar in composition to the feldspars and formed as a secondary weathering product. They are associated with slowly accumulating deep-sea sediments—particularly, brown clays. The two most widespread groups in deep-sea sediments are phillipsite and clinoptilolite.

Phillipsite occurs as elongate, prismatic crystals and is the most important zeolite in deep-sea sediments. It may comprise 50 percent of the carbonate-free sediment in regions of very slow sedimentation. Pacific phillipsite is associated with iron and manganese oxides, montmorillonite clays, palagonite, and other volcanic debris. Murray and Renard suggested in 1891 that phillipsite forms from the alteration (weathering) of volcanic debris on the ocean floor. In particular, phillipsite may form from the submarine alteration of palagonite, which itself is an alteration product of submarine volcanic rocks, occurring as a brown to yellow interstitial material in pillow lavas.

Clinoptilolite occurs in all three major oceans but is the most common zeolite in Atlantic sediments [Biscaye, 1965]. It is generally considered that clinoptilolite forms as a result of the alteration of acidic volcanic materials, especially rhyolite or volcanic glass but also opaline silica. Hence it may originate from both volcanic and nonvolcanic sources.

Marine Barite

Barite ($BaSO_4$) is widespread in deep-sea sediments in crystalline or microcrystalline phases or as replacement material of fecal pellets. Its concentration averages 1 percent, but it can form up to 10 percent by weight of the calcium carbonate-free fraction in some sediments [Cronan, 1974]. It is formed either by accumulations from submarine hydrothermal activity or from biogenic material. An association between barium and organic production is well established [Arrhenius, 1952].

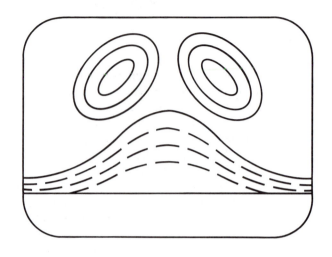

fifteen

Geological Effects

of Bottom Currents:

Motion and Commotion

In this way, it would seem inevitable that the surface waters of the northern and southern frigid zones must, sooner or later, find their way to the bottom of the rest of the ocean; and there accumulate to a thickness dependent on the rate at which they absorb heat from the crust of the earth below, and from the surface water above.

Thomas H. Huxley

THE GEOLOGIC RECORD OF BOTTOM CURRENTS

Introduction

During the late nineteenth century, curiosity about the deep ocean created oceanography. Initially, controversy raged over whether life could even exist at the great pressures, low temperatures, and in the complete darkness of the deep ocean; the *Challenger* expedition was a direct result of these controversies. This expedition rapidly dispelled the notion of a lifeless deep abyss, but it took nearly 100 years more to dispel the equally incorrect notion that the deep sea is a tranquil environment. Misled geologists inferred fossil ripple and scour marks found in rocks of all ages to be proof of ancient shallow-water deposits. Even when the DSDP began in 1968, most stratigraphers expected to find complete, unbroken sequences of fossils and sediments in the deep sea. Unbroken sequences have been found to be the exception rather than the rule. In recent years, the erosion and redeposition of sediment on the deep-ocean floor has come to be recognized as a geographically and temporally important process. Differential sediment accumulation patterns and unconformities indicate that the deep-sea depositional record is generally incomplete. These unconformities have been interpreted to be indicative of episodic bottom-current flow, which may result from changes in the climatic conditions or tectonic configurations of the ocean basins. The high probability that bottom-water circulation plays a significant role in the reorganization of deep-sea sediments was first recognized by Georg Wust in the 1930s. He initially correctly inferred relatively strong bottom currents in the South Atlantic, from calculations of potential temperatures and, later, inferred velocities of 10 to 15 cm/sec from dynamic calculations. Direct measurements of deep-sea currents using current meters later confirmed his calculations, but in the meantime geologists remained skeptical about the presence of currents in the deep sea of high enough velocity to create erosion. Few skeptics remained after the ocean floor was extensively photographed. Bottom camera stations normally contain between 10 and 30 photographs, with each frame representing a 2-by-3m view of the sea floor. The effect of ocean currents and other depositional processes on the morphology of the sea floor has been demonstrated and the criteria for the identification and interpretation of morphologic features have been provided. Current-generated bottom features range from subtle smoothing of the sea floor and gentle deflection of organisms through lineations, sediment tails, and ripples to current scour and pavement swept clean of sediment.

Understanding the ways in which various forms of the ocean bed have developed provides information on the nature and structure of the **benthic boundary layer** of the oceans. The benthic boundary layer differs from the water at higher levels because of the proximity of the seabed. Assessing the dynamics and history of bed-form development leads to a measure of how important and how fast deep currents have been in the past and how they shape the sediment surface at present. It should be noted that major surface features of the ocean floor may not reflect characteristics of modern abyssal circulation, as there is evidence that extensive ridges were formed in the past when bottom currents were more active. Although these ridges are relict features, they will be treated in this chapter on the sediment-surface features of the modern sea floor.

Most of the deep-sea floor is swept by rather slow (less than 2 cm/sec) currents, which transport cold, dense bottom waters from the polar sources to other parts of the ocean basins, where bottom water moves slowly upward and replaces surface waters. However, direct measurements of currents in certain sectors of the abyss exhibit velocities up to 40 cm/sec at mid-ocean depths. Velocities of at least 15 to 20 cm/sec can persist to within 0.5 m of the ocean floor. Superimposed on the mean velocities may be tidal oscillations of several centimeters per second, although these are still poorly understood. Where a current is obstructed by a seamount, the velocities may be locally increased by as much as a factor of two.

Abyssal circulation (see Chapter 8) is controlled by four principal factors: formation of bottom waters in source regions, deep-ocean topography, interocean connections and passages, and the earth's rotation. The density-driven flows seek the deepest routes throughout the oceans. These routes are mostly governed by the interaction of configuration of the ocean floor and the forces set up by the spin of the earth. The currents tend to lie along the west side of basins through which they flow. This is the simple effect of the Coriolis effect (see Chapter 8), guiding currents along bathymetric contours. These are called **contour currents**. However, a current may not follow the contour if there are other forces, such as potential density difference and the reaction to other moving water masses. Nevertheless, the geological effects of bottom currents are most conspicuous along the western margins of the major oceans. In the Atlantic, the northward flow of dense AABW has a major effect on sediment patterns in the Southwest Atlantic, while in the Northwest Atlantic, the southward flow of NADW along the eastern side of North America is almost equally as strong. These currents scour and transport sediment to the south, although the Gulf Stream may extend deep enough to complicate the pattern. Similar, bottom-water effects exist on the western side of the Pacific and Indian oceans. Strong currents are also associated with oceanic circulation through passages between continents and islands, such as the Drake Passage, Straits of Gibraltar, Florida Strait, and Yucatan Channel and the channels through Macquarie ridge south of New Zealand. The oceanographically important channels are the *gateways* between different oceans. The areas of greatest interest are those where the largest transfer of water occurs; these are also the areas that exhibit the greatest geological effects from bottom-water flow.

METHODS OF STUDY

The features of the ocean floor formed by current action range in size from millimeters to hundreds of kilometers. Such a range creates difficulties in resolving the features' characteristics, so a wide variety of methods, from bottom photography to various seismic reflection methods, must be used. The history of bottom water has also been studied, using various sedimentary and micropaleontological studies, each of which is briefly summarized here.

Visual and Geophysical Methods

Bottom photographs

Direct observation of the ocean floor to study **microtopography** requires use of a submersible, remote-control television, or an automated underwater camera. Ripple and scour marks and rock outcrops were first photographed by Heezen in the late 1940s in the Atlantic and by Menard in the early 1950s in the Pacific. They gathered the first direct evidence of strong currents in the deep sea. Interpretation techniques were later developed by Heezen and Hollister using photographs from virtually every known ocean environment. The majority of photographs on topographic high points in the ocean, such as seamounts and escarpments, and in deep-sea channels show conspicuous evidence of bottom-current activity. The current effects seen in bottom photographs include attached benthonic organisms bending in the flow, dispersal of mud clouds stirred up by the camera and current lineations, scour marks, and ripples in the sediment. Certain filter-feeding benthonic organisms, such as gorgonians and crinoids, have developed a planar shape suitable for the interception of maximum quantities of water, and they orient themselves perpendicular to the current. Thus the orientation of organisms bending to bottom currents is indicative of current direction. Bottom currents produce several characteristic features recognizable in bottom photographs (Figs. 15-1 and 15-2). These can be ranked by the strength of the currents. At lowest current

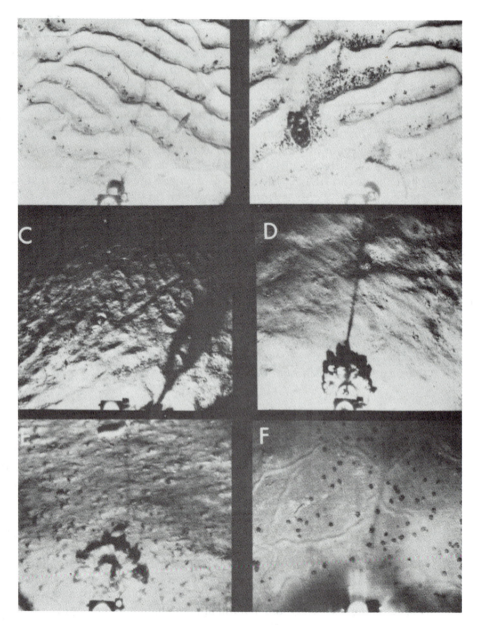

Figure 15–1 Bottom photographs from the Southeast Indian Ocean that indicate strong to very strong bottom currents. (A) Short-crested asymmetrical ripple marks with a few manganese micronodules in troughs; Holothurian seen at right middle (2750 m). (B) Short-crested asymmetrical ripple marks with concentrations of manganese micronodules and sand in troughs (2750 m). (C) Ripple marks (4574 m). (D) Current-formed sediment lineations (4241 m). (E) Sediment tails formed behind partially buried manganese nodules; Holothurian at upper center (2974 m). (F) Common manganese nodules on sandy bottom with distinct bioturbation (4302 m). (From J. P. Kennett and N.D. Watkins, *Geol. Soc. Amer. Bull.*, vol. 87, p. 333, 1976, courtesy The Geological Society of America)

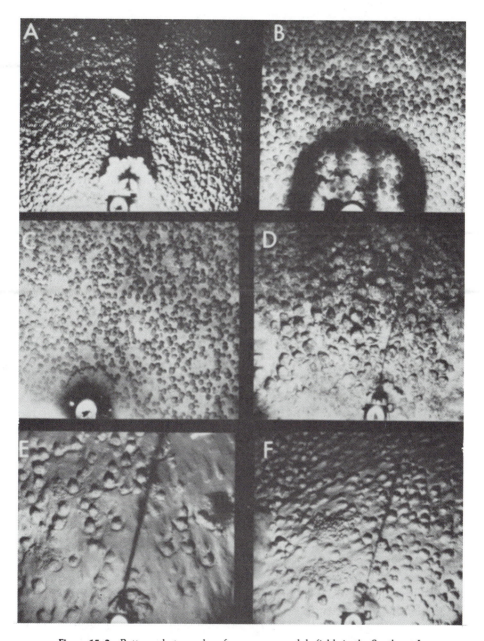

Figure 15-2 Bottom photography of manganese nodule fields in the Southeast Indian Ocean. All frames considered to indicate very strong bottom currents. (A-D) shows manganese nodule pavement with extremely tightly packed nodules. (A) Photograph from the Southeast Indian Ocean manganese nodule pavement; Holothurian near center (A) 4178 m; (D) 4541 m. (B,C) Very abundant spherical manganese nodules within sandy ooze sediment (B) 4428 m; (C) 4207 m. (E,F) Abundant manganese nodules partially buried by sediment and forming distinct tails. Moating has formed around individual nodules in (E) (E) 4003m; (F) 3798 m. (From J. P. Kennett and N. D. Watkins, *Geol. Soc. Amer. Bull.*, vol. 87, p. 334, 1976, courtesy The Geological Society of America)

velocities, muddy sea bottoms characteristically have abundant, distinct, faunal tracks, which become erased with low bottom-current velocities until the bottom develops a smooth appearance; however, evidence of life in the form of partly erased tracks may still be abundant. As velocities are increased, surfaces develop elongate sedimentary deposits called **streamers,** behind positive features like lumps or feces (Fig. 15–1(E)). Further velocity increase may cause ripple marks to develop (Fig. 15–1(A–C)). Some ripples exhibit winnowed sand, gravel, and manganese micronodules in their troughs (Fig. 15–1(A),(B)). Increasing current velocities are sometimes associated with increased numbers of manganese nodules (Fig. 15–2) which lie on a sediment bottom. Invariably, where sediment separates individual nodules, strong evidence of associated moderate-to-high current velocities exists, either the smoothing of sediment (Fig. 15–2(C)), the development of crag and tail (Figs. 15–1(E) and 15–2(E), (F)), or scoured moats (Fig. 15–2(E)).

Currents may also be recorded by a lack of sediment cover on rocks or from the character of sediment lodged in cracks and crevasses. Current directions are often determined using compass bearings on dispersed mud clouds photographed by a bottom camera and by orientated animals and sediment features, such as ripples and streamers. Current directions determined by the first two methods are only momentary indicators of currents. While persistent flows are necessary to create sediment features, these features are not easily erased by short-term fluctuations in currents either, unless the fluctuations are particularly strong. The orientation of microtopographic features can be highly variable in the same region, probably being influenced by very local topographic features superimposed on broad regional patterns.

Seismic-reflection profiles

Low-frequency echoes   The surface of the ocean floor is studied by the way in which it reflects sound, or its **echo character.** The seismic character of the ocean floor is discerned by the character of low frequency echoes, which penetrate deeply into the sediment column but yield only general information on the upper surface layers. High-frequency echoes provide detailed information on the seismic character of the upper part of the sequence, but penetration is shallow. Seismic records in the deep ocean clearly show a considerable range of topographic features either formed or modified by bottom currents, including extensive, long ridges in the North Atlantic constructed from sediments transported by bottom currents.

Sediment deposits formed by bottom currents are not uniformly distributed in relation to basement topography, but are often accumulated as positive elongate structures parallel to bottom-current directions and lying beneath the margins of the sediment-transporting current. The upper surface of the sediment is often wavy, with a typical wavelength of 2 km and an amplitude of 50 m. Where the sediment bodies encounter basement highs, a marginal moat or channel typically forms on one or more sides and sediments dip and thin toward the obstacle. Conversely, sediment can be banked up on the other side. The acoustic character of sediments can vary significantly from *acoustically well stratified* to *acoustically poorly stratified.* In the North American basin, preglacial deposits are acoustically poorly stratified in comparison with glacial-age, bottom-current deposits, which are well stratified. In general, areas under the influence of strong bottom currents exhibit a lack of pelagic draping and ponding and are instead smeared and incised by the action of the bottom currents.

High-frequency echoes   Characteristics of echo-sounding profiles recorded by conventional, wide-beam profilers can be used to distinguish certain properties of sea-floor relief. These short-ping (less than 5 ms), high-frequency (3.5–12 kHz) echograms have been used as a basis for understanding sediment processes on the sea floor. The types of echoes from sediment sequences are classified according to three criteria: coherence of the echo return (whether the echo consists of a single return or several closely spaced returns); the presence or absence of side echoes (the presence of hyperbolic echoes from topographic features below the resolution of the echo sounder); and wavelength, height, spacing, and regularity of any sea-floor relief. Hyperbolic echoes are often associated with topograhic features formed by bottom currents. A sequence of giant sediment waves revealed by such seismic-reflection profiling is shown in Fig. 15-3. Sediment waves are also known as *lower continental rise hills, abyssal antidunes, giant ripples,* and large *mud waves* and are well known from most continental margins where they are associated with contour currents. A series of hyperbolic echoes elongated parallel to topographic contours is interpreted as a number of current-formed furrows.

Features that lie on the deep-sea floor, from a few meters to a few kilometers in scale, are difficult to determine because they are between the range of echo sounders and bottom photography. These difficulties have been largely overcome by the development of a deeply towed instrument package created by the Marine Physical Laboratory, Scripps Institution of Oceanography, called *Deep Tow*. This package is towed at a height of 10 to 100 m above the sea floor at speeds of 2-4

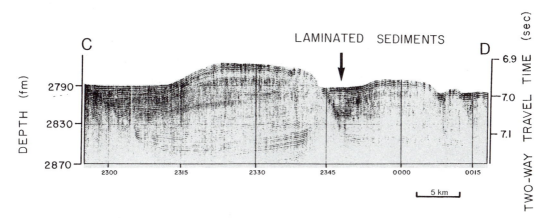

Figure 15-3 High resolution minisparker seismic profile of uppermost part of sediment column (200 m) from the Northwest Atlantic Ocean, showing irregular asymmetric hills with troughs partially filled with bedded sediments. (Courtesy U. S. Geological Survey, Woods Hole, Massachusetts)

km/hr and is carried within an acoustic transponder net. The system has a side scanning sonar, narrow-beam (4°) echo sounder, 4-kHz subbottom profiler, stereo photography, and continuous temperature measurement. A pair of side-looking sonars, operating at 110 kHz, record acoustic backscattering properties of the sea floor up to about 500 m from the vehicle. Because this system can determine sea-floor features varying in scale from centimeters to kilometers, it has been successfully employed for mapping patches of abyssal bed forms in considerable detail and relating these characteristics to bottom currents and sediment cores [Lonsdale and Speiss, 1977].

Sedimentary Methods

Sediment mapping

A useful approach to the study of the effects of bottom currents and their history is the regional mapping of surface-sediment ages and sedimentation rates. Erosion by bottom currents often exposes older sediments; the mapping of surface-sediment ages identifies regions of active erosion (Fig. 15–4). Sedimentation rates are also affected by the redistribution of sediments by bottom currents on the ocean

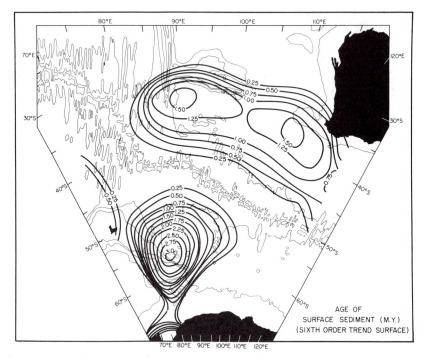

Figure 15–4 Map of trend surface analysis (6th order) of age in millions of years of surface sediment in the Southeast Indian Ocean. Note two broad areas of surface sediment erosion. (From J. P. Kennett and N. D. Watkins, 1975)

floor. Sedimentation rates in cores can be mapped, exhibiting patterns of bottom-current distributions. The use of combined paleomagnetic and micropaleontological dating of suites of cores over large areas is the most effective approach.

Sediment particle-size analysis

As currents scour sediment the particle-size distribution is changed and recorded in textural parameters, such as *mean grain size, skewness,* and *sorting*. An increase in current velocity will increase the mean grain size and cause more positive skewness and better sorting. Using these properties it is possible to identify horizons in cores that have been exposed to increased bottom current action. Such horizons may be subtle and would go undetected without the use of such quantitative approaches. The range of silt sizes has been used for particle-size analysis instead of the complete range of sizes, due to difficulties caused by the use of several other methods needed to determine the particle-size distributions of the sand, silt, and clay-size fractions. The relationship between current velocity and particle size is still poorly known and thus the subject of intense investigation. Sand-size foraminiferal tests may be eroded by currents with velocities greater than 15 cm/sec. Selective removal of fine particles can decrease the proportion of calcareous nannofossils relative to the larger foraminifera. Subsequent deposition of the winnowed fine material leads to the accumulation of distinctive fine-grained sediment. Sediment particles may therefore indicate changes in bottom-water velocity.

Changes in skewness values also reflect the dynamics of sediment deposition. A skewness value more negative than -0.2 indicates a depositional environment influenced by pelagic settling, while values more positive than -0.2 reflect influence by bottom currents [Huang and Watkins, 1977].

Mapping of hiatuses

The distribution of **hiatuses,** or unconformities, are readily determined by conventional biostratigraphic and paleomagnetic dating of sediment sequences. Most hiatuses are distinguishable in cores by any or all of the following: a sudden change in the stratigraphic ranges of microfossils, a lithologic change, or a paleomagnetic reversal. An unconformity may be recognized by the presence of a zone of manganese nodules or micronodules, greater amounts of coarse debris, or subtle changes in bedding features revealed by X-ray radiography. However, a lack of these characteristics does not exclude the possibility of a hiatus in the sequence. In general, hiatuses provide only the extreme limits of the bottom-current velocities creating the features. The full range of variations in current velocities creating a hiatus cannot be determined because the sediment, and hence the criteria for evaluating current velocity, is completely removed. Detection of disconformities spanning periods of times shorter than about 0.2 m.y. requires special techniques. The analysis of sediment skewness is of value here (Fig. 15-5). An increase in bottom-water velocity changes the skewness from negative to positive and decreases

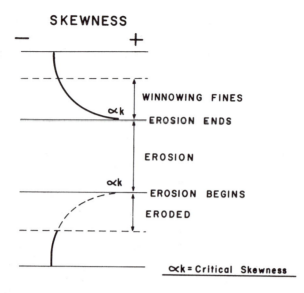

SKEWNESS

WINNOWING FINES

∝k

EROSION ENDS

EROSION

∝k

EROSION BEGINS

ERODED

∝k = Critical Skewness

Figure 15–5 A model relating sediment skewness to erosion by bottom water. An increase in bottom-water velocity changes the skew from negative to positive. The sedimentation rate decreases until a critical velocity is reached (sediment skew αk) when active erosion begins, removing that sediment deposit during the inital current increase (dashed lines). Erosion ends when the bottom current decreases (skew dropping back to αk), so that the final sampled core will feature a sharp discontinuity in skew (solid curves). (From T. C. Huang and N. D. Watkins, *Marine Geology*, vol. 23, p. 128, 1977)

the sediment rate; when critical velocity is reached, active erosion begins removing the sediment deposited during the increase in current velocity (Fig. 15–5). The sampled core will then feature a discontinuity, from consistent negative to declining positive skewness, in association with a hiatus. The critical skewness, marking the point at which erosion begins, will vary between cores because the critical bottom-current velocity required to initiate erosion will vary depending on sediment type.

Diatom transport

Displaced high-latitude diatoms have long been used to describe the path of spreading bottom waters from the Antarctic. These diatoms can be displaced a considerable distance from their original source, agreeing with the spreading paths of Antarctic bottom water. The method serves as a general criterion for studying the circulation patterns of deep waters, including transport through narrow channels in oceanic ridges. Antarctic diatoms become entrained in newly formed Antarctic bottom water. As this water mass spreads northward, the diatoms are transported and deposited on the sea floor, where they serve as a more-or-less permanent record of the passage of this water mass. Using this method, Antarctic diatoms have been traced as far north as 30°N in the Atlantic and also as far north as the equatorial regions of the Pacific [Booth and Burckle, 1976].

Paleomagnetic fabric

The paleomagnetic fabric of deep-sea cores has been used for determining the relative magnitude and direction of bottom currents [Ellwood and Ledbetter, 1979]. The alignment efficiency of magnetic grains is related to areas of inferred current velocities, based on relationships with other sedimentary parameters. This alignment is detected by using particular magnetic measurements (**anisotropy of magnetic susceptibility,** or AMS). Measurements of California beach sand has shown that magnetic particle orientations reflect the alignment of grains previously determined independently by standard sedimentologic measurements. In the Vema Channel, in the Southwest Atlantic, bottom currents are known to be particularly strong, and the strongest magnetic alignments and coarsest particles are found in the axis of the channel. The larger magnetic grains in these sediments are oriented parallel to depth contours of the channel, suggesting alignment by currents. Areas marked by active bottom-sediment reworking by benthonic organisms should not retain such orientation for long and the methods are probably only useful on ocean floors with little animal life.

Benthonic foraminiferal distributions

Another way to determine distribution patterns of abyssal water masses and of fluctuations in abyssal water masses is the use of the benthonic fauna. There is a strong correlation between elements of the benthonic fauna and particular abyssal water masses [Streeter, 1973; Schnitker, 1974; Lohmann, 1978; Corliss, 1979]; these associations can be used to infer changes in the distribution patterns, through time, of bottom waters. Foraminifera are usually the only components of the benthos that can be recovered in sufficient numbers from deep-sea sediments to be useful in this approach. Ostracoda are also valuable in some areas.

The modern deep-sea environment is much more uniform than the shallow-marine environment. Deep-sea salinity is usually less than 35‰ in temperature ranges between 1° to 4° C at depths greater than 2000 m; dissolved oxygen ranges from less than 1 to more than 6 ml//l. Major changes occur at these depths in the degree of undersaturation of calcium carbonate, with both the lysocline and the CCD effecting calcareous assemblages. Despite the relatively small physical differences between different deep-water masses, the benthonic foraminifera have become sensitive to small environmental differences in adapting to these conditions. Recent studies of modern deep-water benthonic foraminifera have shown that in the North Atlantic Ocean, particular deep-water species are associated with, and mark, specific deep-water masses. An analysis of the distribution of frequently occurring foraminiferal species has led to the recognition of three consistently recurring faunal assemblages. The association of these faunal assemblages with three discrete deep-water masses can be inferred from their distribution. A widely occurring faunal assemblage, dominated by the species *Epistominella umbonifera,* occurs in areas associated with AABW (greater than or equal to 1.5–2°C). The second faunal assemblage, marked by abundant *Epistominella exigua* and *Planulina wuellerstorfi,* is present in areas associated with the Arctic bottom water (ABW) (less than or equal to 2°–3°C). At about 40°N latitude, *E. umbonifera* and *E. exigua* faunas lose their identities, and samples from this area, the area of confluence and mixing of the two bottom-water types, have an intermediate character. Because the temperature differences between water masses are minor, possibly the slightly higher salinity and dissolved oxygen content of the ABW or differences in alkalinity or nutrient content are sufficient to bring about the observed faunal differences. Samples widely scattered on the continental rise and slope of the mid-Atlantic ridge contain faunas in which *Uvigerina peregrina, Globocassidulina subglobosa* and *Hoeglundina elegans,* or *Nummoloculina irregularis* and *Cibicides kullenbergi* are variously dominant. These faunas are associated with the NADW (about 2°C–4°C).

From these observations it may be concluded that the fauna are definitely not bathymetrically controlled, although their distribution in places may roughly parallel bathymetry. Also, temperature, oxygen, and salinity individually do not appear to control the distributions of the deep-sea benthonic foraminifera. Instead, other factors in combination with temperature may be more helpful in explaining the distribution.

Because *E. umbonifera* so readily marks the AABW, it can be used to trace movement of this water through the deep-ocean basins. The most important implication of these studies is that the benthonic foraminifera can be used to examine the distribution of abyssal water masses through geologic time. During the Quaternary these changes were of considerable magnitude and were ocean-wide. However, because of the substantial taxonomic turnover among benthonic organisms during the middle Miocene, the ecological understanding of modern deep-water benthonic foraminifera cannot be simply extrapolated back to the earlier Cenozoic. Any understanding of environmental preferences of benthonic foraminifera before the middle Miocene is based on indirect evidence, such as their paleobiogeography and paleobathymetry and comparison with isotopic and sedimentological parameters.

EROSION, TRANSPORTATION, AND DEPOSITION BY BOTTOM CURRENTS

Factors Controlling the Cycle

Whether erosion, nondeposition, or sediment accumulation occurs in the deep sea is determined by the dynamic balance between the rate at which sediment is supplied to and removed from the sea floor. The rate of sediment supply is determined by the rate of biogenic productivity and terrigenous input, and the rate of removal is determined by the rate of bottom flow and the corrosiveness of bottom waters to biogenic sediments. Continued exposure to bottom water, through nondeposition, would lead to greater dissolution of biogenic sediment. Removal of sediment by dissolution is a form of erosion.

Small-scale features generally result from slow rates of water flow and develop either perpendicular (ripples) or parallel (lineations and scour and tail features) to currents. As the speed of flow increases, transverse structures develop, usually as megaripples and sand waves. Net flow in one direction will give rise to asymmetry in cross sections—gentler slopes on the upstream side and steeper slopes on the downstream side. Oscillations cause symmetrical cross sections; as flow rates increase, parallel structures, such as banks, develop. The nature of bed forms produced will also depend on the rate of supply of sediments to the area, although this is poorly known.

It is difficult to determine precisely the critical bottom-water velocities required to erode, transport, and deposit sediments in the oceans. Although much has been learned about the dynamics of sediment deposition of nearshore marine environments, little is known about deep-sea sediments. The critical traction transport velocities for deep-sea sediment particles, if determined, could provide estimates of minimum velocities required to form ripples, scour marks, current lineations, and other current features. Some laboratory flume experiments have been conducted on pelagic biogenic sediments, and it has been found that the critical velocity for the erosion of carbonate ooze ranges from about 15 to 35 cm/sec. Heezen and Hollister [1964] and Postma [1967] summarized several studies which examined the traction transport of sediments in streams (Fig. 15-6) [Flood, 1978]. The results serve primarily as a first approximation, because the sea-floor environment and deep-sea sediments differ drastically from those of rivers and streams. Erosion of unconsolidated clay and silt grains requires bottom-current velocities between 10 and 20 cm/sec and velocities of about 20 to 40 cm/sec in the case of sand-size material (Fig. 15-6). However, for consolidated sediments the velocities need to be increased considerably [Flood, 1978; Gardner, 1977; Hollister and others, 1978; McCave, 1978]. Once in motion, very low current velocities will maintain the finer sediments in suspension (Fig. 15 6). These estimates imply that currents of considerable velocity occur over wide areas of the ocean floor, because erosion of deep-sea sediments is so widespread.

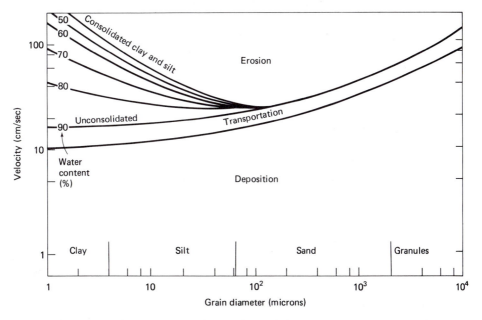

Figure 15-6 Bottom-current velocities (cm/sec) for erosion, transportation, and sedimentation of sediments according to grain size. Velocity measured 15 cm above the bottom. Note effects of consolidation upon commencement of erosion in clays and silts. (After Postma, 1967)

Geological Effects of Bottom Currents: Motion and Commotion   **517**

In general, the formation of ripples depends on the sediment characteristics and the velocity of bottom currents. For sediment densities of about 1.5 g/cm³ (about the density of foraminiferal oozes) and particle sizes of 200 μm, ripples will form undercurrent velocities of 20 to 40 cm/sec. Much more work is required to determine accurately the current velocities required to initiate deep-sea erosion and maintain sediment transportation.

In some locations, deep-ocean currents reflect the combined influences of the dissipation of *tidal energy* and the net transport of bottom water from one oceanic region to another. Deep-ocean current velocities are periodic: Both semidiurnal and monthly tidal variations have been recorded. Oscillatory tidal currents are known to have velocities up to 17 cm/sec immediately above the sea floor on seamounts. This process may operate continuously, maintaining the sharpness of ripple-mark crests and smoothing the tracks and trails of benthonic organisms. Tidal energy is dissipated in the oceans at a rate of 2.7×10^{19} erg/sec, a large proportion of which is dissipated on the continental margins and in shallow seas. The mechanisms controlling tidal currents in the deep ocean today and their geological importance are poorly understood, but it seems that, at least in shallow regions, they play an important role in the formation of microtopography. Their effects would be particularly important in combination with other currents, which singularly may not reach sufficient velocities to bring particles to suspension, although they are able to accomplish this in combination. Thus they may be very important in enhancing the erosion of the shallow-water sea floor. During glacial episodes, when sea level was about 130 m lower than present sea level, many of the world's continental shelves and shallow seas were exposed and no longer were available for the dissipation of tidal energy. Although the potential effects of this process are unclear, greater tidal energy dissipation would have occurred in the open ocean regions, including the continental margins.

Erosional Features

Erosional features are widespread throughout the ocean basins and range in scale from millimeters to hundreds of kilometers (Fig. 15–7). These features are concentrated at the axis of deep-flowing currents, whereas depositional features (Fig. 15–7) tend to occur at the margins of such currents. Thus the greatest amount of erosion and the largest erosional features are concentrated in ocean gateways, interbasinal channels, and other bottom-water pathways. The gentlest erosional features may be subtly indicated by an absence of burrows, mounds, and tracks. Small-scale topographic features create regions where current flow becomes intensified locally; in these regions both chemical dissolution and mechanical erosion proceed more rapidly than in more protected regions nearby. Larger features include scour and extensive moats formed at the margins of topographic irregularities, such as seamounts, knolls, certain ridges and escarpments, and extensive unconformities.

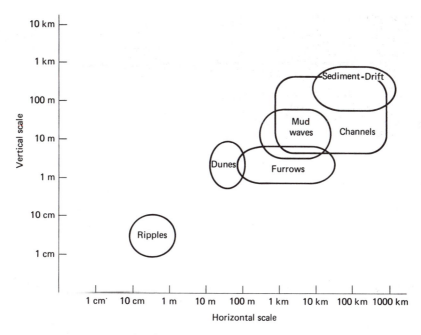

Figure 15-7 Horizontal and vertical scales of ocean bottom features. (From Report of Ocean Crustal Dynamics Committee of JOI Inc., 1979)

Furrows

One of the most spectacular and, perhaps, the most important erosional bed forms of the deep-sea floor are the **furrows,** or erosional troughs, up to several kilometers in length and a few meters in width, regularly spaced at intervals of 10 to 100 m and from 1 to 20 m deep (Fig. 15-8) [Flood, 1978]. They extend parallel to the flow direction and may curve to conform to the local contours. Furrows are often associated with hyperbolic echoes and thus are below the limit of the ship's echo-sounding resolution. Furrowed topography can be distinguished by the spacing of hyperbolic echoes. Although the topographic form can be studied using the hyperbola, furrows are best studied with detailed near-bottom, side-looking sonar and carefully positioned bottom cameras. Many furrows have ripple-covered sides and flat floors that may reflect a lithologic control on the morphology. Furrows seem to develop well in fine-grained, cohesive sediment.

The major problem in explaining the origin of furrows is their long and narrow shape, but it seems that they are formed by narrow threads of high-velocity currents separated by broad bands of much lower current velocities. The velocities necessary to create furrows are unknown but seem to be formed by current ribbons with velocities greater than 10 cm/sec. There is some geological evidence that furrows are relicts from periods of high-bottom current flow. These have since become areas of overall deposition, but the furrows are maintained because their topography continues to concentrate any bottom currents existing in the area.

Geological Effects of Bottom Currents: Motion and Commotion   **519**

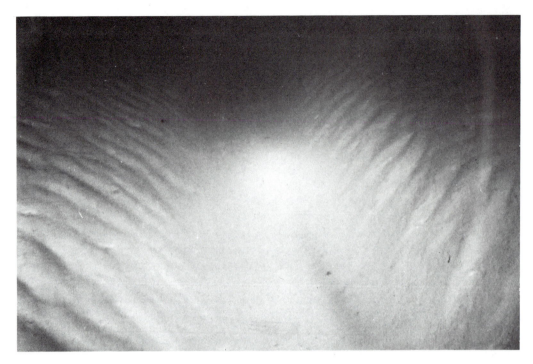

Figure 15-8 Oblique bottom photograph of a furrow. Note its asymmetry. Flat furrow floor is about 1 m in width. (From R. D. Flood, doctoral dissertation, 1978)

Measured currents of up to 16 cm/sec are associated with larger furrows [Wimbush, personal communication], but the actual formation of the furrows may have required much greater bottom-current velocity. This discovery of such narrow, elongate erosional features was unexpected and poses important questions about the circulation in the deep sea.

Moats and marginal channels

Seismic-reflection profiles have shown the widespread occurrence of large scale scour features around topographic features throughout the ocean basins [Davies and Laughton, 1972]. These include **moats** and **marginal channels,** which are formed by concentration of currents due to topographically high features (Figs. 15-9 and 15-10). The depth of the moats and channels is determined by the duration of the current creating the erosion, the accessibility of sediment sources, and the presence of resistant stratigraphic horizons, which limit further erosion. Marginal channels are associated with western boundary currents in both the Pacific and the Atlantic. Active erosional channels may initially be created by faulting, with subsequent erosional processes reshaping topography so that prominent escarpments are no longer present. Tectonic processes can, therefore, facilitate sediment erosion.

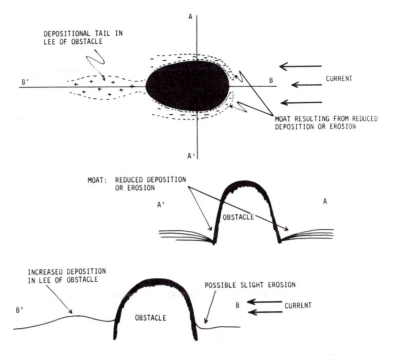

Figure 15-9 Schematic diagrams showing the formation of moats by bottom currents. (From T. A. Davies and A. Laughton, 1972)

Hiatuses

Erosion causes unconformities, or hiatuses. Hiatuses are caused either by nondeposition or, more commonly, by erosion and are associated with the flow path of bottom waters. Hiatuses caused by erosion are most likely to occur where the flow is most intensified, such as areas swept by western boundary currents or near topographic highs or constrictions. Hiatuses can cover any amount of geological time, up to tens of millions of years. The DSDP has demonstrated the widespread occurrence of hiatuses; this is really a paradox, as the deep-ocean basins are the ultimate sediment catchments. Surface-sediment studies reveal very extensive regional hiatuses, such as one documented in the central equatorial Pacific by Johnson [1972]. This hiatus exhibits erosion of several hundred meters of sediment, with sediments up to Eocene age having been removed. Other extensive hiatuses occur in the region of the Southern Ocean. Everything that is washed from the continents is finally carried to the ocean bottom. Any change in the productivity of surface water, the rate of bottom-water flow, or the chemical character of bottom water can affect the balance between accumulation and erosion and hence the distribution of hiatuses. Drastic changes in the flow paths and sources of bottom waters have been created by major tectonic adjustments, which have changed the patterns of hiatus occurrence. There are several possible explanations for the occurrence of each hiatus in the sedimentary record, but the difficulty is to determine that genesis. The range of possible explanations for hiatus formation has been sum-

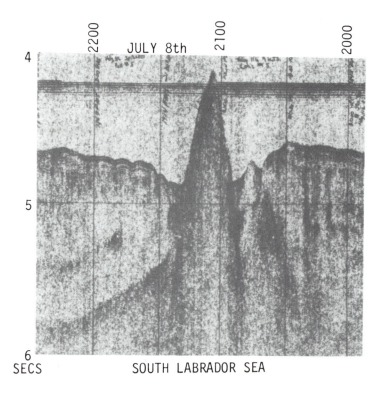

Figure 15-10 Seismic reflection profile in North Atlantic showing moat around seamount. (From T. A. Davies and A. Laughton, 1972)

marized by van Andel and others [1975] and Moore and others [1978] and is illustrated in Fig. 15-11. If each hiatus results from a single erosional event, the time of maximum extent can be determined; the time of erosion cessation can be determined if the overlying material has not been subsequently eroded by a secondary erosional event. The time of initiation of an erosional event cannot be precisely determined; however, regional synchronism and the age of an individual unconformity does suggest that the time of initiation is close to the age determined for the underlying sediment.

The rate of erosion is determined not only by bottom flow, but by the corrosiveness of bottom waters on biogenic sediments as well. In general, young bottom water is stripped of dissolved silica and so is more corrosive to siliceous tests than older bottom water, which gradually becomes more enriched in silica as these tests dissolve. The effect is opposite for calcite, with older waters being richer in CO_2 and hence more corrosive to calcium carbonate (see Chapter 14). Thus biogenic dissolution plays a crucial part in the formation of hiatuses. Although the

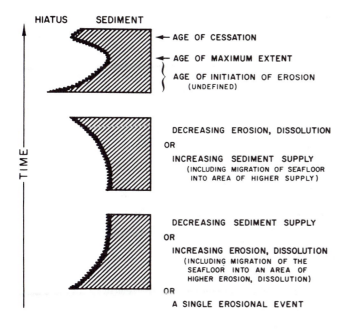

HIATUS   SEDIMENT

← AGE OF CESSATION

← AGE OF MAXIMUM EXTENT

AGE OF INITIATION OF EROSION
(UNDEFINED)

DECREASING EROSION, DISSOLUTION

OR

INCREASING SEDIMENT SUPPLY
(INCLUDING MIGRATION OF SEAFLOOR
INTO AREA OF HIGHER SUPPLY)

DECREASING SEDIMENT SUPPLY

OR

INCREASING EROSION, DISSOLUTION
(INCLUDING MIGRATION OF THE
SEAFLOOR INTO AN AREA OF
HIGHER EROSION, DISSOLUTION)

OR

A SINGLE EROSIONAL EVENT

TIME

Figure 15–11 Diagrammatic cross section of hiatuses with possible causes. Age of maximum extent and age of cessation can be defined; age of initiation cannot. (From T. C. Moore, Jr. et al., *Micropaleontology*, vol. 24, pp. 115, 118, 1978)

number of hiatuses in deep-sea sediment sections is striking, they are minor contributors to the recycling of biogenic materials. Estimates of dissolution of calcium carbonate and opaline silica in the water column and on the sea floor indicate that at least 80 percent of the carbonate and 96 percent of the silica is dissolved before it is incorporated into the stratigraphic record. Thus if hiatuses in this record result from the destruction of biogenic components, the average contribution toward the recycling of these components must be only another 10 percent of the carbonate supply and 2 percent of the silica supply. Hiatus formation is a major criterion for understanding oceanic circulation in the past; the historical implications are discussed in Chapters 17 through 19.

Manganese nodule pavements

It is well known that manganese nodules often occur in areas where the rate of sediment deposition is low enough to ensure that accreting manganese does not become covered (see Chapter 14). The rate of sediment deposition is particularly low in the area underlying the circum-Antarctic current [Kennett and Watkins, 1975]. In the Australasian sector of the Southern Ocean, all extensive fields of manganese nodules are associated with erosion of deep-sea sediments, and even more restricted occurrences of manganese nodules are associated with ocean-floor features indicating bottom-water activity. However, all manganese nodule fields are not associated with areas of active bottom currents (see Chapter 14); neither are all areas of active bottom erosion, like the Kerguelen and Campbell plateaus in the Southern Hemisphere, associated with widespread manganese nodule development. While it is clear that erosion enhances manganese nodule development, such dynamic activity by itself is not sufficient to produce widespread manganese nodule growth. Other factors of importance are the chemical environment and the proximity of element sources.

Transportation Processes

Bottom photographs of certain parts of the oceans show clouds of sediment particles close to the ocean floor. These clouds show transportation of sediment in bottom waters. Low concentrations of suspended materials are generally found throughout the water column, but a gradual increase in particulates is found from 500 to 1700 m above the bottom and a very large increase in particulate matter occurs within 50 to 200 m of the bottom. Transportation of sediments in the deep ocean takes place in two different, but intergrading, bodies: in conventional, high-velocity bottom currents in which suspensions can be extremely concentrated at least for short periods of time, and in generally more diffuse **nepheloid layers** (cloudy layers) of much lower concentrations (50 to 100 $\mu g/l$). Nepheloid layers are longer lived bodies of suspended sediments that may reach heights of several hundred meters above the ocean floor. They were discovered by Jerlov in 1953. These two processes, which certainly act in unison, laterally transport sediment in the deep sea. High-velocity bottom currents, close to the benthic boundary layers, stir up the sea floor periodically and transport even the coarser materials over short distances before redeposition; finer particles, injected into the nepheloid layer, can be transported over much greater distances.

The extent of lateral transport of deep-sea sediments was first realized during a study of the geographic distribution of distinct lithologies in the Northwest Atlantic Ocean. In 1961, Ericson and others found that brick-red clays occur in interglacial episodes of piston cores from the continental rise south of Cabot Strait between Newfoundland and Nova Scotia, but not in cores east of this boundary. It was later suggested that the red sediments were derived from late Paleozoic redbeds of New Brunswick and Nova Scotia and that bottom currents associated with the western boundary current transported this material southward. The thickness percentages of these red sediments decrease southward, from greater than 50 percent on the continental rise off Cabot Strait to less than 10 percent off Cape Hatteras, and these materials have even been detected on sediments adjacent to the Bahamas.

Erosion of material may take place very quickly, but deposition involving the entire sediment load is a very slow process. Bottom photographs show sediment clouds moving along the bottom at concentrations greater than those cited for the abyssal region as a whole. It probably takes only a few days to deposit much of the coarse material, but much longer periods of time are required for deposition of the finer materials from the nepheloid layer.

Nepheloid layers were first extensively surveyed and named by Ewing and Thorndike [1965] using an **optical nephelometer** in waters from 2000 to 4500 m off the east coast of North America. The nephelometer is a meter for measuring light scattering in situ. It is a photographic instrument that continually profiles the intensity of white-light scattering through the water column down to any oceanic depth. Determining the degree of light scattering is complex and is dependent upon particle concentration, size, shape, and composition, as well as other factors. Concentrations range from 10 to 100 mg/*l*. The suspended material in filtered water samples from areas of known light scattering are a control for the amounts of suspended materials carried in nepheloid layers. Because light-scattering properties result from many factors, calibrations are required for different water masses.

Water bottles have been used to collect suspended particulate matter in the ocean for mass and compositional analyses. Particulate concentrations are high in oceanic surface waters of high biological activity. Decomposition and remineralization rates are rapid above the seasonal thermocline. Particulate concentrations decrease at middle depth, but, as the sea floor is approached, levels may increase to those measured near the surface, largely from an increase in resuspension of sediments from the sea floor.

Biscaye and Eittreim [1977] developed a model of particulate concentrations in the oceans (Fig. 15–12). They plotted particulate concentrations in the clear intermediate waters and in the more turbid surface and bottom waters. **Clear water** is

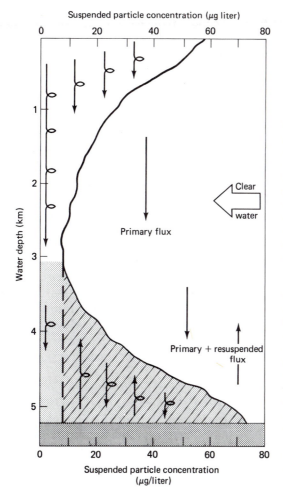

Figure 15–12 Typical nephelometer profile from an area with a strong nepheloid layer. The minimum in light scattering (suspended particulate concentration), which is at the top of the stippled area, is called the *clear-water minimum*. The clear-water minimum is defined as the upper limit of the nepheloid layer and all suspended matter below it as the gross particulate standing crop in units of g/cm^2 (stippled area). The model shown here schematically assumes that all particles falling from above the clear-water minimum have come from the surface layers. The suspended matter below the clear water which is in excess of the clear-water concentrations is defined as the net particulate standing crop (diagonally barred area), and it is assumed that this represents particles mixed upward or injected "upstream" and/or "upslope." The primary flux (F_p) represents the remnant of the surface-water particulate load during its downward transit. (After P. E. Biscaye and S. L. Eittreim, 1977 and Gardner, 1977)

a broad minimum in concentration and light scattering occurring at various intermediate water depths. Concentrations in clear water were as much as one to two orders of magnitude lower than those in surface water; suspended particulate load in this region reflects the downward transport of particles from surface waters. Nearer the bottom, particulate concentrations increase again due to resuspension, vertical mixing, and advection of sediments from the sea floor.

Vertical fluxes within the nepheloid layer are caused by primary material (that which settles from the upper water column) and resuspended particles (resuspension flux) and can be separated by determining the primary flux. If the net standing crop of suspended particles in the nepheloid layer is determined, gross estimates can be made for the residence time of resuspended particles in the nepheloid layer, assuming a steady state and uniform deposition and erosion. Analyses of sediment traps are also used to estimate the residence time for particles resuspended in the nepheloid layer [Gardner, 1977]. This residence time ranges from several days to weeks in the lowest 15 m of the water column and weeks to months in the lowest 100 m. These relatively short residence times indicate a rapid exchange between the surface sediment and the nepheloid layer, so for particles to be carried long distances in the nepheloid layer, they will frequently be deposited on the bottom and then resuspended.

Nepheloid layers are widespread in the world oceans and are intimately related to potential temperatures of bottom waters; thus they indicate the individual bottom waters. Biscaye and Eittreim [1977] mapped the suspended-particle load in the bottom nepheloid layer of the Atlantic (Fig. 15–13); this map shows that concentrations exceed those found in clear waters at intermediate depths. Most of the sediments in the nepheloid layer occur in a strip along the strong western boundary currents. Suspended materials are in much lower concentrations in other parts of the Atlantic Ocean. Similar work has also shown high concentrations associated with western boundary current activity in the Indian Ocean. The nepheloid layer is not a passive body of water with particles remaining in suspension indefinitely; it is maintained by constant resuspension and deposition of sediment, without which the nepheloid layer could only last a few months. The dynamic nature of these layers makes it possible for particles to be resuspended many times, being carried far from their origin, before being finally buried. The recycling of particles near the sea floor may increase dissolution of siliceous and carbonate material. Nepheloid layers are probably long-lived phenomena, because even though erosion can take place quickly, redeposition required much longer periods of time. Injection of particles into the nepheloid layer may occur at various topographic levels on the ocean floor, which can produce a nepheloid layer hundreds of meters thick and probably accounts for major oscillations in suspended load at different levels in the water column.

Most of the material in nepheloid layers consists of clay particles originally derived from land. Pak and others [1971], Carder and others [1971], and Zaneveld and others [1974] have examined the character of particles that make up nepheloid layers in the Pacific. They found that 90 percent of the total scattering is produced by particles ranging in diameter from 0.5 to 8.5 μm. There are several possible mechanisms for injecting this material into the water column. The primary mechanism is the erosion of the seabed by bottom currents. Other mechanisms in-

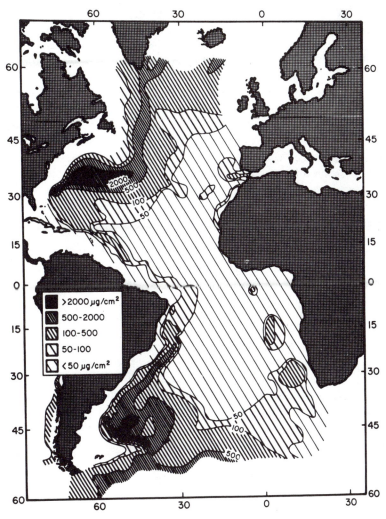

Figure 15-13 Distribution of suspended material in the nepheloid layer in the Atlantic Ocean. Material settling from above into the layer has been subtracted. (From I. N. McCave (after Biscaye and Eittreim, 1977), *Oceanus*, vol. 21, no. 1, 1978)

clude the injection of the fine material by turbidity currents, fine sediments moving down submarine canyons, and glacial flour being directly injected into the bottom waters at their sources around Antarctica. Information on the settling velocities of particles from nepheloid layers is insufficient.

Depositional Features

The ocean floor abounds with sedimentary depositional features created by bottom currents. Variation in sediment thicknesses is considerable over a few kilometers. Because the supply of pelagic material to the sea floor is uniform over great distances, irregular sediment-distribution patterns must be caused mostly by processes resulting from bottom-current action. Varying sediment thicknesses are associated with topographic irregularities, such as seamounts, knolls, ridges, and escarpments, and are attributed to streamlining and accelerating water masses as they flow around the obstructions. Differential sediment accumulations are more pronounced in the Atlantic than the Pacific, possibly due to greater dissolution of sediment involved in bottom transport in the Pacific Ocean.

There are complete gradations between the features of the abyssal landscape (Fig. 15–14), resulting from an intimately related cycle of erosion and deposition. Bed forms range in size from regular sinusoidal mud waves, several kilometers between crests and tens of meters high, to very small lineations and ripples, centimeters in size. The smaller features are short lived, on the order of days to weeks, while the huge sediment drifts are built by long-term bottom currents sweeping a region for hundreds of thousands or millions of years. The most dramatic current-produced abyssal bed forms occur on the western side of the ocean basins, the most prominent of which occur in the Atlantic Ocean. Very large supplies of terrigenous sediments have been available for reworking by bottom currents in the Atlantic Ocean, especially in northern parts, which underwent periodic major glacial activity.

There are several types of depositional features on the deep-ocean floor, including current lineations, ripples, sand and mud waves, sand dunes, and major ridges (Fig. 15–7). Current lineations tend to be depositional rather than erosional and result from deposition on the lee sides of solid objects, such as rocks, nodules, or clasts. Crag and tail features are common for abyssal microtopography and provide information on current directions. Wave and current ripples are best studied from bottom photographs, but medium-scale bed forms like sand waves (dunes) and mud waves are best studied from high-resolution, side-looking sonar records. The larger scale corrugations, with wavelengths of 1 to 5 km, have been mapped with echosounders.

Ripples

Ripples are small-scale, regular alternations of depositional ridges a few centimeters apart. They may be symmetric or asymmetric or transverse or longitudinal to flow; they are formed by currents or waves on relatively unconsolidated sand, silt, or mud. Wavelengths range from about 10 cm to a meter (Fig. 15–7) and amplitudes range from barely perceptible to 20 cm or more. Ripples occur at all depths in the ocean, from the beach to the abyss. The dynamics of deep-sea ripples formed by sand are well understood because of their similarity to shallow-water forms. Less is known about ripples formed in mud, which tend to occur in deeper water. Where ripples are formed in noncohesive sediment, they usually occur in well-sorted foraminiferal oozes and are known to form as currents travel at low rates (15–20 cm/sec). Ripples may be produced in noncohesive sediments as fine as 10 μ, at speeds of 30 cm/sec. Ripples may also form by wave action on shallow seamounts and ridges.

Figure 15-14 The effects of increasing currents on the ocean floor. (From *The Face of the Deep* by Bruce C. Heezen and Charles D. Hollister. Copyright © 1971 by Oxford University Press, Inc. Reprinted by permission.)

Sediment waves and dunes

As ripples become larger they grade into features called **sediment waves** (Fig. 15-15), formed in both sand and mud with wavelengths ranging from about 10 m to 10 km (Fig. 15-7). **Mud waves** are regular, large-scale ridges of cohesive muddy sediment (Fig. 15-15), occurring commonly down-current from high sediment input regions like submarine canyons. These are very common features on many continental rises.

Sand waves, or **dunes,** are the large-scale counterparts of sand ripples with typical lateral dimensions of 10 to 100 m (Fig. 15-7). All known abyssal dunes are transverse to flow with downstream slip faces; only isolated dunes are aligned in rows parallel to the flow. These features are best seen on side-looking sonar records. An excellent example, described by Lonsdale and Malfait [1974], is a major dune field on the Carnegie ridge in the eastern equatorial Pacific (Fig. 15-16). This field includes isolated abyssal barchans arranged in down-current rows, just as subaerial desert barchans often are. Also present are transverse dunes in regular trains, with many transitional forms. Most of the barchans are between 10 and 100 m wide between the horns and in length from dune apex to the horn tips (median length is 20 m). Their salient characteristics are gentle upstream slopes and avalanching slip faces with inclinations, photogrammetrically determined, between 25° and 30°; they are close to the angle of repose for the cohesionless sand of which they are composed. The dunes are inferred to move down the flat floor of an erosional valley that has been excavated through calcareous ooze to a resistant, manganese-encrusted chalk stratum. Geologic and hydrographic data suggest that these sand waves migrate episodically, probably during spillover of dense water across the Carnegie ridge, which evidently occurs at intervals of several years or more. After an interval of 30 months, individual dunes could still be recognized in the same relative position indicating little, if any, migration during that interval.

The nature of the gradation between ripples and dunes is not fully understood, and the transition velocity can only be determined empirically. Flume

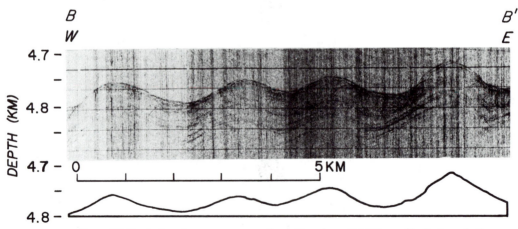

Figure 15-15 Sediment waves on ocean floor. Top shows 3.5 kHz profile; bottom, bathymetric profile. (From R. D. Flood, doctoral dissertation, 1978)

530 Geological Effects of Bottom Currents: Motion and Commotion

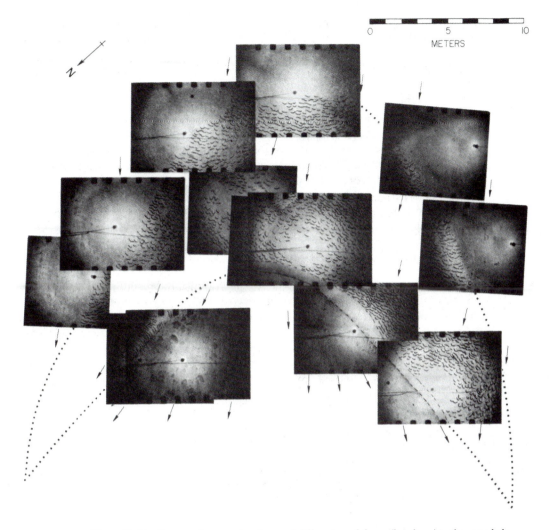

Figure 15-16 Bottom photographs of several different sand dunes (barchans) in the same belt arranged as a composite "photomosaic" of a typical barchan. Arrows show sand transport paths inferred from the orientation of ripples on the dune surface and of sand streamers on the adjacent pavement. (From P. Londsdale and B. Malfait, *Geol. Soc. Amer. Bull.*, vol. 85, p. 1706, 1974, courtesy The Geological Society of America)

studies suggest that currents of 40 to 50 cm/sec are eventually required to build dunes of quartz sand with diameter 100 μm. The threshold for motion of low-density foraminiferal sand in flumes is lower (about 20 cm/sec, measured 1 m above the bed) [Southard and others, 1971; Lonsdale and others, 1972]. Lonsdale and Malfait [1974] suggested that the dunes on the Carnegie ridge formed at velocities greater than 30 ± 10 cm/sec.

Abyssal sand dunes are so difficult to detect with conventional techniques that their numbers have been greatly underestimated. Because of their slight amplitude, they are usually imperceptible on echo-sounder records and are frequently unrecognized when partially captured in the narrow field of bottom photographs. Their distribution is restricted to areas where the currents are strong (probably more than 30 cm/sec) and where the superficial sediment is uncohesive sand. Strong, deep-sea currents can create a sandy bed by winnowing, especially if the sediment contains abundant foraminiferal tests. However, many of the fastest bottom currents are of AABW and, except at high southern latitudes, this water mass generally impinges on sediment (below the calcite compensation depth) that has such a dearth of coarse components that a sand bed cannot be formed. In these circumstances, sand dunes are not observed even where the flow is locally accelerated in topographic constrictions. Dunes, unlike mud waves, are probably also rare on continental rises because of the rapid influx of cohesive terrigenous lutites. However, where western boundary currents are particularly strong, a superficial bed of winnowed sand, with ripples and dunes, can occur.

Depositional ridges

A major step in understanding the magnitude of bottom currents was the discovery of large, elongate ridges composed of thick piles of sediments, such as the Feni, Gardar, and Erik ridges in the North Atlantic (Fig. 15-17). These **sediment drifts** (or ridges) are acoustically poorly stratified features, elongated parallel to the bottom-current flow; they are often mantled with "mud waves" having wavelengths of 1 to 2 km and amplitudes of a few tens of meters. The ridges themselves are often hundreds of kilometers long and tens of kilometers wide and

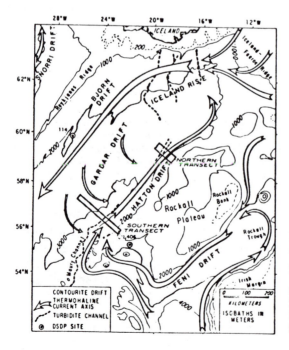

Figure 15-17 Location of Feni, Hatton, and Bjorn sediment drifts in the North Atlantic south of Iceland, in relation to bottom-current axes. (From I. N. McCave and others, *J. of Sed. Petrol.*, vol. 50, 1980, reprinted by permission of the Society of Economic Paleontologists and Mineralogists)

vary from sharply peaked to slightly convex. Sediment accumulation rates are generally high, with known measured rates of at least 12 cm per 1000 years. These ridges are formed by deposition from contour currents, called contourites by some workers [Hollister and Heezen, 1972]. Well-sorted sands or silts are formed in well-defined layers from 0.1 to 10 cm thick, with sharp upper and lower contacts and often with cross bedding. The sediment deposition forming these elongate ridges results from the interactions between different currents (such as the Gulf Stream and the Western Boundary Undercurrent, or WBUC) or between active bottom currents and relatively static water (Fig. 15–18). Sediment ridges form parallel to these currents, not under the center of a current but at the margin of the flow (Fig. 15–18). Under some conditions, where the current is not banked against a steep topography, interaction between the current and the adjacent stationary water may occur on both sides, developing twin ridges. These large sediment drift deposits are of considerable age, much of their formation having occurred during glacial inter-

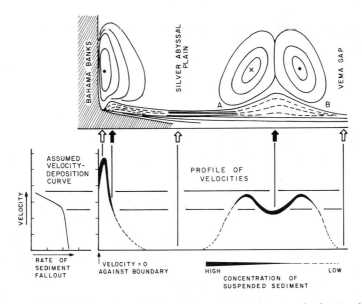

Figure 15–18 A schematic model for the formation of depositonal ridges on the ocean floor. Shown is modern bottom flow regime, concentration of transported sediment, and depositional conditions across the Caicos Outer Ridge and Greater Antilles Outer Ridge. Dots represent currents flowing toward the reader; X's are currents flowing away. Maximum deposition occurs where two criteria are met: (1) high concentrations of suspended sediment in the flow, and (2) current speeds dropping into the critical velocity range for deposition. At lower left is a schematic representation of the relationship between current velocity and rate of sediment fallout. The curve assumes a rapid increase in fallout over a relatively short velocity range, although this feature is not critical to the model. The curve would be shifted to higher or lower velocities for coarser and finer grain sizes of suspended sediment, respectively. Open and filled arrows indicate zones of slow and rapid deposition, respectively. (From B. E. Tucholke and J. I. Ewing, *Geol. Soc. Amer. Bull.*, vol. 85, p. 1800, 1974, courtesy The Geological Society of America)

vals; hence they are relict features, and measured current velocities associated may not reflect the conditions under which major formation took place.

North Atlantic sediment drifts

The importance of currents in modifying and controlling sedimentation processes in the deep basin of the western North Atlantic has been demonstrated in numerous studies [Heezen and Hollister, 1964; Hollister and Heezen, 1972; Laine and Hollister, 1981; Davies and Laughton, 1972; Flood, 1978]. The synthesis of diverse lines of evidence, such as seismic profiling records, sediment composition, suspended matter distribution, and bottom-current characteristics, has revealed many sediment drifts in this region, and the distribution patterns show a general correlation with the spreading direction of dense bottom water. Bottom water is derived mostly from the Norwegian Sea and spills over ridges between Iceland, Scotland, and Greenland (see Chapter 8). This water source first became available due to major tectonic reorganization of sea-floor spreading on the Reykjanes ridge during the middle Cenozoic (early Miocene, 18 Ma) when most drifts began to form. The bottom current trends to the right due to Coriolis effect and follows the regional bathymetric contours (Fig. 15–17). Initially, flow is confined to the topographic contours of a series of basins and ridges in high latitudes of the North Atlantic (Fig. 15–17) and later flows south along the continental rise of eastern North America as the WBUC. Studies using tracers have demonstrated that sediment has been transported by the WBUC from the area off New England and the Maritime Provinces of Canada at least as far south as the Blake-Bahama outer ridge [Hollister and others, 1978]. There are some indications from sediment-distribution patterns that sediments may be transported even further southward to the Caicos outer ridge and the Greater Antilles outer ridge.

After the bottom water flows over the Iceland-Faeroes ridge, it travels southwest along the eastern side of the Rockall Bank, producing the Feni Drift (Fig. 15–17) [McCave and others, 1980]. The deep current continues around the toe of the Rockall Plateau, hugging the contours but moving to the right at every opportunity. The current swings northward, this time on the western side of the Rockall Plateau, depositing the Hatton Drift. At the head of the basin south of Iceland, the bottom current is enhanced by bottom waters spilling through the Iceland-Faeroes Channel. The current is then redirected southward along the eastern side of the Reykjanes ridge, forming the Bjorn Drift (Fig. 15–17) and later the Gardar Drift.

The Norwegian Sea overflow current seems to lose most of its load of suspended sediment on the Gardar Drift before moving through a conduit in the mid-Atlantic ridge, called the Charlie-Gibbs fracture zone. As soon as the bottom water enters the Northwest Atlantic basin, it follows the topographic contours of the western side of the Reykjanes ridge northward through the Irminger basin. At the northern end of this basin, the currents are invigorated by addition of new bottom waters, which then continue their flow around Greenland, depositing the Erik ridge, continuing through the Labrador Sea, and sweeping southward as the WBUC. The Western Boundary Undercurrent is a mixture of some fresher water from the Labrador basin and saltier North Atlantic basin water. This current plays

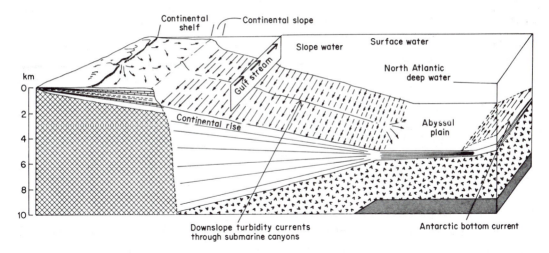

Figure 15-19 Shaping of the continental rise of eastern North America by contour currents. (After B. C. Heezen et al., *Science*, vol. 152, 1966, pp. 502–508, copyright © 1966 by the American Association for the Advancement of Science)

a major role in molding the lower continental rise of North America (Fig. 15-19). Antarctic bottom water entering the system via the Vema gap–southern Bermuda rise, flows around the Hatteras abyssal plain and joins the WBUC at about the latitude of Cape Hatteras. It seems that a large amount of North American sediment, swept from the continent over several million years, has been incorporated into this flow and has become modeled into a feature resembling a continental rise—the Blake-Bahama outer ridge—which is located seaward of the Blake Plateau about 700 km east of Florida.

This ridge is a major topographic feature over 800 km long and 400 km wide, with an average thickness of 2 km. The Blake-Bahama outer ridge is bounded on the east by the Hatteras abyssal plain and on the north by the continental rise of North America. Instead of forming along the continental margin, the ridge has formed as a separate topographic protrusion into the Atlantic. It seems likely that this has resulted from the interaction between the northward-flowing Gulf Stream and the southward-flowing WBUC. The Gulf Stream, although dominantly a surface current, extends deep enough in places to erode and redistribute sediment and create significant bedforms. When the current returns to the continental margin along the steep Blake escarpment after flowing around the toe of the Blake-Bahama outer ridge, it again flows southward at an increased rate (greater than 30 cm/sec), transporting considerable amounts of fine carbonate sediment from the Bahama Banks and depositing the Caicos outer ridge. The last of the large North Atlantic sedimentary deposits formed from the WBUC is the Greater Antilles outer ridge. Deep-sea drilling on the ridge suggests that the ridge was formed by bottom currents during the last 10 m.y.

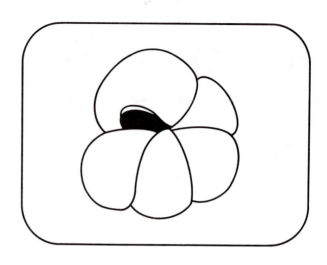

sixteen
Oceanic Microfossils

The microscopic organisms are very inferior in individual energy to lions and elephants, but in their united influences they are far more important than all of these animals.

C. G. Ehrenberg, 1862; translated by Mrs. Bury.

The vast majority of planktonic and benthonic organisms in the ocean have only soft parts and therefore have no chance of preservation. A few groups, however, do have hard parts capable of fossilization. These are very important because they are the major contributors of material to deep-sea sediments and because oceanic paleoenvironmental history has been derived largely from the study of these groups. The marine geologist is therefore concerned only with those groups that do leave a fossil record.

Oceanic microorganisms that contribute to deep-sea sediments are floating, or planktonic, forms and bottom-dwelling, or benthonic, forms. The planktonic forms, both phytoplankton and zooplankton, are the most important in the formation of deep-sea sediment and are significantly different in their ecological functions. The benthonic forms are not normally important as sediment builders except on parts of the continental shelf; they do, however, provide vital information about the benthonic environment and its associated water masses.

Most oceanic microfossil groups construct their shells of calcite (or aragonite) or opaline silica. Those groups that construct their shells of other materials, such as complex organic substances or cemented sandgrains, are relatively minor as sediment builders and as sources of information about the marine paleoenvironment.

The calcareous oceanic microplankton are the foraminiferida and the coccolithophorida, with small contributions from other groups, such as Mollusca and Ostracoda. The siliceous oceanic microplankton are the diatoms and the radiolaria. The marine benthonic groups important in sediments at depths greater than the continental shelf are the foraminifera and ostracoda. Although some groups, such as the Mollusca and ahermatypic corals do occur at great depths, they are too rare in the fossil record to be of much value to the marine geologist.

CALCAREOUS MICROFOSSILS

Foraminifera

The foraminifera (Class, Sarcodina; Order, Foraminiferida) are an extremely diverse group of marine, shell-bearing protozoans with several benthonic families and few planktonic families. They normally range in size from 50 to 400 μm, but they may grow larger. Although the benthonic forms are taxonomically more diverse than planktonic forms, there are fewer individuals per species. The tests of foraminifera are constructed of calcium carbonate except in one major suborder (Textularinae) of benthonic foraminifera, which construct their tests of cemented grains of sand and—in one small group—aragonite. The function of the test is probably multiple, as pointed out by Marzalek and others [1969], including protection from predation and unfavorable environmental conditions, control of buoyancy, sinking, and even stabilizing the animal if buildup of light-weight lipids and fats occurs.

Foraminifera are considered marine, although they are important in marginal marine waters of both low and high salinities. Foraminifera have been studied more extensively than any other group of oceanic microfossils and make up 2.5 percent of all known animal species since the Cambrian. About 38,000 types have been named, with approximately 10,000 others still present in the seas [Boltovskoy and Wright, 1976]; however, the actual number is much lower because many have been described more than once. A group known as the larger foraminifera are, as the name suggests, much larger than normal (up to 16 mm) and are somewhat different ecologically. They occupy shallow, warm seas and require different methods for their study, including thin-section work to study critical internal structures. Although the larger foraminifera are important for correlating tropical shallow-water marine sequences, they are unimportant in studies of deep-sea sediments and so will not be considered further.

The foraminifera are of great importance geologically and oceanographically because of their high diversity and their occurrence in all marine environments (from marginal marine to deep basins and in surface or near-surface waters) at all latitudes and within all oceans. The group is well studied and taxonomically well known. Taxonomic work now is not much concerned with the description of new forms, but with regrouping entities previously oversplit by zealous taxonomists. Their relative abundance in marine sediments has been instrumental in placing marine paleoecology on a firm quantitative basis in many environments. Originally foraminifera were used stratigraphically in late Mesozoic and Cenozoic rocks and later were intensively studied to assist oil exploration. Benthonic foraminifera were studied almost exclusively during these early investigations, but since the late 1950s efforts have been concentrated on the planktonic forms. The use of foraminifera in geologic and paleoceanographic analysis is closely related to their basic habitat—planktonic versus benthonic. Because the characteristics of these two groups are very different, each will be described separately.

Planktonic Foraminifera

Planktonic foraminifera first developed during the late Jurassic as small, simple, *Globigerina*-like forms. These apparently insignificant forms showed dramatic evolutionary radiation throughout the Cretaceous and have been important components of the oceanic plankton ever since. More than 40 genera and at least 400 species have evolved during the last 130 m.y. During this time the group has undergone major changes in evolution; extinction and diversity has led them to be widely used in correlating and dating rock sequences.

In the present-day oceans there are about 30 species (Fig. 16–1) which are grouped into two families: the Globigerinidae (spinose forms) and the Globorotaliidae (nonspinose forms). Their past and present stratigraphic distribution, along with aspects of their test morphology, are well known. Little is known about the biology and ecology of the living organisms. Plankton tows show that this group lives mainly in the euphotic zone and that the few deep-water species probably spend their earlier life stages in near-surface waters. This is as expected, because most food resources occur in the upper 200 m of the water column. They live in marine waters of normal salinity and are very rare in brackish waters. They are uncommon in most nearshore shallow waters, except for insular regions and regions of low freshwater runoff, such as Baja California.

Planktonic foraminiferal test morphology has been of wide variety during the Cretaceous and Cenozoic, including biserial, planispiral, trochospiral, streptospiral, and enveloping stages (Figs. 16–1 and 16–2). Major evolutionary trends have acted upon the morphology, resulting in morphotypes of differing importance. For example, biserial forms were prevalent during early evolutionary development in the Cretaceous, were drastically reduced during the global biotic crisis at the end of the Cretaceous, and today are a single, small, rather insignificant species.

Flotation and vertical distribution

The nonmotile planktonic foraminifera must be buoyant. The test of planktonic foraminifera is composed of calcite having a density about 2.7 times that of seawater, and experimental evidence based on the settling rates of empty tests in seawater indicates sinking velocities in the range of 0.3 to 2.3 cm/sec. In order to remain buoyant all plankton are small. This increases their specific surface area, thus increasing frictional drag, resulting in increased buoyancy. Their globose tests and their high surface relief also increase surface area. The group conserves the amount of material in the test by being more porous than most benthonic foraminifera. Approximately one-half of the present-day forms (family, Globigerinidae) possess spines, which radiate out from the test surface (Fig. 16–2). These serve as an additional support for the protoplasm. Some spine systems are up to five times the diameter of the chamber from which they grow. In some species spines are flattened

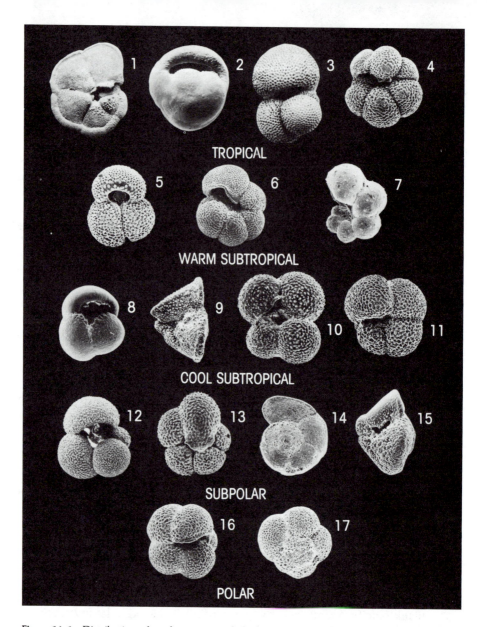

Figure 16-1 Distribution of modern species of planktonic foraminifera:

Tropical forms: (1) *Globorotalia menardii,* × 54; (2) *Pulleniatina obliquiloculata,* × 110; (3) *Globigerinoides sacculifer,* × 148; (4) *Neogloboquadrina dutertrei,* (form with umbilical teeth), × 81.

Warm subtropical forms: (5) *Globigerinoides ruber,* × 102: (6) *Neogloboquadrina dutertrei,* × 72; (7) *Hastigerina pelagica,* × (60).

Cool subtropical forms: (8) *Globorotalia inflata,* × 60; (9) *Globorotalia truncatulinoides,* × 86; (10) *Globigerina falconensis,* × 120; (11) *Neogloboquadrina pachyderma,* (dextral coiling), × 138.

Subpolar forms: (12) *Globigerina bulloides,* × 91; (13) *Globigerina quinqueloba,* × 180; (14) *Globorotalia truncatulinoides* × 60; (15) *Globorotalia truncatulinoides,* × 89.

Polar forms: (16) *Neogloboquadrina pachyderma,* (sinistral coiling), × 141; (17) *Neogloboquadrina pachyderma,* (sinistral coiling), × 152.

or have different flanges when viewed in cross section, which increases surface area (Fig. 16–2). Several species develop relatively large apertures or numerous smaller apertures, which further reduce shell weight.

The viscosity of water increases as temperature decreases. The viscosity of water at 0°C is twice that at 25°C. Viscosity therefore increases with both depth and latitude. Because planktonic foraminiferal tests will sink twice as rapidly in warmer waters, they must be more lightly constructed in these environments. Therefore tropical planktonic foraminifera tend to have thinner shells, higher porosity, more exaggerated apertures, and distinctive, well-developed spines. Planktonic foraminifera also may increase buoyancy by including gas vacuoles, lipids, and other fatty substances within the protoplasm. Thus the combination of both test and protoplasm has a specific gravity approximately the same as the density of seawater (about 1.026 g/cm^3).

Certain species of planktonic foraminifera need not only to maintain buoyancy but also to maintain certain positions within the water column. Many species prefer particular depth habitats (depth stratification), and extensive vertical migration occurs within the life cycles of certain species. These migrating groups must be capable of sinking. Since this vertical migration is irreversible, it often occurs late in the ontogeny of many species, including most species of *Globorotalia*. Individuals in near-surface waters are usually smaller than those from greater depths. Some planktonic foraminifera also undergo diurnal and other oscillating vertical migration. Emiliani [1954] first postulated that different species prefer different depth habitats based on isotopic temperature measurements on tests from the sediment. Because the maximum density of protoplasm only slightly exceeds the density of seawater, effective sinking is not possible without the test. Many forms of planktonic foraminifera, particularly *Globorotalia*, increase their test densities by forming a calcite crust (secondary calcification); the deepest-dwelling morphotypes have the thickest tests. Thus the planktonic foraminifera control their buoyancy by the specific gravity of the protoplasm, the calcite (by test growth and secondary calcification), and by absolute size of the test. Planktonic foraminifera are not aimless drifters in the water column. They can regulate their buoyancy, allowing them to exploit particular environmental niches. The density of seawater is a major factor in the distribution of foraminifera. Planktonic foraminifera are reliable indicators of seawater temperature because of the direct correlation between seawater density and seawater temperature.

Emiliani [1971], Bé [1977], and other workers have shown vertical distribution in Recent planktonic foraminifera. Bé [1977] recognizes shallow, intermediate, and deep dwellers including the following:

SHALLOW: Most spinose species, including the species of *Globigerinoides* and some *Globigerina* (Fig. 16–2).

INTERMEDIATE: (50–100 m). Spinose species: *Globigerina bulloides, Hastigerina pelagica, Orbulina universa, Globigerinella aequilateralis, Globigerina calida;* Nonspinose species: *Pulleniatina*

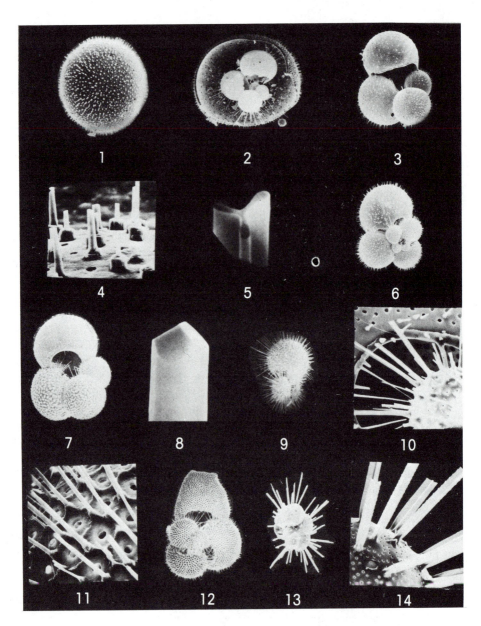

Figure 16–2 Spinose planktonic foraminifera captured living in the water column:

Orbulina universa: This species is distinguished by a juvenile trochospiral stage and an adult spherical stage.

(1) Spherical chamber, ×220; (2) A dissected spherical test showing internal trochospiral test, ×120; (3) Juvenile trochospiral form, ventral view, showing the umbilical primary aperture and a small sutural aperture, ×200; (4) Oblique view of the surface showing spines with terraced spinal shoulders, ×3500; (5) Tip of a triradiate spine, ×23,200; (6) Juvenile trochospiral form, dorsal view, showing sutural supplementary apertures, ×180.

obliquiloculata, Neogloboquadrina dutertrei, Candeina nitida, Globigerinita glutinata.

DEEP: *Globorotalia, Neogloboquadrina pachyderma, Sphaeroidinella dehiscens.*

Productivity and sinking

Little is known about what planktonic foraminifera eat. Several are in symbiotic association with zooxanthellae, which probably provide part or all of the food through photosynthesis. Most planktonic foraminifera are probably herbivorous, feeding on algae; however, one species is known to be carnivorous, preying on other plankton as large as copepods [Bé, 1977]. The planktonic foraminifera, *Globigerina bulloides* and *Globorotalia truncatulinoides,* are reported to have multiple food requirements [Lee and others, 1966; Tolderlund and Bé, 1971]. The distribution of planktonic foraminifera, like other zooplankton, appears to be governed mainly by the availability of food. Surface waters with high nutrient content generally support the largest populations [Bradshaw, 1959; Parker, 1960] except in areas greater than 60° latitude, where light intensity is reduced greatly during the year.

Estimates of the concentration of planktonic foraminifera vary significantly depending on the mesh size used in capturing them. Reported concentrations may vary between 10^{-3} and 10^3 per cubic meter [Berger, 1971]. Many estimates $(10/m^3)$ are much too low because many small specimens are lost through coarse mesh sizes. Life assemblages of planktonic foraminifera have been captured in net tows, by some pumps, and now by scuba divers with bottles. It is important to understand the life cycle because foraminifera vacate their tests when reproducing, thereby contributing to the sediment during reproduction and after death. Virtually nothing is known of the life cycle of planktonic foraminifera because of little success with culturing. Experiments in the Santa Barbara basin area, California, indicate life spans of from 2 weeks to 1 or 2 months [Berger and Soutar, 1967; Berger, 1971A].

Within each oceanic realm, concentrations seem to be at least ten times greater in the fertile high-latitude coastal and equatorial regions of upwelling than in the

---

Globigerina bulloides
(7) Ventral view showing spines in umbilical region, × 200; (8) Tip of a single spine showing crystal surfaces, × 30,000.

Globigerinella aequilateralis
(10) Surface of early chambers and part of aperture showing triradiate, round, and triangular spines, × 1000.

Globigerinoides sacculifer
(9) Side view showing spines, × 180; (11) Surface of ultimate chamber showing spines and newly formed ridges, × 1800; (12) Dorsal view showing supplementary apertures, × 140.

Hastigerina pelagica: This species is somewhat similar to *G. aequilateralis* except it has distinct spines.
(13) Oblique view showing distribution of spines, × 100; (14) Surface of ultimate chamber showing spines and spinal banks, × 700.

(Courtesy G. Vilks, Bedford Institute of Oceanography)

more sluggish central and mid-latitude regions. In general, calculated values of foraminiferal standing stock parallels the range of phosphate concentrations and total plankton standing stock. The highest known concentrations (North Pacific) are about 1000 times greater than the lowest ones (Sargasso Sea). Total abundance in the Atlantic appears to be about half of that for the Indo-Pacific.

Sinking velocities are important to paleoecology because lateral drift during sinking can effect the comparison of overlying assemblages with surface sediment assemblages; also, if sinking is very slow, dissolution can occur (see Chapter 14). This drift depends on how rapidly empty foraminiferal tests sink. Early laboratory experiments suggested that tests sink at a rate of 2 cm/sec, but these estimates were on relatively thick-shelled tests in a nonturbulent environment. Berger [1971B] considers these rates to be about ten times too high. Nevertheless, most foraminifera would sink to the bottom after only a few tens of days, making sinking rates unimportant in paleoecological interpretations.

Modern biogeography

As a result of the *Challenger* expedition, Murray [1897] first recognized the temperature control in the distribution of many species, the polar-tropical diversity gradient, and individual species distribution. Now general outlines of biogeographic patterns have become quite firm.

Planktonic foraminifera are very sensitive markers of water masses. Their distributions in Recent surface sediments have been the subject of most distribution research, but some classical studies have examined them in the water masses. The environmental factors usually discussed to account for distribution patterns are those that are easily measured: temperature, salinity, food supply, oxygen, and pH. The first studies, done by Bradshaw [1959], distinguished clear faunal water-mass relations. Even within small areas, sharp boundaries between water masses are clearly distinguished by well-defined discontinuities in the fauna (Fig. 16-3). This is demonstrated by the distribution in the Northwest Pacific of the warm Kuroshio and cool Oyashio currents. Of the various environmental parameters controlling the distribution of planktonic foraminifera, temperature is generally considered the most important, although with some species other factors, such as salinity, are of equal or greater importance. It has been found [Jones, 1967] that relatively small variations in the measured environment exert major control in the population dynamics of groups. Temperature variations as small as 2°C and salinity variations as little as 0.2 to 0.5 ppm exert primary control over the abundance of many species. For instance, the Equatorial Undercurrent (marked by high salinities) is enriched in *Neogloboquadrina dutertrei* (a euryhaline form; Fig. 16-4). This suggests that this species is strongly controlled by small variations in salinity. However, because temperature and salinity—as well as a number of other conservative water properties—are closely related, it is often difficult to differentiate which of these factors is in control. For example, cold-water species tend to be associated with low salinities because of the relationship of low temperature with low salinities. Thus water-mass definitions of distribution are more appropriately used than are single

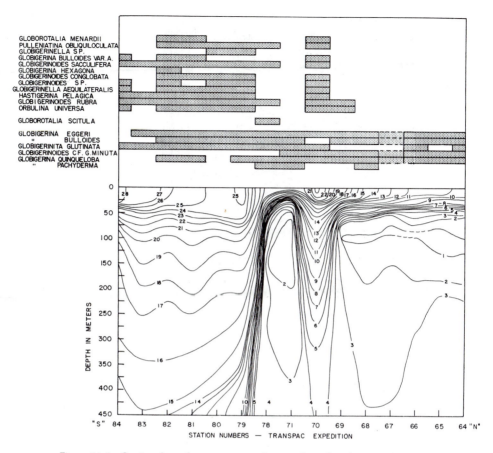

GLOBOROTALIA MENARDII
PULLENIATINA OBLIQUILOCULATA
GLOBIGERINELLA S.P.
GLOBIGERINA BULLOIDES VAR.A.
GLOBIGERINOIDES SACCULIFERA
GLOBIGERINA HEXAGONA
GLOBIGERINOIDES CONGLOBATA
GLOBIGERINOIDES S.P.
GLOBIGERINELLA AEQUILATERALIS
HASTIGERINA PELAGICA
GLOBIGERINOIDES RUBRA
ORBULINA UNIVERSA

GLOBOROTALIA SCITULA

GLOBIGERINA EGGERI
      "   BULLOIDES
GLOBIGERINITA GLUTINATA
GLOBIGERINOIDES CF. G.MINUTA
GLOBIGERINA QUINQUELOBA
      "   PACHYDERMA

DEPTH IN METERS

"S" 84 83 82 81 80 79 78 71 70 69 68 67 66 65 64 "N"
STATION NUMBERS — TRANSPAC EXPEDITION

Figure 16-3   Section through upper water column at boundary between the warm
Kuroshio and the cold Oyashio currents showing the effects upon planktonic
foraminiferal distributions in surface waters. (From J. S. Bradshaw, *Cushman Foundation for Foraminiferal Research*, vol. 10, p. 95, 1959)

parameters. The planktonic foraminifera act as "fingerprints" of pattern changes
in these water masses.

Planktonic foraminifera are distributed in the world oceans in roughly a
bipolar manner with distinct gradations occurring between the polar and equatorial
assemblages. The latitudinal changes exhibited by planktonic foraminifera which
need consideration are as follows:

1. Diversity of the total assemblage.

2. Taxonomic difference (presence or absence of species at different latitudes).

3. Changes in percent frequency of total abundance.

4. Intraspecific morphological variation, which is either phenotypic variation

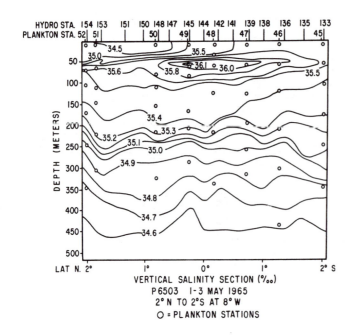

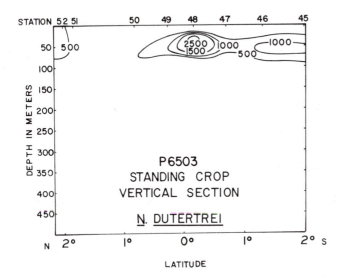

Figure 16-4 Relationship between salinity in upper water column and standing crop of *Neogloboquadrina dutertrei* in the Gulf of Guinea. (From J. I. Jones, *Micropaleontology* vol. 13, pp. 496, 498, 1967)

or subspeciation and acts upon such features as coiling direction, general shape, and test thickness.

Highest diversity occurs in the tropical regions, where over 25 species are represented. Diversity decreases with increasing latitude to only one species in true Arctic and Antarctic waters (Fig. 16-5). The changing species distributions and

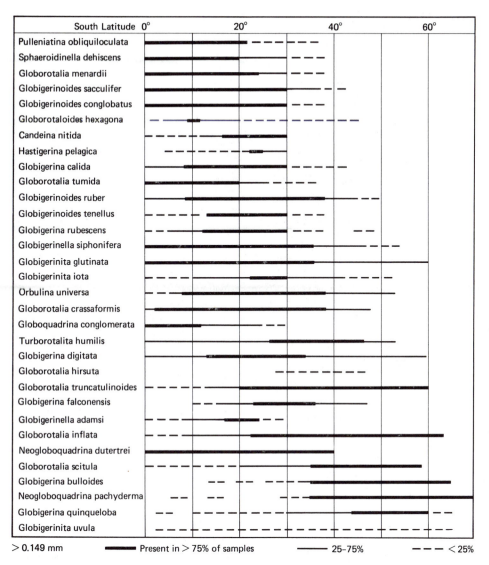

Figure 16-5 Latitudinal distribution of planktonic foraminifera in South Pacific. (After F. Parker and W. H. Berger, 1971)

abundances enable distinct faunal groupings to be defined [Bé, 1959; Parker, 1971]. In general, five distinct groups determined by water mass can be differentiated between the tropics and polar areas (temperature ranges are approximate).

1. Tropical: 24–30°C.

2. Warm subtropical: 20–24°C.

3. Cool subtropical: 12–20°C.

4. Subpolar: 5–12°C.

5. Polar (Antarctic or Arctic): 0–5°C.

The boundaries between faunas and subfaunas often correspond to distinct physical oceanographic features.

1. The tropical-subtropical faunal boundary with the tropical convergence.

2. The cool-to-warm subtropical boundary at 20°C.

3. The subpolar-to-cool subtropical boundary at the subtropical or sub-antarctic convergence.

4. The polar-to-subantarctic, with the Antarctic convergence or polar front.

TROPICAL SPECIES INCLUDE:	*Globorotalia menardii* (or *cultrata*) *Globorotalia tumida* *Pulleniatina obliquiloculata* *Sphaeroidinella dehiscens* *Neogloboquadrina dutertrei* (toothed form)
SPECIES IMPORTANT IN BOTH TROPICAL AND WARM SUBTROPICAL WATERS INCLUDE:	*Globigerinoides ruber* *Globigerinoides sacculifer*
WARM SUBTROPICAL SPECIES INCLUDE:	*Hastigerina pelagica* *Globorotalia hirsuta* *Neogloboquadrina dutertrei* (nontoothed form)
COOL SUBTROPICAL SPECIES INCLUDE:	*Globorotalia inflata* *Globorotalia truncatulinoides* *Globigerina falconensis* *Neogloboquadrina pachyderma* (dextral coiling)
SUBPOLAR SPECIES INCLUDE:	*Globigerina bulloides* *Globigerina quinqueloba* *Neogloboquadrina pachyderma* (sinistral coiling)
POLAR SPECIES IS:	*Neogloboquadrina pachyderma* (sinistral coiling)

There are several species that are distinctive of one water mass but often range into other water masses. Furthermore, differences exist in ranges between oceans and between hemispheres. In the North Atlantic, Imbrie and Kipp [1971] and Kipp [1976] found that factor analysis of surface sediments yielded 6 assemblage groupings: polar, subpolar, transitional, subtropical, tropical, and a gyre margin assemblage consisting of certain tropical species. The assemblage distribution (Fig. 16-6) follows the Gulf Stream transport into higher North Atlantic latitudes and

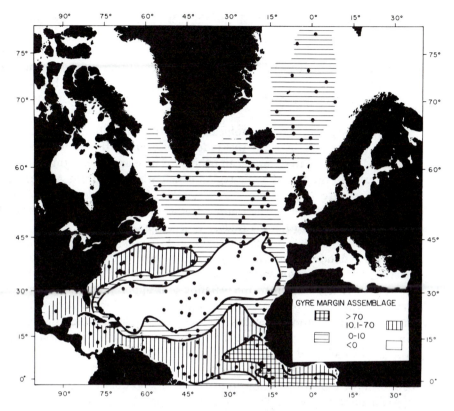

Figure 16-6 Distribution of the gyre margin planktonic foraminiferal assemblage in surface sediments of the North Atlantic. Values plotted are varimax factor scores. (From N. G. Kipp, in *Investigation of Late Quaternary Paleoceanography and Paleoclimatology,* ed. R. M. Cline and J. D. Hays, GSA Memoir 145, p. 27, 1976, courtesy The Geological Society of America)

ends in the area where the surface expression of the Gulf Stream markedly weakens. Cifelli [1976] also observed close relationships between marginal assemblages and clockwise gyral circulation as expressed in patterns of the Gulf Stream and Canary Current, which cross latitudinal lines. Distinct cross-latitudinal transport also occurs in the Northwest Pacific with the Kuroshio Current. Cooler faunas are transported into relatively low latitudes by California and Canary currents in the Northern Hemisphere and by the Peru–Chile and Angola currents in the Southern Hemisphere. Thus the plankton assemblages closely mirror the asymmetric surface current system of the oceans.

Although bipolarity in the planktonic foraminiferal assemblages of the oceans is striking, there are several departures. The most important is the distribution differences of *Globorotalia truncatulinoides* between the two hemispheres. In the Southern Hemisphere the species is well represented (as a distinct phenotype) in subantarctic waters with temperatures as low as 5°C. In the Northern Hemisphere the species is confined to water with temperatures greater than 10°C. Stratigraphic studies show that the species adapted to subantarctic conditions only 200 to 300,000 years ago but has not adapted in the Northern Hemisphere. Thus biogeographically there has been important genetic isolation during the Quaternary between bipolar cool-water populations despite bipolar expansions of colder water masses toward the equator. This genetic isolation is also exemplified by the phenotypes of the polar species *Neogloboquadrina pachyderma,* which are morphologically distinct between Antarctic and Arctic species and possibly between subarctic and subantarctic species.

There are also biogeographically important distributional differences in the oceans, particularly between the Indo-Pacific and Atlantic tropical assemblages separated by the Central American isthmus. There are at least three tropical species which occur in the Indo-Pacific region which are absent from Atlantic faunas [Bé, 1967]. They are *Globorotaloides hexagona, Globoquadrina conglomerata,* and *Globigerinella adamsi.* Stratigraphic studies indicate that the Central American isthmus developed during the middle to late Pliocene (about 3 Ma) [Saito, 1976A; Keigwin, 1976], thus isolating the tropical assemblages. *G. hexagona* disappeared from the Atlantic about 90,000 years ago, near the end of the last interglacial episode, and *G. conglomerata* disappeared during the Pliocene or early Pleistocene. *G. adamsi* is not known in the Atlantic and apparently is a rather recent Indo-Pacific form that evolved from *G. aequilateralis* in Pacific isolation, even though the latter is known in all oceans [Parker, 1965]. Despite the disappearance of *G. conglomerata* from the Atlantic, its abundance in Pacific plankton suggests that it is still a well-adapted form.

Patterns of foraminiferal distribution are known to be disturbed in areas of coastal upwelling. In addition to affecting the productivity of the plankton, such upwelling also influences the taxonomic composition of the assemblage. Upwelling can expel certain species characteristic of surface layers and favor more opportunistic species. Furthermore, such upwelling is known to mix cool and warm assemblages not normally occurring together in open ocean conditions and to allow certain species more characteristic of warmer regions to be brought in by countercurrents and upwelled into relatively cool surface waters.

Benthonic Foraminifera

Unlike the planktonic forms, the benthonic foraminifera live at or near the sediment-water interface. Most are mobile, while sessile forms may be attached temporarily by pseudopodia or permanently by cementation. Benthonic foraminifera are easily differentiated from planktonic foraminifera in deep-sea sediments by differences in shape and wall-surface texture. The benthonic foraminifera occur in brackish to normal marine habitats and live at all depths. They are found at all latitudes, although the biogeographic provinciality is, in part, latitudinally determined and highest diversities are found in tropical areas. The diversity gradient between polar and tropical regions can be complicated by local environmental variations. Antarctic faunas are much more diverse than Arctic faunas and similarities between polar faunas are small, particularly at the species level.

The benthonic foraminifera are subdivided into two general groups: the "smaller" benthonics—20 μm to 300 μm—and the "larger" benthonics or foraminifera—up to 16 mm. The larger foraminifera are quite distinct taxonomically, occur mainly in shallow tropical seas, and are of little use in marine geological work. The smaller foraminifera are taxonomically more diverse than the planktonic foraminifera, but in the deep sea the number of individuals per species are much fewer. On the continental shelves, where planktonic foraminifera are much less frequent, the number of individual benthonic foraminifera can be relatively high. They are abundant enough in most marine sediments to be valuable paleoenvironmentally and, in some cases, for stratigraphic control. Recent benthonic foraminifera are taxonomically well known, although much oversplitting has occurred.

The oldest known benthonic foraminifera are pseudochitinous and agglutinated forms of Cambrian age. Microgranular calcareous forms are important throughout the Paleozoic. The forerunners of the perforate hyaline forms, which were so important during the Cenozoic, first appeared during the late Paleozoic.

The biology of benthonic foraminifera is better understood than that of planktonic forms, because they are easier to culture. The life cycles of several species are well known. However, the majority of this information is on shallow, nearshore species and may not be typical of the deep-water forms important in ocean-basin paleoecology.

Importance of benthonic foraminifera in marine geology

The distribution patterns of neritic to abyssal benthonic foraminifera in the ocean basins can provide a variety of valuable paleoenvironmental information.

Paleoreconstructions of benthonic foraminiferal distributions suggest that distinct faunas may be associated with different deep-water masses. Deep-water benthonic foraminifera have shown remarkable evolutionary stability in their morphology over the past 14 m.y. (since the middle Miocene). Because their ecological evolution has been so conservative, we should be able to extrapolate knowledge of their modern environmental preferences back to the middle Miocene and thus reconstruct the history of deep-sea circulation in a three-dimensional framework. The only alternative in deep-sea paleoceanography is oxygen isotope analysis (see Chapter 17). Studies of benthonic foraminiferal changes during the late Quaternary (the last 150,000 years) have already shown that there have been major distributional changes in deep-sea assemblages. These changes are partly in phase with climatic changes believed to have caused them.

A most useful contribution of post-middle Miocene benthonic foraminiferal information is as a paleontological criterion for identification of samples deposited within AABW. This criterion provides a means, independent of both isotopic and sedimentologic evidence, for recognizing the evolution of AABW and delineating its areal extent in the past.

The tests of benthonic foraminifera provide material from the deep sea for oxygen and carbon isotopic measurements. This has provided data about bottom-water history and distribution and the thermal-haline structure of the oceans.

Distribution patterns may indicate transport with sediment downslope as it has long been recognized that benthonic foraminifera are transported to deeper waters with sediments.

The bathymetric distribution of benthonic foraminifera in the bathyal and abyssal zones ($\pm$ 500 m) is well enough known to provide tectonic information. Foraminifera can provide information on the depth of sediment deposited on mid-ocean ridges to substantiate "back-tracked" paleodepth estimates, the history of subsidence (or uplift) on aseismic ridges, and the tectonic history of continental margins.

Paleodepth information provided by benthonic foraminifera in the shallower sequences of the continental shelf and upper continental slope also can indicate eustatic sea-level history. Also, faunas of nearshore to the outer continental shelf group into distinctive assemblages, which are useful markers for a diversity of shallow-water sedimentary environments.

Morphology and classification

The test of benthonic foraminifera ranges in complexity from extremely simple, single-chambered forms to complex multichambered forms with intricate internal structures. The most conspicuous differences concern the test wall, which may be constructed of a variety of materials including membranous or pseudochitinous material, agglutinated or arenaceous material (which is constructed of cemented sedimentary particles), secreted calcite or aragonite of differing microstructure, or opaline silica, secreted by certain rare deep-sea forms. Foraminifera with pseudochitinous tests are of little value in paleoecological or stratigraphic studies because they are very scarce as fossils, due to their fragile tests. The agglutinated forms use a wide range of materials, including quartz and other mineral grains, carbonate fragments, and other biogenic material, such as sponge spicules. Some species are extremely selective in regard to the type and range of size of material incorporated, while others use a wide variety of materials. Certain deep-sea forms incorporate calcareous biogenic material, which provides a smooth surface texture. Others are remarkable in that they construct their tests of the coccoliths of a single species of calcareous nannoplankton. The sedimentary particles are variously cemented by organic cement, calcium carbonate, carbonate iron, or a combination of the three. Forms living in water highly undersaturated in calcium carbonate use a noncalcareous cement. Such agglutinated forms are important in polar and abyssal areas.

Modern calcareous or aragonitic benthonic foraminifera have been classified as three basic types [Loeblich and Tappan, 1964].

1. Porcellaneous tests have opaque walls with a distinct porcellaneous luster, which results from random crystal orientation in much of the test wall. Pores are lacking.

2. Hyaline calcareous tests have perforate, radial walls. In such walls, the small prisms of calcite are primarily oriented with the axis perpendicular to the test surface. In polarized light it may produce a black cross and colored rings.

3. Hyaline calcareous tests have perforate granular walls, which show a granular structure when viewed in polarized light. Most construct their tests of calcite, although some forms are constructed of aragonite.

Subdivisions using radial and granular test walls may be oversimplistic and artificial but, nevertheless, have not yet been substantially modified.

Chamber forms and arrangements are extremely varied. Test shapes include forms that are globular, spherical, ovate, pyriform, tubular, hemispherical, radial elongate, angular, fistulose, and biconvex, among others [Loeblich and Tappan, 1964]. Chamber arrangements include forms that are rectilinear, arcuate, planispiral (evolute or involute), trochospiral, streptospiral, milioline, uniserial, biserial, triserial, or multiserial [Loeblich and Tappan, 1964].

Virtually all foraminifera possess one or more apertures that differ significantly in shape and position and are often modified by accessory structures, such as a lip, tooth, or flap. Some forms have internal canal systems of tubular cavities within the shell material. Ornamentation is often beautiful, with ribs, keels, spines, and other features.

The classification of benthonic foraminifera is relatively stable, primarily because of the compilations and taxonomic studies of Loeblich and Tappan [1964; 1974]. The foraminifera belong to one of the few living animal groups for which classification has been largely based on test morphology. Most studies of living individuals (restricted to a few shallow-water species) and features (such as reproduction) have shown that classification based on hard parts closely approaches a natural one. On the other hand, alternation of asexual and sexual generations within the life cycle can produce a dimorphism reflected in megalospheric and microspheric tests. These are marked by differing proloculus (initial chamber) size or even by two distinctly different generations. In some instances this dimorphism is known to cut across existing classifications [Arnold, 1964].

According to the latest and most widely used classification [Loeblich and Tappan, 1964], test-wall composition and microstructure are important in distinguishing suborders and superfamilies. Within suborders, the unilocular or multilocular nature is important in the lower groups. Next in importance are the apertural characters, then chamber form and arrangement, and finally whether the taxa are free living or attached. Chamber form and arrangement are of less importance in classification because taxa of similar form have entirely different wall structures and have developed independent lineages by parallel evolution.

The three suborders of benthonic foraminifera of value in marine paleoecological studies are the

TEXTULARIINA: agglutinated forms;

MILIOLIINA: porcellaneous forms;

ROTALIINA: hyaline calcareous perforate forms.

The last group includes both calcareous and aragonitic forms, along with a large number of important superfamilies, which display great diversity of form and environmental tolerance.

Distribution

The distribution patterns of benthonic foraminifera are complex, because they are controlled by a large number of physical characteristics other than temperature and salinity. Benthonic foraminiferal distributions show a strong correlation with depth, because this factor controls many other environmental parameters, such as light availability, nutrient concentrations, temperature, salinity, pressure, and oxygen and carbon dioxide content. Lower bathyal and abyssal foraminiferal assemblages are associated with particular water masses, which migrate over different water depths through the ocean basins. Hence depth and faunal-assemblage relationships are only general. Most studies of benthonic foraminifera have insufficient ecological data to determine the primary causes of the distribution patterns, but as with living planktonic foraminifera, the distribution of benthonic foraminifera can be very patchy in terms of species distribution, diversity, and standing crop. Such differences are caused by small changes and are usually unimportant except when conducting detailed distribution studies. A useful review of the ecology of benthonic foraminifera is given by Boltovskoy and Wright [1976].

Although there are many limiting factors in any single area, only one or two environmental parameters will reach the critical limit for individual species. In different areas of the sea floor, the distribution of individual species may be controlled by different limiting factors [Murray, 1973]. Also, a species may be able to survive in an environment in which it is unable to reproduce, hence affecting paleoecological conclusions.

Varying degrees of importance have been placed on individual ecological parameters. Greiner [1970] argued, based on observed distribution patterns of agglutinated (noncalcareous) and calcareous forms, that the saturation level of calcite in the ocean is a major controlling factor. However, once the level of calcite availability is reached, it is no longer a limiting factor [Murray, 1973]. Depth as a limiting factor has been stressed by many because of the depth-related trends of observed benthonic foraminifera. There was interest in establishing paleodepth relations of marine sequences in order to understand the evolution of sedimentary basins. Unfortunately, few were concerned with the actual environmental factors controlling the general depth relationships. More recent studies have stressed the need to understand as many environmental factors as might control distributions [Lohmann, 1978; Corliss, 1979].

Benthonic foraminifera display clear biogeographic distribution patterns in response to latitudinal changes. Shallow-water faunas have resulted from interoceanic and intraoceanic isolation. The distribution of several distinct provinces has been summarized in Boltovskoy and Wright [1976]. They distinguish five major benthonic foraminiferal provinces circumscribing South America. On a worldwide basis, the major shallow-water faunal provinces are controlled mainly by temperature changes [Murray, 1973].

Depth biotopes

The majority of past studies dealing with benthonic foraminiferal distributions have been concerned with the establishment of depth sequences or zonations. Most species do exhibit a preference for certain water depths and are generally viewed as if they were distributed in bathymetric zones, even though the zones may be at different depths in various regions. Nevertheless, most workers would agree that it is easy to distinguish between faunas, or **biotopes,** from nearshore areas, different depths on the continental shelf, the continental slope, and deep-ocean basins. The limits of these depth zones are related to the natural physical and chemical zones of the ocean bottom. As a result, such biotopes are more clearly defined in shallow-water areas, where natural boundaries are more conspicuous and more abrupt. The changing nature of these faunas is also evident in geological sequences. Fossil benthonic foraminiferal assemblages have been widely used for paleobathymetric reconstructions of sedimentary basins since the 1930s [Natland, 1957]. The approach, with its limitations, has been quite sound. The primary problem of benthonic foraminiferal depth zonation is that the causes of the patterns are not clearly known. Funnell [1967] considered that depth itself is probably not a limiting factor and that species distributions are controlled by the other environmental parameters. These environmental factors vary in water depth, so there are no absolute depths inhabited by the species.

General features of different bathymetric zones have been described by many workers and have been summarized by Boltovskoy and Wright [1976] as follows:

INTERTIDAL ZONE:
Foraminiferal tests in this zone are often flattened and firmly attached to substrate, in response to the normally high-energy environment. Examples: *Discorbis* and attached forms of *Cibicides.*

Tests also are often strong with thickened walls. Examples: *Ammonia beccarii, Elphidium.*

Many species reside here that have particular shapes, the benefits of which are not yet understood. Example: *Buliminella.*

INNER NERITIC (SHELF) ZONE (0-30 M, FIG. 16-7):
The taxonomic composition and forms are often similar to the littoral zone and the most common genera include *Elphidium, Ammonia, Quinqueloculina,* numerous other miliolid forms, and *Poroeponides.*

MIDDLE NERITIC (SHELF) ZONE (30-100 M):
Diversity increases relative to the inner neritic zone.

Agglutinated species are not yet important, but those present have relatively simple morphologies and generally lack a complex interior structure. Examples: *Textularia, Trochammina,* and *Reophax.*

The most common genera include *Ammonia, Elphidium, Quinqueloculina, Triloculina, Spiroculina, Discorbis, Bulminella,* and *Buccella.*

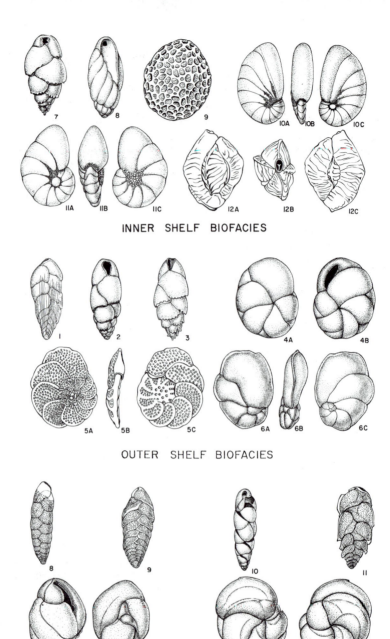

INNER SHELF BIOFACIES

OUTER SHELF BIOFACIES

UPPER BATHYAL BIOFACIES

Figure 16-7 Representative benthonic foraminifera typical of inner and outer shelf biofacies and upper bathyal biofacies in the Gulf of California:

Inner shelf forms

(7) *Bulimina marginata* d'Orbigny var. ×115; (8) *Buliminella elegantissima* (d'Orbigny), ×188; (9) *Gypsina vesicularis* (Parker and Jones), ×67;

Shallow tropical waters are marked by numerous additional forms, which include *Amphistegina, Peneroplis, Archaias,* and *Heterostegina.*

OUTER NERITIC (SHELF) ZONE (100–130 M, FIG. 16–7): Diversity continues to increase relative to shallower waters.

Taxonomic composition still has much in common with the shallow neritic zone, but there are some important differences.

Hyaline calcareous species increase at the expense of porcellaneous forms. Examples: *Lagenids, Buliminids,* and *Cibicids.*

Typical nearshore euryhaline forms, such as *Ammonia beccarii,* have mostly disappeared.

The most common genera include *Cassidulina* with sharp edges and fairly compressed tests, *Cibicides, Nonionella, Uvigerina, Fursenkoina,* and *Pullenia,* with relatively compressed tests compared with deeper water forms.

Agglutinated taxa include forms with complex interiors.

UPPER AND MIDDLE BATHYAL ZONE (130–1000 M; FIGS. 16–7 AND 16–8): Increased species diversity continues relative to the outer neritic zone.

Common genera include *Bolivina, Uvigerina, Cassidulina* with more globular tests, *Gyroidina, Bulimina, Pullenia,* and *Cibicides.*

The nodosariids become much more diverse.

Among the porcellaneous forms, biloculine forms—especially *Pyrgo*—become important.

---

(10) *Nonionella basispinata* (Cushman and Moyer), × 80; (11) *Nonionella atlantica* Cushman × 135; (12) *Quinqueloculina catalinensis* Natland, × 47.

Outer shelf forms
(1) *Bolivina acutula* Bandy, × 113; (2) *Bulimina denudata* Cushman and Parker, × 96; (3) *Bulimina marginata* d'Orbigny var., × 90; (4) *Cassidulina minuta* Cushman, × 225; (5) *Planulina ornata* (d'Orbigny), × 75; (6) *Cancris auricula* (Fichtel and Moll), × 80.

Upper bathyal forms
(8) *Bolivina seminuda* Cushman, × 50; (9) *Bolivina spissa* Cushman, × 43; (10) *Buliminella exilis* (Brady) var. *tenuata* Cushman, × 50; (11) *Loxostomum pseudobeyrichi* (Cushman), × 45; (12) *Cassidulinoides cornuta* (Cushman), × 85; (13) *Cassidulina delicata* Cushman, × 122.

(From O. L. Bandy, *Micropaleontology,* vol. 7, pp. 23, 24, 25, 1961)

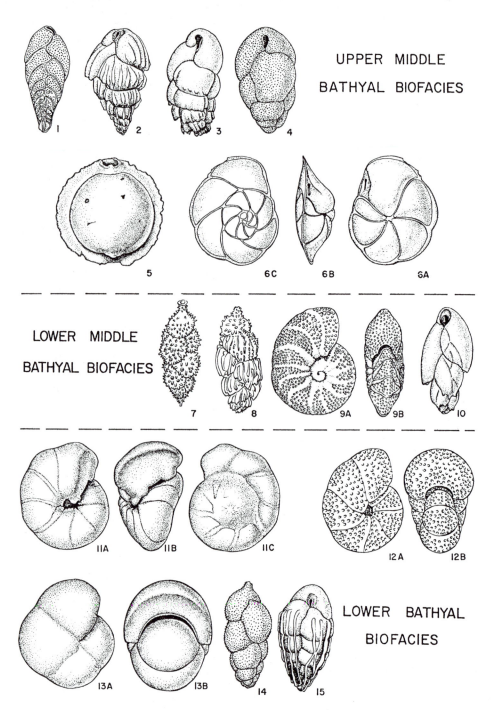

UPPER MIDDLE BATHYAL BIOFACIES

LOWER MIDDLE BATHYAL BIOFACIES

LOWER BATHYAL BIOFACIES

Figure 16–8 Representative benthonic foraminifera typical of middle and lower bathyal biofacies.

Upper middle bathyal forms
(1) *Bolivina argentea* Cushman, ×59; (2) *Bulimina striata* d'Orbigny var. *mexicana* Cushman, ×79; (3) *Bulimina* sp., ×43; (4) *Bulimina affinis* d'Orbigny, ×50;

Test size begins to increase in several groups, and surface ornamentations begin to be more complex, even elegant.

Lower bathyal zone (1000–3000 m; Fig. 16–8): Diversity in general appears to decrease in these assemblages. Abundance is decreased in the benthonic foraminifera as a whole.

The distribution of these forms is widespread, and many species can be regarded as cosmopolitan.

Agglutinated species become an increasingly important component of the assemblage. Calcareous forms become less important below approximately 3000 to 4000 m.

Common genera include *Oridorsalis, Stilostomella, Pleurostomella, Melonis, Gyroidina, Globocassidulina, Cibicides, Epistominella, Pyrgo,* and *Eggerella.*

Abyssal zone (3000–5000 m): Agglutinated forms are particularly important as a result of calcium carbonate dissolution. Calcareous forms are much depleted.

Common genera include *Bathysiphon, Cyclammina, Haplophragmoides, Rhabdammina,* and *Cribrostomoides.*

Benthonic foraminifera are rare, although reduced sedimentation tends to concentrate those that are present.

Ostracoda

The ostracods are very specialized crustaceans with bivalved carapaces (Fig. 16-9) in which the soft parts of the animal live, ostensibly for defense against predators and crushing by a mobile substrate. The group occurs in fresh, brackish,

---

(5) *Pyrgo murrhina* (Schwager), ×43; (6) *Epistominella smithi* (R. E. and K. C. Stewart), ×150.

Lower middle bathyal forms
(7) *Uvigerina hispida* Schwager, ×49; (8) *Uvigerina peregrina* Cushman var. *dirupta* Todd, ×43; (9) *Melonis barleeanus* (Williamson), ×79; (10) *Virgulina spinosa* Heron-Allen and Earland, ×135.

Lower bathyal forms
(11) *Gyroidina soldanii* d'Orbigny, ×47; (12) *Nonion pompiliodes* (Fichtel and Moll, ×104; (13) *Pullenia bulloides* (d'Orbigny), ×150; (14) *Uvigerina senticosa* Cushman, ×57; (15) *Bulimina rostrata* Brady, ×113.

(From O. L. Bandy, *Micropaleontology,* vol. 7, pp. 23, 24, 25, 1961)

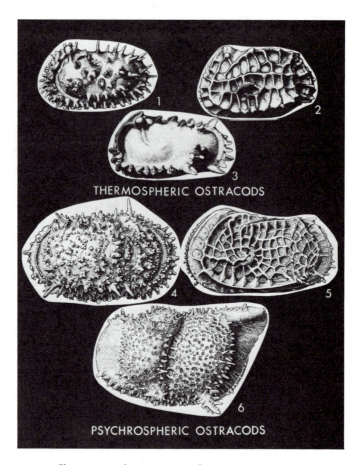

Figure 16–9   Characteristic deep-sea ostracods.

Thermospheric forms
(1) *"Cythere" scutigera,* ×80; (2) *Bradleya andamanae,* ×80; (3) *Pterygocythereis ceratoptera,* ×80.

Psychrospheric forms
(4) *"Cythere" acanthoderma,* ×100; (5) *Bradleya dictyon,* ×100; (6) *Bythoceratina scaberrima,* ×100. Thermospheric forms are found in shallower, warmer waters, whereas psychrospheric forms are associated with deeper colder water masses.

(Courtesy R. Benson, Smithsonian Institution)

and normal marine waters and most are benthonic. Their small size (0.5–2 mm) is relatively large compared with most other microfossils preserved in deep-sea sediments. The carapace appears to be constructed of chitinous-rich calcite with up to 5 percent $MgCO_3$ [Milliman, 1974; Cadot and others, 1972], so that is the part which is commonly fossilized. Planktonic ostracods have an organic-walled carapace that is not preserved in deep-sea sediments. Like other arthropods, the ostracods grow by a series of steps associated with molting (**ecdyses**). The younger carapaces, or **instars,** may be preserved in the sediment in association with the fully grown morphotypes. Sexual dimorphism is often common.

The two valves are joined by transverse adductor muscles, which leave attachment scars that are useful in classification. Hinges between the valves range from single-valve overlap to teeth and socket arrangements. Ornamentation is highly variable (Fig. 16–9); additional interior pores and features are important in taxonomic differentiation. Classification is based on characteristics of the carapace, which varies during ontogeny and shows sexually related differences. Such features include size, outline, convexity of valves, hinge characteristics, and ornamentation and surface texture. In marine and marginal marine environments, males may be rarer than females, so a large number of species have been described only from female populations.

Ostracods are long-ranging geologically; they have had strong representation since the early Ordovician and were important stratigraphically in the Paleozoic. The marine forms have been classified in over 30 families, mostly living in shallow oceanic areas.

Strong relationships, particularly in size and strength, exist between the environment and the architecture of the ostracod carapace. Shallower forms have fewer spines, are more robust, have coarser ornamental reticulation, and have thicker and smaller valves. Deeper forms have morphologically more complex carapaces with intricate ornamentation (Fig. 16–9).

Distribution

Ostracods inhabit a wide range of environments. Species diversity increases with increased salinity and decreases in species dominance. The benthonic ostracods have no known pelagic larval stages, so they are always closely linked to their benthonic environment. They have no active means of locomotion and are passively dispersed.

Although the present deep-sea ostracods are very sensitive to temperature and other environmental changes, they are cosmopolitan [Benson, 1975]. Deep-sea forms are not diverse, consisting of only about 50 to 60 well-defined species—fewer total taxa globally than are found living in many single, local, shallow-water environments. Population density is also low in the deep sea, being only a fraction of that of shallow-water environments.

Benson [1972] has distinguished two major global faunas (Fig. 16–9):

1. The **psychrospheric** assemblage occurs in cold-deep waters (greater than or equal to 500 m). It may extend to shallow areas in high latitudes or in upwelling regions.

2. The **thermospheric** assemblage occurs in the upper, less-dense, warmer (more than 10°C) layers of the water column.

These two assemblages are separated by the mixing zone.

The psychrospheric assemblage contains most of the bathyal and all of the abyssal faunas. There are two major groupings within the psychrospheric faunas, which are separated at lower latitudes at approximately 4 to 6°C, around 2000 m. The psychrospheric assemblage contains individuals that are usually larger than average, with adults often exceeding 1 mm in length. Shell walls are thin, with a noticeable increase in surface-to-volume ratio. To compensate for less shell material, deep-water ostracods have developed structures that add strength to the carapace, including increased ornamentation. The animals are blind: Eye tubercles are missing. Ornate forms range from delicately reticulate or densely spinose forms to those with thin, smooth carapaces. The deep-sea fauna is believed to be a relict original fauna found in warm conditions in the early Cenozoic.

The thermospheric fauna is much different and very diverse compared with the deeper dwelling psychrospheric faunas. Many of the families have never resided at great oceanic depths.

Importance of Ostracoda

Although ostracods are rare in comparison with other preserved microfossils, they are very sensitive to environmental change and respond rapidly to moderate variations in depth and salinity. This high state of adaptive development and many available characters make the ostracods useful as paleoecologic tools. The distributions of faunas associated with cold, deep and warm, shallow-water masses is a valuable characteristic. The evolution of these assemblages provides important information on the evolution of these water masses.

The distribution of the psychrospheric assemblage is important in determining the degree of deep-water connections between the open ocean and certain marginal seas, particularly the Mediterranean. The entry and maintenance of the global psychrospheric assemblage within a series of partially landlocked basins require free access to the open oceans at considerable depths. Present-day sill depths at the western end of the Mediterranean are too shallow to allow penetration of the global psychrospheric ostracod assemblage into the Mediterranean. The psychrospheric assemblage existed in the Mediterranean until the earliest Quaternary, indicating deep-water connection with the Atlantic [Benson, 1972]. Their disappearance reflects further oceanic isolation of the Mediterranean and the latest phase in the disintegration of the Tethys Sea.

The depth-related ostracods are also important paleodepth indicators. Their occurrence in freshwater to marine deposits makes them useful in stratigraphic subdivision and correlation over these wide-ranging environments.

Pteropods and Heteropods

Pteropods and heteropods are planktonic gastropods (Fig. 16–10), belonging to the suborder Euthecosomata, that have fragile aragonitic shells. They are common and widespread in the surface plankton of the world oceans. There are approximately 38 species of pteropods and 17 species of heteropods; they secrete shells up to 28 mm long, but usually range between 0.3 and 10 mm. Shell shapes vary, including trochospiral, biconvex or discoid spiral, and elongate cones (Fig. 16–10). The gastropod foot has been adapted to wings, which provide locomotion (hence the common name *sea butterflies*) and assistance in feeding [Bé and Gilmer, 1977].

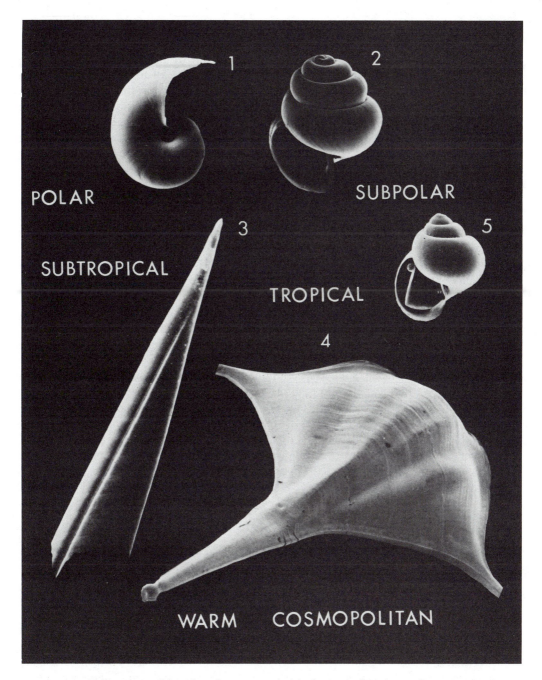

Figure 16-10 Selected modern pteropods (planktonic mollusca) according to water mass distribution.

(1) *Limacina helicina*, ×54—polar (Arctic-Antarctic); (2) *Limacina retroversa*, ×54—subpolar; (3) *Styliola subula*, ×27—subtropical; (4) *Diacria trispinosa*, ×20—cosmopolitan (warm water) (5) *Limacina trochiformis*, ×56—tropical.

(Courtesy A. W. H. Bé, Lamont-Doherty Geological Observatory)

Most species live within the upper few hundred meters of the water column, but there are some bathypelagic species living at depths greater than 500 m.

Although the geological range of pteropods is from Eocene to Recent (some doubtful Mesozoic records exist [Chen, 1968]), almost all fossils are restricted to the Quaternary. They are very rare in pre-Quaternary sediments because the aragonitic shells are less stable and more susceptible to dissolution than the calcite shells of foraminifera and calcareous nannofossils. Although pteropods are found in surface sediments of most water masses, they are mostly restricted to tropical areas in depths less than 2500 m (aragonite compensation depth; see Chapter 14) as compared to 4700 m in the North Atlantic for foraminifera and coccoliths.

Like other planktonic organisms, the pteropods and heteropods show distinct water-mass preferences (Fig. 16–10). Most species occur in the tropics and subtropics, while a single species is known from the Arctic and two species are known from the Antarctic [Bé and Gilmer, 1977]. Five examples are shown in Figure 16–10. Biogeographical differences also exist between the Indo-Pacific and Atlantic faunas; the Red Sea (Indo-Pacific province) and Mediterranean (Atlantic province) faunas are distinctly different. Biogeographic differences even exist between the eastern and western basins of the Mediterranean [Bé and Gilmer, 1977].

The planktonic gastropods are not very useful as biostratigraphic tools because of their restriction to the Quaternary. But they have been useful in Quaternary paleoclimatic investigations, especially in more enclosed or marginal seas, such as the Mediterranean, Caribbean, and Gulf of Mexico, where preservation is good. For instance, Chen [1968] has determined climatic oscillations in Mediterranean and Red sea faunas based on quantitatively determined changes in cold and warm assemblages. In the Caribbean and Gulf of Mexico, the abundance of planktonic gastropods is greater during glacial episodes. This may reflect a deepening of the aragonite-compensation depth and, therefore, less-widespread dissolution of aragonite shells. Few investigations have been conducted on pteropod paleoecology, so their paleoenvironmental potential has not been realized.

Calcareous Nannofossils

The coccolithophorida are a group of biflagellate, unicellular, golden brown algae belonging to the class Haptophyceae. They are included in the Haptophyceae because some possess a sensor, or **haptonema,** between the more flexible flagella. Coccolithophorida differ from other members of the class, and from other algae, in their ability to form calcareous surface scales, or bodies, known as **coccoliths** (Fig. 16–11). Each species possesses its own characteristic coccolith morphology on which taxonomic identification is based. They generally range from 2 to 10 μm in diameter. Each organism (entire cell) is called a *coccosphere,* which is usually spherical (Fig. 16–11) and ranges in diameter from 2 to 25 μm. Each coccosphere contains 10 to 150 coccoliths on its surface, with an average of 20 [Black, 1963; Honjo, 1976]. The coccoliths are usually not well enough articulated to remain intact in sediment, so only individual coccoliths are found in sediments.

The coccolithophorids form an important component of the marine

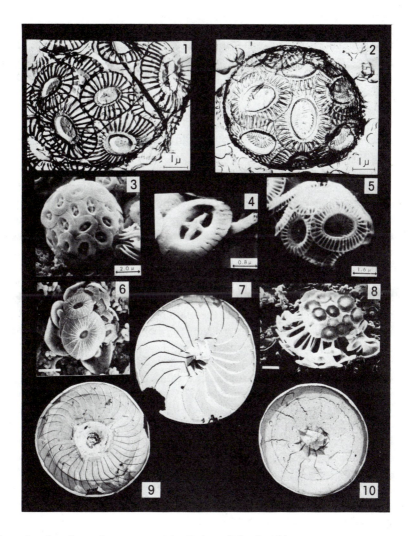

Figure 16–11 Examples of modern calcareous nannofossils (coccolithophorids).

Electron microscope views of coccosphere: (1) *Emiliania huxleyi* × 7000 (cold-water form showing solid proximal shield and pore covering). (A. McIntyre and A. W. H. Bé, 1967). (2) *Emiliania huxleyi* × ~ 6500 (warm-water form with open grill covering the central pore and T-shaped elements in both shields). (A. McIntyre and A. W. H. Bé, 1967)

Scanning electron microscope views of coccosphere: (3) *Cruciplacolithus neohelis;* coccosphere of about 50 tightly interlocking coccoliths. (4) *C. neophelis;* view of individual coccolith showing cruciform structure which is unique among modern forms, but characteristic of Eocene and older forms. (5) *Emiliania huxleyi;* coccosphere of 12 interlocking coccoliths. (6) *Umbellosphaera* cf. *tenuis;* two types of coccoliths can be seen: macro and micro; bar is 3 μm. (From Okada and S. Honjo, 1970). (8) *Discosphaera tubifera;* showing tubercule coccoliths; bar is 3 μm. (Okada and Honjo, 1970). (7) (9) (10) *Cycloccolithus leptopora;* note phenotypic variation exhibited in size differences and in the number of elements used in the construction of the coccolith. Numbers 7 and 9 are predominantly transitional water types; number 10 is a warmer form marked also by smaller size, related to the warm Benguela current flow in the South Atlantic. Figure 7, × 9200; Figure 9, × 6000; Figure 10, × 15,300. (Figures 3, 4, and 5 courtesy W. Dudley, University of Hawaii; Figures 7, 9 and 10 courtesy B. Molfino and A. McIntyre, Lamont-Doherty Geological Observatory)

phytoplankton in subpolar and warmer waters. Together with diatoms and dinoflagellates, they form the bulk of marine phytoplankton. To the deep-sea paleoecologist the group is very important because it, along with an extinct and probably related group—the discoasters—forms a large fraction of recent and fossil sediments. They are the most important and diverse constituent of common chalk and were particularly widespread in the Cretaceous. In the Atlantic most of the fine-grained carbonate is believed to be composed of coccolith debris forming 5 to more than 20 percent of the total sediment [McIntyre and Bé, 1967]. Furthermore, in some modern enclosed basins, such as the Mediterranean, coccoliths still dominate the carbonate component.

Coccoliths were first discovered in 1836 by Ehrenberg, who first thought they were inorganic concretions. In 1858 Huxley discovered them in deep-sea muds and, because of their resemblance to the green algae *Protococcus,* named them coccoliths. In 1861 Sorby proved that the coccoliths were part of an organism when he discovered an intact coccosphere, and Lohmann, in the early twentieth century, recognized them as primary producers in the oceans.

Because of their minute size, the study of coccoliths is difficult even using a light microscope. Progress was slow until modern studies used the transmission-electron microscope and, later, the scanning-electron microscope (SEM). These two instruments have revolutionized the study of this group. Their extremely small size is a major advantage, allowing studies of large assemblages in very small sediment samples, which require a minimum of preparation. Furthermore, the preservation of nannofossils is usually superior to other calcareous microfossils. Each coccolith is enclosed in an organic membrane [Chave, 1965] that helps prevent dissolution, perhaps analogous to that believed to be present in diatom frustules [Lewin, 1961]. This, in combination with rapid evolutionary diversification, has made them extremely valuable biostratigraphic tools widely used in the study of deep-sea sediments, particularly in connection with the DSDP. Indeed, the rapidity of age determinations of core samples aboard the *Glomar Challenger* is due to the use of calcareous nannofossils. Much is now known about the taxonomy and distribution of living and fossil coccolithophorids, but little is known about the living organism because their sizes and oceanic habitats make this subject difficult to study. There may be as many as 150 species [Loeblich and Tappan, 1966], but only about 16 species are relatively abundant and possess preservable skeletal elements [Gaarder, 1971; McIntyre and Bé, 1967]. The oldest reported coccoliths are of early Jurassic age and all major groups evolved by the early Cretaceous. A comprehensive review of the coccolithophorids is provided by Tappan [1980].

Living forms

The coccolithophorids are photoautotrophic algae and primary producers in the oceans. Some forms of coccoliths, however, have a heterotrophic capability, possessing the ability to ingest organic molecules. Laboratory studies show that growth can be stimulated by organic material. Although coccolithophorids are frequently observed at depths of several hundred meters in the sea, where no light is available for photosynthesis, there is, as yet, no evidence to show that these forms grow satisfactorily in the nonilluminated zone of the ocean.

Data about life cycles are mostly from cultures of nearshore or eurythermal ocean forms; culturing of stenothermal oceanic types has been unsuccessful. Reproductive cycles may vary, but most are believed to multiply by binary fission in a motile or nonmotile coccolith-bearing stage. In some species, generations alternate, while other species produce benthonic states in which the coccoliths are absent and the coccosphere is naked. In division, some species divide the coccoliths equally between the daughter cells; in others the coccoliths are lost preceding reproduction and the daughter cells grow their own.

Coccoliths

Coccoliths grow inside the cell, probably being secreted by the golgi apparatus. They are then extruded onto the surface of the cell, where they remain, coated with an organic membrane. Here they may or may not interlock, or form one or two layers. The function of the coccoliths is unknown, but they may be a defense mechanism, focus light on the cell or act as a shield for light, be a by-product of either respiration or photosynthesis, or a flotation apparatus to maintain required levels in the water column. Some sinking occurs during ontogeny, possibly in response to nutrient depletion in surface waters [Smayda, 1970], and descending speeds may be controlled by using coccoliths as ballast for the cell. In laboratory culture, some species drop their coccoliths when in nonoptimum conditions, while others die with their coccoliths intact. Because coccolithophorids live in the upper 150 m of the water column, they are important as recorders of paleoenvironmental changes that occur at these levels in the ocean.

The coccoliths are of low magnesium calcite. Most are formed of minute rhombohedral crystals, although a few have hexagonal prisms. The shape of the crystals may be influenced by the temperature at which the cells are growing. Some of the forms of Jurassic age have very elaborate crystal morphologies, but the elaborate forms generally have shorter geologic ranges. If anything, modern forms have adopted simpler crystallographic growth characters.

The morphology of coccoliths forms the basis for classification of the Coccolithophoridae. This is an unnatural and unsatisfactory classification because in many species each cell carries two, or even three, distinct types of coccoliths (Fig. 16-11). However, coccolith morphology of given types is remarkably constant within any given species. Also, the life cycles within an individual species can embrace two widely different coccolithophorid types, with both large nonmotile cells with relatively few but heavier coccoliths and smaller, motile cells with numerous, delicate coccoliths. In living species these characteristics can be sorted out, but in extinct species different coccoliths from a single species are normally regarded as a distinct species, thus providing an unnatural classification.

Diversity and modern biogeography

As with other groups, the calcareous nannofossils generally are most diverse in warm seas, with decreasing diversities toward the higher latitudes. They are absent in Arctic and Antarctic waters. The total diversity of living coccolithophorids is greater than for modern planktonic foraminifera, but the total diversity in surface sediments is somewhat less due to preservational characteristics. Modern marine calcareous nannoplankton can, as a group, tolerate temperatures from 0 to 34°C, most species preferring temperatures from 9 to 13°C. A few freshwater species exist, and some species are euryhaline (for instance, *E. huxleyi*), tolerating salinity ranges from 15 to 40‰. *E. huxleyi* is the most pervasive modern species. The distribution of this species clearly delineates the subpolar-polar boundary. Regional or local variations in nutrient supply, salinity, and turbidity are at least as important as temperature in determining the floral composition. As with most plankton, the group displays distinct seasonal fluctuations.

Studies of both the plankton [Honjo and Okada, 1974] and surface sediments [McIntyre and others, 1970; Geitzenauer and others, 1976] indicate well-defined floral assemblages closely associated with water masses. Colder assemblages show an increasing dominance of more robust types. Factor analysis of Pacific surface sediments show six coccolith assemblages or factors. The two major assemblages are one dominated by *Gephyrocapsa oceanica* (Fig. 16–12; Assemblage A) and one by *Cycloccolithina leptoporus* (variety C; Fig. 16–13). The former is a large species, abundant in the Pacific equatorial water mass and the marginal seas of the western Pacific, and is associated with the warm Kuroshio Current. The latter is characteristic of higher latitude areas of the central regions. This cool assemblage is carried to lower latitudes by the Peru Current system and to the eastern part of the basin by the South Pacific central water circulation.

The scheme of McIntyre and others [1970] distinguishes the following assemblages in the Pacific and serves to illustrate distribution patterns.

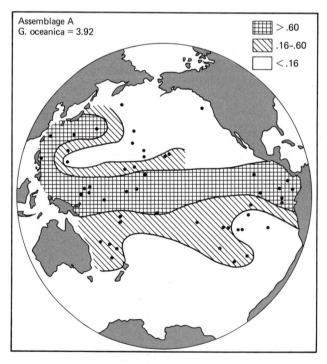

Figure 16–12 Distribution of Assemblage A (*Gephyrocapsa oceanica* assemblage) in surface sediments of the Pacific. Numbers are varimax factor values. (After K. R. Geitzenauer, M. B. Roche, and A. McIntyre, 1976)

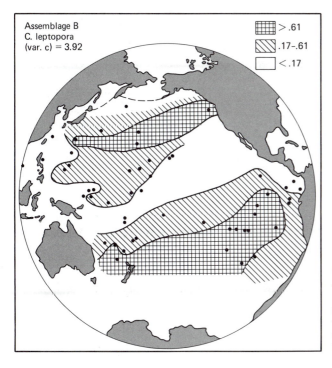

Assemblage B
C. leptopora
(var. c) = 3.92

▦ > .61
▧ .17-.61
☐ < .17

Figure 16-13 Distribution of Assemblage B (*Cylcoccolithina leptopora* var. C assemblage) in surface sediments of the Pacific. Numbers are varimax factor values. (After K. R. Geitzenauer, M. B. Roche, and A. McIntyre, 1976)

SUBARCTIC: (0°–6°C): *E. huxleyi;* cold-water variety almost exclusively present.

TRANSITIONAL: (6°–14°C): *G. caribbeanica, C. leptoporus,* variety C, *C. pelagicus,* and *E. huxleyi,* cold-water variety.

SUBTROPICAL: (14–21°C): addition of *G. ericsonii, R. stylifera, D. tubifera,* and *V. tenuis.*

TROPICAL: (21°C): *U. irregularis; G. oceanica* dominant with *C. leptoporus,* variety B, and other forms in reduced numbers.

Depth distribution

Water-column studies of living coccolithophorids [Honjo and Okada, 1974] reveal that biogeographic patterns of surface waters do not extend to levels deeper than 125 m because species assemblages are significantly different in the lower photic zone (Fig. 16-14). Furthermore, the average assemblage of species in the water column may not be that of the surface water, as there are conspicuous vertical distribution differences; maximum concentrations of coccospheres may occur at deeper depths, particularly in the central zone. In the equatorial regions, species distribution is more vertically homogeneous, so surface assemblages are similar to total assemblages.

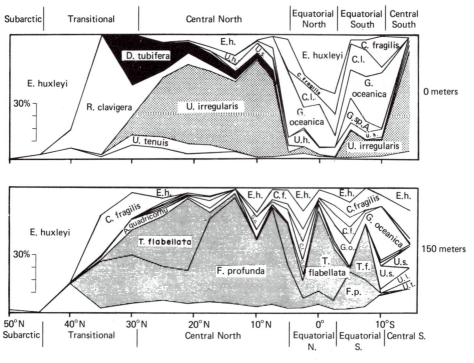

Figure 16-14 Changes of species in coccolithophorid assemblages along north-south traverse in the North Pacific at the surface and 150 m water depth. (After Honjo and Okada, 1974)

Although the calcareous nannoplankton are confined to the photic zone close to the ocean's surface, they clearly show vertical species stratification. This can be important paleoecologically because changes in conditions at various depths in the upper part of the water column may preferentially concentrate species of different depth habitats, thereby effecting distributions in deep-sea sediments.

Production rates

Laboratory cultures show that coccolithophorids are among the fastest-growing planktonic algae, with up to 2.25 divisions per day. Field studies by Honjo [1976] yield more conservative turnover rates of 4 to 10 days in temperate to tropical waters. Production rates in the equatorial zone appear to be relatively constant throughout the year, whereas other areas fluctuate seasonally. Measured standing crops range from 10^3 to 10^4 individuals per liter of seawater, but counts of more than 10^5 individuals per liter may occur in near surface waters of the subarctic zone (upper 30 m) [Honjo, 1976]. In equatorial areas, maximum reported populations are between 50 and 100 m, while at intermediate latitudes the lower-standing crops are more evenly distributed between 0 and 150 m. Although surface populations are greater at high latitudes, coccolith-rich sediments are generally restricted to lower latitudes [Black, 1965].

Sinking and dissolution

The majority of coccolithophorids are grazed by zooplankton and aggregated into fecal pellets [Smayda, 1970; Honjo, 1976]. Studies using the SEM have shown that coccoliths are generally not dissolved or crushed, because fecal pellets are covered by pellicles, which protect the contents from immediate chemical dissolution. Those that are released are rapidly dissolved. SEM studies have shown little evidence of selective dissolution of coccoliths in suspension, probably because of the organic coatings.

Deposition of coccoliths on the deep-sea floor is mainly via the rapid sinking of fecal pellets (see Chapter 14). Honjo [1976] estimated that in the equatorial Pacific, about 92 percent of coccoliths produced in the photic zone reach the deep-sea floor. This accelerated sinking (about 150 m/day) assures that the taxa in surface waters arrive at the ocean floor in similar biogeographic patterns. Usually, differences in susceptibility to dissolution produce marked differences between the assemblages in the plankton and those in the surface sediments.

Most dissolution of calcareous nannofossils occurs at the ocean floor and the coccolith abundance in sediments is strongly controlled by the rate of dissolution of individual species. For instance, in the North Atlantic at depths below 5000 m, *E. huxleyi, C. leptoporus, C. oceanica,* and *U. mirabilis* form almost 100 percent of the flora [McIntyre and McIntyre, 1971], and with increasing dissolution certain species become almost indistinguishable from others. Heavier, overlapped coccolith elements are the best at surviving dissolution. Because species that live in cold water tend to be constructed of heavy elements, this will bias the sedimentary assemblage and affect paleoecological interpretations as poorly preserved faunas will have a colder water aspect (Fig. 16–15) [Berger, 1975]. A preselection for hardy types provides a relative high abundance of resistant forms on the sea floor even in well-preserved assemblages. This makes it difficult to define a coccolith lysocline [Berger, 1973B].

Discoasters

This is an important, star-shaped group of calcareous nannofossils. The last discoasters became extinct near the Pliocene-Quaternary boundary, about 1.8 Ma. All are placed within one genus, *Discoaster,* and range in diameter between 6 and 25 μm. Discoasters have tabular crystalline forms and are more coarsely constructed than the coccoliths. This makes them more dissolution resistant and therefore more useful in biostratigraphic studies.

The discoasters first appeared in the late Paleocene and displayed rapid evolutionary radiation during most of the Cenozoic, so they are stratigraphically valuable. Most forms are tropical and warm subtropical, so their presence in high numbers is an indicator of warm surface water. Although there are no known modern representatives, their affinities are similar to the coccoliths. Forms from the early Cenozoic are fairly solid, but forms from later in the Cenozoic tend to have less calcite in their structure. A reduction in calcite is also observed in the coccoliths through the Tertiary. This evolutionary pattern has not yet been adequately explained.

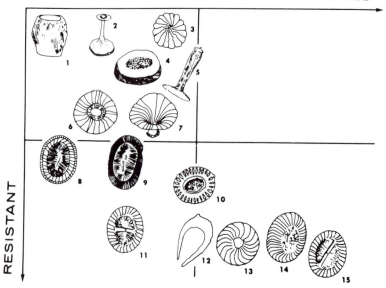

Figure 16-15 Calcareous nanofossils as indicators of paleotemperature and preservational environment. Note that poorly preserved floras appear to have a cold-water character. Coccoliths: (1) *Scyphosphaera* sp.; (2) *Discosphaera tubifera*; (3) *Cycloccolithina fragilis*; (4) *Pontosphaera* sp.; (5) *Rhabdosphaera clavigera*; (6) *Cyclolithella annual*; (7) Umbellosphaera sp.; (8) *Syracosphaera pulchra*; (9) *Helicopontosphaera kamptneri*; (10) *Emiliania huxleyi*; (11) *Gephyrocapsa oceanica*; (12) *Ceratolithus cristalus*; (13) *Cyclococcolithina leptopora*; (14) *Coccolithus pelagicus*; (15) *Gephyrocapsa caribbeanica*. (From W. H. Berger and P. H. Roth, *Reviews of Geophysics and Space Physics*, vol. 13, p. 564, 1975, copyrighted by the American Geophysical Union)

Paleoceanographic value of calcareous nannofossils

Because the distribution of the calcareous nannoplankton is closely related to water masses, they are of value in paleoclimatic-paleoceanographic studies. The earliest studies to employ calcareous nannofossils as paleoclimatic indicators were those of McIntyre [1967] for the Quaternary in the North Atlantic and Geitzenauer [1969] for the Quaternary in subantarctic regions. Both studies showed large-scale mass migrations of surface water related to paleoclimatic change.

The calcareous nannofossils evolved much more rapidly than did the planktonic foraminifera through the Cenozoic. Haq [1978] has estimated that coccoliths evolved at the rate of 1 species every 30,000 years and the discoasters at the rate of 1 every 50,000 years. These rapid evolutionary rates make them much less effective as Tertiary paleoclimatic indicators. Modern assemblages have little application as paleoenvironmental indicators in the Tertiary. Nevertheless, they have been used to good advantage by Haq and Lohmann [1976] and Haq and others [1977] toward the understanding of the early Tertiary paleoceanographic record. This has been accomplished by first mapping their distribution patterns for different past times to determine their latitudinal distributions. They found that latitudinal differences have existed during most of the early Cenozoic, with the exception of the earliest Paleocene (65–64 Ma). During this interval, which followed a period of massive extinction of marine plankton in the latest Cretaceous, only a few ecologically robust remnants survived; these had wide latitudinal distributions. Thus, owing to the wide geographic range of taxa and the low total diversities, the earliest Paleocene shows little or no latitudinal provinciality.

During the Cenozoic, there were major shifts of assemblages through space and time, as well as evolutionary changes that involved the appearance of new dominant groups and disappearance of old ones, which changed the nature of the assemblages. The temporal oscillations in the assemblages are interpreted as being largely caused by major climatic fluctuations. Four marked cooling episodes are recorded within the early Cenozoic: middle Paleocene (60–58 Ma), middle Eocene (46–43 Ma), the earliest Oligocene (37–35 Ma), and the middle Oligocene (32–28 Ma). A particularly marked warming episode occurred during the late Paleocene to early Eocene (54–51 Ma).

SILICEOUS MICROFOSSILS

Radiolaria

The radiolaria (Class, Actinopodea; subclass, Radiolaria) are a diverse group of planktonic sarcodinid protozoans that construct an elaborate biogenic siliceous or strontium sulfate skeleton. They are geologically important because they are an abundant and diverse microfossil group in deep-sea sediments. More than 7000 species have been described, although this number may be inflated by taxonomic splitting.

There are three major groups of radiolaria: two have skeletal material that does not preserve well, so they are of little paleoecological and biostratigraphic value. They are the Order Acantharia (strontium sulfate skeletons) and the Order Tripylea (siliceous-organic skeletons). The acantharians are, nevertheless, important members of the zooplankton and are almost certainly important in strontium geochemistry of the oceans. The third group, Order Polycystina, has crystallographically amorphous opaline silica ($SiO_2 \cdot nH_2O$) skeletons and is an important component of marine microfossils. The polycystine radiolarians are further divided into two major suborders: the Spumellaria, which are spherical forms, and the Nassellaria, which are ring or cap-shaped. The vast majority of radiolaria are single, although some colonial forms are known.

There are about 300 modern species of Radiolaria [W. Riedel, personal communication] and they are the most diverse group of oceanic microfossils. In terms of skeletal complexity, radiolaria are unrivaled among the pelagic fossil-forming groups. Many taxonomic problems remain unsolved. The majority of living and fossil radiolarians are not satisfactorily known taxonomically and many described species have been heavily oversplit. Unlike the diatoms, there are no freshwater or benthonic forms.

The soft body (protoplasm) of radiolaria may be divided into two parts—a dense intracapsular (endoplasm) and extracapsular protoplasm (ectoplasm)—by a central capsule which is made of pseudochitin [Haeckel, 1887]. A radiating mesh of pseudopodia (axopods) projects from the organism. The axopods each contain a rod (axone), which begins in the central cytoplasm and consists of longitudinal tubular structures called microtubules. The ultrastructural characteristics of the axopodial apparatus and the position of the nucleus varies between different radiolarian groups and should be important in radiolarian taxonomy [Petrushevskaya, 1975].

Although the biology of the polycystine radiolarians is largely unknown, distributional information on living forms and their fossil remains in surface sediments [Nigrini, 1967; Sachs, 1973; Moore, 1973a; Lozano and Hays, 1976] suggests that they are in many ways similar to planktonic foraminifera. Most radiolaria, like planktonic foraminifera, inhabit the near-surface waters (50–200 m) and can be grouped into assemblages that have geographic boundaries similar to those of the surface-water masses. Skeletons range in size from 50 to 400 μm, and some species are known to have symbiotic zooxanthellae. Radiolarian reproduction includes both asexual binary fission and sexual reproduction involving flagellated gametes [Haeckel, 1887].

Some radiolarian species live at water depths of several thousand meters. Radiolaria do not frequent nearshore waters, as do planktonic foraminifera, so they are rare or absent in most uplifted marine sequences. Little is known about the skeletal formation in radiolaria, although existing skeletal parts show thickening throughout the life span. The known geological range of radiolaria is from the Cambrian to the present day.

Importance of radiolaria in marine geology

Several characteristics of the radiolaria make them a useful group in paleoecological and biostratigraphical studies.

1. They are often abundant in deep-sea sediments, particularly in regions where siliceous oozes are deposited.

2. They have robust, opaline tests, which are often well preserved in locations where other microfossils are either badly corroded or absent.

3. They are a comparatively diverse group; therefore quantitative analysis of the radiolarians can be an accurate measure of faunal-environmental relationships. This is particularly important in the polar regions. There are about 30 Antarctic species, which make them by far the most diverse microfossil group.

4. Because of their morphological diversity and skeletal complexity, radiolarians are great recorders of evolutionary history. However, evolutionary trends are difficult to understand because the biological and functional aspects of skeletons in modern forms are not yet well known.

Morphology and classification

Although this group has a wide range of morphological characteristics, some general characteristics need to be mentioned. Most of the present classification was assembled by Haeckel [1862; 1887] at a time when virtually nothing was known about the biology, stratigraphy, or phylogeny of the group. These early classifications were artificial and based on morphological subdivisions that cut across natural generic and species groupings. The variability within a single species was considered to justify generic or species differentiation. This classification is being revised, based on phylogenetic relationships by Riedel, Petrushevskaya, Nigrini, and other workers. One of the major problems of working with radiolarians is to distinguish successfully between the more stable and the more variable morphologic elements.

The classification of radiolarians is based mainly on hard parts. Recent biological investigations have indicated several major types of cytoplasmic characteristics, especially in connection with the axopodial form and position of the nucleus. These are important taxonomically and will enter into future classification schemes [Petrushevskaya, 1975].

Spumellarians These polycystine forms are spherical, oval, or discoidal. Sphaerid forms (Fig. 16-16) may have more than one shell in concentric arrangement, interconnected by radial beams. The inside of the shell is called the **medullary shell** and the main shell is the **cortical shell.** Surrounding the cortical shell may be an extra cortical shell, which is not always preserved. Various types of spumellarians are illustrated in Fig. 16-16. Of the spumellarians, the Collosphaeridae and the Orosphaeridae are potentially useful in paleoecological work.

Collosphaeridae are colonial forms (Fig. 16-16) that become dissociated in sediments. They are spherical or subspherical and often irregular in outline. The shell wall resembles a smooth plate perforated by pores. The colonies contain several morphotypes. Collosphaeridae are relatively warm-water forms in modern seas and are absent at latitudes higher than 40°.

The Orosphaeridae (Fig. 16-16) are a highly characteristic group with robust skeletons of either lattice shell or two concentric shells. They have conspicuous radial spines, which are usually circular in cross section. The forms are fairly large (1-2 mm) and are commonly found as fragments.

Nassellarians The most common and conspicuous members of the nassellarians are bell- or cap-shaped, with one or more segments uniserially arranged (Fig. 16-17). This order also includes ring-shaped and bilobed members, which have obvious homologous skeletal elements. During development, the cephalis and thorax form first, followed by later segments. These are divided internally by fairly complete septa, marked by constrictions on the exterior. In addition to the test lattice, internal parts of the skeleton—a primary, or sagittal, ring and a basal tripod—are important. The tripod extends throughout most of the length of the skeleton.

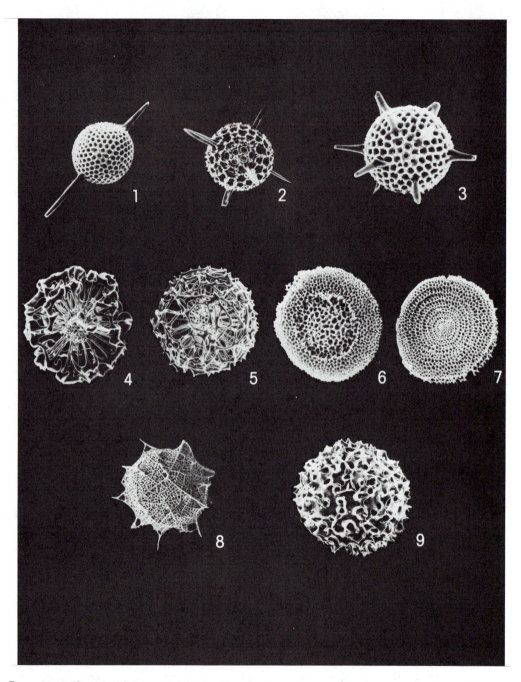

Figure 16-16 Scanning electron micrographs showing several major groups of radiolaria. Note major shape differences that occur in spherical-discoid forms.

(1) *Stylosphaerid*, ×242; (2) *Cubosphaerid*, ×274; (3) *Astrosphaerid*, ×352; (4 and 5) *Actinomina*, ×143 & ×187; (6) *Cyclodiscid*, ×225; (7) *Phacodiscid*, ×203; (8) Orosphaerid, ×37; (9) Collosphaerid, ×110.

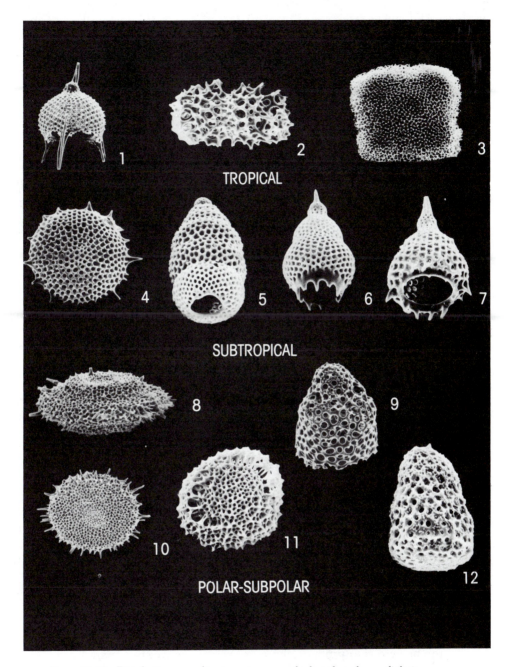

Figure 16–17   Distribution according to water mass of selected modern radiolaria.

Tropical: (1) *Pterocanium praetextum*, ×245; (2) *Ommatartus tetrathalanias*, ×307; (3) *Spongaster tetras*, ×249.

Subtropical: (4) *Phacodiscid*, ×249; (5) *Sticocyrtis* sp., ×297; (6) *Lamprocyclas maritalis* (cool), ×248; (7) *Lamrocyclas maritalis* (warm), ×297.

Polar-subpolar: (8) *Spongotrochus glacialis*, ×269; (9) *Antarctissa strelkovi*, ×265; (10) *Spongotrochus glacialis*, ×242; (11) *Lithelius nautiloides*, ×344; (12) *Antarctissa denticulata*, ×292.

Spines in nassellarians are generally not hollow, although they may be spongy. Pore shape varies and the pores are generally hexagonally arranged.

Ecological aspects

Little is known about the ecology of modern radiolaria. Life cycles of about 1 to 3 months have been proposed for shallow-water forms [Casey and others, 1971; Berger, 1976]. A study within the water column indicates that numerical abundances of radiolaria and planktonic foraminifera are roughly the same [Cifelli and Sachs, 1966] and the highest standing crops reported occur from 150 to 400 m in tropical to Antarctic waters. In tropical to temperate waters, radiolarian abundances decrease rapidly below 500 m [Casey, 1971], but in Antarctic waters they dominate the protozoan plankton at great depths [Petrushevskaya, 1966; 1967].

Radiolaria have definite depth preferences. Surface-water forms are more delicate, spinose, and ornate, while deeper water forms are larger with fewer, shorter spines and smaller pores with laterally compressed tests [Haecker, 1907]. These are clearly adaptations for maintenance of buoyancy at desired water-column levels. Shape differences between radiolaria are also believed to be closely related to flotation, although little is known about this and few speculations have been made on adaptive morphology. Juvenile forms are usually found at the shallowest depth within the depth range inhabited by the species.

Diversity and modern biogeography

Biogeographic study of modern radiolaria is partly impeded by limitations of the surface sedimentary record because radiolaria are notably absent from large areas of sediments in the ocean basins due to a lack of preservation.

Like the other planktonic groups, the radiolaria are most diverse in the tropical areas and least diverse in polar areas. Also, radiolaria are more diverse than planktonic foraminifera. In general the nassellarians and spumellarians are equally abundant. Radiolarian faunas occur in distinct provinces, closely related to water masses (Fig. 16–18). Various biogeographic schemes, proposed by a number of workers, differ only in detail [Casey, 1971; Nigrini, 1970]. According to Hays [1965], polar or bipolar species have more restricted distributions than warmer water forms. In areas that might be marginal to radiolarian assemblages, such as the severely cold and seasonally ice-covered embayments of Antarctica (the Ross and Weddell seas) and the highly saline tropical regions of the Pacific, diversity decreases and spumellarians become more prevalent. Several lists of radiolarian species characteristic of various water masses have been presented. These are far too numerous to produce, but a few representative species are listed.

ANTARCTIC: *Actinomma antarctic, Lithelius nautiloides, Triospyris antarctica, Theocalyptra davisiana, Antarctissa denticulata,* and *Antarctissa stelkovi.*

SUBANTARCTIC: *Lamprocyclas maritalis, Axoprunum staurax-*

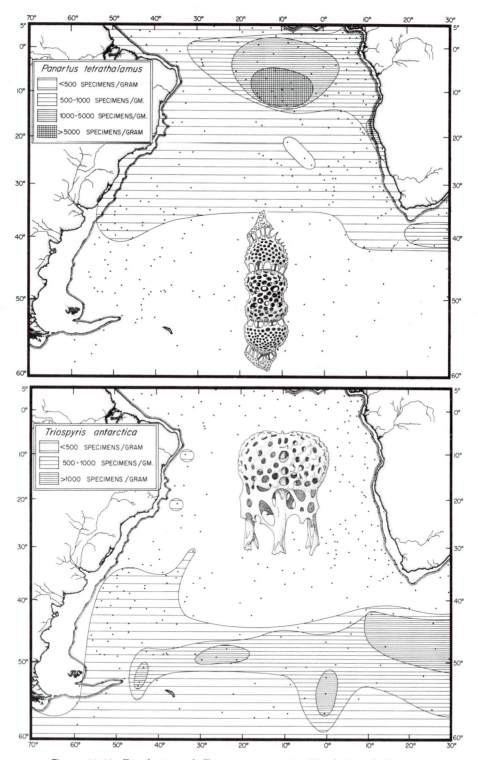

Figure 16–18 Distribution of *Triospyris antarctica* (Haecker) and *Panartus tetrathalamus* (Haeckel) in South Atlantic surface sediments. Abundance refers to specimen number per gram of carbonate-free sediment. (From R. M. Goll and K. R. Bjørklund, *Micropaleontology*, vol. 20, pp. 58, 63, 1974)

onium, Stichopilium annulatum, and *Acti-nomma leptodermum.*

TEMPERATE: *Androcyclas gamphonycha, Lithomitra clevei,* and *Theocalyptra craspedota.*

SUBTROPICAL: *Phorticium pylonium,* and *Cornutella bimar-ginata.*

TROPICAL: *Euchitonia elegans, Botryocyrtis scutum, Pterocanium praetextum, Eucyrtidium hexa-gonatum,* and *Pterocanium trilobum.*

More recently, quantitative studies have been conducted on radiolarian assemblages based on counts of assemblages (greater than 62 μm). Studies, such as those by Goll and Bjorklund [1971] and Lozano and Hays [1976], show quantitative changes with changing latitude in individual species frequencies (Fig. 16–19). Radiolarian assemblages from various areas have also been analyzed using factor analysis. In the Southern Ocean, Lozano and Hays [1976] distinguished three main factors which compare well with the distribution of water masses (Fig. 16–19).

1. *Subtropical.* This assemblage is composed primarily of *Spongotrochus glacialis,* with significant contributions from a species of *Ommatodiscus* and *Lithelius minor.*

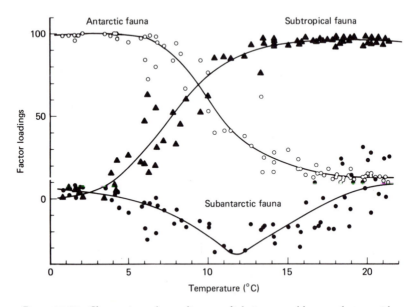

Figure 16-19 Changes in surface sediment radiolarian assemblages or factors with changing sea-surface February temperatures in the Southern Ocean. Dominance is expressed as factor loadings × 100. Note changing relations among three dominant groupings. (After J. A. Lozano and J. D. Hays, 1976)

2. *Antarctic.* This assemblage is dominated by *Antarctissa strelkovi* and *Antarctissa denticulata.* Petrushevskaya [1971] and Keany [1973] have shown that *A. strelkovi* is more prevalent.

3. *Subantarctic.* This assemblage is dominated by *S. glacialis, Ommatodiscus, Theocorythium trachelium, Lithelius minor,* and *Lamprocyclas maritalis.* This factor is not dominated in any one location, but attains maximum negative values under subantarctic waters.

Phenotypic variation is very conspicuous in radiolaria. For example, distinct morphological variation occurs between different water masses for single species [Haeckel, 1887; Popovsky, 1908]. Hays [1965] discovered that Antarctic morphotypes tend to have thicker, heavier tests, while subantarctic forms are lighter and more delicate. The reason for this is unknown but might be due to lower concentrations of silica in subantarctic surface waters, slower rates of reproduction south of the Antarctic convergence, which would allow more time for skeletal construction, or a phenotypic response to different water densities to maintain desired water-mass levels.

Dissolution and reworking

Despite the high degree of undersaturation of silica in the ocean, siliceous skeletons are constructed, although the skeletal elements are rapidly destroyed after organic decomposition. Radiolarian preservation varies significantly and might be related to several factors, including surface-water productivity. Increased siliceous microfossil productivity appears to enhance preservation in underlying surface sediment [Johnson, 1974]. Other preservational factors include vertical temperature distribution and dissolved silica concentration [Berger, 1968], as well as sedimentation rates [Riedel, 1959].

Naturally, solid forms are preferentially concentrated and are more characteristic of deeper water. This can have a great effect on any paleoecological interpretations. Like their calcareous counterparts, the radiolaria species show different susceptibilities to dissolution, which can be ranked [Johnson, 1974]. This ranking shows the relative degree of dissolution a sample has undergone. Collosphaerid and orosphaerid fragments are fairly resistant to dissolution, while nasselarians show a wide range of susceptibility to dissolution [Johnson, 1974]. This dissolution of assemblages creates major differences between the faunas within the plankton and those in the underlying surface sediments.

Diatoms

Diatoms are solitary or colonial algae of the Class Bacillariophyceae within the Chrysophyta that secrete minute bivalved tests or frustules of opaline silica. They are important in marine stratigraphy and paleoecology. The group is a major primary producer in the biosphere, comprising more than 70 percent—and sometimes over 90 percent—of the suspended silica in the water column [Lisitzin, 1972]. Radiolaria are second in importance in the production of siliceous biogenic material. As elements of the phytoplankton, diatoms are confined to the photic zone in marine and freshwaters and may even occur in soils where light is available. They show a wide range of habitat adaptation occurring as planktonic or benthonic, attached or motile.

Diatoms are classified as either centric or pennate forms (Fig. 16–20). Centric diatoms have spherical to cylindrical tests with radial or concentric sculpture; most planktonic diatoms belong in this group. The pennate diatoms are spindle-, wedge-, or rod-shaped and are bilaterally symmetrical around a median line as are bird feathers, from which the name is derived. The pennate forms tend to be more heavily silicified and hence fossilize more easily. Most benthonic diatoms belong in this group, except in Antarctica, where planktonic pennate forms are unusually diverse. Diatom frustules vary in size between 2 μm to 2 mm, although most forms range between 10 and 100 μm. Chainlike colonies can be of considerable length.

Diatoms are quite diverse because they tolerate wide ranges of environments. About 600 living and fossil genera and 20,000 living and fossil species have been described. Of the latter, approximately 10,000 are known to be living today [Wornardt, 1969].

Frustules are especially important in marine sediments that underlie water masses of high fertility, particularly in high-latitude areas and areas of coastal upwelling. They can also be important to the stratigrapher and paleoecologist studying deep-water sediments, which lack calcium carbonate. In tropical areas, radiolaria generally outnumber diatoms in the sediments. However, in relatively shallow depths and in areas of high rates of sediment accumulation, diatom preservation is favored and may be more abundant.

The oldest known diatoms are probably of Jurassic age. They are well represented during the Cretaceous and underwent a major evolutionary radiation in the middle Cretaceous. The ancestors were probably naked or only weakly silicified and therefore were not fossilized. The centric diatoms preceded the pennate forms, occurring exclusively throughout the Cretaceous to early Paleocene. Pennate forms first appeared in the late Paleocene and have shown major evolutionary radiation during the Tertiary. The centric forms have been evolutionarily conservative [Tappan and Loeblich, 1973; Tappan, 1980].

Biological aspects

The living forms and life cycles of diatoms are the best studied of all the microfossils. Most work has been carried out on coastal eurytopic forms, which may not be typical of oceanic plankton forms; therefore this must be considered when interpreting the deep-sea sedimentary record.

The protoplasm contains chromatophores for photosynthesis, which give color to the cell and to the water at times of production blooms. Cells also contain fat globules, which may assist in flotation and be used in food storage. Buoyancy is probably also aided by highly vacuolated cytoplasm. Frustules are enclosed in an organic membrane, which may delay dissolution. They have two valves, which fit together like halves of a pillbox. The "box" is called the **hypotheca** and the "lid" is the **epitheca**. These are connected by a thin bank called a **girdle**. Valves are almost always separated in fossil form.

Figure 16–20 A selection of marine diatoms showing range of morphologies and water mass relations.

Colonial form: (1) *Skeletonema costatum,* ×5000.

Cosmopolitan forms: (2) *Nitzschia bicapitata,* ×7800; (3) *Thalassiosira eccentrica,* ×2300.

Temperate or tropical forms: (4) *Thalassiosira* sp.; ×2500; (5) *Asteromphalus hepactis,* ×600; (6) *Planktoniella sol,* ×800.

Polar forms: (7) *Nitzschia cylindrus,* ×2800; (8) *Nitzschia kerguelensis,* ×4000; (9) *Thalassiosira gracilis* var. *expecta* ×4000; (10) *Thalassiosira* cf. *lentiginosa,* ×1200.

(Photographs 1 and 4 to 10 courtesy P. Hargraves, Graduate School of Oceanography, University of Rhode Island; 2 and 3 courtesy G. Hasle, University of Oslo)

Diatom growth can be very rapid, with populations capable of doubling in about 1 day. Reproduction is by simple cell-division and by auxospore formation. Following cell division, each daughter cell retains one valve and then grows a new one to form the pair. Thus one daughter cell is the same size as the parent, while the other is smaller. This leads to a steady decrease in average size of successive generations. If size continued to decrease, it would clearly be detrimental to the species; hence diatoms have added another stage to the life cycle to return successive generations to the original full size. This stage is called the *auxospore* and results from sexual fusion, autogamy, or from purely vegetative processes. Diatoms in the auxospore stage are about three times as large as the mother cell. Following auxospore formation, simple cell-division begins again. As the successive generations become smaller, changes also may occur in basic geometry and ornamentation of the cells; these are significant in taxonomic differentiation. Diatom life cycles may also include a resting spore stage. The spores are specialized cells that form singly, in pairs, or in chains of four during unfavorable growth conditions. Open ocean forms generally do not contain a resting spore stage in their life cycles.

Morphology and taxonomy

Many diatoms have regularly repeated perforations or areolae through the frustule and are organized in radial (centric diatoms) or striated (pennate diatoms) forms. Furthermore, pennate diatoms possess a cleft: If it runs parallel to the opical axis, it is called a raphe; on a smooth area, it is called a pseudoraphe. Most pennate forms from benthonic habitats bear raphes, which facilitate movement on surfaces.

The auxospore's structure is different than the normal frustule. Resting spores also differ; they are heavily silicified and robust and have two valves, usually without a girdle. Because they are thicker, these forms have a better chance of being preserved in sediment.

The taxonomy of diatoms in most instances is based on the morphology of the frustule, and recent advances in electron microscopy have revealed specialized structures which have required extensive reclassification within the group. Phylogenetic and evolutionary aspects, as well as life-cycle characteristics, require more consideration. Commonly used hard-part characteristics include the outline of the valve plane (circularity, angularity), the structure of areolation, the division of the valve plane into different sections, nature of processes, spines, colony formation, and many others.

Importance of diatoms in marine geology

During the past 10 years, diatoms have become increasingly important for Cenozoic biostratigraphy, and recent work has shown them to be invaluable as paleoceanographic tools. They are valuable for many reasons. For example, diatoms are powerful tools for biostratigraphic subdivision in marine sequences from equatorial to polar regions [Burckle, 1972]. Resulting biostratigraphic zonations are important in deep-sea sedimentary sequences, which lack calcareous microfossils, and in high-latitude areas, marked by poor calcareous microfossil assemblages. Also, because freshwater and marine diatoms are so highly distinctive, they provide a rapid method for differentiating sediments formed in these environments. Paleoecological studies of Quaternary diatom assemblages have shown that like all other planktonic microfossil groups, the diatoms readily distinguish near-surface cold and warm water oscillations.

It is generally accepted that most diatom-bearing sediments are directly related to the productivity of the overlying waters. Diatomaceous sediment is thus most conspicuous in regions of upwelling waters and in highly productive coastal regions. If diatom productivity is reduced or terminated in such areas, the underlying sediment will become barren in diatoms or will contain only a few dissolution resistant forms. In the eastern equatorial Pacific, diatomaceous sediments are important within 10° to 15° north and south of the equator in response to upwelling in the equatorial region. However, in the central gyre areas where upwelling does not occur, diatoms are virtually absent. The quantitative distribution of diatoms, therefore, should reflect the distribution of surface productivity. In conjunction with the previous relationships, it has been found that highly productive regions are also marked by specific diatom assemblages. Thus the Peru–Chile Current and the upwelling region off the Southeast Arabian Peninsula are both marked by their own specific assemblages. Many of the important elements of these assemblages can be traced back to the middle Miocene, so distributions of these assemblages should indicate the initiation of strong upwelling during the Cenozoic. In addition to being clearly responsive to temperature, diatoms also exhibit a strong biogeographic provincialism. Very few species are bipolar. Thus specific assemblages mark the North Pacific, the equatorial region, and the Southern Ocean. This has been generally true since the late Miocene, when the modern flora essentially developed. Diatom studies, therefore, offer important information about Cenozoic development of water masses and the evolution of the various floral provinces.

Also of potential value is the application of diatoms to the study of the distribution of sea ice through time. Many species of high-latitude diatoms either live attached to the underside of sea ice or are ice-related forms, found most abundantly in leads around the sea-ice edge. In the Arctic, at least 24 species of diatoms are usually found growing on the underside of sea ice [Usachev, 1949]. For the Antarctic region, L. Burckle [personal communication] has found that several species, while not exclusively ice-attached forms, are found most abundantly in or near pack ice. The first appearance of such forms may indicate when sea ice first developed, and distribution changes may indicate fluctuations in the extent of sea ice.

Because of their small size, lightness, and hydrodynamic properties, diatoms are readily transported by oceanic bottom currents. They have been used as tracers of the flow of bottom waters from high- to low-latitude regions. Burckle and Hays [1966] found that the Antarctic diatom species *Eucampia balaustium* is currently being transported northward into the Argentine Basin by Antarctic bottom water. Such transportation is so efficient that Antarctic and subantarctic diatoms have been reported as far north as 10°S in the Atlantic (Burckle and Stanton, 1975). Some freshwater diatoms are readily transported by the wind to the ocean, where they are deposited in marine sediments. Their location reveals the history of wind-blown lake sediments from arid areas, such as the Sahara, and records the history of continental aridity.

Ecological aspects

The distribution of marine diatoms is governed by essentially the same variables that effect their abundance. It is difficult to determine the most important parameters affecting their distribution and standing crop. Varying degrees of light intensity clearly affect the depth and latitudinal distribution of diatoms. Within the water column, there is a depth zonation in diatom distribution. For instance, *Planktoniella sol* is known to live at greater depths within the photic zone, and a number of Arctic diatoms appear to have low light requirements and so are capable of active growth on the underside of sea ice.

Temperature and salinity are other critical factors controlling latitudinal changes in floral assemblages. Warm-water forms are found sporadically in waters colder than the normal temperature ranges and it appears that some growth is possible at temperatures much lower than optimal, while a modest increase above the optimal temperature range frequently prohibits growth [Paasche, 1968; Hasle, 1976]. Availability of critical nutrients also effects distribution, particularly the productivity of diatom assemblages. In certain nutrient-rich areas (such as upwelling zones) massive occurrences of diatoms, which even cause the discoloration of the ocean, have been reported. The most important nutrients controlling productivity are phosphorous, nitrate, and silica. In zones rich in these three, huge colonies have been reported [Schrader, 1973; Hasle, 1976].

Low concentrations of silica may selectively influence the species composition of diatom assemblages. Laboratory studies have shown that changing concentrations of silica can affect species growth differentially [Kilham, 1971; Paasche, 1973]. In general, however, it appears that silica concentrations in oceanic waters do not limit species growth because the organisms can convert even an insignificant fraction of dissolved silica into biogenic silica [Lisitzin, 1972].

For a species to survive in Antarctic waters, it must be able to tolerate very low temperatures, low light intensity during part of the year, and high and fairly constant salinities [Smayda, 1958]. Diatoms have succeeded in adapting to these conditions and have relatively diverse floras. In contrast, calcareous nannoplankton have no present-day representatives in these waters and dinoflagellates have only a few species present.

Productivity in polar regions is limited during the winter by low light intensities, while in more stable environments (such as the central Pacific), productivity is limited during the summer by the nutrient supply [Venrich, 1971]. Diatom associations are also known to change seasonally, and these changes depend somewhat on the nature of the water mass. Diatom groups that occupy more stable environments appear to have relatively permanent associations, dissociating into smaller groups during times of rapid growth. This dissociation is a response to local environmental differences. In the subarctic, a generalized winter group of diatom species separates into three groups during the summer (Fig. 16–21) [Venrich, 1971]. Diatom groups in areas of greater environmental instability show strong seasonal oscillations, each having to redevelop from cells introduced from adjacent environments [Venrich, 1971].

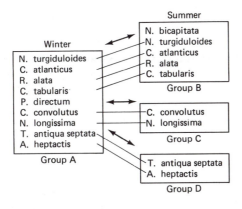

Figure 16-21 Temporal fluctuations of diatom species associations between winter and summer in the subarctic Pacific. (From E. L. Venrich, 1971)

Distribution

Various schemes for biogeographic subdivision of diatom assemblages have been proposed. For example, Kanaya and Koizumi [1966] distinguished seven diatom assemblages in the North Pacific that are related to distinct water masses and Jouse and others [1971] recognized seven general Pacific biogeographic provinces from the Arctic to Antarctic. Several species have bipolar distributions, which has led some workers to speculate that this resulted from transportation of resting spores by bottom waters between the high-latitude regions.

Many diatoms may have very large geographic distributions, implying wide environmental tolerances [Venrich, 1971]. Some of the smallest marine planktonic diatoms appear to have a cosmopolitan distribution [Hasle, 1976]. Following is a representative grouping of diatom species that mark different water masses [Jouse and others, 1971].

ANTARCTIC: The Antarctic assemblage includes forms associated with continental-shelf and -slope areas, as well as open-ocean conditions; *Fragilariopsis curta, Eucampia balaustium,* and *Coscinodiscus oculoides.*

SUBANTARCTIC: *Fragilariopsis antarctica* and *Coscinodiscus lentiginosus.* These are robust forms that preserve well.

SUBTROPICAL: *Thalassionema nitzschiodes, Thalassionema decipiens, Coscinodiscus radiatus,* and *Roperia tesselata.*

TROPICAL: *Coscinodiscus nodulifer* and *Nitzschia marina.*

EQUATORIAL: *Asteromphalus imbricatus* and *Coscinodiscus africanus.*

Morphotypic variation

Like all other groups of microfossils, the diatoms change morphologically with changing environmental conditions. For instance, warm-water diatoms in general tend to be large and spherical, with thin walls, while cold-water diatoms tend to be smaller, with thicker frustules and large appendages. This appears to be a common trend in all planktonic microfossil groups. Intraspecific variation is linked with environmental parameters. For example, rapidly dividing cells usually form thinner frustules.

Dissolution

Diatom assemblages in deep-sea sediments are significantly changed by dissolution, at levels higher than those for radiolaria and silicoflagellates. Studies of sediment suspensions indicate that frustules are rapidly destroyed in the upper 100 m of the water column, with the rate of dissolution decreasing greatly with depth. As a result of dissolution, species diversity decreases with depth and thicker-shelled forms increase in importance. There are several factors that control the differential dissolution rate between diatom species, including the changing specific surface area of frustules (strongly controlled by microporosity of the shell), the relative thickness of the frustule wall, and the changing degree of hydration of the silica [Hurd and Theyer, 1975].

Silicoflagellates

The silicoflagellates are single-celled, flagellated marine plankton with an internal skeleton constructed of opaline silica. They are tubular-shaped and range in size from 10 to 100 μm. They are widespread in oceanic sediment, and so are of value in paleoecological and biostratigraphic studies. Silicoflagellates are regarded as both animals, within the Protozoa, and plants, within the Chrysophyta, because the cytoplasm surrounding the internal skeleton contains yellow discoid chromatophores. They have one flagellum. Because they are algae, they are confined to the photic zone. Laboratory culture studies show that a generation is about 2 days [Van Valkenburg and Norris, 1970].

The skeleton may be a simple ring, ellipse or triangle, but it is often more complex, consisting of two rings or polygons joined by a series of rods (Fig. 16-22). That skeletons exhibit a wide range of variation is very evident in neritic sediments; however, open-ocean conditions are more equable so populations are rather homogeneous over wide areas [Poelchau, 1974]. Silicoflagellate skeletal morphology aids flotation by both maintaining the greatest diameter perpendicular to the sinking direction and by the construction of a minimal framework of hollow tubular rods [Lipps, 1970A].

Silicoflagellate taxonomy is based entirely on skeletal characters, yet the skeleton of modern silicoflagellates in culture vary widely within single populations [Van Valkenburg and Norris, 1970]. As a result, this group is not systematized, due to disagreement on the importance of characters used in establishing the taxa.

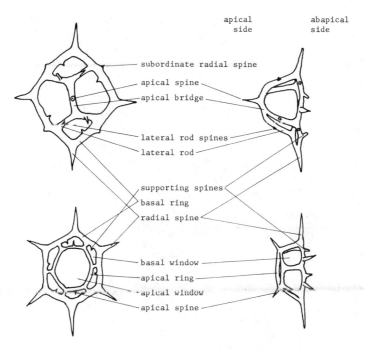

apical
side

abapical
side

subordinate radial spine

apical spine

apical bridge

lateral rod spines

lateral rod

supporting spines

basal ring

radial spine

basal window

apical ring

apical window

apical spine

Figure 16-22 Morphology of the silicoflagellate skeleton and terminology of its parts. Upper form is *Dictyocha*, lower form is *Distephanus*. (From H. S. Poelchau, *Micropaleontology*, vol. 22, p. 164, 1976)

Distribution and paleoecology

Approximately 58 living species are known [Tappan and Loeblich, 1973], while only two genera are very important in Recent sediments: *Dictyocha* and *Distephanus* (Fig. 16-23). Although silicoflagellates occur in sediments in all parts of the ocean, they are seldom abundant and are rarely a major component in marine sediments, except for the siliceous oozes of late Miocene to early Pliocene age on the Falkland Plateau [Stadum and Burckle, 1973]. Gemeinhardt [1934] showed biogeographic provinciality among the silicoflagellates, thereby establishing their potential for paleoecological work. In the South Atlantic *Dictyocha* was found in low and middle latitudes and *Distephanus* in higher latitudes. Quantitative studies of modern silicoflagellates have been performed [Poelchau, 1974], and most species correlate with surface-water temperature. In the North Pacific, *Distephanus octangulatus* and *D. speculum* occur predominantly in cold-water areas, while *Dictyocha messanensis* is more typical of warm water.

Several workers have conducted paleoenvironmental studies in the Cenozoic by using changes in the relationship between warm-water *Dictyocha* and cold-water *Distephanus* [Mandra and Mandra, 1972; Jendrezejewski and Zarillo, 1972; Ciesielski and Weaver, 1974].

The silicoflagellates appeared somewhat later than the diatoms, appearing in

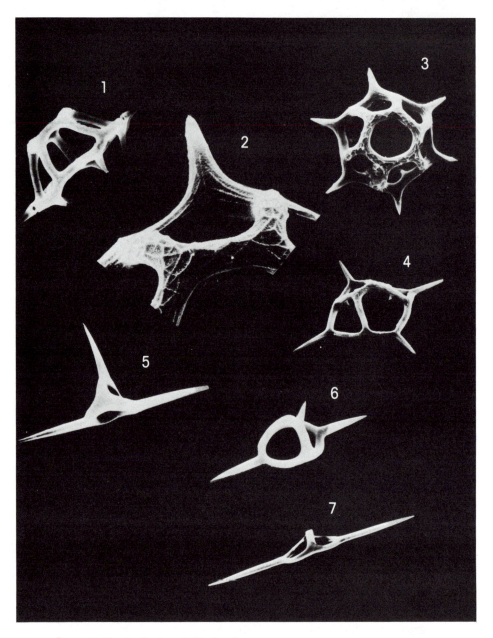

Figure 16-23 A selection of silicoflagellates.

Modern forms
(1) *Distephanus*, apical window view; (2) *Distephanus*, apical bridge view; (3) *Distephanus*, radial spine; surface structure includes nannodivides and nonnovalleys; (4) *Dictyocha*, oblique view.

Late Eocene form
(5 and 6) *Naviculopsis* with apical bridge spine; (7) *Naviculopsis* with no apical bridge spine. All forms × 1500.

(Photographs courtesy York T. Mandra, San Francisco State University)

590   Oceanic Microfossils

the middle Cretaceous. The group is more diverse in the Tertiary than in the Cretaceous and Quaternary and is particularly diverse in the Miocene [Lipps, 1970B; Tappan and Loeblich, 1973].

Importance of silicoflagellates in marine geology

The constant presence of silicoflagellates in siliceous-rich sediments makes them of some use both stratigraphically and paleoecologically. The stratigraphic resolution of silicoflagellate zones is moderate. The average length of a zone is 5 m.y., and zones range from 1 to 9 m.y. [Sanfilippo and others, 1973]. Higher resolution should be achieved with more work, but parallel studies on paleoenvironmental and preservational factors are needed. Paleoenvironmental differences between different genera and species will provide additional criteria for studies of paleoceanography. This will be useful in areas where calcareous fossils are absent.

IV
Ocean History

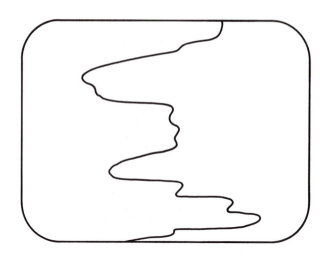

seventeen
Approaches
to Paleoceanography

The hypothesis which we accept ought to explain phenomena which we have observed. But they ought to do more than this: our hypotheses ought to foretell phenomena which have not yet been observed.

William Whewell

THE NATURE OF PALEOCEANOGRAPHY

Definition of Paleoceanography

Global tectonics forms a fundamental basis for our study and understanding of the evolution of the earth's lithosphere, its morphological and physical character, and its motion through time. Global tectonics also provides a framework for understanding the environmental evolution of the earth, its physical geography (for instance, the Appalachian Mountains [Dewey and Bird, 1970]), the oceans [Heirtzler and others, 1968] and oceanographic and climatic conditions [van Andel and others, 1975; Kennett and others, 1975; Valentine, 1973]. The movement of the solid crustal boundaries is the major cause for large changes in climate. For example, the presence of a circumglobal tropical ocean has been linked with relatively warm global climates. The closing of such tropical oceanic connections, together with the opening of seaways at high latitude, are thought to be associated with the dramatic cooling and formation of continental glaciers in polar regions [Kennett and others, 1975; van Andel and others, 1975; Hayes and Frakes, 1975]. The isolation of ocean basins at high latitudes appears to be associated with the development of continental glaciations [Kennett and others, 1975; Hayes and Frakes, 1975]. The study of the development of the ocean systems is called **paleoceanography.**

Paleoceanography includes studies of the development of surface circulation, the history and effects of bottom-water circulation patterns, planktonic and benthonic biogeographic development and rates of organic evolution, the history of biogenic productivity and its effects on sediment distribution, and the history of calcium carbonate and silica deposition and dissolution. None of these disciplines is easily separated from the others.

Paleoceanography and related global climatic evolution have been the major controlling forces on the development of marine biogeography. Biogeography is concerned with the geographic distribution of organisms throughout the earth. Such investigations attempt to show spatial relations between organisms, to explain these relations, and to determine the evolutionary pathways that produced the biogeographic patterns. The history of development of the oceans and global climates are critical to our understanding of the present-day marine environment. The means by which the oceans and climate have changed in response to changing boundary conditions gives insight not only into the possible natural extremes of the earth's climate, but also into the basic behavior of the air-ocean-ice system.

The Deep-Sea Sediment Record

Early concepts of deep-sea sedimentation presumed that the deep sea was a tranquil, unchanging environment with extensive, unbroken sedimentary records. These assumptions were based on the incorrect notion that the earth is static. Now it is known that there have been many changes in the physical and chemical state of the oceans throughout the earth's history. The deep-sea sedimentary record offers a more complete record of the earth's history since the middle Mesozoic. Deep-sea cores are extremely useful because data can be extracted from them and compared globally and because they contain records generally more continuous than those located on land. The physical, chemical, and paleontologic characters of marine-sediment sequences record many important features of the ancient ocean, atmosphere, and **cryosphere,** or glacial world. Each core contains a multichannel record permitting reconstructions of numerous aspects of oceanic and climatic variations. For instance, the shells of myriad planktonic species, each adapted to its own peculiar set of oceanographic conditions, are major components of deep-sea sediments. Changes in global ice volume are reflected in the oxygen isotope ratios of foraminiferal tests. Changes in sediment character reflect a host of varying environmental conditions, such as those related to sediment rafting in icebergs, continental erosion, and biological productivity. The covariance of these indicators (both intercore and intracore) provides input basic to understanding the cause of oceanic and climatic changes.

Objectives of Paleoceanography

Rapid recent progress in Cenozoic paleoceanography has followed technical and conceptual breakthroughs in several areas. First, and above all else, has been the spectacular success in acquiring sediment samples from the deep sea by drill and piston cores. The second advance is the development of biostratigraphic schemes that have been chronologically calibrated and globally correlated to provide potential for analysis of time series and stratigraphic framework for constructing pale-oceanographic charts. Third, plate tectonics provides the required basic framework for reconstructing paleogeography. Fourth, numerous paleontologic, geochemical, and mineralogical techniques are now available for application to a wide range of paleoenvironmental reconstructions, including paleoclimatology. The methods and techniques available for the study of paleoceanography all require the determination of conditions in the oceans in three dimensions. Horizontal *and* vertical distributions of parameters are equally important, and certain critical areas that form "gateways" (enabling or denying particular circulation or migration opportunities) are especially significant. When integrated, these methods constitute a monitoring system capable of observing the following:

1. The distribution of surface water masses.

2. The average position of major currents and fronts.

3. The regional pattern of sea-surface temperatures.

4. Changes in the global ice volume and sea-ice extent.

5. Changes in the vertical thermal structure.

6. Changes in bottom currents and bottom water masses.

7. Changes in the intensity and position of trade winds.

8. Paleobiogeographic changes that are manifested by all of the above.

Horizontal and vertical gradients can be measured from the floral, faunal, physical sedimentological, and geochemical patterns derived from analyzing the sediments. The tools for delineating these patterns include stable isotopic compositions, census data on the microfossil assemblages, morphometric and taxonomic data on both the planktonic and benthonic microfossils, as well as organic compositions and trace metal contents of the sediments. Stable isotopic parameters include the ratio of O^{18} to O^{16} and the ratio of C^{13} to C^{12} (see Chapter 3). The microfossil trends define gradients in the water properties and reveal the displacement of particular water masses.

Sedimentological patterns on the sea floor are associated with the following patterns.

1. Dissolution and basin to basin fractionation, both for opaline silica and calcium carbonate.

2. Upwelling and productivity, primarily for opaline silica, but also for calcium carbonate.

3. Surface circulation, derived largely from microfossils but also based on distribution of ice-rafted material.

4. Deep-water circulation, indicated by benthonic microfossils, contour current deposits, spatial pattern of hiatuses, and by geochemical fractionation patterns.

5. Atmospheric circulation, primarily associated with distribution of continental detritus, such as quartz and illite.

Because all these patterns are controlled by processes, changes in distributions with time can be interpreted in terms of changes in the controlling processes. In this way, changes in the distribution of sedimentary components can be used as another tool in deciphering the changes in oceanic and atmospheric circulation and the effects of these changes on global climate.

Paleoceanography differs from modern physical oceanography in three fundamental ways. First, estimates of physical parameters are made by the various indirect approaches mentioned above, rather than by the direct measurements possible by physical oceanographers. Second, the manner in which the sedimentary record is formed limits this paleoenvironmental monitoring system to recording conditions averaged over hundreds to thousands of years. However, the record can usually be sampled on nearly the same time scale as that associated with the residence time of deep waters; that is every 1000 to 2000 years. Third, the time dimension extends from minutes to decades (as in physical oceanography) to centuries, millenia, and millions of years. The extension of the time span allows the study of the oceans' past characters, as well as the evolutionary steps toward the development of our modern oceans.

GENERAL SCENARIO OF LATE PHANEROZOIC PALEOCEANOGRAPHIC CHANGE

Oceanic circulation and global climates have undergone a dynamic evolution throughout the late Phanerozoic. This has resulted primarily from lithospheric evolution due to plate tectonics and has fundamentally changed both the spatial relationships between land and sea and the ocean basins themselves. Recently, progress has been made in our understanding of this paleoceanographic evolution. A number of paleoenvironmental syntheses of existing DSDP data have been completed for particular regions and data sets [for example, Berggren and Hollister, 1974; 1977; van Andel and others, 1975; 1977; Moore and others, 1978; Kennett, 1977], especially for the Cenozoic. A preliminary framework of late Phanerozoic global paleoceanographic change has been established, providing a basis for future investigations and presenting a unified view of global paleoenvironmental history, stratigraphy, and geochemistry. This is one of the most important legacies of the first decade of deep-ocean drilling.

At the beginning of the Mesozoic, all of the present continents were joined together as a single supercontinent, Pangaea. Pangaea was surrounded by the superocean Panthalassa, which stretched from pole to pole and around more than half the circumference of the earth. During the subsequent 200 m.y., this ocean fragmented to form the Atlantic, Arctic, Indian, and Southern oceans, leaving the Pacific a scaled-down version of the original superocean. In the early Mesozoic the two major components of the supercontinent (Gondwanaland and Laurasia) became separated by a triangular reentrant of Panthalassa and the Tethys Sea. Circumtropical circulation developed through the Tethys Seaway; longitudinal seaways were created when the Americas rifted apart from Africa and Europe. In the late Jurassic to early Cretaceous (about 130 Ma), fragmentation of Gondwanaland began as India separated from western Australia and opened the Indian Ocean. The Tethys Seaway, after its gradual diminution, became our Mediterranean Sea. Associated northward movement of the southern continents allowed a circum-Antarctic circulation pattern to develop. Because of the intimate relationships between ocean, atmosphere, and land, these paleoceanographic changes have been a major driving force in the evolutionary development of the whole paleoenvironmental system; this has determined the evolution of marine biota ranging from the protozoans to whales.

Fragmentation of Pangaea led to fundamental changes in the distribution of active and passive continental margins. Initially, the edges of Pangaea were almost entirely dominated by active, subducting margins. Since the early Mesozoic, a reorganization of the global pattern of plate tectonics has led to the formation of rifted margins and the transformation of active margins into passive margins.

Global climates have also undergone dynamic changes, the most important having occurred between the various geologic epochs and periods. Warm, equable Mesozoic climates were marked by an absence of pronounced latitudinal temperature gradients, combined with a relatively homogeneous oceanic thermal structure. Together they created a sediment regime of low energy. Sea levels were high, leading to the deposition of extensive, shallow sequences of carbonate sediment. Low environmental gradients and restricted circulation in the narrow, but expanding, deep seaways led to the development of stagnant conditions and deposition of anoxic sediments.

The Mesozoic conditions were subsequently replaced by cold oceans with bottom waters close to freezing that created strong environmental gradients between tropics and poles. The global climatic changes did not proceed uniformly. They were marked by discrete and sudden coolings, especially during the Cenozoic. These seem to have resulted from marked changes in the geometry of the ocean basins and associated changes in ocean circulation. Cause-and-effect relationships almost certainly exist between changes in ocean circulation and global climates. One of the most prominent changes took place near the Eocene-Oligocene boundary (38 Ma) when the cold, deep water of the oceans developed following the onset of ice formation around Antarctica. Subsequent coolings occurred during the middle Miocene (14 Ma), when most of the ice accumulated in the Antarctic ice cap, and in the late Pliocene (3 Ma), when Northern Hemisphere ice sheets began to develop, perhaps in response to the closing of the Central American Seaway (Panama). Present-day patterns have been established only recently and have undergone numerous, significant modifications due to repeated climatic oscillations associated with the glacial-interglacial episodes. This steplike progression of climatic evolution resulted from the solid earth's changing crustal topography and geography, features which form the boundaries of the earth's fluid spheres.

Thus the historical development of the world ocean-current system is controlled by two major factors: the evolution of oceanic basins through plate tectonic processes and deep-sea topography and the climatic and glacial evolution of the earth, particularly in the polar regions. Three major elements are involved in the evolution of Cenozoic paleoceanography. The first of these is the diminishing role of oceanic circulation in the equatorial areas, resulting from the closure of the Tethys Seaway, the construction of the Indonesian area, related to northward movement of Australia, and the bridging of the Central American connections. The second major element is the development of circum-Antarctic circulation, resulting primarily from the separation of Australia from Antarctica and the opening of the Drake Passage south of South America. The third major element is the changing character of the bottom-water circulation of the world ocean. All three elements, together, control the broad pattern of deep-sea sediments during the Cenozoic, including those resulting from changing loci of high biogenic productivity, from deep-sea erosion and the formation of widespread disconformities by bottom-current activity, and from chemical changes controlling calcium carbonate and opaline silica dissolution.

First Requirements

Before it is possible to make paleoceanographic reconstructions using data from individual sample sites, a number of basic requirements have to be met. The following basic parameters for each site are needed for paleoceanographic reconstructions: For each site the migration track of the site (**site backtracking**), as well as its subsidence history, has to be determined. A standardized lithologic description, correlative stratigraphy, and a chronologic framework for the sequence must be determined. Also, rates of deposition for bulk sediments and for biogenic components must be known. In addition, before any quantitative paleoceanographic estimates are made, it is necessary for the taxonomy of the most abundant microfossil forms being used for the paleoceanographic reconstructions to be reasonably well known.

Paleoreconstructions

The shape of the ocean basins, their lateral boundaries, the continental positions, and the absolute and relative positions of individual sample sites are all affected by plate motion. The movement of individual sites is particularly important for biogeographic reconstructions when such motion has taken place across latitudinal lines. The consideration of depth changes associated with sea-floor spreading is also important to the interpretation of the isotopic, faunal, floral, and lithologic records. The migration path of each site with respect to the spin axis of the earth, as well as the change with time of the configuration and position of the ocean basin itself and the evolving bathymetry, must be determined.

A simple relation has been found to exist between depth and age of the oceanic crust [Sclater and others, 1971; see Chapter 5]. It has been shown that this relation is caused by cooling of the oceanic crust as it moves away from a spreading center [Sclater and Francheteau, 1970]. Thus, to construct charts of the past bathymetry of our ocean, it is necessary only to know the history of sea-floor spreading of the ocean floor and the relationship between depth and age. The first series of paleobathymetric charts were constructed by Sclater and McKenzie [1973] for the South Atlantic Ocean. Since then an attempt has been made to chart the Indian Ocean [Sclater and others, 1977A] and to draw a more detailed chart for the whole Atlantic Ocean [Sclater and others, 1977B].

Paleobathymetric analysis is now a well-established and useful technique in the oceans. Paleobathymetric charts can allow the gross, sediment-free paleobathymetry of the oceanic crust younger than 50 Ma to be mapped within ± 400 m over approximately 80 percent of the ocean floor. Assuming that models for aseismic ridges and other anomalously shallow areas are correct, coverage to such precision (± 400 m) can be extended to more than 90 percent of the floor of the oceans.

Microfossil taxonomy

The intensive, quantitative use of pelagic microfossils for biogeographic and paleoceanographic reconstructions requires a working taxonomy. An initial phase in studies of global paleoenvironments, and one that must continue, has been the discovery, description, and classification of fossil organisms and the study of their evolutionary sequences. This precedes the next step, that of determining their habits and habitats for use in paleoecology. The characters of the four major groups (foraminifera, calcareous nannofossils, radiolaria, and diatoms) used in paleoceanography and oceanic biostratigraphy are outlined in Chapter 16. The four major groups vary sufficiently in their paleobiological and preservational aspects to provide different kinds of useful information. Furthermore, abundant assemblages of carbonate and siliceous forms are often mutually exclusive because of preservational differences, so comprehensive paleoceanographic reconstructions over large areas of the oceans require the use of all four groups. Taxonomic knowledge varies considerably for each of the microfossil groups. The taxonomy and phylogenetic relationships of planktonic foraminifera and calcareous nannofossils are well enough known to carry out quantitative analyses on the majority of the fauna and flora. Radiolarian taxonomy is less well established, but it is advancing rapidly and most of the abundant Cenozoic elements are becoming taxonomically better known. Diatom taxonomy still requires much investigation. Paleoceanographic analyses are rapidly advancing for each of the various groups, using quantitative changes in the assemblages over most parts of the ocean basins.

Stratigraphic correlations

In order to reconstruct the biogeography and paleoceanography for certain times past and to understand global paleoceanographic history, it is necessary to be able to correlate each site accurately. For spatial reconstructions, this correlation has to be accurate enough that fluctuations in the assemblages within the interval in question are smaller than the differences between sites. In the past, almost all correlations between Tertiary sequences were carried out using planktonic microfossils. Other approaches are oxygen isotopic stratigraphy and paleomagnetic stratigraphy (see Chapter 3). These have been particularly important in correlating Quaternary sequences. In fact, resolution from correlations using oxygen isotopic stratigraphy may never be much improved upon in the Quaternary!

As a result of the efforts of many scientists involved with the DSDP, the Cenozoic and Cretaceous biostratigraphies of foraminifera, radiolaria, calcareous nannofossils, and diatoms have been greatly improved in recent years. Biostratigraphic schemes based on planktonic foraminifera have been usefully applied for a longer time, but they have improved significantly by the study of the deep-sea sequences. These biostratigraphic schemes are based on extinctions and first appearances of individual species. Correlation and calibration of Cenozoic planktonic microfossil zonations have rapidly advanced over the past decade, with major summaries presented by Berggren [1969; 1972], Berggren and van Couvering [1974, 1978] and Srinivasan and Kennett [1981]. Low-latitude Cenozoic zonal schemes for calcareous and siliceous microfossils are now standardized and, when combined with multiple microfossil datum planes, offer a biostratigraphic resolution averaging 1 to 0.5 m.y., as documented by Berggren. Until recently, a similar level of biostratigraphic resolution was not possible in middle- and higher-latitude Neogene sequences due to the low diversity of calcareous plankton in those regions, lack of knowledge about high-latitude siliceous planktonic microfossils, and difficulties in extending low-latitude zonal schemes into cooler-water provinces. This situation has changed markedly as low- through high-latitude diatom zonations have been established [Burckle, 1972].

Because major differences exist between microfossil assemblages in high and low latitudes and because the stratigraphic ranges of microfossils can vary markedly, even between what are inferred to be adjacent water masses, there are major problems with widespread biostratigraphic correlations across the Tertiary paleoceans. Quaternary correlation is aided greatly by oxygen isotopic and paleomagnetic stratigraphy.

Paleomagnetic stratigraphy, which has been fundamental in the correlation and dating of Pliocene and Quaternary deep-sea piston cores, has been of little value in drilled sequences because of mechanical drilling disturbance. Now the hydraulic piston core is providing mechanically undisturbed sequences which produce high-quality paleomagnetic polarity records. This is a major step in the detailed correlation and chronology of unlithified sequences.

Chronological framework

Quantitative paleoceanography requires a chronological and stratigraphic framework to determine rates of oceanographic change. Established chronological frameworks, in relation to the standard planktonic biostratigraphy and schemes, are discussed in Chapter 3 [Berggren, 1972; Berggren and van Couvering, 1974]. In order to resolve the most conspicuous paleoceanographic changes through the late Phanerozoic, a temporal resolution of about 1 to 2 m.y. is required. Such resolution can be obtained over most areas of the ocean, except where core recovery has been poor or where microfossil assemblages are poorly preserved. The available chronology for the Cenozoic has already been widely used to make sedimentation-rate calculations at individual sites and over wider areas. This chronology seems to be a satisfactory basis for rate change studies within the oceans. Chronological frameworks are, however, only as accurate as the quality and amount of radiometric data used in their establishment, so they must be updated periodically as new data become available. Intercalibration of biostratigraphies and chronology are poorer for the Cretaceous, so determination of rates of paleoceanographic change are much more difficult in this interval.

BASIC APPROACHES TO PALEOCEANOGRAPHIC MAPPING

Time Series and Time Slices

There are two basic approaches used to study changes in the global air-ocean-ice system: changes in the system as a function of time at various locations (**time series**) and spatial oceanographic patterns that existed at selected times in the past (**time slices**). The methods are different, but they complement one another. Without time-slice studies it would be difficult to determine how one area was related to another and how this relationship changed with time. Without time-series studies in different locations, it would be difficult to determine the character of changes in the ocean from one stage of its evolution to the next. Many time-series analyses have been carried out for all parts of the Cenozoic and Cretaceous. These are easier to complete because they can be conducted independently of series carried out in other areas. Time-slice work requires much more effort, because correlations between sequences must be very well resolved to guarantee that the same interval of time is being examined at each location. This level of stratigraphic correlation normally requires the integration of numerous biostratigraphic and stable isotopic criteria. Once the isochronous surface has been established, it is relatively easy to map biogeographic, geochemical, sedimentological, and paleotemperature parameters on a global basis. Other than the modern ocean, the best known time slice is the last glacial world 18,000 years ago, studied by the CLIMAP (Climate Long Range Investigation Mapping and Planning) group [CLIMAP, 1976]. This time slice reveals the character of surface-water circulation 18,000 years ago.

Surface-Water Gradients and Patterns

The ocean is a multicomponent system with regard to advective and convective flow. The most energetic components of this system are the surface-water masses, which play the major role in energy exchange and transport in the oceans. Nevertheless, surface waters are but a thin interface between the atmosphere and the vast bulk of the hydrosphere which lies beneath. Because surface circulation is so critical in the transportation of energy from low to high latitudes, it is important to establish both latitudinal and meridional gradients in these surface waters. Changes in oceanic heat storage and transport are responsible for changes in the earth's climate. Oceanographic patterns related to these processes are reflected in biogeographic patterns. Variations in surface-water conditions, such as shifting water-mass boundaries, temperature variations or upwelling rates, are recorded by changing microfossil assemblages in deep-sea sediments. The relative importance of individual water masses is established by determining proportions of individual microfossil assemblages or forms. A second major tool used in establishing surface-water gradient patterns is oxygen isotopic analyses of planktonic microfossils. The oxygen isotopic composition of planktonic foraminifera, in part, varies in response to surface-water temperatures in the ocean.

Surface-water gradients are determined by analyzing transects of deep-sea sediments across oceans or oceanic boundary zones. Such zones include the eastern and western boundary currents, which show sharp hydrographic and faunal contrasts (for instance, the Peru Current with the nearby regions of the central gyres). These boundary currents interrupt the zonal flow and are important features in transporting heat poleward, in determining patterns of productivity, and in the interaction of deep and surface waters. Of particular interest in the study of boundary currents are changes in their width, the sharpness of their gradients, and the shape of their decay oceanward. Changing patterns of surface-water gradients associated with the boundary currents provides insight into past strength and patterns of upwelling. Sharp gradients that occur in the open oceans mark regions of convergence or divergence that separate major water masses. These boundaries lead to partial isolation of the individual water-mass packages and, depending on the sharpness of their definition, indicate the intensity of mixing and zonal transport. Shifts in the location of these zones indicate changes in past patterns of water masses that may be associated with climatic change or tectonically caused changes in the boundaries of the oceans.

In certain areas, north-south, surface-water gradient changes between extreme oceanic conditions are significantly amplified. For instance, the Northeast Atlantic has exhibited particularly large surface-water gradient changes that extend over 10 to 20° of latitude between interglacial and glacial times [McIntyre, 1967; McIntyre and others, 1976]. Figure 17–1 shows the differences in surface-water isotherm maps for this region between the present day and the last glaciation 18,000 years ago. Polar water in the eastern North Atlantic retreated to 65°N (Iceland) during interglacials and expanded southward to beyond 45°N during glacial maxima. During glacial maxima a latitudinally orientated frontal system developed along 45°N, with extreme gradients similar to those in the Southern Ocean. Subpolar and transitional waters were essentially eliminated by being compressed into a narrow band between the clockwise subtropical gyre and the counterclockwise polar gyre. This pattern minimized northward heat transport in the surface waters [Kipp, 1976]. A much smaller change in surface-water gradients occurred in the subtropical gyres between glacial and interglacial episodes, reflecting their greater stability.

Surface-Water and Assemblage Relationships

The value of planktonic marine microfossils for paleoclimatic-paleoceanographic analysis was inherent in Murray's [1897] early recognition that patterns of these shell-forming groups on the deep-sea floor reflect their living distributions in

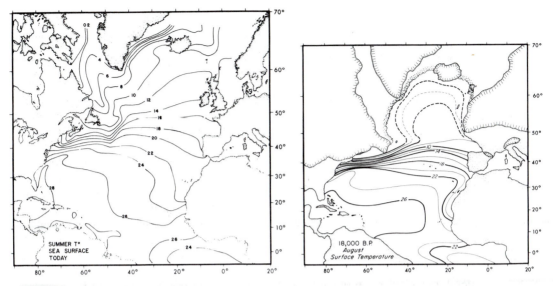

Figure 17-1 Surface-water isotherm maps for August today and August 18,000 years ago in the North Atlantic. (After A. McIntyre and others, 1976, courtesy The Geological Society of America)

the overlying water masses and the positions of surface isotherms (see Chapter 16). Thirty-eight years later, Schott [1935] published a pioneering paper in which he clearly demonstrated that the percentage of frequency variations of temperature-sensitive planktonic foraminifera offered a powerful tool for determining Quaternary climatic events. The following two decades saw reinforcement, extension, and evolution of these early insights, progressing from qualitative identifications of subjectively defined "warm" versus "cool" planktonic assemblages in gravity cores [Phleger, 1939; Arrhenius, 1952] to semiquantitative analysis of selected species in longer piston cores [Ericson and Wollin, 1956, 1964].

These initial qualitative analyses provided important, concrete evidence of the validity of micropaleontologic analysis for paleoceanographic interpretation within Quaternary sequences. However, more sophisticated methods of study, including ratio manipulations based on quantitative analysis of entire assemblages [Lidz, 1966] demonstrated that Quaternary paleotemperatures comparable to those derived by oxygen isotope analysis [Emiliani, 1964] could be estimated directly from micropaleontologic data. Parallel advances in quantitative study of biogeographic distributions of living planktonic foraminifera, calcareous nannofossils, diatoms, and radiolarians, as related to water-mass characteristics, together with the advent of a more-precise Quaternary chronology through application of paleomagnetic stratigraphy and increasing availability of long-piston cores provided the stimulus and framework for detailed study of Quaternary paleoceanographic oscillations [Hays and others, 1969; Ruddiman, 1971]. These studies have culminated in determining quantitative relations between Quaternary planktonic assemblages and isotherm patterns through the use of transfer-function analysis. This technique uses factor and regression analyses of quantitative assemblage data and direct comparison of living distributions of planktonic foraminifera to temperature [Imbrie and Kipp, 1971; Hecht, 1973]. The result of this research was a major advance in our understanding of Quaternary paleoceanographic changes in relation to glacial-interglacial oscillations. Planktonic foraminifera continue to be the prime tools used in these analyses because their modern distributional patterns are well known and because of the low diversity at any particular time in the world ocean (maximum of about 30 species in low-latitude areas grading to monospecific populations in polar areas). Similar quantitative analyses and statistical manipulations of other groups, including calcareous nannoplankton [McIntyre and others, 1976] and radiolaria [Sachs, 1973] are yielding valuable results in Quaternary sequences.

Most paleoceanographic reconstructions are dependent in large measure on the confidence with which fossil plankton can be associated with water-mass boundaries. Studies of the ecology and biogeography of living planktonic foraminifera over the last several years have firmly established distributional patterns for this group in the world ocean [Bradshaw, 1959; Bé and Tolderlund, 1971]. Species ranges, as well as assemblages of species, are clearly linked to individual water masses, and some species, such as *Globorotalia inflata,* reach optimum development within oceanic frontal systems. This provides a method for examining water-mass boundaries. Both qualitative and quantitative analyses of these patterns have shown that planktonic foraminiferal assemblages are excellent tracers of tropical, subtropical, temperate (transitional), subpolar, and polar water masses; it is this general model that has formed the basis for transform-function analyses of Quaternary paleoclimates [Imbrie and Kipp, 1971]. Data on modern distributions of other shelled planktonic groups show similar water-mass–assemblage relationships. Census data on biogeographic trends during the Quaternary provides a workable basis for establishing paleoceanographic patterns and even maps of absolute temperature trends [CLIMAP, 1976].

The task becomes much more difficult when dealing with older faunas, in which most of the elements do not exist in present-day oceans. Nevertheless, planktonic assemblages of the past oceans have also been related to individual water masses. Thus, even without a thorough understanding of their biology, oceanic microfossils can be used to map the oceans of the past. An important observation is that the gross biogeographic distributions and diversities of Neogene and Paleogene microfossil assemblages are such that similar associations of fossil taxa are readily discernable throughout the Cenozoic. These express similar water-mass–temperature control, despite differences in taxonomic composition, changing ocean configurations, and major differences in isotherm positions [Cifelli, 1969; Haq, 1973].

Great similarities between modern and pre-Quaternary biogeographic distributions are inherent in correlating planktonic foraminiferal morphotypes with latitude and surface isotherms. Highest diversity assemblages, containing relatively exotic morphotypes, have inhabited tropical water throughout most of the Cenozoic, while low-diversity assemblages of simple globose morphotypes have dominated middle and high latitude regions [Cifelli, 1969]. This was not the case during the early Paleocene and Oligocene, when relatively low-diversity assemblages and simple morphotypes dominated all latitudes due to different paleoceanographic conditions. These trends, together with recurrent species associations, assist in assigning extinct species to appropriate high-, middle-, and low-latitude planktonic assemblages and in the process, to effectively overcome one of the primary problems facing pre-Quaternary paleoceanography—determining the environmental significance of the fossils [Berger and Roth, 1975].

Statistical techniques have since been developed to objectively select paleoceanographically significant assemblages of extinct Paleogene calcareous nannoplankton through use of Q-mode varimax factor and oblique factor analyses [Haq and Lohmann, 1976]. These allow differentiation and mapping of low-, middle-, and high-latitude nannofloras. These same techniques can be applied to any group for which appropriate census data exist.

Bottom-Water Circulation

Approaches have also been developed to study the history of bottom-water change and distribution. These approaches are more difficult than those used in the study of surface-water patterns. We have seen from the above discussion that surface-water gradients change steadily from region to region, and that these changes in gradients can be very rapid. We have also seen that the changes are well marked by gradients in planktonic microfossils. Bottom-water patterns are unlike those of surface waters (see Chapter 8), in that minimal changes occur in their characters over wide areas of the oceans once individual bottom-water masses have formed. The physical character and distribution of bottom waters is governed largely by conditions in their areas of formation (primarily at polar latitudes), by the bathymetry of the oceans, and interocean connections that control the bottom-water flow patterns. The chemical nature of bottom waters changes gradually and is dependent on the amount of time elapsed since formation and the rates of supply and regeneration of the nutrient (nonconservative) elements. Thus the same deep water mass may cover vast areas of the sea floor and show very gentle gradients in the character of the faunas associated with it.

Studies of deep-sea sediments have shown that the character of the bottom waters of the oceans has evolved as a result of polar glacial evolution. Hays and Pitman [1973] suggested that the extensive epicontinental seas of the late Mesozoic and the absence of polar glaciations might have resulted in the production of deep and bottom water in areas distant from the poles, perhaps producing a much more fragmented pattern of deep circulation. On a much shorter time scale, changes in the distribution of oceanic bottom waters have occurred in response to changing polar glacial conditions between interglacial and glacial episodes during the late Cenozoic.

The most important technique employed to study bottom-water history is one that establishes relationships between benthonic foraminiferal or ostracode assemblages and individual bottom-water types. Schnitker [1974] and Streeter [1973] pioneered this approach for the deep sea by discovering that North Atlantic deep water and Antarctic bottom water in the North Atlantic are marked by distinctive benthonic foraminiferal assemblages. Furthermore, downcore studies of the benthonic fauna indicate that major changes have taken place in the deep water masses and that they are very different in character from those of the surface water indicators. These initial investigations have developed into four-dimensional studies of bottom water using benthonic foraminiferal assemblages at individual time slices in traverses of cores at different abyssal and lower-bathyal water depths [Lohmann, 1978]. One of the major constraints in studying the history of bottom waters is that abyssal bottom waters tend to be highly undersaturated in calcite, so benthonic foraminiferal assemblages are usually lost or severely altered by dissolution. As a result, waters at intermediate ocean depths provide much more information on changes in the deep ocean than do deeper waters.

Another approach in the study of bottom-water circulation is to examine the distribution of hiatuses in deep-sea cores throughout the oceans. As bottom waters are channeled by abyssal topography, changes in topography affect their distribution patterns and the patterns of unconformities and sedimentation rates. For instance Kennett and others [1972] inferred that a major change had occurred in the distribution of bottom waters in the Tasman Sea region during the Cenozoic in the Southwest Pacific because Paleogene deep-sea sequences are disrupted by several unconformities caused by northward-flowing bottom water through the region. Less disrupted Neogene sequences indicate that these bottom currents were later diverted to other areas of the South Pacific.

Vertical Oceanographic Gradients

Vertical oceanographic gradients reflect the stratification of the oceans and are associated with depth-specific sedimentary facies, microfossils, and stable isotopes. The gradients of greatest interest are in the thermohaline pycnocline, the oxygen-minimum zone, and the zone of carbonate deposition. The first two are rarely intersected by the sea floor under normal conditions; their record in sediments must be sought in special regions. Vertical oceanographic gradients are important for the general description of an oceanic environment. Distinct oceanic states have very different vertical gradients, ranging in the extreme from highly stratified anaerobic (**stagnant**) oceans such as those that occurred at particular times in the Mesozoic to the highly mixed aerobic (**dynamic**) oceans of the late Cenozoic. Most studies involving paleoceanography of the deep waters have concentrated on determining the history of bottom-water temperature change throughout the Cretaceous and Cenozoic rather than on reconstructing vertical environmental gradients through deep and intermediate waters for discrete intervals of geologic time.

The principal record for shallow- and intermediate-water structure is from cores taken on anomalously shallow crust. Suitable areas are ancient ridge flanks, deep-sea plateaus, aseismic ridges and other suitably elevated areas, subsiding offshore platforms, and continental margin junctions.

The most suitable slopes for such analyses are located on tectonically simple crustal blocks with minimal sedimentary redeposition or erosion. Furthermore, most topographic elevations show evidence of subsidence with time. The use of this record requires the ability to reconstruct the subsidence history of each site [Detrick and others, 1977], which is a difficult task. Given an accurate paleodepth reconstruction of a subsiding slope, it is possible to extract vertical gradients from the sediments originally deposited on this slope at particular times in the past (Fig. 17-2). Three sites are shown at three successive times and their cores record the particular environmental conditions at successively greater water depths. The oldest sites have sampled three water masses over the slope. Since the slope is subsiding, each site records parameters related to increasing water depths. Vertical oscillations in water masses are also recorded in such core traverses. Rates of subsidence are more rapid in earlier stages of crustal cooling; thus closely spaced sites are required on the steeper, younger slopes in order to study the history of vertical gradients.

The most important vertical, physical gradients in the oceans are changes in temperature, salinity, and density. There are also major chemical gradients, such as those reflected in the preservation of calcite and of organic matter. All of these parameters reflect the influences of different water masses. These water masses can be identified by the analysis of calcite dissolution, isotopic paleotemperatures, microfossil sequences, organic carbon content, and other parameters.

Oxygen isotopic profiles

In tropical regions, surface waters are like an envelope of warm water resting on an ocean of cold water. Temperature decreases with increasing depth but most of the change takes place in the upper few hundred meters. Temperature at depths greater than 1000 m in the present day are very stable and vary little (about 2°C globally). The level in the ocean where temperature changes occur most rapidly is

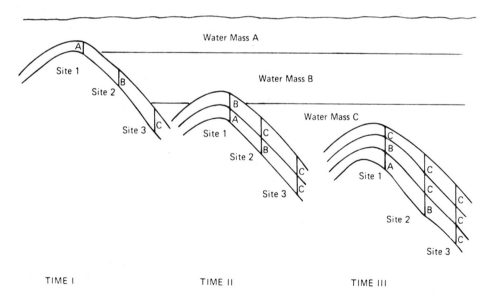

Figure 17-2 Schematic figure of subsiding oceanic ridge at three time intervals showing the influence of three vertically stable water-mass layers on the sediments being deposited on the ridge.

the thermocline (see Chapter 8). Its depth varies latitudinally and seasonally and is well developed in tropical regions, but it is only weakly developed or absent in polar areas. Oxygen isotopic measurements made on both planktonic foraminifera, that lived at different water depths and benthonic foraminifera from different ocean depths can provide a vertical structural picture of the oceans at selected time intervals. It is possible, for instance, to determine paleotemperature gradients for near-surface and bottom waters. The gradients give valuable insights into the relationships between temperature changes in surface waters at low and intermediate latitudes and the temperature histories of the high-latitude and deep-ocean waters [Savin and others, 1975]. For instance, Savin and others [1975] used isotopic gradients to determine that surface- and bottom-water temperature covaried until about the middle Miocene. Before the middle Miocene the surface-to-bottom temperature gradient was about 60 percent of that after the middle Miocene, implying that the thermocline was less clearly defined.

The thermal structure of the oceans for specific times past can also be determined by analysis of oxygen isotopes of benthonic foraminifera over wide depths. For instance, Keigwin and others [1979] examined the vertical structure of the Pacific during the early Pliocene (3 to 5 Ma) in depths from 1 to 4.5 km (Fig. 17–3). The average early Pliocene $\delta ^{18}O$ values of benthonic foraminifera at all sites increase with increasing water depths (Fig. 17–3), as expected for equilibrium precipitation of calcite in waters of increasingly cold temperatures. The greatest enrichment in $\delta ^{18}O$ occurs between about 1 and 2 km. Below about 3 km, $\delta ^{18}O$ appears relatively stable throughout the tropical and temperate Pacific Ocean. When the ancient $\delta ^{18}O$ data are plotted against modern temperature at the depth and location of each site, they provide a line with a slope similar to that of the line for equilibrium precipitation of calcite, $-0.26/ml/°C$. This suggests that the temperature gradient of the Pacific Ocean during the early Pliocene was similar to today's gradient.

It is also possible to construct realistic vertical temperature profiles for the upper few hundred meters of the oceanic water column using the $\delta ^{18}O$ values of a

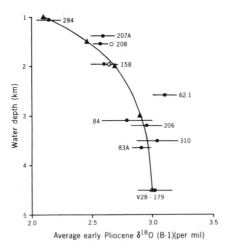

Figure 17-3 Average early Pliocene $\delta^{18}O$ plotted against water depth. Samples are DSDP sites of early Pliocene age. Error bars are ± 1 standard deviation. (From L. D. Keigwin, M. L. Bender, and J. P. Kennett, 1979)

number of types of foraminifera that inhabited different parts of the water column. Planktonic foraminifera have been depth-stratified in the water column since Cretaceous time [Emiliani, 1954; Douglas and Savin, 1973]. There has also been a consistent relationship between shell morphology of a species and its depth habitat. Species with globigerine morphology have consistently occupied shallower depths than have species with globorotalid morphology. Most planktonic species are confined to the upper 400 m of the water column [Bé and Tolderlund, 1971]; thus vertical paleotemperature profiles based on isotopic analyses of these forms are restricted to the upper 400 m and must assume that the depth distribution of the form used is known or can be approximated.

Vertical separation of planktonic foraminiferal species in the water column is most developed in tropical regions. The large temperature and salinity variation in the upper few hundred meters of tropical water produces a wide range of water densities for the foraminifera. A larger number of environmental niches exist in the tropical oceans. This probably contributes significantly to the higher taxonomic diversity at these latitudes.

In polar regions there is little vertical variation of temperature or salinity in the upper few hundred meters of the ocean. Planktonic foraminifera are neither abundant nor diverse in high latitudes; they are concentrated in the near-surface layer [Bé, 1960; Berger, 1969; Bé and Tolderlund, 1971]. The distinct depth zonation characteristic of warm-water regions is lacking because of the smaller number of depth-density habitats in polar seas [Douglas and Savin, 1978]. The upper 400 m of ancient temperature-depth curves have been defined using the isotopic compositions of planktonic foraminifera [Douglas and Savin, 1978]. Planktonic species are grouped as shallow, intermediate, or deep dwelling. In each case the warmest isotopic temperature obtained from a species is taken to be the temperature of shallow water (20–50 m). The coldest isotopic temperature obtained from a species is assumed to be a deep-water (200–400 m) temperature. Intermediate isotopic temperatures are taken to be water temperatures at intermediate depths (100–200 m).

REGIONS OF SPECIAL PALEOCEANOGRAPHIC IMPORTANCE

The present ocean-circulation system in the world has evolved in response to topographic changes in the ocean basins, the changing position of continents, and changes in the earth's climatic characteristics, especially in the polar regions. Certain regions of the earth's surface have played more important roles in controlling paleoceanographic evolution because of their strategic locations. Table 17–1 indicates several regions, which can be subdivided into those which record vertical gradients or horizontal gradients or are located close to *gateways,* or conduits, for current flow either between oceans or between different basins. Such regions include those areas that are or have been avenues of major interoceanic communication. These include the Pacific-Atlantic ocean connections at the Drake Passage and the now-extinct Middle American Seaway, as well as Indian-Pacific ocean connec-

TABLE 17-1 OPTIMAL AREAS FOR STUDYING VERTICAL
AND HORIZONTAL GRADIENTS OF WATER PROPERTIES
AND DYNAMIC CHANGES IN THESE PROPERTIES THROUGH TIME

	Vertical	Horizontal	"Gateways"
Deep sea	Ridge flanks Plateaus Guyots and other aseismic elevated features	Frontal regions Convergences and divergences	
			On flanks and in basins downstream from passage
Marginal	Offshore margin monitors Subsiding offshore platforms Intersections of aseismic ridges with continental margins Deep-sea margin transition	Eastern boundary currents Western boundary currents Antarctic deep currents	

tions, such as the Indonesian and Tasmanian seaways. Certain deep-sea passages, such as the Vema Passage, the Tonga-Samoa Passage, the Broken-Ridge Naturiste Plateau Passage, and several western Indian Ocean pathways of present-day bottom waters, are critical in interbasinal bottom-water circulation. Other important sites are those that have been source areas for bottom waters, marginal areas which have exhibited significant upwelling, and those ocean basins surrounded by continents (Fig. 17–4). Additionally, there are sites adjacent to deserts that have monitored the development of these climatic regimes.

Gateways

Gateways are relatively narrow passages that control surface or deep-water circulations. They are important in changing both the distributions of surface-water masses and the vertical structure of the oceans. The Gibraltar sill is the gateway in the modern ocean that controls the character of circulation in and out of the Mediterranean Sea and is instrumental in the formation of Mediterranean deep water, which enters and spreads out over large areas of the Atlantic Ocean. Examples of particularly important paleocirculation changes related to the opening or closing of gateways or seaways are the opening of the Tasmanian Seaway south of the Tasman rise in the Australian region, which was a vital component in circum-Antarctic current evolution and the opening of the Drake Passage Seaway south of South America, also vital in circum-Antarctic current evolution. The final closing of the Tethys Seaway between the Indian and Atlantic oceans in the middle Cenozoic, the increased isolation of the tropical Indian Ocean from the tropical Pacific by the closing of the Indonesian Seaway during the Cenozoic, and the closing of the Central American Seaway in the late Cenozoic were all crucial to paleocirculation changes at low latitudes.

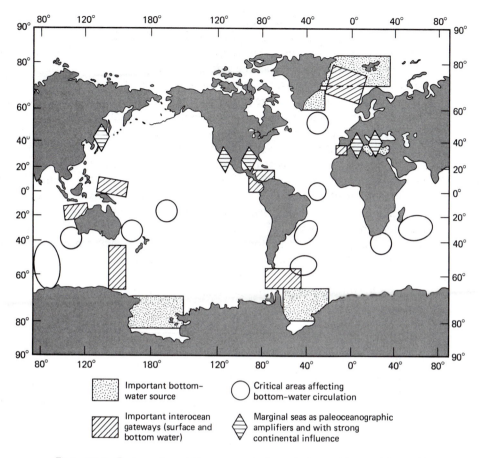

Figure 17-4 Regions of special importance in the evolutionary history of the world ocean.

Legend:

- ⬚ Important bottom-water source
- ⬭ Critical areas affecting bottom-water circulation
- ▨ Important interocean gateways (surface and bottom water)
- ⧖ Marginal seas as paleoceanographic amplifiers and with strong continental influence

The closing of gateways leads to fractionation of waters between oceans and differing geochemical and sedimentary regimes. Resulting oceanic isolation can also lead to increasing biogeographic provincialism. Because geochemical sedimentary and biogeographic parameters are so sensitive to the degree of interchange between oceans, they are good monitors of any such changes. Geophysical reconstructions provide a basic framework within which to construct paleoceanographic models, but they are often more limited in providing the exact timing because of the structural complexities that often exist in gateway areas. The paleoceanographic effects of changes at gateways are best studied not directly at the passage, where strong currents through the passage will destroy the record, but on either side and particularly *downstream* from the passage, where sediments accumulate even during strong circulation phases. Studies must be based on transects downstream in the depositional lee of the gateway—one transect for disturbances in the horizontal gradients and a second for disturbances in the vertical gradients.

Important gateways also exist for deep bottom-water flow. A number of gateways, such as the Drake Passage Seaway and the Tasmanian Seaway, result from separating continental blocks, which affect the entire water column—including bottom waters. However, many gateways develop through deep topographic features, such as the mid-ocean ridges or aseismic ridges. These are commonly called **channels.** Channels for interbasinal flow of deeper oceanic waters are usually related to fracture zones through the mid-ocean ridge systems, while those through aseismic ridges are often related to critical phases in their subsidence history. The evolution of the deep-ocean channel system is controlled primarily by the distribution of bottom water through the ocean basins. Figure 17–5 shows subsidence curves for a number of channels that have been important in controlling bottom-current flow in the South Atlantic. For instance, the Vema Channel is a major conduit of Antarctic bottom-water flow to the Northwest Atlantic basins. This channel only subsided below 4000 m near the beginning of the Cenozoic, and thus has long influenced deep bottom-water flow in this region. Nevertheless, as the upper surface of certain bottom waters probably has never been much shallower than about 4500 m (for example, Antarctic bottom water), a channel at 4000 m will have a blocking effect. Certain deep-sea drilled sections have been placed at strategic locations both within and near the ends of such conduits to examine the history of bottom-water flow.

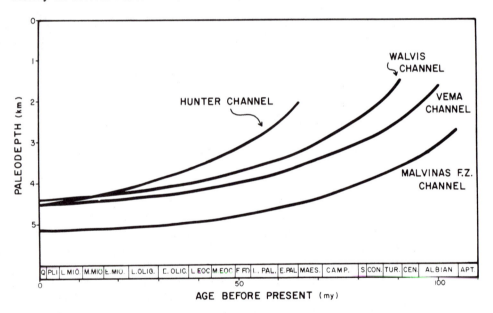

Figure 17-5 Subsidence curves with increasing age of critical passageways in the South Atlantic controlling bottom and intermediate water-mass circulation. Malvinas Fracture Zone Channel connects the Southern Ocean with the Argentine Basin; the Vema and Hunter channels connect the Argentine and Brasil basins; the Walvis Channel connects the Cape Basin with the Angola Basin. Subsidence is based upon estimated age in the deepest part of the channel and on the assumption that the channels formed early in the history of the ridges. (From Tj. H. van Andel, J. Thiede, J. G. Sclater, and W. W. Hay, *J. Geol.*, vol. 85, pp. 651–98, 1977, reprinted with permission of the University of Chicago Press)

Source Areas for Bottom Waters

The most important sources of bottom waters in the present-day oceans are in the polar regions, especially Antarctica and the North Atlantic. Within the Antarctic, the Weddell and Ross seas are the most important areas where bottom waters are formed (see Chapter 8). Because the production of these bottom waters is intimately related to glacial evolution and sea-ice production, an understanding of the climatic evolution in these areas is fundamental. Much progress has been made toward that end (see Chapter 19); but as the polar areas are logistically the most difficult to work, much remains to be understood. There are a number of approaches used in developing a **paleoglacial,** or **cryospheric evolution,** including oxygen isotopic measurements as a record of ice-volume change (see Chapter 3) and the character and distribution of abyssal versus lower-bathyal benthonic foraminiferal assemblages in Cenozoic sequences close to the bottom-water source areas (these assemblages serve as traces of particular bottom-water types). The history of ice-rafted debris is an indication of polar glacial history and the geometry of the flow of high-latitude surface water (see Chapter 13), and certain planktonic microfossil forms, such as the radiolarian *Cycladophora davisiana,* increase in importance at times of sea-ice formation [Hays and others, 1976].

Zones of Upwelling

The bulk accumulation rates of siliceous and calcareous tests can give estimates of productivity and rates of dissolution at the sea floor that are not strongly influenced by evolutionary changes in the biota [van Andel and others, 1975]. These estimates can be related to changes in the character of surface and deep-water circulation. In particular, opal dissolves more readily in chemically "young" waters (low in dissolved silica) and calcite dissolves more readily in chemically "old" waters (rich in dissolved CO_2 [Heath and others, 1976]; see Chapter 14).

The feasibility of this approach—that is, the determination of the bulk carbonate and opaline silica contents of deep-sea sediments and conversion of these values to rates of accumulation using biostratigraphic age estimates and GRAPE (Gamma Ray Attenuation Porosity Evaluator) bulk-density measurement—was demonstrated for deep-drilled cores from the equatorial Pacific by van Andel and others [1975] and Leinen [1976]. These studies revealed large variations in the vertical gradient of carbonate dissolution, apparently related to the evolution of sources of NADW and AABW, as well as the productivity contrast between the equatorial zone and adjacent central water masses.

The study of biogenic sedimentary sequences at or near zones of oceanic upwelling holds much promise in assisting with a general understanding of changes in the vertical water-mass structure through time, the degree of oceanic turnover, and oceanic circulation in general. Present areas of upwelling and related high-biogenic productivity include the Antarctic convergence, the equatorial divergence and various eastern boundary currents, including the Peru-Chile Current and the

Benguela Current. Upwelling is also associated with monsoonal climatic conditions in the Northwest Indian Ocean. Much progress has been made towards a better understanding of the history of biogenic productivity associated with equatorial divergence [van Andel and others, 1975] and Antarctic convergence [Hayes and Frakes, 1975; Kennett and others, 1975; Tucholke and others, 1976].

Marginal Basins as Paleoenvironmental Amplifiers

Certain marginal marine basins are semi-isolated from the world ocean because of narrow gateways or shallow sill depths. These tend to amplify certain paleoenvironmental events of global significance. For instance, the Mediterranean Basin became isolated in the latest Miocene, about 6 Ma, and caused a salinity crisis within the basin (see Chapter 19). This isolation seems to have been caused by a drop in sea level, which may have been partly due to increased ice-volume growth in the Antarctic [Ryan and others, 1973]. Thus an event (increased polar glaciation) occurring on another part of the earth is amplified as a salinity crisis in the Mediterranean Basin. Such changes in sea level can cause environmental changes in marginal basins at widely separated locations. The Sea of Japan and the Gulf of California are good examples. Burckle [1976] reported a freshwater diatom assemblage of late Miocene age from the central part of the Japan Sea. This suggests that the Japan Sea was completely isolated from the world ocean at some time in the late Miocene and became a freshwater lake. This event seems to have occurred at the same time as the isolation of the Mediterranean Basin. Also, late Miocene evaporitic deposits from the southern Imperial Valley, California, represent filled-in portions of the northern Gulf of California. These deposits may also have resulted from isolation of the depositional basins.

Desiccation of a large ocean basin isolated from the world ocean should cause a transgression elsewhere. The present-day Mediterranean, with a volume of 3.8×10^6 km^3 [Menard and Smith, 1966], contains 0.28 percent of the volume of the world's oceans. Consequently, emptying of the present basin, similar to the drying proposed for the late Miocene, would raise the sea level of the remaining world ocean by about 10 m [Berger and Winterer, 1974]. Larger effects require larger basins, such as those formed during the early stages of separation of South America and Africa in the Mesozoic.

Both the Mediterranean Sea and the Gulf of Mexico are almost surrounded by continents as well. Thus the deposition of Cenozoic sediment sequences is strongly influenced by the climate and topography of the surrounding continents. The high rate of evaporation in the Mediterranean allowed the desiccation of the Mediterranean in the late Miocene. In contrast, at times during the late Quaternary, the Gulf of Mexico received enormous supplies of fresh waters from the melting ice sheets to the north via the Mississippi River. Large influxes of the meltwater, which spread out over the surface of the Gulf, are recorded by $\delta\ ^{18}O$ signals in planktonic foraminiferal tests [Kennett and Shackleton, 1975; Emiliani and others, 1978]. The meltwater signals amplify and monitor the process of ice-sheet destruction in the

Northern Hemisphere. Other paleoclimatic signals also are not truly oceanic, but reflect the influence of the surrounding continents.

COMPLICATING FACTORS AND LIMITATIONS

The deep-sea sedimentary record, either during or after sedimentation, can be affected by a number of processes that complicate interpretations of the stratigraphic record, including microfossil dissolution and the creation of unconformities. These reduce the value of sequences providing certain types of information, but provide data about other paleoenvironmental processes. For instance, dissolution of biogenic sediments can alter, and eventually destroy, the very assemblages that provide data on paleotemperatures or ocean circulation. However, patterns of dissolution reflect the character of the geochemical environment of present and past oceans, basin-to-basin geochemical fractionation, and the age and general circulation of bottom waters.

Likewise, in a variety of topographic settings, deep-sea sequences in all oceans are punctuated by unconformities of varying importance. The extent of sea-floor erosion has been fully appreciated only since deep-sea drilling. This process has intensely disrupted the multichannel record of many deep-sea sequences. These unconformities are formed by, and provide information on, the erosive flow of density-driven bottom currents derived primarily from the polar regions. Nevertheless, unconformities are primarily negative stratigraphic features; much larger amounts of information are provided by continuous sequences. The paleoceanographer's task has been made much more difficult because of the vast extent of deep-sea erosion through time.

Another group of limitations only reduce our ability to reconstruct the history of ancient oceans. These are limitations due to site location, poor core recovery, poor core quality, bioturbation, and sediment diagenesis.

Site Location

A relatively dense network of piston cores has been obtained from the ocean basins (except the Arctic Ocean). The high latitudes from the southern Indian Ocean and the South Pacific are relatively poorly represented. This network of cores obtained has been vital for late Quaternary paleoceanography, including time-slice mapping [CLIMAP, 1976]. The logistics of deep-sea drilling have provided much greater constraint on the development of more ancient paleoceanography. Ocean-floor spreading has further confined successively older sequences toward the continental margins, reducing paleogeographic coverage and decreasing potential for comprehensive paleoceanographic mapping with increasing age. The oldest oceanic rocks penetrated by the drill bit are only from the late Jurassic. Although over 500 sites have now been drilled throughout the ocean basins (except the Arctic Ocean) this is still only 1 site per 1,000,000 km^2! The majority (about 75 percent) of these have been drilled in the Northern Hemisphere, the global continental sphere! Many of these sites from the early stages of the DSDP were selected for reconnaissance purposes: for determining the age of oceanic basement, age of sediments, and general sediment character. Relatively few sites were selected for examining specific paleoceanographic problems, although the suite of existing sites has provided enough information to establish our modern paleoceanographic models. Detailed paleoceanographic studies require well-selected traverses or groups of sites to delineate the various gradients that have been previously discussed.

Core Recovery

The length of piston cores has generally been confined to the late Quaternary, but piston cores from areas of low sedimentation have allowed important studies of the early Quaternary and Pliocene to be done. The proportion of core recovered during drilling plays a major role in stratigraphic resolution. Moore [1972] compiled statistics showing that core recovery not only varies widely, but is totally inadequate at a very large number of sites. During the early stages of deep-sea drilling, site selection and coring programs were dominated by tectonic objectives; the coring record is often poor and of much less value for the reconstruction of the depositional and paleoceanographic history of a region. The later phases of the DSDP and of IPOD provided much more continuous coring, but core recovery and quality still left much to be desired. The hydraulic piston corer has provided about 80 to 90 percent core recovery of excellent quality—an enormous step for future paleoceanographic studies. Cenozoic paleoceanographic analyses have benefited and much of our knowledge has been derived from a relatively small number of carefully located and continuously cored sites.

Core Quality

Piston cores have generally provided sequences of good quality. However, drilling technology has its limits. Young, unconsolidated sediments have been particularly prone to drilling disturbance, which decreases as consolidation increases. As a result, deep-drilled sequences from the late Cenozoic are often mixed, reducing their potential stratigraphic resolution. The presence of chert layers in older sequences, especially the middle Paleogene to late Cretaceous, has reduced core recovery because drilling through even thin layers requires washing for bit lubrication and chip removal in the hole. Core distortion and technologically caused coring gaps can be remedied only by advances in technology. The hydraulic piston core is a great improvement for collecting relatively unconsolidated sediments of late Cenozoic age and provides cores of quality equal to those of piston cores.

Bioturbation

Our understanding of paleoceanography is tied to quantitative data related to sediment age and composition, fossil composition, and geochemistry, including stable isotopes. These data permit the use of rate-of-change information and only rate-change information permits paleoceanographic inference [van Andel and others, 1975]. The resolution at which paleoceanographic processes can be examined is strongly controlled by **bioturbation,** the process of sediment mixing by benthonic organisms. Bioturbation effectively reduces the time resolution of deepsea cores and smooths the record of paleoenvironmental change. Loss of stratigraphic resolution because of bioturbation is directly proportional to sedimentation rates. Because sedimentation rates below 2 cm per 1000 years characterize a high proportion of the ocean area, the paleoenvironmental history defined by changes in oxygen isotopes and microfossil assemblages is a smoothed representation of changes that actually occurred over time periods of less than a few thousand years. Fortunately, several areas exist where sedimentation rates are sufficiently high to provide paleoenvironmental records with a time resolution less than the mixing time of the ocean [Shackleton, 1977]. Furthermore, there are a few basins such as Santa Barbara Basin, California [Pisias, 1978], Guayamas Basin, Gulf of California [Schrader and others, 1980] and the Orca Basin, Gulf of Mexico [Kennett and Penrose, 1978] marked by anaerobic bottom environments that lack benthonic organisms and hence also lack sediment mixing. This factor, combined with high sedimentation rates, has enabled extraction of paleoenvironmental history at resolutions of less than 10 years for geological periods of several thousand years. Estimates of the stratigraphic thickness over which sediment is redistributed by burrowing organisms have been made by examining particle distribution deposited instantaneously (isochronously), such as a volcanic ash layer from a single volcanic eruption or a microtektite layer. Glass [1969] found that microtektites from a layer deposited in the middle Quaternary throughout the Indian Ocean have been redistributed about 90 cm upward and about 20 cm downward in these sequences. This implies mixing of Indian Ocean sediment ranging in age from 33,000 to 320,000 years (averaging 120,000 years). However, *intense* bioturbation does not normally penetrate much deeper than about 10 cm in the sediment column. On the continental shelf, bioturbation extends to much greater depths, perhaps up to 4 m. This results from greater infaunal densities and larger, deeper burrowing organisms. Where anoxic conditions occur, sediments are commonly laminated and nonmottled and contain high-resolution historical information.

Quantitative models for biological mixing rates have been developed by Berger and Heath [1968] and Guinasso and Schink [1975]. Guinasso and Schink [1975] have described vertical mixing in deep-sea sediments in terms of a time-dependent model that involves the rate of mixing, the depth to which mixing takes place, and the sedimentation rate. Berger and others [1977] have also developed an *unmixing* equation, in order to reconstruct a stratigraphic signal as it might have appeared had there been no benthonic mixing to distort it.

Sediment Diagenesis

The DSDP has provided sequences thick enough and old enough to allow the study of diagenesis and lithification of sediments. The term *diagenesis,* introduced in the middle 1800s by von Gümbel, refers to the process of lithification of sediment following deposition. Diagenesis of sediments includes all those processes that act upon these materials after their initial deposition but before elevated temperatures and pressures metamorphose minerals and structures. Diagenetic processes transform loose sediment into a cohesive rock. These processes include **gravitational compaction, grain interpenetration,** and **cementation.** Diagenesis has an impact on paleoceanography because it alters the original character of the sediment and microfossils, rendering paleoenvironmental interpretations difficult. Information about postdepositional environments is provided at the expense of information about original environments of deposition.

Sediments of uniform composition, porosity, and other physical properties change discontinuously down a sequence [Davies and Supko, 1973]; each abrupt change often reflects a visible change in lithification, or diagenetic, state. However, these abrupt changes do not occur everywhere at the same depths, suggesting that the physical set of properties of sediments are complexly controlled.

The general trend toward increasing lithification with length of time and depth of burial is interrupted by local reversals in lithification. Thus soft plastic oozes can occur below stiff, pliable chalks, and the latter can occur beneath more dense limestones [Schlanger and Douglas, 1974]. These variations are due to a complex of geochemical processes which are still poorly understood.

PRIMARY TOOLS

Lithology

The first step in paleoceanographic analysis is the determination of lithological sequence. Lithological change in marine sequences is critical to the understanding of paleoenvironmental history involving interrelations among the total oceanic geochemical system—the biosphere, the vertical water-mass structure, continental sources, the atmosphere, the cryosphere, and postdepositional environments and plate motions. Models of oceanic sediments have concentrated on the biogenic components: opaline, silica, calcium carbonate, and organic carbon. The sedimentary record has been used to make quantitative inferences about past climatic conditions and geochemical history, including temperature, circulation, and stratification. Certain other lithological components, such as ice-rafted debris and windblown sands, directly reflect changes in continental climate.

The sediments recovered during the DSDP form the basis of paleoceanography. Because of the uniform manner in which samples from the project have been handled, they now constitute the most homogeneous body of data available on sedimentary rocks. The methods by which cores are described are provided in each of the DSDP initial reports. A lithological data bank has been developed [Davies and others, 1977] that allows machine-screening of the raw data with the appropriate classification, so that standardized descriptions can be obtained rapidly and accurately.

Rates of deposition have been calculated for each site using paleontologically determined ages. Age-depth diagrams are provided in the initial reports and are also obtainable from a biostratigraphic data bank. But these require careful screening for possible hiatuses and biostratigraphic conflicts. Because sedimentation is strongly influenced by compaction, age, diagenesis, and the nature of the initial sediment, it is more valuable to convert data to accumulation rates in g/cm² per 1000 years, using wet-bulk densities and porosities, derived from GRAPE measurements (using sonic velocities) made immediately after each core was collected [van Andel and others, 1975]. Accumulation rates so obtained have yielded consistent results in diverse applications [van Andel and others, 1975; Stakes and Leinen, 1976]. Using concentrations of different components, the accumulation rates can be apportioned to various sediment phases. The use of rates instead of concentrations of the various components permits an entirely new set of inferences to be made.

Biogenic debris is volumetrically the most important constituent in pelagic sediments. The accumulation rate of biogenic debris is determined both by the rate at which it is supplied to the sea floor and by the rate at which it is dissolved by corrosive bottom waters. Assuming steady-stage conditions in the ocean, the dissolution rate is controlled by the rate at which the shell-building material is supplied to the ocean by rivers. Based on studies of the preservation of calcareous shells (for example, Berger [1971B]) and the distribution of calcium carbonate concentrations in marine sediments [Berger and others, 1976; Biscaye and others, 1976], its distribution is largely controlled by spatial differences in the rate of calcite dissolution. Changes in the distribution pattern of carbonate, at least in the Quaternary, were caused by global or regional changes in the dissolution rate (for example, Broecker [1971]; Berger, [1971B]). Recent studies of fluctuations in the CCD [Berger and Winterer, 1974; van Andel, 1975] and dissolution gradients [van Andel and others, 1975; Heath and others, 1976] suggest that major changes in carbonate distribution during the Cenozoic are associated with changes in the carbonate dissolution gradient in the deep sea. The general Cenozoic patterns of fluctuations in the CCD [van Andel, 1975] are similar in the various ocean basins and are likely to have been caused by a global mechanism (such as a change in the hypsometry of the ocean basins or a change in supply of carbonate to the oceans). There are, however, ocean-to-ocean differences in this general pattern. These differences are likely to be caused by basin-to-basin fractionation of carbonate [Berger, 1970A] and can be used to learn more about the history of interbasinal exchange of deep and surface waters. Furthermore, by knowing the dissolution gradient within an ocean basin, estimates of the relative differences in carbonate productivity in different regions in that basin can be determined from regional differences in the CCD [Heath and others, 1976].

Although the basic controls on the distribution of biogenic opal in deep-sea sediments are the same as those on carbonate, the resulting distribution patterns are quite different. There is no opal-compensation depth below which siliceous tests are totally dissolved and basin-to-basin fractionation has the opposite effect on opal distribution [Berger, 1970A]. The distribution of opal in modern marine sediments is very similar to oceanic patterns of productivity [Heath, 1974], so, unlike calcite, the supply of opal to the sea floor appears to be the primary factor controlling its distribution. Although the patterns are similar, the relationship between opal accumulation rates and primary productivity remains qualitative. It is feasible, however, to use changes in opal accumulation rates as a measure of changes in productivity within a region. A global picture is necessary to interpret regional changes. A summary of the changing relative abundances of biogenic carbonate and opal for each of the major ocean basins is shown in Fig. 17–6 [T. Moore, personal communication]. The patterns reflect the complex interactions of basin-to-basin fractionation, biogenic productivity, and preservation history, as described above.

Ocean-Sediment Reservoirs and Fluxes

The relationships between the various processes of recent oceanic sedimentation are summarized in Fig. 17–7 (clastic sedimentation) and Fig. 17–8 (biogenic and other pelagic sedimentation). Clastic sediments are derived from the land (dissolved detrital and volcanic materials), with a tiny cosmic fraction. These

	Antarctic		Tropical Pacific		North Pacific		North Atlantic	
	carb	silica	carb	silica	carb	silica	carb	silica
Quaternary and Pliocene	L	H	M to L	M to H	L	H	H	L
Late Miocene	L	H	L	M to H	L	M	H	L
Middle Miocene	L	H	H	M to H	L	L to M	H	L
Early Miocene	M	M	M	M	L	L	M	M
Late Oligocene	H	L	M to H	L	M	L	L	M
Early Oligocene	H	L	M to H	M	M	L	L	M
Middle to Late Eocene	L	L	L	H	L	L	L	M
Early Eocene and Paleocene	?	?	L	L	?	L	L	L

Figure 17-6 Generalized patterns of the changing relative abundance of biogenic calcium carbonate and silica in selected ocean basins through the Cenozoic: H = high; M = moderate; L = low. These changing patterns between the basins reflect changes in dissolution, upwelling, and basin-to-basin geochemical fractionation. Assignments are highly generalized because large differences in abundance of biogenic material can occur within individual basins for a large number of reasons. (Courtesy T. Moore, Jr.)

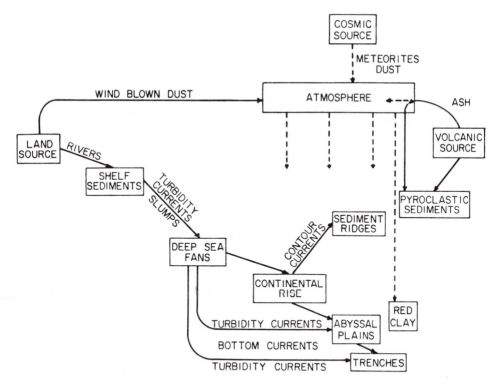

Figure 17-7 Processes of clastic sedimentation in the oceans. (From W. W. Hay in *Studies in Paleo-oceanography*, p. 2, 1974, reproduced with permission of the Society of Economic Paleontologists and Mineralogists)

materials can be captured and deposited at various stages during their transportation to the deep sea (see Chapter 13). Most material is trapped on shelves and continental margins, including fans, but a significant proportion is transported by turbidity currents to the deeper parts of the ocean and deposited in abyssal plains and trenches. At all stages, bottom currents can redistribute the materials (see Chapter 15). Wind also transports clastic materials to the deep sea.

Pelagic sediments are dominated by the biogenic fraction, which varies considerably between latitudes, from tropical pteropod oozes to high-latitude diatom oozes. Glacial-marine sediments, although of clastic character, are of pelagic origin. The nonbiological pelagic component consists largely of windblown clays and ice-rafted debris. Attempts have been made, particularly by Broecker [1974] and Hay and Southam [1977], to estimate the budgets of the major components of ocean sediments and their relationships with the ultimate terrigenous source.

Hay and Southam [1977] summarize oceanic sedimentation as dominated by the interaction between the surface ocean and the deep ocean, separated by the thermocline (Fig. 17-8). These two reservoirs communicate through vertical mixing, resulting from the downwelling of surface water and the upwelling of deep water and through the sinking of particulate matter produced by organisms in the surface ocean. The primary source matter is dissolved, and particulate material is carried in

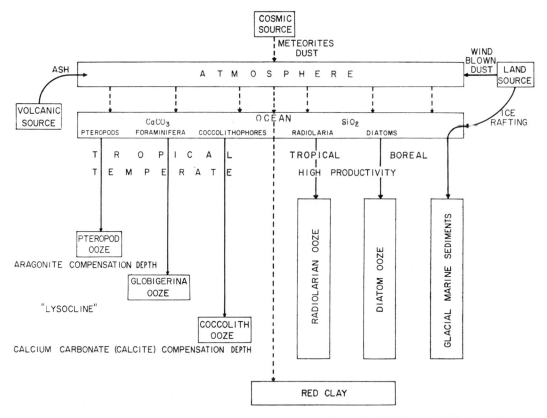

Figure 17-8 Pelagic sedimentation in the oceans. (From W. W. Hay, in *Studies in Paleooceanography*, p. 3, 1974, reproduced with permission of the Society of Economic Paleontologists and Mineralogists)

rivers. The primary sink is biogenic particulate matter, which settles out as sediment throughout the ocean basins.

The fundamental sedimentary processes and their interrelationships are fairly well known, with the remaining problems concerning accurate estimates of the amounts of material involved. Knowledge of amounts of source material is not only related to data on dissolved loads of rivers and atmospheric contributions, but to dynamic changes as well. For instance, large amounts of sediment can be locked up temporarily behind dams, such as those created by the continental shelves during high stands of sea level. This reduces the influx of material to the deep ocean. Because the modern ocean is of interglacial mode, fluxes of terrigenous detritus to the deep ocean are considered to be much lower and atypical of conditions that prevailed through much of the last few million years.

Garrels and Mackenzie [1971] reviewed and discussed modern sediment transport into the ocean (by means of rivers, groundwater, glaciers, and atmospheric dust). They ranked the present annual fluxes as mechanical loads of rivers (183×10^{14} g), dissolved loads of rivers (39×10^{14} g), mechanical loads of glaciers (20 × 10^{14} g), dissolved loads of ground water (4 × 10^{14} g), erosion on coast (2.5 × 10^{14} g of solids), and atmospheric dust loads (0.6 × 10^{14} g). In their discussion they pointed out that the basic controlling factors are land area, precipitation and climate, slopes, character of coarse material, and type of weathering. The dissolved load of the rivers is proportional to the area of the continent and independent of elevation of the continent; the mechanical load of the rivers is proportional to the area of the continent and increases exponentially with the average elevation of the continent. The dissolved load of the rivers has probably not changed appreciably through geologic time. However, significant changes in the ocean-sediment budget must have resulted from the changing elevation of continents. Sea-level changes resulting from alternating glacial and interglacial episodes should have had a major influence on continental elevations and the mechanical loads of rivers. During low stands of sea level (glaciations), the mechanical erosion rate of continents and the loads of rivers would have approximately doubled. There is strong evidence supporting this dramatic increase in rate. Rates of erosion by rivers were higher in nonglaciated areas and glacial erosion in the areas of the continental ice sheets (North America and Europe) also seems to have been higher. Also, much less sediment was stored on the continental shelves because lowered sea levels enabled rivers to transport their loads to the outer edge of the continental shelf for more immediate transportation down the continental slopes.

Paleoceanographic changes have been examined using changing rates of sedimentation in the major ocean basins [Davies and others, 1977; Davies and Worsley, 1981] based on average sedimentation rates in DSDP sites. Average sedimentation rates (total sediment) were prepared for a number of Cenozoic time intervals (Fig. 17–9). Although problems exist in the compilation of this kind of data, it is clear from Fig. 17–9 that fluctuations have occurred in the rate of sediment accumulation in each of the major ocean basins. In general, these fluctuations appear to be globally synchronous (with some clear exceptions), reflecting large-scale paleoceanographic changes [Davies and others, 1977]. Because DSDP sites are overwhelmingly biased in favor of oceanic pelagic sediments, the total sediment curves show predominantly pelagic sedimentation. Several Cenozoic sedimentation episodes are apparent (Fig. 17–9). During the Paleocene to early Eocene, the rates of sedimentation were very low, but were high during the middle Eocene. Again, during the late Eocene to middle Oligocene, the rates were low. Middle Oligocene to Recent sedimentation rates vary somewhat but high rates mark the middle to late Oligocene, and the late Cenozoic. Thus the oceans have alternated between periods of generally high and low sedimentation [Davies and others, 1977; Davies and Worsley, 1981]. These broad trends are highly significant paleoceanographically. Davies and Worsley [1981] suggest that these changes in sedimentation rates correlate with the global sea-level fluctuations postulated by Vail and others [1977]. High accumulation rates occur at times of low sea level, and vice versa. As pointed out by Rona [1973A], a correlation between sea level and sedimentation rate suggests that at times of high sea level, material eroded from the continents is trapped on the continental shelves. During times of low sea level, the shelf sediments are exposed, chemical erosion occurs on the continents, and the resulting products are then flushed to the deep sea, increasing sedimentation rates.

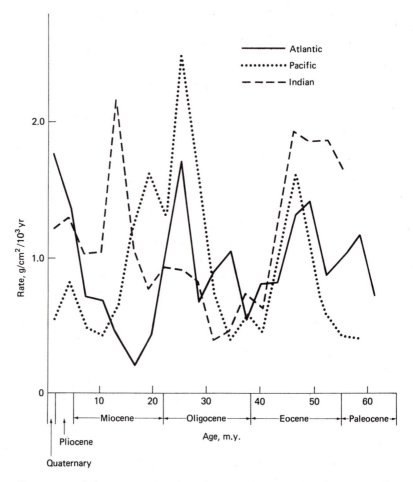

Figure 17-9 Sedimentation rates through time in the major ocean basins normalized with respect to stratigraphic distribution of unconformities and recovery rate differences in DSDP sites. (After T. A. Davies, 1981)

Because of the limited drainage areas of the Pacific basin and because the deep Pacific is surrounded by trenches, transport of continental detritus is reduced. Davies and Worsley [1981] propose that the correlation between sea level and sedimentation rate represents mainly biogenic precipitates, rather than terrigenous debris. In this model, high sea levels allow higher rates of biogenic precipitation on the shelves, thus starving the ocean basins, whereas during low sea levels, dissolved river loads reach the deep sea.

The nonbiogenic mineralogical composition of deep-sea sediments has also shown change through time that reflects changing dominance of continental or volcanic sources. From the middle Eocene (45 Ma) to the late Miocene (10 Ma), Pacific sediments were dominated by montmorillonite-rich and quartz- and pyroxene-poor sediments. Beginning in the late Miocene chlorite, kaolinite, and pyroxene were supplied in greater quantities, and in the late Cenozoic (Pliocene and Quaternary) quartz- and illite-rich sediments began to dominate [Heath, 1969]. These compositional differences reflect earlier Cenozoic volcanic sources derived from active Pacific volcanism, which was replaced by increasing amounts of continentally derived sediments in the late Cenozoic. The dominance of continental sources in deep-sea nonbiogenic sediments is a geologically recent phenomena and reflects a broad spectrum of global changes related to glacial development, climatic cooling, sea-level lowering and high-amplitude oscillations, and increased tectonism and mountain building.

Plate Stratigraphy

For deep-sea stratigraphy, plate tectonics implies sedimentation on horizontally moving plates, which are slowly subsiding as they cool [Berger and Winterer, 1974]. We have already considered the techniques that allow accurate paleoreconstruction of the ocean floor and of backtracking of depositional sites through time along a generalized subsidence curve (see Chapter 5).

Sedimentation occurring with sea-floor spreading gives rise to a particular set of facies relationships that do not occur in continental sequences. The interpretation of the complex geometrical relations among oceanic sediments governed by the rules of sea-floor spreading has been called **plate stratigraphy** [Berger and Winterer, 1974]. The principal elements involved in plate stratigraphy are shown in Fig. 17–10. The depth of the CCD is usually between 4 and 5 km and the average elevation of mid-ocean ridges is between 2.5 and 3 km. Thus the upper flanks of mid-ocean ridges accumulate carbonate, while clay or siliceous ooze are deposited further down the flanks (Fig. 17–10) below the major facies boundary, or "snow line." The resulting stratigraphic record shows basalt, overlain by a thin layer of metal-enriched basal sediments, overlain by carbonate which, in turn, is overlain by a thinner layer of clay or siliceous sediments (or both). The subsurface boundary

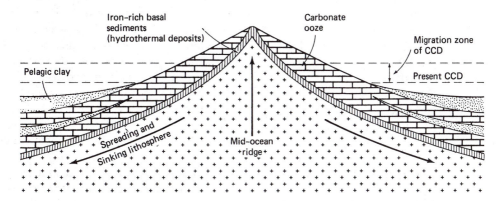

Figure 17–10 Sediment facies related to plate stratigraphy. Fluctuating CCD levels produced interfingering layers of pelagic clay and carbonate ooze. (After W. H. Berger and E. L. Winterer, 1974; and T. A. Davies and D. Gorsline, 1976)

between the two carbonate and clay layers is broadly time-transgressive and constitutes the fossil trace of the ancient carbonate line [Berger and Winterer, 1974]. In the Pacific Ocean this boundary ranges in age from early Cretaceous in the west to late Cenozoic in the east. New sea floor from the mid-ocean ridge must subside about 2 km to reach the CCD. This process takes about 30 to 35 m.y. Carbonate accumulation usually ranges between 4 and 20 m per million years, so 30 Ma would allow between 120 and 600 m of carbonate sedimentation [Berger and Winterer, 1974]. These estimates generally agree with the range of thickness of carbonate in drilled sections.

Some drilled sites show several alternations between carbonate and clay. Such a sequence is produced by a fluctuating CCD. Thus the facies relationships that result in the sea are largely a product of interrelations between at least two kinds of vertical movements: tectonic subsidence and variation in the CCD.

It is important to understand how paleodepth reconstructions use the CCD in DSDP sites, because this contributes to our knowledge of CCD fluctuations through time within the different ocean basins. Site 137 from the North Atlantic Ocean is a good example [Berger and Winterer, 1974]. The basement is 105 Ma (early Cenomanian), the site is presently at a water depth of 5360 m, and sediment thickness is 400 m. The amount of subsidence for 105 Ma is about 3160 m. Therefore the original paleodepth was 2400 m (5360 − 3160 m + 200 m, or ½ × 400 m). Similar calculations of subsidence age show that carbonate sediments accumulated down to a depth of about 3500 m, so the CCD was about 3500 at 90 Ma (late Cenomanian to early Turonian). After passing through the CCD, the site ceased accumulating carbonate, and clay deposition commenced. However, this was a time of marine transgression and sea level was about 300 m higher. The water depth of the CCD was therefore 3800 m, which is rather shallow but within the present range of values [Berger and Winterer, 1974].

One of the major elements of the tectonic evolution of the oceans, which has played a major role on depositional history, is the sea-floor subsidence away from the mid-ocean ridge. A second major influence has been the migration of the sea floor beneath major oceanographic features—for example, the northward migration of the Pacific plate across the equator [van Andel and others, 1975]. Winterer [1973] considers that one of the principal results of the DSDP has been to show the progressive northward shift of biogenic sediments deposited during the last 40 Ma beneath the biologically productive equatorial belt. We have already seen (Chapter 8) that the equator is marked by high rates of biogenic productivity. This is due either to upwelling at the equatorial divergence or because the Cromwell Undercurrent increases instability in the water column, or both. Because of the Coriolis effect, this upwelling and the resulting increased biogenic productivity should have remained fixed on the equator throughout geological history. Biogenic sedimentation, therefore, has remained higher at the equator. For instance, Hays and others [1969], calculated that accumulation rates during the Brunhes paleomagnetic epoch (since 0.7×10^6 years ago) have been at a maximum at the equator. Beneath the region, within 2° of the equator, rates of biogenic accumulation are often several times as rapid as areas only 5° away. However, the northward motion of the Pacific plate produces a northward displacement of a thick equatorial sediment bulge.

Seismic reflection surveys [Ewing and others, 1968] showed maximum thickness of the bulge 4° north of the equator. The drilling results confirm the seismic picture and show the details of this northward movement epoch by epoch back to the early Cenozoic [Winterer, 1973]. This is one of the most striking examples of the principle that facies relationships, tied to latitudinal zonation, allow us to determine latitudinal migration of the sea floor by their progressive displacement with time [Berger and Winterer, 1974]. The equatorial crossing can be recorded in different ways depending on the depth at which the site was situated at the time of crossing and on its situation with respect to the CCD. If a site passes the equator while remaining above the CCD, an increase in the rate of carbonate sedimentation will result. If the site was below the CCD during crossing, the sediments will record higher rates of siliceous biogenic sedimentation. In this case, the crossing is marked by either radiolarian ooze or abundant chert [Lancelot and Larson, 1975].

Van Andel and others [1975] have examined detailed changes in lithology and thickness in Cenozoic sediments that have at some time traversed the equator in the Pacific. A north-south traverse through the region (Fig. 17-11) is dominated since

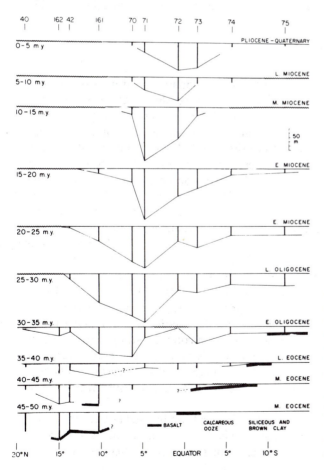

Figure 17-11 Sediment thickness variations and general lithology along north-south equatorial Pacific traverse (140° W) for separate 5 m.y. intervals in the Cenozoic. Numbers at top are DSDP sites. Wavy lines mark erosion; heavy dashed line is basement. (From Tj. H. van Andel, G. R. Heath, and T. C. Moore, Jr., *Cenozoic History and Paleoceanography of the Central Equatorial Pacific Ocean*, GSA Memoir 143, p. 27, 1975, courtesy The Geological Society of America)

at least the middle Eocene by the equatorial zone of maximum sedimentation. Each of the time intervals is marked by a well-defined and centrally located thickness maximum, which thins rapidly to the north and south (Fig. 17–12). Each interval is marked by a carbonate sediment bulge, which grades laterally into siliceous ooze and clay. The locus of maximum sedimentation for each time interval is displaced progressively northward with increasing age. A total displacement of 15° has occurred since the middle Eocene. During the last 5 m.y. the zone of maximum thickness has been located squarely on the equator. Since 20 to 25 Ma (late Oligocene), it moved 5° to the north.

Plate motion also produces other types of facies changes. For instance the volume of windblown detritus varies (as measured by eolian quartz) as the Pacific plate moves north through the wind belts, such as the zone of westerlies [Leinin and Heath, 1981]. Also, a wedge of volcanogenic sediments is found within 1000 km of the island arcs of the Northwest Pacific. This wedge steadily thickens toward the Pacific margin, which is the source of the windblown volcanogenic sediments. The shape and extent of the wedge is a function of two counterbalancing effects: the amount and extent of distribution of volcanic debris and the westward movement to ultimate subduction of the Pacific plate [Heezen and others, 1973].

Oxygen Isotopic Methods

The use of oxygen isotopic data from carbonate microfossils (naturally occurring variations in the relative abundances of stable oxygen isotopes) has been previously discussed. In Chapter 3 we considered the use of oxygen isotope variation for stratigraphic correlation in marine sequences containing carbonate microfossils. It was shown that, since the middle Miocene (about 14 Ma) when the earth began to develop polar ice sheets, most of the change recorded in oxygen isotopes in the marine record has resulted from oceanic compositional changes due to ice-volume fluctuations at high latitudes. Furthermore, earlier in this chapter it was shown that oxygen isotopes provide a valuable approach to learning about the vertical water-mass structure, the magnitude of surface-to-bottom gradients, and the degree of water-mass stratification. It needs to be emphasized that oxygen isotopes are not only important for stratigraphic correlation and for providing information on paleoenviromental gradients, but also for providing a paleotemperature message when certain conditions are met. This seems ironic because the early pioneering work of Emiliani emphasized the temperature part of the signal rather than the part caused by ice-volume fluctuations. The oxygen isotope approach contributes to our understanding of paleotemperature history over most of geologic history and is a primary tool in determining paleotemperature gradients and paleocirculation in the oceans. Most oxygen isotopic measurements have been made on foraminifera, although a few have been carried out on calcareous nannofossils, mollusks, and corals.

Whether or not the oxygen isotopic record can be used as a tool for estimating absolute paleotemperatures depends on whether the material being analyzed is from a time of ice-free conditions on the earth or from a time when ice sheets existed in

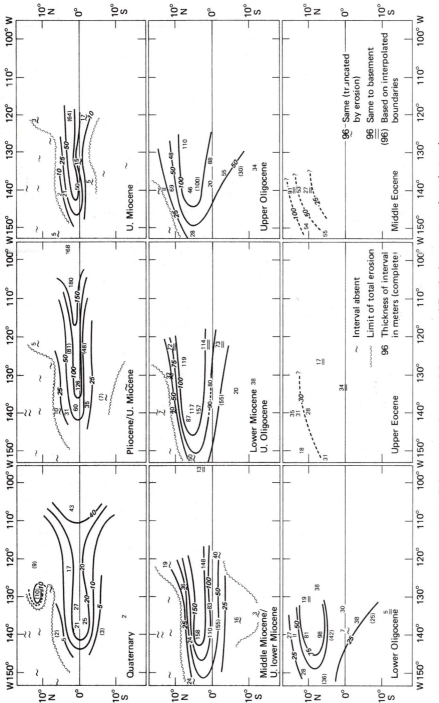

Figure 17-12 Sediment isopach maps of the equatorial Pacific for nine time slices during the Cenozoic. Heavy dashed lines on Eocene maps represent very uncertain isopachs. (From Tj. H. van Andel, G. R. Heath, and T. C. Moore, Jr., *Cenozoic History and Paleoceanography of the Central Equatorial Pacific Ocean*, GSA Memoir 143, p. 28, 1975, courtesy The Geological Society of America)

the polar areas. The development of ice sheets in the polar regions (Antarctica) in the middle Miocene complicated the temperature message that the oxygen isotopes provide by changing the isotopic composition of sea water. The isotopic record of foraminifera during pre-Miocene times is simply interpreted by most workers in terms of changing ocean temperatures. In the Quaternary oceans, where abyssal temperatures are believed to be approximately constant, the change in the δ ^{18}O of benthonic foraminifera serves as an approximate index of ice volume and provides a basis for making estimates of surface-water isotopic paleotemperatures based on planktonic foraminifera. During intermediate times, when both ice volumes and abyssal paleotemperatures varied, it may be impossible to make estimates of absolute temperatures, but at all times sea-surface temperatures may be computed relative to bottom-water values and relative surface-water temperatures may be estimated with changing latitude. Thus, as long as it is known that abyssal paleotemperatures have not varied, the oxygen isotopic analysis of planktonic foraminifera can measure temperature. The greatest difficulty is to conclusively demonstrate that bottom-water temperatures have not varied. The temperature is probably constant if the benthonic foraminiferal assemblage has not changed significantly. We can summarize by stating that if calcium carbonate is deposited in isotopic equilibrium with water (for example, ocean water), then the ratio of O^{18} to O^{16} of the carbonate differs from that of the water. The amount by which it differs is a function solely of the temperature. Hence, if we measure the ratio of O^{18} to O^{16} in a carbonate fossil and the ratio of O^{18} to O^{16} in the water in which it grew, and if we know (or assume) that the carbonate were deposited in isotopic equilibrium with the water, we can calculate the temperature at which the carbonate formed. In practice, most of the problems encountered in estimating temperatures of $CaCO_3$ formation using this method are the uncertainty about the O^{18} to O^{16} ratio of the water in which the carbonate precipitated, the precipitation of $CaCO_3$ not being in isotopic equilibrium with the water, and the diagenetic alteration of the isotopic composition of the carbonate after its formation. These problems have been reviewed by Berger and Gardner [1975], Savin [1977], and Hecht [1976]. The analytical precision in measuring oxygen isotope ratios is about $\pm 0.1\%$, corresponding to a change in temperature of only 0.5°C. Thus the limiting factor in obtaining paleoclimatic information is not analytical uncertainty, but our ability to meet the given criteria. In addition to global changes in oxygen isotopic composition of the ocean due to ice-volume fluctuations, there is local variability in the isotopic composition of near-surface waters due to the balance of evaporation and precipitation over the oceans [Craig and Gordon, 1965].

A third complicating factor in oxygen isotopic analyses is that shell secretion may not be in isotopic equilibrium with seawater. For example, echninoderms and corals show nonequilibrium effects, which Urey [1947] called *vital effects*. Many species of foraminifera also show vital effects, especially the benthonic forms [Smith and Emiliani, 1966; Duplessy, 1972; Shackleton, 1974]. Carbon isotopes may also be in disequilibrium. Fortunately, the degree of departure from isotopic equilibrium among species seems to have been genetically controlled. Thus intercalibrations between species are possible, including those in equilibrium.

A fourth complication in interpreting δ ^{18}O values of planktonic foraminifera is the isotopic temperature record. Temperature is recorded at the depth of growth of the species not at surface-water temperature. As discussed earlier in this chapter, ideally surface temperatures can be approximated from the isotopic temperature of the shallowest-dwelling species in a sample. A relative depth ranking can be applied and information provided on vertical oceanographic structure. Figure 17–13 compares oxygen isotope records of a shallow- and a deep-dwelling planktonic foraminifer through the late Quaternary.

In agreement with Emiliani's pioneering studies, oxygen isotopic analyses of planktonic and benthonic foraminifera from deep-sea cores have generally revealed the same trends of quasiperiodic fluctuations [Duplessy, 1978]. However, because of the various constraints imposed by the method, a precise interpretation of these fluctuations is still controversial.

The climatic changes from glacial to interglacial times have been remarkably constant over the last 0.7×10^6 years, producing similar minima (Fig. 17–14). Also, this temperature record can be correlated with other climatically regulated parameters, such as sea-level change and microfossil oscillations. The close correlation of these events and their cyclicity provides striking evidence for the *near* synchroneity of climatic change, as measured in different regions, and for the inherent stability in the mechanism producing the Quaternary climatic changes.

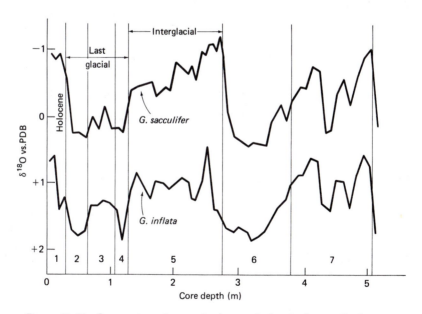

Figure 17–13 Oxygen isotopic record of two planktonic foraminiferal species through the late Quaternary (approximately 260,000 years) in the southern Indian Ocean. Pleistocene stage numbers are shown at bottom. (After J. C. Duplessy, 1978)

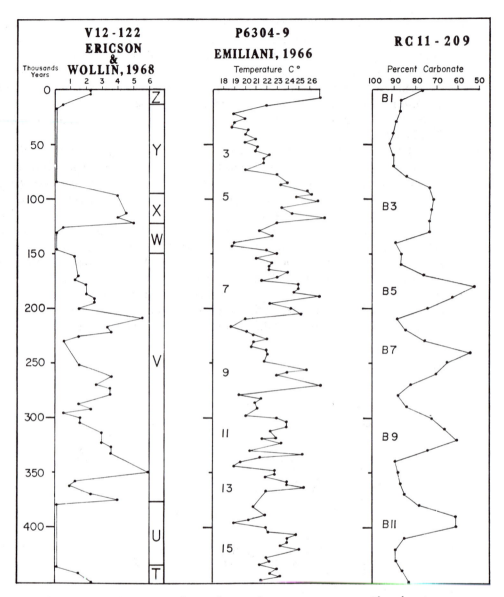

Figure 17-14 Comparison of two relative paleotemperature curves with carbonate curves from an equatorial Pacific core. Curve at left is based on oscillations in percent of the tropical planktonic foraminiferal species *Globorotalia menardii*. (From J. D. Hays and others, *Geol. Soc. Amer. Bull.* vol. 80, p. 1507, 1969, courtesy The Geological Society of America)

Microfossil Methods

Of the two primary methods of determining changes in the ocean system—biotic and isotopic—the biotic method shows change in broad distributional patterns of species associated with major current systems and water masses, but these changes are difficult to quantify. Nevertheless, the recent application of new concepts and approaches in micropaleontology has led to refinements in paleoceanography. Because the marine ecosystem is a complex network of relationships between organisms and environmental parameters, multivariate analyses have been employed to examine simultaneously the variations and interactions of many factors.

Uniformitarianism

In paleoceanographic analysis of the Quaternary, uniformitarianism allows comparison between modern environments and their associated taxa and environments that prevailed at the time of deposition of the microfossils. Implicit in most studies of this type is that there has been no evolution in the taxa or in the relationships between assemblages and their environments. This assumption is generally accepted for the Quaternary, because insufficient time has elapsed for much biotic evolution to have taken place. However, statistically extracting estimated paleotemperatures through one-to-one relationships between living and fossil species and their environments becomes increasingly difficult through time because of the extinction and evolution of species. Since evolution may be a response to environmental stress or change, some shift in environmental tolerance in the taxa is probably involved. Furthermore, even when there is no conspicuous morphologic change in the hard parts of the microfossil, the organism still may have undergone evolution. As species become extinct or migrate to new regions, the existing species probably adjust their environmental tolerances if only because of changes in biotic competition. Despite these constraints pre-Quaternary paleoceanography is making enormous advances; paleoceanographic history is being established, although details on the relative magnitude of temperature changes cannot yet be determined.

Group selection

At any one location the major picture of oceanographic change might be provided by only one of the microfossil groups. But which group does one choose? The biology of the plants is better known; and, because of their need for light, the plants must respond to changes in the surface water. However, they provide little information on the deeper waters. Comparison between the phytoplankton and the zooplankton is useful. Selection of major zooplankton and phytoplankton microfossils, one of each group with calcareous and siliceous tests, allows a global coverage of biotic indicators, with both plant and animal kingdoms represented.

Of the zooplankton microfossils, the foraminifera are the most widely used and, taxonomically, the best-known group (see Chapter 16). Although the biology of the polycystine radiolarians is largely unknown, distributional information on both living forms (for example [Cifelli and Sachs, 1966]) and their fossil remains in surface sediments [Nigrini, 1967; Sachs, 1973; Moore, 1973; Lozano and Hays, 1976] suggest that they are, in many ways, similar to planktonic foraminifera (see Chapter 16). Most radiolaria and planktonic foraminifera inhabit the near-surface waters (50–200 m) and both can be grouped into assemblages with geographic boundaries similar to those of the surface-water masses. There are, however, two characteristics of the radiolarians that make them a particularly useful group in the suite of microfossils to be employed: they have robust opaline tests, which are often well preserved at locations where other microfossils are either badly corroded or totally absent, and they are a comparatively diverse group. So, a quantitative analysis of the radiolarians can provide a measure of faunal-environmental relationships that are more statistically stable than those of the less-diverse groups. Both of these factors have caused the radiolaria to be particularly important in reconstructing Quaternary temperature patterns of the Pacific and Antarctic oceans.

Of the phytoplankton microfossils, the calcareous nannoplankton (coccoliths and discoasters; see Chapter 16) have been used most extensively for paleoceanographic work. Diatoms have largely been of stratigraphic importance, but have proved useful in some Quaternary paleoceanographic studies. The calcareous nannofossils have fairly well-defined taxonomies and phylogenies and are less diverse than most other microfossil groups. Table 17–2 compares and contrasts a number of characteristics of microfossil groups that are relevant to their usefulness in paleoceanographic studies.

Quantitative, semiquantitative, and qualitative information

Micropaleontological data are of three types: qualitative, semiquantitative, and quantitative. **Qualitative information** is that which simply records the presence or absence of a species and has been almost completely generated for stratigraphic and chronological reasons. Although this is the largest body of paleontologic information, its helpfulness in paleoceanographic reconstructions is limited. Plotting the estimated or actual changes in species abundance with depth in a core (Fig. 17–14) or in the ratio of several species is the simplest form of data presentation, but—by definition—it fails to take into account the entire assemblage.

Semiquantitative data are those in which estimates of species abundance frequencies have been established without actual counts being made. The primary advantage is that these data can be produced in a fraction of the time required for quantitative data by rapid scanning of the slide. The taxa present are ranked according to relative abundance, adding little extra effort to qualitative data reporting. Also semiquantitative data are amenable to statistical treatment and, if accurate, can be used to produce biogeographic patterns that are equivalent to those based on quantitative data [Sancetta, 1979B].

Quantitative data are naturally the most objective, detailed, precise, and reproducible, but it takes considerably longer to generate than semiquantitative data. For instance, frequency counts of foraminiferal species are usually made of splits of about 300 random specimens, which can take from 15 minutes in some well-preserved samples to several hours if the preservation is poor or the taxonomy is not well known.

Quantitative approaches can be classified into five main lines of inquiry. First is the analysis of variations in generic and species diversities through time based on analogy with the modern high-to-low, equator-to-pole gradients extant among

TABLE 17-2 COMPARISON OF CHARACTERISTICS OF MAJOR MICRO-
FOSSIL GROUPS USED IN MARINE GEOLOGICAL WORK

	Foraminifera	Radiolaria	Calcareous nannoplankton	Diatoms
1. Taxonomy	+ [a]	− [c]	+	−
2. General diversity	✔ [b]	+	−	✔
3. Diversity sufficiently high for polar Cenozoic paleoceanographic studies	✔	+	−	+
4. Diversity sufficiently high for subpolar Cenozoic paleoceanographic studies	+	+	+	+
5. Biostratigraphy known	+	+	+	✔
6. Biological controls known	✔	−	✔	+
7. Modern vertical and geographic distribution known	+	−	+	✔
8. Species and assemblage patterns match surface water masses	+	+	+	+
9. Morphologic variation related to environmental change	+	−	✔	−
10. Tests resistant to dissolution	−	+	✔	✔
11. Census data can provide data on original assemblages	✔	+	✔	✔
12. Tests resistant to lateral displacement (winnowing)	+	✔	−	−
13. Commonly found over wide areas in Cenozoic sediments	+	+	+	✔
14. Relative simplicity of counting	+	+	+	+
15. Tests suitable for isotopic measurements	+	−	+	✔

[a] + = relatively high values, or well known
[b] ✔ = moderately high values, or only partially known
[c] − = relatively low values, or poorly known

calcareous plankton in the modern ocean [Haq, 1973]. This probably results from increased environmental stability toward the equator and an increase in the degree of stratification of the upper water column, providing additional environmental niches for exploitation. Significant variations in planktonic diversity (taxonomic frequency) and in evolutionary rates (frequency of originations and extinctions) have occurred through the Cenozoic [Cifelli, 1969; Berggren, 1969; Haq, 1973]. This implies that the evolution of these groups has been strongly influenced by ecological stresses produced by dynamic changes in the thermal structure of the world oceans. Evolutionary radiations are clearly linked to times of reduced environmental stress and climatic warmings; extinctions closely follow drastic changes in the ecologic balance. The most conspicuous changes in the late Phanerozoic occurred at the Cretaceous-Tertiary boundary and the Eocene-Oligocene boundary.

Second is quantitative analysis of relative frequencies of species (Fig. 17–15), total faunal analysis of planktonic assemblages (for foraminifera, see [Ingle, 1967; Kennett and Watkins, 1974; Poore and Berggren, 1975]), and analysis of frequencies and ratios of temperature sensitive species (for diatoms and radiolaria, see

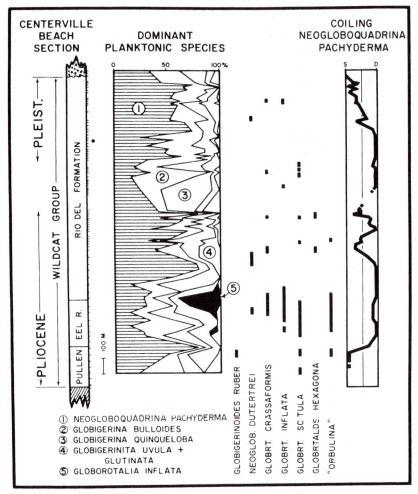

Figure 17-15 Quantitative variation of planktonic foraminifera within the Pliocene-Pleistocene Centerville Beach section, California. Note the appearance of a transitional water-mass assemblage characterized by *Globorotalia inflata* and *G. puncticulata* in the middle Pliocene representing a major warm event within the California Current and expressed elsewhere in the North Pacific. Also shown are coiling direction shifts in *Neogloboquadrina pachyderma*: S-axis is 100 percent sinistral; D-axis is 100 percent dextral. (Courtesy J. Ingle, Jr.)

[Kanaya, 1959; Koizumi, 1968; Casey, 1972; Schrader, 1973; Barron, 1973; Keany and Kennett, 1972]). For example, in the subpolar and temperate regions, changes in the percentages of the single foraminiferal species with polar water-mass preferences—*Neogloboquadrina pachyderma,* sinistral-coiling—are excellent first-order indicators of oceanic sensitivity to climatic change (Fig. 17–15). High percentages indicate low temperatures (glacial climates), and low percentages signify higher temperatures (interglacial climates).

A more useful approach is to combine all species that are typical of a modern set of environmental conditions into a diagnostic assemblage and present, in a composite plot, the simultaneous changes in abundances of all such assemblages (for instance [Kennett, 1970; McIntyre and others, 1976]). This procedure produces several partially interdependent curves, each of which is part of the total paleoclimatic information. Other workers have combined this type of data from a number of species into a single paleoclimatic curve (Fig. 17–16).

Phleger and others [1953] presented single, idealized paleoclimatic curves and considered each counted sample as indicative of middle, high, or low latitudes. Ruddiman [1971] combined warm and cold foraminiferal species into a single paleoclimatic curve, but did not make gradational distinctions within the warm and cold faunal groups among species whose temperature preferences are known to vary

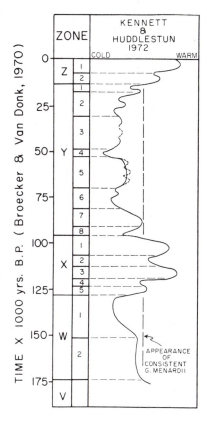

Figure 17–16 Late Pleistocene paleoclimatic curve and planktonic foraminiferal zones for the western Gulf of Mexico. The paleoclimate curve is based on species frequency oscillations in the fauna (greater than 175 μm). (From J. P. Kennett and P. Huddlestun, *Quat. Res.* vol. 2, p. 386, 1972)

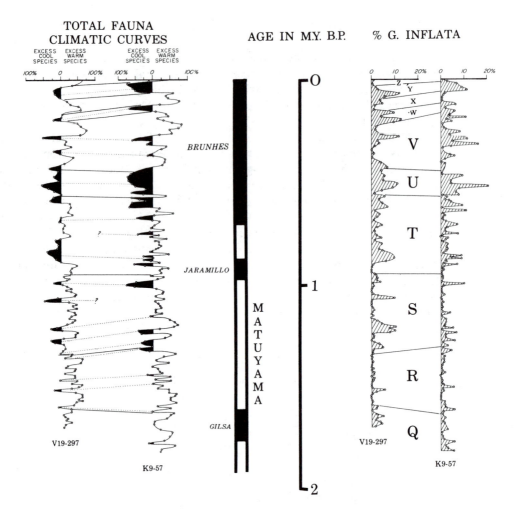

Figure 17–17 Total fauna paleoclimatic curves for South Atlantic cores plotted against paleomagnetic stratigraphy. Zones in which cool nonequatorial planktonic foraminiferal species outnumber warm tropical forms are marked by darkened zones. The most prominent cold zones occur above the Jaramillo Normal Event. (From W. Ruddiman, *Geol. Soc. Amer. Bull.*, vol. 82, p. 294, 1971, courtesy The Geological Society of America)

in the modern oceans (Fig. 17–17). McIntyre and others [1976] and Ingle [1977] used present-day foraminiferal and coccolith environmental preferences to reconstruct past geographic movements of particular water masses with the latitudinal extent of the environmental shifts serving as a quantitative measure of paleoclimatic change (Fig. 17–18) [Ingle, 1977; Ruddiman, 1977B].

The third main line of inquiry is the quantitative analysis of frequency variations of temperature-sensitive morphologic traits (ecophenotypes) of selected planktonic foraminifera based on similar clines established for modern popula-

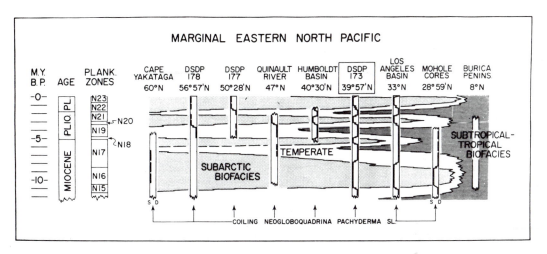

Figure 17-18 Schematic representation of major late Neogene oscillations of temperature sensitive planktonic foraminiferal biofacies within the California Current system and adjacent Alaskan Gyre. Note that north–south traverse extends from 60° N to 8° N. Exclusively sinistral coiling populations of *Neogloboquadrina pachyderma* are interpreted as evidence of surface temperatures lower than 10°C, whereas dextral populations are interpreted as representing temperatures higher than 15°C. The leading edge of the subtropical-tropical biofacies corresponds to the 20°C isotherm. (From J. Ingle, 1977)

tions, including coiling-ratio patterns [Bandy, 1960; Ingle, 1967; Kennett, 1967; 1968; Srinivasan and Kennett, 1974], changes in shape, size, and porosity, and many other morphological variables. For instance, mean test size of *Orbulina universa*, a spherical planktonic foraminifer, was observed by Bé and others [1973] to vary with latitude in the Indian Ocean. They concluded that a divergent correlation exists between *Orbulina* test size and water temperature and density. A particularly strong latitudinal gradient in shell size is observed in the middle latitudes, probably in response to the proximity of the subtropical convergence zone in which subtropical and subantarctic waters mix.

Fourth is the mathematical analysis of planktonic biofacies through factor analysis and related techniques capable of objectively defining paleoecologically meaningful groupings of extant or extinct species (for example [Haq and Lohmann, 1976]; see Fig. 17–19). Lastly, the paleoecological transfer functions allow computation of past conditions using modern calibrated microfossil-environmental relationships. This approach has become so important in Quaternary paleoceanography that it is discussed in more detail on pages 642 through 646.

All of these techniques give quantitative and paleoceanographically meaningful parameters that can be used to establish faunal and floral trends either vertically within individual sequences, forming time series, horizontally within a given unit or "slice" of time, generating areal patterns, or within both temporal and geographic frameworks. However, all but simple diversity analysis require counting of standardized individual faunal elements or statistically defined key elements.

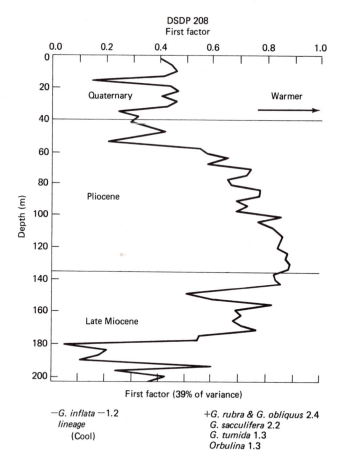

DSDP 208
First factor

Quaternary

Warmer

Pliocene

Late Miocene

Depth (m)

First factor (39% of variance)

−G. inflata −1.2
lineage
(Cool)

+G. rubra & G. obliquus 2.4
G. sacculifera 2.2
G. tumida 1.3
Orbulina 1.3
G.cf. conoidea 1.3
(Warm)

Figure 17-19 Factor analysis of planktonic foraminiferal assemblages in the late Cenozoic DSDP Site 208. Assemblage groupings (factors) are shown at bottom.

Transfer functions

The major problem in using modern foraminiferal species as standards of reference for Quaternary paleoclimatology is the interrelationship of temperature, salinity and productivity in the oceans [Ruddiman, 1977B]. These factors in combination have produced Quaternary microfossil assemblages. Most areas show particular associations of environmental parameters. For instance, areas of cold water, often exhibit low salinities and high fertility. Thus it is difficult to extract independent estimates of paleotemperature, paleosalinity, or productivity from deep-sea microfossil assemblages [Ruddiman, 1977B]. This problem has, in part, been overcome by the design of a factor analysis–transfer function technique that is used to reconstruct an estimate of an individual paleoenvironmental parameter by using surface-sediment assemblages and a modern atlas of oceanic parameters as a predictive data base. These are **paleoecological transfer functions,** which are empirically derived equations for calculating quantitative estimates of past oceanic or atmospheric conditions from paleontological data [Imbrie and Kipp, 1971]. The relationships in these functions are based on the spatial correlations between modern climatic data and census data for surface-sediment assemblages, which are used as calibration sets [Sachs and others, 1977]. Each equation is derived by standard multivariate-regression techniques. The use of such numerical techniques in micropaleontology became feasible with advances in computer technology and the assembly of large data sets. From a theoretical standpoint, the Imbrie-Kipp technique is a more comprehensive technique than any other conceived at this time. In their original study, Imbrie and Kipp [1971] used planktonic foraminiferal counts to derive the transfer functions for estimating past sea-surface temperatures and salinity. Five species groups were recognized in North Atlantic surface sediments: polar, subpolar, subtropical, tropical, and gyre margin. The faunal assemblages form geographically coherent patterns and emphasize ecologically important characteristics of each data set. Paleoclimatic estimates for a Caribbean Quaternary core using these equations are shown in Fig. 17–20. When a large set of cores is examined, a three-dimensional framework of paleoceanographic oscillations, such as that shown in Fig. 17–21, develops. This approach has been extended to the global ocean by the CLIMAP group in order to construct maps of sea-surface temperature on the 18,000 years B.P. glacial maximum datum [CLIMAP, 1976].

There are four principal steps in paleotemperature analysis using the Imbrie-Kipp technique [Hutson, 1978] (Fig. 17–22). The first step is to describe faunal assemblages in core-top data set using a factor analysis. The second step is the derivation of a transfer function using multiple regression. The third step is resolution of downcore data into the same faunal assemblages described by the analysis of the core-top data:

$$U_{dc} = F_{dc}V$$

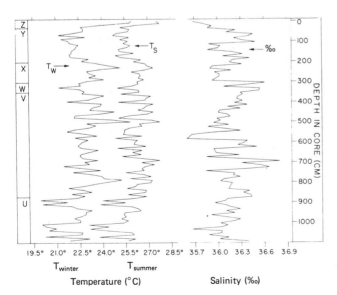

Figure 17–20 Paleotemperature (T winter and T summer) and paleosalinity estimates for surface waters in Caribbean core based on paleoecological equations described in text. Planktonic foraminiferal zones are shown at left. (After J. Imbrie and N. G. Kipp, in *The Late Cenozoic Glacial Ages*, ed. K. K. Turekian, p. 118, 1971, reprinted with permission of Yale University Press)

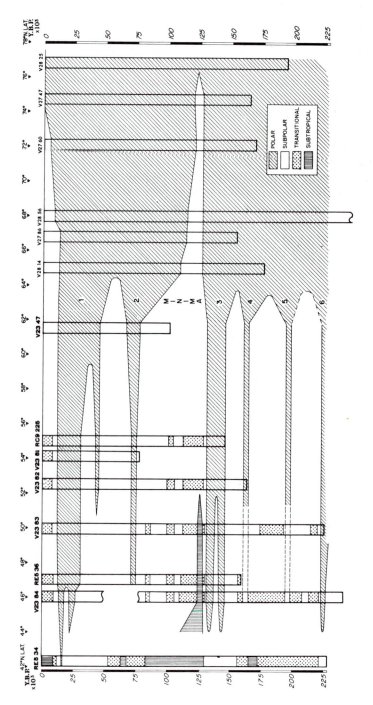

Figure 17-21 Paleoceanographic oscillations with latitude in the Norwegian Sea and northern North Atlantic during the late Quaternary. Plotted are planktonic foraminiferal-coccolith assemblages specific to particular water masses. Note that subpolar faunas have penetrated into Norwegian Sea only twice during the last 150,000 years: at present and about 120,000 years ago. (From T. B. Kellogg, in *Investigation of Late Quaternary Paleoceanography and Paleoclimatology*, ed. R. M. Cline and J. D. Hays, GSA Memoir 145, p. 105, 1976, courtesy The Geological Society of America)

644 Approaches to Paleoceanography

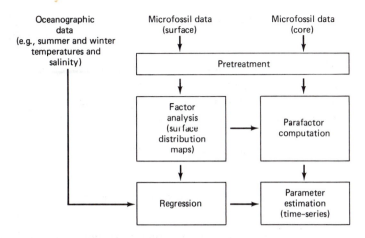

Figure 17-22 Principal steps in transforming oceanic microfossil assemblage data into temperature or salinity estimates.

where U_{dc} represents the estimated faunal assemblages downcore, F_{dc} is the downcore census data, and V is the assemblage description matrix. The final step is the application of the transfer function (T) to the faunal assemblages downcore (U_{dc}) to estimate paleotemperature (E_{dc}):

$$E_{dc} = U_{dc}T$$

To summarize, paleoecological transfer functions share the following characteristics [Sachs and others, 1977]:

1. They produce calibrated quantitative estimates of some parameter of a past environment, such as seasonal or monthly air or ocean-surface temperature.

2. They use algebraic methods to formulate these estimates.

3. The algorithms rely on multivariate computational techniques to analyze multicomponent fossil data.

4. The functions are calibrated from an adequate set of modern distributional data.

5. The calibrated function is then applied to older samples to estimate environmental parameters for past times.

Furthermore, the accuracy of transfer functions is a measure of how closely they estimate actual temperatures. This will depend on how closely the following basic assumptions, which underlie the use of transfer functions [Sachs and others, 1977], are met.

The multivariate approach is assumed to be preferable to estimating past conditions using variations of a single species. Although this seems reasonable for most ecosystems, it has not been rigorously demonstrated. Multivariate estimates may be buffered from some types of depositional or ecological anomalies that selectively affect single species, and the use of these methods has allowed the discovery of such anomalies.

Biological responses must be systematically related to the physical attributes of the environment, and the physical parameters, such as temperature, should either be important in controlling the biota or at least related to other controlling parameters. Noncorrelated parameters are readily detected with all transfer function approaches used.

The following simple relationships are assumed to exist between past and present conditions.

1. Evolutionary change of the fossil species has been negligible, so ecological roles are constant.

2. Climatic or oceanographic conditions fall within the range of modern calibration data.

3. Preservational conditions have been relatively constant through time.

If conditions within the downcore data set are not within the calibration set, then transfer functions based on modern surface-sediment data will be in error. These are known as **no-analogue** situations. An example is exhibited by changes in the radiolarian *Cycladophora davisiana* in Antarctic Quaternary sediments. This species is highly abundant about 18,000 years B.P. in the Antarctic [Hays and others, 1976]. In the modern ocean, this form is abundant only in the Okhotsk Sea, an area of extensive seasonal sea-ice formation [Robertson, 1975], so the abundance maximum 18,000 years B.P. in the Antarctic is interpreted by Hays and others [1976] to reflect times of extensive sea-ice development. There is, however, no modern analogue of high abundances of this species in the Antarctic.

Hutson [1976] has recognized two classes of no-analogue conditions: **environmental** and **biological**. Environmental no-analogue situations can be the result of artifact assemblages produced by chemical dissolution. Biological no-analogue situations may arise from incomplete sampling of modern environments or from modern environments not including particular conditions of the past. Also, evolution may cause biological no-analogue situations.

Time-Series Paleoceanographic Signals

The problem of forcing functions

We have now examined the various isotopic, micropaleontological, and sedimentological techniques that provide us with the needed paleoenvironmental information, including time-slices of the ocean. Finally, we need to examine the various ways by which time-series data are employed to better understand the nature of paleoceanographic change and the forcing functions which lead to and control this change. Climatic, geochemical, and biological changes through geologic time are believed to be, in large part, reactions to forcing functions, such as **astronomic, atmospheric,** and **telluric** (for instance, continental drift and ocean-basin evolution) causes. In order to explain these changes, the mixture of forcing functions, as well as the internal mechanisms of response of the ocean-atmosphere-biosphere system, must be identified. Solar energy runs the ocean, so any change in solar energy reaching the ocean will result in change of oceanic circulation. Important components of the response mechanisms are storage of heat and carbon and the transfer of heat and carbon from one reservoir to another. There are three primary processes that change the oceanic energy input.

1. *Orbital variations.* The amount of solar radiation striking the earth's upper atmosphere at any given latitude and season is fixed by three elements of the earth's orbit around the sun: the **eccentricity,** the **obliquity of the ecliptic,** and the **longitude of perihelion** with respect to the moving vernal point (Fig. 17–23). Each of these orbital elements is a quasi-periodic function of time. The geometries of past and future orbits, originally calculated by Milankovitch in 1941, are known as the **Milankovitch orbital perturbations.** The most recent and accurate calculations are Berger's [1978]. The Milankovitch orbital perturbations are now considered a major forcing function in the glacial to interglacial oscillations that marked the Quaternary period.

2. *Solar variations.* Changes in the surface convective patterns of the sun are visibly observed by counting the number of sunspots. Counts of sunspots for the last 300 years reveal marked changes in the sun's behavior. In addition to an 11-year cycle, sunspots may nearly completely disappear during extended periods, such as between A.D. 1650 and 1710. Solar variability translates into changes in solar-wind magnetic properties and, in turn, into cosmic-ray intensity, so ^{14}C production rates change. Past atmospheric ^{14}C levels are derived from tree-ring ^{14}C analyses. These show two periodicities of about 130 and 200 years [Stuiver and Quay, 1980]. A less precise, 8000-year-long record of ^{14}C levels in bristlecone pine shows periodicities of about 500, 900, and 2400 years [Suess, 1980]. It is possible that these longer-term ^{14}C variations reflect climatically related changes in carbon-reservoir exchange rates. Little work of this type has yet been carried out using deep-sea cores, mostly because of inadequate stratigraphic resolution due to bioturbation and low sedimentation rates.

3. *Albedo changes.* Albedo changes have been caused by both telluric and atmospheric changes. The telluric changes result from latitudinal drift of continents. This creates climatic change, resulting in changing desertification and in ice-sheet formation and hence in changes in the earth's albedo, or solar, reflectivity. The importance of the longer term changes in continent-to-ocean relations in creating major global paleoclimatic changes are discussed in Chapter 19.

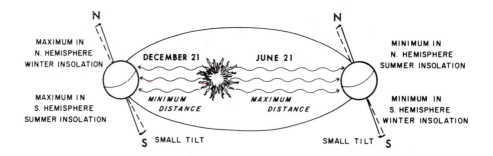

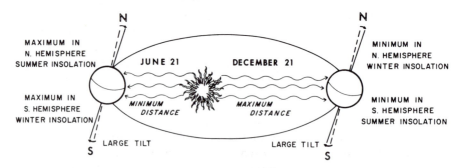

Figure 17-23   Astronomical configurations of the earth in relation to the sun that
may (a) enhance ice growth in the Northern Hemisphere; (b) melt ice in the Northern Hemisphere. (Courtesy W. Ruddiman and A. McIntyre)

Atmospheric changes in the earth's albedo relate largely to loading of volcanic
dust. Bryson and Goodman [1980] theorize that historical volcanic eruptions have
had a measureable short-term effect on climate. On a much longer time scale, Kennett and Thunell [1977] have shown major changes in volcanicity over the last
several million years, and Kennett [1980] and Bray [1979] have suggested relations
between major climatic episodes during the last 20 m.y. A direct link between these
changes in volcanicity and the late Cenozoic evolution of major ice sheets has also
been suggested [Bray, 1979]. On an intermediate time scale (10^3 years to 10^6 years)
no compelling relationships have yet been demonstrated between large volcanic
eruptions and oceanographic or climatic change. Yet marine sedimentary records
indicate many major eruptions over the past few million years [Ninkovich and
Donn, 1975; Ninkovich and Shackleton, 1975].

The various kinds of analytical approaches used in paleoceanographic
research include **event analysis, event correlation, phase-shift analysis,** and **spectral
analysis.**

1. *Event analysis.* The geologic record over various time scales is marked by paleoceanographic-paleoclimatic events that are transitional between one paleoceanographic mode and another. These are **transient events.** In several cases these events are irreversible paleoenvironmental changes reflecting evolutionary stages in global oceanographic and climatic history. Others are temporary reversible events. An example of an irreversible transient event is that which occurred at Eocene-Oligocene boundary [Kennett and Shackleton, 1976]; bottom waters cooled suddenly and several other changes, such as a drop in the CCD, occurred (see Chapter 19). Apparently the character of the deep water changed considerably due to cooling and sinking of bottom water off Antarctica. Bottom water has not since returned to its previous warmth, so this event is the transition through a **paleoceanographic threshold.** Examples of reversible transient events are those that occurred in the Quaternary from an interglacial to a glacial state. Event analysis involves the study of these transitional environmental states. The rate of the change evinces the size of the reservoir involved and the inertia of the system. The character of the transition may indicate a monotonic shift or the existence of unstable oscillations through the change from one oceanographic state to the next (for example, the middle-Miocene buildup of the Antarctic ice sheet).

2. *Event correlation.* Another dimension is added to paleoceanographic analysis by the correlation and comparison of events recorded in stratigraphic sections from widely separated locations. This provides insight about the magnitude of a particular change and its ramifications on other parts of the oceanic-climatic system. Correlations may include latitudinal (high versus low), cross current (inside versus outside of gyre), downcurrent (eastern versus western boundary currents), interbasin (such as North Atlantic versus North Pacific versus Mediterranean) and shallow-deep (paleoenvironmental gradients with depth).

3. *Phase-shift analysis of event sequencing.* The chronology of change recorded within and between different areas provides insight into the sequence of oceanographic history, and in turn separates the primary driving forces from later repercussions (the cause-and-effect chain). Unless sequences of changes are known, the dynamics remain obscure. Because the sediments are multichannel recorders, changes in different parts of the air-sea ice system can be studied in the same sedimentary section. Examples of signals in which phase shifts are known include paleotemperatures of surface and bottom waters, oxygenation, carbonate dissolution, and oxygen isotopic composition of the oceans due to changes in the cryosphere. Often only a few hundred to a few thousand years separate the signals, so undisturbed sections with high stratigraphic resolution are necessary.

4. *Spectral analysis.* Power-spectral, or frequency-amplitude, analysis is a major tool for examining the dynamics of change in the air-sea-ice system. Functions driving the changes have been revealed, such as astronomical forc-

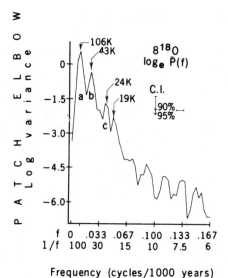

Figure 17–24 Spectra of climatic variations (in δ¹⁸O) in subantarctic piston cores. High resolution spectra expressed as the natural log of the variance as a function of frequency (cycles per thousand years). Prominent spectral peaks are labeled a, b, and c. (From J. D. Hays and others, *Science*, vol. 194, p. 1127, 1976, copyright © 1976 by the American Association for the Advancement of Science)

ing (Milankovitch perturbations) of glacial-interglacial fluctuations, Prominent spectral peaks in time series from deep-sea cores [Hays and others, 1976] occur at 23,000 and 41,000 years (Fig. 17–24). Concentrations of variance in ice volume and paleoclimatic data are predicted at these periods by a simple version of the theory of astronomical forcing. These spectral peaks vary greatly in amplitude from area to area. The discovery of such frequencies in the Quaternary have also provided a new tool for the refinement of our time scale, in that time-scale adjustment is possible by "tuning" the time series to the astronomical frequencies.

eighteen
Paleoceanographic and Sediment History of the Ocean Basins

Come wander with me, she said,
Into regions yet untrod;
And read what is still unread,
In the manuscripts of God.

Henry Wadsworth Longfellow

The evolution of the oceanic environment, its sedimentary history, and paleobio-geography have been well studied, due to the DSDP. In Chapter 6 we examined the tectonic evolution of each of the major ocean basins. This chapter contains a summary of the sedimentary and paleoceanographic evolution of the Pacific, North Atlantic, South Atlantic, and Indian oceans. These summaries are generalized and simplified from several sources, most based on material gathered by the DSDP. Each of the initial reports of the project represents summaries of the geological evolution of a particular region and contains enormous amounts of detailed information that is lost in the summaries that follow. In addition to the initial reports, there are a few other valuable summaries of much larger oceanic regions.

The sedimentary and environmental evolution of a particular region or ocean contains two sets of information: local and global. However, because an ocean works as one geochemical and circulatory machine, a local change may have global effects. Global patterns evolving with paleoclimatic change provide the necessary context for understanding the oceanographic history of an ocean. Overprinted on this global pattern are special characteristics of a region, resulting mainly from its physiographic evolution [van Andel and others, 1975]. The local overprint dominates the marine environment during the early life of an ocean basin because tectonic barriers partially or completely isolate early basins from the global mainstream. As the oceans widen during maturity and tectonic barriers are reduced, the environmental characteristics become more like those of the world ocean. Thus each description of the environmental history of an individual ocean contains both information that is unique to that region and that is a response to change in the global ocean. The most important global effects are summarized in more detail in Chapter 19.

PACIFIC OCEAN

The Pacific Ocean exhibits a different tectonic history from the Atlantic and Indian oceans, which resulted in a unique sedimentary history. These differences derive largely from the following factors.

1. Other oceans have grown through the Mesozoic and Cenozoic at the expense of the Pacific. During the early Mesozoic, the Pacific was considerably larger than today, represented by the super or global ocean, Panthalassa. Since then the Pacific has become more restricted, but it still represents the largest global ocean in which large areas are remote from terrigenous influences.

2. Active margins, related trenches and marginal seas almost completely surround the Pacific. As a result, terrigenous material is trapped in marginal areas, leaving vast areas of the Pacific basin exposed only to the effects of pelagic sedimentation, which consists largely of biogenic material and small amounts of eolean dust. Terrigenous clastic sedimentation of more than local significance occurs in the northeast. Changes in sedimentation of the Pacific Ocean would, therefore, seem to be a reliable indication of the overall behavior of the oceanic part of the earth's geochemical system [Worsley and Davies, 1979]. Thus the sedimentary history of the Pacific is little influenced by local-basin or continental evolution and records phenomena that have affected the oceans as a whole.

3. The sedimentary history of the Pacific Ocean is steadily being destroyed in the surrounding subduction zones compared to other oceans, where the record has been steadily accumulating from oceanic inception to the present day. Thus, although the Pacific Ocean is oldest, there is no good record older than the middle Mesozoic within its basins. Part of this older record is found in the accretionary prisms of the surrounding active margins. These offer the only means of obtaining information about Panthalassa.

4. Although the Pacific Ocean has been substantially reduced in size by the creation of the Atlantic and Indian oceans, it was originally so immense that it has retained much of its original shape. Circulation patterns thus have not changed drastically, such as in the much younger oceans. The sedimentary patterns of the new oceans were strongly affected by the evolution of paleocirculation as their basins developed, especially early in their histories. In contrast, for more than 200 m.y. the Pacific has been marked by open-ocean circulation, characterized by large gyres in both Northern and Southern hemispheres (Fig. 18–1).

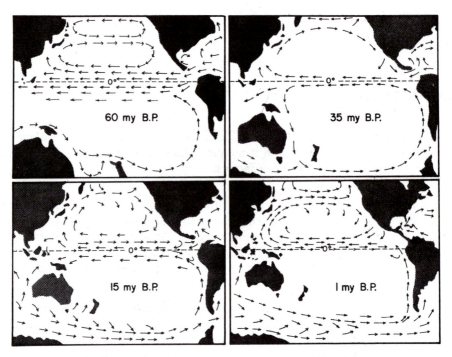

Figure 18–1 Postulated changes in the surface circulation of the Pacific Ocean as the continental configurations changed with time. (From Tj. H. van Andel, 1979)

5. Large areas of the western and southwestern Pacific have undergone complex tectonic histories as marginal seas developed. In these areas, the history of Cenozoic sedimentation is complex.

6. The sedimentary records of the Mesozoic and Cenozoic over large areas of the Pacific basin reflect the motion of the Pacific plate with respect to the oceanographic environments (see Chapter 17). The tectonic evolution involves northward motion of the plate, gradual subsidence with increasing age of the flanks of the ancestral East Pacific rise, and changes in height, shape, and position of the rise [van Andel and others, 1975]. The primary paleoceanographic changes include variations in plankton productivity and changes in the nature and flow of bottom waters, which are partly related to the history of Antarctic glaciation. These waters enter the deep Pacific from the Antarctic, increasing dissolution and, in some cases, physically eroding the sediments, thereby further modifying the pattern of accumulation. Sedimentation rates decrease rapidly with increasing depth (and age) of the basement as a result of carbonate dissolution [van Andel and others, 1975]. Furthermore, the productivity of biogenic material sharply decreases away from the equatorial zone. Chert layers of early Cretaceous age, now located in the far Northwest Pacific, have been interpreted by Lancelot and Larson [1975] to represent previous siliceous oozes deposited while crossing the equator.

The sedimentary history of the Northwest Pacific reflects the interplay of plate motion, the CCD, and equatorial productivity. All DSDP sites reveal the same relatively simple sedimentary history: initial deposition of carbonate ooze at shallow depths on the mid-ocean ridge during some part of the Cretaceous; followed by deposition of clays or siliceous biogenic sediments (or both) as the ridge flank spread permanently beneath the CCD; increased siliceous biogenic productivity as a DSDP site moved northward across the high biogenic-productivity zone of the equator; followed by slowly deposited clays in the central gyre regions. Late Cenozoic volcanic ash deposits in part reflect the movement of the plate closer to volcanic sources in the areas of the Northwest Pacific.

The most important review of Pacific paleoceanographic and sedimentary history is by van Andel and others [1975]; this review deals primarily with the equatorial region, but it also provides concepts that apply to wide areas of the Pacific basin. Other more general reviews include Fischer and others [1970] and Worsley and Davies [1979]. Regional reviews include Scholl and Creager [1973], von Huene and Kulm [1973], and Lancelot and Larson [1975] for the North Pacific; Ingle [1973] for the marginal Northeast Pacific; Karig [1975] for the West Pacific; and Burns and Andrews [1973] and Packham and Terrill [1975] for the Southwest Pacific.

Mesozoic History

Ancient gyres of Panthalassa

The Mesozoic paleoceanography of the Pacific is poorly known because most of the remaining record is now confined to the far northwestern sector. It has been inferred, however, that the combination of large size, evidence of climatic homogeneity during the Mesozoic, and large heat capacity of the oceans produced uniform and uneventful histories compared to the Cenozoic [Berggren and Hollister, 1977]. Surface circulation was probably similar to the modern Pacific basin (Fig. 18-1), with the two main gyres in the Northern Hemisphere: the anticyclonic subtropical and the cyclonic subarctic. Cretaceous analogues of the Kuroshio, Oyashio, and California currents were probably already well developed [Luyendyk and others, 1972]. In the Southern Hemisphere, two well-developed gyres would also have existed since an absence of the Circum-Antarctic Current prevented interference with the development of an inferred cyclonic subantarctic gyre. A broad westward-flowing equatorial current would have extended into the northern Indian Ocean, because Australia was still attached to Antarctica. The long equatorial transit would have transported very warm waters to the western Pacific and northern Indian oceans. Since Antarctica was not yet heavily glaciated, low surface-water temperature gradients would have existed between low and high latitudes. In general, the equatorial regions remained tropical and stable throughout the late Mesozoic and Cenozoic, with distinct gradients developing during the middle Cenozoic. The vigor of the equatorial circulation would also have been enhanced by unrestricted flow between the equatorial Pacific and Atlantic oceans, through the Central American Seaway. This connection continues until the late Pliocene (3 Ma). From the late Cretaceous to the Paleocene (80 to 60 Ma), the Tasman basin began to develop by sea-floor spreading, separating New Zealand from Australia [Hayes and Ringis, 1973]. At that time the region between the Tasman spreading center and New Zealand would have been part of the Pacific plate, rather than the India plate as later. This was the first of several basins to develop in the Southwest Pacific. During the Cenozoic, there was no effective paleoceanographic connection between the Pacific and Arctic oceans. The Bering Straits represent only a very shallow-water seaway between the Arctic and Pacific, which became exposed above sea level during Quaternary and perhaps during earlier low stands of sea level.

Cenozoic History

Early Cenozoic evolution

Australia remained part of Gondwanaland through the Paleocene and early Eocene, until 53 Ma [Weissel and Hayes, 1972]. Biogeographic evidence indicates the presence of warm subtropical conditions as far north as 60°N on both western and eastern margins of the Pacific and in New Zealand and Australia and only slightly cooler temperatures (warm temperate) adjacent to Antarctica [Berggren and Hollister, 1977]. Carbonate oozes were deposited at high latitudes in both the North and South Pacific, reflecting these warmer conditions. During the Eocene, the equatorial belt was a broad, ill-defined zone with low accumulation rates. Terrigenous sedimentation was very high in the Gulf of Alaska, beginning 50 Ma [Scholl and Creager, 1973]. This continued until 30 Ma, forming the Aleutian abyssal plain.

During the middle Eocene, Australia began to move northward; the volcanic arcs and marginal seas in the Southwest Pacific began to form a few million years later. Sedimentation in most of these basins has been pelagic due to isolation from terrigenous sources, although major input has also been from volcanogenic components.

Most Eocene sedimentation was continuous, though low, throughout much of the Pacific basin. Hiatuses occurred at specific times, such as at the Paleocene-Eocene boundary. The early Eocene is nonfossiliferous over broad areas of the deeper Pacific basin. Carbonate sedimentation was restricted to areas shallower than 2500 m; extensive carbonate dissolution resulted in low rates of carbonate accumulation. The broad equatorial zone of biological productivity became restricted to a much narrower zone by the early Oligocene. Carbonate sedimentation continued on shallower parts of the mid-ocean ridge in the equatorial areas, but by the late Eocene to early Oligocene the Emperor Seamounts in the Northwest Pacific ceased to receive carbonate deposition, suggesting a sharpening of latitudinal dissolution gradients [Worsley and Davies, 1979]. Carbonate sedimentation continued during the late Eocene in some shallow areas adjacent to Antarctica.

Eocene-Oligocene boundary event

About 38 Ma, at the end of the Eocene, the CCD dropped abruptly by as much as 1500 m to depths that have not since fluctuated greatly (see page 689). In the equatorial Pacific, the CCD reached about 5000 m by the late Oligocene and temporarily rose again to 4400 m during the middle Miocene to late Miocene. The CCD drop at the Eocene-Oligocene boundary increased carbonate sedimentation over wide areas of the Pacific; the equatorial carbonate belt widened and accumulation rates increased [van Andel and others, 1975]. This event is thought to be associated with the first significant development of sea ice around Antarctica, which indicates a major cooling and increased turnover of bottom waters (see Chapter 19). The climatic change in Antarctica was perhaps triggered by the initial opening of the seaway south of Tasmania to surface-water currents [Kennett, 1978].

During the early Oligocene, a hiatus minimum occurred only in the eastern tropical Pacific, but in the western Pacific, this interval is represented by an extensive hiatus [Kennett and others, 1972; 1975; Rona, 1973,b]. This probably was associated with the major cooling of Antarctica and increased production of Antarctic bottom water. This, in turn, led to increased erosion by northward flowing deep bottom currents, which followed newly created fracture zones as far north as equatorial latitudes, a process which has continued throughout the Cenozoic and was enhanced during the Quaternary [Johnson and Johnson, 1970; Berggren and Hollister, 1977].

Oligocene carbonate accumulations

By the early Oligocene, the deep equatorial passage north of Australia was closed due to extensive tectonism along the northern margin of Australia [Veevers, 1969] and in the western Pacific [Moberly, 1972]. Thus the developing Circum-Antarctic Current south of Australia was matched by destruction of the Tethys Seaway north of Australia. The Caroline basin in the Philippine Sea developed in the early to middle Oligocene. In the Southwest Pacific, the Caledonia and the eastern part of the Fiji Basin also formed, and obduction of the sea floor occurred at the island of New Caledonia. The developing circum-Antarctic circulation profoundly affected sediment patterns over almost the entire South Pacific (see Chapter 19). Indeed, Oligocene conditions differed markedly in the world ocean and heralded the onset of modern circulation patterns. During the Oligocene, carbonate sedimentation continued over the present day Antarctic convergence. The character of sedimentation in the Oligocene south of this remains largely unknown.

The period from 33 to 26 Ma marks a transition to a new regime, where dissolution rates at depths began to increase again. However, this effect was compensated until 26 Ma by the increased rate of supply and perhaps by depression of the lysocline [van Andel and others, 1975; Worsley and Davies, 1979]. A broad equatorial carbonate zone resulted. By the late Oligocene, there were high rates of carbonate sedimentation over a wide area, suggesting a time of maximum carbonate accumulation in the deep Pacific [Worsley and Davies, 1979]. By the middle Oligocene (30 Ma), the Gulf of Alaska ceased to be a site of significant terrigenous sediment accumulation. A major pulse of volcanism occurred from the late Oligocene to the middle Miocene (30–12 Ma) in the Bonin-Mariana-Yap volcanic arc and in Northwest Japan, peaking in the early Miocene. Also the Parece Vela–Shikoku Basin opened rapidly between the late Oligocene and early Miocene [Karig, 1975]. By the late Oligocene, a deep passage was established south of Tasmania [Kennett and others, 1975] and the Drake Passage possibly opened [Barker and Burrell, 1976], permitting development of a full system of circum-Antarctic deep current. During the Mesozoic, seaways may have existed at the southern end of the Antarctic Peninsula as a result of motion between East and West Antarctica [Duncan, 1981], but the paleoceanographic significance of such possible seaways is not clear. During the early Cenozoic, shallow-water marine connections must have existed between the South Pacific and Atlantic oceans through the East Antarctic region, which would have provided important biogeographic connections. No evidence exists for the development of any deep-water seaways until the middle Cenozoic (see Chapter 19).

Neogene siliceous sedimentation

As Antarctic circulation developed during the Neogene, the siliceous biogenic belt expanded northward. During the early Miocene, in the South Pacific sector, the sequences exhibit well-developed oscillations between siliceous and calcareous sediments. By 15 Ma, major growth of the Antarctic ice sheet had occurred. This led to further increases in carbonate-sediment dissolution rates as the CCD shoaled, the equatorial belt of carbonate sediments became narrower, and erosion once more became widespread. The middle Miocene coincides with a peak in hiatus abundance over much of the Pacific, although maximum sediment accumulation rates occurred near the equator, probably because of increased upwelling and fertility [van Andel and others, 1975].

During the latter part of middle Miocene and late Miocene, a dramatic increase in diatom productivity over large areas of the Pacific heralded worldwide climatic deterioration related to increased upwelling and oceanic turnover. Diatomaceous sediments were deposited throughout the continental margins of the west coast of North America and Japan [Ingle, 1973]. In the Antarctic region, siliceous sediments completely replaced carbonate deposits, and rates of biogenic sedimentation began to increase steadily toward a peak in the Quaternary. Likewise, at high latitudes of the North Pacific, upwelling also increased beginning 13 Ma [Creager and others, 1973]. This is reflected on the guyots at the northern end of the Emperor seamount chain by a transition from predominately clay sediments to diatom oozes having high sedimentation rates (8 m/m.y.). Warm water, middle-Miocene assemblages in Japan and California were replaced in the late Miocene by cool-temperate and subarctic assemblages. Alaskan vegetation reflects a parallel terrestrial cooling. Toward the end of the late Miocene, glaciation became evident onshore and in nearshore sediments.

Terrigenous sediments became important again in the Northwest Pacific, reflecting increased tectonism of the areas surrounding the Gulf of Alaska [von Huene and Kulm, 1973]. Volcanogenic components also increased in the deep basins during the middle Miocene as explosive volcanism increased in intensity [Kennett and others, 1977; Hein and others, 1978].

In the Southwest Pacific, tectonism continued in the marginal basins and certain adjacent landmasses. Turbidite deposition commenced in the Coral Sea basin during the early part of late Miocene, reflecting the uplift of the mountains of New Guinea. This tectonism has continued through to the present day, resulting in deposition of 2000 m of turbidites [Burns and others, 1973]. The development of the South Fiji basin would have carried the volcanic arc, the Lau ridge, and the islands of Fiji to the east. The Fiji Plateau was added to the Pacific plate as a result of a shift in position of the plate boundary.

Pliocene-Quaternary glaciations and neotectonism

Pliocene-Pleistocene sedimentation in the Pacific is marked by increased terrigenous input around the margins, increased volcanogenic input, increased eolian input, enlargement of the ice-rafting and biogenic siliceous provinces at high latitudes in both hemispheres, and increased rates of deposition of siliceous biogenic sedimentation.

Activated tectonism is reflected by orogenesis in New Zealand, Japan, California, and western South America. The Mariana Trough has been opening at rates of at least 10 cm/year during the last 2 to 3 m.y. [Karig, 1975]. On the west coast of North America, marginal basins—including the California borderland—accumulated thick, rapidly deposited Pliocene-Quaternary terrigenous sediments over the predominantly diatomaceous interval as a result of tectonic uplift and erosion. In the North Pacific, terrigenous sediment input also increased during the late Pliocene. The abyssal plains in this region record Quaternary sedimentation rates of 60 m/m.y. This is related to pronounced uplift of the coastal regions of southern and southwestern Alaska and to increased glaciation. Explosive volcanism also intensified in association with the increased tectonism.

The most important change in circulation during the latest Cenozoic was the closure of the seaway across the Isthmus of Panama—the final element in the destruction of the equatorial circumglobal passage. This occurred in the late Pliocene about 3 Ma and appears to have coincided with the onset of ice-sheet formation in the Northern Hemisphere. In the North Pacific, there was significant southward displacement of surface isotherms as much as 20 to 30° of latitude during glacial episodes, mirroring similar fluctuations of the polar water mass in the North Atlantic. Upwelling continued to increase in high latitudes of the North and South Pacific, producing the highest rates of siliceous biogenic productivity for the entire Cenozoic.

NORTH ATLANTIC

The main North Atlantic basin was part of the circumglobal Tethys Sea from its early middle-Jurassic formation to the late Cretaceous, at which time the South Atlantic became wide enough to form a connection with the Southern Ocean for exchange of surface and deep-water masses. The opening and evolution of the Atlantic from a narrow, latitudinal, tropical sea to a wide, longitudinal basin during the last 160 to 180 m.y. (Figs. 6-7 and 6-8) has played a major role in paleoceanographic history. The early Atlantic Ocean was a critical component of equatorial circulation through the Tethys Seaway; the modern Atlantic is critical to interpolar water-mass circulation because it is the only ocean connecting both polar hydrospheres.

Sedimentation in the Atlantic occurs within a tectonic framework of an expanding ocean basin bordered by passive margins, which have continuously subsided and accumulated large sedimentary prisms. Freshwater discharge into the Atlantic constitutes drainage from over one-half of the earth's land area and results in a high rate of terrigenous influx and deposition.

A number of important syntheses have been carried out on the sedimentary history of various parts of the North Atlantic Ocean. The history of the surrounding continental margins is discussed in Chapter 11. This chapter concentrates on the stages of the basin's paleoenvironmental evolution based on the valuable contributions of Thiede [1979], Lancelot and Seibold [1978], Arthur [1979], Tucholke and Vogt [1979], and Cita and Ryan [1979] and other workers. Deep-sea drilling in the different sectors of the North Atlantic has shown that, although every basin has its own depositional history, imprints of major paleoceanographic events are preserved over very large regions and form the basis of North Atlantic paleoenvironmental evolution. Regional differences in continental climates, in paleocirculation associated with the North Atlantic gyre, and in basinal depths have created the basin-to-basin differences. For instance the northeastern North Atlantic received little terrigenous sediment input during the last 130 to 140 m.y. because wide, epicontinental seas in Europe acted as centers for the deposit of terrigenous sediment before this reached the ocean. The sediments in the northeastern North Atlantic are therefore largely calcareous. The southeastern basin received much less carbonate sediment than the southwestern basin during the Mesozoic, and carbonate sedimentation has increased here only during the last 20 m.y. Asymmetry in the erosive and transporting capacity of currents between the western and eastern North Atlantic have created major differences in the depositional regime. These asymmetries are enhanced by the abyssal plains, which are much larger in the western North Atlantic than in its eastern basins.

Mesozoic History

Early evaporites and carbonates (late Jurassic to early Cretaceous)

Mesozoic sequences can now be correlated with some precision between the western and eastern North Atlantic. The sedimentary sequences exhibit wide lateral distribution, lithological similarity, and reasonable time synchronism. Jansa and others [1978] show schematic lithostratigraphic cross sections for these basins in Figures 18-2 to 18-4. A succession of seven Mesozoic lithostratigraphic units are recognized in ascending order, as follows: Oxfordian greenish gray limestone, Oxfordian-Kimmeridgian reddish brown argillaceous limestone, Tithonian-

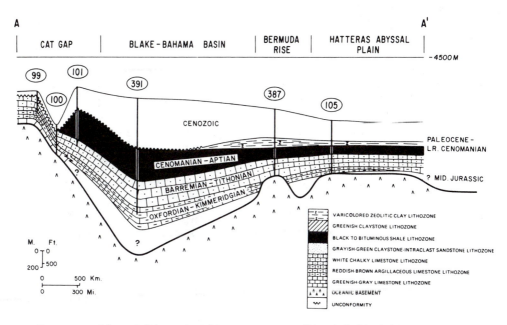

Figure 18-2 Schematic lithostratigraphic cross section parallel with the North American continental margin in the western North Atlantic. DSDP sites are shown. (From Jansa, Gardner, and Dean, 1977)

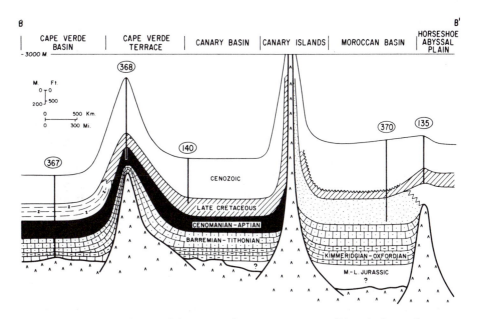

Figure 18-3 Schematic lithostratigraphic cross section parallel with the northwestern African continental margin in the eastern North Atlantic. Legend for lithologies as in Fig. 18-2. (From Jansa, Gardner, and Dean, 1977)

Paleoceanographic and Sediment History of the Ocean Basins   **661**

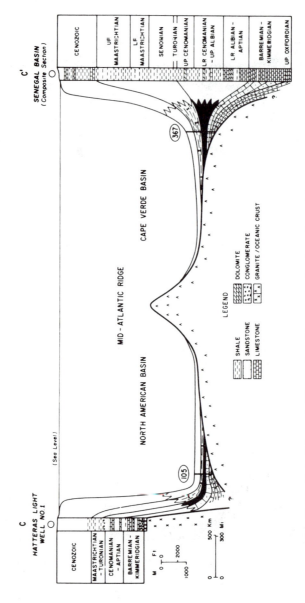

Figure 18-4 Schematic lithostratigraphic cross section from Senegal Basin to Cape Hatteras across the central North Atlantic. Legend for lithologies as in Fig. 18-2. (From Jansa, Gardner, and Dean, 1977)

Hauterivian white chalk, Aptian-Cenomanian black bituminous shale, Valanginian-Albian greenish gray claystone with turbidites, late Cretaceous greenish colored claystone, and late Cenomanian-Maestrichtian to Paleocene varicolored zeolitic clay [Jansa and others, 1978]. These sequences provide the information on North Atlantic paleoceanographic evolution during the Mesozoic.

The earliest stages of development of the North Atlantic during the middle Jurassic are not well understood. Evaporites in the Gulf of Mexico and along the West African and North American continental margins indicate troughs that were isolated from the world ocean by shallow sills. The timing of salt deposition off West Africa and North America is difficult to determine, but probably is early or middle Jurassic in age [Thiede, 1979]. Western Europe was at that time a series of epicontinental seas and land areas [Ziegler, 1975; 1977].

The oldest part of the main North Atlantic basin opened between the late Jurassic and early Cretaceous. It was closed to the north and south, but was probably connected in the southwest to the Pacific and in the northeast to the ancient Tethys Sea [Thiede, 1979]. The sediments deposited during this early stage of the North Atlantic resemble Tethyan Mesozoic lithofacies [Bernoulli, 1972]. Specifically, the pelagic lithofacies provide evidence of a complex late Jurassic and Cretaceous history of circulation, oxygenation, and contribution of organic matter. Resulting sediments consisted of multicolored clays and limestones and of black shales.

The oldest sedimentary rocks recovered by the DSDP in the North American basin are red argillaceous limestones no older than Oxfordian (late Jurassic). Mineralogy and sedimentary structures suggest well-ventilated conditions in deep waters during this early stage of the basin's evolution. Deposition of these limestones took place in a deep, bathyal environment, near but above the CCD. Limestone deposition persisted over wide areas through to the end of the Neocomian. During this interval, the presence of dark, laminated shales indicates brief anoxia of deeper waters, representing the first stage in development towards extensive anoxic conditions.

Black shales (early to middle Cretaceous)

By the early Cretaceous, well-developed anoxic conditions had become widespread, especially between the Barremian and the middle Albian. Anoxic or near-anoxic conditions in the water column were periodic in most of the North Atlantic. At the same time, rudist-coral reefs flourished along the continental margin as far north as Nova Scotia. Black shales were deposited with a pronounced terrigenous imprint including sand, silt, and organic matter derived from land plants. These sequences exhibit significant spatial and temporal variations in sedimentation rate, amount of terrigenous detritus, and amounts and sources of organic matter. Such variations are related to differences in paleodepth, proximity to deltaic influx, and local sea-surface fertility [Arthur, 1979]. Most black-shale sections are characterized by rhythmic sedimentation. Interbedded dark, typically laminated, organic-rich mudstone are lighter colored, and bioturbated, sometimes more calcareous, layers occur with a period of about 50,000 years [Arthur, 1979]. Such rhythmic sedimentation suggests a cyclic climatic control of oxygen depletion, input of terrigenous sediment and organic matter, sea-surface productivity, and carbonate preservation. Although bottom waters became sluggish, the widespread distribution of Tethyan biota throughout the early and middle Cretaceous seaway [Berggren and Hollister, 1974] indicates that barriers to bottom-water circulation did not block circumglobal surface water exchange between the Atlantic, Pacific, and Indian oceans.

Oxygen deficiency intensified and reached a maximum between late Albian and Cenomanian (95–90 Ma), with highest organic-carbon deposition (more than 10 percent) consisting of organic carbon at paleodepths of 3 to 5 km. The black shales are, however, interbedded with oxidized pelagic carbonates and radiolarian sands. This implies overturn of nutrients and increase in sea-surface productivity over much of the North Atlantic with a distinct waning of terrigenous input.

The cause of the anoxic conditions in which the black shales formed is a major and controversial problem. Most workers consider stagnation of the basins to be caused by the formation of circum-Atlantic deep circulation barriers. However, physical restriction was not the only cause, and a combination of factors, including the following, influenced changes in the amount of organic matter preserved in Cretaceous sediments [Arthur, 1979]:

1. Stable salinity stratification due to sinking of dense, saline waters derived from evaporative shelf settings (such as the Gulf of Mexico region and the Florida-Bahamas platform).

2. Mixing of these saline waters with old, already oxygen-depleted deep waters derived from adjoining ocean basins along an equatorial track.

3. Sluggish surface-water circulation and relatively poor exchange of gases between surface waters and atmosphere due to equable climate and high surface temperatures.

4. Possible periodic enhancement of stable salinity stratification by extensive outflow of low-salinity surface waters from delta systems.

All these factors could have contributed to ineffective oxygenation of deep water masses. In addition, a large volume of terrigenous organic matter was fed into the North Atlantic from numerous coastal deltaic complexes. Oxidation of this terrigenous material, coupled with that of marine organic matter derived from phytoplankton productivity, further exhausted the low oxygen content of deep Atlantic waters [Arthur, 1979]. Some workers believe that the sluggish bottom-water circulation and black-shale formation are not just an Atlantic phenomenon, but appear to be global [Schlanger and Jenkyns, 1976; Thiede, 1979]. Hence they may be ultimately caused by large-scale climatic fluctuations rather than local physiographic control.

Reoxygenation and carbonate depletion (late Cretaceous)

The beginning of deposition of vivid, multicolored clay and red clay during the Cenomanian (about 90 Ma) marks the return to continuous oxygenated conditions in deep waters throughout most of the North Atlantic. Several potential new bottom-water sources could have caused the deep circulation rejuvenation. Exchange of deep water with the South Atlantic is one of the most likely causes. Post-Cenomanian sediments are marked by widespread hiatuses, lower rates of sedimentation, or both. Sea-floor erosion resulted from increased bottom-water circulation, while a Cenomanian sea-level rise led to entrapment of terrigenous sediments behind barrier reefs in shallow-shelf seas, reducing sedimentation rates in the deep basins [Arthur, 1979]. The CCD was at a very shallow level until the Maestrichtian and calcium-carbonate deposition was reduced in the multicolored clays. The shallow CCD was caused by high sea levels, favoring accumulations of large amounts of carbonates in the epicontinental seas (for example, the late Cretaceous chalks of northern Europe). This tended to deplete the deep ocean floor of carbonate [Berger, 1970A; Berger and Winterer, 1974].

By Aptian times (110 Ma), a shallow and narrow seaway connected the North and South Atlantic. Hence the Atlantic became a north-south trending ocean basin connected to the Pacific and Tethyan realms. The basin remained closed to the north because the pathway through the Norwegian and Greenland seas had yet to open. Oceanographically, the North Atlantic of the late Cretaceous might have resembled today's northern Indian Ocean with its characteristic mid-water oxygen minimum [Thiede, 1979]. During the Santonian to Maestrichtian (about 80–70 Ma), a deep-water connection between the North and South Atlantic had been firmly established [van Andel and others, 1977; Thiede, 1979].

Latest Cretaceous carbonate sedimentation

During the latest Cretaceous (middle Maestrichtian; 68 Ma) a sharp depression occurred in the CCD to more than about 5 km. This event resulted in basin-wide chalk deposition of Maestrichtian age overlying the carbonate-poor, late-Cretaceous clay deposits. The sudden deepening of the CCD predated the mass plankton extinctions that occurred at the Cretaceous-Tertiary boundary (65 Ma) by several million years. At the end of the Mesozoic, the North Atlantic comprised an ocean basin several thousand kilometers wide. The floors of both eastern and western basins had subsided to a water depth of more than 5 km [Sclater and others, 1977].

Cenozoic History

The major Cenozoic paleoceanographic events that affected sedimentation in the North Atlantic include the breakup and widening of the Norwegian and Greenland seas during the Paleocene to Eocene (Fig. 18–5); the establishment of a deep-water connection to the Arctic Ocean during the Oligocene to Miocene; the

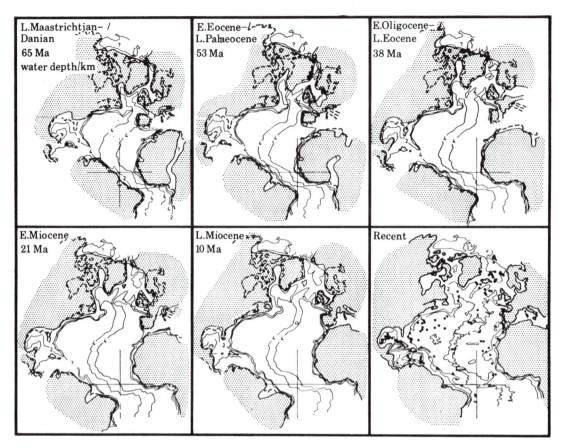

Figure 18-5 Paleobathymetric and paleogeographic evolution of the North Atlantic, the Norwegian-Greenland seas, and their epicontinental seas during the last 65 Ma. The dots in the Recent ocean represent DSDP sites drilled between 1968 and 1977. (From J. Thiede, *Philosophical Transactions of the Royal Society of London*, vol. 294, p. 179, 1980)

closure of the circumglobal Tethys Seaway between Europe and Africa and between the Americas (Fig. 18–5); and major changes of the paleogeography of epicontinental seas in western Europe, western Africa, and in the southern part of North America [Thiede, 1980].

The elusive Paleocene

During the early Paleocene, the North Atlantic consisted of eastern and western basins divided by the mid-ocean ridge (Fig. 18–5). The deep basins extended northward as narrow appendages into the Labrador Sea, which had opened during the late Cretaceous [Laughton, 1975], and into the Rockall Trough, which had developed during the late Jurassic to early Cretaceous [Roberts, 1975]. More is known about the tectonic history than the sedimentary history during the Paleocene, because the Paleocene is poorly represented in deep-sea sequences. Its widespread absence is partly due to formation of unconformities. Paleocene sediments, where present, are largely represented by clays and terrigenous sediments. Increased rates of sedimentation occur during the Paleocene at some locations compared with the underlying multicolored clays. This probably reflects an influx of terrigenous sediments from North America. By late Cretaceous to early Paleocene, continental debris probably topped the shelf-reef barriers along the margin, and sea-level regression allowed sediment dispersal directly to the deep basins, thus developing an early continental rise [Tucholke and Vogt, 1979].

During the late Paleocene to early Eocene, the Labrador basin continued to widen. Spreading began in the Norwegian and Greenland seas, generating the first deep-marine troughs between Scandinavia and Greenland. Late Paleocene to early Eocene marine faunas on Svalbard [Livsic, 1974] indicate the formation, for the first time, of a shallow-water seaway between the Arctic Ocean and the North Atlantic via the Norwegian and Greenland seas [Thiede, 1980].

Paleogene siliceous sedimentation

The Eocene marks a dramatic change in the character of North Atlantic sediment history, because it was at this time that the deposition of siliceous biogenic sediments began. Related formation of cherts gave rise to prominent seismic reflectors, referred to as Horizon A, occurring over much of the North Atlantic (see Chapter 2). The cause of the increased siliceous sedimentation is uncertain. The intermediate level of the CCD certainly enhanced the relative abundance of siliceous debris by reducing carbonate accumulation, but it is also likely that the deep water was generally more silica-rich, allowing preservation of more siliceous microfossils and volcanic debris. The developing abyssal circulation resulting from the opening of the Arctic gateway 50 Ma may have been responsible for increased deposition and preservation of siliceous sediments. This would have stimulated upwelling of nutrient-rich water and enhanced productivity [Berggren and Hollister, 1974], with cooler water originating in an area of active early-Eocene volcanism [Tucholke and Vogt, 1979]. Berggren and Hollister [1974] considered that strong equatorial currents associated with the Tethys enhanced biogenic siliceous productivity in low-latitude areas. Chert layers are less widespread in the deep basins of the western North Atlantic compared with the eastern sector. Their absence was probably caused by erosion of the siliceous sediments by the stronger western boundary currents, before they had time to be transformed into hard chert layers [Lancelot and Seibold, 1978]. Thus Horizon A, which is generally correlated with Eocene cherts in many parts of the world, often corresponds with a hiatus in the North American basin.

Further events in deep circulation

During the Oligocene to earliest Miocene, the abyssal circulation intensified because of global climatic cooling [Shackleton and Kennett, 1975] and active bottom-water circulation from the Norwegian and Greenland seas. A major unconformity encompassing much of the Oligocene, and even older deposits, was created on the continental rise. This erosional unconformity separates hemipelagic clays of lower to middle Miocene from underlying sediments as old as the early Cretaceous.

A major lowering of sea level within the Oligocene (about 30 Ma) [Vail and others, 1977] is not recorded in the basins by any increase in volume of terrigenous deposits. The precise timing of the erosional events are unknown, but they occurred between the late Eocene and early to middle Miocene [Tucholke and Vogt, 1979].

The probable source of the bottom water generating the early Western Boundary Undercurrent was in the subpolar regions of the northern Atlantic Ocean (the Labrador and Norwegian-Greenland seas). The high latitudes were still probably subtropical, so the bottom waters generated there were not as cold or dense as the modern Norwegian Sea overflow or Labrador seawater. Potential bottom-water influx from the Pacific was blocked by the Antilles (and perhaps Panamanian barriers) in the Caribbean, and there was no significant bottom-water contribution from the Tethys [Berggren and Hollister, 1974]. Antarctic bottom water probably did not flow past the Rio Grande rise into the North Atlantic in significant quantities until the Miocene, although the early NADW may have flowed south into the South Atlantic as early as the Eocene [McCoy and Zimmerman, 1977; Tucholke and Vogt, 1979].

Neogene eastern and western basin contrasts

During the Neogene, a sharp change occurred in the depositional regime of the North Atlantic related to further changes in bottom-water circulation. During the middle Miocene there was subsidence of the main platform of the aseismic Iceland-Faeroe ridge and its deep channel (Fig. 18–5). This allowed deep water from the Norwegian and Greenland seas to enter the main North Atlantic basin and the NABW was created, further invigorating bottom circulation [Schnitker, 1980]. Development of this circulation is significant because it currently serves to fractionate chemically the waters of the world ocean. As a result, a disproportionate amount of the world's carbonate sedimentation occurs in the Atlantic and a correspondingly disproportionate accumulation of siliceous sediments occurs in the Pacific. The CCD deepened in the Atlantic at this time, producing more widespread carbonate sediments in the eastern basins. Any increase in carbonate sediment deposition in the western basins was highly diluted by transport of terrigenous debris by the current.

The western basins along the North American margins were dominated by hemipelagic deposition of gray green terrigenous mud through the Neogene and continues to the present. These sediments were laid down under the influence of strong bottom currents derived from high latitudes of the North Atlantic. The thickest accumulations began in the Miocene and were responsible for building the foundations of the continental rises off North America. Mass-flow deposits from the continental shelves and slopes have resulted in massive deposits of late-Cenozoic turbidite sequences forming the abyssal plains.

The Neogene evolution of the western North Atlantic basins differs greatly from the eastern basins, which are sediment starved because of the arid African climate. Aridity in the Sahara region seems to have begun or increased during the early Miocene [Sarnthein and Walger, 1974]. Because terrigenous sedimentation was minimal, carbonate sedimentation was enhanced in the basins. The contrasts between terrigenous, rapid sedimentation in the western basins with carbonate, slower sedimentation in the eastern basins has continued to the present.

By the late Miocene the North Atlantic and the Norwegian seas had attained their approximate present size, shape, and depth (Fig. 18–5). The pathways of the circumequatorial seaway were virtually closed [Thiede, 1980]. The Mediterranean dried out for a brief period during the latest Miocene, forming evaporitic deposits. However, circulation patterns were relatively stable during the Miocene and sediment deposition exhibited no drastic changes in the deeper basins. This changed suddenly again in the Pliocene as a result of two events, which represented the final stages in the paleoceanographic evolution of the North Atlantic to an essentially modern circulation pattern [Berggren and Hollister, 1977]. The elevation of the Isthmus of Panama (about 3 Ma) finally severed the marine connection and faunal interchange between the Atlantic and Pacific oceans, which had persisted since the middle Mesozoic. The deflection of the westward-flowing North Equatorial Current contributed to a more vigorous Gulf Stream. This event was followed shortly afterwards (about 3 Ma) by the initiation of Northern Hemisphere continental glaciation which, by 2.5 Ma, had already developed a sizeable polar ice sheet [Shackleton and Kennett, 1975b]. The cold Labrador Current was formed at this time as a significant water mass and displaced the Gulf Stream to its present position south of latitude 45°N [Berggren and Hollister, 1977]. The North American ice sheet eroded enormous amounts of continental debris, which were transported to the ocean and rapidly accumulated as the continental rise and abyssal plains. Sediment ice-rafting became an important process at high latitudes.

The sedimentary evolution of the Atlantic basin, therefore, has been controlled by the tectonic opening and closing of a number of gateways to surface and bottom waters, as the ocean developed from a narrow, latitudinal, tropical sea to a wide, meridional basin. Thiede [1979B] summarized the sequence of evolving paleoceanographic regimes and the principal tectonic events controlling gateways; see Table 18–1.

SOUTH ATLANTIC

The South Atlantic Ocean began to form during the early Cretaceous, about 130 Ma, and is thus somewhat younger than the North Atlantic. The South Atlantic was completely separate from the North Atlantic during its early evolution. Even after an interoceanic connection was established during the late Cretaceous, much of South Atlantic sedimentary history remained local, superimposed upon changes that have affected all of the ocean basins. Like the North Atlantic, topographic subdivisions led to regional differences in sediment character. The mid-Atlantic ridge and the Rio Grande rise–Walvis ridge system divide the South Atlantic into several distinct sedimentary provinces. Although often breached, the ridge systems form effective barriers to bottom currents and associated sediment transportation and have even led to differences in the depth of the CCD. As in other ocean basins, a considerable part of South Atlantic history has either never been recorded or has

TABLE 18–1 IMPORTANT PALEOTECTONIC AND PALEOCEANOGRAPHIC EVENTS IN THE NORTH ATLANTIC DURING THE MESOZOIC AND CENOZOIC[a]

Age (Ma)	General character	Important epicontinental seas	Opening and closure of pathways for surface water circulation	Opening and closure of pathways for deep water circulation
10–30	Part of a longitudinal ocean system connecting the polar deep-water environments of both hemispheres. Final subsidence of Iceland–Faeroe ridge and disconnection from circumequatorial current system during late Miocene.		Closure of connections to Pacific and Mediterranean. Subsidence of shallowest parts of Iceland–Faeroe ridge.	Closure of pathways to Pacific and Mediterranean. Opening of gap between Spitzbergen and Greenland.
50		Trans-European Sea	Opening of Norwegian and Greenland seas. Surface water exchange with Arctic Ocean.	
65		Trans-Saharan Sea		Opening of deep connection to the South Atlantic.
70–90	Part of a North-south trending Atlantic Ocean, but virtually closed in the north.			
88		Trans-North American Sea		
110			Opening of connection to the South Atlantic.	

670

TABLE 18-1 (continued)

Age (Ma)	General character	Important epicontinental seas	Opening and closure of pathways for surface water circulation	Opening and closure of pathways for deep water circulation
120	East-west trending central North Atlantic basin as part of the Tethys Ocean.			
130				Pathway into the Tethys Ocean.
140		Circum-Fennoscandian Sea		Pathway through the Caribbean in the Pacific.
150			Establishment of marine connection between Gulf of Mexico/ Caribbean land Pacific to NW-European epicontinental seas and Tethys Ocean.	
160	Isolated, narrow marine basin with a restricted depositional environment [evaporites]: Gulf of Mexico, W. African continental margin due to spillover from Pacific, North American continental margin due to spillover from Tethys.			
170				
180				
190				
200				

a From Thiede, 1980.

been subsequently destroyed by erosion or dissolution. Van Andel and others [1977] estimated that at least 40 percent of aggregate time is occupied by hiatuses in South Atlantic sequences.

There are several reviews about South Atlantic paleoceanographic and sedimentary history. These include van Andel and others [1977] and McCoy and Zimmerman [1977]. Other reviews of more regional nature include Supko and Perch-Nielson [1977], Barker, Dalziel and others [1977B], Melguen [1978], and Natland [1978]. The paleogeographic and bathymetric history of the South Atlantic have been reconstructed by Sclater and others [1977B] and Rabinowitz and La Brecque [1979].

Mesozoic History

Early history

Initial rifting in the South Atlantic probably began during the late Jurassic, about 180 Ma. Ocean-floor spreading began much later however, during the early Cretaceous (130 to 140 Ma). The first widespread marine transgression occurred during the late Valanginian over the Agulhas Bank. At this stage, the transgressing marine environment was restricted to the south of the Torres arch–Walvis ridge complex [Rabinowitz and La Brecque, 1979]. North of this complex, thick lacustrine deposits of early Cretaceous to Aptian age are found bordering the margins of eastern Brazil, Gabon, and Angola, indicating that the Torres arch–Walvis ridge formed an effective barrier to northward ocean penetration. Continued early-Cretaceous transgression eventually formed a linear seaway much like the Red Sea, with a maximum depth of 2 km (Fig. 18–6). With the exception of a few aseismic ridges, the entire basin was created by normal sea-floor spreading and has steadily and systematically widened and deepened [van Andel and others, 1977]. Four basins have formed, separated meridionally by the mid-ocean ridge and latitudinally by the Walvis ridge and Rio Grande rise, which at that time were shallower than 1 km. Ridges shallower than 2 km, located in the northern and southern parts, formed effective barriers between the North Atlantic and the Southern Ocean until about 100 Ma [van Andel and others, 1977].

The sedimentary history of the very early stages of South Atlantic development is poorly understood. The oldest marine sediments are of Jurassic age, from the eastern Falkland Plateau, where marine sand and siltstones overlie continental sediments containing lignites. These probably reflect an inland sea from middle to late Jurassic [Barker, Dalziel, and others, 1977B], which later became oceanic. Terrigenous sediments were important during the early stages when the ocean was shallow and narrow and where most areas of the ocean floor were close to continental sources. Carbonate sediments outline the shallower topographic features [McCoy and Zimmerman, 1977]. Oceanic sediments deposited in the earliest rift of the central basin have not been sampled and little is known of pre-Aptian and Aptian pelagic environments. They are, however, assumed to be continental and lacustrine in character [Franks and Nairn, 1973]. Timing of the earliest marine transgressions is not well determined [Nairn and Stehli, 1973].

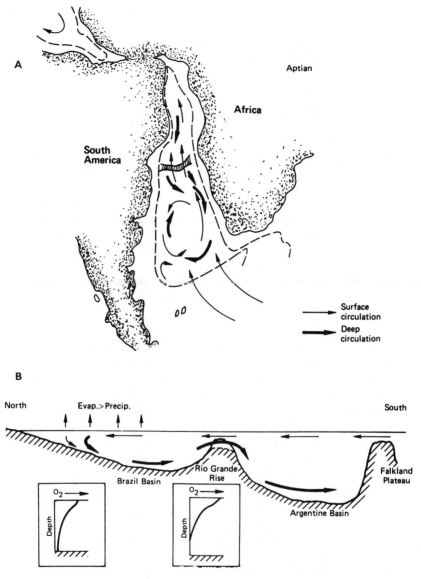

Figure 18-6 Aptian (middle Cretaceous) paleocirculation of the South Atlantic. Cross-hachured stripe in A is the ancient Rio Grande Rise. In B inferred meridional circulation and schematic oxygen profiles are shown. (From F. W. McCoy and H. B. Zimmerman, 1977)

Evaporitic sediment deposition

During most of the Cretaceous, the newly opened basins of the South Atlantic were supplied with ocean water only from the south. Apart from one narrow channel of water stretching to high latitudes (Fig. 18-6), the northern reaches of the

South Atlantic were entirely landlocked in most directions for thousands of kilometers [Natland, 1978]. During the early Cretaceous (Aptian), the eastern tip of the Falkland Plateau (Ewing Bank) had not cleared South Africa, but a narrow channel existed between the Falkland and Agulhas margins (Fig. 18–6). Seawater had access to the South Atlantic only by way of the Agulhas fracture zone [Francheteau and Le Pichon, 1972; Dingle and Scrutton, 1974]. The northern end of the South Atlantic was located within the tropics and the southern end at cool, temperate latitudes. Thus the waters of the Angola and Brazil basins should have been considerably warmer, more saline, and denser than waters of the Cape and Argentine basins. Circulation at this time may have behaved similarly to that between the Red Sea and the Gulf of Aden (Fig. 18–6) [Natland, 1978]. North of the Falkland Plateau sill were four basins. Two southern basins (Cape and Argentine) were divided by the mid-ocean ridge and were further separated from narrow northern basins (the Angola and Brazil) by another west-east sill formed by the Rio Grande rise–Walvis ridge (Fig. 18–6). Thus almost complete isolation existed between the Angola and Brazil basins and the Cape and Argentine basins. Only small amounts of seawater passed north of the obstruction through narrow sills. Landlocked on three sides and blocked by the Rio Grande rise–Walvis ridge, evaporation occurred in excess of precipitation in the arid, equatorial northern basin, resulting in the deposition of massive salt layers (Fig. 18–7) [McCoy and Zimmerman, 1977]. Up to 2000 m of evaporitic sediment deposits are centered at 105 Ma (late Aptian). These deposits currently lie beneath the continental margins of both Brazil and Angola (Fig. 18–7) [Pautot and others, 1973], where the seaward boundaries of the associated diapiric salt fields mark the split by subsequent ocean-floor spreading [Leyden and others, 1976]. Salt deposition lasted only for a few million years and terminated in the early Albian as the South Atlantic circulation became less restricted.

During the 5 m.y. represented by the Aptian stage, an anoxic water mass persisted in the Cape and Argentine basins, while the Angola and Brazil basins were accumulating salts. Very thick deposits of black shales were deposited, intercalated within massive sandstone rich in plant debris [Melguen, 1978]. Shales and sandstones filled the Cape Basin at high rates (60 m/m.y.). Thus, although restricted conditions existed at this time in the Cape Basin, climatic conditions were not warm enough to form evaporitic sediments [Bolli, Ryan, and others, 1978]. A shallow sill at the Agulhas fracture zone maintained these restricted conditions until the end of the early Cretaceous, when anoxic conditions in the Cape Basin ceased.

Middle-to-late Cretaceous anoxic conditions

In the Cape Basin, sedimentation continued throughout the late Cretaceous with the deposition of shales under aerobic conditions but with a shallow CCD [Bolli, Ryan, and others, [1978]. The presence of the Rio Grande rise–Walvis ridge, however, continued to restrict circulation to the north during the Albian to the Coniacian-Santonian (middle-to-late Cretaceous). This led to the deposition of black shales overlying the evaporites of the northern basins. Stagnant conditions were interspersed however, with periods of oxygenation of bottom waters and the

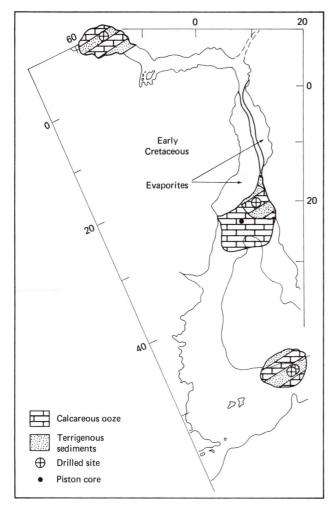

Legend:
- Calcareous ooze
- Terrigenous sediments
- Drilled site
- Piston core

Figure 18-7 Sediment patterns in the South Atlantic during its earliest opening phase in the early Cretaceous. (After F. W. McCoy and H. B. Zimmerman, 1977)

deposition of limestones [Melguen, 1978]. The black shales exhibit lower organic content and the deposits tend to be more pelagic, reflecting developing open-ocean conditions.

North-south connection: 90 Ma

During the Turonian, about 20 m.y. following the deposition of evaporites, the continuing enlargement of the Atlantic eventually led to the permanent establishment of the communication between the South and North Atlantic (Fig. 18-8). This resulted in the disappearance of stagnant bottom waters and the installation of open-ocean conditions during the late Cretaceous. Sedimentation in the deep basins has since been almost entirely pelagic, continuous, and in an increasingly deepening environment as subsidence proceeded. More rapid overturn of nutrients associated with a more vigorous circulation led to increased biogenic

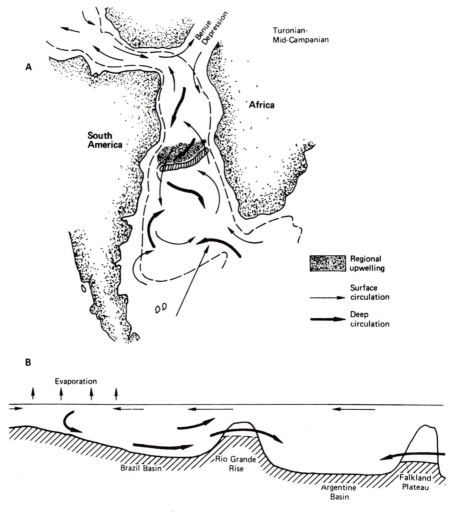

Figure 18-8 Turonian (late Cretaceous) paleocirculation of the South Atlantic. Cross-hachured stripe in A is Rio Grande Rise. In B inferred meridional circulation is shown. (From F. W. McCoy and H. B. Zimmerman, 1977)

productivity and to the deposition of chalks in shallower regions (Fig. 18-9). During the late Cretaceous, terrigenous sediment remained important in the western sector until the end of the Cretaceous (Fig. 18-9), when the terrigenous contribution abruptly decreased. In the eastern sector, the terrigenous influence steadily decreased during the late Cretaceous, reflecting a significant decline in the denudation and erosion of the African continent as the climate became less humid.

As the barriers gradually deepened, increasingly deep circulation developed between individual basins and between the North and South Atlantic. Deeper water circulation between the North and South Atlantic was established by the late Cretaceous (Campanian-Maestrichtian) and by this time passages with depth above 4 km may have connected the Southern Ocean with the North Atlantic through

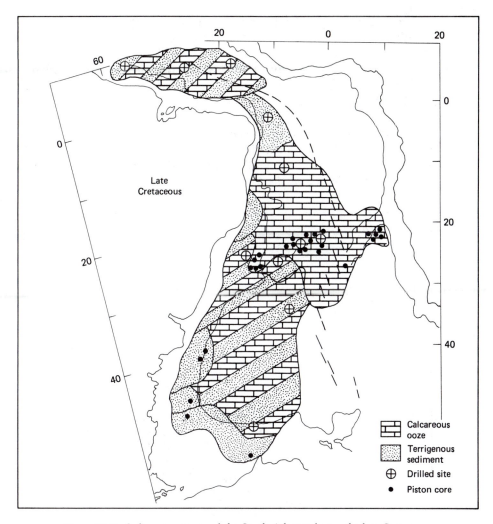

Figure 18-9 Sediment patterns of the South Atlantic during the late Cretaceous. (After F. W. McCoy and H. B. Zimmerman, 1977)

several gaps [van Andel and others, 1977]. However, the subsiding ridges still remained as effective barriers to flow of bottom waters, and deep bottom-water circulation did not develop until the early Cenozoic [Berggren and Hollister, 1977].

Cenozoic History

The end of the Mesozoic (latest Maestrichtian) in the South Atlantic sequence is marked by a major hiatus, or drastic reduction in sedimentation rates, as in other parts of the world ocean. Slower sedimentation resulted partly from a general decrease of terrigenous input, but over wider areas, it was due to a worldwide rise

of the CCD [Worsley, 1974]. Paleocene sediments are usually missing in an unconformity, but where present, they are represented by pelagic clays and carbonates in
the deeper basins. Carbonate sediments continued to outline the Rio Grande
rise–Walvis ridge system (Fig. 18-10), which remained as major east-west barriers
to deep circulation. The first appearance of siliceous biogenic sediments on the
Falkland Plateau and the Rio Grande rise (Fig. 18-10) may have resulted from local
increases in biogenic productivity or may reflect the first, relatively minor signs of
high-latitude upwelling related to climatic cooling. The Paleocene, in many sites in
the South Atlantic, terminates as another hiatus centered near the Paleocene-
Eocene boundary, as in many other parts of the world ocean.

By the beginning of the Cenozoic, the South Atlantic was a wide basin (Fig.

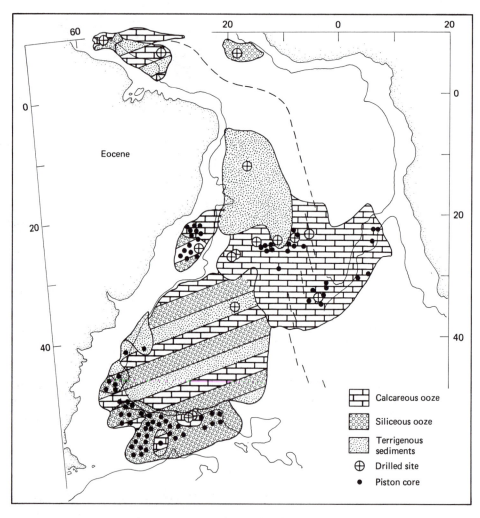

Figure 18-10 Sediment patterns of the South Atlantic during the Eocene. (After
F. W. McCoy and H. B. Zimmerman, 1977)

18–11) and was an integral part of the world ocean. From the early Cenozoic onward, the sedimentary history of the South Atlantic is linked with that of Pacific and Indian oceans [van Andel and others, 1977]. Global processes began to dominate the sedimentary history of both the South and North Atlantic. Such processes include those that control the CCD, the position of the lysocline, surface-water productivity, and the hiatus occurrences, reflecting erosion and dissolution by bottom-water flow [van Andel and others, 1977]. From the early Cenozoic to the present day, climatic evolution in the polar regions has controlled bottom-water movement in the Atlantic Ocean. When related coherent patterns of vigorous deep-

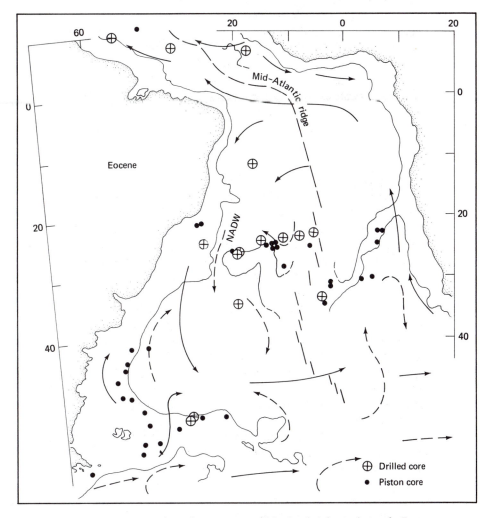

Figure 18–11 Inferred circulation patterns of the South Atlantic during the Eocene. Surface circulation shown as heavy arrows; bottom circulation shown as dashed arrows. (After F. W. McCoy and H. B. Zimmerman, 1977)

water flow in the world ocean developed in the early Cenozoic, a northward route through the South Atlantic was probably not quite established (Fig. 18-11). Surface-water connections were, however, well developed.

Beginnings of siliceous biogenic sedimentation: Eocene

During the Eocene, deposition of terrigenous sediments decreased in the South Atlantic (Fig. 18-10). Despite subsidence of increasingly larger areas of the sea floor below CCD levels, there seems to be no corresponding increase in fine terrigenous sediments as a residual deposit [McCoy and Zimmerman, 1977]. Furthermore, sedimentation rates did not decrease because of progressively important siliceous biogenic sediments related to Antarctic glacial development (Fig. 18-10). In northern South Atlantic areas, calcareous clays were partially replaced by biogenic siliceous sediments. Sedimentation rates of noncalcareous biogenic materials also increased in the Argentine Basin toward the end of the Eocene (Fig. 18-10). These represent accumulations of Antarctic drift sediments—diatomaceous clays transported northward by AABW [van Andel and others, 1977]. This northward transportation was possible because several channels had developed through the fracture zone system associated with the Falkland Plateau system.

Bottom-water activity in middle Cenozoic

As in many other parts of the world ocean, a hiatus at the Eocene-Oligocene boundary is well developed in the South Atlantic [Supko and Perch-Nielsen,1977]. This hiatus is present or inferred in all western Atlantic sites, on the Rio Grande rise, in the Brazil and Angola basins, on the Falkland Plateau, on the São Paulo Plateau, and probably in the Argentine Basin. Although sedimentation is continuous across this interval in Cape Basin sequences, carbonate dissolution increased. This hiatus marks the beginning of a more vigorous deep-water circulation related to cooling of surface waters around Antarctica [Shackleton and Kennett, 1975; Kennett and others, 1975]. The hiatus also includes much of the Oligocene in the Brazil Basin, São Paulo Plateau, and Angola Basin, where it ended in the latest Oligocene or even in the earliest Miocene [Supko and Perch-Nielsen, 1977].

The increase in overturn of the Cenozoic Ocean that began at the end of the Eocene probably increased surface-water fertility in the South Atlantic and caused a rapid drop in the CCD from 3 km in the late Eocene to 4 km in the early Oligocene, and carbonate sediments became dominant [van Andel and others, 1977]. The CCD remained deep until the middle Miocene (about 14 Ma) when it rose sharply to 3.2 km in response to further major paleoceanographic changes in the polar areas [Berger and Von Rad, 1972]. Since then, the CCD has gradually fallen to its present-day depth. During the Oligocene, a complete disappearance of a carbonate component occurred in the Argentine Basin due to subsidence of the basin floor below the CCD and the intrusion of cold AABW northward as far as the Rio Grande rise [McCoy and Zimmerman, 1977]. Despite a major Oligocene regression and a change from warm Eocene climates to cooler Oligocene climates, there is no evidence of an increase in terrigenous sediment influx in the South Atlantic [McCoy and Zimmerman, 1977].

Miocene terrigenous and biosiliceous sedimentation

Two major changes in sedimentation took place during the Miocene in the South Atlantic. These include a renewed influx of terrigenous sediments in the western basins and an intensification of siliceous biogenic sedimentation at high southern latitudes. Carbonate sedimentation began to reestablish its importance in the late middle Miocene as the CCD began to deepen again to its modern depths. First-order topographic features from the equator to the Walvis ridge remained well defined by their cover of carbonate sediments.

An influx of terrigenous sediments into the Argentine and Brazil basins resulted from tectonism in South America. The South American watersheds are placed asymmetrically with the larger drainage centers fronting the Atlantic, so that the volume of terrigenous sediment entering the South Atlantic is disproportionately large [van Andel and others, 1977]. The Amazon became a large and well-established river system by the late Miocene [Damuth and Kumar, 1975]. The Miocene sediment influx is the largest contribution of material from South America during the Cenozoic [McCoy and Zimmerman, 1977] and corresponds to pronounced sediment input elsewhere in the oceans [Davies and Supko, 1973].

Siliceous biogenic sediments expanded northward during the Miocene as the Antarctic water mass developed. By the late Miocene, the Antarctic convergence had migrated to about its present-day position (55°S). As biogenic productivity increased in response to increased upwelling, siliceous biogenic sedimentation rates also increased.

Continued subsidence of the aseismic ridge systems developed additional channels for mid- and deep-water communication between the various basins of the Atlantic (Fig. 17-5). The Hunter Channel, the axis of which lies at about 3700 m, is located between the Rio Grande rise and the mid-Atlantic ridge. The subsidence of this channel to nearly this depth by the early Miocene was important because it then served as a major conduit for the southward flow of NADW. Deep dense AABW was apparently still prevented from flowing to the North Atlantic through the Vema Channel [McCoy and Zimmerman, 1977]. The Falkland Plateau was eroded by vigorous bottom currents associated with the circum-Antarctic Current, which was now well developed, and intensified Southern Ocean currents related to increasing glaciation.

The Pliocene is marked by patterns of sediment distribution similar to the modern South Atlantic. Patterns established during the Miocene lasted through the latest Cenozoic, although the rates of some processes intensified. For example, terrigenous sedimentation rates increased along the east coast of South America from the São Paulo Plateau northward. Coarse, terrigenous material in the Brazil Basin reflects this influx [McCoy and Zimmerman, 1977]. Influx from the Congo River was particularly marked (Melguen, 1978). Much of this terrigenous material was emplaced by turbidity currents flowing down many submarine canyons. Major abyssal plains, such as the Argentine abyssal plain, continued to develop, with the northward-flowing AABW adding a major sediment component. These waters also sculptured extensive areas of the plains and continental margins. The Vema Channel had probably subsided sufficiently to allow northward penetration of AABW to the North Atlantic. Biogenic productivity increased in response to enhanced upwelling along the continental margin extending along the African margin from Walvis Bay to Luanda [Melguen, 1978].

INDIAN OCEAN

The Indian Ocean is the youngest of the three major ocean basins being formed by the separation of Africa, Antarctica, India, and Australia. Its formation began with the breakup of Gondwanaland in the Mesozoic. Because few trenches bound the Indian Ocean, its tectonic history is easily deciphered from mapped fracture zones and marine magnetic anomalies. This history has been outlined in Chapter 6. Norton and Sclater [1979] have identified seven major tectonic events.

1. First break between East and West Gondwanaland in the late Triassic–early Jurassic with initial motion along transform faults parallel to the present east coast of Africa.

2. Early Cretaceous separation of Africa and South America and possibly simultaneous separation between India and Australia-Antarctica.

3. Cessation of motion between Africa and Madagascar.

4. Break between India and Madagascar in the late Cretaceous.

5. Paleocene reorganization in the northwest Indian Ocean, when migration occurred between the Seychelles and India.

6. Eocene separation between Australia and Antarctica, with Australia joining the Indian plate.

7. India's collision with Asia in the early Eocene and subsequent commencement of spreading on the central Indian ridge and later opening of Drake Passage in the middle Cenozoic.

We are concerned here with the history of sedimentation of the Indian Ocean during the late Phanerozoic. Carbonate sedimentation has always played a dominant role, and early stages in the ocean's evolution were marked by more widespread deposition of carbonate because a higher percentage of the ocean floor was above 4000 m and thus above the CCD. As the size of the ocean basins increased and increasing proportions of the ocean floor became deeper, carbonate sedimentation, in general, became more restricted. Estimations by Sclater and others [1971] of the surface area covered by carbonate are 60 percent in the early Cretaceous, 50 percent in the Eocene, 40 percent in the Oligocene, and 60 percent at present. Fluctuations in the importance of carbonate have also depended on the changing level of the CCD [van Andel, 1975; Sclater and others, 1977A]. During the late Jurassic (140 Ma), the CCD was only at about 3.5 km; it became shallower (about 3 km) during the middle Cretaceous (about 100 Ma) and then dropped to 4 km at the end of the Cretaceous. It then remained level before another drop to 4.5 to 5 km, which occurred either once or twice between the late Oligocene and the middle Miocene. It seems that the CCD curve for the Indian Ocean follows closely that of the Pacific but contrasts with the Atlantic [van Andel, 1975; McGowran, 1978]. A similarity with the Pacific and contrast with the Atlantic would support Berger's [1970A] contrast of the Atlantic ("estuarine" deep-water outflow) to Pacific and Indian ("antiestuarine" or "lagoonal" deep-water inflow).

Mesozoic History

The sedimentary history of the Indian Ocean has been reviewed by Davies and Kidd [1977] and Kidd and Davies [1978]. Most of the following summary is extracted from their work. Early Mesozoic sediments have been encountered at very few localities. The late Jurassic has been cored at only one location (DSDP Site 261) northwest of Australia, where sediments are dominated by clays. The separation of India from Australia during the early Cretaceous produced a narrow, restricted basin, or series of basins, in which fine-grained terrigenous sediments were deposited. The presence of significant amounts of pyrite and organic material suggest deposition under anoxic conditions. To the north, siliceous plankton productivity was high in open-ocean conditions.

During the late Cretaceous (70 Ma), sedimentation of detrital and pelagic clays continued in small, developing basins (Fig. 18–12), while carbonates were deposited in the central part of the ocean and on the ridges. Detrital clays rich in pyrite and organic carbon were deposited due to restricted circulation in the semi-isolated Mozambique Basin and off Madagascar. The ocean at this time was bisected by an almost continuous topographic barrier formed by Ninetyeast and Broken ridges between India and Antarctica. The southern end of Ninetyeast ridge, at this time, was marked by volcanic activity, while shallow-water pelagic carbonates were deposited on its northern end.

Cenozoic History

By the early Cenozoic, more open-ocean conditions began to prevail (Fig. 18–13), but tectonic activity and bottom-water erosion created a patchy and discontinuous record, especially in the Eocene and early Oligocene. Earliest Paleocene sediments are rarely identified because of the hiatus at the Cretaceous-Tertiary boundary. Sedimentation in the later Paleocene was dominantly calcareous (with diagenetic cherts) on many of the ridges and shallow plateaus, while pelagic clays were deposited in the deep basins. Terrigenous sediments were deposited off East Africa and the east coast of India.

During the Paleogene, surface-circulation patterns were different from those of the modern Indian Ocean. Because of the southern position of Australia, a major western prolongation of the South Equatorial Pacific Current would have occurred to the north of Australia [Frakes and Kemp, 1972]. The more southern position of India would, at that time, have blocked westward flow of this equatorial

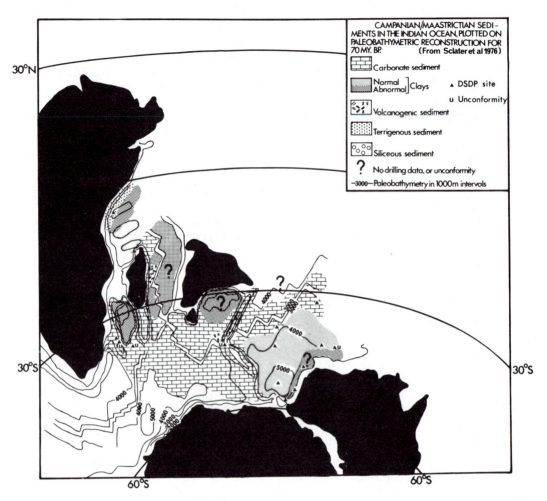

Figure 18–12 Patterns of sediment distribution in the Indian Ocean at 70 Ma (late Cretaceous). (From T. A. Davies and R. B. Kidd, in *Indian Ocean Geology and Biostratigraphy Studies following DSDP Legs 22–29*, p. 45, 1977, copyrighted by the American Geophysical Union)

current system to the western Indian Ocean. The flow would have been deflected into higher latitudes. This current may have winnowed sediments on the Madagascar and Mozambique ridges [Le Claire, 1974].

Early Eocene sedimentation continued much as during the late Cretaceous and Paleocene. An increase in bottom-water erosion occurred at the Paleocene-Eocene boundary, causing an extremely widespread hiatus [McGowran, 1978]. The Wharton Basin, fringed to the north by carbonate platforms associated with the Ninetyeast ridge and a former mid-ocean ridge, was steadily deepening, while remaining open to the Tethys Sea and the Pacific Ocean. Carbonate sediments dominated in the western Indian Ocean in association with the mid-ocean ridge and extensive shallow seas associated with the Chagos-Laccadive ridge and the

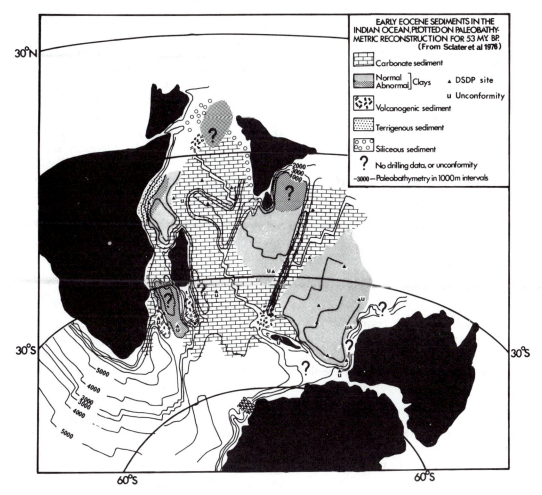

Figure 18-13 Patterns of sediment distribution in the Indian Ocean at 53 Ma (late early Eocene). (From T. A. Davies and R. B. Kidd, in *Indian Ocean Geology and Biostratigraphy Studies following DSDP Legs 22–29*, p. 43, 1977, copyrighted by the American Geophysical Union)

Mascarene Plateau. The northwestern connection to the Tethys Sea continued to become constricted.

The succession on Ninetyeast ridge during the Paleocene to early Eocene suggests deposition in very shallow water, and at one site (DSDP Site 214) nonmarine sediments were found. At Site 214, the oldest shallow-marine sediments are about 58 Ma and the onset of oceanic sedimentation occurred about 50 Ma. Volcanic sedimentary and paleontological data all indicate rapid sinking of Ninetyeast ridge, together with a rapid northward movement during latest Cretaceous and earliest Tertiary times. Pollen assemblages in the early Tertiary nonmarine sediments at Site 214 are similar to assemblages in southern Australia and New Zealand and have been interpreted by Kemp and Harris [1975] to reflect a cool-to-warm temperate climate with high rainfall. Northward movement of the Ninetyeast ridge has transferred this site to present-day tropical latitudes.

By the early Eocene, India had collided with Asia and completed the process of the evolution of the eastern Tethys into the Indian Ocean. This event created the first major influx of terrigenous sediments into the northern Arabian Sea, presumably as a result of the closure of the Indus Trough and the beginning of uplift of the Himalayas. The Indus and Bengal fans began to accumulate.

The early Oligocene is the poorest part of the Paleogene record because of widespread unconformities. Hiatuses occur all around the margins, on most shallow ridges and plateaus, and in the deeper basins. They were created by intensification of bottom currents and of carbonate dissolution and are pronounced in the western Indian Ocean, presumably because of even greater intensification of the western boundary undercurrents. The unconformities seem to be centered in the early Oligocene and end by late Oligocene to early Miocene time. They are not specific to the Indian Ocean, but are widespread throughout the oceans and were created by global intensification of bottom currents.

By the early Oligocene, the Indian Ocean closely resembled its present-day configuration with three distinct regions (Fig. 18-14): an almost totally enclosed northwest region; a central region split by the inverted "Y" of the spreading ridges; and an eastern region dominated by the Wharton Basin and with extensive communication to the Pacific [Kidd and Davies, 1978; Davies and Kidd, 1977]. The topographic resemblance of the Oligocene Indian Ocean to that of the present day led to a similar pattern of sedimentation. Maturity of the ocean basins created more extensive development of the pelagic clays than during any previous period. These are represented by thin, unfossiliferous sequences that accumulated at very slow rates. Also during the Oligocene, an extensive input of terrigenous sediments began on the northern submarine fans, especially the Bengal fan to the northwest.

During the Miocene, there were few major changes in sedimentation throughout the Indian Ocean. Of primary importance, however, was the development of extensive biogenic siliceous sedimentation in the Antarctic sector, as in other parts of the Southern Ocean. A less dramatic increase in biogenic siliceous sedimentation occurred in the equatorial zone of high productivity in the Northwest Indian Ocean. Clay sediments reached their maximum geographic extent in the early Miocene, when the deep basins were at their maximum extent prior to the most recent drop of the CCD to its present depth range [Kidd and Davies, 1978]. As the CCD level dropped, carbonate accumulations became more widespread. Terrigenous sedimentation continued to increase in importance, with increases in sedimentation rates related to increased elevation of the Himalayas, the formation of the Zambesi fan, and continued progradation of the East African margin. Miocene to Recent subduction of the Indian plate under the Eurasian plate created the Sunda–Java Trench. Associated volcanism in the Indonesian Arc caused extensive buildup of volcanogenic sediments during the Pliocene and Quaternary.

Present-day patterns of sedimentation are shown in Fig. 18-15. Carbonate sedimentation has assumed its earlier importance as in the Mesozoic and earliest Cenozoic because of a drop in the CCD to about 4500 m. The CCD in fact may reach 5000 m in the equatorial zone, especially in the western Indian Ocean and

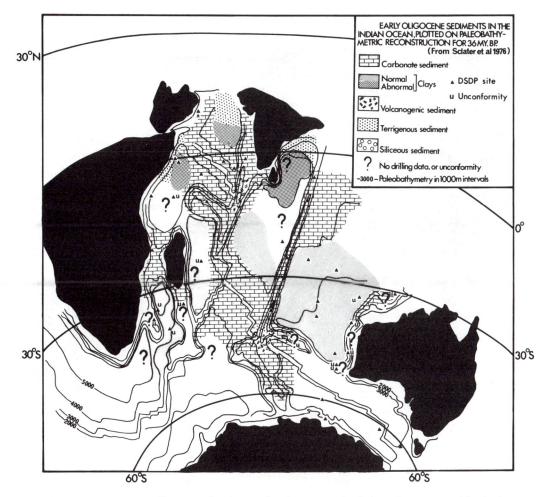

Figure 18-14 Patterns of sediment distribution in the Indian Ocean at 36 Ma (early Oligocene). (From T. A. Davies and R. B. Kidd, in *Indian Ocean Geology and Biostratigraphy Studies following DSDP Legs 22–29*, p. 41, 1977, copyrighted by the American Geophysical Union)

shoals toward the Antarctic, reaching less than 4000 m in the area south of 50°S. Siliceous sediments are poorly represented in low latitudes of the Indian Ocean compared with the Pacific Ocean. Nevertheless, they do occur in the equatorial region of the northeast sector and to the south in the Southern Ocean. Deposition is either minimal or nonexistent in several deeper basins swept by well-developed, northward-flowing bottom currents, such as the Wharton basin, the southern Mascarene basin, and parts of the Crozet and Australian-Antarctic basins. Extensive manganese nodule pavements have developed in the Southwest Indian Ocean [Kennett and Watkins, 1975] as a result of these geostrophic currents. Thick accumulations of terrigenous sediments continue to be deposited in association with the outflows of some of the world's largest rivers in the northern and western parts

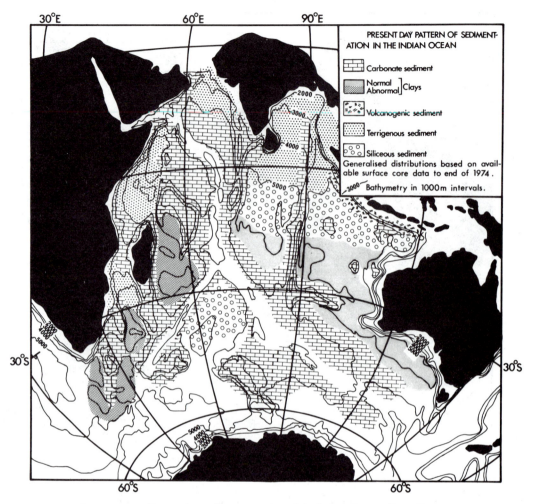

Figure 18-15 Present-day sediment patterns of the Indian Ocean. (From T. A. Davies and R. B. Kidd, in *Indian Ocean Geology and Biostratigraphy Studies following DSDP Legs 22-29*, p. 81, 1977, copyrighted by the American Geophysical Union)

Within the map:

PRESENT DAY PATTERN OF SEDIMENT-ATION IN THE INDIAN OCEAN

Carbonate sediment

Normal] Clays
Abnormal]

Volcanogenic sediment

Terrigenous sediment

Siliceous sediment

Generalised distributions based on available surface core data to end of 1974.

3000 Bathymetry in 1000m intervals.

of the Indian Ocean. No such deposits occur to the east because of Australia's aridity.

OCEANIC HISTORY OF CALCIUM CARBONATE COMPENSATION DEPTH (CCD)

The level of the CCD has changed dramatically through the late Mesozoic to Cenozoic (Fig. 18-16). Oscillations of up to about 2000 m are recorded. These fluctuations in the CCD [Berger and Winterer, 1974; van Andel, 1975] suggest that major changes in carbonate distribution during the Cenozoic are associated with

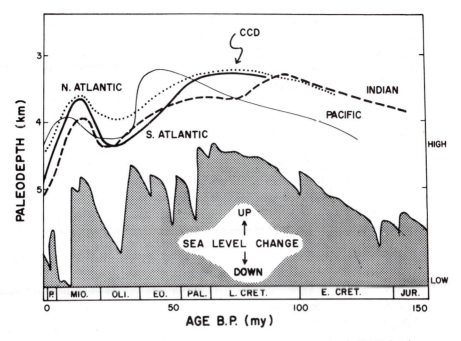

Figure 18-16 Temporal changes in the calcite compensation depth (CCD) for the Indian, Pacific, and Atlantic oceans for the last 150 m.y. (after van Andel, 1975), and eustatic sea-level changes (after Vail and others, 1977). (After Tj. H. van Andel, 1979)

changes in the carbonate dissolution gradient in the deep sea. The general Cenozoic pattern of fluctuations in the CCD [van Andel, 1975] are similar within the different ocean basins and therefore are probably caused by a global mechanism (such as a change in the hypsometry of the ocean basins or a change in supply of carbonate to the oceans). There are, however, ocean-to-ocean differences in this general pattern. These differences may be caused by basin-to-basin fractionation of carbonate [Berger, 1970A] and help determine the history of interbasin exchange of deep and surface waters. Long-term CCD trends have important global implications for paleoclimatic and paleoceanographic history.

Little is known about pre-Cenozoic fluctuations of the CCD. During the late Jurassic, a major shift seems to have occurred in carbonate deposition to the deep sea as the major carbonate-producing planktonic microfossil groups evolved (planktonic foraminiferal; calcareous nannofossils). However, the CCD remained quite shallow (about 3 km) through the Cretaceous [van Andel and others, 1975; Berger, 1976]. At times the CCD was so shallow that only small areas of the ocean floor received carbonate sediments.

During the Eocene, the CCD remained shallow, being close to 3200 m in the Pacific and 3600 m in the Indian Ocean (Fig. 18-16). Toward the end of the Paleogene, the CCD gradually began to deepen in the Atlantic and Indian oceans. About 38 Ma, at the Eocene-Oligocene boundary, it dropped abruptly to about 4500 m in the Pacific region and more gradually in the Indian and Atlantic oceans.

The CCD remained deep during the Oligocene to the early Miocene, until it began to rise to a Neogene shallow peak about 10 to 15 Ma. Following this short-lived early Neogene peak, the CCD descended to its present position (about 4500 to about 4900 m), the deepest ever attained [van Andel, 1975]. In the equatorial Pacific, the CCD behaved in a similar but more extreme fashion. One of the most pronounced differences between the CCD curves of the ocean basins occurred during the middle Miocene. At this time, when the CCD was shallowing in the Indo-Pacific region, it deepened in the North Atlantic, causing more widespread carbonate deposition [Davies and Worsley, 1979].

The global extent of these features suggests a global cause and supports the explanation of Berger [1970A] and Berger and Winterer [1974] that they result from partitioning of carbonate deposition between shallow and deep seas. This suggestion is supported by the parallelism between the CCD curve and changes in eustatic sea level given by Vail and others [1977] (Fig. 18-16). During global marine transgressions, carbonate precipitation was stimulated on the shallow continental shelves and inland seas and became unavailable to the open-ocean cycle. Shelf-sea carbonates subtracted carbonate from the oceans. As a result, the CCD became very shallow to maintain a balance between input and output of carbonate and therefore effectively decreased the area of deposition in the deep oceans. This occurred during the late Mesozoic to Eocene and during the middle Miocene. Conversely, during times of marine regression, abundant carbonate was injected into the oceanic cycle, deepening the CCD. This occurred during the Oligocene and the latest Cenozoic.

Although the sea level–CCD oscillations are generally similar throughout the Cenozoic, they differ in detail, and thus the CCD level must respond to other factors as well as sea level [Arthur, 1979]. For example, changes in CCD seem to be out of phase with each other during the middle to late Miocene in the North Atlantic, South Atlantic, and Pacific (Fig. 18-16). These differences are probably tied to the changes in bottom-water circulation and ocean fractionation between the oceans. These changes, in turn, are due to the progressive glacial evolution of Antarctica and the Arctic through the Cenozoic and to changes in interocean connections. For example, the drastic drop in the CCD at the base of the Oligocene (Fig. 18-16) is not matched by a large sea-level regression [Vail and Hardenbol, 1979], but is probably caused by fundamental changes in the character and turnover of oceanic bottom waters [Kennett, 1978; van Andel and others, 1975]. During significant sea-ice formation around Antarctica, cold AABW began to form, stimulating thermohaline circulation throughout the ocean basins. This decreased the CO_2 buildup in the deep-ocean basins and led to lowering of the CCD. Likewise, the difference in the CCD curve of the North Atlantic and the Pacific (Fig. 18-16) coincided with the development of the NADW as an important component of global bottom-water circulation.

The deepening position of the CCD during the late Cenozoic (last 10 m.y.) corresponds with increasingly rapid rates of sedimentation, which, in the Quaternary, have been almost double the rate of any other Cenozoic interval [Worsley and Davies, 1979]. This reflects lowered sea levels and greater clastic sediment supply from an increasingly glacial world. Superimposed on this broad late Cenozoic drop have been more minor, but nevertheless conspicuous, changes in the CCD as sea level has risen and fallen during interglacial and glacial episodes. Arrhenius [1952] was the first to investigate fluctuating rates of carbonate in the deep sea during the Quaternary. He found distinct oscillations in calcium carbonate content in deep-sea Quaternary cores from the eastern equatorial Pacific. High and low carbonate stages alternate in an orderly fashion, while this cyclicity is also reflected in concentration fluctuations of other sediment components, such as planktonic and benthonic foraminifera, radiolarians, and diatoms. Arrhenius [1952] suggested that glacial strata are rich and interglacial strata are poor in carbonate, a correlation confirmed by the stratigraphy of Hays and others[1969]. Therefore the terms *glacial* and *interglacial* can be used interchangeably with *high-carbonate* and *low-carbonate* episodes in the Pacific (Fig. 18–17). Arrhenius proposed that the differences in carbonate content are due to a greatly increased production rate of calcareous shells during glacial episodes. Berger [1973B] showed, however, that the carbonate cycles were caused by dissolution: the CCD rose during interglacials and fell during glacials, an observation now accepted for much of the Pacific Ocean. Conditions are less simple in the equatorial Atlantic, where it has been shown [Gardner, 1975] that dissolution of carbonate was greater during glacial episodes and less during interglacial episodes (Fig. 18–17). The difference in the equatorial Atlantic seems due to an increase in dissolution effects by corrosive bottom waters during glacial episodes, overriding any change due to sea level.

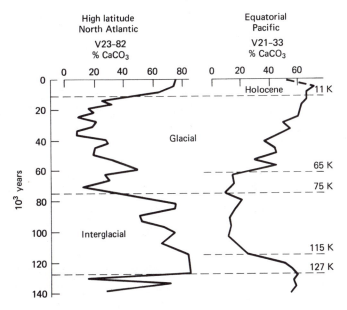

Figure 18-17 Typical percent CaCO₃ curves of the high-latitude North Atlantic (after McIntyre and others, 1972) and of the equatorial Pacific for the late Quaternary. In the North Atlantic, the variations indicate primarily climate-induced changes in the production of CaCO₃. In the equatorial Pacific the variations result primarily from fluctuations in the intensity of carbonate dissolution. Note that changes in dissolution occur after changes in production. (After B. Luz and N. J. Shackleton, 1975)

Any changes in the shape of the ocean basins or in the oceanic circulation patterns can have major regional, or even global, effects on the balance between sediment accumulation and removal. If removal outstrips accumulation, hiatuses may form. Hiatuses are widespread in all facies of the oceans—from the continental shelf to abyssal depths—and prove that dynamic changes have occurred in the oceans over time. The temporal and spatial distribution of hiatuses has been summarized for the Cenozoic deep-drilled sequences by Moore and others [1978]. Several general observations can be made concerning the widespread hiatuses that occur in deep ocean sediments.

1. The distribution of hiatuses has changed over time.

2. Hiatuses occur over a wide depth range in the oceans.

3. The pattern of occurrences for hiatuses in the late Cenozoic follows paths of bottom-water flow and suggests logical bottom-water flow paths for early geologic times within plate-tectonic models.

4. Hiatuses are much more developed on the western side of the ocean basins where strong, western boundary currents prevail. The causes of individual hiatus formation seem to have ceased operating sooner in the western than in the eastern parts of ocean basins.

5. Several major hiatuses in different ocean basins are almost perfectly synchronized.

6. The timing of major changes in the pattern of hiatuses tends to match times of major changes in the configuration of ocean basins, which may have created fundamental changes in circulation.

Figure 18–18 is a plot of the ages from a set of deep-drilled sequences in the Southwest Pacific region. The early Cenozoic is broken up by numerous hiatuses. Two appear to be regional: at the Paleocene-Eocene boundary and the Eocene-Oligocene boundary. Sediment sequences younger than the Oligocene are much more continuous. Compilations of unconformities on an ocean-wide basis by Moore and others [1978] (Fig. 18–19) show the proportion of the stratigraphic record represented by hiatuses. Generally, more than half of the stratigraphic record in deposits older than the Oligocene and as much as 90 percent of the record of early Paleocene deposits is missing. The increasing proportion of hiatuses with age may result from the greater likelihood that an older sedimentary deposit will have suffered some destruction of the section. For all ocean basins, maxima in the abundance of hiatuses occurred at the Cenozoic-Mesozoic boundary, in the late Eocene (spanning the Eocene-Oligocene boundary), and in the middle Miocene to late Miocene. Fewer hiatuses are found in the middle Eocene, early Miocene to middle Miocene, and the Quaternary. It may be significant that the stratigraphic

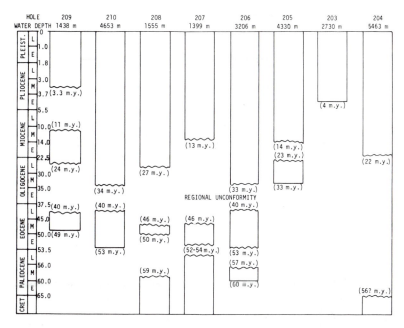

Figure 18-18 Sediment ages in DSDP cores from the Southwest Pacific showing position of unconformities. Note regional unconformity centered within the early Oligocene. (From J. P. Kennett and others, 1975)

boundaries of the Paleocene and Eocene are marked by maxima in hiatus abundance, while Miocene boundaries are marked by minima.

The increase in glacial detritus (both windborne and waterborne) supplied to the sea floor may be responsible for the Quaternary minimum in hiatus abundance. Or perhaps the coarse resolution of stratigraphies used could be responsible, as well as the fact that a hiatus may now be in the process of forming. The most common duration of a hiatus found in the DSDP sites is 4 m.y., which means that the peak at 3 Ma might be just the start of another major phase of widespread hiatus formation. Certainly silica and carbonate are being dissolved on the sea floor, and lateral movement and reworking of sediments is occurring. Also, when more detailed stratigraphies are applied to Quaternary sections, short hiatuses are found in some areas of the ocean, although only global changes in hiatus occurrence markedly affect the average global curve (Fig. 18–19). Thus, in the Atlantic and Indian oceans, a distinct hiatus maximum is found at about 15 m.y., but on the global average curve (Fig. 18–19), this is counteracted by a minimum in hiatus abundance in the eastern Pacific and Southern oceans.

The most prevalent and well-defined maxima of hiatus abundance probably shows major changes in oceanic configuration, affecting not only bottom-water circulation, but also circulation of surface and intermediate waters. For instance, the peak in hiatus abundance at the Cretaceous-Tertiary boundary coincides with a time of rapid opening for the Atlantic and possibly the closure of free circulation

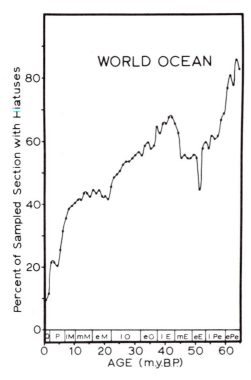

Figure 18–19 Proportion of sampled section represented by hiatuses (unconformities) as a function of time in the world ocean. (From T. C. Moore, Jr., Tj. H. van Andel, C. Sancetta, and N. Pisias, *Micropaleontology*, vol. 24, p. 118, 1978)

between the Arctic and the Pacific. Also, the development of the Tasmanian Seaway is associated with the Eocene-Oligocene boundary peak in hiatus abundance.

nineteen
Global Paleoceanographic
Evolution: Critical Events
in Ocean History

We shall not cease from exploration
And the end of all our exploring
Will be to arrive where we started
And know the place for the first time.

T. S. Eliot

INTRODUCTION

In the last chapter we briefly examined the paleoceanographic evolution and associated sedimentary history of individual ocean basins. In this chapter we shall summarize paleoceanographic evolution on the global scale. Paleoceanography is a very young scientific endeavor [Schopf, 1980], and the summary provided here represents but a beginning of this work. Many of the ideas presented are speculative, and many of the problems remain unsolved. Nevertheless, the sequence of changes provides a framework for future studies of the paleoenvironmental evolution of the oceans. It is clear that in many respects the global ocean acts as a single entity: Changes occurring in a single region can affect the overall condition of the ocean. Other local perturbations may have little or no effect on the ocean as a whole.

In Chapter 17, a general scenario of the history of late Phanerozoic paleo-ceanographic change was outlined. One of the most pronounced characteristics of this change was a gradual cooling in the world's high-latitude oceans and land-masses. During much of the Cretaceous period, warm, equable climates existed with low equator-to-pole temperature gradients. At that time, mean annual temperatures must have been near or above freezing point at the poles. These conditions were gradually replaced through the Cenozoic by a cooling of the higher latitudes, and ultimately by the highly variable, glacial-mode climatic conditions that mark the Quaternary period. On the time scale during which this change has occurred (about 100 m.y.), the most dramatic associated factors have been the changing geographic distribution of landmasses and changes in the dimensions of the land surface caused by eustatic sea-level fluctuations. It is apparent from a qualitative viewpoint that these paleogeographic changes must have been a major cause for climatic change. The unequal heating of the earth, coupled with its rotation, are the driving forces for circulation in the atmosphere and oceans. Heating varies with latitude, but it also exhibits important longitudinal gradients in relation to the distribution of land and sea. The Southern Hemisphere Ocean, for instance, is highly zonal in character compared to the Northern Hemisphere Ocean because its flow path is not interrupted by continents.

A major problem of paleoceanography and paleoclimatology is to evaluate how much climatic change is caused by paleogeographic alteration and to determine the extent to which it is responsible for the pronounced global cooling of the Cenozoic [Barron and Thompson, 1980]. A second major problem is to determine which elements of the climatic system (the lithosphere, cryosphere, hydrosphere, and atmosphere) are the principal forcing functions in creating the changes. That is, what are the mechanisms by which climatic change is affected? Different workers place varying stress upon different components. Donn and Shaw [1977] were among the first workers to examine quantitatively the effects on climate of changing land and sea distributions due to continental drift. They considered that the decline in late Phanerozoic temperatures resulted from increased continentality of the Northern Hemisphere, which had the effect of increasing the albedo (or solar reflectivity) at the earth's surface, in addition to leading to the buildup of snow and ice as the continents moved into higher latitudes. Because of contrasts of albedo between ocean and land and between vegetated and arid land, paleogeographic changes modify the planetary albedo. The initial results of an analysis of surface albedo as a function of land-sea fraction and incoming solar radiation indicates that paleogeography has a large potential for changing the heat budget, producing a *maximum* change of 2 percent in surface albedo when no atmospheric feedbacks are considered. It is especially significant that the major change in land area between 100 Ma and the present occurs as an increase in the subtropical regions (Fig. 19-1).

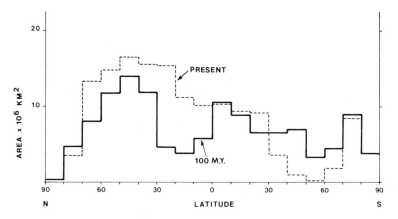

Figure 19-1   The distribution of land area at 100 Ma (middle Cretaceous) and at present. (From E. J. Barron and S. L. Thompson, 1980)

The arid regions of the subtropics have a high albedo and because the low latitudes receive a large solar influx, this change in paleogeography is an important component of the global cooling trend [Barron and Thompson, 1980].

Figure 19-2 shows a comparison of global albedo distribution (assuming 50 percent cloud cover) between the present day and the Cretaceous. The geographic variations in incoming and outgoing solar energy are balanced by adjusting the radiative processes by oceanic and atmospheric transportation. This is the basic assumption of simple energy-balance climate models in which the energy influx at a particular latitude must be balanced by the flux out of the area by water vapor in the atmosphere and by heat transported by the atmosphere and ocean. As the geography changed through time, this transport has changed by modifying the amplitude of the seasonal temperature cycle in the equator-to-pole gradients. Thus, in order to understand paleoceanographic evolution, it is necessary to consider changes in surface albedo and also those factors that modify heat transport around the globe. These clearly include changes in both atmospheric and oceanic circulation, as well as continental evolution. For instance, large-scale crustal warping (**epeirogeny**) and mountain building (**orogeny**) have also influenced paleoclimatic evolution [Damon, 1968]. During times of maximum transgression of epicontinental seas onto the continents, mild, uniform climates prevailed; during times of maximum regression, cooler, more varied climates prevailed. Because orogeny continued during both transgressive and regressive cycles, it is not by itself sufficient to cause an onset of global cooling or warming. On the other hand, continental glaciation can develop when an orogenic phase coincides with regression of continental

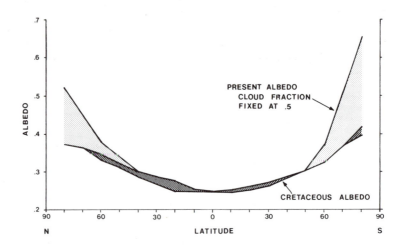

Figure 19-2 Comparison of latitudinal albedo gradients between the present day and Cretaceous. The graph is separated into two components: (a) that fraction of the change resulting from the assumed temperature contrast between the Cretaceous and the present (light shading); (b) that fraction resulting only from changes in the land-ocean proportions at the various latitudes. (From E. J. Barron and S. L. Thompson, 1980)

seas, such as during the late Cenozoic, the late Paleozoic, and the late Precambrian [Damon, 1968].

There may have been a general cyclicity in paleoceanographic change during the last 200 m.y. Two major oceanographic modes or states have alternated—an **oligotaxic** mode and a **polytaxic** mode [Fischer and Arthur, 1977]. The oligotaxic mode is illustrated by Quaternary oceanic conditions, which are marked by cool waters and relatively steep environmental gradients between the equator and the poles and in the water column. The CCD is relatively deep at these times. Environmental extremes create an increase in circulation rates and more oxygenation of ocean waters. In addition, because bottom currents are more vigorous, hiatuses tend to be more widespread. Oligotaxic modes are also marked by increased rates of extinction of planktonic and nektonic organisms, reduced complexity of biotic communities, and reduced biotic diversity on a global scale. It may be of great significance that oligotaxic modes are often associated with prolonged periods of sea-level regression.

In contrast, the polytaxic mode is marked by more diverse and complex biotic communities. Sea-surface temperatures were warmer at high latitudes than the Quaternary and thus gentler temperature gradients existed between the equator and the poles and in the water column. As a result, rates of oceanic circulation are reduced, including the vigor of deep-sea currents. The oxygen-minimum layer was intensified and expanded, creating widespread sedimentation of organic-rich sediments. The CCD was shallower. Polytaxic modes have normally been associated with marine transgression.

Fischer and Arthur [1977] considered that oligotaxy recurs rhythmically at intervals of 32 m.y. The episodicity of distinct polytaxic modes are less well defined. Oligotaxic episodes were centered at the following times: 222 m.y. (Permotriassic boundary); 190 m.y. (Triassic-Jurassic boundary); 158 m.y. (Bathonian-Callovian boundary); 126 m.y. (early Neocomian); 94 m.y. (Cenomanian); 62 m.y. (latest Cretaceous–earliest Tertiary); 30 m.y. (mid-Oligocene); and during the last 2 m.y. (Quaternary). Before the work of Fischer and Arthur [1977], other workers had recognized cycles that roughly matched their suggested periodicity. Dorman [1968] suggested a 30–m.y. temperature cycle in the oxygen isotopic record, while Damon [1971] suggested a 36–m.y. periodicity for sea-level regressions. The recognition of such cycles may be an oversimplification, but the observations have assisted in recognizing distinct oceanic modes in the geologic past. If such cycles do exist in the geologic record, a major problem concerns their ultimate cause, because it is not apparent that they are caused by plate-tectonic evolution, although this must influence the expression of such modes. Perhaps a major cause has been changes in sea level. Higher sea levels have tended to ameliorate climates because of water's relatively high thermal inertia; the development of intracontinental seaways provided additional avenues for marine circulation and the reduction in albedo as discussed earlier. The ultimate cause of sea-level change is, however, still not satisfactorily explained (see Chapter 9), although Fischer and Arthur [1977] speculated that global tectonic processes may have influenced sea levels by changing the earth's configuration and may simultaneously have affected the atmosphere—and therefore climates—by volcanism.

PALEOZOIC PALEOCEANOGRAPHY:
ANCIENT CONTINENTAL BITS AND PIECES

The distribution of continents and oceans during much of the Paleozoic and earlier times is far from clear. It is generally accepted, however, that plate-tectonic processes were operating during the Paleozoic and probably in some form even during the Proterozoic as well [Burke and Dewey, 1973]. Continental drift was a continuing process and global geography constantly changed. Maps of the Paleozoic world are quite unlike those of modern times. A milestone toward an understanding of Paleozoic continental and oceanic distributions was a series of maps based largely upon paleomagnetic reconstructions [Smith and others, 1973]. Later maps have become more refined by the addition of new paleomagnetic data, as well as of paleoclimatic information [Zonenshayn and Gorodnitskiy, 1977], of biogeographic data [Ziegler and others, 1977; 1979], and of tectonic data [Morel and Irving, 1978].

The plate-tectonic history of the Paleozoic is largely that of drift of continents initially disseminated in low latitudes, then gradually moving together to form a continuous barrier from pole to pole. Thus geographic relationships existed in the Paleozoic that have not been repeated since. Ziegler and others [1979] have produced a set of paleogeographic reconstructions for a number of selected intervals during the Paleozoic. They have recognized six major paleocontinents differing in position, size, and shape from those that developed during the Mesozoic and Cenozoic. These have been called Gondwanaland, Laurentia, Baltica, Siberia, Kazakhstania, and China. Laurentia, for example, consisted of nuclear North America, in addition to Greenland, Scotland, Spitzbergen, and parts of eastern Siberia, but not Florida and Avalonia (Nova Scotia and Newfoundland). Gondwanaland, the largest and best-known paleocontinent, consisted of Antarctica, Australia, Africa, South America, India, Arabia, and Madagasca, as well as New Zealand, Florida, southern and central Europe, Turkey, Iran, Afghanistan, and Tibet. Together these made up more than 50 percent of the continental area [Scotese and others, 1979]. The elements that made up the other Paleozoic continents and their changes were described by Ziegler and others [1979]. Gondwanaland formed near the end of the Precambrian and remained nearly intact during the Paleozoic and early Mesozoic. However, during the late Paleozoic and early Triassic, continental fragments—such as Tibet-Iran-Turkey and Malaysia—were rifted from the northern parts of Gondwanaland eventually to become parts of Asia, Europe, and even North America. The paleoenvironmental and paleontological history associated with these continental blocks has thus been complex and variously related either to Gondwanaland or the northern continents.

By the early Carboniferous, coal swamps and glaciers had become well developed and, together with the evaporitic and carbonate deposits, provide valuable criteria for testing the accuracy of the paleogeographic reconstructions. In general, the coal deposits were located in low-latitude areas on the windward side of the dispersed continental masses. Tillites occur at high paleolatitudes in South American and South African parts of Gondwanaland on the western side of the supercontinent, where westerlies provided the necessary precipitation for ice accumulation. Carbonates and evaporites are confined to low-latitude regions. Continental configurations of the Permian were not significantly different from those of the late Carboniferous; however, paleoclimatic conditions began to change drastically, limiting coal deposits to temperate areas and restricting late Paleozoic ice sheets to eastern Australia by the late Permian.

LATE PALEOZOIC SUPEROCEAN AND SUPERCONTINENT

By 240 Ma, in the late Permian, the continents were largely grouped together as Pangaea (see Chapters 4 and 6). As a result, the world ocean, Panthalassa, was truly enormous, spanning the globe from pole to pole and extending for 300° of longitude at the equator [Scotese and others, 1979; Bambach and others, 1980], twice the width of the modern Pacific Ocean (Fig. 19-3). Given this size and

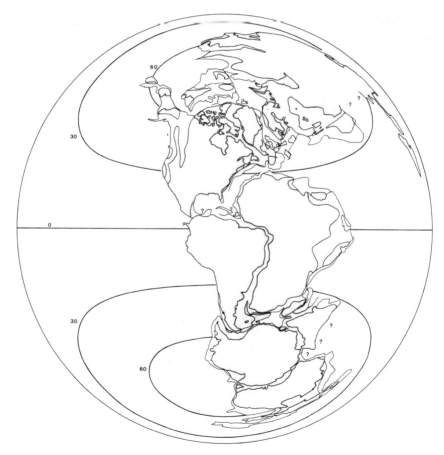

Figure 19-3 Paleogeographic reconstruction at 160 Ma (Jurassic). (From Barron, Sloan, and Harrison, *Paleo. Paleo. Paleo.*, vol. 30, pp. 17–40, 1980)

geometry, it is probable that the paleocirculation in Panthalassa was much more simple than that of the modern ocean. Enormous, single-circulatory gyres would have existed in both the Northern Hemisphere and the Southern Hemisphere. Equatorial currents driven by the trade winds would have flowed uninterrupted around 85 percent of the earth's circumference. East-to-west paleoceanographic extremes were probably never larger than during the time when Panthalassa exhibited its largest dimensions. The western margin of this superocean must have been very warm in contrast to its eastern sector. In contrast, it is probable that rather low thermal contrasts existed between the equator and the poles, forming rather sluggish meridional circulation. Little biotic provincialization existed in the marine assemblages, but distinct tropical and bipolar (boreal) realms existed [Arkell, 1956].

The separation of Pangaea into Laurasia and Gondwanaland during the middle Jurassic to late Jurassic (Fig. 19–3) formed the Tethys Seaway, which allowed the westward equatorial current of Panthalassa to become a globe-encircling flow. The few late Jurassic sediments available from the oceans indicate sedimentation generally comparable to that of the modern low-latitude oceans, with calcareous and siliceous sediments (chalks, limestones, and cherts) in the Pacific and clays, marls, and siliceous sediments that accumulated under oxidizing bottom conditions in the Atlantic Gulf.

TRANSITION FROM A SUPEROCEAN TO FRAGMENTED OCEAN

During the most recent quarter billion years of geologic time, the most fundamental change in global geography has been the progressive dismemberment of Pangaea, the formation of the Atlantic and Indian oceans, and the collision of some of the fragments of Pangaea to form the Afro-Eurasian landmass. Separate circulation systems unfolded in each of the developing ocean basins, although the various components of the world-ocean system nearly always remained in partial communication.

During the early Mesozoic, oceanic circulation was highly restricted in the beginning stages of opening of the Atlantic Ocean (see Chapter 8). Partial barriers occurred between several shallow Atlantic basins, and connections with the world ocean were limited. Conditions were suitable for evaporitic deposition, which occurred in the North Atlantic and Gulf of Mexico during the late Triassic to middle Jurassic. In the South Atlantic, which began to open later than the North Atlantic, evaporite sequences were deposited in the early Cretaceous. The only connection that the South Atlantic had with the Indian-Pacific Ocean at that time was across a shallow barrier formed by the Falkland Plateau.

By the Cretaceous, a two-gyre system had been established in the North Pacific and a single major gyre in the widening North Atlantic [Luyendyk and others, 1972]. Paleoceanographic conditions of the Cretaceous Ocean were considerably different than the modern ocean. The fundamental difference was its general warmth and equability. More efficient global heat absorption resulted from lower albedo on a global scale because of high sea level and an absence of ice at high latitudes. Oxygen isotopic data indicate low thermal gradients [Emiliani, 1961; Bowen, 1966; Saito and van Donk, 1974; Douglas and Savin, 1975]; deep oceanic water may not have been much colder than surface waters, although few data exist. If vertical temperature gradients were very small, the density of oceanic waters would have been controlled more by salinity than by temperature differences. The low thermal gradients and resulting sluggishness of ocean mixing would have allowed a greater chemical fractionation within the oceans. Slow renewal of oxygenated waters in parts of the deep ocean led, at times, to the preservation of large amounts of organic carbon in the sedimentary record.

Because no ice existed at the poles, albedo would have been rather low, providing high heat retention, especially during summers when total daily radiation is relatively high in polar areas. This would have caused increased seasonality at high latitudes. Conditions in the Cretaceous Ocean differed drastically from those of the modern ocean, which is more highly thermally stratified, consists of a warm surface layer and an underlying, enormous, cold, deep layer, and exhibits much more vigorous meridional deep-water circulation. In the modern ocean, deep waters are mainly replenished in the cold polar areas, which are ice covered and thus exhibit high albedo and lower seasonality.

The low environmental gradients of the Cretaceous permitted broad distribution of planktonic assemblages, but provinciality was markedly increasing in shallow-water marine assemblages due to increased oceanic compartmentalization [Valentine, 1973]. Paleotemperatures were probably warmer during the middle Cretaceous than any later time. Oxygen isotopic data indicate a general temperature decline since about 100 Ma [Douglas and Savin, 1975].

MESOZOIC OXYGEN-DEFICIENT OCEANS

One of the most striking differences between the Mesozoic and modern oceans is the presence in the Mesozoic of episodes when the oceans were relatively depleted in oxygen, so that deposition of organic matter became widespread. These intervals are the **oceanic anoxic events** and are marked by deposition of sequences of black, organic-rich (1–30%), laminated muds and shales, separated by well-oxygenated biogenic sedimentary sequences (see Chapter 14). During the oceanic anoxic events, carbonaceous sediments were deposited over a wide range of paleotectonic and paleogeographic settings within the oceans. They have been found in the North and South Atlantic oceans, the Indian Ocean, and even on aseismic ridges in the Pacific Ocean (Fig. 19-4). Widespread deposits occur in the late Aptian, Albian, parts of the Cenomanian to Turonian, and to a lesser extent, Coniacian to Santonian [Schlanger and Jenkyns, 1976]. Widespread anoxia in the oceans and epicontinental seas at these times is also supported by lack of benthonic microfaunas and depleted macrofaunal associations [Kauffman, 1967]. These times are also marked

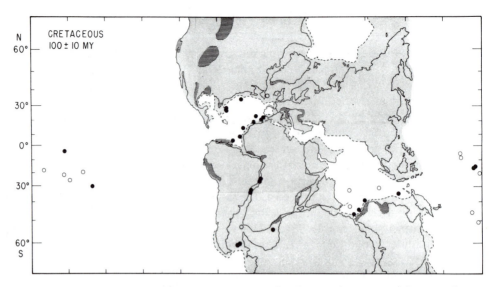

Figure 19-4 Occurrence of black, carbonaceous-rich sediments of 100 m.y. age (Aptian–Albian Stages) in drilled sites shown by black circles. Sites lacking such sediments are shown as open circles. Occurrences of black, organic-rich sediments are shown on shaded continents. (From A. G. Fischer and M. A. Arthur, Secular Variations in the Pelagic Realm, Society of Economic Paleontologists and Mineralogists, Spec. Publ. No. 25, p. 32, 1977)

by little or no erosion by bottom currents. Episodes of anoxia are also globally widespread at older times, including parts of the Ordovician, Silurian, and Devonian and during the early Jurassic [Hallam, 1975; Berry and Wilde, 1978].

The widespread occurrence of the carbonaceous deposits has led numbers of workers to speculate on their origin and possible global significance [Schlanger and Jenkyns, 1976; Ryan and Cita, 1977; Fischer and Arthur, 1977] (see Chapter 14). Other workers have stressed hypotheses related to conditions within individual, partly isolated ocean basins [Lancelot and others, 1972; McCoy and Zimmerman, 1977; Thiede and van Andel, 1977; Natland, 1978]. There is, however, no consensus as to the exact mechanisms involved in the creation of the oceanic anoxic events. However, it is clear that the primary prerequisite for the preservation of unusual amounts of organic matter in deep-sea sediments is stable water-mass stratification. This is normally accomplished by increasing the density stratification by a strong influx of heavy (very saline, very cold, or both) water, which isolates deep waters from overlying shallower waters. Similarly, a large influx of freshwater or very warm water will isolate the upper ocean layer from deeper water masses [Thierstein and Berger, 1979]. The buildup of organic carbon in the oceans is also enhanced by increased rates of input and burial of organic material of either marine or terrigenous origin.

Development of oxygen deficiencies in the global ocean during the early and middle Cretaceous is probably due to high temperatures and low-latitude temperature gradients, consequent low solubility of oxygen, and to the periodic stabilizing effect of high salinity contrasts between surface and bottom waters [Arthur, 1979]. The two most widely discussed hypotheses employed to account for anoxia are an expansion of the oceanic oxygen minimum layer [Schlanger and Jenkyns, 1976; Fischer and Arthur, 1977; Thiede and van Andel, 1977] and stagnation of all the ocean basins [Ryan and Cita, 1977; Natland, 1978; Arthur and Natland, 1979]. Those supporting the first theory believe that the carbonaceous deposits are restricted to a paleodepth range of several hundred meters below the surface to 3000 m in the South Atlantic and formed where the sea floor intersects a strong oxygen-minimum layer [van Andel, 1975]. Thus a completely stagnant basin is not required. Thierstein and Berger [1979] also believe that abyssal anaerobism may have been amplified (with increased water-column stability) by the injection of brines derived from the South Atlantic evaporitic basins during its early opening stages (Aptian stage). These brine injections reinforced haline stratification in a global ocean already highly salinity stratified. Possibly of additional fundamental importance in the formation of oceanic anoxic events is their apparent correlation with periods of late but not peak marine transgressions and early regression. Fischer and Arthur [1977] have correlated the widespread Aptian-Albian black clays (Fig. 19–5) with polytaxic episodes of global proportions, marked by maximum biotic diversity and warmer and more uniform oceanic temperatures. Conversely, oxygenated deep-ocean intervals were marked by low carbon accumulations such as during the Campanian to Maestrichtian times. These times were associated with global climatic cooling, eustatic lowering of sea level, epicontinental regression, and the stimulation of oceanic circulation—the oligotaxic episodes of global proportions [Fischer and Arthur, 1977]. However, the polytaxic-oligotaxic cycles do not by themselves adequately explain the variable stratigraphic extent of the black clays in the various ocean basins (Fig. 19–5). Tucholke and Vogt [1979] believe instead that their stratigraphic distribution is related primarily to tectonically controlled paleocirculation history of individual basins.

From the early part of late Cretaceous (about 90 Ma) onward, the deposition of anoxic sediments became rare. It is possible that reduced oceanic anoxia resulted from an increase in open-ocean circulation resulting from the steadily enlarging ocean basins, perhaps a reduction in sea level, and increased stimulation of oceanic circulation due to climatic cooling episodes. Whatever the cause it seems that the present abundance of highly oxidized red sediment is not typical of much of the Cretaceous period.

MESOZOIC ORIGINS OF BIOGENIC SEDIMENT BUILDERS

Of major concern in marine geology is the timing of origins of the primary biogenic sediment-building planktonic microplankton. The character of oceanic sedimentation changed drastically when each of the four major groups developed: planktonic foraminifera, calcareous nannofossils, diatoms, and radiolarians. One of the most distinctive characteristics of the Cretaceous oceans was the major evolutionary radiation within all of these groups, creating fundamental changes in the structure

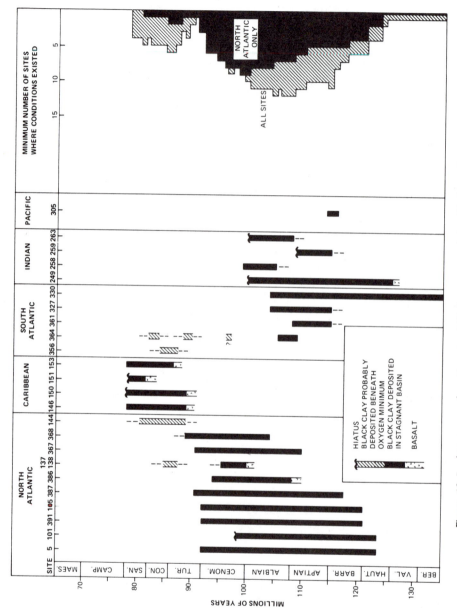

Figure 19-5 Age range of sections in drilled cores containing black, carbonaceous-rich sediments. Black bars represent inferred deposition in anoxic basins; shaded bars represent inferred deposition, probably in oxygen minimum zone at intermediate depths as diagnosed by Thiede and van Andel [1977]. Dashed lines indicate uncertain limits on extent of black clays. (From B. Tucholke and P. Vogt, 1979)

of pelagic communities. Rootstocks of several important microfossil groups had already developed during the Jurassic and formed the basis for this evolution. The ancestry of most modern biotas is to be found within these radiations. During the Triassic, microplankton diversity was low. This was followed in the Jurassic by the initial development of coccolithophorids and planktonic foraminifera [Lipps, 1970B]. These groups continued to diversify into the Cretaceous, at which time they exhibited large evolutionary radiations. At that time, the coccolithophorids and planktonic foraminifera became productive enough to become the important sediment builders over wide areas of the ocean. Diatoms and silicoflagellates first appeared somewhat later, during the late Cretaceous, adding another two siliceous-biogenic sedimentary components to oceanic sediments. This had been previously represented only by the radiolaria, which have an uninterrupted fossil record from the Ordovician period. Thus the middle Mesozoic to late Mesozoic is both the time of development and expansion of the deep-sea carbonate sedimentary facies and the expansion, slightly later, of the siliceous-biogenic facies as the diatoms began to radiate. The Cretaceous, for the first time, marks the dual distribution of carbonate and siliceous biogenic sediments over large oceanic areas.

During the Cretaceous the ground rules of preservation of calcareous microfossils in deep-sea sediments were also different from the rules of cold Cenozoic oceans. The chemistry of the Cretaceous oceans is unfamiliar [Thierstein and Berger, 1978]. We have already seen that the oxygen content was often much lower in the Cretaceous oceans compared with the Cenozoic. Small changes in oxygen levels should have had pronounced effects upon preservation, which is observed in Cretaceous deep-sea sequences [Thierstein and Berger, 1978].

Zooplankton diversity generally correlates with that of the phytoplankton in the modern ocean, and these two groups have had similar geologic histories. The late Paleozoic and early Mesozoic radiolaria exhibit a decrease in diversity followed by an evolutionary burst during the Jurassic and Cretaceous [Tappan and Loeblich, 1973]. This was followed by a diversity decrease during the Cenozoic (Fig. 19–6). Like the radiolaria, the benthonic foraminifera have had a long geologic history from the Cambrian. However, it was not until the Jurassic that the first clear planktonic foraminifera appeared, at about the same time as other pelagic groups began to radiate. The planktonic foraminifera exhibit maximum diversity during the late Cretaceous, followed by a sharp drop at the Cretaceous-Tertiary boundary (Fig. 19–6). Diversity increased through the Paleogene, but still fell short of the late Cretaceous maximum [Tappan and Loeblich, 1973]. During times of highest diversity a wide range of morphologies are found, including simple coiled forms, species with keels, and radially elongate chambers and apertural modifications of various types. During times of lower diversity a reduced range of morphological forms is found.

It is well established that coccoliths appeared by the early Jurassic (Fig. 19–6). The reported occurrence of coccoliths for pre-Jurassic sediments is questionable because the forms discovered may not be related to coccoliths at all [Gartner, 1977]. Whether the appearance of coccoliths marks the Triassic or Jurassic boundary is not clear. Of particular importance, however, was the major evolutionary expansion of the calcareous nannofossils during the late Cretaceous, giving rise to

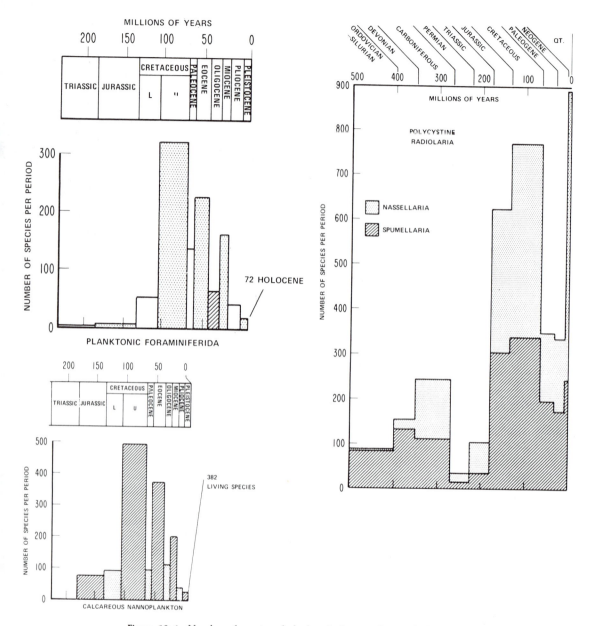

Figure 19-6 Number of species of planktonic foraminifera, radiolaria, and calcareous nannofossils within specific intervals of geologic time. (From H. Tappan and A. R. Loeblich, Jr., *Earth-Science Rev.*, vol. 9, pp. 207–240, 1973)

widespread chalk and marl deposition over many continental margins, as well as over the deep-ocean basins. Chalks have never been so widespread as during the late Cretaceous. Early Cretaceous chalk facies are less common but more widely distributed than Jurassic species [Tappan and Loeblich, 1973].

The oldest record of diatoms is apparently of Jurassic age, but the Jurassic and Cretaceous records include only about 50 species, all of which are centric forms [Tappan and Loeblich, 1973]. These underwent an explosive development during the late Cretaceous and early Cenozoic, which has continued to the present. The pennate diatoms first appeared in small number in the Paleocene and expanded to a peak in diversity during the Miocene and Pliocene. The oceanic siliceous-sediment record was affected mostly by evolution within the centric diatoms, because these are planktonic in habit (see Chapter 16). The pennate diatoms are, in contrast, mostly benthonic and neritic in habit, and hence their appearance in the early Cenozoic may have had an important effect upon the oceanic-silica budget by providing a shallow-water silica sink. Silicoflagellates first appeared somewhat later than the centric diatoms, being first known from the middle Cretaceous. They exhibit a major expansion during the late Cretaceous, although their diversities were higher at times during the Cenozoic.

It can be observed in Fig. 19-6 that three of the major sediment building groups, the coccolithophorids, planktonic foraminifera, and radiolaria, all exhibit highest diversities during the late Cretaceous, giving rise, in particular, to conspicuous and widespread calcareous sediment (with cherts) of this age throughout the earth. The Cretaceous was the age of expansion of the oceanic fossil microplankton. This would have had major affects upon the geochemistry of the oceans. For example, before these groups became important in the open ocean, a high proportion of carbonate must have been deposited in shallow seas, accounting for the high proportion of limestones among older rocks of the continental areas. Since then, the deep-ocean basins have represented enormous sinks for calcium carbonate deposition.

A fundamental question arising from these observations is why so many important sediment-building planktonic microfossil groups expanded so rapidly in the Cretaceous. What were the causes of this evolution? We have seen that most groups appeared much earlier as rootstocks in the Jurassic and perhaps even earlier but they did not show major evolutionary radiations until the Cretaceous, when we can infer conditions were favorable for this to occur. But why were the Cretaceous marine environments so favorable? There are probably several reasons. First, the Cretaceous was marked by the active development of new ocean basins: the North and South Atlantic and the Indian Ocean. This resulted in large-scale isolation of segments of formerly associated biotas and competition among formerly isolated biotas. Biotic provinciality began to increase, resulting in evolutionary radiations. Yet, while interoceanic provinciality was developing, the warm, stable climatic conditions provided additional opportunities for evolutionary radiations. Related, greater vertical partitioning of the water column (greater stratification) provided additional ecologic niches, thus promoting evolutionary radiations. In stable environments, species develop strategies that promote more complete utilization of the habitat. The environments usually have low nutrient levels, and interspecific competition for these can be intense [Lipps, 1970B]. Such competition can promote ecologic segregation and specialization of species. Such conditions probably existed during the Cretaceous. The high sea levels typical of the Cretaceous also probably contributed to an increase in evolution of the marine planktonic microfossils by helping create widespread, equable, moist climates, broadening thermal gradients and thus promoting greater climatic stability, and by increasing the primary ecospace of the upper pelagic zone of the open ocean [Kauffman, 1979]. The spread of epicontinental seas caused equable climates, apparently resulting in the extension of the potential ecospace for the growth and speciation of nannoplankton. During the Cretaceous, there were at least 10 distinct global transgressions and regressions, although sea level was generally high. Times of lower sea level are marked by cooler temperatures, intensification of temperature gradients, and greater vertical mixing of the water column, which in turn reduced the number of ecological niches and created lower pelagic diversity [Fischer and Arthur, 1977]. It seems then that major radiations are controlled by the degree of depth stratification. The well-known correlation between planktonic microfossil diversity and temperature [Berggren, 1969; Stehli and others, 1972; Haq, 1973] may result from changes in oceanic stratification rather than to simple temperature change. Nevertheless the high temperatures of the Cretaceous seas would have encouraged calcium carbonate precipitation in a larger range of marine organisms. However, the problem of the evolutionary radiation of so many important planktonic microfossil groups during the late Mesozoic still has not been explained satisfactorily in detail and continues to be a challenging problem.

MASS EXTINCTIONS AT THE END OF CRETACEOUS

During the period of 570 million years for which abundant fossil remains are available, there have been five great crises of living organisms during which many groups died out. Of these, only the most recent (at 65 Ma, which marks the Cretaceous-Tertiary boundary), is still preserved in the open-ocean sedimentary record. Extinction of many groups of shallow and planktonic marine organisms and a number of taxa of land invertebrates and plants at the boundary between the Maestrichtian and Danian stages remains among the great unsolved puzzles in the history of the earth and ranks high in the list of important problems remaining for study about oceanic evolution.

The extinctions followed a late Cretaceous period of warm, equable climates, high biotic diversity (perhaps highest in the late Phanerozoic), and widespread chalk deposition in transgressive seas that were well ventilated.

The Cretaceous-Tertiary boundary is marked by the extinction of the marine reptiles, the flying reptiles, and both orders of dinosaurs, ammonites, and numerous families of scleractinian corals, bivalves (such as the inoceramids and rudistids), gastropods, and echinoids. In addition, the coccolithophorids, the planktonic foraminifera, and the belemnites suffered nearly complete extinction, with only a few species surviving the crisis. Many genera of the larger benthonic foraminifera and radiolaria also disappeared. Many of these groups became extinct within a very brief period of time, at or near the apex of their evolutionary radiation. On the other hand, a number of groups were little affected, including many groups of land plants, crocodiles, snakes, mammals, numerous invertebrate groups, the freshwater organisms, and benthonic animals from deep-water areas. Russell [1979] estimated that about half of the living genera became extinct during this crisis. However, extinction seems to have been particularly severe in the pelagic communities and in the reef setting. The crisis created that subdivision of time called the Tertiary period—the age of the mammals, birds, and angiosperms.

One important aspect of Mesozoic-Tertiary faunal change is the continuity across the boundary of benthonic foraminifera living in environments ranging from bathyal to abyssal, with little change observed from Campanian to the top of the Paleocene. This is in contrast to the drastic changes exhibited by planktonic foraminifera and shallower living benthonic foraminifera. The planktonic foraminifera became so impoverished that they are represented in sediments immediately overlying the Cretaceous-Tertiary boundary by a peculiar, dwarfed fauna, assigned in some areas to the *Globigerina eugubina* zone (Pla) of earliest Tertiary age. The length of this zone is probably less than 0.5 m.y. [Tucholke and Vogt, 1979].

The biotic crisis is usually accompanied by changes in the character of sediments at the boundary, such as the widespread occurrence of clays. Also, hiatuses occur at the boundary in most sections, representing a period of time of essentially no biogenic deposition, increased carbonate dissolution, or erosion due to intensified deep-sea circulation. The length of time of nondeposition varies from place to place, but the record seems interrupted in almost every location examined. A pronounced sea-level regression is associated with the boundary [Vail and others, 1977].

The crisis at the end of the Cretaceous is generally assumed to have occurred instantly and rapidly around the earth. Paleomagnetic evidence also suggests that there was at least a 0.5 m.y. time lag between extinction of organisms in the oceans [Alvarez and others, 1977] and those on land [Butler and others, 1977]. The last occurrence of dinosaurs measured in terrestrial sections is within the upper part of a normal polarity zone that correlates with Anomaly 29 of the magnetic time scale; after a barren interval, this is followed by the first occurrence of Paleocene mammal fossils within the succeeding reversed-polarity zone (between Anomalies 28 and 29). In the marine section at Gubbio, Italy, the boundary is instead placed at the reversed interval between Anomalies 29 and 30 [Alvarez and others, 1977].

Many hypotheses have been proposed to explain the extinctions at the Cretaceous-Tertiary boundary, although no consensus yet exists. Theories involve both terrestrial and extraterrestrial causes, although most models rely on forms of nutrient depletion or a cosmic event. These include an increase in cosmic radiation [Schindewolf, 1954] and, more specifically, the appearance of a supernova near our solar system [Russell, 1979]; increase in toxic trace-element concentrations [Cloud, 1959]; drastic changes in sea level [Newell, 1962]; decline in oceanic fertility [Bramlette, 1965]; periodicity in phytoplankton productivity and its effect on oxygen levels [Tappan, 1968]; changes in the earth's magnetic field [Simpson, 1966]; CO_2 crises and upward excursions of the CCD to surface waters [Worsley, 1974]; volcanic events and metal poisoning [Vogt, 1972B]; flooding of the ocean surface by freshwater from the Arctic [Gartner and Keany, 1978; Thierstein and Berger, 1978]; or to climatic changes. Most of these theories are insufficient by themselves to explain the range of changes involved in the crisis, but the majority have neither been disapproved nor strongly supported [Percival and Fischer, 1977].

It has become clear, however, that environmental changes occurred at the end of the Cretaceous that affected all or nearly all of the biosphere in a number of ways. It is also clear that climatic changes were relatively minor and temporary and led to no apparent permanent global effects. Contrary to previous suggestions [Saito and Van Donk, 1974; Savin and others, 1975] no dramatic cooling of ocean surface waters occurred at the boundary. If anything, it is possible that surface-water to deep-water temperatures rose several degrees on a global basis across the boundary [Boersma and others, 1979; Thierstein and Berger, 1978] (Fig. 19-7). This temperature rise was short-lived (Fig. 19-7). Could such a temperature rise affect the changes in the wide spectrum of the earth's living organisms? If the change did have an effect it may have been because the late Cretaceous climatic conditions had been so stable that any stress of a new climatic regime, however slight, would have resulted in widespread extinctions. In this way the changes at the Cretaceous-Tertiary boundary differ from those that occurred during the Quaternary—the extinctions were profound within relatively minor paleoceanographic changes and independent of any trend. These observations make gradualistic explanations unattractive and thus, in some respects, catastrophic causes appear more plausible.

Catastrophic causes have received stimulation by the discovery of increased noble element concentrations found in clays at the Cretaceous-Tertiary boundary by Alvarez and others [1980] and Smit and Hertogen [1980]. Platinum and related transition metals are depleted in the earth's crust relative to their cosmic abundances. Concentrations of iridium in sediments at the boundary of up to 160 times more than the background level (although little such data exist) in several areas are considered by Alvarez and others [1980] to have been the result of a great influx of extraterrestrial material. Their scenario proposed a postimpact distribution of pulverized meteoric rock in the earth's atmosphere for several years, resulting in both the heating and darkening of the atmosphere. The resulting suppression of photosynthesis would have had an immediate detrimental effect on the biota, as observed in the paleontological record. According to Emiliani [1980], an Apollo object 10 km in diameter would introduce about 4×10^{30} ergs of energy into the earth's atmosphere in a few seconds, which, if trapped, would be sufficient to raise the temperature of the top 50 m of the ocean and that of the lower troposphere by 5 to 10°C. Such a temperature rise would be lethal to much of the earth's biota in the tropics, but would be less important in the deep sea and at higher latitudes where regional temperatures are lower. Particularly affected would be those organisms most sensitive to high temperature [Emiliani, 1980].

The environmental and climatic effects of such a meteorite impact remain controversial. For instance, Hickey [1981 and personal communication to R. Kerr, 1980] believes that the character of land plant extinction is opposite to that expected for dust blocking and sudden heat rise, because a much higher percentage of plants at higher Northern Hemisphere latitudes become extinct compared with the tropics, where extinction was less distinct. If the entire globe were shrouded by dust, the less hardy tropical plants should have suffered the most, rather than the least. Catastrophic theories are further weakened by the possibility that the terrestrial ex-

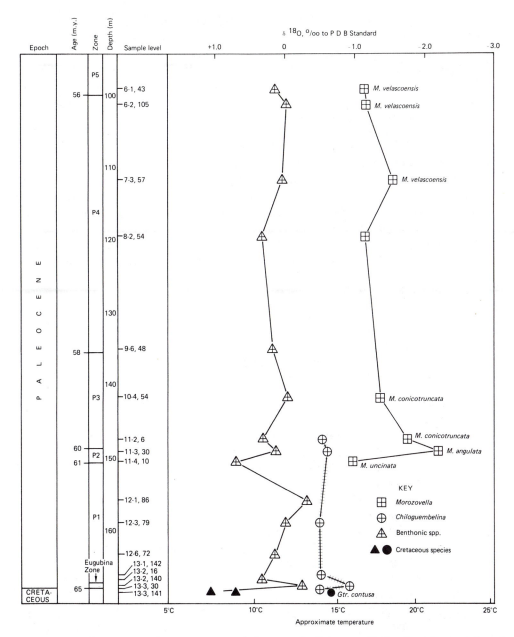

Figure 19-7 Oxygen isotopic records across the Cretaceous-Tertiary boundary and through the Paleocene in DSDP Site 384 in the western North Atlantic. Note shift toward lighter δ¹⁸0 across the boundary in the benthonic foraminiferal species. (From A. Boersma, N. J. Shackleton, and Q. Given in B. E. Tucholke, P. R. Vogt, et al., *Initial Reports of the Deep Sea Drilling Project*, 43, Washington, D.C.: U. S. Govt. Printing Office, pp. 695–717, 1979.)

tinction occurred somewhat later than the marine extinctions. To test further if various habitats were indeed affected sequentially, further paleomagnetic stratigraphy is required for the correlation of continental and oceanic sedimentary sequences within an accurate time framework. In the final analysis, it may be that the Cretaceous-Tertiary boundary crisis was a combination of gradual paleoenvironmental changes, which sequentially affected a series of changes perhaps beginning with a drop in sea level, but upon which some superimposed catastrophic event amplifed the change.

Whatever the cause of this crisis, it seems that of the calcareous nannofossils and planktonic foraminifera, only a few species survived; these appear to have been environmental generalists constituting only a minor proportion of Cretaceous assemblages. These few species were critical because they provided all of the genetic material needed for the relatively rapid, new, evolutionary radiations that took place at the beginning of the Tertiary. By the middle Paleocene a diverse and flourishing planktonic foraminiferal and coccolithophorid community had been reestablished. Diversity continued to increase gradually and reached a peak in the middle Eocene corresponding to generally warm surface-ocean water at all latitudes [Savin and others, 1975].

CENOZOIC PALEOCLIMATIC HISTORY: TRANSITION FROM WARM TO COLD OCEANS

The Tertiary is readily differentiated from the Cretaceous because of the differences in faunas and floras, resulting largely from the brief biotic crisis at the Cretaceous-Tertiary boundary. Another major difference is that of climatic change. The Tertiary exhibits a long cooling interval from the warm climates and low thermal gradients typical of the late Mesozoic oceans to the cold oceans and glacial climates of the late Tertiary and Quaternary. A major question is the cause of this change. How are these transitions related to changes in the ocean boundaries? We have seen in Chapters 6 and 18 that the present world-ocean circulation system has undergone fundamental changes through time as a result of topographic evolution of the ocean basins and the changing position of the continents. Three major elements of paleoceanographic change can be distinguished for the Cenozoic [Kennett, 1977]. The first of these is related to the development of the circum-Antarctic circulation system as southern landmasses moved aside, creating unrestricted latitudinal flow; the second is a generally reciprocal breakup of equatorial and lower-latitude circulation as landmasses moved across or developed in these regions; the third element is related to the development and history of bottom waters of the oceans in response to climatic and glacial events at high latitudes. Two of the three major elements involved in global Cenozoic paleoceanographic evolution are directly related to events occurring at high latitudes, particularly in the Southern Hemisphere. The evolution of the Southern Ocean itself has had a fundamental influence on the development of Antarctic glaciation, which in turn has affected the direction of global climatic evolution, including the north polar region.

During the Cenozoic, the once virtually continuous equatorial seaway became progressively segmented by the closure of the Tethys, the northward migration of Australia, and finally by the closure of the Central American Seaway. This process had substantial effects on the width and intensity of the equatorial current system and the associated upwelling, nutrient budgets, biological productivity, and faunal and floral communities. Of particular importance has been the paleoceanographic transition that occurred in the Tethyan region, which was under a compressional regime throughout the Cenozoic. In the segment from the Mediterranean through to the Himalayas, the oceanic areas were eliminated, and the mobile belts folded, thickened, and raised into mountain chains. At the eastern and western ends of the Mediterranean, the compression has been limited, and the system in these areas has remained mainly oceanic with only minor orogeny. The Mediterranean itself is in an intermediate stage of development. Concurrently with these changes, rifting essentially unrelated to the Tethys has intersected the belt in various directions: The opening of the Atlantic rifted away the western (Caribbean) end of the original Tethys Seaway, the Indian Ocean cut into the eastern portion, and the Red Sea split the Arabian corner from Africa [Fischer, 1975].

The destruction of equatorial circulation during the Cenozoic coincided, in general, with the development of the circum-Antarctic circulation system as southern landmasses moved away from the continental mass, which at the beginning of the Tertiary still consisted of Antarctica, Australia, and South America (Fig. 19–8). Changing boundary conditions in this region included the opening of the Tasmanian Seaway, the opening of the Drake Passage, and the development of the Kerguelen Plateau. The formation and later development of the Circum-Antarctic Current had the effect of thermally isolating Antarctica by decoupling the warmer subtropical gyres from the colder subantarctic and Antarctic gyres. This, in turn, led to the development of increased Antarctic glaciation and later ice-sheet formation, a climatic regime which itself had a profound effect on the environmental evolution, and thus the biogeography, at high southern latitudes. As we shall see later, environmental characteristics resulting from the development of this climatic regime included extensive seasonal sea-ice production, the cooling of waters surrounding the continent, and meterologically forced upwelling of nutrient-rich intermediate waters, which had a profound effect on biogenic productivity in the Southern Ocean. The thermal barriers in high southern latitudes represented by the Antarctic convergence and the Subtropical convergence also came to represent major biogeographic barriers, profoundly affecting the distribution of planktonic organisms in this region. The development of the Circum-Antarctic Current also had the effect of removing biogeographic provinciality previously occurring within different sectors of the Southern Ocean, resulting from former land barriers [Kennett, 1978]. Major elements involved in the Cenozoic paleoceanographic evolution are summarized in Table 19–1.

Cenozoic paleoclimatic evolution is distinguished by three major characteristics.

1. The temperature decreased.

2. The cooling trend was largely a high-latitude phenomenon; the tropics were little affected.

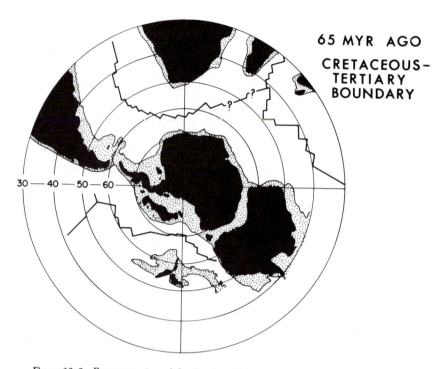

30 — 40 — 50 — 60

Figure 19-8 Reconstruction of the Southern Ocean 65 m.y. ago, about the time of the Cretaceous/Tertiary boundary. Antarctica and Australia are joined as a single continent; South America and Antarctica are joined at the position of the present-day Drake Passage. The southern tip of India is visible in the southern Indian Ocean. Spreading ridges and connecting fracture zones are shown as jagged lines. Reconstructions are compiled from those produced for different sectors by Weissel and others [1977] and Sclater and others [1977a, b]. (From J. P. Kennett, *Marine Micropaleontology*, vol. 13, pp. 301–345, 1978)

3. The climatic cooling was not a smooth, gradual change, but consisted of a series of rather rapid, steplike transitions from one climatic state to another. These steps largely represented changing threshold levels from water to ice in a number of high-latitude regions. The climatic thresholds probably occurred because of positive feedback upon the attainment of a particular stage of snow and ice development, which changed the global albedo [Berger and others, 1981].

Figure 19-9 shows the character of Cenozoic paleoclimatic evolution based on oxygen isotopic evidence. Cooling events were quite rapid. Dramatic declines occur from the early Eocene to middle Eocene and across the Eocene-Oligocene boundary, the middle Miocene, and the late Pliocene. A number of warming periods have occurred, especially at times during the Eocene and during the late early Miocene to middle Miocene. The significance of each of these events is discussed in detail later.

TABLE 19-1 MAJOR ELEMENTS OF PALEOCEANOGRAPHIC CHANGE DURING THE CENOZOIC

A. Diminishing equatorial circulation.

   1. Destruction of Tethyan Seaway (early to middle Cenozoic).
   2. Blocking of Indonesian Seaway (middle to late Cenozoic).
   3. Bridging of Central American Seaway (Pliocene).

B. Development of circum-Antarctic circulation.

   1. Opening of Australian-Antarctic Seaway (earliest Oligocene).
   2. Development of Kerguelen Plateau.
   3. Opening of Drake Passage (Oligocene).

C. Development of psychrospheric circulation; linked with polar glacial evolution (earliest Oligocene).

In essence, the Tertiary can be subdivided into two principal climatic modes: during the Paleogene and the Neogene. The Paleogene interval (65–23 Ma) represents an intermediate stage between the late Mesozoic and Neogene, marked by changing thermal patterns in the ocean and a transition from a predominantly latitudinal, warm-water circulation at all depths to a predominately meridional, thermohaline, cold-water circulation. The Paleogene is also intermediate between the Cretaceous nonglacial mode and the Neogene glacial mode of the earth. The

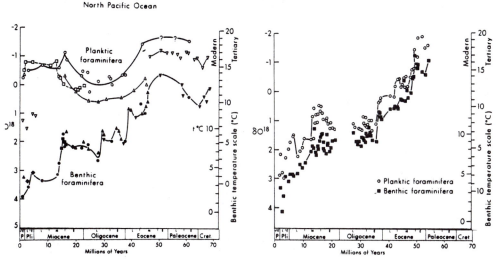

Figure 19-9 Oxygen isotopic trends for foraminifera through the Tertiary low latitudes (left) and high latitudes (right). Left: Isotopic paleotemperature data for planktonic foraminifera (open symbols) and benthonic foraminifera (closed symbols) primarily from the North Pacific (after Savin, 1977). Right: Isotopic paleotemperature analyses of planktonic and benthonic foraminifera from the subantarctic Pacific by Shackleton and Kennett [1975]. The Modern and Tertiary benthonic temperature scales were calculated assuming water $\delta^{18}O$ values of -0.08 and -1.00 per mil, respectively. (After Savin, 1977)

earth's glaciation was initiated during the early Tertiary, certainly by the Oligocene, and perhaps in limited form even earlier, although glaciation during the Eocene was probably very local and thus of little global significance. Ice cover became more important during the Oligocene; but ice sheets and ice shelves on a scale, such as exist today, did not develop until middle Miocene times. The climatic thresholds passed during the Cenozoic seem to be linked to the development of ice and snow at high latitudes. As this developed it reflected much of the incoming radiation back to space, thus providing a feedback to the climatic regime that assisted in maintaining the new paleoclimatic state.

THE PALEOCENE AND EOCENE:
PRELUDE TO A GLACIAL EARTH

During the Paleocene and Eocene, the major avenues for interoceanic circulation remained at equatorial and low-latitude regions with unrestricted connections in the Indo-Pacific north of Australia and the Atlantic-Pacific via the Middle American Seaway (Fig. 19-10). The Indian Ocean must have been relatively warm throughout, as equatorial currents flowing into this ocean from the east would have flowed uninterrupted across the entire Indo-Pacific equatorial zone, becoming increasingly warmer due to the long residence time [Frakes and Kemp, 1972]. Middle-latitude, interocean circulation existed between the Indian and Atlantic oceans south of Africa and in the Northern Hemisphere via the Tethys Seaway, which was becoming highly restricted during this time (Fig. 19-10).

Expansion of the Atlantic and Indian oceans continued to restrict the size of

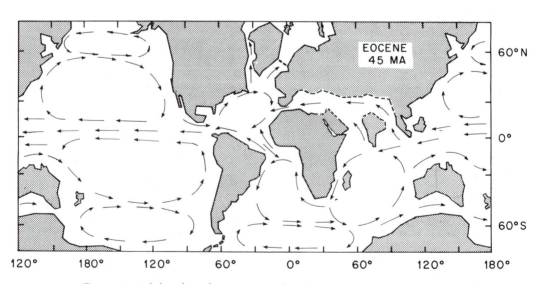

Figure 19-10 Inferred circulation patterns of surface waters at 45 Ma (middle Eocene). (From B. U. Haq, *Oceanologica Acta*, vol. 4, in press, 1981)

the Pacific Ocean. Rotation of North America toward Asia had closed the Arctic-Pacific connection during the late Cretaceous [Smith and Briden, 1977]. This eliminated any previous source of bottom waters from high latitudes in the North Pacific [Schnitker, 1980], and cool water from the south became increasingly important. North Atlantic avenues to the Arctic remained closed. As in the present ocean, partial isolation of bottom waters between the oceans created geochemical differences and led to interoceanic geochemical fractionation. As a result, differences developed in the history of the CCD; in the Pacific, the CCD shoaled continuously through the Paleocene and Eocene, while at the same time deepening in the Atlantic and Indian oceans. Paleoclimates during much of the Paleocene and Eocene remained relatively warm and equable, with low pole-to-equator temperature gradients. Nevertheless, temperatures decreased through much of this interval (Fig. 19–9), leading to greater latitudinal climatic differentiation. This, in turn, led to an intensification of oceanic circulation and increased biogenic productivity at low and middle latitudes, which created widespread biogenic siliceous sediments.

During the Paleocene (65–55 Ma), the Antarctic continent was in a polar position (Fig. 19–8), although no significant glaciation existed on the continent; also, Australia and Antarctica were joined. Sea-surface temperatures were relatively high (subantarctic about 18°C) [Shackleton and Kennett, 1975]. Deep waters were also warm (about 16°C at about 1000 m) because of high surface-water temperatures adjacent to Antarctica. The inferred surface paleocirculation during the middle Eocene is shown in Fig. 19–10 [Haq, 1981]. Circulation became restricted at this time in the northern Indian Ocean as India approached the Asian continent, although the tropical Tethys Current continued to flow westward through the narrow northern passage and the triangular reentrant west of India (Fig. 19–10). In the North Atlantic, *minor* surface water exchange may have commenced with the Arctic [Talwani and Eldholm, 1977] and the Labrador Passage may have allowed relatively warm water to be transported into the Arctic Basin.

During the late early Eocene to middle Eocene (about 53 Ma), Australia began to drift northward from Antarctica, creating an ocean between the two continents; this ocean has continued to increase in size (Fig. 19–11) [Weissel and Hayes, 1972]. This was the only major continental separation to take place entirely within the Cenozoic (Fig. 19–12); it had profound consequence upon global circulation, climates, and biotic evolution. The deep basin forming to the southwest of the South Tasman rise during the Eocene as a result of this spreading received fine-grained, poorly sorted detrital sediments with little biogenic material and high organic carbon content, reflecting highly restricted deep-water circulation and a lack of any deep circum-Antarctic current flow [Andrews and others, 1975]. Circum-Antarctic current flow was blocked by continental masses associated with the present-day South Tasman rise (which linked Australia with Antarctica) and the Drake Passage, which continued throughout the Eocene, although rather shallow-water marine connections of outer neritic–uppermost bathyal water depths (about 100–300 m) formed over the South Tasman rise during the late Eocene (about 40 Ma) [Kennett and others, 1975]. This shallow-water marine connection probably produced the first direct communications between shallow-water and planktonic

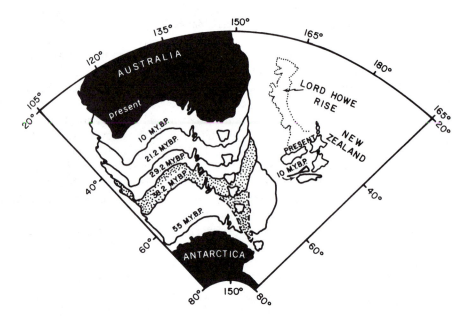

Figure 19-11 Successive positions of Australia relative to Antarctica as Australia moved northward during the Cenozoic. The position of Australia at 38 m.y. ago (Eocene–Oligocene boundary) is shown (stippled) to include the South Tasman rise, which is of continental crust and which prevented the development of deep circum-Antarctic circulation well after spreading commenced. (Modified after Weissel and Hayes, 1972; from J. P. Kennett, 1977)

marine organisms of the southern Indian and Pacific oceans [Kennett, 1978]. In the Southwest Pacific, surface-water temperatures in the subantarctic were about 20°C in the early Eocene, but dropped to 12 to 14°C by the middle Eocene and as low as 10°C by the late Eocene [Shackleton and Kennett, 1975]. Bottom waters were still relatively warm (about 7°C), reflecting moderately high surface temperatures adja-

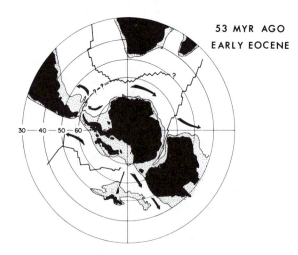

53 MYR AGO
EARLY EOCENE

Figure 19-12 Reconstruction of the Southern Ocean and suggested bottom-water circulation during the early Eocene 53 m.y. ago. A spreading ridge has formed between Australia and Antarctica heralding the beginning of northward drift of Australia. India has now moved northward off the map projection. Spreading ridges and connecting fracture zones are shown as jagged lines. Dashed line in South Atlantic represents fracture zone. Reconstructions are compiled from those produced for different sectors by Weissel and others [1977] and Sclater and others [1977a, b]. (From J. P. Kennett, *Marine Micropaleontology*, vol. 3, pp. 301–345, 1978)

cent to the Antarctic continent. These also decreased through the Eocene, similar to subantarctic surface-water temperatures. Temperatures on at least parts of the continent were warm enough during the Eocene to support a cool temperate vegetation (*Nothofagus* flora), a conclusion based on palynological evidence [Cranwell and others, 1960; McIntyre and Wilson, 1966]. Paleontological evidence from Eocene sequences elsewhere in the world shows that warm humid conditions were prevalent [Wolfe and Hopkins, 1967; Frakes and Kemp, 1973].

Evidence for glaciation during the Eocene [Geitzenauer and others, 1968; Margolis and Kennett, 1971; Le Masurier, 1972] is not conclusive and remains to be confirmed by further deep-sea drilling off Antarctica. Assuming that the relatively warm bottom-water temperatures do indeed reflect those of surface water around the coast of Antarctica, extensive Antarctic glaciation at sea level and ice-shelf development is inconceivable during the Eocene. However, if glaciation did occur, it may have been confined to alpine glaciers in elevated coastal areas, such as West Antarctica. During the late Eocene (about 38 Ma), calcareous biogenic sediments were deposited adjacent to the continent, and siliceous biogenic sedimentation is largely unknown from the region [Hayes and Frakes, 1975; Kennett and others, 1975].

Paleoclimatic conditions were by no means stable during the Paleocene and Eocene. Paleobiogeographic changes of oceanic microfossils [Haq and Lohmann, 1976] and changes in terrestrial vegetation [Wolfe, 1978] indicate a number of warm and cool intervals through this interval. After an important cooling event in the middle Paleocene, a late Paleocene warming trend culminated in a period of peak warming during the early Eocene to middle Eocene (53–49 Ma), probably the warmest interval of the Cenozoic [Haq and others, 1977]. This coincided with a widespread transgression [Vail and others, 1977], which apparently expanded the ecospace for marine organisms. Diversities of microfossils were particularly high at this time for the Cenozoic, reflecting a peak in evolutionary radiation. Surprisingly warm vertebrate faunas lived on Ellesmere Island, north of Baffin Bay in the Canadian Arctic, indicating minimum temperatures of 10 to 12°C during the winters [Estes and Hutchinson, 1980; McKenna, 1980]. Furthermore, during the Paleocene through Eocene, a broad-leaved evergreen forest extended north of latitude 60°N in the Arctic [Wolfe, 1978]. Wolfe has speculated that the existence of these floras at such high latitudes can be explained only by invoking more prolonged light at these northern latitudes. This, in turn, can be accomplished only by a significantly smaller inclination of the earth's rotational axis (at least 15° less during the middle Eocene than now!). Such a conclusion, if correct, implies fundamentally different paleoenvironmental conditions of the earth than today.

THE TERMINAL EOCENE EVENT

We have already seen that major changes in oceanic circulation generally developed rapidly and were separated by long periods of relative stability. Within the Cenozoic, the most significant paleoceanographic change and related crisis in global biota occurred about 38 Ma at the Eocene-Oligocene boundary. This has been called the **terminal Eocene event** [Wolfe, 1978]. At this time the abyssal ocean filled up with cold water, establishing the psychrosphere, hence its significance [Benson, 1975]. Oxygen isotopic changes in deep-sea benthonic foraminifera from the subantarctic [Shackleton and Kennett, 1975; Kennett and Shackleton, 1976; Keigwin, 1980], the tropical Pacific region [Savin and others, 1975; Keigwin, 1980], and the shallow and deep North Atlantic [Buchardt, 1978; Vergnaud-Grazzini and others, 1978] indicate that bottom temperatures decreased rapidly by about 4 or 5°C (Fig. 19-13). This temperature reduction was calculated by Kennett and Shackleton [1976] to have occurred within 10^5 years, which is remarkably abrupt for preglacial

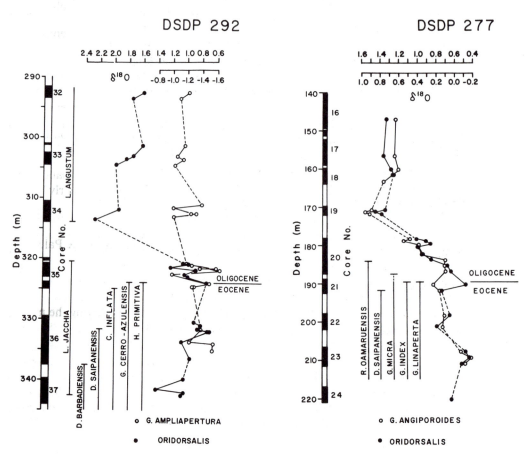

Figure 19-13 Oxygen isotopic changes in benthonic foraminifera (closed circles) and planktonic foraminifera (open circles) over Eocene-Oligocene boundary in DSDP Site 277 (subantarctic) and Site 292 (tropical Pacific). Stratigraphic ranges of selected significant microfossils also are shown. Note that in the subantarctic site both the planktonic and benthonic species exhibit oxygen isotopic shifts at the boundary, while in the tropics only the benthonic foraminifera exhibit a change. Upper horizontal $\delta^{18}O$ scale is for benthonic foraminifera; lower scale is for planktonic foraminifera. (From L. D. Keigwin, 1980, reprinted by permission from *Nature*, vol. 287, pp. 722-25, copyright © 1980 Macmillan Journals, Ltd.)

Tertiary times, and is considered to represent the time when large-scale freezing conditions developed at sea level around Antarctica, forming the first significant sea ice. It is inferred that at this time Antarctic bottom water began to be produced, that bottom water temperatures fell to levels close to (but still above) present-day values, and that thermohaline circulation much like that in the present ocean developed. A major plunge occurred in the CCD in all oceans at the boundary (Fig. 19–14).

The terminal Eocene event did not cause such severe widespread changes in the biota as did the terminal Cretaceous event. Nevertheless, the deep-sea benthonic assemblages were more drastically affected at the end of the Eocene than at the end of the Cretaceous. Large changes in the benthonic foraminifera occur in the assemblages over a wide stratigraphic interval straddling the boundary [Corliss, 1979B]. These differences in the way the benthonic fauna respond to change reflect

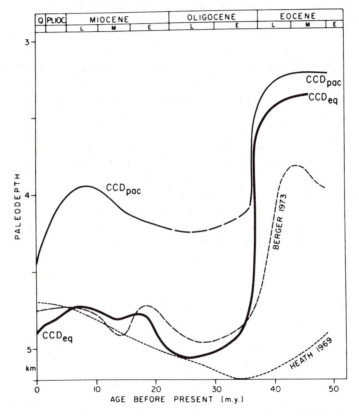

Figure 19-14 Changes in calcium carbonate compensation depth (CCD) during the Cenozoic for the equatorial zone between latitudes 3° N and 3° S (CCDeq) and for the Pacific north of latitude 4° N and south of latitude 4°S (CCDpac) are shown [van Andel and others, 1975]. These data are compared with CCD curves of Heath [1969] and Berger [1973]. Note major plunge of all CCD curves near Eocene–Oligocene boundary. (From Tj. H. van Andel, G. R. Heath, and T. C. Moore, Jr., GSA Memoir 143, p. 47, 1975, courtesy The Geological Society of America)

Global Paleoceanographic Evolution: Critical Events in Ocean History   **723**

the differences in the paleoenvironmental changes occurring during the two crises—specifically, major cooling and glacial development occurred at high latitudes at the end of the Eocene, but not at the end of the Cretaceous. The terminal Eocene event, in fact, was largely in response to critical development of high-latitude glaciation and climatic cooling. On the other hand, as we have seen, the fundamental driving force for the crisis at the Cretaceous-Tertiary boundary was not climatic change.

The suddenness of the Eocene termination is striking. Strong positive feedback was probably involved, presumably from albedo change due to an increase in snow and ice over parts of Antarctica and adjacent oceans [Berger and others, 1981]. The cooling also coincided with sea-level regression [Vail and others, 1977] over the continental margins, as exhibited in areas such as the Gulf Coast, Europe, and Australia.

Why did the first significant sea-ice apparently form at high latitudes 38 m.y. ago and why did the first glacial conditions appear so much later than the time when Antarctica first moved into a south polar position during the Mesozoic? It is not clear what geological event produced the conditions necessary to cross this climatic threshold, but it certainly was partly due to increasing isolation of Antarctica from Australia, the development of an ocean between these two continents, and particularly the establishment of the first surface-water connections between the southern Indian and Pacific oceans, over the South Tasman rise (Tasmanian Seaway) [Kennett, 1977].

As soon as the Tasmanian Seaway opened (Fig. 19–15), relatively cool surface waters from high latitudes of the southern Indian Ocean would have been transported into the Ross Sea embayment and the narrow waterway between East and West Antarctica (Fig. 19–15). Previous to this event, the Ross Sea region would have been largely under the influence of the relatively warm East Australian current. The influence of the new cool water in the Ross Sea area may thus have triggered freezing, sea-ice formation, and the production of cold bottom waters near the beginning of the Oligocene.

Decrease in Antarctic surface-water temperatures also produced major changes in the high-latitude planktonic foraminiferal faunas, which at the beginning of the Oligocene assumed characteristics typical of the present-day assemblage; these include very low diversity and relatively simple morphology [Kaneps, 1975]. These high-latitude changes coincided with more widespread planktonic microfossil extinctions and reduction in diversity at middle and even low latitudes. These changes produced the distinctive Oligocene assemblages of the oligotaxic mode.

The Eocene-Oligocene boundary also marks the most profound event of the Tertiary terrestrial floras. According to Wolfe [1978], the vegetation changed drastically in the middle-to-high latitudes of the Northern Hemisphere. Within a geologically short period of time, areas that had been occupied by broad-leaved evergreen forests became occupied by temperate, broad-leaved deciduous forest of particularly low diversity. A major decline in mean annual temperature occurred—about 12 to 13°C at latitude 60° in Alaska and about 10 to 11°C at latitude 45° in the Pacific Northwest. Just as profound, however, was the shift in temperature equability: In the Pacific Northwest, for example, the mean annual

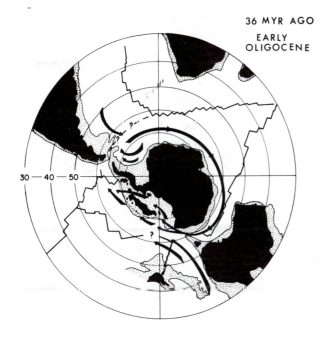

36 MYR AGO
EARLY
OLIGOCENE

30 — 40 — 50

Figure 19-15 Reconstruction of the Southern Ocean and suggested surface-water circulation during the earliest Oligocene 36 m.y. ago. A substantial ocean has now formed between Australia and Antarctica, although the southward extension of the continental South Tasman rise and Tasmania continued to block deep circum-Antarctic flow between these two continents. The Drake Passage remains closed between South America and Antarctica. Spreading ridges and connecting fracture zones are shown as jagged lines. A shallow, surface-water connection is by this time established over the South Tasman rise, leading possibly to cooling and sea-ice formation in the Ross Sea region. Reconstructions are compiled from those produced for different sectors by Weissel and others [1977] and Sclater and others [1977a, b]. (From J. P. Kennett, *Marine Micropaleontology*, vol. 3, pp. 301–345, 1978)

range of temperature, which had been at least as low as 3 to 5°C in the middle Eocene, must have been at least 21°C and probably as high as 25°C in the Oligocene [Wolfe, 1971; 1978].

OLIGOCENE OLIGOTAXIC OCEANS

During the Oligocene, a number of paleoceanographic changes occurred in the world oceans, which were to transform finally global circulation patterns from the Cretaceous-Eocene mode to the present-day mode. Three principal events occurred.

1. The eastern Tethys became almost completely closed by the early Oligocene, thereby severely restricting the westward flow of the Tethys Current (Fig. 19-16) and equatorial circulation.

2. Australia continued its northward movement away from Antarctica, so that by the early part of late Oligocene, the trailing edge of the continent, the South Tasman rise, cleared Victoria Land, Antarctica. The initiation of the Circum-Antarctic Current dates from this point, isolating Antarctica within a ring of cold water. This was a factor in reducing the efficiency of meridional heat transport from the equator to the pole, thus increasing the temperature gradient between these regions [Kemp, 1975].

3. The Drake Passage between South America and Antarctica opened. Evidence from magnetic anomalies indicates that this occurred during the

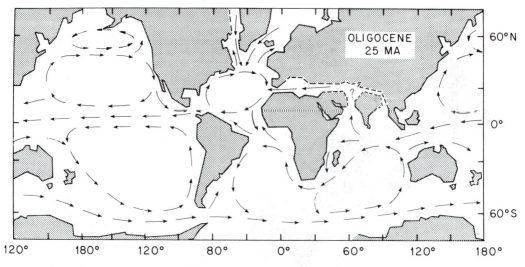

Figure 19-16 Inferred circulation patterns of surface waters at 25 Ma (late Oligocene). (From B. U. Haq, *Oceanologica Acta*, vol. 4, in press, 1981)

Oligocene, but it is not clear exactly when. The most commonly quoted date is 30 Ma [Barker and Burrell, 1977]; however, more recent reconstructions indicate a shallow-water connection by the latest Eocene to earliest Oligocene [Norton and Sclater, 1979]. Differences in the oxygen isotopic records between the Pacific and Atlantic oceans, however, indicate that deep-water connections had not been established through the Drake Passage during much of the Oligocene. Boesma and Shackleton [1977] have interpreted oxygen isotopic records in such a way that during the Oligocene, the deep water of the Atlantic was a few degrees warmer than that of the Pacific. It thus seems that closure to deep water at the Drake Passage prevented cold water penetrating this way into the Atlantic. The flow of bottom water into the Atlantic was a long one, with mixing and warming along the way.

The development of the circum-Antarctic Current would have effectively eliminated previous cyclonic gyres that may have existed at high latitudes in the Southern Hemisphere (Fig. 19-16). The development of the circum-Antarctic circulation caused major reorganization of circulation and sedimentation patterns in the Southern Hemisphere. The deep Circum-Antarctic Current began to erode bottom sediments south of Tasmania. Bottom-water activity in the northern Tasman Sea diminished and was redirected to form a deep western boundary-current flow east of New Zealand [Kennett, 1977], (Fig. 19-17). These deep-sea circulatory patterns established in the late Oligocene have persisted to the present, although the overall intensity of circulation has fluctuated. Calcareous biogenic sediments continued to be deposited close to the continent, while siliceous biogenic sediments, although expanding in importance, remained restricted and inconspicuous in distribution except in the southwest area [Hayes and Frakes, 1975; Tucholke and others, 1976; Barker, Dalziel, and others, 1977B].

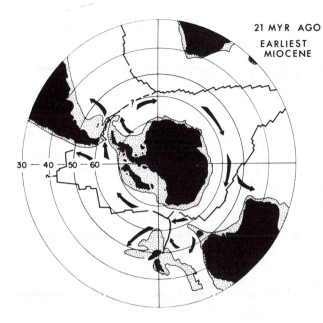

Figure 19-17 Reconstruction of the Southern Ocean and suggested bottom-water at the time of the Paleogene–Neogene boundary 21 m.y. ago. Australia and the South Tasman rise are now well separated from Antarctica, and the Drake Passage is open. The movement of previously obstructing landmasses has, by this time, allowed the formation of the circum-Antarctic water-mass system, and bottom-water transportation south of the South Tasman rise and through the Drake Passage is feasible. Spreading ridges and connecting fracture zones are shown as jagged lines. Reconstructions are completed from those produced for different sectors by Weissel and others [1977] and Sclater and others [1977a, b]. (From J. P. Kennett, *Marine Micropaleontology*, vol. 3, pp. 301–345, 1978)

The conspicuous climatic cooling at the beginning of the Oligocene also produced widespread glacial conditions throughout Antarctica. Yet oxygen isotopic evidence indicates that, although glaciation was widespread in the Oligocene, no ice sheets had yet developed on Antarctica [Shackleton and Kennett, 1975; Savin and others, 1975]. Subantarctic surface-water temperatures had fallen to about 7°C by the early Oligocene, a value similar to those of the present day (Fig. 19-9). Antarctic surface-water temperatures are also inferred to have reached values close to but probably still slightly warmer than those of the present day. A variety of paleoenvironmental evidence throughout the world indicates relatively cool global climates [Edwards, 1968; Wolfe and Hopkins, 1967; Crowell and Frakes, 1970; Hornibrook, 1971; Frakes and Kemp, 1973]. An abundance of fossil whales and penguins in Oligocene marine sequences of New Zealand gives an impression of coolness [Fleming, 1962; 1975]. With the available evidence, it is difficult to translate these low Oligocene temperatures into the extent of Antarctic ice cover; there is no clear evidence of ice-rafting of substantial quantities of glacial debris for Antarctica until the late Oligocene (about 25 Ma) [Hayes and Frakes, 1975]. Even then, ice-rafted sediment is uncommon and restricted to sequences closely adjacent to Antarctica. Although no strong evidence of major glaciation exists prior to the late Oligocene, the record is difficult to interpret with available data, and the possibility of extensive older glaciation exists. Nevertheless it is doubtful that any *large* ice accumulations occurred anywhere on the earth prior to the late Oligocene or even middle Miocene (see below) although this has been disputed by Matthews and Poore [1980].

Support for only partial Antarctic glaciation comes from paleobotanical evidence. According to Kemp [1975], vegetation may have persisted in the Ross Sea region until the late Oligocene. The pollen spectrum suggests a shrub or tree cover of low diversity, with *Nothofagus* and Podocarps dominant. These floras seem to have been eliminated from Antarctica by the early Miocene, if not earlier, due to the severe climatic conditions.

These paleoenvironmental changes had a drastic effect on the biogeographic evolution of planktonic faunas and floras in the world ocean. At the beginning of the Oligocene, the distinctive Antarctic biogeographic provinciality characteristic of the present day began to develop. Relatively diverse late-Eocene, planktonic-foraminiferal assemblages from areas adjacent to Antarctica were replaced by low-diversity, sparse faunas of early Oligocene age and younger. It is not clear how rapidly these changes took place at high latitudes because of inadequate marine sections, but it can be predicted, from the available oxygen isotopic evidence, that the faunal and floral changes were abrupt. In the subantarctic areas, microfossil changes, although distinct, seemed to have occurred more gradually across the Eocene-Oligocene boundary. Well-defined paleontological changes have been reported to occur almost everywhere in the world ocean near the Eocene-Oligocene boundary. Among planktonic-foraminiferal assemblages, diversity is lowest for the entire Cenozoic during the early to middle Oligocene (producing an oligotaxic ocean), after which it increased again through the late Oligocene. These trends reflect the early-Oligocene biogeographic crisis, followed by the gradual replacement of forms characteristic of the Paleogene by typical Neogene taxa [Berggren, 1969].

In southern Australia, warm echinoids were replaced by cold echinoids [Foster, 1974]. Jenkins [1974] described major changes in planktonic-foraminiferal assemblages from the early part of late Eocene, marked by high-diversity faunas containing more complex forms, such as *Globigerapsis* and *Hantkenina,* which were replaced by lower-diversity assemblages of simpler, cooler water forms of *Globigerina* and nonkeeled *Globorotalia*. These changes form a gradation toward the so-called twilight zone, representing an interval of time during which earlier Paleogene assemblages disappeared and faunas of modern aspects had not yet appeared [Cifelli, 1969]. This left Oligocene planktonic foraminiferal faunas of remarkable simplicity and superficially similar to those of the early Paleocene (Danian) because of dominance by simple globigerinid forms. The dominance by *Globigerina* persisted through much of the Oligocene, although these forms continued to diversify conservatively. Biostratigraphic subdivision of the Oligocene is largely based on these elements, although low diversity has allowed relatively few biostratigraphic subdivisions to be established compared with the remainder of the Cenozoic [Cifelli, 1969; Berggren, 1969]. The Antarctic diatom assemblage also underwent drastic changes involving many taxa close to the Eocene-Oligocene boundary [Hajos, 1976].

As discussed by Kennett [1978], the Oligocene also marks very important developments in the marine vertebrate faunas. The earliest known baleen whales (Mysticeti) are of middle-Oligocene age in New Zealand marine sequences [Fordyce, 1977]. Fordyce [1977] postulated that the development of filter feeding in some cetacea, particularly the *Mauicetus,* was induced by increased biogenic productivity in the Southern Ocean during the Oligocene. This seems possible because of the establishment of the Circum-Antarctic Current at this time, establishing a fundamentally different circulation pattern and setting the stage for later developments. Nevertheless, the deep-sea sediments in most sectors still reflect rather low biogenic productivity during the Oligocene, although siliceous biogenic oozes began to form in the South Atlantic sector during the Oligocene [Barker, Dalziel and others, 1977B]. It seems clear, however, that the successful development of filter feeding in whales during the Oligocene must have been in response to suitably high levels of biogenic productivity at least in some parts of the Southern Hemisphere and perhaps even in coastal areas that are marked by oceanic upwelling. In general, pre-Oligocene toothed whales (Archaeoceti) are morphologically different from post-Oligocene forms (Mysticeti and Odontoceti—modern toothed forms), suggesting a radiation of alternative adaptive types [Lipps and Mitchell, 1976]. However, the Oligocene is marked by rather low diversities in whales [Lipps and Mitchell, 1976], which probably reflect rather low productivity. Alternatively, the markedly different oceanographic conditions during the Oligocene, which reduced diversity in all marine groups, also affected the whales [Gaskin, 1976], probably through a reduction in trophic resources [Lipps and Mitchell, 1976]. Although the first filter-feeding whales are known from the Oligocene, it is important to note that their major evolutionary radiation (increase in diversity), as well as that of the toothed forms (Odontoceti), did not seem to commence until the early Miocene [Gaskin, 1976; Lipps and Mitchell, 1976]. This is expected because the early Miocene clearly marks the time of increased siliceous biogenic productivity (increased oceanic upwelling) in the Southern Ocean, as inferred from increased rates of biogenic sedimentation [Kennett and others, 1975]. Thus the trophic resources were provided for the evolutionary explosion which followed in the middle and late Cenozoic. The success of the Mysticeti in the late Cenozoic, particularly in terms of large physical size, reflects the efficiency of filter feeding [Fordyce, 1977], especially as upwelling and biogenic productivity continued to increase in the Southern Ocean region, providing the required increase in trophic resources for this successful evolutionary radiation.

NEOGENE OCEANS—INTO THE GLACIAL MODE

By the early Miocene, about 22 m.y. ago, the ocean basins had essentially assumed their modern shapes, if not the same proportions [Schnitker, 1980]. The development of the Circum-Antarctic Current during the Oligocene created the thermal isolation of Antarctica by decoupling the warm subtropical gyres from the cold subpolar gyres. This thermal isolation of Antarctica led to increasing glaciation and eventually to the development of the Antarctic ice sheets in the middle Miocene and the further expansion of sea ice. However, in the early Miocene there seem to have been no ice sheets; the Arctic was probably ice free and the Antarctic, while possessing extensive ice cover, had not developed a significant ice sheet. Increased glaciation of Antarctica and development of the Circum-Antarctic Current led to the expansion of the Antarctic water mass and perhaps the initiation of the Antarctic convergence during the early Miocene. As a result, the high-latitude siliceous biogenic sedimentary province began a northward expansion at the expense of carbonate-sediment distributions. Sedimentation rates began to increase at high latitudes. The early Miocene also marks the establishment of a permanent steep temperature gradient between the polar and tropical regions related to the development of polar-to-tropical, surface water-mass belts [Kennett, 1977]. Since then, these belts have largely retained their identities during middle and late Cenozoic climatic changes but have oscillated latitudinally. Intensification at global climatic gradients, in turn, led to increased vigor in oceanic circulation involving both vertical and horizontal transport. The surface temperature gradient was about 15°C [Savin and others, 1975], as was the vertical gradient, with bottom temperatures about 5°C and tropical surface temperatures of 20°C [Shackleton and Kennett, 1975; Savin and others, 1975; Sancetta, 1979A].

During the early Miocene, ice-rafted sediments are recorded in Southeast Pacific deep-drilled sequences adjacent to Antarctica. A slight temporary warming of surface waters occurred through the early Miocene, although this is superimposed upon a general trend of continued glaciation. This warming seemed to cause a temporary reduction in flow of AABW in equatorial Pacific sequences, as inferred by van Andel and others [1975] on the basis of fewer hiatuses, a shoaling of the CCD, decreasing width of the equatorial, carbonate-rich zone, and other features resulting from CO_2 buildup and increased carbonate dissolution. This temperature increase is also reported in other regions of the world on the basis of a variety of paleontological evidence [Hornibrook, 1971].

The establishment of circum-Antarctic circulation and related high-latitude water masses, the development of extensive polar glaciations, and the ultimate elimination of low-latitude interocean circulation makes the Neogene paleoceanographically and paleoclimatically distinct from the Paleogene. The initiation of Antarctic ice-sheet formation and the onset of Northern Hemisphere glaciation in the late Pliocene were both critical stages in global environmental evolution during the late Cenozoic. These steps ultimately determined both the overall physical, chemical, and biological characteristics of the oceans as we know them today and readied the earth for the remarkable cyclic, glacial-interglacial conditions that have marked the latest Cenozoic. The exact sequence of the events involved in this evolution and the determination of the hierarchy of interactions between boundary changes, ice-sheet development, surface circulation, and deep circulation still remain to be determined. It seems clear, however, that there were three principal Neogene paleoceanographic events that led to modern oceanic conditions; one of these occurred in the Antarctic region and the two others occurred in the Northern Hemisphere.

1. The opening of the Drake Passage to *deep-water* flow.

2. The subsidence of the Iceland-Greenland-Faeroe ridge, allowing relatively efficient exchange of cold Arctic water with the Atlantic Ocean, and thus the world ocean.

3. The continued restriction and final closure of the Tethys Seaway, first between Asia and Africa and finally between North and South America.

We have already seen that the Drake Passage became open to surface flow sometime during the Oligocene. At about 22 m.y. ago, near the Oligocene-Miocene boundary, this gateway seems to have opened to deep-water flow [Barker and Burrell, 1976], judging from changes in sediment distribution patterns and hiatuses in the Southwest Atlantic and Southeast Pacific [Craddock and Hollister, 1976; Ciesielski and Wise, 1977]. This event amplified circum-Antarctic flow, increasing the thermal isolation of Antarctica.

North of Australia, the Tethys Seaway continued to be restricted throughout the Miocene by the northward drift of Australia, separating the equatorial Indian Ocean from the equatorial Pacific. This closed off previous deep-water communications through the region. The collision of Eurasia and Africa during the late early Miocene (Burdigalian Stage) about 18 m.y. ago is indicated by the exchange of land vertebrates between the two continents [Berggren and van Couvering, 1974]. The closure of this part of the Tethys must have created an evaporation basin at about 30°N, which may have served as a source of warm, saline water to the North Atlantic like the present-day Mediterranean Sea [Schnitker, 1980].

One of the most conspicuous changes in the Neogene Ocean has been the increasing importance of siliceous biogenic productivity and sedimentation at high latitudes during the Neogene. Antarctic waters have long been noted for their prolific surface-biogenic productivity. Present surface fertility results from the upwelling of nutrient-rich intermediate water south of the Antarctic convergence (see Chapter 14). In the present ocean the Antarctic convergence sharply coincides with the deposition of calcareous biogenic oozes to the north and siliceous biogenic oozes to the south, resulting from the rapid temperature change in surface waters, which creates large differences in the planktonic biota. This observation has been used by several workers studying Cenozoic sedimentary sequences in the Atlantic as an index of prior positions of the Antarctic convergence [Kennett and others, 1975; Hayes and Frakes, 1975; Tucholke and others, 1976]. Furthermore, skeletal opal—in the form of diatom, radiolaria, and silicoflagellate tests—provides a dependable index of surface-water productivity (see Chapter 14). Antarctic surface water became more conducive to siliceous biogenic productivity with the progressive thermal isolation of Antarctica. As the cold Antarctic water mass expanded northward, so did the siliceous biogenic province. As upwelling increased south of the Antarctic convergence, so did the rate of siliceous biogenic sedimentation. Variations in rates of biogenic productivity are apparently largely related to climatic oscillations, which alter the strength of the westerly winds that drive surface water away from Antarctica and hence control the amount of upwelling of nutrient-rich waters [Kennett, 1977; Brewster, 1977; 1980]. Colder climates thus create increased upwelling in the Antarctic region.

Studies of the temporal distribution of diatomaceous sediment (Fig. 19-18) have shown that the first such sediments were deposited closely adjacent to Antarctica and that the northern limits of this province have expanded northward during the Neogene (Fig. 19-18), creating diachronous biogenic sedimentary facies [Kemp and others, 1975; Tucholke and others, 1976]. Biogenic sedimentation rates, as monitored largely at DSDP Site 278 on the present-day Antarctic convergence,

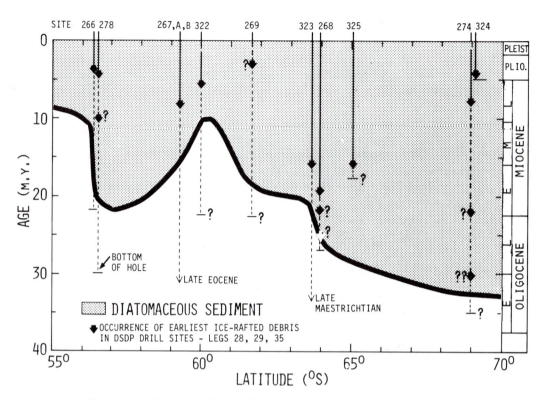

Figure 19-18 Temporal and geographic distribution of diatomaceous sediment and ice-rafted debris in the Southern Ocean DSDP sections drilled during Legs 28, 29, and 35. Note diachronous latitudinal distribution. (From Tucholke, Hollister, Weaver, and Vennum, 1976)

show major changes during the Neogene [Kennett and others, 1975; Brewster, 1977; 1980]. The Paleogene is marked by a fairly constant low level of productivity (apart from a questionable middle-Eocene peak). At Site 278 the Oligocene is represented by carbonate oozes. No siliceous oozes had yet been deposited. During the early Miocene, siliceous sedimentation commenced in this area; however, low sedimentation rates suggest that upwelling was sluggish but beginning to increase in northern Antarctic waters as the Antarctic convergence developed. By the middle Miocene alternations of siliceous-rich and calcareous-rich biogenic sediments in DSDP Site 278 almost certainly record minor latitudinal fluctuations of the Antarctic convergence.

Brewster [1977; 1980] has distinguished two principal episodes of enhanced siliceous productivity in the Antarctic during the Neogene (Fig. 19-19). One centered in the earliest Miocene about 22 Ma and the other in the late Cenozoic from 5 Ma to the present day. During the late Cenozoic, after an initial increase in opal accumulation from 8 to 5 Ma, opal accumulation rates exhibit a fourfold increase to 1100 g/cm^2/m.y. at about 5 Ma, followed by a temporary decrease to 666 g/cm^2/m.y. about 3 Ma. Since then, opal accumulation rates have increased to the

ANTARCTIC   CENTRAL PACIFIC

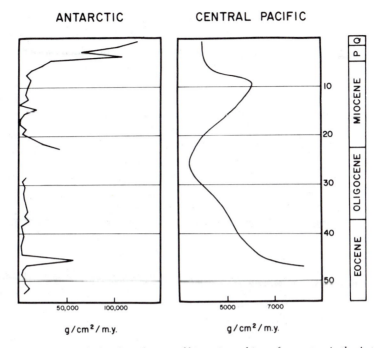

g/cm²/m.y.   g/cm²/m.y.

Figure 19-19   Calculated production of biogenic opal in surface waters in the Antarctic and central Pacific through the Cenozoic, assuming a constant rate of
dissolution through time. Pacific data are from Leinen [1977]. (From N. A.
Brewster, *Geol. Soc. Amer. Bull.* vol. 91, p. 345, 1980, courtesy The Geological
Society of America)

present Cenozoic maximum of 1240 g/cm²/m.y. [Brewster, 1980]. The particularly
high rates in the Quaternary are assumed to be related to higher oceanic turnover in
response to the late Cenozoic glacial development [Kennett and others, 1975].

Since the Oligocene, times of increasing productivity in the Antarctic correspond to periods of decreasing productivity in the central equatorial Pacific (Fig.
19-19). This is tied to the changing paleoceanographic patterns during the
Cenozoic. The efficiency of the biological cycling of silica has progressed in the
Antarctic to the extent that much of the silica assimilation and accumulation has
been transferred to the Antarctic at the expense of other productive oceanic regions,
such as the central equatorial Pacific [Brewster, 1977].

In addition to the large oceanic sedimentological changes that began near the
base of the Neogene, major permanent change also began in the character of
oceanic faunas and floras.

The Paleogene-Neogene transition is marked in the world ocean by a major
biogeographic change involving both the development of distinct and permanent
latitudinal belts of planktonic assemblages from the tropics to the poles and the
development of permanently high faunal-floral diversity gradients between the
tropical and polar areas [Kennett, 1978]. Oceanic circulation during the Paleogene
was also marked at times by the presence of these distinct latitudinal planktonic

belts and high-diversity gradients (such as during much of the Eocene). At other times, however, especially during the Oligocene and the Paleocene, faunal and floral gradients were less distinct, species had much wider distributions, and latitudinal differences in planktonic assemblages were much less conspicuous. During the Neogene, this water-mass-related biogeographic provinciality largely retained its identity despite major climatic coolings. Thus fundamental patterns of water-mass distributions characteristic of the present-day oceans became established during the early Neogene, although many changes have since occurred in the geographic extent of the water masses, their relative movements, and the intensity of oceanic processes, thereby affecting planktonic biogeography.

Evolutionary radiations were renewed in the early Miocene in response to these new paleoceanographic settings. For instance, the siliceous planktonic groups became extremely important in the Southern Ocean at the beginning of the Neogene as the Antarctic water mass developed and siliceous biogenic productivity increased. The Miocene marks the beginning of rapid evolutionary change in Antarctic radiolarian assemblages. Strong endemism in Antarctic radiolarian assemblages resulted and modern radiolarian characteristics began to develop as the ancestors of modern forms began to evolve [Petrushevskaya, 1975]. These changes became progressively more pronounced through the Miocene. Planktonic foraminifera began new Neogene radiations throughout the oceans. Also, the disappearance of larger penguins by the early Miocene probably resulted from competition and predation from toothed whales and seals which, as we have seen, began to evolve rapidly at the beginning of the Miocene [Stonehouse, 1969].

ANTARCTIC ICE-SHEET FORMATION: MIDDLE MIOCENE

The middle Miocene represents the next crucial stage in the development of global climates, for at this time (about 14 Ma) most of the Antarctic ice sheet formed [Shackleton and Kennett, 1975; Savin and others, 1975]. This event is marked by a sharp increase in the $\delta^{18}O$ values of calcareous plankton and of benthonic foraminifera (Fig. 19-9). This increase certainly reflects in part a major period of ice-sheet growth, as well as a drop in surface temperatures on the Antarctic coast. Shackleton and Kennett [1975A] concluded that by the late Miocene, bottom temperatures were near freezing and that the East Antarctic ice sheet had achieved approximately its present dimensions, which are limited by the coastline of the continent. On the other hand, Savin and others [1975] suggested that the data did not allow differentiation between ice buildup and decrease in temperature, although they believed that by the early part of late Miocene a large Antarctic ice sheet had formed. Furthermore, comparison of the oxygen isotopic records between high and low latitudes (Fig. 19-9) suggests that during the middle-Miocene event, the planetary temperature gradient markedly steepened and temperatures at high and low latitudes became much less closely coupled. The existence of a widespread ice sheet from the middle Miocene is confirmed by the presence of common and persistent ice-rafted sediments around the Antarctic continent from that time. Ice-rafted sediments in older sediments are much less common and are restricted to areas closely adjacent to the continent. The first reported ice-rafted debris in DSDP Site 278 on the Antarctic convergence in the Southwest Pacific occurred in the middle Miocene [Margolis, 1975, Tucholke and others, 1976; Craddock and Hollister, 1976]. It is generally believed that the East Antarctic ice sheet (by far the largest on Antarctica) has existed in essentially its present form since the middle Miocene and has not undergone any significant reduction in size.

The formation of the East Antarctic ice sheet in the middle Miocene seemed to have led to a further decrease in oceanic bottom temperatures supporting the contention of Savin and others [1975] that part of the oxygen isotopic change resulted from temperature change, as well as from the buildup of ice. Evidence for this comes from the large changes at that time of deep-sea benthonic foraminiferal assemblages. Coincident with the isotopic change was the replacement of numerous species originating in the Oligocene or earlier assemblages composed of taxa that dominate the late Cenozoic and modern deep-sea environment [Woodruff and Douglas, 1981]. This change was both distinctive and rapid, which is unusual for deep-sea benthonic foraminiferal assemblages as they tend to change rather slowly. A temperature decrease of bottom waters took place from 5° to 2°C [Schnitker, 1980].

The number of hiatuses also abruptly increased in the equatorial Pacific Ocean about 15 m.y. ago, probably due to an increase in corrosiveness of bottom water resulting from the development of this new AABW upon the formation of the Antarctic ice sheet [van Andel and others, 1976]. Also this might account for a coeval shoaling of the CCD in the equatorial Pacific and the development of a narrower equatorial carbonate zone [van Andel and others, 1975]. Atmospheric circulation probably also increased in strength and is reflected by a slight increase in the rate of biogenic sedimentation in the Antarctic [Brewster, 1980].

Since the development of the East Antarctic ice sheet, global climates have at no time returned to the previous warmth that occurred during the early Miocene and the early part of middle Miocene (a climax for the Neogene). This cooling episode seemed to have created major change in the character of terrestrial plants, perhaps over wide areas. Tropical forests over parts of East Africa were replaced at this time by the first woodland-grassland habitats, probably because of the development of drier climates [Andrews and van Couvering, 1975]. Mammalian evolution responded by the first large-scale development of grazing (as compared with browsing) mammals. Perhaps also of great significance was the first known appearance of apparently bipedal primates, *Ramapithecus* [Pilbeam, 1972], the evolution of which may have been triggered by the changing habitats in East Africa and perhaps part of Asia. Because *Ramapithecus* is considered to be a strong possibility in the primate lineage that ultimately led to *Homo sapiens,* the global climatic changes related to the development of the East Antarctic ice sheet may have indirectly influenced human evolution.

The exact cause of development of the East Antarctic ice sheet remains unexplained. Why did it take 20 m.y. for the Antarctic ice sheet to form when the continent had been well isolated by surrounding oceans in the Oligocene and when conditions seemed to have been favorable for major glacial development well before the middle Miocene? Why did the Antarctic ice sheet form at a time of globally warm climate? The answer may well lie with plate-tectonic events that took place not around Antarctica, but in the northern North Atlantic [Moore and others, 1978; Schnitker, 1980]. There the Iceland ridge had effectively isolated the Arctic Ocean from the Atlantic since the middle Eocene (Fig. 19-20), when sea-floor spreading had been initiated between Greenland and Scandinavia. By the latest Oligocene the first segments of the Iceland ridge had begun to subside beneath sea level [Vogt, 1972; Schrader and others, 1976; Talwani and Eldholm, 1977]. This early subsidence of the ridge (Fig. 19-20) allowed North Atlantic surface waters to flow into the Norwegian Sea for the first time, but did not allow the passage of any deep-water flow. However the surface water that flowed into the Norwegian Sea was probably warm and of high salinity, characteristics which were inherited from very salty, warm waters flowing from the nearby Mediterranean Sea. This water cooled and formed the Norwegian Sea which, although probably not very cold, was highly saline and thus dense and flowed into the North Atlantic as a new bottom water (Fig. 19-21). The flow of this Norwegian overflow increased bottom-water circulation in the North Atlantic and created major changes in the patterns of sedimentation [Tucholke and Vogt, 1979]. In the North Atlantic a clear change occurred in the depositional style near the beginning of the Neogene along the deeper part of the continental margin, reflecting the influence of these new abyssal currents. The Neogene sediments often show the effects of current-controlled deposition compared with older sediments which were ponded, draped, or deposited within submarine fans [Tucholke and Vogt, 1979].

By the middle Miocene, the Iceland ridge had sunk below sea level (Fig. 19-20) linking the Arctic Ocean as a heat sink to the Atlantic and thus the world ocean. This created fundamental changes in global deep-water circulation

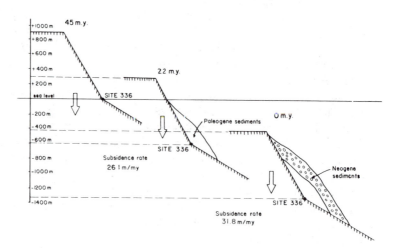

Figure 19-20 Subsidence history of the top of the Iceland–Faeroe ridge and of DSDP Site 336 from 45 Ma to the present. (From T. H. Nilsen and D. R. Kerr, *Geological Magazine*, vol. 115, p. 174, 1978)

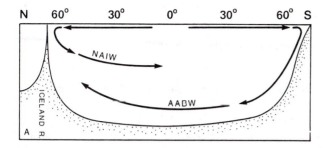

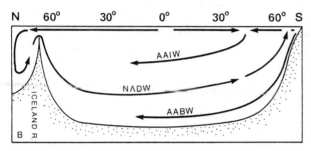

Figure 19-21 Cross sections showing inferred general paleocirculation patterns of the Atlantic Ocean. (A) Pre-Miocene with emergent Iceland-Faeroe ridge; (B) post mid-Miocene, with submerged Iceland–Faeroe ridge and Norwegian Sea overflow. (From D. Schnitker, 1980; reprinted by permission from *Nature*, vol. 284, pp. 615–616. Copyright © 1980 Macmillan Journals, Ltd.)

[Schnitker, 1980]. To replace increasing volumes of outflowing cold, dense bottom waters, increasing amounts of surface waters were drawn to the north. The gradual closing of the Central American Seaway during the late Cenozoic also probably assisted by intensifying the northward flow of warm waters by way of the Gulf Stream. By the middle Miocene the outflow of this NADW was sufficiently large to affect the water structure of the oceans of the Southern Hemisphere. According to Schnitker [1980] the NADW then flowed the entire length of the Atlantic Ocean and inserted itself from below into the Circum-Antarctic Current system as an intermediate water mass (Fig. 19-21). The insertion of NADW destabilized the vertical structure of this system so that the warm, saline North Atlantic water began to upwell south of the Antarctic convergence, replacing the previous upwelling of cold, low-salinity local waters from shallower depths (Fig. 19-21). The upwelling water was now relatively warm and contained heat that was converted to latent heat by evaporation. High evaporation rates thus supplied the required moisture to the Antarctica to build the ice sheet. Thus, although Antarctica had been cold enough to form an ice sheet for at least 25 m.y. before the middle Miocene, no significant ice buildup had formed until then because of insufficient precipitation [Schnitker, 1980]. This theory is supported by the fact that the ice sheet formed largely during a period of increased surface-water temperatures at high southern latitudes [Kennett, 1977]. The middle Miocene increase in NADW production further stimulated bottom-water flow and began to form conspicuous sediment drift accumulations of reworked sediments in the North Atlantic [Tucholke and Ewing, 1974] (see Chapter 15).

THE TERMINAL MIOCENE EVENT:
GLOBAL COOLING
AND THE MEDITERRANEAN SALINITY CRISIS

Since the middle Miocene, the earth's climatic regime has been in a glacial mode. During this entire interval, the East Antarctic ice sheet has been a permanent feature, although oxygen isotopic evidence indicates some changes in volume. Global climates were cool during much of the late Miocene and ice-rafting and the siliceous biogenic sediment belt continued northward expansions. Ice-rafted debris became conspicuous for the first time in the Falkland plateau region in the late Miocene. Also at this time there was a distinct and rapid northward movement (300 km) in the Antarctic siliceous biogenic sediment belts, as well as a further increase in opal production [Kemp and others, 1975; Tucholke and others, 1976]. This event is equated with a rapid northward movement of the Antarctic convergence and is related to a widespread latest Miocene cooling about 6.5 to 5.0 Ma. A large body of paleontological evidence indicates that during the latest Miocene, very cool water masses expanded over New Zealand, California, and other regions and terrestrial climates became cooler. Strong microfossil evidence suggests a substantial cooling episode during the latest Miocene [Bandy, 1967; Ingle, 1967; Wolfe and Hopkins, 1967; Barron, 1973; Kennett and Vella, 1975; Kennett and Watkins, 1976], including the Kapitean stage of New Zealand [Kennett, 1967]. A distinct decrease in precipitation also occurred over much of Australia, in addition to severe climatic cooling [Kemp, 1978].

This cooling episode may have been related to an expansion of the Antarctic ice sheet, although the data are conflicting. For instance, the oxygen isotopic data of Shackleton and Kennett [1975] support significant glacial expansion of the Antarctic ice sheet, supporting the results of continental glacial geology [Denton and others, 1971; Mayewski, 1975]. On the other hand, isotopic data of Keigwin [1979] from the equatorial Pacific indicate no single large ice buildup restricted to the latest Miocene, but several through the late Miocene. Support of at least some Antarctic ice expansion can be inferred from the discovery by Mercer [1978] of glacial tills in southern South America of latest Miocene age (6.75 to 5 Ma). These represent the earliest known tills from this area and reflect glacial expansion beyond the Andean Mountain front. Mercer [1978] considered that this glacial expansion may reflect the initial growth of the West Antarctic ice sheet to the south of South America. The West Antarctic ice sheet is largely marine based, set within an archipelago now mostly overridden by the ice sheet. The growth of such an ice sheet requires the accumulation of cold ice from the beginning, which requires colder summers (perhaps as much as 10°C colder) than for the buildup of the land-based East Antarctic ice sheet [Mercer, 1978]. This requirement of colder conditions suggests a possible later development than the East Antarctic ice sheet, which formed largely during the middle Miocene. The required cooler conditions needed for development of the West Antarctic ice sheet may thus have occurred during the latest Miocene.

Perhaps the apparent absence of a large global oxygen isotopic signal heralding further major ice accumulation [Keigwin, 1979] in association with widespread surface-water cooling in the latest Miocene can be accounted for by the expansion of the West Antarctic ice cap and associated ice shelves. This large glacial expansion does not involve accumulation of large amounts of ice on the continents [Ciesielski and others, 1981]. Another aspect of glacial importance is that the West Antarctic ice sheet is inherently unstable and can survive only so long as its grounded portion is buttressed by fringing ice shelves—particularly the enormous Ross and Filchner-Ronne ice shelves. If this is so, it follows that the ice shelves must have formed before or at the same time as the West Antarctic ice sheet.

During the latest Miocene, abundant evidence exists for intensification of oceanic circulation. This would have resulted from higher temperature gradients between the polar and equatorial regions, increasing wind velocities, and enhanced upwelling. In addition to the increase in siliceous biological productivity in the Southern Ocean [Kennett, 1977; Brewster, 1977], there was also similar increase in upwelling in the equatorial Pacific [Leinen, 1976]. Furthermore, widespread diatomaceous sediments were deposited around the Pacific rim [Ingle, 1973; Ingle and Garrison, 1977]. Dramatic increases in carbonate biogenic sedimentation rates in Indonesia have been attributed to increased upwelling related to latest Miocene [van Gorsel and Troelstra, 1981]. Sustained upwelling also began in the Benguela current system off West Africa, based on sedimentological and paleontological evidence [Siesser, 1980]. Also a particularly intense acceleration of the Gulf Stream system occurred during the latest Miocene, as indicated by erosion on the Blake Plateau [Kaneps, 1979]. Phosphorite deposition increased at this time in many shallower water marine sequences [Carter, 1978], including Southeast Australia, the Aghulas Bank, the Campbell Plateau south of New Zealand, the Chatham Rise east of New Zealand, and Florida; this can be attributed to increased upwelling and biological productivity.

The latest Miocene is marked by three other important events.

1. A carbon shift occurring almost simultaneously with the onset of the global cooling.

2. A major sea-level regression.

3. The isolation of the Mediterranean Basin, leading to an enormous salt buildup.

These are discussed in turn as follows: The latest Miocene carbon isotope shift is a marked semipermanent shift in ^{13}C to ^{12}C ratios in the tests of foraminifera [Keigwin, 1979; Bender and Keigwin, 1979; Keigwin and Shackleton, 1980], which has been paleomagnetically dated by Loutit and Kennett at 6.3 Ma. The shift is toward lighter $\delta^{13}C$ values by about 0.8‰ and is considered to reflect a geologically instantaneous change in the rate of turnover of oceanic circulation [Bender and Keigwin, 1979]. Because of its extremely widespread occurrence throughout the oceanic sediment sequences [Haq and others, 1980; Vincent and others, 1980] and its age, it is almost certainly tied to the paleoceanographic changes associated with the latest Miocene glacial episode.

Increased glacial activity of Antarctica has also been cited as the cause of a major lowering of sea level during latest Miocene [Kennett, 1967; Vail and others, 1977]. Shackleton and Kennett [1975] calculated that an isotopic change of about 0.5 per mil and an isotopic composition for the removed water of 150 per mil would cause a glacioeustatic lowering of sea level of about 40 m. This assumes that the ice was all stored above sea level, which is not the case during the buildup of much of the West Antarctic ice sheet. A latest Miocene regressive phase is indeed widespread throughout much of the shallower marine areas of the oceans. Among the best-dated sections are those facing the eastern North Atlantic. In the Andalusian stratotype of southern Spain, Berggren and Haq [1976] documented a drop in sea level of about 50 m during the latest Miocene. Other areas where unconformities or shallowing of facies are well documented in the latest Miocene include New Zealand [Kennett, 1967], Australia [Carter, 1978], and Fiji [Adams and others, 1979], as well as Florida, the Atlantic coastal plain of North America [Adams and others, 1977], and many other areas. Paleodepth interpretations in the New Zealand marine sections [Kennett, 1967] indicated that depths of deposition were in general greater before the latest Miocene (Kapitean stage) regression than following it in the early Pliocene. If such depth differences resulted from glacioeustatic causes, then it follows that much of the inferred ice that built up on Antarctica remained during a return to warmer climatic conditions in early Pliocene [Shackleton and Kennett, 1975]. There is even some evidence that the glacioeustatic lowering of sea level during the late Miocene may have caused isolation of the Sea of Japan from the Pacific, perhaps turning it into a freshwater lake [Burckle and Akiba, 1978].

The late-Miocene glacioeustatic fall in sea level may also have at least partly caused one of the most spectacular geologic events of the entire Cenozoic—the isolation of the Mediterranean Sea from the world ocean during the Messinian stage. The Messinian stage represents the time elapsed between the end of the Tortonian stage and the beginning of the Pliocene (Tabianian stage) and has been dated at between about 6.2 and 5 Ma (Fig. 19–22) [Benson, 1976]. During this time the Mediterranean was transformed into a series of large inland lakes (*Lago Mare*), in which was precipitated a thick and extensive (1×10^6 km^3) sequence of evaporites including gypsum, halite, and other salts.

The modern Mediterranean Sea has a water volume of 3.7×10^6 km^3. The excess evaporation is 3.3×10^3 km^3/yr. If the Straits of Gibraltar were closed today, the present Mediterranean would dry up in about 1000 years! The late-Miocene evaporites are up to 2 or 3 km thick in places; one basin of Mediterranean seawater could not have supplied enough salt for such thick deposits. If 70 m of salt is deposited each time the basin was desiccated, it follows that the Mediterranean must have been desiccated about 40 times during the latest Miocene [Ryan, 1973], withdrawing a total of about 6 percent of the salt from the world ocean.

Because the event seems to have created an almost complete marine biotic sterilization, it has been called the **salinity crisis** [Cita, 1976]. There is no general consensus about what actually created the final isolation of the Mediterranean Basin in the latest Miocene. The final connection with the Atlantic is believed to have been the Betic Strait, entering into the Mediterranean from Andalusia, Spain. Although the closing of this connection—the *Iberian Portal*—is related to large-

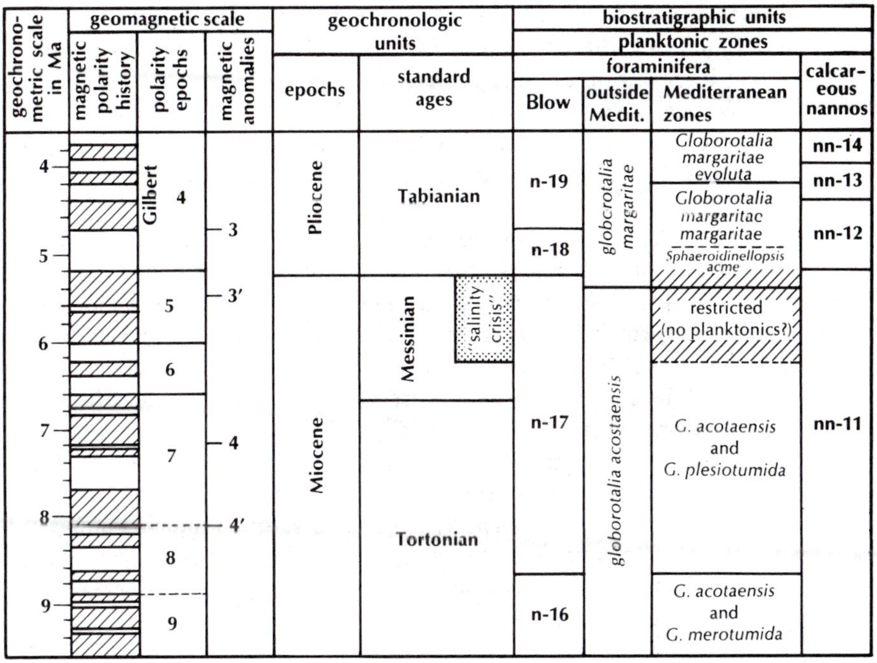

geochrono-metric scale in Ma	geomagnetic scale			geochronologic units		biostratigraphic units			
	magnetic polarity history	polarity epochs	magnetic anomalies	epochs	standard ages	planktonic zones — foraminifera			calcar-eous nannos
						Blow	outside Medit.	Mediterranean zones	
4	(Gilbert)	Gilbert 4	3	Pliocene	Tabianian	n-19	globorotalia margaritae	Globorotalia margaritae evoluta	nn-14
									nn-13
5						n-18		Globorotalia margaritae margaritae	nn-12
								Sphaeroidinellopsis acme	
		5	3'	Miocene	Messinian "salinity crisis"			restricted (no planktonics?)	
6		6				n-17	globorotalia acostaensis	G. acotaensis and G. plesiotumida	nn-11
7		7	4		Tortonian				
8		8	4'						
9		9				n-16		G. acotaensis and G. merotumida	

Figure 19-22 Late Miocene-early Pliocene stratigraphy in relation to the Mediterranean "salinity crisis" (modified from Ryan et al., 1974; Berggren and van Couvering, 1974; Stainforth et al., 1975; Cita, 1976). (From L. A. Smith, 1977)

scale plate movements between Africa and southern Europe, its final severing may have resulted from the lowering of sea level in the latest Miocene [Adams and others, 1977]. The synchronism between the beginning of the salinity crisis and the global cooling and sea-level fall as dated in the Southern Hemisphere [Loutit and Kennett, 1979] strongly suggests that the isolation was triggered by glacioeustatic change.

The history of desiccation of the isolated Mediterranean basins is complicated, but major evaporite deposition occurred in two main phases [Hsu and others, 1977]. Various theories have been proposed to account for the buildup of evaporitic sediments, all of which require a restricted, shallow portal. Major arguments have centered on whether the Mediterranean was a shallow basin during the latest Miocene or a deep basin, as in the present day, and whether sea level was maintained at or far below that of the world ocean. It is generally believed that the Messinian evaporites were formed in shallow-water conditions [Cita, 1976]. Such a conclusion is based upon (1) evidence of subaerial exposure of the stratigraphically lowermost evaporitic rocks, and (2) erosional surfaces developed beneath the main salt layer on the Mediterranean margins, which extend toward the center of the major basins [Ryan and Cita, 1978]. For example, major rivers flowing into the Mediterranean, such as the Nile and Rhone, have buried gorges incised for hundreds of meters below sea level and extending well upstream. At Aswan, which is 800 km from the mouth of the Nile River, a gorge filled with Pliocene-Quaternary sediments extends 200 m below the present level of the Mediterranean. These observations indicate that the level of the sea in the Mediterranean was well below its present level during the late Miocene.

We have seen earlier that the volume of the Mediterranean evaporite, calculated at 1×10^6 km^3, lowered the salinity of the global ocean by 6 percent, or a mean reduction in oceanic salinity of 2‰. This decrease in salinity should have had important paleoclimatic repercussions [Ryan and others, 1974] because it would have raised the freezing point of seawater, allowing sea ice to form at slightly higher temperatures. This would, in turn, have increased the albedo of the earth and hence stimulated Antarctic ice growth during latest Miocene. The hypothesis of Ryan and others [1974] further suggests that the salinity crisis of the Mediterranean was the cause, not the effect, of the global cooling and glacial expansion of Antarctica during latest Miocene. However, a more likely scenario of paleoclimatic cause and effect is that the Antarctic ice sheets expanded sufficiently to lower sea level enough to isolate the Mediterranean Basin, after which paleoenvironmental feedback occurred which at least temporarily encouraged global cooling.

All of this changed abruptly about 5 m.y. ago at the beginning of the Pliocene when the Mediterranean Basin was permanently filled again. Normal marine conditions were restored and deep-sea carbonate oozes and hemipelagic clays were once again deposited [van Couvering and others, 1976].

THE LATE PLIOCENE EVENT: ONSET OF THE ICE AGE

By the early Pliocene, environmental conditions in the Southern Ocean began to approach closely those of the Quaternary. Higher rates of siliceous biogenic productivity than during the Miocene suggest further intensification in oceanic circulation, but it still was conspicuously lower than the maximum levels that occurred during the Quaternary [Kennett and others, 1975; Tucholke and others, 1976; Brewster, 1980]. Antarctic glaciation itself may have been less extensive than during the late Miocene. An early-Pliocene ice retreat [Shackleton and Kennett, 1975; Berggren and Haq, 1976] resulted in subsequent marine transgression. Geological evidence on the Antarctic continent also indicates a conspicuous decrease in the volume of Antarctic ice following an increased glacial episode dated at greater than 4.2 Ma [Mayewski, 1975]. Pliocene ice-rafted sediment limits were further south than during the Quaternary, suggesting less extensive glaciation [Kennett and others, 1975]. Global climatic conditions were generally warmer during the early and middle Pliocene, compared with the late Miocene and the following late Pliocene to Quaternary.

With the firm establishment of the Antarctic convergence by the Pliocene, calcareous nannofossils were excluded from Antarctic waters and coccolith deposition ceased in this region. In contrast, siliceous microfossil groups are very well established by the early Pliocene. The radiolarian sequence is marked by rapid evolutionary changes. The base of the Pliocene is marked by the first appearance of the most important modern Antarctic forms, including *Antarctissa*.

It was in the late Pliocene—about 3 m.y. ago—that the most significant Pliocene geologic event occurred with the formation of Northern Hemisphere ice sheets. It is clear that the formation of these ice sheets occurred much later than those on Antarctica and represents the next major global climatic threshold. Since then, major oscillations have continued in Northern Hemisphere ice sheets, forming the classic Quaternary glacial and interglacial episodes [Emiliani, 1954; Shackleton and Opdyke, 1973]. The strongest evidence for this ice accumulation is a paleomagnetically dated oxygen isotopic change of about 0.4‰ dated at 3.2 Ma (just below the Mammoth event; see Fig. 19–23) [Shackleton and Opdyke, 1977]. This date is in substantial agreement with the inception of ice-rafting observed in drilled cores in the North Atlantic [Berggren, 1972; Poore, 1981]. From 2.5 to 1.8 Ma, during the latest Pliocene, glacial isotopic excursions of about 1.0‰ began (Fig. 19–24). During the Quaternary, isotopic differences between glacial and interglacial episodes are about 1.60‰ (Fig. 19–24). Thus by the late Pliocene there is clear evidence that glaciations of a magnitude of at least two-thirds that of the late-Pleistocene glacial maxima were occurring. This also was probably the scale of glaciation through the early Pleistocene [Shackleton and Opdyke, 1977].

In addition to the changes in oxygen isotopes, other strong evidence exists for the development of the Northern Hemisphere ice sheets during the late Pliocene. This includes glacial deposits between 2.7 and 3.1 Ma in the Sierra Nevada [Curray, 1966] and evidence of extensive ice sheets overlying basalt dated about 3.1 m.y. in Iceland [McDougall and Wensink, 1966]. Distinct surface water-mass cooling in the Southern Hemisphere is associated with the ice buildup during the late Pliocene [Kennett and others, 1971; 1979]. In the Mediterranean Sea, Thunell [1979] found

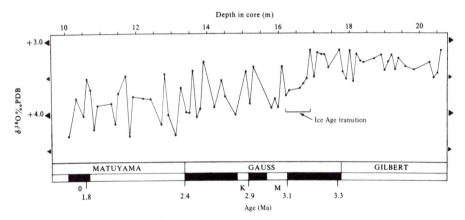

Figure 19-23 Oxygen isotopic variations in core V28–179 through the middle and late Pliocene plotted against paleomagnetic stratigraphy. Note the shift in oxygen isotopic values in the early Gauss paleomagnetic epoch considered to represent an ice-age transition. In effect this represents the initial development of Northern Hemisphere ice caps. Magnetic events are indicated by 0 (Olduvai), K (Kaena), and M (Mammoth). Isotopic measurements were made on the benthonic foraminifera *Globocassidulina subglobosa*. (From N. J. Shackleton and N. Opdyke, 1977; reprinted by permission from *Nature*, vol. 270, pp. 216–219. Copyright © 1977 Macmillan Journals Ltd.)

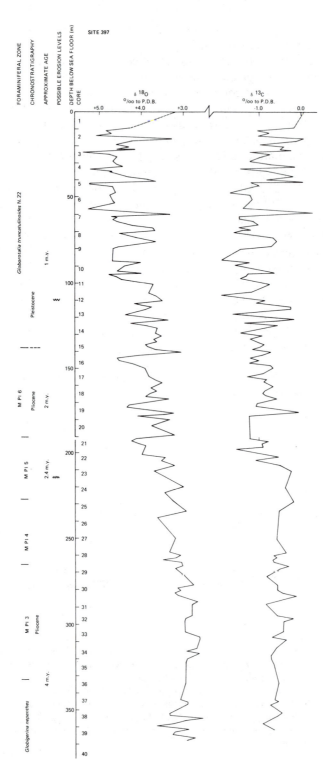

Figure 19-24 Oxygen and carbon isotopic record for the Pliocene and Quaternary in DSDP Site 397 in the Atlantic off Northwest Africa. Note steps in oxygen isotopic record at about 210 m, 130 m, and 60 m, reflecting successive increases in ice accumulation on the continents. (From N. J. Shackleton and M. B. Cita, 1979)

quantitative foraminiferal evidence for a cooling beginning about 3.2 Ma. Isotopic results from this same sequence suggest a surface-water cooling at that time, followed by an ice-volume increase between about 3.0 and 3.1 Ma [Keigwin and Thunell, 1979]. In the North Pacific, ice-rafted debris has been identified in cores as old as 2.4 m.y. [Kennett and others, 1971]. It must be stressed that ice-rafted debris has been found at a number of high northern-latitude regions prior to 3.2 m.y. ago in the early Pliocene, but it is much more scattered and usually occurs at very high latitudes [Warnke and Hansen, 1977; Margolis and Herman, 1980]. This probably represents montane glaciation rather than the development of continental ice sheets. It also seems that the Arctic Ocean has been continually ice covered, at least since the middle Pliocene [Clark, 1971].

It is difficult to explain the relatively sudden onset of Northern Hemisphere ice-sheet formation in the late Pliocene nearly 10 m.y. later than that of Antarctica, but the explanation is probably related to an intensified late-Cenozoic orogenic phase [Hamilton, 1965] and further changes in oceanic circulation patterns, especially those resulting from the final closure of the Middle American Seaway [Hamilton, 1965; Keigwin, 1976]. This separation of the North Atlantic from the Pacific may have increased northward transport of warm Gulf Stream waters and significantly increased precipitation in high latitudes. Saito [1976A] and Keigwin [1978] date the final elevation of the Central American Isthmus between 3.1 and 3.5 m.y., which coincides with Northern Hemisphere ice buildup. That the Gulf Stream was amplified at this time is consistent with observations of sediment patterns made from the Yucatan Channel between Mexico and Cuba. Prior to 3.4 Ma, the sediments are unwinnowed and a drastic change in the depositional regime at about that time is attributed to increased transport of the Gulf Stream [Brunner, 1978]. The effect of an intensified Gulf Stream system would be to provide an increased moisture supply from a warmer subpolar North Atlantic. A vigorous meridional atmospheric circulation directed along the strong surface thermal gradient off the east coast of North America is the regime most compatible with the process of rapid development of the Laurentide ice sheet over North America (Fig. 19-25) [Ruddiman and others, 1980].

THE QUATERNARY GLACIAL FINALE

We have seen that although severe cooling and glaciation began in the Northern Hemisphere during the late Miocene [Ingle, 1967; Bandy, 1967; Wolfe and Hopkins, 1967], the development of ice sheets on North America and Scandinavia did not begin until the late Pliocene, about 3 m.y. ago. A critical global climatic threshold was passed at that time; following this, the earth has exhibited glacial oscillations, reflecting times of ice sheet accumulation and melting in the Northern Hemisphere. Thus Northern Hemisphere ice-sheet fluctuations began during the Pliocene rather than in the Quaternary.

The base of the Quaternary (Pliocene-Pleistocene boundary) is defined stratigraphically in the boundary stratotype located in Le Castella, southern Italy

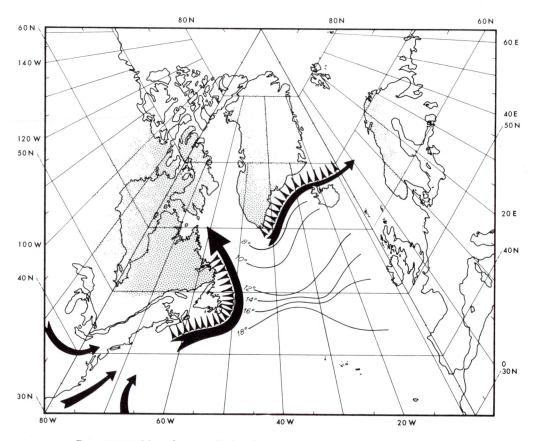

Figure 19-25 Major features of inferred atmospheric circulation patterns in the North Atlantic 75,000 years ago, a time of intermediate ice growth in the Northern Hemisphere. Ice cover shown by stippling. Estimated summer sea-surface temperatures shown for North Atlantic. Main tracks of cyclonic storms shown by heavy arrows, and major thermal fronts and albedo boundaries of the earth's surface indicated by thin black wedges. (From W. F. Ruddiman et al., *Quaternary Res.*, vol. 13, p. 58, 1980)

(see Chapter 3). This level is dated paleomagnetically at about 1.6 Ma and is at, or slightly younger than, the top of the Olduvai normal paleomagnetic event. The Quaternary was originally established as that "time distinguished by severe climatal conditions throughout the great part of the northern hemisphere" [Forbes, 1846]. This statement remains an accurate generalization of the Quaternary period. A considerable body of paleontological and oxygen isotopic evidence clearly shows that the most obvious difference between the Quaternary and the Tertiary is that glaciations were more severe and extensive, particularly on the Northern Hemisphere continents but probably also on Antarctica and by extremes recorded in global climatic oscillations involving glacial-to-interglacial changes. Furthermore, during the last million years, the basic climate of the earth has been glacial: More than 90 percent of the time, extensive land areas were covered by ice, and cooler air and ocean temperatures prevailed. Periodically, this glacial regime has been interrupted by intervals of relative warmth (for example, the Holocene). In essence, the Quaternary has been marked by the development and destruction of Northern Hemisphere ice sheets, producing the classical Quaternary glacial and interglacial episodes. The degree of change that the Antarctic ice sheets underwent during Quaternary glacial to interglacial oscillations is still largely unknown, although there is some evidence of significant changes in volume of the East Antarctic ice cap [Mayewski, 1975]. The most striking difference in the Southern Hemisphere was the much greater extent of sea ice [CLIMAP, 1976]. It seems, however, that global climatic effects upon the oceanic biosphere were driven more by glacial changes in the Arctic region than by those in the Antarctic. There is every reason to believe that the present interglacial episode will be short-lived and that the earth will return to the dominant, glacial regime.

In the relatively short Quaternary time span (about 1.6 m.y.), enormous changes have occurred on the earth.

1. As many as 30 glacial episodes, each of which has been related to extensive ice development at high and middle latitudes in the Northern Hemisphere and which drastically altered the biogeography of both terrestrial and marine organisms [Berggren and van Couvering, 1979].

2. Repeated large-scale latitudinal displacements of climatic zones by as much as 20 to 30°.

3. Large scale fluctuations in oceanic-circulation patterns.

4. Sea-level oscillations with a range of up to about 100 m (see Chapter 9).

5. Large-scale transportation of terrigenous sediments to the deep-ocean basins due to generally lower sea level and glacial erosion (see Chapter 13).

6. Considerable oscillations in oceanic-biogenic productivity and the rate of supply of biogenic sediments to the deep sea floor, as well as biogenic sediment dissolution in the deep-ocean basins. Also, an increase in the rate of deposition of nonbiogenic pelagic sediments, some of which were derived from arid and semiarid continental regions and transported to the open ocean by intensified winds.

7. Much of the evolution of intelligence of *Homo sapiens* and all of the related cultural evolution.

Furthermore, the increased glaciation of the Quaternary stimulated both vertical and horizontal oceanic circulation, as reflected by increased biogenic productivity in certain oceanic regions, including the Southern Ocean. In this area, siliceous biogenic productivity continued to increase through the late Pliocene and Quaternary, as indicated by increased siliceous biogenic sedimentation rates. Maximum levels for the entire Cenozoic are recorded in the late Quaternary [Kennett and others, 1975]. As a result the filter-feeding whales reached maximum size in the history of their evolution. Further northward, increases in ice-rafted limits also occurred, so that these reach a maximum for the entire Cenozoic [Margolis, 1975;

Hayes and Frakes, 1975]. Widespread erosion on the ocean floor during the Quaternary also probably resulted from related intensification of deep oceanic circulation [Kennett and Watkins, 1976], through increased AABW production, intensified flow of the Circum-Antarctic Current, and steepening of latitudinal temperature gradients in both the Northern and Southern hemispheres [Watkins and Kennett, 1972; Kennett and Watkins, 1976].

What effects did the Quaternary, glacially related paleoceanographic changes have on global marine biogeography? The severity of climatic change during the Quaternary might itself suggest a major biogeographic response. However, although extensive movements of marine planktonic provinces are evident during the Quaternary, no biotic crisis, as measured by the number of extinctions in planktonic microfossils, is evident in the marine record. Paleoenvironmental conditions during the Quaternary created little new stress on the planktonic biota. This should not be surprising, since the glacial, climatic, and paleoceanographic development took place over a long period of time, commencing in the early Tertiary. Also, Quaternary changes do not represent the creation of fundamentally new conditions, but rather geographic shifts in the boundaries between water masses. Paleoceanographic and climatic changes of greater biogeographic importance took place near the Eocene-Oligocene boundary and during the middle and late Miocene [Kennett, 1978].

The nature of glacial episodes within the Quaternary is now much better known, largely as a result of efforts by workers in the CLIMAP Project. The first major global paleoclimatic experiment completed by the CLIMAP group simulates global climate at a time that continental glaciers reached their maximum extent in the last ice age, approximately 18,000 years ago. In order to simulate the atmosphere of 18,000 years before present, four boundary conditions had to be reconstructed: (1) the geography of the continents; (2) the albedo of land and ice surfaces; (3) the extent and elevation of permanent ice; and (4) the sea-surface temperature pattern of the world ocean 18,000 years ago, using regression analysis of temperature-sensitive planktonic microfossils (see Chapter 17). In addition to the well-known occurrence of immense Northern Hemisphere ice sheets, which reached up to 3 km in thickness, this glacial episode was also marked by a greater extent of sea ice around Antarctica, lowered sea level (of at least 85 m), and an increase in grasslands, steppes, and deserts at the expense of forests. The oceans during such a glacial episode are marked by considerable steepening of thermal gradients along polar frontal systems and equatorward displacement of polar frontal systems (Fig. 19–26); general cooling of most surface waters with a global average decrease of 2 to 3°C; increased cooling and upwelling along equatorial divergences in the Pacific and Atlantic; cooler waters extending toward the equator along the western coasts of Africa, Australia, and South America (Fig. 19–26); and near-stable positions and temperatures of the central gyres in the subtropics. Overall, the entire ocean contained a more energetic circulation system [CLIMAP, 1976].

In the Northern Hemisphere the large shift in the path of Gulf Stream waters caused the Atlantic to exhibit the largest change. During major climatic changes, the North Atlantic polar front moved as a line hinged in the western Atlantic south to southeast of Newfoundland, sweeping out an arc larger than 45° and increasing

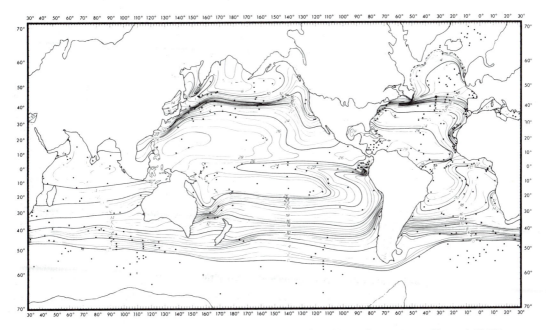

Figure 19-26 Sea-surface temperatures for Northern Hemisphere summer (August) 18,000 years ago as mapped by the CLIMAP Project. Contour intervals are 1°C for isotherms. Continental outlines represent a sea-level lowering of 85 m. The locations of all cores used in the sea-surface temperature reconstruction are shown by black dots. Continental ice extent shown for Northern Hemisphere. (Used with permission of CLIMAP)

in angle slightly to the northeast toward the European mainland [Ruddiman and McIntyre, 1977]. Its position is a critical monitor of whether warm saline NADW flowed into the Northeast Atlantic and Southeast Norwegian Sea or turned southeastward and was contained within the subtropical gyre. Only during peak interglacial portions of the last 800,000 years (as at present) did North Atlantic water penetrate into the Labrador Sea and west of Iceland [Ruddiman and McIntyre, 1977].

Oxygen isotopic and paleontological evidence has indicated that both amplitude and frequency of climatic change have increased through the late Pliocene and Quaternary, reaching maximum values during the late Quaternary from about 1 m.y. ago to the present day (Fig. 19-24). Several steps can be recognized, the latest of which has been paleomagnetically dated at 0.9 Ma at the Jaramillo paleomagnetic event [Shackleton and Opdyke, 1976]. Since then, about 11 glacial episodes have occurred (stages 22 to 1), and these represent glaciations of greatest extent in the Northern Hemisphere.

Since the work of Schott [1935], it has been recognized that deep-sea sediment sequences have been affected by the global climatic-interglacial cycles. In general, the major oxygen isotopic signals of the last several hundred thousand years exhibit an erratic saw-toothed geometry [Broecker and van Donk, 1970]; a gradual cooling led to maximum glaciation, generally ending at the same level of ice development and then followed by a rapid deglaciation or warming (terminations), culminating in a peak warm period. The trend toward increasing glaciation can be easily explained by feedback from increasing albedo, but the rapid periods of deglaciation are difficult to explain.

The question of the causal mechanisms of these paleoclimatic cycles has received considerable attention. The most widely accepted theory is that of Milankovitch [1938], who proposed that the orientation of the earth's axis of rotation and the shape of its orbit affect the distribution of sun's radiation (insolation) received over the earth's surface. Strong support for the theory was obtained from the deep-sea sediment record by Hays and others [1976]. They carried out power-spectrum analysis of the last half million years and showed that the climatic curve is dominated by fluctuations with period-lengths close to 100,000 years (50 percent of variance), 40,000 years (25 percent of variance), and 20,000 years (10 percent of variance) [Hays and others, 1976]. These peaks correspond to the dominant periods of the earth's solar orbit as follows [Hays and others, 1976]:

1. 42,000 year. Same period as variations in the obliquity of the earth's axis. Obliquity is the tilt of the earth's axis away from the plane of the earth's orbit. It is now 23.5°, but it slowly varies between 22.1 and 24.5°, completing one cycle every 41,000 years.

2. 23,000 year. Same period as the quasi-periodic precession index. The precession is the change in the orientation of the earth's axis. At this time the axis points toward the North Star. Precession determines at what point on the earth's elliptical orbit winter and summer occur. Warmer winters result when the earth is close to the sun.

The major problem of the Milankovitch theory is the unexplained dominance of the 100,000-year cycle. The only connection between this cycle and orbital variations is through the inferred insignificant effect of variations in the eccentricity or ellipticity of the earth's orbit. The eccentricity cycle causes at most a 0.1 percent change in insolation of the earth. Thus large changes in climate are linked only to small changes in the orbital characteristics. Other major problems concerning the Quaternary climatic cycles involve the means by which orbital variations are translated into regional climatic changes, the mechanism of transport of water from the oceans needed to build the ice sheets, and the problem of rapid deglaciation.

Orbital variations also do not explain the steplike increases in the glacial state of the earth during the Pliocene to Quaternary (Fig. 19-24). In fact, no single theory adequately explains all of the characteristics of late-Cenozoic climatic history. Many theories exist and a number of other processes are sure to have modulated climatic change. Such theories include variations in heat output from the sun, the concentration of interstellar dust, changes in the earth's magnetic field, surging of ice sheets, resulting from their possible mechanical instability, albedo changes resulting from sea-level variation, and the amount of volcanic dust injected into the atmosphere. Kennett and Thunell [1975; 1977], for instance, discovered that the Quaternary is marked by a conspicuous increase in the amount of explosive volcanism producing volcanic ash. The correlation of increased explosive volcanism during the Quaternary with rapid deterioration in climate suggests a link between the two.

Whatever the cause of Quaternary climatic cycles, a very intriguing aspect has been their modulation of many of the earth's processes and characteristics, including oceanic circulation, ocean chemistry, cycles in carbonate and siliceous biogenic productivity and dissolution, sea-level change, and terrigenous sedimentation in the oceans and biogeography.

References

ADAMS, C. G., and others [1977]. "The Messinian Salinity Crisis and Evidence of Late Miocene Eustatic Changes in the World Ocean," *Nature* 269: 383–86.

ADAMS, C. G., RODDA, P., and KITELEY, R. J. [1979]. "The Extinction of the Foraminiferal Genus *Lepidocyclina* and the Miocene/Pliocene Boundary Problem in Fiji," *Mar. Micropaleo.* 4: 319–39.

ALDRICH, L. T., and WETHERILL, G. W. [1958]. "Geochronology by Radioactive Decay," *Ann. Rev. Nucl. Sci.* 8: 257–98.

ALMAGOR, G., and GARFUNKEL, Z. [1979]. "Submarine Slumping in Continental Margin of Israel and Northern Sinai," *Am. Assoc. Petrol. Geol. Bull.* 63: 324–40.

ALVAREZ, L. W., ALVAREZ, W., ASARO, F., and MICHEL, H. V. [1980]. "Extraterrestrial Cause for the Cretaceous-Tertiary Extinction," *Science* 208: 1095–1108.

ALVAREZ, W., and others [1977]. "Upper Cretaceous-Paleocene Magnetic Stratigraphy at Gubbio, Italy. V. Type section for the Late Cretaceous-Paleocene Geomagnetic Reversal Time Scale," *Geol. Soc. Am. Bull.* 88: 383–89.

AMPFERER, O. [1906]. "Über das Bewegungsbild von Faltengebirge: Austria," *Geol. Bundesanst., Jahrb.* 56: 539–622.

ANDERSON, D. L. [1967]. "Latest information from seismic observations." In *The Earth's Mantle*, pp. 355–420, ed. T. F. Gaskill. New York: Academic Press.

ANDERSON, J. B. [1972]. "Nearshore Glacial-Marine Deposition from Modern Sediments of the Weddell Sea," *Nature Phys. Sci.* 240: 189–192.

ANDERSON, J. B., CLARK, H. C., and WEAVER, F. M. [1977]. "Sediments and Sediment Processes on High Latitude Continental Shelves." Ninth Annual Offshore Technology Conference, Houston, Texas.

ANDERSON, R. N., LANGSETH, M. G., and SCLATER, J. G. [1977]. "The Mechanism of Heat Transfer through the Floor of the Indian Ocean," *J. Geophys. Res.* 82: 3391–4409.

ANDREWS, P. B., and others [1975]. "Synthesis Sediments of the Southwest Pacific Oceans, Southwest Indian Ocean and South Tasman Sea," in *Initial Reports of the Deep Sea Drilling Project,* Vol. 29, p. 1147, ed. J. P. Kennett, R. E. Houtz, and others, Washington, D. C.: U.S. Govt. Printing Office.

ANDREWS, P. B., and VAN COUVERING, J. H. [1975]. "Palaeoenvironments in the East African Miocene," in *Approaches to Primate Paleobiology,* ed. F. S. Szalay, pp. 62–103, Basel, Switzerland: Karger.

ANIKOUCHINE, W. A., and STERNBERG, R. W. [1973]. *The World Ocean, An Introduction to Oceanography.* Englewood Cliffs, N.J.: Prentice-Hall, Inc.

ARGAND, E. [1924]. "La Tectonique de l'Asie," XIIIe Congr. Geol. Intern., Brussels, Compt. Rend., pp. 171–372.

ARKELL, W. J. [1956]. *Jurassic Geology of the World.* Edinburgh: Oliver and Boyd, Ltd.

ARNOLD, Z. M. [1964]. "Biological Observations on the Foraminifera *Spiroloculina Hyalina Schulze,*" *University of California Publications in Zoology 72.*

ARRHENIUS, G. [1952]. "Sediment Cores from the East Pacific," *Swedish Deep Sea Expedition Rep. No. 5* Götenborg, pp. 1–227.

ARTHUR, M. A. [1979]. "North Atlantic Cretaceous Black Shales: Record at Site 398 and a Brief Comparison with Other Occurrences," in *Initial Reports of the Deep Sea Drilling Project,* vol. 47, pt. 2, pp. 719–52, ed. J. L. Sibuet, W. B. F. Ryan, and others, Washington D.C.: U.S. Government Printing Office.

ARTHUR, M. A., and NATLAND, J. H. [1979]. "Carbonaceous Sediments in the North and South Atlantic: The Role of Salinity in Stable Stratification of Early Cretaceous Basins," in *Maurice Ewing Series 3,* pp. 375–401, ed. M. Talwani, W. Hay, and W. B. F. Ryan. American Geophysical Union.

ATWATER, T. [1970]. "Implications of Plate Tectonics for the Cenozoic Tectonic Evolution for Western North America," *Geol. Soc. Am. Bull.,* 81: 3513–66.

ATWATER, T., and MOLNAR, P. [1973]. Relative Motion of the Pacific and North American Plates Deduced from Sea-Floor Spreading in the Atlantic, Indian and South Pacific Oceans," in *Proceedings of the Conference on Tectonic Problems of the San Andreas Fault System,* pp. 136–148, Stanford, Calif: Stanford Univ. Publs., Geol. Seis., Vol. 13.

AUMENTO, F. [1967]. "Magnetic Evolution of the Mid-Atlantic Ridge," *Earth and Planet. Sci. Letts.,* 2: 225.

BAKER, G. [1959]. "Fossil Opal-Phytoliths and Phytolith Nomenclature," *Australian J. Sci.,* 21: 305–306.

BALLARD, R. D., HOLCOMB, R. T., and VAN ANDEL, Tj. H. [1979]. "The Galapagos Rift at 86°W, 3, Sheet flows, Collapse Pits, and Lava Lakes of the Rift Valley," *J. Geophys. Res.,* 84: 5407–22.

BALLARD, R. D., and UCHUPI, E. [1975]. "Triassic Rift Structure in Gulf of Maine," *Am. Assoc. Petrol. Geol. Bull.,* 59: 1041–72.

BALLARD, R. D., and VAN ANDEL, Tj. H. [1977]. "Morphology and Tectonics of the Inner Rift Valley at Lat. 36°50′N on the Mid-Atlantic Ridge," *Geol. Soc. Am. Bull.,* 88: 507–30.

BALLARD, R. D., VAN ANDEL, Tj. H., and HOLCOMB, R. T. [1981]. "The Galapagos Rift at 86°W: 5. Variations in Volcanism, Structure, and hydrothermal Activity Along a 30 km. Segment of the Rift Valley," *J. Geophys. Res.* (in press).

BALLY, A. W. [1975]. "A geodynamic scenario for hydrocarbon occurrences," *Proceedings of the Ninth World Petroleum Congress,* Tokyo 2(Geology), Paper TD-1, pp. 33–44.

BALLY, A. W. [1976]. "Canada's Passive Continental Margins—A review," *Mar. Geophys. Res.,* 2: 327–40.

BALLY, A. W., [1979]. *Continental Margins, Geological and Geophysical Research Needs and Problems.* A. W. Bally, Chairman, and other panel members. Washington, D.C.: National Academy of Sciences.

BAMBACH, R. K., SCOTESE, C. R., and ZIEGLER, A. M. [1980]. "Before Pangea: The Geographies of the Paleozoic World," *Am. Scientist* 68: 26–38.

BANDY, O. L. [1960]. The Geologic Significance of Coiling Ratios in the Foraminifer *Globigerina pachyderma* (Ehrenberg)," *J. Paleont.* 34: 671–81.

BANDY, O. L. [1961]. "Distribution of Foraminifera, Radiolaria and Diatoms in Sediments of the Gulf of California," *Micropaleo.* 7: 1–26.

BANDY, O. L. [1967]. "Foraminiferal Definition of the Boundaries of the Pleistocene in Southern California," in *Progress in Oceanography,* vol. 4., ed. M. Sears, New York: Pergamon Press.

BANNER, F. T., and BLOW, W. H. [1965]. "Progress in the Planktonic Foraminiferal Biostratigraphy of the Neogene," *Nature* 208: 1164–66.

BARAZANGI, M., and DORMAN, J. [1969]. "World seismicity maps compiled from ESSA, Coast and Geodetic Survey, Epicenter Data, 1961–1967," *Seismol. Soc. Am. Bull.* 59: 369–80.

BARAZANGI, M., and ISACKS, B. L. [1976]. "Spatial Distribution of Earthquakes and Subduction of the Nazca Plate beneath South America," *Geology* 4: 686–92.

BARKER, P. F., and BURRELL, J. [1976]. "The Opening of Drake Passage," in *Proceedings of the Joint Oceanographic Assembly,* p. 103. Rome: Food and Agricultural Organization of the United Nations.

BARKER, P. F., and BURRELL, J. [1977]. "The Opening of Drake Passage," *Mar. Geol.* 25: 15–34.

BARKER, P. F., DALZIEL, I. W. D., and others [1977A]. *Initial Reports of the Deep Sea Drilling Project,* vol. 36. Washington, D.C.: U.S. Government Printing Office.

BARKER, P. F., DALZIEL, I. W. D., and others [1977B]. "Evolution of the Southwestern Atlantic Ocean Basin: Results of Leg 36, Deep Sea Drilling Projects," in *Initial Reports of the Deep Sea Drilling Project,* leg 36, pp. 993–1014, ed. P. F. Barker, I. W. D. Dalziel, and others, Washington, D.C.: U.S. Government Printing Office.

BARRON, E. J., HARRISON, C. G. A., and HAY, W. W. [1978]. "A Revised Reconstruction of the Southern Continents," *Trans. Am. Geophys. Union* 59: 436–49.

BARRON, E. J., SLOAN, J. L., II, and HARRISON, C. G. A. [1980]. "Potential Significance of Land-Sea Distribution and Surface Albedo Variations as a Climatic Forcing Factor, 180 m.y. to the Present," *Paleo. Paleo. Paleo.* 30: 17–40.

BARRON, E. J., and THOMPSON, S. L. [1980]. "Paleogeography as Climatic Forcing Factor," *Birliner Geowissenschaftliche Abhandlungen, Reihe A/Band* 19: 19–21.

BARRON, J. A. [1973]. Late Miocene–Early Pliocene Paleotemperatures for California from Marine Diatom Evidence," *Paleo. Paleo. Paleo.* 14: 277–91.

BAUMGARTNER, T. R., and VAN ANDEL, Tj. H. [1971]. "Diapirs of the Continental Margin of Angola, Africa," *Geol. Soc. Am. Bull.* 82: 793–802.

BÉ, A. W. H., [1959]. "Ecology of Recent Planktonic Foraminifera, Part 1, Areal Distribution in the Western North Atlantic," *Micropaleo.* 5: 77–100.

BÉ, A. W. H. [1960]. "Some Observations on Arctic Planktonic Foraminifera," *Contrib. Cushman Found. Foram. Res.* 11: 64–68.

BÉ, A. W. H. [1967]. "Foraminifera, Families: *Globigerinidae* and *Globorotaliidae*, Fiche 108," in *Fiches d'Identification du Zooplancton*, Redige par J. H. Fraser. Charlottenlund, Denmark: Cons. Internal. Explor. Mer.

BÉ, A. W. H. [1977]. "An Ecological, Zoogeographic and Taxonomic Review of Recent Planktonic Foraminifera," in *Oceanic Micropaleontology*, ed. A. T. S. Ramsay, vol. 1, pp. 1–88. New York: Academic Press.

BÉ, A. W. H., and GILMER, R. W. [1977]. "A Zoogeographic and Taxonomic Review of *Euthocosomatous Pteropoda*," in *Oceanic Micropaleontology* vol. 1, pp. 733–96, ed. A. T. S. Ramsay. New York: Academic Press.

BÉ, A. W. H., HARRISON, S. M., and LOTT, L. [1973]. "*Orbulina universa* in the Indian Ocean," *Micropaleo.* 19: 150–92.

BÉ, A. W. H., and TOLDERLUND, D. S. [1971]. Distribution and Ecology of Living Planktonic Foraminifera in the Surface Waters of the Atlantic and Indian Oceans," in *Micropaleontology of the Oceans*, pp. 105–49, ed. B. M. Funnell and W. R. Riedel. London: Cambridge University Press.

BECKINSALE, R. D., and GALE, N. H. [1969]. "A Reappraisal of the Decay Constants and Branching Ratio of ^{40}K," *Earth and Planet. Sci. Letts.* 6: 289–94.

BENDER, M. L., and KEIGWIN, L. D. [1979]. "Speculations about the Upper Miocene Change in Abyssal Pacific Dissolved Bicarbonate $\delta^{13}C$," *Earth and Planet. Sci. Letts.* 45: 383–93.

BENIOFF, H. [1949]. "Seismic Evidence for the Fault Origin of Oceanic Deeps," *Geol. Soc. Am. Bull.* 60: 1837–56.

BENSON, R. H. [1972]. "Ostracodes as Indicators of Threshold Depth in the Mediterranean during the Pliocene," in *The Mediterranean Sea: A natural Sedimentation Laboratory*, pp. 63–73. ed. D. J. Stanley. Stroudsburg, Pennsylvania: Dowden, Hutchinson and Ross.

BENSON, R. H. [1975]. "The Origin of the Psychrosphere as Recorded in Changes of Deep-Sea Ostracode Assemblages," *Lethaia* 8: 69–83.

BENSON, R. H. [1976]. "Changes in the Ostracodes of the Mediterranean with the Messinian Salinity Crisis," *Paleo. Paleo. Paleo.* 20: 147–70.

BERGER, A. L. [1978]. Long-term Variations of Caloric Insolation Resulting from the Earth's Orbital Elements," *Quat. Res.* 9: 139–167.

BERGER, W. H. [1968]. "Planktonic Foraminifera: Selective Solution and Paleoclimatic Interpretation," *Deep Sea Res.* 15: 31–43.

BERGER, W. H. [1969]. "Ecological Patterns and Living Planktonic Foraminifera," *Deep Sea Res.* 16: 1–24.

BERGER, W. H. [1970A]. "Biogenous Deep-Sea Sediments: Fractionation by Deep-Sea Circulation," *Geol. Soc. Am. Bull.* 81: 1385–1402.

BERGER, W. H. [1970B]. "Planktonic Foraminifera in Selective Solution and the Lysocline," *Mar. Geol.* 8: 111–83.

BERGER, W. H. [1971A]. "Planktonic Foraminifera: Sediment Production in an Oceanic Front," *J. Foram. Res.* 1: 178–86.

BERGER, W. H. [1971B]. "Sedimentation of Planktonic Foraminifera," *Mar. Geol.* 11: 325–58.

BERGER, W. H. [1973A]. "Cenozoic Sedimentation in the Eastern Tropical Pacific," *Geol. Soc. Am. Bull.* 84: 1941–54.

BERGER, W. H. [1973B]. "Deep-Sea Carbonates: Pleistocene Dissolution Cycles," *J. Foram. Res.* 3: 187–95.

BERGER, W. H. [1974]. "Deep-Sea Sedimentation." In *The Geology of Continental Margins*, pp. 213–241, ed. C. A. Burk and C. D. Drake. New York: Springer-Verlag.

BERGER, W. H. [1975]. "Dissolution of Deep-Sea Carbonates: An Introduction," in *Spec. Publ. No. 13*, ed. W. V. Sliter, A. W. H. Bé, and W. H. Berger. Lawrence, Kansas: Cushman Foundation for Foraminiferal Research.

BERGER, W. H. [1976]. "Biogenous Deep-Sea Sediments: Production, Preservation, and Interpretation," in *Treatise on Chemical Oceanography* vol. 5, pp. 265–388, ed. J. P. Riley and R. Chester. New York: Academic Press.

BERGER, W. H., ADELSECK, C. G., and MAYER, L. A. [1976]. "Distribution of Carbonate in Surface Sediments of the Pacific Ocean," *J. Geophys. Res.* 81: 2617–27.

BERGER, W. H., BÉ, A. W. H., and SLITER, W. V. [1975]. "Dissolution of Deep-Sea Carbonates: An Introduction," in *Spec. Publ. No. 13*, ed. W. V. Sliter, A. W. H. Bé, and W. H. Berger. Lawrence, Kansas: Cushman Foundation for Foraminiferal Research.

BERGER, W. H., and GARDNER, J. V. [1975]. "On the Determination of Pleistocene Temperatures from Planktonic Foraminifera," *J. Foram. Res.* 5: 102–113.

BERGER, W. H., and HEATH, G. R. [1968]. "Vertical Mixing in Pelagic Sediments," *J. Mar. Res.* 26: 134–43.

BERGER, W. H., JOHNSON, R. F., and KILLINGLEY, J. S. [1977]. "Unmixing of the Deep-Sea Record and the Deglacial Meltwater Spike," *Nature* 269: 661–63.

BERGER, W. H., and ROTH, P. H. [1975]. "Oceanic Micropalentology: Progress and Prospect," *Reviews of Geophys. and Space Phys.* 13: 561–85.

BERGER, W. H., and SOUTAR, A. [1967]. "Planktonic Foraminifera: Field Experiment on Production Rate," *Science* 156: 1495–97.

BERGER, W. H., and THIERSTEIN, H. [1979]. "On Phanerozoic Mass Extinctions," *Naturwessenschaften* 66: 46–47.

BERGER, W. H., VINCENT, E., and THIERSTEIN, H. R. [1981]. "The Deep-Sea Record: Major Steps in Cenozoic Ocean Evolution," in *The Sea* vol. 8, ed. C. Emiliani (in press).

BERGER, W. H., and VON RAD, U. [1972]. "Cretaceous and Cenozoic Sediments from the Atlantic Ocean," in *Initial Reports of the Deep Sea Drilling Project,* vol. 14, pp. 787–954, ed. D. E. Hayes, A. C. Pimm, and others. Washington, D.C.: U.S. Government Printing Office.

BERGER, W. H., and WINTERER, E. L. [1974]. "Plate Stratigraphy and the Fluctuating Carbonate Line," in *Pelagic Sediments: On Land and Under the Sea* Pub. No. 1, pp. 11–98, ed. K. J. Hsü and H. C. Jenkyns. Blackwell, Oxford: International Association of Sedimentologists.

BERGGREN, W. A. [1969]. "Rates of Evolution in Some Cenozoic Planktonic Foraminifera," *Micropaleo.* 15: 351–65.

BERGGREN, W. A. [1972]. "Late Pliocene-Pleistocene Glaciation," in *Initial Reports of the Deep Sea Drilling Project* vol. 12, pp. 953–963, ed. A. S. Laughton, W. A. Berggren, and others. Washington, D.C.: U.S. Government Printing Office.

BERGGREN, W. A., and HAQ, B. [1976]. "The Andalusian Stage (Late Miocene): Biostratigraphy, Biochronology and Paleoecology," *Paleo. Paleo. Paleo.* 20: 67–129.

BERGGREN, W. A., and HOLLISTER, C. D. [1974]. "Paleogeography, Paleobiogeography and the History of Circulation in the Atlantic Ocean," in *Studies in Paleo-Oceanography,* pp. 126–186, ed. W. W. Hay. Society of Economic Paleontologists and Mineralogists, Special Publication 20.

BERGGREN, W. A., and HOLLISTER, C. D. [1977]. "Plate Tectonics and Paleocirculation-Commotion in the Ocean," *Tectonophys* 38: 11–48.

BERGGREN, W. A., PHILLIPS, J. D., BERTELS, A., and WALL, D. [1967]. "Late Pliocene-Pleistocene Stratigraphy in Deep Sea Cores from the South-Central North Atlantic," *Nature* 216: 253–55.

BERGGREN, W. A., and VAN COUVERING, J. A. [1974]. "The Late Neogene Biostratigraphy, Geochronology, and Paleoclimatology of the Last 15 m.y." *Paleo. Paleo. Paleo.* 16: 1–216.

BERGGREN, W. A., and VAN COUVERING, J. A. [1978]. "Biochronology," in *Contributions to the Geologic Time Scale,* pp. 39–56, ed. G. V. Cohee, M. F. Glaessner, and H. D. Hedberg. American Association of Petroleum Geologists.

BERGGREN, W. A., and VAN COUVERING, J. A. [1979]. "Quaternary," in *Treatise on Invertebrate Paleontology,* pp. A505–A543, ed. R. A. Robinson and C. Teichert, Lawrence, Kansas: Geological Society of America.

BERNOULLI, D. [1972]. "North Atlantic and Mediterranean Mesozoic Facies: A Comparison," in *Initial Reports of the Deep Sea Drilling Project* Vol. 22, ed. C. D. Hollister, J. I. Ewing, and others. Washington, D.C.: U.S. Government Printing Office.

BERRY, W. B. A., and WILDE, P. [1978]. "Progressive Ventilation of the Oceans—An Explanation for the Distribution of the Lower Paleozoic Black Shales," *Am. J. Sci.* 278: 257–75.

BEZRUKOV, P. L., ed. [1970]. "Sedimentation in the Pacific Ocean," in *The Pacific Ocean,* vol. 6, 2 pts. Moscow: Izdat, Nauk.

BIBEE, L. D., SHOR, G. G., JR., and LU, R. S. [1980]. "Inter-Arc Spreading on the Mariana Trough," *Mar. Geol.* 35: 183–97.

BIGGS, R. B. [1978]. "Coastal Bays," in *Coastal Sedimentary Environments,* pp. 69–100, ed. R. A. Davis, Jr. New York: Springer-Verlag.

BIRD, J. M., and DEWEY, J. F. [1970]. "Lithosphere Plate Continental Margin Tectonics and the Evolution of Appalachian Orogen," *Geol. Soc. Am. Bull.* 81: 1031–60.

BISCAYE, P. E. [1965]. "Mineralogy and Sedimentation of Recent Deep-Sea Clay in the Atlantic Ocean and Adjacent Seas and Oceans," *Geol. Soc. Am. Bull.* 76: 803–32.

BISCAYE, P. E., and EITTREIM, S. L. [1974]. "Variations in Benthic Boundary Layer Phenomena: Nepheloid Layer in the North American Basin, in *Suspended Solids in Water,* pp. 227–60, ed. R. J. Gibb. New York: Plenum.

BISCAYE, P. E., and EITTREIM, S. L. [1977]. "Suspended Particulate Loads and Transports in the Nepheloid Layer of the Abyssal Atlantic Ocean," *Mar. Geol.* 23: 155–72.

BISCAYE, P. E., KOLLA, V., and TUREKIAN, K. K. [1976]. "Distribution of Calcium Carbonate in Surface Sediments of the Atlantic Ocean," *J. Geophys. Res.* 81: 2595–2603.

BLACK, M. [1963]. "The Fine Structure of the Mineral Parts of Coccolithophoridae," *Proc. Linn. Soc.* (London), 174: 41–46.

BLACK, M. [1965]. "Coccoliths," *Endeavour* 24: 131–37.

BLACKWELDER, B. W., PILKEY, O. H., and HOWARD, J. D. [1979]. "Late Wisconsinan Sea Levels on the Southeast U.S. Atlantic Shelf Based on In-Place Shoreline Indicators," *Science* 204: 618–20.

BLATT, H., MIDDLETON, G. V., and MURRAY, R. C. [1972]. *Origin of Sedimentary rocks.* Englewood Cliffs, N.J.: Prentice-Hall, Inc.

BLOOM, A. L. [1967]. "Pleistocene Shorelines: A New Test of Isostasy," *Geol. Soc. Am. Bull.* 78: 1477–94.

BLOOM, A. L. [1978]. *Geomorphology, A Systematic Analysis of Late Cenozoic Land Forms*. Englewood Cliffs, N.J.: Prentice-Hall, Inc.

BLOOM, A. L. and others [1974]. "Quaternary Sea Level Fluctuations on a Tectonic Coast: New ^{230}Th/^{234}U Dates from the Huon Peninsula, New Guinea," *Quat. Res.* 4: 185–205.

BODVARSSON, G., and WALKER, G. P. L. [1964]. "Crustal Drift in Iceland," *Geophys. J. Roy. Astron. Soc.* 8: 285–300.

BOERSMA, A., and SHACKLETON, N. [1977]. "Tertiary Oxygen and Carbon Isotope Stratigraphy, Site 357 (Mid-Latitude South Atlantic)," in *Initial Reports of the Deep Sea Drilling Project,* vol. 39, pp. 911–924, ed. P. R. Supko, K. Perch-Nielson, and others. Washington, D.C.: U.S. Government Printing Office.

BOERSMA, A., SHACKLETON, N., HALL, M., and GIVEN, Q. [1979]. "Carbon and Oxygen Isotope Records at Deep Sea Drilling Project Site 384 (North Atlantic) and Some Paleocene Paleotemperatures and Carbon Isotope Variations in the Atlantic Ocean," in *Initial Reports of the Deep Sea Drilling Project,* vol. 43, pp. 695–717, ed. B. E. Tucholke, P. R. Vogt, and others. Washington, D.C.: U.S. Government Printing Office.

BOLLI, H. M. [1957]. Planktonic Foraminifera from the Oligocene-Miocene Cipero and Lengua Formations of Trinidad, B. W. I.," *U.S. Natl. Mus. Bull.* 215: 97–123.

BOLLI, H. M., RYAN, W. B. F., and others [1978]. *Initial Reports of the Deep Sea Drilling Project,* vol. 40. Washington, D. C.: U. S. Govt. Printing Office.

BOLTOVSKOY, E., and WRIGHT, R. [1976]. *Recent Foraminifera,* The Hague: Dr. W. Junk. b. v., publisher.

BONATTI, E., and FISHER, D. E. [1971]. "Oceanic Basalts: Chemistry Versus Distance from Oceanic Ridge," *Earth. and Planet. Sci. Letts.* 2: 307–11.

BOOTH, J., and BURCKLE, L. H. [1976]. "Displaced Antarctic Diatoms in the Southwestern and Central Pacific," *Pacific Geol.* 2: 99–108.

BOSTRÖM, K. [1970]. Geochemical Evidence to Ocean Floor Spreading in South Atlantic Ocean," *Nature* 227: 1041.

BOSTRÖM, K., and PETERSON, M. N. A. [1965]. Precipitates from Hydrothermal Exhalations on the East Pacific Rise," *Econ. Geol.* 61: 1258–65.

BOTT, M. H. P. [1971]. *The Interior of the Earth.* New York: St. Martin's Press.

BOUMA, A. H. [1962]. *Sedimentology of Some Flysch Deposits: A Graphic Approach to Facies Interpretation.* Amsterdam: Elsevier Publishing Company.

BOWEN, D. Q. [1978]. *Quaternary Geology, A Stratigraphic Framework for Multidisciplinary Work.* New York: Pergamon Press.

BOWEN, N. L. [1928]. *The Evolution of the Igneous Rocks.* Princeton, N.J.: Princeton University Press.

BOWEN, R. [1966]. *Paleotemperature Analysis.* Amsterdam: Elsevier Scientific Publishing Company.

BRADSHAW, J. S. [1959]. "Ecology of Living Planktonic Foraminifera in the North and Equatorial Pacific Ocean," *Contrib. Cushman Found. Foram. Res.* 10: 25–64.

BRAMLETTE, M. N. [1965]. "Massive Extinctions in the Biota at the End of Mesozoic Time," *Science* 148: 1696–99.

BRAMLETTE, M. N., and BRADLEY, W. H. [1940]. "Lithology and geologic interpretations, Parts 1 and 2, in *Geology and Biology of North Atlantic Deep-Sea Cores.* U.S Geological Survey Professional Paper 196: 1–24.

BRAY, J. R. [1979]. "Neogene Explosive Volcanicity, Temperature and Glaciation," *Nature* 282: 603–605.

BRENNINKMEYER, B. [1976]. "Sand Fountains in the Surf Zone," in *Beach and Nearshore Sedimentation,* pp. 69–91, ed. R. A. Davis and R. L. Ethington. Society of Economic Paleontologists and Mineralogists, Special Publication 24.

BREWSTER, N. A. [1977]. *"Cenozoic Biogenic Silica Sedimentation in the Antarctic Ocean, based on Two Deep Sea Drilling Sites."* (Unpublished thesis, Oregon State University) 98 pages.

BREWSTER, N. A. [1980]. "Cenozoic biogenic silica sedimentation in the Antarctic Ocean, Based on Two Deep Sea Drilling Project Sites," *Geol. Soc. Am. Bull.* 91: 337–47.

BRIDEN, J. C., DREWRY, D. J., and SMITH, A. G. [1974]. "Phanerozoic Equal-Area World Maps," *J. Geol.* 82: 555–574.

BROECKER, W. S. [1971]. "Calcite Accumulation Rates and Glacial to Interglacial Changes in Oceanic Mixing," in *The Late Cenozoic Glacial Ages,* pp. 239–65, ed. K. K. Turekian. New Haven, Connecticut: Yale University Press.

BROECKER, W. S. [1974]. *Chemical Oceanography.* New York: Harcourt, Brace, Jovanovich, Inc.

BROECKER, W. S., and TAKAHASHI, T. [1978]. The Relationship Between Lysocline Depth and *in situ* Carbonate Ion Concentration," *Deep Sea Res.* 25: 65–95.

BROECKER, W. S., TUREKIAN, K. K., and HEEZEN, B. [1958]. "The Relation of Deep-Sea Sedimentation Rates to Variations in Climate," *Am. J. Sci.* 256: 503–517.

BROECKER, W. S., and VAN DONK, J. [1970]. "Insolation Changes, Ice Volumes, and the O^{18} Record in Deep-Sea Cores," *Reviews of Geophys. and Space Phys.* 8: 169–97.

BROOKFIELD, M. [1971]. "Periodicity of Orogeny," *Earth and Planet. Sci. Letts.* 12: 419–24.

BRUNNER, C. A. [1978]. "Late Neogene and Quaternary Paleoceanography and biostratigraphy of the Gulf of Mexico." (Unpublished doctoral thesis, University of Rhode Island) 341 pages.

BRYAN, W. B., and MOORE, J. G. [1977]. Compositional Variations of Young Basalts in the Mid-Atlantic Ridge Rift Valley near Lat. 36°40'N," *Geol. Soc. Am. Bull.* 88: 556–70.

BRYAN, W., THOMPSON, G., and FREY, F. [1979]. "Petrologic Character of the Atlantic Crust from DSDP and IPOD Drill Sites," in *Maurice Ewing Series 2*, pp. 273–84, ed. M. Talwani, C. G. Harrison and D. E. Hayes. Washington, D.C.: American Geophysical Union.

BRYANT, W. R., ANTOINE, J. W., EWING, M., and JONES, B. [1968]. "Structure of Mexican Continental Shelf and Slope, Gulf of Mexico," *Am. Assoc. Petrol. Geol. Bull.* 52: 1204–28.

BRYSON, R. A., and GOODMAN, B. M. [1980]. "Volcanic Activity and Climatic Changes," *Science* 207: 1041–44.

BUCHARDT, B. [1978]. Oxygen Isotope Paleotemperatures from the Tertiary Period in the North Sea Area," *Nature* 275: 121–23.

BUFFLER, R. T., WATKINS, J. S., and DILLON, W. P. [1979]. "Geology of the Offshore Southeast Georgia Embayment, U.S. Atlantic Continental Margin, Based on Multichannel Seismic Reflection Profiles," in *Geological and Geophysical Investigation of Continental Margins*, ed. J. S. Watkins, L. Montadert, and P. W. Dickerson. Amer. Assoc. Petrol. Geol. Memoir 29.

BULLARD, E. C., EVERETT, J. E., and SMITH, A. G. [1965]. "The Fit of the Continents around the Atlantic," in *A Symposium on Continental Drift*, A258, pp. 41–51, ed. P. M. S. Blackett, E. Bullard, and S. K. Runcorn. Philosophical Transactions, Royal Society of London.

BULLARD, F. M. [1976]. *Volcanoes of the Earth*. Austin, Texas: University of Texas Press.

BURCKLE, L. H. [1971]. "Correlation of Late Cenozoic Marine Sections in Japan and the Equatorial Pacific," *Trans. Proc. Paleont. Soc. Japan, N. S.* 82: 117–28.

BURCKLE, L. H. [1972]. "Late Cenozoic Planktonic Diatom Zones from the Eastern Equatorial Pacific," *Beih. Zwr Nova Hedwegia*, Helf, 39: 217–46.

BURCKLE, L. H. [1976]. "Pleistocene Changes in the Antarctic Bottom Water Flow through the Vema Channel," *EOS Trans. Am. Geophys. Union* 57: 257.

BURCKLE, L. H., AND AKIBA, F. [1978]. "Implications of Late Neogene Freshwater Sediment in the Sea of Japan," *Geology* 6: 123–27.

BURCKLE, L. H., and HAYS, J. D. [1966]. "Tertiary Sediments on Falkland Platform and Argentine Continental Slope," abstract, *Am. Assoc. Petrol. Geol. Bull.* 50: 607.

BURCKLE, L. H., and STANTON, D. [1975]. "Distribution of Displaced Antarctic Diatoms in the Argentina Basin," *Nova Hedwegia* 53: 283–91.

BUREK, P. T. [1970]. "Magnetic Reversals: Their Application to Stratigraphic Problems," *Am. Assoc. Petrol. Geol. Bull.* 54: 1120–39.

BURKE, K. [1979]. "The Edge of the Ocean: An Introduction," *Oceanus*, 22: 3–11.

BURKE, K., and DEWEY, J. F. [1973]. "An Outline of Precambrian Plate Development," in *Implications of Continental Drift to the Earth Sciences*, ed. D. H. Tarling and S. K. Runcorn, vol. 2, pp. 1035–45. New York: Academic Press.

BURKE, K., and WILSON, J. T. [1972]. "Is the African Plate Stationary?" *Nature* 239: 387–90.

BURNS, R. E., and ANDREWS, J. E. [1973]. "Regional Aspects of Deep Sea Drilling in the Southwest Pacific," in *Initial Reports of the Deep Sea Drilling Project*, vol. 21, pp. 897–906, ed. R. F. Burns, J. E. Andrews, Washington, D.C.: U.S. Government Printing Office.

BURNS, R. E., ANDREWS, J. E., and others [1973]. *Initial Reports of the Deep Sea Project*, vol. 21. Washington, D.C.: U.S. Government Printing Office.

BUTLER, R. F., LINDSAY, E. H., JACOBS, L. L., and JOHNSON, N. M. [1977]. "Magnetostratigraphy of the Cretaceous-Tertiary Boundary in the San Juan Basin, New Mexico," *Nature* 267: 318–23.

CADOT, H. M., SCHMUS, W. R., and VAN KAESLER, R. L. [1972]. "Magnesium in Calcite of Marine Ostracoda," *Geol. Soc. Am. Bull.* 83: 3519–22.

CALDER, N. [1974]. *Restless Earth*. British Broadcasting Corporation.

CALLAHAN, J. E. [1971]. "Velocity Structure and Flux of the Antarctic Circumpolar Current of South Australia," *J. Geophys. Res.* 76: 5859–70.

CALVERT, S. E. [1971]. "Nature of Silica Phases in Deep-Sea Cherts of the North Atlantic," *Nature* 234:133–134 (London).

CALVERT, S. E. [1974]. "Deposition and Diagnosis of Silica in Marine Sediments," in *Pelagic Sediments on Land and Under the Sea*, pp. 273–99, ed. K. J. Hsü and H. Jenkyns. Special Publication 1. Oxford: Blackwell.

CANN, J. R. [1974]. "A Model for Oceanic Crustal Structure Development," *Geophys. J. R. Astr. Soc.* 39: 169–87.

CARDER, K. L., BEARDSLEY, G. F., and PAK, H. [1971]. "Particle Size Distributions in the Eastern Equatorial Pacific," *J. Geophys. Res.* 76: 5070–77.

CAREY, S. W. [1958]. "A Tectonic Approach to Continental Drift," in *Continental Drift, A Symposium*, pp. 177–355, ed. S. W. Carey. Hobart: University of Tasmania.

CARTER, A. N. [1978]. "Phosphatic Nodule Beds in Victoria and the Late Miocene-Pliocene Eustatic Event," *Nature* 276: 258–59.

CASEY, R. E. [1971]. Radiolarians as Indicators of Past and Present Water Masses," in *The Micropaleontology of the Oceans*, pp. 151–160, ed. B. M. Funnell and W. R. Riedel. London: Cambridge University Press.

CASEY, R. E. [1972]. Neogene Radiolarian Biostratigraphy and Paleotemperatures. The Experimental Mohole, Antarctic Core. E14-8," *Paleo. Paleo. Paleo.* 12: 115–30.

CASEY, R. E., PARTRIDGE, T. M., and SLOAN, J. R. [1971]. Radiolarian Life Spans, Mortality Rates, and Seasonality Gained from Recent Sediment and Plankton Samples," in *Proc. II Plankt. Conf. 2,* ed. A. Famnacci, pp. 159–65.

CHAMLEY, H. [1979]. "North Atlantic Clay Sedimentation and Paleoenvironment since the Late Jurassic," in *Maurice Ewing Series 3,* ed. N. Talwani, W. Hay, and W. B. F. Ryan. Washington, D.C.: American Geophysical Union.

CHAPPELL, J. M. A. [1974]. "Geology of Coral Terraces, Huon Peninsula, New Guinea; A Study of Quaternary Tectonic Movements and Sea Level Changes," *Geol. Soc. Am. Bull.* 85: 553–70.

CHAPPELL, J. M. A. [1975]. "On Possible Relationships between Upper Quaternary Glaciations, Geomagnetism and Volcanism," *Earth and Planet. Sci. Letts.* 26: 370–76.

CHASE, C. G. [1972]. "The N-Plate Problem of Plate Tectonics," *Geophys. J. R. Astr. Soc.* 29: 117.

CHAVE, K. E. [1965]. "Carbonates: Association with Organic Matter in Surface Seawater," *Science* 148: 1723–24.

CHAVE, K. E., and SUESS, E. [1970]. "Calcium Carbonate Saturation in Seawater: Effects of Organic Matter," *Limnol. and Oceanog.* 15: 633–37.

CHEN, C. S. [1965]. "The Regional Lithostratigraphic Analysis of Paleocene and Eocene Rocks of Florida," *Florida Geol. Survey Bull.* 45: 105.

CHEN, C. [1968]. "Zoogeography of the Cosomatous Pteropods in the West Antarctic Ocean," *Nautilus* 81: 94–101.

CHRISTENSEN, N. I., and SALISBURY, M. H. [1975]. "Structure and Constitution of the Lower Oceanic Crust," *Rev. Geophys. Space Phys.* 13: 57–86.

CHRISTOFFEL, D. A., and FALCONER, R. H. K. [1972]. "Marine Magnetic Measurements in the Southwest Pacific Ocean and the Identification of New Tectonic Features: Antarctic Oceanology II: The Australian–New Zealand Sector," in *Antarctic Research Series 19,* pp. 197–209, ed. D. E. Hayes. Washington, D.C.: American Geophysical Union.

CHUBB, L. J. [1957]. The Pattern of Some Pacific Island Chains," *Geol. Mag. Great Britain* 94: 221–28.

CIESIELSKI, P. F., LEDBETTER, M. T., and ELLWOOD, B. B. [1981]. "The Development of Antarctic Glaciation and the Neogene Paleoenvironment of the Maurice Ewing Bank," *Mar. Geol.* (in press).

CIESIELSKI, P. F., and WEAVER, F. M. [1974]. "Early Pliocene Temperature Changes in the Antarctic Seas," *Geology* 2: 511–16.

CIESIELSKI, P. F., and WISE, S. W., JR. [1977]. "Geologic History of the Maurice Ewing Bank of the Falkland Plateau (Southwest Atlantic Sector of the Southern Ocean) Based on Piston and Drill Cores," *Mar. Geol.* 25: 175–207.

CIFELLI, R. [1969]. "Radiation of the Cenozoic Planktonic Foraminifera," *Syst. Zool.* 18: 154–68.

CIFELLI, R. [1976]. "Evolution of Ocean Climate and the Record of Planktonic Foraminifera," *Nature* 264: 431–32.

CIFELLI, R., and SACHS, K. N. [1966]. "Abundance Relationship of Planktonic Foraminifera and Radiolaria," in *Deep-Sea Research and Oceanographic Abstracts* 13: 731–53. New York: Pergamon Press.

CITA, M. B. [1975]. "The Miocene/Pliocene boundary: History and Definition," in *Late Neogene Epoch Boundaries,* Spec. Pub. 1, pp. 1–30, ed. T. Saito and L. H. Burckle. New York: Micropaleontology Press.

CITA, M. B. [1976]. "Early Pliocene Paleoenvironment after the Messinian Salinity Crisis," VI. *African Micropaleontol. Colloq.,* Tunis, 1974.

CITA, M. B., and RYAN, W. B. F. [1979]. "Late Neogene Paleoenvironment: Interpretation of the Evolution of the Ocean Paleoenvironment," in *Initial Reports of the Deep Sea Drilling Projects,* vol. 47, ed. U. von Rad, W. B. F. Ryan, and others. Washington, D.C.: U.S. Government Printing Office.

CLAQUE, D. A., DALRYMPLE, G. B., and MOBERLY, R. [1975]. "Petrography and K-Ar Ages of Dredged Volcanic Rocks from the Western Hawaiian Ridge and the Southern Emperor Seamount Chain," *Geol. Soc. Am. Bull.* 86: 991–98.

CLAQUE, D. A., and JARRARD, R. D. [1973]. Pacific Plate Motion Deduced from the Hawaiian-Emperor Chain," *Geol. Soc. Am. Bull.* 84: 1135–54.

CLARK, D. L. [1971]. "Arctic Ocean Ice Cover and Its Late Cenozoic History," *Geol. Soc. Am. Bull.* 82: 3313–24.

CLARK, J. A., and LINGLE, C. S. [1979]. "Predicted Relative Sea-Level Changes (18,000 Years B.P. to Present) Caused by Late-Glacial Retreat of the Antarctic Ice Sheet," *Quat. Res.* 11: 279–98.

CLARK, S. P., and RINGWOOD, A. E. [1964]. "Density Distribution and Constitution of the Mantle," *Rev. Geophys.* 2: 35–88.

CLIMAP Project Members [1976]. "The surface of the Iceage Earth," *Science* 191: 1131–37.

CLOUD, P. E., JR. [1959]. "Paleoecology—Retrospect and Prospect," *J. Paleont.* 33: 926–62.

COLE, J. W., and LEWIS, K. B. [1981]. "Evolution of the Taupo-Hikurangi Subduction System," *Tectonophysics* 72: 1–22.

COOPER, A. K., MARLOW, M. S., and SCHOLL, D. W. [1976]. "Mesozoic Magnetic Lineations in the Bering Sea Marginal Basin," *J. Geophys. Res.* 81: 1916–34.

CORLISS, B. H. [1979A]. "Recent Deep-Sea Benthonic Foraminiferal Distributions in the Southeast Indian Ocean: Inferred Bottom-Water Routes and Ecological Implications," *Mar. Geol.* 31: 115–38.

CORLISS, B. H. [1979B]. "Response of Deep-Sea Benthonic Foraminifera to Development of the Psychrosphere Near the Eocene/Oligocene Boundary," *Nature* 282: 63–65.

CORLISS, J. B., and others [1979]. "Submarine Thermal Springs on the Galapagos Rift," *Science* 203: 1073–83.

COWAN, D. S., and SILLING, R. M. [1978]. "A Dynamic, Scaled Model of Accretion at Trenches and Its Implications for the Tectonic Evolution of Subduction Complexes," *J. Geophys. Res.* 83, 5389–96.

Cox, A. [1969]. "Geomagnetic reversals," *Science* 163: 237–44.

Cox, A. ed., [1973]. *Plate Tectonics and Geomagnetic Reversals*. San Francisco: W. H. Freeman and Company.

Cox, A., Dalrymple, G. B., and Doell, R. R. [1967]. "Reversals of the Earth's Magnetic Field," *Scientific American* 216: 44–54.

Cox, K. E., Bell, J. D., and Parkhurst, R. J. [1979]. *The Interpretation of Igneous Rocks.* London: George Allens and Unwin.

Craddock, C., and Hollister, C. D. [1976]. "Geologic Evolution of the Southeast Pacific Basin," in *Initial Reports of the Deep Sea Drilling Project,* vol. 35, p. 723, ed. C. Craddock, C. D. Hollister, and others. Washington, D.C.: U.S. Government Printing Office.

Craig, H. [1957]. "Isotopic Standards for Carbon and Oxygen and Correction Factors for Mass-Spectrometric Analysis of Carbon Dioxide," *Geochim. et Cosmochim. Acta* 12: 133–49.

Craig, H., and Gordon, L. I. [1965]. "Deuterium and Oxygen-18 Variations in the Ocean and the Marine Atmosphere," *Proc. Spoleto Conf. on Stable Isotopes in Oceanographic Studies and Paleotemperatures* 2: 1–87.

Cranwell, L. M., Harrington, H. J., and Speden, I. G. [1960]. "Lower Tertiary Microfossils from McMurdo Sound, Antarctica," *Nature* 186: 700.

Creager, J. S., Scholl, D. W., and others [1973]. *Initial Reports of the Deep-Sea Drilling Project,* vol. 19. Washington, D.C.: U.S. Government Printing Office.

Cronan, D. S. [1974]. "Authigenic Minerals in Deep-Sea Sediments," in *The Sea, Marine Chemistry,* vol. 5, pp. 491–526, ed. E. D. Goldberg. New York: Wiley-Interscience.

Cronan, D. S. [1975]. "Manganese Nodules and Other Ferro-Manganese Oxide Deposits from the Atlantic Ocean," *J. Geophys. Res.* 80: 3831–37.

Cronan, D. S. [1976]. "Manganese Nodules and Other Ferro-Manganese Oxide Deposits," in *Chemical Oceanography,* vol. 5, pp. 217–63, ed. J. P. Riley and R. Chester. New York: Academic Press.

Cronan, D. S. [1977]. "Deep-Sea Nodules: Distribution and Geochemistry," in *Marine Manganese Deposits,* pp. 11–44, ed. G. P. Glasby. Amsterdam: Elsevier.

Crowell, J. C. [1974]. "Origin of Late Cenozoic Basins in Southern California," in *Tectonics and Sedimentation,* pp. 190–204, ed. W. R. Dickinson. SEPM Special Publication 22.

Crowell, J. C., and Frakes, L. A. [1970]. "Phanerozoic Glaciation and the Causes of Ice Ages," *Am. J. Sci.* 268: 193.

Curray, J. R. [1964]. "Transgressions and regression," in *Papers in Marine Geology,* Shepard Commemorative Volume, pp. 175–203, ed. R. C. Miller. New York: Macmillan.

Curray, J. R. [1965]. "Late Quaternary History, Continental Shelves of the United States," in *The Quaternary of the United States,* pp. 723–735, ed. H. E.

Wright, Jr. and D. G. Frey. Princeton, New Jersey: Princeton University Press.

Curray, J. R. [1969]. "Shore Zone Sand Bodies: Barriers, Cheniers, and Beach Ridges," in *The New Concepts of Continental Margin Sedimentation: Application to the Geological Record,* ed. D. J. Stanley, JC11-1–JC11-19. Washington, D.C.: American Geological Institute.

Curray, J. R. [1969]. "History of Continental Shelves," in *The New Concepts of Continental Margin Sedimentation: Application to the Geological Record,* pp. JC 6-1–JC-6-7, ed. D. J. Stanley. Washington, D.C.: American Geological Institute.

Curray, J. R. [1977]. "Modes of Emplacement of Prospective Hydrocarbon Reservoir Rocks of Outer Continental Marine Environments," in *Geology of Continental Margins,* pp. E1–E14, Course Notes, Series 5, AAPG Continuing Education.

Curray, J. R., Emmel, F. J., Moore, D. G., and Raitt, R. W. [1981]. "Structure, Tectonics and Geological History of the Northeastern Indian Ocean," in *Ocean Basins and Margins, The Indian Ocean,* vol. 6, ed. A. E. M. Nairn and F. G. Stehli. New York: Plenum Press.

Curray, J. R., and Moore, D. G. [1971]. "Growth of the Bengal Deep-Sea Fan and Denudation of the Himalayas," *Geol. Soc. Am. Bull.* 82: 565–72.

Curray, R. P. [1966]. "Glaciation about 3,000,000 years ago in the Sierra Nevada," *Science* 154: 770–71.

Damon, P. E. [1968]. "The Relationship between Terrestrial Factors and Climate," *Meterol. Mono.* 8: 106–11.

Damon, P. E. [1971]. "The Relationship between Late Cenozoic Volcanism and Tectonism and Orogenic-Epeirogenic Periodicity," in *Late Cenozoic Glacial Ages,* pp. 15–36, ed. K. K. Turekian. New Haven, Connecticut: Yale University Press.

Damuth, J. E., and Kumar, N. [1975]. "Amazon Cone: Morphology, Sediments, Age, and Growth Pattern," *Geol. Soc. Am. Bull.* 86: 863–78.

Dana, J. D. [1885]. "Origin of Coral Reefs and Islands," Series 3, *Am. J. Sci.* 30: 89–105, 169–91.

Darwin, C. [1842]. *The Structure and Distribution of Coral Reefs.* London: Smith, Elder. (Reprinted in 1962 by University of California Press, Berkeley–Los Angeles, California.)

Darwin, C. [1846]. "An Account of the Fine Dust Which Often Falls on Vessels in the Atlantic Ocean," *Q. J. Geol. Soc. London* 2: 26–30.

Davies, T. A., and Gorsline, D. S. [1976]. "Oceanic Sediments and Sedimentary Processes," in *Chemical Oceanography,* Part 5 (2nd ed.), pp. 1–80, ed. J. P. Riley and R. Chester. New York: Academic Press.

Davies, T. A., Hay, W. W., Southam, J. R., and Worsley, T. R. [1977]. "Estimates of Cenozoic Oceanic Sedimentation Rates," *Science* 197: 53–55.

Davies, T. A., and Kidd, R. B. [1977]. "Sedimentation in the Indian Ocean Through Time, in *Indian Ocean Geology and Biostratigraphy,* pp. 61–85, eds. J. R. Heirtzler and others. Washington, D.C.: American Geophysical Union.

DAVIES, T. A., and LAUGHTON, A. S. [1972]. "Sedimentary Processes in the North Atlantic," in *Initial Reports of the Deep Sea Drilling Project,* vol. 12, pp. 905–34, ed. A. S. Laughton, W. A. Berggren, and others. Washington, D.C.: U.S. Government Printing Office.

DAVIES, T. A., MUSICH, L. F., and WOODBURY, P. B. [1977]. "Automated Classification of Deep-Sea Sediments," *J. Sediment. Petrol.* 47: 650–56.

DAVIES, T. A., and SUPKO, P. R. [1973]. "Oceanic Sediments and Their Diagenesis: Some Examples from Deep Sea Drilling," *J. Sediment. Petrol.* 43: 381–90.

DAVIES, T. A., and WORSLEY, T. R. [1981]. "Paleoenvironmental Implications of Oceanic Carbonate Sedimentation Rates," in *SEPM Spec. Pub.* (in press), ed. R. Douglas and E. Winterer.

DAVIS, R. A., JR. [1978]. "Beach and Nearshore Zone," in *Coastal Sedimentary Environments.,* ed. R. A. Davies, Jr. New York: Springer-Verlag.

DAVIS, R. A., JR., and FOX, W. T. [1972]. "Coastal Processes and Nearshore Sand Bars," *J. Sediment. Petrol.* 42: 401–412.

DE BEAUMONT, E. [1845]. *Leçons de Geologie Pratique: 7 me Leçon-Levées de sables etgalets* pp. 223–252. Paris.

DEFANT, S. [1961]. *Physical Oceanography* vol. 1. New York: Pergamon Press.

DEGENS, E. T., and ROSS, D. A. [1972]. "Chronology of the Black Sea over the Past 25,000 Years," *Chem. Geol.* 20: 1–16.

deMASTER, D. J. [1979]. "The Marine Budgets of Silica and ^{32}Si." (Unpublished doctoral dissertation, Yale University).

DENTON, G. H., ARMSTRONG, R. L., and STUIVER, M. [1971]. "The Late Cenozoic Glacial History of Antarctica," in *The Late Cenozoic Glacial Ages,* pp. 267–306, ed. K. K. Turekian. New Haven, Connecticut: Yale University Press.

DETRICK, R. S., SCLATER, J. G., and THIEDE, J. [1977]. "The Subsidence of Aseismic Ridges," *Earth and Planet. Sci. Letts.* 34: 185–96.

DEWEY, J. F. [1972]. "Plate tectonics," *Scientific American* 266: 56–68.

DEWEY, J. F. [1980]. "Collisional Zones," in *Internationales Alfred-Wegener Symposium, Kurzfassungen der Beitrage,* p. 35, ed. U. Dornsiepen und V. Haak. Berlin: Verlag von Dietrich Reimer.

DEWEY, J. F., and BIRD, J. M. [1970]. "Mountain Belts and the New Global Tectonics," *J. Geophys. Res.* 75: 2625–47.

DICKINSON, K. A., BERRYHILL, H. L., and HOLMES, C. W. [1972]. "Criteria for Recognizing Ancient Barrier Coastlines," in *Recognition of Ancient Sedimentary Environments,* pp. 192–214, ed. J. K. Rigby and W. H. Hamblin. SEPM Spec. Pub. 16.

DICKINSON, W. R. [1973]. "Widths of Modern Arc-Trench Gaps Proportional to Past Duration of Igneous Activity in Associated Magnetic Arcs," *J. Geophys. Res.* 78: 3378–89.

DICKINSON, W. R. [1974]. "Plate Tectonics and Sedimentation," in *Tectonics and Sedimentation,* pp. 1–27, ed. W. R. Dickinson. SEPM Spec. Pub. 22.

DICKINSON, W. R. [1977]. "Tectono-Stratigraphic Evolution of Subduction-Controlled Sediment Assemblages," in *Maurice Ewing Series 1,* pp. 33–40, Washington, D.C.: American Geophysical Union.

DICKINSON, W. R., and HATHERTON, T. [1967]. "Andesitic Volcanism and Seismicity around the Pacific," *Science* 157: 801–803.

DICKINSON, W. R., and SEELY, D. R. [1979]. "Structure and Stratigraphy of Forearc Regions," *Am. Assoc. Petrol. Geol. Bull.* 63: 2–31.

DIETRICH, G. [1963]. *General Oceanography.* New York: John Wiley.

DIETZ, R. S. [1952]. "Geomorphic Evolution of Continental Terrace, Continental Shelf and Slope," *Am. Assoc. Petrol. Geol. Bull.* 36: 1802–19.

DIETZ, R. S. [1961]. "Continent and Ocean Basin Evolution by Spreading of the Sea Floor," *Nature* 190: 854–57.

DIETZ, R. S., and HOLDEN, J. C. [1970]. "Reconstruction of Pangaea: Breakup and Dispersion of Continents, Permian to Present," *J. Geophys. Res.* 75: 4939–56.

DILL, R. F. [1969]. "Submerged Barrier Reefs on the Continental Slope North of Darwin, Australia," Abstracts with Programs for 1969, pt. 7, pp. 264–266. Boulder, Colorado: Geological Society of America.

DILLON, W. P., and OLDALE, R. N. [1979]. "Late Quaternary Sea-Level Curve: Reinterpretation Based on Glaciotectonic Influence," *Geology* 6: 56–60.

DILLON, W. P., PAULL, C. K., BUFFLER, R. T., and FALL, J. P. [1978]. "Structure and Development of the Southeast Georgia Embayment and Northern Blake Plateau: Preliminary Analysis," in *Geological and Geophysical Investigations,* pp. 27–41, American Association Petroleum Geologists Memoir 29.

DINGLE, R. V., and SCRUTTON, R. A. [1974]. "Continental Breakup and the Development of Post-Paleozoic Sedimentary Basins around Southern Africa, *Geol. Soc. Am. Bull.* 85: 1467–74.

DOBRIN, M. B. [1976]. *Introduction to Geophysical Prospecting.* New York: McGraw-Hill.

DOLAN, R., HAYDEN, B., and LINS, H. [1980]. "Barrier Islands," *Am. Sci.* 68: 16–25.

DONN, W. L., and SHAW, D. M. [1977]. "Model of Climate Evolution based on Continental Drift and Polar Wandering," *Geol. Soc. Am. Bull.* 88: 390–396.

DORMAN, F. H. [1968]. "Some Australian Oxygen Isotope Temperatures and a Theory for a 30-Million-Year World Temperature Cycle," *J. Geol.* 76: 297–313.

DOTT, R. H., JR., and BATTEN, R. L. [1971]. *Evolution of the Earth.* New York: McGraw-Hill.

DOUGLAS, R. G., and SAVIN, S. M. [1973]. "Oxygen and Carbon Isotope Analysis of Cretaceous and Tertiary Foraminifera from the Central North Pacific," in *Initial Reports of the Deep Sea Drilling Project,* vol. 17, pp. 591–605, ed. E. L. Winterer, J. I. Ewing, and others. Washington, D.C.: U.S. Government Printing Office.

DOUGLAS, R. G., and SAVIN, S. M. [1975]. "Oxygen and Carbon Isotope Analyses of Tertiary and Cretaceous Microfossils from Shatsky Rise and Other Sites in the North Pacific Ocean," in *Initial Reports of the Deep Sea Drilling Project,* vol. 32, pp. 509–20, ed. R. L. Larson, R. Moberly, and others. Washington, D.C.: U.S. Government Printing Office.

DOUGLAS, R. G., and SAVIN, S. M. [1978]. "Oxygen Isotopic Evidence for the Depth Stratification of Tertiary and Cretaceous Planktonic Foraminifera," *Mar. Micropaleo* 3: 175–96.

DRAKE, C. L., and BURK, C. A. [1974]. "Geological Significance of Continental Margins," in *The Geology of Continental Margins,* pp. 3–10, ed. C. A. Burk and C. L. Drake. New York: Springer-Verlag.

DRAKE, C. L., EWING, M., and SUTTON, G. H. [1959]. "Continental Margins and Geosynclines; the East Coast of North America North of Cape Hatteras," in *Physics and Chemistry of the Earth,* vol. 3, pp. 110–98. New York: Pergamon Press.

DRAKE, C. L., IMBRIE, J., KNAUSS, J. A., and TUREKIAN, K. K. [1978]. *Oceanography.* New York: Holt, Rinehart & Winston.

DUNCAN, R. A. [1981]. "Hotspots in the Southern Oceans—An Absolute Frame of Reference for Motion of the Gondwana Continents," *Tectonophysics* 74: 29–42.

DUPLESSY, J. C. [1972]. "*La Geochimie des Isotopes Stables du* Carbone dans la Mer." (Unpublished doctoral thesis, University of Paris, France).

DUPLESSY, J. C. [1978]. "Isotope Studies," in *Climatic Change,* ed. J. Gribbin. Cambridge: Cambridge University Press.

DUTOIT, A. L. [1937]. *Our Wandering Continents.* Edinburgh: Oliver and Boyd.

EATON, G. P. [1964]. "Windborne Volcanic Ash. A Possible Index to Polar Wandering," *J. Geol.* 72: 1–35.

EDMOND, J. M. [1974]. "On the Dissolution of Carbonate and Silicate in the Deep Ocean," *Deep-Sea Res.* 21: 455–80.

EDMOND, J. M. [1980]. "Ridge Crest Hot Springs: The Story So Far." *EOS* 61: 129–131. Washington, D.C.: American Geophysical Union.

EDMOND, J. M., and others [1979]. "On the Formation of Metal-Rich Deposits at Ridge Crests," *Earth and Planet. Sci. Letts.* 46: 19–30.

EDWARDS, A. R. [1968]. "The Calcareous Nannoplankton Evidence for New Zealand Tertiary Marine Climate," *Tuatara* 16: 26.

EDWARDS, A. R. [1979]. "Classification of Marine Paleoenvironments," *Geol. Soc. N. Z. Newsletter* 48: 18.

EKMAN, V. W. [1902]. "On Jordrotationens Inverkan På Vindströmmar I. Hafvet," *Nyt. Mag. f. Naturvid.,* 40, Kristiania.

ELASSER, W. M. [1971]. "Sea-Floor Spreading as Thermal Convection," *J. Geophys. Res.* 76: 1101–12.

ELLIOTT, T. [1979]. "Clastic Shorelines." In *Sedimentary Environments and Facies,* pp. 143–77, ed. H. G. Reading. New York: Elsevier.

ELLIS, D. B., and MOORE, T. C., JR. [1973]. "Calcium Carbonate, Opal and Quartz in Holocene Pelagic Sediments and the Calcite Compensation Level in the South Atlantic Ocean," *J. Mar. Res.* 31: 210–27.

ELLWOOD, B. B., and LEDBETTER, M. T. [1979]. "Paleocurrent Indicators in Deep-Sea Sediment," *Science* 203: 1335–37.

EMBLEY, R. W. [1975]. "Studies of Deep-Sea Sedimentation Processes using High-Frequency Seismic Data," (Unpublished doctoral dissertation, Columbia University), 334 pages.

EMERY, K. O. [1960]. *The Sea off Southern California. A Modern Habitat of Petroleum.* New York: John Wiley.

EMERY, K. O. [1968]. "Relict Sediments on Continental Shelves of World," *Am. Assoc. Petrol. Geol. Bull.* 52: 445–464.

EMERY, K. O. [1969]. "The Continental Shelves," *Scientific American* 221: 106–22.

EMERY, K. O. [1977]. "Structure and Stratigraphy of Divergent Continental Margins," in *Geology of Continental Margins,* pp. B1–B20, Short Course, American Association of Petroleum Geologists.

EMERY, K. O. [1980]. "Continental Margins—Classification and Petroleum Prospects," *Am. Assoc. Petrol. Geol. Bull.* 64: 297–315.

EMERY, K. O., and GARRISON, L. E. [1967]. "Sea Levels 7,000 to 20,000 Years Ago," *Science* 157: 684–87.

EMERY, K. O., and SKINNER, B. J. [1977]. "Mineral Deposits of the Deep-Ocean Floor," *Mar. Mining* 1: 1–71.

EMERY, K. O., and UCHUPI, E. [1972]. "Western North Atlantic Ocean: Topography, Rocks, Structure, Water, Life and Sediments," *Am. Assoc. Petrol. Geol. Memoir* No. 17.

EMERY, K. O., and others [1970]. "Continental Rise off Eastern North America," *Am. Assoc. Pet. Geol. Bull.* 54: 44–108.

EMILIANI, C. [1954]. "Depth Habitats of Some Species of Pelagic Foraminifera as Indicated by Oxygen Isotope Ratios, *Am. J. Sci.* 252: 149–58.

EMILIANI, C. [1961]. "The Temperature Decrease of Surface Sea-Water in High Latitudes and of Abyssal-Hadal Water in Open Oceanic Basins During the Past 75 Million Years," *Deep Sea Res.* 8: 144–147.

EMILIANI, C. [1964]. "Paleotemperature Analysis of the Caribbean Cores A254-BR-C and CP-28," *Geol. Soc. Am. Bull.* 75: 129–44.

EMILIANI, C. [1971]. "The Amplitude of Pleistocene Climatic Cycles at Low Latitudes and the Isotopic Composition of Glacial Ice," in *The Late Cenozoic Glacial Ages,* pp. 183–197, ed. K. K. Turekian. New Haven, Connecticut: Yale University Press.

EMILIANI, C. [1980]. "Death and Renovation at the End of the Mesozoic," *EOS,* 61: 505–506. Washington, D.C.: American Geophysical Union.

EMILIANI, C., and MILLIMAN, J. D. [1966]. "Deep-Sea Sediments and Their Geological Record," *Earth-Sci. Rev.* 1: 105–132.

EMILIANI, C., and ROOTH, C. and STIPP, J. J. [1978]. "The Late Wisconsin Flood into the Gulf of Mexico," *Earth and Planet. Sci. Letts.* 41: 159–62.

ENGEL, A. E. J., ENGEL, C. G., and HAVENS, R. G. [1965]. "Chemical Characteristics of Ocean Basalts and the Upper Miocene," *Geol. Soc. Am. Bull.* 76: 719–734.

ERICSON, D. B., EWING, M., WOLLIN, G., and HEEZEN, B. C. [1961]. "Atlantic Deep-Sea Sediment Cores," *Geol. Soc. Am. Bull.* 72: 193–286.

ERICSON, D. B., and WOLLIN, G. [1956]. "Correlation of Six Cores from the Equatorial Atlantic and the Caribbean," *Deep Sea Res.* 3: 104–25.

ERICSON, D. B., and WOLLIN, G. [1964]. *The Deep and the Past.* New York: Knopf.

ESTES, R., and HUTCHINSON, J. M. [1980]. "Eocene Lower Vertebrates from the Ellesmere Island, Canadian Arctic Archipelago," *Paleo. Paleo. Paleo.* 30: 325–47.

EWING, J. [1963]. "Elementary Theory of Seismic Refraction and Reflection Measurements," in *The Sea,* vol. 3, pp. 3–19, ed. M. N. Hill. New York: Wiley-Interscience.

EWING, J. I., EDGAR, N. T., and ANTOINE, J. [1970]. "Structure of the Gulf of Mexico and Caribbean Sea," in *The Sea,* vol. 4, pt. II, pp. 321–58, ed. A. E. Maxwell. New York: Wiley-Interscience.

EWING, J., and EWING, M. [1967]. "Sediment Distribution on the Mid-Ocean Ridges with Respect to Spreading of the Sea Floor," *Science* 156: 1590–92.

EWING, J., and EWING, M. [1970]. "Seismic Reflections," in *The Sea,* vol. 4, pt. 1, pp. 1–52, ed. A. E. Maxwell. New York: Wiley-Interscience.

EWING, J. I., EWING, M., AITKEN, T., and LUDWIG, W. J. [1968]. "North Pacific Sediment Layers Measured by Seismic Profiling," *Am. Geophys. Union Geophys. Mon.* 12, 147–73.

EWING, J., WORZEL, J. L., EWING, M., and WINDESCH, C. C. [1966]. "Ages of Horizon A and Oldest Atlantic Sediments," *Science* 154: 1125–32.

EWING, M. [1965]. "The Sediments of the Argentine Basin," *Quat. J. Roy. Astr. Soc.* 6: 10–27.

EWING, M., ERICSON, D. B., and HEEZEN, B. C. [1958]. "Sediments and Topography of the Gulf of Mexico," in *Habitat of Oil. A Symposium.,* pp. 975–1053, ed. L. G. Weeks. American Association of Petroleum Geologists.

EWING, M., and HEEZEN, B. C. [1956]. "Some Problems of Antarctic Submarine Geology," *Geophys. Mon.* 1: 75–81.

EWING, M., and HEEZEN, B. C. [1960]. "Continuity of the Mid-Oceanic Ridge and Rift Valley in the Southwestern Indian Ocean Confirmed," *Science* 131: 1677–79.

EWING, M., and LUDWIG, W. J., and EWING, J. I. [1964]. "Sediment Distribution in the Oceans: The Argentine Basin," *J. Geophys. Res.* 71: 1611–36.

EWING, M., SUTTON, G. H., and OFFICER, C. B. [1954]. "Seismic Refraction Measurements in the Atlantic Ocean, Part VI: Typical Deep Stations, North Atlantic Basin," *Bull. Seis. Soc. Am.* 44: 21–38.

EWING, M., and THORNDIKE, E. M. [1965]. Suspended Matter in Deep Ocean Water," *Science* 147: 1291–94.

FAIRBRIDGE, R. W. [1960]. "The Changing Level of the Sea. *Scientific American* 202: 70–79.

FAIRBRIDGE, R. W. [1961]. "Eustatic Changes in Sea Level," *Phys. Chem. Earth* 4: 99–185.

FAIRBRIDGE, R. W., ed. [1966]. *The Encyclopedia of Oceanography: Encyclopedia of Earth Science Series,* vol. 1. New York: Reinhold.

FAURE, G. [1977]. *Principles of Isotope Geology.* New York: John Wiley.

FISCHER, A. G. [1975]. "Tethys," in *Geol. of Italy,* vol. 1, pp. 1–9, ed. C. H. Squyres. Tripoli, Libyan Arab Republic: Earth Sci. Society.

FISCHER, A. G., and ARTHUR, M. A. [1977]. "Secular Variations in the Pelagic Realm," pp. 19–50. SEPM Spec. Pub. No. 25.

FISCHER, A. G., and others [1970]. "Geological History of the Western North Pacific," *Science* 168: 1210–14.

FISHER, O. [1889]. *Physics of the Earth's Crust,* 2nd ed. London: MacMillan.

FISHER, R. V. [1964]. "Maximum Size, Median Diameter, and Sorting of Tephra," *J. Geophys. Res.* 69: 341–55.

FLEMING, C. A. [1962]. "New Zealand Biogeography—A Paleontologist's Approach," *Tuatara* 10: 53–108.

FLEMING, C. A. [1975]. "The Geological History of New Zealand and Its Biota." in *Biogeography and Ecology in New Zealand,* ed. C. Kuschel and W. Junk. The Hague: Dr. W. Junk, b.v., publisher.

FLOOD, R. [1978]. "Studies of Deep Sea Sedimentary Microtopography in the North Atlantic Ocean." (Unpublished doctoral dissertation, Woods Hole Oceanographic Institution and Massachusetts Institute of Technology).

FOLGER, D. W. [1970]. "Wind Transport of Land-Derived Mineral, Biogenic, and Industrial Matter over the North Atlantic," *Deep Sea Res.* 17: 337–52.

FOLGER, D. W. and others [1979]. "Evolution of the Atlantic Continental Margin of the United States," in *Maurice Ewing Series 3,* pp. 87–108, eds. M. Talwani, W. Hay, and W. B. F. Ryan. Washington, D.C.: American Geophysical Union.

FORBES, E. [1846]. "On the Connection Between the Distribution of the Existing Fauna and Flora of the British Isles, and the Geological Changes which Affected Their Area, Especially during the Epoch of the Northern Drift," *Geol. Surv. Gt. Brit. Mem.* 1: 336–42.

FORDYCE, R. E. [1977]. "The development of the Circum-Antarctic Current and the Evolution of the Mysticeti (Mammalia–Cetacea)," *Paleo. Paleo. Paleo.* 21: 265–71.

FORSYTH, D., and UYEDA, S. [1975]. "On the Relative Importance of the Driving Forces of Plate Motion," *Geophys. J. R. Astr. Soc.* 43: 163–200.

FOSTER, R. J. [1974]. "Eocene Echinoids and the Drake Passage," *Nature* 249: 751.

FOX, P. J., and OPDYKE, N. D. [1973]. "Geology of the Oceanic Crust: Magnetic Properties of Oceanic Rocks," *J. Geophys. Res.* 78: 5139–54.

FRAKES, L. A., and KEMP, E. M. [1972]. "Tertiary Climate-Influence of Continental Positions," *Nature* 240: 97–99.

FRAKES, L. A., and KEMP, E. M. [1973]. "Paleogene Continental Positions and Evolution of Climate," in *Implications of Continental Drift to the Earth Sciences,* vol. 1, p. 539, ed. D. H. Tarling and S. K. Runcorn. New York: Academic Press.

FRANCHETEAU, J., and LE PICHON, X. [1972]. "Marginal Fracture Zones as Structural Framework of Continental Margins in South Atlantic Ocean." *Am. Assoc. Petrol. Geol. Bull.* 56: 991–1007.

FRANCHETEAU, J., SCLATER, J. G., and MENARD, H. W. [1970]. "Pattern of Relative Motion from Fracture Zone and Spreading Rate Data in the Northeastern Pacific," *Nature* 219: 1328–33.

FRANCIS, T. J. G., and SHOR, G. G., JR. [1966]. "Seismic Refraction Measurements from the Northwest Indian Ocean," *J. Geophys. Res.* 71: 427–49.

FRANKS, S., and NAIRN, A. E. M. [1973]. "The Equatorial Marginal Basins of West Africa," in *The Ocean Basins and Margins,* vol. 1, *The South Atlantic,* pp. 301–50, ed. A. E. M. Nairn and F. G. Stehli. New York: Plenum Press.

FREY, F. A., BRYAN, W. B., and THOMPSON, G. [1974]. "Atlantic Ocean Floor: Geochemistry and Petrology of Basalts from Legs 2 and 3 of the Deep Sea Drilling Project," *J. Geophys. Res.* 79: 5507–27.

FRIEDMAN, G. M., and SANDERS, J. F. [1978]. *Principles of Sedimentology.* New York: John Wiley.

FUNNELL, B. M. [1967]. "Foraminifera and Radiolaria as Depth Indications in the Marine Environment," *Mar. Geol.* 15: 333–47.

FUSOD (The Future of Scientific Ocean Drilling) [1977]. "Report of the Subcommittee of the JOIDES Executive Committee," Woods Hole, Massachusetts, March, 1977.

GAARDER, K. [1971]. "Comments on the Distribution of Coccolithophorids in the Oceans," in *The Micropaleontology of Oceans,* pp. 97–102, ed. B. M. Funnell and W. R. Riedel. London: Cambridge University Press.

GANSSER, A. [1964]. *Geology of the Himalayas.* New York: Wiley-Interscience.

GARDNER, J. V. [1975]. "Late Pleistocene Carbonate Dissolution Cycles in the Eastern Equatorial Atlantic," *Cushman Found. Foram. Res. Spec. Pub.* 13: 129–141.

GARDNER, W. D. [1977]. Fluxes, Dynamics, and Chemistry of Particulates in the Ocean. (Unpublished doctoral dissertation, Woods Hole Oceanographic Institution and Massachusetts Institute of Technology.)

GARRELS, R. M., and MAC KENZIE, F. T. [1971]. *Evolution of Sedimentary Rocks.* New York: W. W. Norton and Company.

GARRISON, L. E., and MC MASTER, R. L. [1966]. "Sediments and Geomorphology of the Continental Shelf Off Southern New England," *Mar. Geol.* 4: 273–89.

GARTNER, S. [1977]. "Nannofossils and Biostratigraphy: An Overview," *Earth Sci. Rev.* 13: 227–50.

GARTNER, S., and KEANY, J. [1978]. "The Terminal Cretaceous Event: A Geologic Problem with an Oceanographic Solution," *Geology* 6: 708–12.

GASKIN, D. E. [1976]. "The Evolution, Zoogeography and Ecology of Cetacea," *Oceanog. Mar. Biol. Ann. Rev.* 14: 247–346.

GEITZENAUER, K. R. [1969]. "Coccoliths as Late Quaternary Palaeoclimatic Indicators in the Subantarctic Pacific Ocean," *Nature* 223: 170–72.

GEITZENAUER, K. R., MARGOLIS, S. V., and EDWARDS, D. S. [1968]. "Evidence Consistent with Eocene Glaciation in a South Pacific Deep-Sea Sedimentary Core," *Earth and Planet. Sci. Letts.* 4: 173.

GEITZENAUER, K. R., ROCHE, M. B., and MCINTYRE, A. [1976]. "Modern Pacific Coccolith Assemblages: Derivation and Application to Late Pleistocene Paleotemperature Analysis," in *Geol. Soc. of Am. Memoir* 145, pp. 423–48, ed. R. M. Cline and J. D. Hays.

GEMEINHARDT, K. [1934]. "Die Silicoflagellaten des Süd-Atlantischen Ozeans," *Wiss. Ergebn. Dtsch. Atlantischen Exped. "Meteor,"* 1925–1927, 12: 274–312.

GIBBS, R. J. [1967]. "The Geochemistry of the Amazon River System: Part 1. The Factors that Control the Salinity and the Composition and Concentration of the Suspended Solids," *Geol. Soc. Am. Bull.* 78: 1203–32.

GIBBS, R. J. [1977]. "Suspended Sediment Transport, and the Turbidity Maximum," in *Estuaries, Geophysics and the Environment,* p. 104, ed. P. E. Abelson and others. Washington, D.C.: National Academy of Science.

GIBSON, T. G., and TOWE, K. M. [1971]. "Eocene Volcanism and the Origin of Horizon A," *Science* 172: 152–54.

GILBERT, G. K. [1885]. "The Topographic Features of Lake Shores." *Fifth Annual Report,* pp. 69–123, U.S. Geological Survey.

GINSBURG, R. N. [1957]. "Early Diagenesis and Lithification of Shallow-Water Carbonate Sediments in South Florida," in *Regional Aspects of Carbonate Deposition,* pp. 80–100, ed. R. J. Leblanc and J. G. Breeting. SEPM Publication 5.

GINSBURG, R. N., ed. [1972]. *South Florida Carbonate Sediments,* in Sedimentation II, p. 72. Coral Gables, Florida: University of Miami.

GLASBY, G. P. [1971]. "The Influence of Aeolian Transport of Dust Particles on Marine Sedimentation in the Southwest Pacific," *J. Roy. Soc. N. Z.* 1: 285–300.

GLASBY, G. P. [1973]. "Manganese Deposits of Variable Composition from North of the Indian-Antarctic Ridge," *Nat. Phy. Sci.* 242: 106–107.

GLASS, B. P. [1967]. "Microtektites in Deep-Sea Sediments," *Nature* 214: 372–74.

GLASS, B. P. [1969]. "Chemical Composition of Ivory Coast Microtektites," *Geochim. Cosmochim. Acta* 33: 1135.

GLASS, B. P. [1972]. "Australasian Microtektites in Deep-Sea Sediments," in *Antarctic Oceanology II, The Australian-New Zealand Sector,* pp. 335–48, Washington, D.C.: American Geophysical Union.

GLASS, B. P., SWINCKI, M. B., and ZWORT, P. A. [1979]. "Australasian, Ivory Coast and North American Tektite Strewnfields: Size, Mass and Correlation with Geomagnetic Reversals and Other Earth Events," *Proc. Lunar Planet. Sci. Conf. 10th,* pp. 2535–45.

GOLL, R. M., and BJÖRKLUND, K. R. [1971]. "Radiolaria in Surface Sediments of the North Atlantic Ocean," *Micropaleo.* 17: 434–457.

GOLL, R. M., and BJÖRKLUND, K. R. [1974]. "Radiolaria in Surface Sediments of the South Atlantic," *Micropaleo.* 20: 38–75.

GOODELL, H. G. and others [1973]. "Marine Sediments of the Southern Oceans," *Antarctic Map Folio Series,* Folio 17, New York: American Geographical Society. 18 pages.

GORDON, A. L. [1971A]. "Comment on the Weddell Sea Produced Antarctic Bottom Water," *J. Geophys. Res.* 76: 5913–14.

GORDON, A. L. [1971B]. "Oceanography of Antarctic Waters," *Antarctic Research Series,* vol. 15, pp. 169–203. Washington, D.C.: American Geophysical Union.

GORDON, A. L. [1971C]. "Recent Physical Oceanographic Studies of Antarctic Waters," in *Research in the Antarctic,* Publ. 93, pp. 609–629, ed. L. O. Quam. Washington, D.C.: American Association for the Advancement of Science.

GORDON, A. L. [1975]. "An Antarctic Oceanographic Section along 170°E," *Deep Sea Res.* 22: 357–77.

GORSLINE, D. S. [1978]. "Anatomy of Margin Basins—Presidential Address," *J. Sediment. Petrol.* 48: 1055–68.

GORSLINE, D. S., and SWIFT, D. J. P., ed. [1977]. *Shelf sediment dynamics: A national overview.* Report of a workshop held in Vail, Colorado, Nov. 2–6, 1976.

GREEN, D. H. [1970]. "The Origin of Basaltic and Nephelinitic Magmas," *Trans. Leicester Lit. Philos. Soc.* 64: 26–54.

GREEN, D. H., and RINGWOOD, A. E. [1967]. "The genesis of basaltic magmas," *Contr. Mineral. Petrol.* 15: 103–90.

GREEN, D. H., and RINGWOOD, A. E. [1968]. "Crystallization of Basalt and Andesite Under High Pressure Hydrous Conditions," *Earth. Planet. Sci. Letts.* 3: 481–489.

GREINER, G. O. G. [1970]. "Distribution of Major Benthonic Foraminiferal Groups on the Gulf of Mexico Continental Shelf," *Micropaleo.* 16: 83–101.

GRIFFIN, J. J., WINDOM, H., and GOLDBERG, E. D. [1968]. "The Distribution of Clay Minerals in the World Oceans," *Deep Sea Res.* 15: 433–59.

GROSS, M. G. [1972]. "Sediment-associated Radionuclides from the Columbia River," in *The Columbia River Estuary and Adjacent Ocean Waters; Bioenvironmental Studies,* pp. 736–754, eds. A. T. Pruter and D. L. Alverson. Seattle: University of Washington Press.

GROSS, M. G. [1977]. *Oceanography—A View of the Earth.* Englewood Cliffs, N.J.: Prentice Hall, Inc.

GROW, J. A. [1973]. "Crustal and Upper Mantle Structure of the Central Aleutian Arc," *Geol. Soc. Am. Bull.* 84: 2169–92.

GROW, J. A., MATTICK, R. E., and SCHLEE, J. S. [1979]. "Multichannel Seismic Depth Sections and Internal Velocities over the Continental Shelf and Upper Continental Slope Between Cape Hatteras and Cape Cod," in *Geological and Geophysical Invest. of Continental Margins,* Amer. Assoc. Petrol. Geol. Memoir 29.

GUILCHER, A. [1958]. *Coastal and Submarine Morphology.* New York: John Wiley.

GUINASSO, N. L., and SCHINK, D. R. [1975]. "Quantitative Estimates of Biological Mixing Rates in Abyssal Sediments," *J. Geophys. Res.* 80: 3032–43.

GUTENBERG, B. [1941]. "Changes in Sea Level, Postglacial Uplift, and Mobility of the Earth's Interior," *Geol. Soc. Am. Bull.* 52: 721–72.

GUTENBERG, B. [1959]. *Physics of the Earth's Interior.* New York: Academic Press.

GUTENBERG, B., and RICHTER, C. F. [1954]. *Seismicity of the Earth and Associated Phenomena* (2nd ed). Princeton, N.J.: Princeton University Press.

HAECKEL, E. [1862]. *Die Radiolarien. Eine Monographie.* Berlin: Reiner.

HAECKEL, E. [1887]. "Report on the Radiolaria collected by H.M.S. *Challenger* during the Years 1873–76," in *Report on the Scientific Results of the Voyage of H.M.S. Challenger,* vol. 18, ed. Sir C. W. Thomson and J. Murray. London: Her Majesty's Government.

HAECKER, V. [1907]. "Altertümliche Spharellarien und Crytellarien aus Grossen Meerestienten." *Archio Protistenk* 10: 114–26.

HAJOS, M. [1976]. "Upper Eocene and Lower Oligocene Diatomaceae, Archaemonadaceae and Silicoflagellatae in Southwestern Pacific Sediments, DSDP Leg 29," in *Initial Reports of the Deep Sea Drilling Project,* vol. 35, pp. 817–83, eds. C. D. Hollister, C. Craddock, and others. Washington, D.C.: U.S. Government Printing Office.

HALL, J. M., and ROBINSON, P. T. [1979]. "Deep Crustal Drilling in the North Atlantic Ocean," *Science* 204: 573–86.

HALLAM, A. [1963]. "Major Epeirogenic and Eustatic Changes since the Cretaceous, and Their Possible Relationships to Crustal Structure," *Am. J. Sci.* 261: 397–423.

HALLAM, A. [1975]. *Jurassic Environments.* Cambridge, England: Cambridge University Press.

HAMILTON, W. [1965]. "Cenozoic Climatic Change and Its Cause," in *Meteoro. Monog.,* vol. 8, ed. J. M. Mitchell, Jr. Boston, Mass.: American Meteorological Society.

HAMPTON, M. A. [1970]. "Subaqueous Debris Flow and Generation of Turbidity Currents." (Unpublished doctoral dissertation, Stanford University.)

HAMPTON, M. A. [1972A]. "The Role of Subaqueous Debris Flows in Generating Turbidity Currents," *J. Sediment. Petrol.* 42: 775-93.

HAMPTON, M. A. [1972B]. "Transport of Ocean Sediments by Debris Flow," Abstract. *Am. Assoc. Petrol. Geol. Bull.* 56: 622.

HAQ, B. U. [1973]. "Transgressions, Climatic Change, and Diversity of Calcareous Nannoplankton," *Mar. Geol.* 15: 25-30.

HAQ, B. U. [1978]. "Calcareous Nannoplankton," in *Introduction to Marine Micropaleontology,* pp. 79-108, ed. B. U. Haq and A. Boersma. Amsterdam: Elsevier.

HAQ, B. U. [1978]. "Silicoflagellates and Ebridians," in *Introduction to Marine Micropaleontology,* pp. 267-76, ed. B. U. Haq and A. Boersma. Amsterdam: Elsevier.

HAQ, B. U. [1981]. Paleogene Paleoceanography—Early Cenozoic Oceans Revisited. *Oceanologica Acta,* 4 (in press).

HAQ, B. U., and LOHMANN, G. P. [1976]. "Early Cenozoic Calcareous Nannoplankton Biogeography of the Atlantic Ocean," *Mar. Micropaleo.* 1, 119-94.

HAQ, B. U., PREMOLI-SILVA, I., and LOHMANN, G. P. [1977]. "Calcareous Planktonic Paleobiogeographic Evidence for Major Climatic Fluctuations in the Early Cenozoic Atlantic Ocean," *J. Geophys. Res.* 82: 3861-76.

HAQ, B. U., and others [1980]. "Late Miocene Marine Carbon-Isotope Shift and Synchroneity of Some Phytoplanktonic Biostratigraphic Events," *Geology* 8: 427-31.

HARDENBOL, J., and BERGGREN, W. A. [1978]. "A New Paleogene Numerical Time Scale," in *Contributions to the Geologic Time Scale,* pp. 213-34, ed., G. V. Cohee, M. F. Glaessner, and H. D. Hedberg. American Association of Petroleum Geologists.

HARRIS, P. [1972]. "The Composition of the Earth," in *Understanding the Earth,* pp. 53-70, ed. I. G. Gass, P. J. Smith, and R. C. L. Wilson. Sussex: The Artemis Press.

HART, S. R., SCHILLING, J. G., and POWELL, J. L. [1973]. "Basalts from Iceland and Along the Reykjanes Ridge: Sr Isotope Geochemistry," *Nature Phys. Sci.* 246: 104-107.

HASLE, G. R. [1976]. "The Biogeography of Some Marine Planktonic Diatoms," *Deep Sea Res.* 23: 319-38.

HATHAWAY, J. C., and others [1979]. "U.S. Geological Survey Core Drilling on the Atlantic Shelf," *Science* 206: 515-527.

HATHERTON, T., and DICKINSON, W. R. [1969]. "The Relationship between Andesitic Volcanism and Seismicity in Indonesia, the Lesser Antilles, and Other Island Arcs," *J. Geophys. Res.* 74: 5301-10.

HAY, W. W. [1974]. "Introduction, Studies in Paleoceanography," pp. 1-6. SEPM Special Publication 20.

HAY, W. W., and SOUTHAM, J. R. [1977]. "Modulation of Marine Sedimentation by the Continental Shelves," in *The Fate of Fossil Fuel CO_2 in the Oceans,* pp. 569-604, ed. N. R. Anderson and A. Malahoff. New York: Plenum.

HAYES, D. E., FRAKES, L. A., and others [1975]. *Initial Reports of the Deep Sea Drilling Project,* Vol. 28. Washington, D.C.: U.S. Government Printing Office.

HAYES, D. E., and FRAKES, L. A. [1975]. "General Synthesis Deep Sea Drilling Project Leg 28," in *Initial Reports of the Deep Sea Drilling Project,* vol. 28, p. 919, ed. D. E. Hayes, L. A. Frakes, and others. Washington, D.C.: U.S. Government Printing Office.

HAYES, D. E., and PITMAN, W. C., III [1973]. "Magnetic Lineations in the North Pacific," in *Geological Investigations in the North Pacific, pp. 291-314, ed. J. D. Hays.* Geological Society of America Memoir 126.

HAYES, D. E., and RINGIS, J. [1973]. "Sea-Floor Spreading in a Marginal Basin: The Tasman Sea," *Nature* 243: 86-88.

HAYES, M. D. [1976]. "Morphology of Sand Accumulation in Estuaries: An Introduction to the Symposium," in *Estuarine Res.,* vol. II2, pp. 3-22, ed. O. L. Gronin. Geology and Engineering.

HAYS, J. D. [1965]. "Radiolaria and Late Tertiary and Quaternary History of the Antarctic Seas," *Biol. Antarctic Seas, Antarctic Res.,* Series 5, 2: 125-184, American Geophysical Union.

HAYS, J. D. [1967]. "Quaternary Sediments of the Antarctic Ocean," in *Progress in Oceanography,* pp. 117-31, ed. M. Sears. New York: Pergamon Press.

HAYS, J. D., IMBRIE, J., and SHACKLETON, N. J. [1976]. "Variations in the Earth's Orbit: Pacemaker of the Ice Ages," *Science,* 194: 1121-32.

HAYS, J. D., and PITMAN, W. C., III [1973]. "Lithospheric Plate Motions, Sea-Level Changes and Climatic and Ecological Consequences," *Nature* 246: 18-22.

HAYS, J. D., SAITO, T., OPDYKE, N. D., and BURCKLE, L. H. [1969]. "Pliocene-Pleistocene Sediments of the Equatorial Pacific, Their Paleomagnetic, Biostratigraphic and Climatic Record," *Geol. Soc. Am. Bull.* 80: 1481-1514.

HEATH, G. R. [1969]. "Carbonate Sedimentation in the Abyssal Equatorial Pacific during the Past 50 million Years," *Geol. Soc. Am. Bull.* 80: 689-694.

HEATH, G. R. [1974]. "Dissolved Silica and Deep Sea Sediments," in *Studies in Paleoceanography,* pp. 77-93, ed. W. W. Hay. SEPM Special Publication 20.

HEATH, G. R., and CULBERSON, C. H. [1970]. "Calcite: Degree of Saturation, Rate of Dissolution, and the Compensation Depth in the Deep Oceans. *Geol. Soc. Am. Bull.* 81: 3157-60.

HEATH, G. R., and MOBERLY, R. L. [1971]. "Cherts from the Western Pacific, Leg 7, Deep Sea Drilling Project," in *Initial Reports of the Deep Sea Drilling Project*, vol. 7, pp. 991–1008, ed. E. L. Winterer, W. R. Riedel, and others. Washington, D.C.: U.S. Government Printing Office.

HEATH, G. R., MOORE, T. C., JR., and DAUPHIN, J. P. [1976]. "Late Quaternary Accumulation Rates of Opal, Quartz, Organic Carbon, and Calcium Carbonate in the Cascadia Basin Area, Northeast Pacific," *Geol. Soc. Am. Memoir* 145: 393–409.

HECHT, A. D. [1973]. "Faunal and Oxygen Isotopic Paleotemperatures and the Amplitude of Glacial/Interglacial Temperature Changes in the Equatorial Atlantic, Caribbean Sea and Gulf of Mexico," *J. Quat. Res.* 3: 671–90.

HECHT, A. D. [1976]. "The Oxygen Isotopic Record of Foraminifera in Deep-Sea Sediments," in *Foraminifera*, vol. 2, pp. 1–43, ed. R. H. Hedley and C. G. Adams. New York: Academic Press.

HEEZEN, B. C. [1962]. "The Deep-Sea Floor," in *Continental Drift*, pp. 235–288, ed. S. K. Runcorn. New York: Academic Press.

HEEZEN, B. C. [1974]. "Atlantic Type Continental Margins," in *The Geology of Continental Margins*, pp. 13–24, ed. C. A. Burk and C. L. Drake. New York: Springer-Verlag.

HEEZEN, B. C., and EWING, M. [1952]. "Turbidity Currents and Submarine Slumps and the 1929 Grand Banks Earthquake," *Am. J. Sci.* 250, 849–78.

HEEZEN, B. C., and EWING, M. [1963]. "The Mid-Oceanic Ridge," in *The Sea*, vol. 3, pp. 388–410, ed. M. N. Hill. New York: Wiley-Interscience.

HEEZEN, B. C., and HOLLISTER, C. D. [1964]. "Deep-Sea Current Evidence from Abyssal Sediments," *Mar. Geol.* 1: 141–74.

HEEZEN, B. C., HOLLISTER, C. D., and RUDDIMAN, W. F. [1966]. "Shaping of the Continental Rise by Deep Geostrophic Contour Currents," *Science* 152: 502–508.

HEEZEN, B. C., and LAUGHTON, A. S. [1963]. "Abyssal Plains," in *The Sea*, vol. 3, pp. 312–64, ed. M. N. Hill. New York: John Wiley.

HEEZEN, B. C., and others [1973]. "Diachronous Deposits: A Kinematic Interpretation of the Post Jurassic Sedimentary Sequence in the Pacific Plate," *Nature* 241: 25–32.

HEEZEN, B. C., and THARP, M. [1965]. "Tectonic Fabric of the Atlantic and Indian Oceans and Continental Drift," *Phil. Trans. Roy. Soc. London, Series A*, 258: 90–106.

HEEZEN, B. C., THARP, M., and EWING, M. [1959]. "The Floors of the Oceans, 1, The North Atlantic," *Geol. Soc. Am. Spec. Paper* 65: 1–122.

HEIN, J. R., SCHOLL, D. W., and MILLER, J. [1978]. "Episodes of Aleutian Ridge Explosive Volcanism," *Science* 199: 137–141.

HEIRTZLER, J. R. [1979]. "The North Atlantic Ridge: Observational Evidence for Its Generation and Aging," *Ann. Rev. Earth Planet. Sci.* 7: 343–55.

HEIRTZLER, J. R., and others [1968]. "Marine Magnetic Anomalies, Geomagnetic Field Reversals, and Motions of the Ocean Floor and Continents," *J. Geophys. Res.* 73: 2119–36.

HEIRTZLER, J. R., and HAYES, D. E. [1967]. "Magnetic Boundaries in the North Atlantic Ocean," *Science* 157: 185–87.

HEIRTZLER, J. R., LE PICHON, X., and BARON, J. C. [1966]. "Magnetic Anomalies over the Reykjanes Ridge," *Deep Sea Res.* 13: 427–43.

HEKINIAN, B., and others [1980]. "Sulfide Deposits from the East Pacific Rise near 21°N," *Science* 207: 1433–44.

HERMAN, Y. [1972]. "Origin of Deep-Sea Cherts in the North Atlantic," *Nature* 238: 392–93.

HERRON, E. M. [1972]. "Sea-Floor Spreading and the Cenozoic History of the East Central Pacific," *Geol. Soc. Am. Bull.* 83: 1671–92.

HERRON, E. M., DEWEY, J. F., and PITMAN, W. C., III [1974]. "Plate Tectonics Model for the Evolution of the Arctic," *Geology* 2: 377–80.

HESS, H. H. [1946]. "Drowned Ancient Islands of the Pacific Basin," *Am. J. Sci.* 244: 722–91.

HESS, H. H. [1948]. "Major Structural Features of the Western North Pacific," *Geol. Soc. Am. Bull.* 59: 417–46.

HESS, H. H. [1962]. "History of Ocean Basins," in *Petrologic Studies: A Volume in Honor of A. F. Buddington*, pp. 599–620, ed. A. E. J. Engel and others. Boulder, Colorado: Geological Society of America.

HEY, R. N., JOHNSON, G. L., and LOWRIE, A. [1977]. "Recent Plate Motions in the Galapagos Area," *Geol. Soc. Am. Bull.* 8: 1385–1403.

HEY, R. N., and VOGT, P. R. [1977]. "Rise Axis Jumps and Sub-Axial Flow near the Galapagos Hotspot," *Tectonophysics* 37: 41–52.

HICKEY, L. J. [1981]. "Land Plant Evidence Compatible with Gradual, Not Catastrophic Change at the End of the Cretaceous," *Nature* (in press).

HILDE, T. W. C., ISEZAKI, N., and WAGEMAN, J. M. [1976]. "Mesozoic Sea-Floor Spreading in the North Pacific," in *The Geophysics of the Pacific Ocean Basin and its Margin*, pp. 205–226, ed. G. H. Sutton, M. H. Manghnani, and R. Moberly. Washington, D.C.: American Geophysical Union.

HILL, M. N. [1963]. "Single-ship Seismic Refraction Shooting," in *The Sea*, vol. 3, pp. 39–46, ed. M. N. Hill. New York: Wiley-Interscience.

HOGAN, L. G., and others [1978]. "Biostratigraphic and Tectonic Implications of ^{40}Ar-^{39}Ar Dates of Ash Layers from the Northeast Gulf of Alaska," *Geol. Soc. Am. Bull.* 89: 1259–64.

HOLCOMBE, T. L. [1977]. "Ocean bottom features—Terminology and Nomenclature," *Geojournal* 6: 25–48.

HOLEMAN, J. N. [1968]. "The Sediment Yield of Major Rivers of the World," *Water Resources Res.* 4: 737.

HOLLISTER, C. D., FLOOD, R., and MC CAVE, I. N. [1978]. "Plastering and Decorating in the North Atlantic," *Oceanus* 21: 5–13.

HOLLISTER, C. D., and HEEZEN, B. C. [1972]. "Geological Effects of Ocean Bottom Currents," in *Studies in Physical Oceanography,* vol. 2, pp. 37–66, ed. A. L. Gordon. New York: Gordon & Breach.

HOLMES, A. [1929]. "Radioactivity and Earth Movements," *Trans. Geol. Soc. Glasgow* 18: 559–606.

HONJO, S. [1975]. "Dissolution of Suspended Coccoliths in the Deep-Sea Water Column and Sedimentation of Coccolith Ooze," in *Special Pub. No. 13,* pp. 114–128, ed. W. V. Sliter, A. W. H. Bé, and W. H. Berger. Lawrence, Kansas: Cushman Foundation for Foraminiferal Research.

HONJO, S. [1976]. "Coccoliths: Production, Transportation and Sedimentation," *Mar. Micropaleo.* 1: 65–79.

HONJO, S. [1977]. "Biogeography and Provincialism of Living Coccolithophorids in the Pacific Ocean," in *Oceanic Micropaleontology,* vol. 2, pp. 951–972, ed. A. T. S. Ramsay. New York: Academic Press.

HONJO, S., and OKADA, H. [1974]. "Community Structure of Coccolithophores in the Photic Layer of the Mid-Pacific," *Micropaleo.* 20: 209–230.

HORN, D. R., ed. [1972]. *Ferromanganese Deposits on the Ocean Floor.* Washington, D.C.: National Science Foundation.

HORN, D. R., EWING, J. I., and EWING, M. [1972]. "Graded-bed Sequences Emplaced by Turbidity Currents North of 20°N in the Pacific, Atlantic and Mediterranean," *Sedimentology.* 18: 247–275.

HORN, D. R., EWING, M., DELACH, M. N., and HORN, B. M. [1971B]. "Turbidites of the Northeast Pacific," *Sedimentology* 16: 55–69.

HORN, D. R., EWING, M., HORN, B. M., and DELACH, M. N. [1971]. "Turbidites of the Hatteras and Sohm Abyssal Plains, Western North Atlantic," *Mar. Geol.* 11: 287–323.

HORNIBROOK, N. DE B. [1971]. "New Zealand Tertiary Climate," *New Zealand Geol. Survey Report 47,* pp. 1–19.

HORNIBROOK, N. DE B. and EDWARDS, A. R. [1971]. "Integrated Planktonic Foraminiferal and Calcareous Nannoplankton Datum Levels in the New Zealand Cenozoic," in *Proceedings of the Second Planktonic Conference,* pp. 649–57. Rome: Edizoni Technoscienza.

HOUTZ, R., and EWING, J. [1976]. "Upper Crustal Structure as a Function of Plate Age," *J. Geophys. Res.* 81: 2490–98.

HOUTZ, R., EWING, J., and LE PICHON, X. [1968]. "Velocities of Deep-Sea Sediments from Sonobuoy Data," *J. Geophys. Res.* 73: 2615–41.

HOYT, J. H. [1967]. "Barrier Island Formation," *Geol. Soc. Am. Bull.* 78: 1125–36.

HSÜ, K. J. [1977]. "Tectonic Evolution of the Mediterranean Basins," in *The Ocean Basins and Margins,* vol. 4A, ed. A. E. M. Nairn, W. H. Kanes, and F. G. Stehli. New York: Plenum.

HSÜ, K. J., and JENKYNS, H. C., ed. [1974]. *Pelagic Sediments: On Land and Under the Sea.* Special Publication of International Association of Sediment. Oxford: Blackwell Science Publishers.

HSÜ, K. J., and others [1977]. "History of the Mediterranean Salinity Crisis," *Nature* 267: 399–403.

HSÜ, K. J., and RYAN, W. B. F. [1973]. "Deep-Sea Drilling in the Hellenic Trench," *Geol. Soc. Greece Bull.* 10: 80-81

HUANG, T. C., and WATKINS, N. D. [1977]. "Contrasts between the Brunhes and Matuyama Sedimentary Records of Bottom Water Activity in the South Pacific," *Mar. Geol.* 23: 113–132.

HUANG, T. C., WATKINS, N. D., and SHAW, D. M. [1975]. "Atmospherically Transported Volcanic Glass in Deep-Sea Sediments: Volcanism in Sub-Antarctic Latitudes of the South Pacific during Late Pliocene and Pleistocene Time," *Geol. Soc. Am. Bull.* 86: 1305–15.

HUANG, T. C., WATKINS, N. D., and WILSON, L. [1979]. "Deep-Sea Tephra from the Azores during the Past 300,000 Years: Eruptive Cloud Height and Ash Volume Estimates: Summary, Part 1," *Geol. Soc. Am. Bull.* 90: 131–33.

HUMPHRIS, C. C., JR. [1979]. "Salt Movement on Continental Slope, Northern Gulf of Mexico," *Am. Assoc. Petrol. Geol. Bull.* 63: 782–98.

HURD, D. C., and THEYER, F. [1975]. "Changes in the Physical and Chemical Properties of Biogenic Silica from the Central Equatorial Pacific," *Advances in Chemistry Series* 147: 211–230.

HUTSON, W. H. [1976]. "The Ecology and Paleoecology of Indian Ocean Planktonic Foraminifera. (Unpublished doctoral dissertation, Brown University.)

HUTSON, W. H. [1978]. "Application of transfer functions to Indian Ocean Planktonic Foraminifera," *Quat. Res.* 9: 87–112.

HYNDMAN, R. D., and DRURY, M. J. [1976]. "The Physical Properties of Oceanic Basement Rocks from Deep-Drilling on the Mid-Atlantic Ridge," *J. Geophys. Res.* 81: 4042–52.

IMBRIE, J., and KIPP, N. G. [1971]. "A New Micropaleontological Method for Quantitative Paleoclimatology: Application to a Late Pleistocene Caribbean Core," in *The Late Cenozoic Glacial Ages,* pp. 71–181. ed. K. K. Turekian. New Haven, Connecticut: Yale University Press.

IMBRIE, J., VAN DONK, J., and KIPP, N. G. [1973]. "Paleoclimatic Investigation of a Late Pleistocene Caribbean Deep-Sea Core: Comparison of Isotopic and Faunal Methods," *Quat. Res.* 3: 10–38.

INGLE, J. C., JR. [1967]. "Foraminiferal Biofacies Variation and the Miocene-Pliocene Boundary in Southern California," *Am. Paleont. Bull.* 52: 217–394.

INGLE, J. C., JR. [1973]. "Neogene Foraminifera from the Northeastern Pacific Ocean, Leg 18, Deep Sea Drilling Project," in *Initial Reports of the Deep Sea Drilling Project,* vol. 18, pp. 517–567, ed. L. D. Kulm, R. von Huene, and others. Washington, D.C.: U.S. Govt. Printing Office.

INGLE, J. C., JR. [1975]. "Summary of Late Paleogene-Neogene Insular Stratigraphy, Paleobathymetry, and Correlations, Philippine Sea and Sea of Japan region," in *Initial Reports of the Deep Sea Drilling Project,* vol. 31, pp. 837–55, ed. D. E. Karig, J. C. Ingle, Jr., and others. Washington, D.C.: U.S. Govt. Printing Office.

INGLE, J. C., JR. [1977]. "Summary of Late Neogene Planktic Foraminiferal Biofacies, Biostratigraphy, and Paleoceanography of the Marginal North Pacific Ocean," *Proc. First Intern. Congress Pacific Neogene Stratigraphy,* Tokyo, pp. 177–182. Tokyo: Kaiyo Shuppan Company.

INGLE, J. C., JR., and GARRISON, R. E. [1977]. "Origin, Distribution and Diagenesis of Neogene Diatomites Around the North Pacific Rim," *Proc. First Intern. Congress on Pacific Neogene Stratigraphy,* Tokyo, pp. 348–350, Tokyo: Kaiyo Shuppan Company.

INMAN, D. L. [1971]. "Nearshore processes," in *Encyclopedia of Science and Technology,* vol. 9, pp. 26–33, New York: McGraw-Hill.

INMAN, D. L., and CHAMBERLAIN, T. K. [1959]. Tracing Beach Sand Movement with Irradiated Quartz. *J. Geophys. Res.* 64: 41–47.

INMAN, D. L., and FRAUTSCHY, J. D. [1966]. "Littoral Processes and the Development of Shorelines," *Coastal Engineering,* pp. 511–36. Santa Barbara Specialty Conference, California.

INMAN, D. L., and NORDSTROM, C. E. [1971]. "On the Tectonic and Morphologic Classification of Coasts," *J. Geol.* 79: 1–21.

IRVING, E. [1956]. "Paleomagnetic and Palaeoclimatological Aspects of Polar Wandering," *Pure Appl. Geophys.* 33: 23–41.

ISACKS, B. L., and MOLNAR, P. [1971]. "Distribution of Stresses in the Descending Lithosphere from a Global Survey of Focal-Mechanism Solutions of Mantle Earthquakes," *Rev. Geophys. Space Phys.* 9: 103–74.

ISACKS, B. L., OLIVER, J., and SYKES, L. R. [1968]. "Seismology and the New Global Tectonics," *J. Geophys. Res.* 73: 5855–5900.

ITO, K. [1970]. "The Fine Structure of the Basalt-Ecologite Transition," in *Fiftieth Anniversary Symposia,* ed. B. A. Morgan. Mineralogical Society of America Special Paper 3.

ITO, K. [1973]. "Analytic Approach to Estimating Source Rock of Basaltic Magmas: Major Elements," *J. Geophys. Res.* 78: 412–31.

ITO, K., and KENNEDY, G. C. [1967]. "Melting and Phase Relations in a Natural Peridotite to 40 Kilobars," *Am. J. Sci.* 265: 519–538.

JACKSON, E. D., SHAW, H. R., and BARGAR, K. E. [1975]. "Calculated Geochronology and Stress Field Orientations along the Hawaiian Chain," *Earth and Planet. Sci. Letts.* 26: 145–55.

JANSA, L. F., GARDNER, J., and DEAN, W. E. [1978]. "Mesozoic Sequences of the Central North Atlantic," in *Initial Reports of the Deep Sea Drilling Project,* vol. 41, pp. 991–1031, ed. Y. Lancelot, E. Seibold, and others. Washington, D.C.: U.S. Government Printing Office.

JENDRZEJEWSKI, J. P., and ZARILLO, G. A. [1972].

"Late Pleistocene Paleotemperature Oscillations Defined by Silicoflagellate Changes in a Sub-Antarctic Deep-Sea Core," *Deep Sea Res.* 196: 327–29.

JENKINS, D. G. [1967]. "Planktonic Foraminiferal Zones and New Taxa from the Lower Miocene to the Pleistocene of New Zealand," *N. Z. J. Geol. and Geophys.* 10: 1064–78.

JENKINS, D. G. [1974]. "Initiation of the Protocircum-Antarctic Current," *Nature* 252: 371.

JENKYNS, H. C. [1980]. "Tethys: Past and Present," *Proc. Geol. Assoc.* 91: 107–18.

JERLOV, N. G. [1953]. "Particle distribution in the ocean," *Rep. Swed. Deep-Sea Exped.* 3: 73–97.

JERLOV, N. G. [1959]. "Maxima in the Vertical Distribution of Particles in the Sea," *Deep Sea Res.* 15: 173–84.

JOHNSON, B. D., POWELL, C. MC A., and VEEVERS, J. J. [1976]. "Spreading History of the Eastern Indian Ocean, and Greater India's Northward Flight from Antarctica and Australia," *Geol. Soc. Am. Bull.* 87: 1560–66.

JOHNSON, D. A. [1972]. "Ocean-Floor Erosion in the Equatorial Pacific," *Geol. Soc. Am. Bull.* 83: 3121–44.

JOHNSON, D. A., and JOHNSON, T. C. [1970]. "Sediment Redistribution by Bottom Currents in the Central Pacific," *Deep Sea Res.* 17: 157–70.

JOHNSON, D. W. [1919]. *Shore Processes and Shoreline Development.* New York: John Wiley.

JOHNSON, H. D. [1978]. "Shallow Siliciclastic Seas," in *Sedimentary Environments and Facies,* pp. 207–258, ed. H. G. Reading. New York: Elsevier.

JOHNSON, T. C. [1974]. "The Dissolution of Siliceous Microfossils in Surface Sediments of the Eastern Tropical Pacific," *Deep Sea Res.* 21: 851–64.

JONES, J. I. [1967]. "Significance of Distribution of Planktonic Foraminifera in the Equatorial Atlantic Undercurrent," *Micropaleo.* 13: 489–501.

JOUSÉ, A. P., KOZLOVA, O. G., and MUHINA, V. V. [1971]. "Distribution of Diatoms in the Surface Layer of Sediments from the Pacific Ocean," in *The Micropaleontology of the Oceans,* pp. 263–269, ed. B. M. Funnell and W. R. Riedel. London: Cambridge University Press.

KANAYA, T. [1959]. "Miocene Diatom Assemblages from the Onnagaua Formation and Their Distribution on the Correlative Formations in Northeast Japan," *Tohoku Univ. Sci. Rep.,* 2nd ser. *Geology* 30: 130.

KANAYA, T., and KOIZUMI, I. [1966]. "Interpretation of Diatom Thanatocoenoses from the North Pacific Applied to a Study of Core V20–130 (Studies of a Deep-Sea core V20–130 Part 4)," *Tohoku Univ. Sci. Rep.,* 2nd ser. *Geology,* 37: 89–130.

KANEPS, A. G. [1975]. "Cenozoic Planktonic Foraminifera from Antarctic Deep-Sea Sediments, Leg. 28," in *Initial Reports of the Deep Sea Drilling Project,* vol. 28, pp. 573–584, ed. D. E. Hayes, L. A. Frakes and others. Washington, D.C.: U.S. Govt. Printing Office.

KANEPS, A. G. [1979]. "Gulf Stream: Velocity Fluctuations during the Late Cenozoic," *Science* 204: 297–301.

KARIG, D. E. [1970]. "Ridges and Basins of the Tonga-Kermadec Island Arc System," *J. Geophys. Res.* 75: 239.

KARIG, D. E. [1971]. "Origin and Development of Marginal Basins in the Western Pacific," *J. Geophys. Res.* 76: 2542–61.

KARIG, D. E. [1973]. "Comparison of Island Arc-Marginal Basin Complexes in the North-west and South-west Pacific," in *The Western Pacific, Island Arcs, Marginal Seas, Geochemistry*, pp. 355–64, ed. P. J. Coleman. New York: Crane, Russak and Co.

KARIG, D. E. [1974]. "Evolution of Arc Systems in the Western Pacific," *Ann. Rev. Earth Planet. Sci.* 2: 51–75.

KARIG, D. E. [1975]. "Basin genesis in the Philippine Sea," in *Initial Reports of the Deep Sea Drilling Project*, vol. 31, pp. 857–879, ed. D. E. Karig, J. C. Ingle, Jr., and others. Washington, D.C.: U.S. Government Printing Office.

KARIG, D. E., ANDERSON, R. N., and BIBEE, L. D. [1978]. "Characteristics of Back-Arc Spreading in the Mariana Trough," *J. Geophys. Res.* 83: 1213–26.

KARIG, D. E., and MOORE, G. F. [1975]. "Tectonic Complexities in the Bonin Arc System," *Tectonophysics* 27: 97–118.

KARIG, D. E., and SHARMAN, G. F. III [1975]. "Subduction and Accretion in Trenches," *Geol. Soc. Am. Bull.* 86: 377–89.

KAUFFMAN, E. C. [1967]. "Cretaceous Thyasira from the Western Interior of North America," in *Smithson. Misc. Coll.* 152: 1–159.

KAUFFMAN, E. G. [1979]. "Cretaceous," in *Treatise on Invertebrate Paleontology*, pp. A418–A487, ed. R. A. Robinson, and C. Teichert. Lawrence, Kansas: University of Kansas Press, Geological Society of America.

KEANY, J. [1973]. "New Radiolarian Paleoclimatic Index in the Plio-Pleistocene of the Southern Ocean," *Nature* 246: 139–41.

KEANY, J., and KENNETT, J. P. [1972]. "Pliocene–Early Pleistocene Paleoclimatic History Recorded in Antarctic-Subantarctic Deep-Sea Cores," *Deep Sea Res.* 19: 529–48.

KEEN, M. J. [1968], *An Introduction to Marine Geology*. New York: Pergamon Press.

KEIGWIN, L. D., JR. [1976]. "Late Cenozoic Planktonic Foraminiferal Biostratigraphy and Paleoceanography of the Panama Basin," *Micropaleo.* 22: 419–42.

KEIGWIN, L. D., JR. [1978]. "Pliocene Closing of the Isthmus of Panama, Based on Biostratigraphic Evidence from Nearby Pacific Ocean and Caribbean Sea Cores," *Geology* 6: 630–34.

KEIGWIN, L. D., JR. [1979]. "Late Cenozoic Stable Isotope Stratigraphy and Paleoceanography of Deep Sea Drilling Project Sites from the East Equatorial and Central North Pacific Ocean," *Earth and Planet. Sci. Letts.* 45: 361–82.

KEIGWIN, L. D., JR. [1980]. "Palaeoceanographic Change in the Pacific at the Eocene-Oligocene Boundary," *Nature* 287: 722–25.

KEIGWIN, L. D., JR., BENDER, M. L., and KENNETT, J. P. [1979]. "Thermal Structure of the Deep Pacific Ocean in the Early Pliocene," *Science* 205: 1386–88.

KEIGWIN, L. D., JR., and SHACKLETON, N. J. [1980]. "Uppermost Miocene Carbon Isotope Stratigraphy of a Piston Core in the Equatorial Pacific," *Nature* 284: 613–614.

KEIGWIN, L. D., JR., and THUNELL, R. [1979]. "Middle Pliocene Climatic Change in the Western Mediterranean from Faunal and Oxygen Isotopic Trends," *Nature* 282: 294–96.

KELLOGG, T. B. [1976]. "Late Quaternary Climatic Changes: Evidence from Deep-Sea Cores of Norwegian and Greenland Seas," in *The Geol. Soc. Am., Mem. 145*, pp. 77–110, ed. R. M. Cline and J. D. Hays.

KEMP, E. M. [1975]. "Palynology of Leg 28 Drill Sites, Deep Sea Drilling Project," in *Initial Reports of the Deep Sea Drilling Project*, vol. 28, pp. 599–623, ed. D. E. Hayes, L. A. Frakes, and others. Washington, D. C., U. S. Government Printing Office.

KEMP, E. M. [1978]. "Tertiary Climatic Evolution and Vegetation History in the Southeast Indian Ocean Region," *Paleo. Paleo. Paleo.* 24: 169–208.

KEMP, E. M., and BARRETT, P. J. [1975]. "Antarctic Glaciation and Early Tertiary Vegetation," *Nature* 258: 507–508.

KEMP, E. M., FRAKES, L. A., and HAYES, D. E. [1975]. "Paleoclimatic Significance of Diachronous Biogenic Facies, Leg 28, Deep Sea Drilling Project," in *Initial Reports of the Deep Sea Drilling Project*, vol. 28, pp. 909–917, eds. D. E. Hayes, L. A. Frakes, and others. Washington, D.C.: U.S. Government Printing Office.

KEMP, E. M., and HARRIS, W. K. [1975]. "The Vegetation of Tertiary Islands on the Ninetyeast Ridge," *Nature* 258: 303–307.

KENNETT, J. P. [1967]. "Recognition and Correlation of the Kapitean Stage (Upper Miocene, New Zealand)," *N. Z. J. Geol. and Geophys.* 10: 1051–63.

KENNETT, J. P. [1968]. "*Globorotalia truncatulinoides* as a paleoceanographic index," *Science* 159: 1461–63.

KENNETT, J. P. [1970]. "Pleistocene Paleoclimates and Foraminiferal Biostratigraphy in Subantarctic Deep-Sea Cores," *Deep Sea Res.* 17: 125–40.

KENNETT, J. P. [1973]. "Middle and Late Cenozoic Planktonic Foraminiferal Biostratigraphy of the Southwest Pacific—Deep Sea Drilling Project, Leg. 21," in *Initial Reports of the Deep Sea Drilling Project*, vol. 21. pp. 575–637, ed. R. E. Burns, J. E. Andrews, and others. Washington, D.C.: U.S. Government Printing Office.

KENNETT, J. P. [1977]. "Cenozoic Evolution of Antarctic Glaciation, the Circum-Antarctic Ocean, and Their Impact on Global Paleoceanography," *J. Geophys. Res.* 82: 3843–59.

KENNETT, J. P. [1978]. "The Development of Plank-tonic Biogeography in the Southern Ocean during the Cenozoic," *Mar. Micropaleo.* 3: 301–345.

KENNETT, J. P. [1980]. "Paleoceanographic and Bio-geographic Evolution of the Southern Ocean during the Cenozoic, and Cenozoic Microfossil Datums. *Paleo. Paleo. Paleo.* 31: 123–152.

KENNETT, J. P. [1981]. "Marine Tephrochronology," in *The Sea,* vol. 7, ed. C. Emiliani. New York: John Wiley (in press).

KENNETT, J. P., and others [1972]. "Australian-Antarctic Continental Drift, Paleocirculation Changes and Oligocene Deep-Sea Erosion," *Nature Phys. Sci.* 239: 51–55.

KENNETT, J. P., HOUTZ, R. E., and others [1975]. *Initial Reports of the Deep Sea Drilling Project,* vol. 29. Washington, D.C.: U.S. Government Printing Office.

KENNETT, J. P., HOUTZ, R. E., and others [1975]. "Cenozoic paleoceanography in the Southwest Pacific Ocean, Antarctic Glaciation and the Develop-ment of the Circum-Antarctic Current," in *Initial Reports of the Deep Sea Drilling Project,* vol. 29, p. 1155, ed. J. P. Kennett, R. E. Houtz, and others. Washington, D.C.: U. S. Government Printing Office.

KENNETT, J. P., and HUDDLESTUN, P. [1972]. "Abrupt Climatic Change at 90,000 yr. B.P.: Faunal Evidence from Gulf of Mexico Cores," *Quat. Res.* 2: 384–395.

KENNETT, J. P., McBIRNEY, A. R., and THUNELL, R. C. [1977]. "Episodes of Cenozoic Volcanism in the Circum-Pacific Region," *J. Volcanol. Geother-mal Res.* 2: 145–163.

KENNETT, J. P., and PENROSE, N. L. [1978]. "Fossil Holocene Seaweed and Attached Calcareous Polychaetes in an Anoxic Basin, Gulf of Mexico," *Nature* 276: 172–73.

KENNETT, J. P., and SHACKLETON, N. J. [1975]. "Laurentide Ice Sheet Meltwater Recorded in Gulf of Mexico Deep Sea Cores," *Science* 188: 147–150.

KENNETT, J. P., and SHACKLETON, N. J. [1976]. "Ox-ygen Isotopic Evidence for the Development of the Psychrosphere 38 Myr. ago," *Nature* 260: 513–15.

KENNETT, J. P., and others [1979]. "Late Cenozoic Oxygen and Carbon Isotopic History and Volcanic Ash Stratigraphy: Deep Sea Drilling Project Site 284, South Pacific," *Am. J. of Sci.* 279: 52–69.

KENNETT, J. P., and THUNELL, R. C. [1975]. "Global Increase in Quaternary Explosive Volcanism," *Science* 187: 497–503.

KENNETT, J. P., and THUNELL, R. C. [1977a]. "Com-ments on Cenozoic Explosive Volcanism Related to East and Southeast Asian Arcs," in *Maurice Ewing Series 1,* pp. 348–352, ed. M. Talwani and W. C. Pitman, III. Washington, D.C.: American Geophysical Union.

KENNETT, J. P., and THUNELL, R. C. [1977b]. "On Ex-plosive Cenozoic Volcanism and Climatic Implica-tions, *Science* 196: 1231–34.

KENNETT, J. P., and VELLA, P. [1975]. "Late Cenozoic Planktonic Foraminifera and Paleoceanography at Deep Sea Drilling Project Site 284 in the Cool Sub-tropical South Pacific," in *Initial Reports of the Deep Sea Drilling Project,* vol. 29, p. 769, ed. J. P. Kennett, R. E. Houtz, and others. Washington, D.C.: U.S. Government Printing Office.

KENNETT, J. P., and WATKINS, N. D. [1974]. "Late Miocene-Early Pliocene Paleomagnetic Stratigraphy, Paleoclimatology, and Biostratigraphy in New Zealand," *Geol. Soc. Am. Bull.* 85: 1385–98.

KENNETT, J. P., and WATKINS, N. D. [1975]. "Deep-Sea Erosion and Manganese Nodule Development in Southeast Indian Ocean," *Science* 188: 1011–13.

KENNETT, J. P., and WATKINS, N. D. [1976]. "Regional Deep-Sea Dynamic Processes Recorded by Late Cenozoic Sediments of Southeastern Indian Ocean," *Geol. Soc. Am. Bull.* 87: 321–39.

KENNETT, J. P., WATKINS, N. D., and VELLA, P. [1971]. "Paleomagnetic Chronology of Pliocene-Early Pleistocene Climates and the Plio-Pleistocene Boundary in New Zealand," *Science* 171: 276–79.

KENT, D. V., HONNOREY, B. M., OPDYKE, N. D., and Fox, P. J. [1978]. "Magnetic Properties of Dredged Oceanic Gabbros and the Source of Marine Magnetic Anomalies, *Geophys, J. Roy. Astro. Soc.* 55: 513–37.

KERR, R. A. [1980]. "Asteroid Theory of Extinctions Strengthened," *Science* 210: 514–17.

KIDD, R. B., and DAVIES, T. A. [1978]. "Indian Ocean Sediment Distribution since the Late Jurassic," *Mar. Geol.* 26: 49–70.

KILHAM, P. [1971]. "A Hypothesis Concerning Silica and the Fresh Water Planktonic Diatoms," *Limnol. and Oceanog.* 16:10–18.

KIPP, N. G. [1976]. "New Transfer Function for Estimating Past Sea-Surface Conditions From Sea-bed Distribution of Planktonic Foraminiferal Assemblages in the North Atlantic," in *Geol. Soc. Am. Memoir 145,* pp. 3–41, ed. R. M. Cline and J. D. Hays. New York: Geological Society of America.

KIRKPATRICK, R. J., and THE LEG 46 SCIENTIFIC PARTY [1978]. "Leg 46 Cruise Synthesis: the Petrology, Structure and Geologic History at Site 396," in *Initial Reports of the Deep Sea Drilling Project,* vol. 46, pp. 417–423, ed. L. Dmitriev, J. Heirtzler, and others. Washington, D.C.: U.S. Government Printing Office.

KLITGORD, K. D., and BEHRENDT, J. C. [1979]. "Basin structure of the U.S. Atlantic margin," in *Geological and Geophysical Investigation of Continental Margin,* pp. 85–112. American Association of Petroleum Geologists Memoir 29.

KLOOTWIJK, C. T., and PEIRCE, J. W. [1979]. "India's and Australia's Pole Path since the late Mesozoic and the India-Asia Collision," *Nature* 282: 605–607.

KNOPOFF, L. [1969]. "The Upper Mantle of the Earth," *Science* 163: 1277–87.

KOIZUMI, I. [1968]. "Tertiary Diatom Flora of Oga Peninsula, Akita Prefecture, Northeast Japan," *Tohoku Univ. Sci. Rept.,* 2nd ser. Geology, 40: 171–240.

KOLB, C. R., and VAN LOPIK, J. R. [1966]. "Depositional Environments of the Mississippi River Deltaic Plain—Southeastern Louisiana," in *Deltas in Their Geologic Framework*, pp. 17–62, ed. J. L. Shirley and J. A. Ragsdale. Houston, Texas: Houston Geological Soc.

KOLLA, V., BÉ, A. W. H., and BISCAYE, P. E. [1976]. "Calcium Carbonate Distribution in the Surface Sediments of the Indian Ocean," *J. Geophys. Res.* 81: 2605–16.

KOMAR, P. D. [1976]. *Beach Processes and Sedimentation*. Englewood Cliffs, N.J.: Prentice-Hall Inc.

KRAFT, J. C. [1978]. "Coastal stratigraphic sequences," in *Coastal Sedimentary Environments*, pp. 361–384, ed. R. A. Davis, Jr. New York: Springer-Verlag.

KRISHNASWAMI, S. [1976]. "Authigenic Transition Elements in Pacific Pelagic Clays. *Geoch. et. Cosmochimica Acta.* 40: 425–34.

KRISTOFFERSEN, Y. [1977]. "Labrador Sea: A Geophysical Study." (Unpublished doctoral dissertation Columbia University, New York).

KRUMBEIN, W. C. [1936]. "Application of Logarithmic Moments to Size-Frequency Distributions of Sediments," *J. Sedi. Petrol.* 6: 35–47.

KUENEN, PH. H. [1937]. "Experiments in Connection with Daly's Hypothesis on the Formation of Submarine Canyons." *Leidsche Geol. Meded.* 8: 327–35.

KUENEN, PH. H. [1958]. "No Geology Without Marine Geology," *Geol. Rundschau* 47: 1–10.

KUENEN, PH. H., and MIGLIORINI, C. I. [1950]. "Turbidity Currents as a Cause of Graded Bedding," *J. Geol.* 58: 91–127.

KULLENBERG, B. [1947]. "The Piston Core Sampler," *Svenska Hydrog-Biol. Komm. Skr.* 3: 46.

KULM, L. D., and FOWLER, G. A. [1974]. "Oregon Continental Margin Structure and Stratigraphy: A Test of the Imbricate Thrust Model," in *Geology of Continental Margins*, pp. 261–283, ed. C. A. Burk and C. L. Drake. New York: Springer-Verlag.

KULM, L. D., SCHWELLER, W. J., and MASIAS, A. [1977]. "A Preliminary Analysis of the Subduction Processes along the Andean Continental Margin, 6° to 45°S," in *Maurice Ewing Series 1*, pp. 285–301, ed. M. Talwani and W. C. Pitman. Washington, D.C.: American Geophysical Union.

KUMAR, N. [1979]. "Origin of Paired Aseismic Rises: Ceará and Sierra Leone Rises in the Equatorial and the Rio Grande Rise and Walvis Ridge in the South Atlantic," *Mar. Geol.* 30: 175–91.

KUNO, H. [1966]. "Lateral Variation of Basalt Magma Type Across Continental Margins and Island Arcs," *Bull. Volcanologique* 29: 223–33.

KURTZ, D. D., and ANDERSON, J. B. [1979]. "Recognition and Sedimentologic Description of Recent Debris Flow Deposits from the Ross and Weddell Seas, Antarctica," *J. Sed. Petrol.* 49: 1159–70.

KUSHIRO, I. [1968]. "Compositions of Magmas Formed by Partial Zone Melting of the Earth's Upper Mantle," *J. Geophys. Res.* 73: 619–34.

LA BRECQUE, J. L., KENT, D. V., and CANDE, S. C. [1977]. "Revised Magnetic Polarity Time Scale for the Cretaceous and Cenozoic," *Geology* 5: 330–35.

LA BRECQUE, J. L., and RABINOWITZ, P. D. [1977]. "Magnetic Anomalies Bordering the Continental Margin of Argentina," *Map Ser. Cat. 826*, American Association Petroleum Geologists.

LAINE, E. P., and HOLLISTER, C. D. [1981]. "Geological Effects of the Gulf Stream System on the Northern Bermuda Rise," *Mar. Geol.* 39: 277–310

LANCELOT, Y. [1973]. "Chert and Silica Diagenesis in Sediments from the Central Pacific," in *Initial Reports of the Deep Sea Drilling Project*, vol. 17, pp. 377–405, ed. E. L. Winterer, J. I. Ewing, and others. Washington, D.C.: U.S. Government Printing Office.

LANCELOT, Y., and SEIBOLD, E. [1978]. "The Evolution of the Central Northeastern Atlantic—Summary of Results of Deep Sea Drilling Project Leg 41," in *Initial Reports of the Deep Sea Drilling Project*, vol. 41, pp. 1215–45, ed. Y. Lancelot, E. Seibold. Washington, D.C.: U.S. Government Printing Office.

LANCELOT, Y., HATHAWAY, J. C., and HOLLISTER, C. D. [1972]. "Lithology of Sediments from the Western North Atlantic, Deep Sea Drilling Project, Leg 11," in *Initial Reports of the Deep Sea Drilling Project*, vol. 11, pp. 901–949, ed. C. D. Hollister, J. I. Ewing, and others. Washington, D.C.: U.S. Government Printing Office.

LANCELOT, Y., and LARSON, R. L. [1975]. "Sedimentary and Tectonic Evolution of the Northwestern Pacific," in *Initial Reports of the Deep Sea Drilling Project*, vol. 32, pp. 925–939, ed. R. L. Larson, R. Moberly, and others. Washington, D.C.: U.S. Government Printing Office.

LANGSETH, M. C., JR., and VON HERZEN, R. P. [1970]. "Heat Flow through the Floor of the World Oceans," in *The Sea*, vol. 4, pp. 299–345, ed. A. E. Maxwell. New York: Wiley-Interscience.

LARSON, R. L., and CHASE, C. G. [1970]. "Relative Velocities of the Pacific, North American and Cocos Plates in the Middle America Region," *Earth and Planet. Sci. Letts.* 7: 425–428.

LARSON, R. L., and CHASE, C. G. [1972]. "Late Mesozoic Evolution of the Western Pacific Ocean," *Geol. Soc. Am. Bull.* 83: 3627–44.

LARSON, R. L., and LADD, J. W. [1973]. "Evidence for the Opening of the South Atlantic in the Early Cretaceous," *Nature* 246: 227–266.

LARSON, R. L., MENARD, H. W., and SMITH, S. M. [1968]. "Gulf of California: A Result of Sea-Floor Spreading and Transform Faulting," *Science* 161: 781–784.

LARSON, R. L., and PITMAN, W. C., III [1972]. "Worldwide Correlation of Mesozoic Magnetic Anomalies, and Its Implications," *Geol. Soc. Am. Bull.* 83: 3645–61.

LARSON, R. L., and PITMAN, W. C., III [1975]. "Worldwide Correlation of Mesozoic Magnetic Anomalies: Reply," *Geol. Soc. Am. Bull.* 86: 270–72.

LAUGHTON, A. S. [1975]. "Tectonic Evolution of the Northeast Atlantic Ocean, A Review," *Norges Geol. Unders.* 316: 169–93.

LAUGHTON, A. S., and ROBERTS, D. G. [1978]. "Morphology of the Continental Margin," in *Phil. Trans. Roy. Soc. London* 290: 75–85.

LAUGHTON, A. S., and WHITMARSH, R. B., and JONES, M. T. [1970]. "The Evolution of the Gulf of Aden," in *Phil. Trans. Roy. Soc. London,* A., 267: 227–266.

LAWVER, L. A., and WILLIAMS, D. [1979]. "Heat Flow in the Central Gulf of California," *J. Geophys. Res.* 84: 5465–5478.

LE CLAIRE, L. [1974]. "Late Cretaceous and Cenozoic Pelagic Deposits—Paleoenvironment and Paleo-Oceanography of the Central Western Indian Ocean," in *Initial Reports of the Deep Sea Drilling Project,* vol. 25, pp. 481–512, ed. E. S. W. Simpson, R. Schlich, and others. Washington, D.C.: U.S. Government Printing Office.

LEE, J. J., and others [1966]. "Tracer Experiments in Feeding Littoral Foraminifera," *J. Protozool.* 13: 659–70.

LEINEN, M. [1976]. "Biogenic Silica Sedimentation in the Central Equatorial Pacific During the Cenozoic." (Unpublished master's thesis, Oregon State University.)

LEINEN, M., and HEATH, G. R. [1981]. "Sedimentary Indications of Atmospheric Activity in the Northern Hemisphere during the Cenozoic. *Paleo. Paleo. Paleo.* (in press).

LEINEN, M., and STAKES, D. [1979]. "Metal Accumulation Rates in the Central Equatorial Pacific during Cenozoic Time," *Geol. Soc. Am. Bull.* 90: 357–75.

LE MASURIER, W. G. [1972]. "Volcanic Record of Cenozoic Glacial History of Marie Byrd I," in *Antarctic Geology and Geophysics,* ed. R. J. Adie. Oslo: Universitetsforlaget.

LE PICHON, X. [1968]. "Sea Floor Spreading and Continental Drift," *Tectonophysics* 7: 3661–97.

LE PICHON, X., and HAYES, D. E. [1971]. "Marginal Offsets, Fracture Zones and the Early Opening of the South Atlantic," *J. Geophys. Res* 76: 6283–93.

LEWIN, J. C. [1961]. "Dissolution of Silica from Diatom Walls," *Geochim. et. Cosmochim.* 21: 182–98.

LEWIS, K. B. [1971]. "Slumping on a continental slope inclined at 1°–4°," *Sedimentology* 16: 97–110.

LEWIS, K. B. [1980]. "Quaternary Sedimentation on the Hikurangi Oblique-Subduction and Transform Margin, New Zealand," in *Sedimentation in Oblique-slip Mobile Zones,* pp. 171–189, ed. P. F. Ballance and H. G. Reading. Special Publication of the International Association of Sedimentologists, vol. 4.

LEYDEN, R., ASMUS, H., ZEMBRUSCKI, S., and BRYAN, G. [1976]. "South Atlantic Diapiric Structures," *Am. Assoc. Petrol. Geol. Bull.* 60: 196–212.

LEYDEN, R., EWING, M., and SIMPSON, E. S. W. [1971]. "Geophysical Reconnaissance on the African Shelf: Capetown to East London," *Am. Assoc. Petrol. Geol. Bull.* 55: 651–57.

LIDDICOAT, J. C., OPDYKE, N. D., and SMITH, G. I. [1980]. "Palaeomagnetic Polarity in a 930-m Core from Searles Valley, California." *Nature* 286: 22–25.

LIDZ, L. [1966]. "Deep-Sea Pleistocene Biostratigraphy," *Science* 154: 1448–52.

LILLIE, A. T., and BROTHERS, R. N. [1970]. "The Geology of New Caledonia," *N. Z. J. Geol. and Geophys.* 13: 145–83.

LIPPS, J. H. [1970A]. "Ecology and Evolution of Silicoflagellates," *Proc. North Am. Paleont. Convent. Part G.,* pp. 965–993.

LIPPS, J. H. [1970B]. "Plankton Evolution," *Evolution* 24: 1–22.

LIPPS, J. H., and MITCHELL, E. [1976]. "Trophic Model for the Adaptive Radiations and Extinctions of Pelagic Marine Mammals," *Paleobiology* 2: 147–55.

LISITZIN, A. P. [1971]. "Distribution of Carbonate Microfossils in Suspension and in Bottom Sediments," in *The Micropaleontology of the Oceans,* pp. 173–196, ed. B. M. Funnell and W. R. Riedel. London: Cambridge University Press.

LISITZIN, A. P. [1972]. "Sedimentation in the World Oceans," SEPM Spec. Publ. 17: 1–218.

LISITZIN, A. P., BELVAYEV, Y. I., BOGDNOV, Y. A., and BOGOYAVLENSKIY, A. N. [1967]. "Distribution Relationships and Forms of Silicon Suspended in Waters of the World Ocean," *Int. Geol. Rev.* 9: 604–23.

LISTER, C. R. B. [1972]. "On the Thermal Balance of a Midocean Ridge," *Geophys. J. Roy. Astron. Soc.* 39: 535–75.

LIVSIĆ, J. J. [1974]. "Paleogene Deposits and the Platform Structure of Svalbard," *Norsk. Polarinst. Skr.* 159: 51.

LOEBLICH, A. R., JR., and TAPPAN, H. [1964]. "Sarcodina, Chiefly 'Thecamoebians' and Foraminiferida," in *Treatise on Invertebrate Paleontology,* Part C vols. 1–2, ed. R. L. Moore. New York: Geological Society of America.

LOEBLICH, A. R., JR., and TAPPAN, H. [1966]. "Annotated Bibliography of the Calcareous Nannoplankton," *Phycologia* 5: 81–216.

LOEBLICH, A. R., JR., and TAPPAN, H. [1974]. Recent Advances in the Classification of the Foraminifera," in *Foraminifera,* vol. 1, pp. 1–53, eds. R. H. Hedley and C. G. Adams. London and New York: Academic Press.

LOHMANN, G. P. [1978]. "Abyssal Benthonic Foraminifera as Hydrographic Indicators in the Western South Atlantic Ocean," *J. Foram. Res.* 8: 6–34.

LOHMANN, H. [1902]. "Die Coccolithophoridae, eine Monographie der Coccolith en bildenden Flagellaten, Zuglish ein Beitrag Zur Kenntrnis des Mittelmeerauftriets," *Archi. Protistenkunde* 1: 89–165.

LONSDALE, P., and MALFAIT, B. [1974]. "Abyssal Dunes of Foraminiferal Sand on Carnegie Ridge," *Geol. Soc. Am. Bull.* 85: 1697–1712.

LONSDALE, P., and NORMARK, W. R., and NEWMAN, W. A. [1972]. "Sedimentation and Erosion on Horizon Guyot," *Geol. Soc. Am. Bull.* 83: 289–316.

LONSDALE, P., and SPEISS, F. N. [1977]. "Abyssal Bedforms Explored with a Deeply Towed Instrument Package," *Mar. Geol.* 23: 57–75.

LOUTIT, T. S., and KENNETT, J. P. [1979]. "Application of Carbon Isotope Stratigraphy to Late Miocene Shallow Marine Sediments, New Zealand," *Science* 204: 1196–99.

LOWRIE, W. [1974]. "Oceanic Basalt Magnetic Properties and the Vine and Matthews Hypothesis," *J. Geophys.* 40: 513–36.

LOWRIE, W. [1979]. "Geomagnetic Reversals and Ocean Crust Magnetization," in *Maurice Ewing Series 2,* pp. 135–150, ed. M. Talwani, C. G. Harrison, and D. E. Hayes. Washington, D.C.: American Geophysical Union.

LOZANO, J. A., and HAYS, J. D. [1976]. "Relationship of Radiolarian Assemblages to Sediment Types and Physical Oceanography in the Atlantic and Western Indian Ocean Sectors of the Antarctic Ocean," in *Geol. Soc. of Am. Memoir 145,* pp. 303–336, ed. R. M. Cline and J. D. Hays.

LUDWIG, W. T., NATE, J. E., and DRAKE, C. L. [1970]. "Seismic Refraction," in *The Sea,* vol. 4, pp. 53–84, ed. A. E. Maxwell. New York: Wiley-Interscience.

LUYENDYK, B. P. [1970]. "Origin and History of Abyssal Hills in the Northeast Pacific Ocean," *Geol. Soc. Am. Bull.* 81: 2237–60.

LUYENDYK, B. P. [1977]. "Deep Sea Drilling on the Ninetyeast Ridge: Synthesis and a Tectonic Model," in *Indian Ocean Geology and Biostratigraphy,* pp. 165–188, ed. J. R. Heirtzler, and others. Washington, D.C.: American Geophysical Union.

LUYENDYK, B. P., FORSYTH, D., and PHILLIPS, J. D. [1972]. "Experimental Approach to the Paleocirculation of the Oceanic Surface Waters," *Geol. Soc. Am. Bull.* 83: 2649–64.

LUZ, B., and SHACKLETON, N. J. [1975]. "CaCO₃ Solution in the Tropical East Pacific during the Past 130,000 Years," in *Spec. Publ. No. 13,* pp. 142–50, ed. W. V. Sliter, A. W. H. Bé, and W. H. Berger. Lawrence, Kansas: Cushman Foundation for Foraminiferal Research.

MacDONALD, K. C. [1973]. "Near-bottom Thermocline in the Samoan Passage, West Equatorial Pacific," *Nature* 243: 461–62.

MacDONALD, K. C., BECKER, K., SPEISS, F. N., and BALLARD, R. D. [1980]. "Hydrothermal Heat Flux of the 'Black Smoker' Vents on the East Pacific Rise," *Earth and Planet. Sci. Letts.* 48: 1–7.

MacDOUGALL, J. D. [1971]. "Fission Track Dating of Volcanic Glass Shards in Marine Sediments," *Earth and Planet. Sci. Letts.* 10: 403–406.

MacILVAINE, J. C., and ROSS, D. A. [1979]. "Sedimentary Processes on the Continental Slope of New England," *J. Sediment. Petrol.* 49: 563–574.

MacINTYRE, I. G., PILKEY, O. H., and STUCKENRATH, R. [1978]. "Relict Oysters in the United States Atlantic Continental Shelf: A Reconsideration of Their Usefulness in Understanding Late Quaternary Sea-Level History," *Geol. Soc. Am. Bull.* 89: 277–82.

MacINTYRE, A., RUDDIMAN, W. F., and JANTZEN, R. [1972]. "Southward Penetrations of the North Atlantic Polar Front: Faunal and Floral Evidence of Large-scale Surface Water Mass Movements over the Last 225,000 Years," *Deep Sea Res.* 19: 61–77.

MALMGREN, B. A., and KENNETT, J. P. [1981]. "Phyletic Gradualism in a Late Cenozoic Planktonic Foraminiferal Lineage: Deep Sea Drilling Project Site 284, Southwest Pacific," *Paleobiology* (in press).

MANDRA, Y. T., and MANDRA, H. [1972]. "Paleoecology and Taxonomy of Silicoflagellates from an Upper Miocene Diatomite near San Felipe, Baja California, Mexico. *Occ. Pap. Calif. Acad. Sci.* 99: 1–35.

MARGOLIS, S. V. [1975]. "Paleoglacial History of Antarctica Inferred from Analysis of Leg 29 Sediments by Scanning-Electron Microscopy," in *Initial Reports of the Deep Sea Drilling Project,* vol. 29, pp. 1039–48, ed. J. P. Kennett, R. E. Houtz, and others. Washington, D.C.: U.S. Government Printing Office.

MARGOLIS, S. V., and HERMAN, Y. [1980]. "Northern Hemisphere Sea-Ice and Glacial Development in the Late Cenozoic," *Nature* 286: 145–149.

MARGOLIS, S. V., and KENNETT, J. P. [1971]. Cenozoic paleoglacial history of Antarctica recorded in subantarctic deep-sea cores. *Am. J. Sci.* 271: 36.

MARSH, B. D. [1979]. "Island-Arc Volcanism," *Am. Sci.* 67: 161–172.

MARSHALL, P. [1912]. "The Structural Boundary of the Pacific Basin," *Rept. Australasian Assoc. Advan. Sci.* 13: 90–99.

MARSZALEK, D. S. [1969]. "Observations on *Inidia diaphana,* a Marine Foraminifer," *J. Protozool.* 16: 599–611.

MARSZALEK, D. S., WRIGHT, R. C., and HAY, W. W. [1969]. "Function of the Test in Foraminifera," *Trans. Gulf. Coast. Assoc. Geol. Soc.* 19: 341–352.

MARTINI, E., and WORSLEY, T. [1970]. "Standard Neogene Calcareous Nannoplankton Zonation," *Nature* 225: 289–90.

MASON, R. G., and RAFF, A. D. [1961]. "A Magnetic Survey Off the West Coast of North America 32°N to 42°N," *Geol. Soc. Am. Bull.* 72: 1259–65.

MATTHEWS, R. K. [1974]. *Dynamic Stratigraphy.* Englewood Cliffs, N.J.: Prentice-Hall, Inc.

MATTHEWS, R. K., and POORE, R. Z. [1980]. "Tertiary δ¹⁸O Record and Glacioeustatic Sea-Level Fluctuations," *Geology* 8: 501–504.

MATTSON, P. H., and PESSAGNO, E. A. [1971]. "Caribbean Eocene Volcanism and the Extent of Horizon A," *Science* 174: 138–39.

MAXWELL, A. E., and others [1970]. *Initial Reports of the Deep Sea Drilling Project,* vol. 3. Washington, D.C.: U.S. Government Printing Office.

MAYEWSKI, P. A. [1975]. "Glacial Geology and Late Cenozoic History of Trans-Antarctic Mountains, Antarctica," Institute of Polar Studies, Report 56. Columbus, Ohio: Ohio State University.

McBIRNEY, A. R. [1971]. "Thoughts on Some Current Concepts of Orogeny and Volcanism," *Comm. Earth. Sci., Geophys.* 2: 69–76.

McBirney, A. R., and Gass, I. G. [1967]. "Relations of Oceanic Volcanic Rocks of Mid-Oceanic Rises and Heatflow," *Earth. and Planet. Sci. Letts.* 2: 265–76.

McBirney, A. R., and others [1974]. "Episodic Volcanism in the Central Oregon Cascade Region," *Geology* 2: 585–89.

McBirney, A. R., and Williams, H. [1969]. "Geology and Petrology of the Galapagos Islands," *Geol. Soc. Am. Memoir 118.*

McCave, I. N. [1978]. "Sediments in the Abyssal Boundary Layer," *Oceanus* 21: 27–33.

McCave, I. N., Lonsdale, P. F., Hollister, C. D., and Gardner, W. D. [1980]. "Sediment Transport over the Hatton and Gardar Contourite Drifts," *J. Sediment. Petrol.* 50:1049–62.

McCoy, F. W., and Rabinowitz, P. D. [1979]. "The Evolution of the South Atlantic. *Oceanus* 22: 48–51.

McCoy, F. W., and Zimmerman, H. B. [1977]. "A History of Sediment Lithofacies in the South Atlantic Ocean," in *Initial Reports of the Deep Sea Drilling Project,* vol. 39, pp. 1047–79, ed. P. R. Supko, K. Perch-Nielsen, and others. Washington, D.C.: U.S. Govt. Printing Office.

McDougall, I. [1977]. "The Present Status of the Geomagnetic Polarity Time Scale," in *The Earth: Its Origin, Structure and Evolution* (A volume in honor of J. C. Jaeger and A. L. Hales), ed. M. W. McElhinny. New York: Academic Press.

McDougall, I. [1979]. "Age of Shield-building Volcanism of Kaui and Linear Migration of Volcanism in the Hawaiian Island Chain," *Earth and Planet. Sci. Letts.* 46: 31–42.

McDougall, I., and Duncan, R. A. [1980]. "Linear Volcanic Chains—Recording Plate Motions?" *Tectonophysics* 63: 275–95.

McDougall, I., and Wensink, H. [1966]. "Paleomagnetism and Geochronology of the Pliocene-Pleistocene Lauas in Iceland," *Earth and Planet. Sci. Letts.* 1: 232–236.

McGowran, B. [1978]. "Stratigraphic Record of Early Tertiary Oceanic and Continental Events in the Indian Ocean Region," *Mar. Geol.* 26: 1–39.

McIntyre, A. [1967]. "Coccoliths as Palaeoclimatic Indicators of Pleistocene Glaciation," *Science* 158: 1314–17.

McIntyre, A., and Bé, A. W. H. [1967]. "Modern Coccolithophoridae of the Atlantic Ocean-I, Placoliths and Cyrtoliths," *Deep Sea Res.* 14: 561–97.

McIntyre, A., Bé, A. W. H., and Roche, M. B. [1970]. "Modern Pacific Coccolithophorida: A Paleontological Thermometer," *N. Y. Acad. Sci. Trans.* 32: 720–31.

McIntyre, A., and others [1976]. "Glacial North Atlantic 18,000 Years Ago. A CLIMAP Reconstruction," in *Geol. Soc. Am. Memoir 145,* pp. 43–76, ed. R. M. Cline and J. D. Hays.

McIntyre, A., and McIntyre, R. [1971]. "Coccolith Concentrations and Differential Solution in Oceanic Sediments," in *The Micropaleontology of the Oceans,* pp. 253–261, ed. B. M. Funnell and W. R. Riedel. London: Cambridge University Press.

McIntyre, D. J., and Wilson, G. J. [1966]. "Preliminary Palynology of Some Antarctic Tertiary Erratics," *N. Z. J. Bot.* 4: 315.

McKenna, M. C. [1980]. "Eocene Paleolatitudes, Climate and Mammals of Ellesmere Island," *Paleo. Paleo. Paleo.* 30: 349–62.

McKenzie, D. P. [1969]. "Speculations on the Consequences and Causes of Plate Motions," *Geophys. J. R. Astr. Soc.* 18: 1–32.

McKenzie, D. P. [1972]. "Plate Tectonics and Sea Floor Spreading," *Am. Sci.* 60: 425–35.

McKenzie, D. P., and Sclater, J. G. [1968]. "Heat Flow Inside the Island Arcs of the Northwestern Pacific," *J. Geophys. Res.* 73: 3173–79.

McKenzie, D. P., and Sclater, J. G. [1971]. "The Evolution of the Indian Ocean Since the Late Cretaceous," *Geophys. J. R. Astr. Soc.* 25: 437–528.

McKenzie, D. P., and Sclater, J. G. [1973]. "The Evolution of the Indian Ocean," *Sci. Am.* 228: 62–72.

McMaster, R. L., La Chance, T. P., and Ashraf, A. [1970]. "Continental Shelf Geomorphic Features Off Portuguese Guinea, Guinea, and Sierra Leone, (West Africa)," *Mar. Geol.* 9: 203–13.

Melguen, M. [1978]. "Facies Evolution, Carbonate Dissolution Cycles in Sediments from the Eastern South Atlantic (Deep Sea Drilling Project Leg 40) since the Early Cretaceous," in *Initial Reports of the Deep Sea Drilling Project,* vol. 40, pp. 981–1024, ed. H. M. Bolli, W. B. F. Ryan, and others. Washington, D.C.: U.S. Government Printing Office.

Menard, H. W. [1952]. "Deep Ripple Marks in the Sea," *J. Sediment. Petrol.* 22: 3–9.

Menard, H. W. [1953]. "Shear Zones of the Northeastern Pacific Ocean and Anomalous Structural Trends of Western North America" (abstract), *Geol. Soc. Am. Bull.* 64: 1512.

Menard, H. W. [1954]. "Topography of the Northeastern Pacific Sea-Floor" (abstract), *Geol. Soc. Am. Bull.* 65: 1284.

Menard, H. W. [1955]. "Deep-sea Channels, Topography, and Sedimentation," *Am. Assoc. Petrol. Geol. Bull.* 39: 236–255.

Menard, H. W. [1967]. "Sea Floor Spreading, Topography, and the Second Layer," *Science* 157: 923–24.

Menard, H. W. [1969]. "The Deep-Ocean Floor," *Scientific American* 221: 126–45.

Menard, H. W. [1976]. "Time, Chance, and the Origin of Manganese Nodules," *Am. Sci.* 64: 519–29.

Menard, H. W., and Atwater, T. [1968]. "Changes in Direction of Sea Floor Spreading," *Nature* 219: 463.

Menard, H. W., and Chase, T. E. [1970]. "Fracture zones," in *The Sea,* pp. 421–443, ed. A. E. Maxwell. New York: John Wiley.

Menard, H. W., and Dietz, R. S. [1952]. "Mendocino Submarine Escarpment," *J. Geol.* 60: 266–78.

Menard, H. W., and Mammerickx, J. [1967]. "Abyssal Hills, Magnetic Anomalies and the East Pacific Rise," *Earth and Planet. Sci. Letts.* 2: 465–72.

MENARD, H. W., and SMITH, S. M. [1966]. "Hypsometry of Ocean Basin Provinces," *J. Geophys. Res.* 71: 4305-25.

MERCER, J. H. [1978]. "West Antarctic Ice Sheet and CO_2 Greenhouse Effect: A Threat of Disaster," *Nature* 271: 321-325.

MIDDLETON, G. V., and HAMPTON, M. A. [1976]. "Subaqueous Sediment Transport and Deposition by Sediment Gravity Flows," in *Marine Sediment Transport and Environmental Management,* pp. 197-218, ed. D. J. Stanley, and D. J. P. Swift. New York: John Wiley.

MILANKOVITCH, M. [1938]. "Astronomische Mittel Zur Erforschung der Erdgeschichtlichen Klimate," *Handbuch der Geophysik.* 9: 593-698.

MILANKOVITCH, M. [1941]. "Kanron der Erdbestrahlung und Seine Andwendung auf das Eiszeiten Problem. *Serb. Akad. Beogr. Spec. Publ. 132,* translated by the Israel Program for Scientific Translations, Jerusalem, 1969.

MILLIMAN, J. D. [1974]. *Marine Carbonates Pt. 1.* New York: Springer-Verlag.

MILLIMAN, J. D., and EMERY, K. O. [1968]. "Sea Levels during the Past 35,000 Years," *Science* 162: 1121-23.

MILLIMAN, J. D., PILKEY, O. H., and Ross, D. A. [1972]. Sediments of the Continental Margin off the Eastern United States, Maine to Florida, *Geol. Soc. Am. Bull.* 82: 1315-34.

MINSTER, J. B., JORDAN, T. H., MOLNAR, P., and HAINES, E. [1974]. "Numerical Modelling of Instantaneous Plate Tectonics," *Geophys. J. R. Astr. Soc.* 36: 541-76.

MITCHELL, A. H. G., and READING, H. G. [1978]. "Sedimentation and Tectonics," in *Sedimentary Environments and Facies,* pp. 439-476, ed. H. G. Reading. New York: Elsevier.

MITCHUM, R. M., JR. [1976]. "Seismic Stratigraphic Investigation of West Florida Slope, Gulf of Mexico," in *Framework, Facies and Oil-trapping Characteristics of the Upper Cont. Margin,* pp. 193-224, ed. A. H. Bouma, G. T. Moore, and J. M. Coleman. American Association of Petroleum Geologists, Studies in Geol. 7.

MITCHUM, R. M., JR., VAIL, P. R., and THOMPSON, S. III [1977]. "The Depositional Sequence as a Basic Unit for Stratigraphic Analysis." in *Seismic Stratigraphy—Applications to Hydrocarbon Exploration,* pp. 53-62, ed. C. E. Payton. American Association of Petroleum Geologists Memoir 26.

MOBERLY, R. [1972]. "Origin of Lithosphere Behind Island Arcs with Reference to the Western Pacific," in *Studies in Earth and Space Sciences,* pp. 35-56, ed. R. Shagam. Geological Society of America Memoir 132.

MOLNAR, P., and ATWATER, T. [1973]. Relative Motion of Hot Spots in the Mantle. *Nature* 246: 288-91.

MOLNAR, P., and ATWATER, T. [1978]. "Interarc Spreading and Cordilleran Tectonics as Alternates Related to the Age of Subducted Oceanic Lithosphere, *Earth and Planet. Sci. Letts.* 41: 330-40.

MOLNAR, P., ATWATER, T., MAMMERICKX, J., and SMITH, S. M. [1975]. "Magnetic Anomalies Bathymetry and the Tectonic Evolution of the South Pacific Since the Late Cretaceous," *Geophys. J. R. Astr. Soc.* 40: 383-420.

MOORE, D. [1966]. "Deltaic Sedimentation," *Earth Sci. Rev.* 1: 82-104.

MOORE, G. T., STARKE, G. W., BONHAM, L. C., and WOODBURY, H. O. [1978]. "Mississippi Fan, Gulf of Mexico—Physiography, Stratigraphy, and Sedimentation Patterns," in *Framework Facies and Oil-Trapping Characteristics of The Upper Continental Margin,* pp. 155-191, ed. A. H. Bouma, G. T. Moore, and J. C. Coleman. AAPG Studies in Geology 7.

MOORE, J. G., FLEMING, H. S., and PHILLIPS, J. D. [1974]. "Preliminary Model for Extrusion and Rifting at the Axis of the Mid-Atlantic Ridge, 36°48′ North," *Geology* 2: 437-40.

MOORE, T. C., JR. [1972]. "Deep Sea Drilling Project: Successes, Failures, Proposals," *Geotimes* 17: 27-31.

MOORE, T. C., JR. [1973]. "Late Pleistocene-Holocene Oceanographic Changes in the Northeastern Pacific," *Quat. Res.* 3. 99-109.

MOORE, T. C., JR., and HEATH, G. R. [1978]. "Sea Floor Sampling Techniques," *Chem. Oceanog.* 7: 75-126.

MOORE, T. C., VAN ANDEL, Tj. H., SANCETTA, C., and PISIAS, N. [1978]. "Cenozoic Hiatuses in Pelagic Sediments," *Micropaleo.* 24: 113-38.

MOORE, W. S., and VOGT, P. G. [1976]. "Hydrothermal Manganese Crusts from Two Sites near the Galapagos Spreading Axis," *Earth and Planet. Sci. Letts.* 29: 349-56.

MOREL, P., AND IRVING, E. [1978]. "Tentative Paleocontinental Maps for the Early Phanerozoic and Proterozoic," *J. Geol.* 86: 535-61.

MORGAN, W. J. [1968]. "Rises, Trenches, Great Faults and Crustal Blocks," *J. Geophys. Res.* 73: 1959-82.

MORGAN, W. J. [1972]. "Plate Motions and Deep Mantle Convection," in *Geological Society of America Memoir 132,* pp. 7-22, ed. R. Shagam and others.

MORNER, N. A. [1976]. "Eustacy and Geoid Changes," *J. of Geol.* 84: 123-51.

MORNER, N. A. [1980]. "The Northwest European Sea-Level Laboratory and Regional Holocene Eustasy," *Paleo. Paleo. Paleo.* 29: 281-300.

MORSE, J. W. [1974]. "Dissolution Kinetics of Calcium Carbonate in Seawater V. Effects of Natural Inhibitors and the Position of the Chemical Lysocline." *Am. J. Sci.* 274: 638-47.

MORSE, J. W., and BERNER, R. A. [1972]. "Dissolution Kinetics of Calcium Carbonate in Seawater: 11. A Kinetic Origin for the Lysocline," *Am. J. Sci.* 272: 840-51.

MOSBY, H. [1934]. "The Waters of the Atlantic Antarctic Ocean," *Sci. Res. Norwegian Antarctic Exped.* 1927-28, 11: 1-117.

MURRAY, H. W. [1939]. "Submarine Scarp off Cape Mendocino, California," *Bull. Calif. Field Eng.* 13: 27-33.

MURRAY, J. [1897]. "On the Distribution of the Pelagic Foraminifera at the Surface and on the Floor of the Ocean." *Nat. Sci.* 11: 17–27.

MURRAY, J., and RENARD, A. F. [1891]. "Report on Deep-Sea Deposits Based on the Specimens Collected during the Voyage of H.M.S. *Challenger* in the Years 1872–1876," in Challenger Reports, London: Government Printer.

MURRAY, J. W. [1973]. *Distribution and Ecology of Living Benthic Foraminiferids.* New York: Crane, Russak and Co., Inc.

NAIRN, A. E. M., and STEHLI, F. G. [1973]. *The Ocean Basins and Margins. The South Atlantic.* New York: Plenum.

NATLAND, J. H. [1978]. "Composition, Provenance and Diagenesis of Cretaceous Clastic Sediments Drilled on the Atlantic Continental Rise off Southern Africa, Deep Sea Drilling Project Site 361—Implications for the Early Circulation of the South Atlantic," in *Initial Reports of the Deep Sea Drilling Project,* vol. 40, pp. 1025–62. ed. H. M. Bolli, W. B. F. Ryan, and others. Washington, D.C.: U.S. Government Printing Office.

NATLAND, M. L. [1957]. "Paleoecology of West Coast Tertiary Sediments," in *Treatise Mar. Ecol. Paleoecol.,* vol. 2, pp. 543–572, Geological Society of America Memoir 67.

NELSON, B. W., ed. [1972]. *Environmental Framework at Coastal Estuaries.* Geological Society of America Memoir 133.

NELSON, C. H. [1976]. "Late Pleistocene and Holocene Depositional Trends, Processes, and History of Astoria Deep-Sea Fan, Northeast Pacific," *Mar. Geol.* 20: 129–73.

NELSON, C. H., and KULM, L. D. [1973]. "Submarine Fans and Deep-Sea Channels, Part II," in *Turbidites and Deep-Water Sedimentation,* pp. 39–70, SEPM Pacific Series, Short Course notes. Anaheim, California.

NELSON, C. H., KULM, L. D., CARLSON, P. R., and DUNCAN, J. R. [1968]. "Mazama Ash in the Northeastern Pacific," *Science* 161: 47–49.

NEUMAYR, M. [1883]. "Klimatische Zonen Während der Jura-and Kreidezeit," *Denkschr. Mathnaturn. kl.* 47: 277–310.

NEWELL, N. D. [1962]. "Paleontologic Gaps and Geochronology," *J. Paleont.* 36: 592–610.

NIGRINI, C. [1967]. "Radiolaria in Pelagic Sediments from the Indian and Atlantic Oceans," *Bull. of Scripps Inst. of Oceanog.* 11: 1–125.

NIGRINI, C. [1970]. "Radiolarian Assemblages in the North Pacific and Their Application to a Study of Quaternary Sediments in Core V20–130," *Geol. Soc. Am. Memoir 126:* 139–83.

NILSEN, T. H., and KERR, D. R. [1978]. "Paleoclimatic and Paleogeographic Implications of a Lower Tertiary Laterite (latosol) on the Iceland-Faeroe Ridge, North Atlantic Region," *Geol. Magazine* 115: 153–236.

NINKOVITCH, D., and DONN, W. L. [1975]. "Explosive Cenozoic Volcanism and Climatic Interpretations," *Science* 194: 899–906.

NINKOVITCH, D., and SHACKLETON, N. J. [1975]. "Distribution, Stratigraphy Position and Age of Ash Layer 'L,' in the Panama Basin Region," *Earth and Planet. Sci. Letts.* 27: 20–34.

NORMARK, W. R. [1970]. "Growth Patterns of Deep-Sea Fans," *Am. Assoc. Petrol. Geol. Bull.* 54: 2170–95.

NORTON, I. O., and MOLNAR, P. [1977]. "Implications of a Revised Fit between Australia and Antarctica for the Evolution of the Eastern Indian Ocean," *Nature* 267: 338–40.

NORTON, I. O., and SCLATER, J. G. [1979]. "A Model for the Evolution of the Indian Ocean and the Breakup of Gondwanaland," *J. Geophys. Res.* 84: 6803–6830.

OFF, T. [1963]. "Rhythmic Linear Sand Bodies Caused by Tidal Currents," *Am. Assoc. Petrol. Geol. Bull.* 47: 324–41.

O'HARA, M. J. [1965]. "Primary Magmas and the Origin of Basalts," *Scottish J. Geol.* 1: 19–40.

O'HARA, M. J. [1968]. "The Bearing of Phase Equilibria Studies in Synthetic and Natural Systems in the Origin and Evolution of Basic and Ultrabasic Rocks," *Earth Sci. Rev.* 4: 69–133.

O'HARA, M. J. [1973]. "Non-primary Magmas and Dubious Mantle Plume Beneath Iceland," *Nature* 243: 507–508.

OLAUSSON, E. [1961]. "Studies of Deep-Sea Cores," Göteborg, Swedish Deep-Sea Expedition, 1947–1948, Repnt. 8, pp. 353–391.

OLAUSSON, E. [1965]. "Evidence of Climate Changes in North Atlantic Deep-Sea Cores," in *Progress in Oceanography,* pp. 221–54, ed. M. Sears. New York: Pergamon Press.

OLDALE, R. N., and O'HARA, C. J. [1980]. "New Radiocarbon Dates from the Inner Continental Shelf Off Southeastern Massachusetts and a Local Sea-Level-Rise Curve for the Past 12,000 yr," *Geology* 8: 102–106.

OLDHAM, R. D. [1906]. "The Constitution of the Interior of the Earth, as Revealed by Earthquakes," *Q. JL. Geol. Soc. London* 62: 456–75.

OPDYKE, N. D. [1972]. "Paleomagnetism of Deep-Sea Cores," *Reviews of Geophys. and Space Physics* 10: 213–49.

OPDYKE, N. D., BURCKLE, L. H., and TODD, A. [1974]. "The Extension of the Magnetic Time Scale in Sediments of the Central Pacific Ocean," *Earth and Planet. Sci. Letts.* 22: 300–306.

OXBURGH, E. R. [1972]. "Plate Tectonics," in *Understanding the Earth,* pp. 263–85, ed. I. G. Gass, P. J. Smith, and R. C. L. Wilson. Sussex: Artemis Press.

OXBURGH, E. R., and TURCOTTE, D. L. [1968]. "Problem of High Heat Flow and Volcanism Associated with Zones of Descending Mantle Convective Flow," *Nature* 218: 1041–43.

OXBURGH, E. R., and TURCOTTE, D. L. [1970]. "Thermal Structure of Island Arcs." *Geol. Soc. Am. Bull.* 81: 1665-88.

PAASCHE, E. [1968]. "The Effect of Temperature, Light Intensity and Photoperiod on Coccolith Formation," *Limnol. Oceanog.* 13: 178-81.

PAASCHE, E. [1973]. "Silicon and the Ecology of Marine Plankton Diatoms II. Silicate Uptake Kinetics in Five Diatom Species," *Mar. Biol.* 19: 262-69.

PACKHAM, G. H., and FALVEY, D. A. [1971]. "An Hypothesis for the Formation of Marginal Seas in the Western Pacific," *Tectonophysics* 11: 79.

PACKHAM, G. H., and TERRILL, A. [1975]. "Submarine Geology of the South Fiji Basin," in *Initial Reports of the Deep Sea Drilling Project,* vol. 30 pp. 617-645. ed. J. E. Andrews, G. Packham, and others. Washington, D.C.: U.S. Government Printing Office.

PAK, H., ZANEVELD, J. R. V., and BEARDSLEY, G. F., JR. [1971]. "Mie Scattering by Suspended Clay Particles," *J. of Geophys. Res.* 76: 5065-69.

PARKER, F. L. [1960]. "Living Planktonic Foraminifera from the Equatorial and Southeast Pacific," *Tohoku Univ. Sci. Rep.,* 2nd ser., Geology, 4: 71-82.

PARKER, F. L. [1965]. "Irregular Distribution of Planktonic Foraminifera and Stratigraphic Correlation," *Progress in Oceanog.* 3: 267-72.

PARKER, F. L. [1971]. "Distribution of Planktonic Foraminifera in Recent Deep-Sea Sediments," in *The Micropaleontology of the Oceans,* pp. 289-308, ed. B. M. Funnell and W. R. Riedel. London: Cambridge University Press.

PARKER, F. L., and BERGER, W. H. [1971]. "Faunal and Solution Patterns of Foraminifera in Surface Sediments of the South Pacific," *Deep Sea Res.* 18: 73-107.

PARKIN, D. W., DELANY, A. C., and DELANY, A. C. [1967]. "A Search for Airborne Cosmic Dust on Barbados," *Geochim. Cosmochim. Acta* 31: 1311-20.

PARMENTER, C., and FOLGER, D. W. [1974]. "Eolian Biogenic Detritus in Deep Sea Sediments. A Possible Index of Equatorial Ice Age Aridity," *Science* 185: 695-98.

PARSONS, B., and SCLATER, J. G. [1977]. An Analysis of the Variation of Ocean Floor Bathymetry and Heat Flow with Age, *J. Geophys. Res.* 82: 803-27.

PAUTOT, G., RENARD, V., DANIEL, J., and DUPONT, J. [1973]. "Morphology Limits, Origin and Age of Salt Layer along South Atlantic African Margin," *Am. Assoc. Petrol. Geol. Bull.* 57: 1658-71.

PAYTON, C. E., ed. [1977]. *Seismic Stratigraphy—Applications to Hydrocarbon Exploration,* Amer. Assoc. of Petrol. Geol. Memoir 26.

PEIRCE, J. W. [1978]. "The Origin of the Ninetyeast Ridge and the Northward Motion of India, Based on Deep Sea Drilling Project Paleolatitudes." (Unpublished doctoral dissertation, Massachusetts Institute of Technology, Cambridge).

PERCIVAL, S. F., and FISCHER, A. G. [1977]. "Changes in Calcareous Nannoplankton in the Cretaceous-Tertiary Biotic Crisis at Zumaya, Spain," *Evolution Theory* 2: 1-35.

PETERSON, J. J., FOX, P. J., and SCHREIBER, E. [1974]. "Newfoundland Ophiolites and The Geology of the Oceanic Layer," *Nature,* 247: 194-196.

PETERSON, M. N. A. [1966]. "Calcite: Rates of Dissolution in a Vertical Profile in the Central Pacific," *Science* 154: 1542-44.

PETERSON, M. N. A. EDGAR, N. T., and others [1970]. *Initial Reports of the Deep Sea Drilling Project,* vol. 2. Washington, D.C.: U.S. Government Printing Office.

PETRUSHEVSKAYA, M. G. [1966]. "Radiolyaric v Planktone 1 v Donnykh Osadkakh," in *Geokhimiya Kremnesema,* pp. 219-45. Moscow: Nauka.

PETRUSHEVSKAYA, M. G. [1967]. "Nassellaria Antarkticheskoe Oblasti (Antarctic Spumelline and Nasseline Radiolarians), *Issled. Fauny Morei 4 (12) (Rez. biol. Issled. sov. antarkt. Eksped. 1955-58)* 3: 5-186.

PETRUSHEVSKAYA, M. G. [1971]. "Spumellarian and Nassellarina Radiolaria in the Plankton and Bottom Sediments of the Central Pacific," in *The Micropaleontology of the Oceans,* pp. 309-18, ed. B. M. Funnell and W. R. Riedel. London: Cambridge University Press.

PETRUSHEVSKAYA, M. G. [1975]. "Cenozoic Radiolarians of the Antarctic, Leg 29, Deep-Sea Drilling Project," in *Initial Reports of the Deep Sea Drilling Project,* vol. 29, pp. 541-76, eds. J. P. Kennett, R. E. Houtz, and others. Washington, D.C.: U.S. Government Printing Office.

PHILIPPI, E. [1910]. "Die Grundproben der Deutschen Südpolar-Expedition, 1901-1903." *Berlin, G. Reimer, Bd. II, Heft 6, Geog. Geol.:* 411-616.

PHILIPPI, G. T. [1965]. "On the Depth, Time and Mechanism of Petroleum Generation," *Geochimica et Cosmochimica Acta* 20: 1021-49.

PHILLIPS, J. D., and FORSYTH, D. [1972]. "Plate Tectonics, Paleomagnetism, and the Opening of the Atlantic," *Geol. Soc. Am. Bull.* 83: 1579-1600.

PHLEGER, F. B. [1939]. "Foraminifera of Submarine Cores from the Continental Slope," *Geol. Soc. Am. Bull.* 50: 1395-1422.

PHLEGER, F. B. [1969]. "Some General Features of Coastal Lagoons," in *Mem. Simp. Intern. Lagunas Costeras,* pp. 5-26, ed. A. A. Castanaris and F. B. Phleger. Mexico City: UNAM-UNESCO.

PHLEGER, F. B., PARKER, F. L., and PEIRSON, J. F. [1953]. "North Atlantic Foraminifera," in *Rep. Swed. Deep-Sea Exped.* 1947-1948, 7: 122.

PILBEAM, D. R. [1972]. "Evolutionary Changes in Hominoid Dentition Through Geological Time." in *Calibration of Hominoid Evolution,* pp. 369-80, ed. W. W. Bishop and J. A. Miller. Edinburgh: Scottish Academic Press.

PIMM, A. C., Mc GOWRAN, B., and GARTNER, S. [1974]. "Early Sinking History of the Ninetyeast Ridge, Northwestern Indian Ocean," *Geol. Soc. Am. Bull.* 85: 1219-24.

PIPER, D. J. W., and BRISCO, C. D. [1975]. "Deep-Water Continental Margin Sedimentation, Deep Sea Drilling Project Leg 28, Antarctica," in *Initial Reports of the Deep Sea Drilling Project,* vol. 28, pp. 727-55. ed. D. E. Hayes, L. A. Frakes, and others. Washington, D.C.: U.S. Government Printing Office.

PIPKIN, B. W., GORSLINE, D. S., CASEY, R. E., and HAMMOND, D. E. [1977]. *Laboratory Exercises in Oceanography.* San Francisco: W. H. Freeman and Company.

PISIAS, N. G. [1978]. "Paleoceanography of the Santa Barbara Basin and the California Current During the Last 8,000 Years." (Unpublished Doctoral Dissertation, University of Rhode Island.)

PITMAN, W. C. III [1978]. "Relationship between Eustacy and Stratigraphic Sequences of Passive Margins," *Geol. Soc. Am. Bull.* 89: 1389-1403.

PITMAN, W. C. III [1979]. "The Effect of Eustatic Sea Level Changes on Stratigraphic Sequences at Atlantic Margins," in *Am. Assoc. Petrol. Geol. Memoir 29,* pp. 453-60.

PITMAN, W. C. III, HERRON, E. M., and HEIRTZLER, J. R. [1968]. "Magnetic Anomalies in the Pacific and Sea-Floor Spreading," *J. Geophys. Res.* 73: 2069-85.

PITMAN, W. C. III, LARSON, R. L., and HERRON, E. M. [1974]. "The Age of the Ocean Basins," *Geol. Soc. Am. Map and Chart Series MC-6.*

PITMAN, W. C. III, and TALWANI, M. [1972]. "Sea-floor spreading in the North Atlantic," *Geol. Soc. Am. Bull.* 83: 619-646.

POELCHAU, H. S. [1974]. "Holocene Silicoflagellates of the North Pacific, their Distribution and Use for Paleotemperature Determination." University of California, San Diego).

POORE, R. Z. [1981]. "Temporal and Spatial Distribution of Ice-Rafted Mineral Grains in Pliocene Sediments of the North Atlantic: Implications for Late Cenozoic Climatic History," in *Deep Sea Drilling Project—A Decade of Progress,* ed. R. G. Douglas and E. Winterer. SEPM Special Publication (in press).

POORE, R. Z., and BERGGREN, W. A. [1975]. "Late Cenozoic Planktonic Foraminiferal Biostratigraphy and Paleoclimatology of the Northeastern Atlantic, Deep Sea Drilling Project Site 116," *J. Foram. Res.* 5: 270-93.

POPOVSKY, A. [1908]. "Die Radiolarien der Antarktis (mit Ausnahme der Tripyleen)," *Deutsche Sudpolar-Exped.* 1901-1903, 10 (Zool. vol. 2): 183-305.

POSTMA, H. [1967]. "Sediment Transport and Sedimentation in the Estuarine Environment," in *Estuaries,* Publ. 83, pp. 158-79, ed. G. H. Lauff. Washington, D.C.: American Association for the Advancement of Science.

PRATT, R. M. [1967]. "The Seaward Extension of Submarine Canyons off the Northeast Coast of the United States," *Deep Sea Res.* 14: 409-20.

PRATT, R. M. [1968]. "Atlantic Continental Shelf and Slope of the United States—Physiography and Sediments of the Deep-Sea Basin," *U.S. Geol. Surv., Prof. Pap.* 529-B: 44.

PRATT, R. M., and DILL, R. F. [1974]. "Deep Eustatic Terrace Levels: Further Speculations," *Geology* 2: 155-59.

PRITCHARD, D. W. [1967]. "What is an Estuary: Physical Viewpoint," in *Estuaries,* Publ. 83, pp. 3-5, ed. G. D. Lauff. Washington, D.C.: American Association for the Advancement of Science.

PYTKOWICZ, R. M. [1970]. "On the Carbonate Compensation Depth in the Pacific Ocean," *Geochimica et Cosmochimica Acta* 34: 836-39.

RABINOWITZ, P. D. [1974]. "The Boundary between Oceanic and Continental Crust in the Western North Atlantic," in *The Geology of Continental Margins,* pp. 67-84, ed. C. A. Burk and C. L. Drake. New York: Springer-Verlag.

RABINOWITZ, P. D., and LA BRECQUE, J. L. [1976]. "The Isostatic Gravity Anomaly: Key to the Evolution of the Ocean-Continent Boundary at Passive Continental Margins," *Earth and Planet. Sci. Letts.* 35: 145-50.

RABINOWITZ, P. D., and LA BRECQUE, J. L. [1979]. "The Mesozoic South Atlantic Ocean and Evolution of Its Continental Margins," *J. of Geophys. Res.* 84: 5973-6001.

RADCZEWSKI, O. E. [1937]. "Die Mineralfazies der Sedimente des Kapverden Beckens," *Wiss. Ergebn. Atlant. Exped. Meteor* 3: 262-77.

RAITT, R. W. [1956]. "Seismic Refraction Studies of the Pacific Ocean Basin, Part I: Crustal Thickness of the Central Equatorial Pacific," *Geol. Soc. Am. Bull.* 67: 1623-40.

RAITT, R. W. [1963]. "The Crustal Rocks," in *The Sea,* vol. 3, pp. 85-102, ed. M. N. Hill. New York: Wiley-Interscience.

RAMSAY, A. T. S. [1971]. "Occurrence of Biogenous Siliceous Sediments in the Atlantic Ocean," *Nature* 233: 115-17.

RAMSAY, A. T. S. [1974]. "The Distribution of Calcium Carbonate in Deep Sea Sediments," in *Studies in Paleo-oceanography,* pp. 58-76, ed. W. W. Hay. SEPM Special Publication 20.

RAMSAY, A. T. S. [1977]. "Sedimentological Clues to Paleo-oceanography," in *Oceanic Micropaleontology,* vol. 2, pp. 1371-1453, ed. A. T. S. Ramsay. New York: Academic Press.

RAUDKIVI, A. J. [1967]. *Loose Boundary Hydraulics.* New York: Pergamon Press.

REA, D. K., and SCHEIDEGGER, K. F. [1979]. "Eastern Pacific Spreading Rate Fluctuation and Its Relation to Pacific Area Volcanic Episodes," *J. Volcan. Geotherm. Res.,* 5: 135-48.

REINSON, G. E. [1979]. "Facies Models 6. Barrier Island Systems," *Geoscience Canada* 6: 57-74.

REX, R. W., and GOLDBERG, E. D. [1958]. "Quartz Contents of Pelagic Sediments of the Pacific Ocean," *Tellus* 10: 153-59.

REX, R. W., and GOLDBERG, E. D. [1963]. "Insolubles," in *The Sea,* vol. 1, pp. 295–304, ed. M. N. Hill. New York: Wiley-Interscience.

RHODES, J. M., and DUNGAN, M. A. [1979]. "The Evolution of Ocean-Floor Basaltic Magmas," in *Maurice Ewing Series 2,* pp. 262–72, ed. M. Talwani, C. G. Harrison, and D. E. Hayes. Washington, D.C.: American Geophysical Union.

RIEDEL, W. R. [1959]. "Siliceous Organic Remains in Pelagic Sediments," in *Silica in Sediments,* SEPM Spec. Publ. 7.

RIEDEL, W. R., and SANFILIPPO, A. (1971). "Cenozoic Radiolaria from the Western Tropical Pacific, Leg. 7," in *Initial Reports of the Deep Sea Drilling Project,* vol. 7, pp. 1529–1672, ed. E. L. Winterer, W. R. Riedel, and others. Washington, D.C.: U.S. Government Printing Office.

RINGWOOD, A. E. [1969]. "Composition and Evolution of the Upper Mantle," in *The Earth's Crust and Upper Mantle,* ed. P. J. Hart, Geophysical Monograph 13. Washington, D.C.: American Geophysical Union.

RINGWOOD, A. E. [1977]. "Petrogenesis in Island Arc Systems," in *Maurice Ewing Series 1,* pp. 311–24, ed. M. Talwani and W. C. Pitman III. Washington, D.C.: American Geophysical Union.

RINGWOOD, A. E., and GREEN, D. H. [1966]. "Petrological Nature of the Stable Continental Crust," in *The Earth Beneath the Continents,* Publ. 1467, pp. 611–619, ed. J. S. Steinhart and T. J. Smith. Washington, D.C.: National Academy of Sciences, Natural Resources Council.

ROBERTS, D. G. [1975]. "Marine Geology of the Rockall Plateau and Trough," *Phil. Trans. R. Soc. London* A278: 447–509.

ROBERTSON, J. H. [1975]. Glacial to Interglacial Oceanographic Changes in Northwest Pacific, Including a Continuous Record of the Last 400,000 years. (Unpublished doctoral dissertation, Columbia University, New York.)

RONA, P. A. [1973A]. "Relations between Rates of Sediment Accumulation on Continental Shelves, Seafloor Spreading, and Eustacy Inferred from the Central North Atlantic," *Geol. Soc. Am. Bull.* 84: 2851–72.

RONA, P. A. [1973B]. "Worldwide Unconformities in Marine Sediments Related to Eustatic Changes of Sea Level," *Nature Phys. Sci.* 244: 25.

ROSS, D. A., and SCHLEE, J. [1973]. "Shallow Structure and Geologic Development of the Southern Red Sea," *Geol. Soc. Am. Bull.* 84: 3827–48.

RUDDIMAN, W. F. [1971]. "Pleistocene Sedimentation in the Equatorial Atlantic: Stratigraphy and Faunal Paleoclimatology," *Geol. Soc. Am. Bull.* 82: 283–302.

RUDDIMAN, W. F. [1977A]. "Late Quaternary Deposition of Ice-rafted Sand in the Subpolar North Atlantic (lat. 40° to 65°N)," *Geol. Soc. Am. Bull.* 88: 1813–27.

RUDDIMAN, W. F. [1977B]. "Investigations of Quaternary Climate Based on Planktonic Foraminifera," in *Oceanic Micropaleontology,* vol. 1, pp. 100–162, ed. A.T.S. Ramsay. New York: Academic Press.

RUDDIMAN, W. F., and MC INTYRE, A. [1977]. "Late Quaternary Surface Ocean Kinematics and Climatic Change in the High-Latitude North Atlantic," *J. Geophys. Res.* 82: 3877–87.

RUDDIMAN, W. F., MC INTYRE, A., NIEBLER-HUNT, V., and DURAZZI, J. T. [1980]. "Oceanic Evidence for the Mechanism of Rapid Northern Hemisphere Glaciation," *Quat. Res.* 13: 33–64.

RUPKE, N. A. [1978]. "Deep Clastic Seas," in *Sedimentary Environments and Facies,* pp. 372–415, ed. H. G. Reading. New York: Elsevier.

RUSSELL, D. A. [1979]. "The Enigma of the Extinction of Dinosaurs," *Ann. Rev. Earth Planet. Sci.* 7: 163–82.

RUSSELL, K. L. [1968]. "Oceanic Ridges and Eustatic Changes in Sea Level," *Nature* 218: 861–62.

RYAN, W. B. F. [1973.]. "Geodynamic Implications of the Messinian Crisis of Salinity," in *Messinian events in the Mediterranean,* pp. 26–38, ed. C. W. Drooger. North-Holland, Amsterdam: Koninklijke Nederlands Akademie van Wetenschappen.

RYAN, W. B. F. [1972]. "Stratigraphy of Late Quaternary Sediments in the Eastern Mediterranean," in *The Mediterranean Sea: A Natural Sedimentation Laboratory,* pp. 149–169, ed. D. J. Stanley. Stroudsburg, Pennsylvania: Dowden, Hutchinson and Ross.

RYAN, W. B. F., and CITA, M. B. [1978]. "The Nature and Distribution of Messinian Erosional Surfaces—Indicators of a Several-Kilometer-Deep Mediterranean in the Miocene," *Mar. Geol.* 27: 193–230.

RYAN, W. B. F., and CITA, M. B. [1977]. "Ignorance Concerning Episodes of Oceanwide Stagnation," *Mar. Geol.* 23: 197–215.

RYAN, W. B. F., and others [1974]. "A Paleomagnetic Assignment of Neogene Stage Boundaries and the Development of Isochronous Datum Planes between the Mediterranean, the Pacific and Indian Oceans in Order to Investigate the Response of World Ocean to Mediterranean «Salinity Crisis»," *Riv. Ital. Paleont.* 80: 631–88.

RYAN, W. B. F., HSÜ, K. J., and others [1973]. *Initial Reports of the Deep Sea Drilling Project,* vol. 13, pt. 1 and 2, p. 1447, Washington, D.C.: U.S. Government Printing Office.

SACHS, H. M. [1973]. "Late Pleistocene History of the North Pacific: Evidence from a Quantitative Study of Radiolaria in Core V21-173," *Quat. Res.* 3: 89–98.

SACHS, H. M., WEBB, T. III, and CLARK, D. R. [1977]. "Paleoecological Transfer Functions," *Ann. Rev. Earth Planet. Sci.* 5: 159–78.

SAITO, T. [1976A]. "Geologic Significance of Coiling Direction in the Planktonic Foraminifera, 'Pulleniatina'," *Geology* 4: 305–309.

SAITO, T. [1976B]. "Late Cenozoic Planktonic Foraminiferal Datum Levels: the Present State of Knowledge toward Accomplishing Pan-Pacific Stratigraphic Correlation." in *Proc. First Inter.*

Cong. on Pacific Neogene Stratigraphy, Tokyo, pp. 61–80. Tokyo: Kaiyo Shuppan Company.

SAITO, T., and VAN DONK, J. [1974]. "Oxygen and Carbon Isotope Measurements of Late Cretaceous and Early Tertiary Foraminifera," *Micropaleo.* 20: 152–77.

SALISBURY, M. H., and others [1979]. "The Physical State of the Upper Levels of Cretaceous Oceanic Crust from the Results of Logging, Laboratory Studies and the Oblique Seismic Experiment at Deep Sea Drilling Project Sites 417 and 418," in *Maurice Ewing Series 2,* pp. 113–34, ed. M. Talwani, C. G. Harrison, and D. E. Hayes. Washington, D.C.: American Geophysical Union.

SALVATORINI, G., and CITA, M. B. [1979]. "Miocene Foraminifera Stratigraphy, Deep Sea Drilling Project Site 397 (Cape Bojador, North Atlantic)," in *Initial Reports of the Deep Sea Drilling Project,* vol. 47, pt. 1, pp. 317–373, eds. U. von Rad, W. B. F. Ryan, and others. Washington, D.C.: U.S. Government Printing Office.

SANCETTA, C. [1979A]. "Paleogene Pacific Microfossils and Paleoceanography," *Mar. Micropaleo.* 4: 363–98.

SANCETTA, C. [1979B]. "Use of Semiquantitative Microfossil Data for Paleoceanography," *Geology* 7: 88–92.

SANFILIPPO, A., BURCKLE, L. H., MARTINI, E., and RIEDEL, W. R. [1973]. "Radiolarians, Diatoms, Silicoflagellates and Calcareous Nannofossils in the Mediterranean Neogene," *Micropaleo.* 19: 209–34.

SARNTHEIN, M., and WALGER, E. [1974]. "Der Äolische Sandstrom aus der W-Sahara zur Atlantikküste," *Geol. Rdsch.* 63: 1065–87.

SAVIN, S. M. [1977]. "The History of the Earth's Surface Temperature during the Last 100 Million Years," in *Ann. Rev. Earth Planet. Sci.,* vol. 5, pp. 319–355, ed. F. A. Donath, F. G. Stehli, and G. A. Wetherill.

SAVIN, S. M., DOUGLAS, R. G., and STEHLI, F. G. [1975]. "Tertiary Marine Paleotemperatures," *Geol. Soc. Am. Bull.* 86: 1499–1510.

SCHEIDEGGER, K. F., and KULM, L. D. [1975]. "Late Cenozoic Volcanism in the Aleutian Arc: Information from Ash Layers in the Northeastern Gulf of Alaska," *Geol. Soc. Am. Bull.* 86: 1407–11.

SCHILLING, J. G. [1973]. "Iceland Mantle Plume: Geochemical Evidence along the Reykjanes Ridge," *Nature* 242: 565–71.

SCHILLING, J. G., UNNI, C. K., and BENDER, M. L. [1978]. "Origin of Chlorine and Bromine in the Oceans," *Nature* 273: 631–36.

SCHINDEWOLF, O. H. [1954]. "Über die Möglichen Ursachen der Grossen Erdges Chechtlichen Fraunenschnitte," *N. Jb. Geol. Paläont. Monatsh,* pp. 457–65.

SCHLANGER, S. O., and DOUGLAS, R. G. [1974]. "The Pelagic Ooze-Chalk-Limestone Transition and Its Implications for Marine Stratigraphy," in *Pelagic Sediments: on land and under the sea,* vol. 1, pp. 117–148, ed. K. J. Hsu and H. C. Jenkyns. Spec. Publs. Int. Assoc. Sediment. Oxford: Blackwell.

SCHLANGER, S. O., and others [1973]. "Fossil Preservation and Diagenesis of Pelagic Carbonates from the Magellan Rise, Central North Pacific Ocean," in *Initial Reports of the Deep Sea Drilling Project,* pp. 467–527, ed. E. L. Winterer, J. I. Ewing, and others. Washington, D.C.: U.S. Government Printing Office.

SCHLANGER, S. O., and JENKYNS, H. C. [1976]. "Cretaceous Oceanic Anoxic Events: Causes and Consequences." *Geol. Mijnb* 55: 179–84.

SCHLEE, J. S. [1977]. "Stratigraphy and Tertiary Development of the Continental Margin East of Florida," *U.S. Geol. Survey Prof. Pap.,* 581-F.

SCHLEE, J. S., and others [1979]. "The Continental Margins," *Oceanus* 22: 40–62.

SCHLEE, S. [1973]. *The Edge of an Unfamiliar World–A History of Oceanography.* New York: Dutton and Co.

SCHLICH, R. [1975]. "Structure et Age de l'Ocean Indien Occidental," *Mem. Horsserie Soc. Geol. de France,* no. 6.

SCHNITKER, D. [1974]. "West Atlantic Abyssal Circulation during the Past 120,000 Years," *Nature* 248: 385–87.

SCHNITKER, D. [1980]. "Quaternary Deep-Sea Benthic Foraminifers and Bottom Water Masses," *Ann. Rev. Earth Planet. Sci.* 8: 343–70.

SCHOLL, D. W., BUFFINGTON, E. C., and HOPKINS, D. M. [1968]. "Geologic History of the Continental Margin of North America in the Bering Sea," *Mar. Geol.* 6: 297–330.

SCHOLL, D. W., and CREAGER, J. S. [1973]. "Geologic Synthesis of Leg 19 (Deep Sea Drilling Project) Results; Far North Pacific, and Aleutian Ridge, and Bering Sea," in *Initial Reports of the Deep Sea Drilling Project,* vol. 19, pp. 897–913, ed. J. S. Creager, D. W. Scholl, and others. Washington, D.C.: U.S. Government Printing Office.

SCHOLL, D. W., MARLOW, M. S., and COOPER, A. K. [1977]. "Sediment Subduction and Offscraping at Pacific Margins," in *Maurice Ewing Series 1,* pp. 199–210, ed. M. Talwani and W. C. Pitman III. Washington, D.C.: American Geophysical Union.

SCHOPF, T. J. M. [1980]. *Paleoceanography.* Cambridge, Mass.: Harvard University Press.

SCHOTT, W. [1935]. "Die Foraminiferen in dem Aquatorialen Teil des Atlantischen Ozeans." *Wiss. Ergeb., Deut. Atlantischen Exped. Vermuss. Forschungsschiff Meteor,* 1925–1927, 111: 43–134.

SCHRADER, H. J. [1971]. "Fecal Pellets: Role in Sedimentation of Pelagic Diatoms." *Science* 174: 55–57.

SCHRADER, H. J. [1973]. "Cenozoic Diatoms from the Northeast Pacific, Leg. 18," in *Initial Reports of the Deep Sea Drilling Project,* vol. 18, pp. 673–698, ed. L. D. Kulm, R. von Huene, and others. Washington, D.C.: U.S. Government Printing Office.

SCHRADER, H. J., and others [1976]. "Cenozoic biostratigraphy, Physical Stratigraphy and Paleoceanography in the Norwegian-Greenland Sea, Deep Sea Drilling Project Leg 38 Paleontological Synthesis," in *Initial Reports of the Deep Sea Drill-*

ing Project, vol. 38, pp. 1197–1211, ed. M. Talwani, G. Udintsev, and others. Washington, D.C.: U.S. Government Printing Office.

SCHRADER, H., and others [1980]. "Laminated Diatomaceous Sediments from the Guaymas Basin Slope (Central Gulf of California): 250,000 Year Climate Record," *Science* 207: 1207–1209.

SCHUBEL, J. R., and PRITCHARD, D. W. [1972]. "The Estuarine Environment, pt. 1," *Jour. Geol. Education,* vol. 20, pp. 60–68, March 1972, Washington, D.C.: Council on Education in the Geological Sciences.

SCLATER, J. G., ABBOTT, D., and THIEDE, J. [1977A]. "Paleobathymetry and Sediments of the Indian Ocean," in *Indian Ocean Geology and Biostratigraphy,* ed. J. R. Heirtzler and others. Washington, D.C.: American Geophysical Union.

SCLATER, J. G., ANDERSON, R. N., and BELL, M. L. [1971]. "The Elevation of Ridges and the Evolution of the Central Eastern Pacific," *J. Geophys. Res.* 76: 7883–7915.

SCLATER, J. G., and DETRICK, R. [1973]. "Elevation of Midocean Ridges and the Basement Age of JOIDES Deep Sea Drilling Sites," *Geol. Soc. Am. Bull.* 84: 1547–54.

SCLATER, J. G., and FISHER, R. L. [1974]. "Evolution of the East Central Indian Ocean, with Emphasis on the Tectonic Setting of the Ninetyeast Ridge," *Geol. Soc. Am. Bull.* 85: 683–702.

SCLATER, J. G., and FRANCHETEAU, J. [1970]. "The Implications of Terrestrial Heat Flow Observations for Current Tectonic and Geochemical Models of the Crust and Upper Mantle of the Earth," *Geophys. J. R. Astr. Soc.* 20: 509–42.

SCLATER, J. G., HAWKINS, J. W., and MAMMERICKX, J. [1972]. "Crustal Extension between the Tonga and Lau Ridges: Petrologic and Geophysical Evidence," *Geol. Soc. Am. Bull.* 83: 505–18.

SCLATER, J. G., HELLINGER, S., and TAPSCOTT, C. [1977B]. "The Paleobathymetry of the Atlantic Ocean from the Jurassic to the Present," *J. of Geol.* 85: 509–52.

SCLATER, J. G., and MC KENZIE, D. P. [1973]. "The Paleobathymetry of the South Atlantic," *Geol. Soc. Am. Bull.* 84: 3203–16.

SCLATER, J. G., and TAPSCOTT, C. [1979]. "The History of the Atlantic," *Scientific American* 240: 156–74.

SCOTESE, C. R., and others [1979]. "Paleozoic Base Maps," *J. Geol.* 87: 217–68.

SCOTT, M. R., and others [1974]. "Rapidly Accumulating Manganese Deposit from the Median Valley of the Mid-Atlantic Ridge," *Geophys. Res. Letts.* 1: 355–58.

SEELY, D. R., VAIL. P. R., and WALTON, G. G. [1974]. "Trench Slope Model," in *The Geology of Continental Margins,* pp. 249–60, ed. C. A. Burk and C. L. Drake. New York: Springer-Verlag.

SEELY, D. R., and DICKINSON, W. R. [1977]. "Structure and Stratigraphy of Forearc Regions," in *Geology of Continental Margins,* American Association of Petroleum Geologists, Continuing Education Course Note Series 5, C1–C23.

SEIBOLD, E., and HINZ, K. [1974]. "Continental Slope Construction and Destruction, West Africa," in *The Geology of Continental Margins,* pp. 179–96, ed. C. A. Burk and C. L. Drake. New York: Springer-Verlag.

SELLEY, R. C. [1970]. *Ancient Sedimentary Environments.* London: Chapman and Hall, Ltd.

SELLWOOD, B. W. [1978]. "Shallow-Water Carbonate Environments," in *Sedimentary Environments and Facies,* pp. 259–313, ed. H. G. Reading. New York: Elsevier.

SHACKLETON, N. J. [1967]. "Oxygen Isotope Analyses and Paleotemperatures Reassessed," *Nature* 215: 15–17.

SHACKLETON, N. J. [1974]. "Attainment of Isotopic Equilibrium between Ocean Water and the Benthonic Foraminifera Genus *Uvigerina:* Isotopic Changes in the Ocean during the Last Glacial," *C.N.R.S.* Colloque Internationaux 219, Paris.

SHACKLETON, N. J. [1977]. "The Oxygen Isotope Stratigraphic Record of the Late Pleistocene," *Phil. Trans. Roy. Soc. Lond.* 280: 169–82.

SHACKLETON, N. J., and CITA, M. B. [1979]. "Oxygen and Carbon Isotope Stratigraphy of Benthic Foraminifers at Site 397: Detailed History of Climatic Change during the Late Neogene," in *Initial Reports of the Deep Sea Drilling Project,* vol. 47, pp. 433–45, ed. U. von Rad, W. B. F. Ryan, and others. Washington, D.C.: U.S. Government Printing Office.

SHACKLETON, N. J. and KENNETT, J. P. [1975A]. "Paleotemperature History of the Cenozoic and the Initiation of Antarctic Glaciation: Oxygen and Carbon Isotope Analyses in DSDP Sites 277, 279, and 281," in *Inital Reports of the Deep Sea Drilling Project,* vol. 29, pp. 743–755, ed. J. P. Kennett, R. E. Houtz, and others. Washington, D.C.: U.S. Government Printing Office.

SHACKLETON, N. J., and KENNETT, J. P. [1975b]. "Late Cenozoic Oxygen and Carbon Isotopic Changes at Deep Sea Drilling Project Site 284: Implications for Glacial History of the Northern Hemisphere and Antarctica," in *Initial Reports of the Deep Sea Drilling Project,* vol. 29, pp. 801–807, ed. J. P. Kennett, R. E. Houtz, and others. Washington, D.C.: U.S. Government Printing Office.

SHACKLETON, N. J., and OPDYKE, N. D. [1973]. "Oxygen Isotope and Palaeomagnetic Stratigraphy of Equatorial Pacific Core V28-238: Oxygen Isotope Temperatures and Ice Volumes on a 10^5 Year and 10^6 Year Scale," *Quat. Res.* 3: 39–55.

SHACKLETON, N. J., and OPDYKE, N. D. [1976]. "Oxygen Isotope and Palaeomagnetic Stratigraphy of Equatorial Pacific Core V28-239, Late Pliocene to Latest Pleistocene," in *Geol. Soc. Am. Memoir 145,* pp. 449–64, ed. R. M. Cline and J. D. Hays.

SHACKLETON, N. J., and OPDYKE, N. D. [1977]. "Oxygen Isotope and Palaeomagnetic Evidence for Early Northern Hemisphere Glaciation," *Nature* 270: 216–19.

SHAW, A. B. [1964]. *Time in Stratigraphy.* New York: McGraw-Hill.

SHAW, D. M., WATKINS, N. D., and HUANG, T. C. [1974]. "Atmospherically Transported Volcanic Glass in Deep-Sea Sediments: Theoretical Considerations," *J. Geophys. Res.* 79: 3087–94.

SHAW, H. R. [1973]. "Mantle Convection and Volcanic Periodicity in the Pacific: Evidence from Hawaii," *Geol. Soc. Am. Bull.* 84: 1505–26.

SHAW, H. R., and JACKSON, E. D. [1973]. "Linear Island Chains in the Pacific: Result of Thermal Plumes or Gravitational Anchors?" *J. Geophys. Res.* 78: 8634–52.

SHEPARD, F. P. [1948]. *Submarine Geology.* (1st ed). New York: Harper and Row.

SHEPARD, F. P. [1961]. "Deep-Sea Sands," *Internat. Geol. Cong., 21st,* Norden, 1960, Repts., 23: 26–42.

SHEPARD, F. P. [1963]. *Submarine Geology* (2nd ed). New York: Harper and Row.

SHEPARD, F. P. [1973]. *Submarine Geology* (3rd ed). New York: Harper and Row.

SHEPARD, F. P., and DILL, R. F. [1966]. *Submarine Canyons and Other Sea Valleys.* Chicago: Rand McNally and Co.

SHEPARD, F. P., DILL, R. F., and VON RAD, U. [1969]. "Physiography and Sedimentary Processes of La Jolla Submarine Fan and Fan-Valley, California," *Am. Assoc. Petrol Geol. Bull.* 53: 390–420.

SHEPARD, F. P., and EMERY K. O. [1973]. "Congo Submarine Canyon and Fan Valley. *Am. Assoc. Petrol. Geol. Bull.* 57: 1679–91.

SHEPARD, F. P., and INMAN, D. L. [1951]. "Nearshore Circulation," in *Conf. on Coastal Engineering, 1st, Proc.,* pp. 50–59, ed. J. W. Johnson. Berkeley, California: Council of Wave Resources.

SHEPARD, F. P., MC LOUGHLIN, P. A., MARSHALL, N. F., and SULLIVAN, G. G. [1977]. Current-Meter Recordings of Low-Speed Turbidity Currents," *Geology* 5: 297–301.

SHERIDAN, R. E., and ENOS, P. [1979]. "Stratigraphic Evolution of the Blake Plateau after a Decade of Scientific Drilling," in *Maurice Ewing Series 3,* pp. 109–122, ed. M. Talwani, W. Hay, and W. B. F. Ryan. Washington, D.C.: American Geophysical Union.

SHOR, G. G., JR. [1963]. "Refraction and Reflection Techniques and Procedure," in *The Sea,* vol. 3, pp. 20–38, ed. M. N. Hill. New York: Wiley-Interscience.

SIESSER, W. G. [1978]. "Age of Phosphorites on the South African Continental Margins," *Mar. Geol.* 26: M17–M28.

SIESSER, W. G. [1980]. "Late Miocene Origin of the Benguela Upwelling System Off Northern Namibia," *Science* 208: 283–85.

SILVA, A., and HOLLISTER, C. D. [1973]. "Geotechnical Properties of Ocean Sediments Recovered with Giant Piston Core, 1, in Gulf of Maine," *J. Geophys. Res.* 78: 3597–3616.

SIMPSON, J. F. [1966]. "Evolutionary Pulsations and Geomagnetic Polarity," *Geol. Soc. Am. Bull.* 77: 227–34.

SLEEP, N. H., and BIEHLER, S. [1970]. "Topography and Tectonics at the Intersections of Fracture with Central Rifts," *J. Geophys. Res.* 75: 2748–52.

SLEEP, N. H., and TOKSOZ, M. N. [1971]. "Evolution of the Marginal Basins," *Nature* 233: 548.

SLOSS, L. L. [1963]. "Sequences in the Cratonic Interior of North America," *Geol. Soc. Am. Bull.* 74: 93–114.

SMAYDA, T. J. [1958]. "Biogeographical Studies of Marine Phytoplankton," *Oikos* 9: 158–91.

SMAYDA, T. J. [1970]. "The Suspension and Sinking of Phytoplankton in the Sea," *Oceanogr. Mar. Biol. Ann. Rev.* 8: 353–414.

SMIT, J., and HERTOGEN, J. [1980]. "An Extraterrestrial Event at the Cretaceous-Tertiary Boundary," *Nature* 285: 198–200.

SMITH, L. A. [1977]. "Messinian Event," *Geotimes* 22: 20–23.

SMITH, A. G., and BRIDEN, J. C. [1977]. *Mesozoic and Cenozoic Paleocontinental Maps.* New York: Cambridge University Press.

SMITH, A. G., BRIDEN, J. C., and DREWRY, G. E. [1973]. "Phanerozoic World Maps," in *Organisms and Continents through time,* pp. 1–42, ed. N. F. Hughes. Paleontological Association Special Papers in Paleontology 12.

SMITH, P. B., and EMILIANI, C. [1966]. "Oxygen-Isotope Analysis of Recent Tropical Pacific Benthic Foraminifera," *Science* 160: 1335–36.

SNIDER-PELLEGRINI, A. [1858]. *La Création et ses Mystères Dévoiles.* Paris: Frank and Dentu.

SOREM, R. K., and FOSTER, A. R. [1972]. "Internal Structure of Manganese Nodules and Implications in Beneficiation," in *Ferromanganese Deposits on the Ocean Floor,* pp. 167–82, ed. D. R. Horn. Washington, D.C.: National Science Foundation.

SOUTHARD, J. B., YOUNG, R. A., and HOLLISTER, C. D. [1971]. "Experimental Erosion of Fine Abyssal Sediment," *J. Geophys. Res.* 76: 5903–5909.

SPROLL, W. P., and DIETZ, R. S. [1969]. "Morphological Continental Drift Fit of Australia and Antarctica," *Nature* 222: 345–48.

SRINIVASAN, M. S., and KENNETT, J. P. [1974]. "Secondary Calcification of the Planktonic Foraminifer *Neogloboquadrina pachyderma* as a Climatic Index," *Science* 186: 630–32.

SRINIVASAN, M. S., and KENNETT, J. P. [1976]. "Evolution and Phenotypic Variation in the Late Cenozoic *Neogloboquadrina dutertrei* Plexus," in *Progress in Micropaleontology,* pp. 329–55, ed. Y. Takayanagi and T. Saito. Micropaleontology Press Special Publication 2. New York: American Museum of Natural History.

SRINIVASAN, M. S., and KENNETT, J. P. [1981]. "A Review of Planktonic Foraminiferal Biostratigraphy: Applications in the Equatorial and South Pacific," *Soc. Econ. Paleon. Mineral Spec. Publ.* (in press).

STADUM, C. J., and BURCKLE, L. H. [1973]. "A Silicoflagellate Ooze from the East Falkland Plateau," *Micropaleontology* 19: 104–109.

STAKES, D., and LEINEN, M. [1976]. "Metal Accumulation Rates in the Central Equatorial Pacific during the Cenozoic," *Trans. Am. Geophys. Union* 57: 269.

STEEN-MCINTYRE, V. [1977]. *A Manual of Tephro-chronology.* Idaho Springs, Colorado.

STEHLI, F. G., DOUGLAS R. G., and KAFESCIOGLU, I. A. [1972]. "Models for the Evolution of Planktonic Foraminifera," in *Models in Paleobiology,* pp. 116–29, ed. T. J. M. Schopf, San Francisco: Freeman, Cooper, and Co.

STETSON, H. C. [1939]. "Summary of Sedimentary Conditions on the Continental Shelf Off the East Coast of the United States," in *Recent Marine Sediments,* pp. 230–44, ed. P. D. Trask. American Association of Petroleum Geologists.

STEWART, R. J. [1976]. "Turbidites of the Aleutian Abyssal Plain: Mineralogy, Provenance, and Constraints for Cenozoic Motion of the Pacific Plate," *Geol. Soc. Am. Bull.* 87: 793–808.

STILLE, H. [1924]. *Grundfragen der vergleichenden Tektonik.* Berlin: Borntraeger.

STODDART, D. R. [1969]. "Ecology and Morphology of Recent Coral Reefs," *Cambridge Phil. Soc. Biol. Rev.* 44: 433–98.

STOFFA, P. L. and others [1980]. "Mantle Reflections beneath the Crustal Zone of the East Pacific Rise from Multi-Channel Seismic Data." *Mar. Geol.* 35: 83–97.

STOFFA, P., and TALWANI, M. [1978]. "Exploring the Crust beneath the Oceans," in *Lamont-Doherty Geol. Observ. Yearbook.* New York: Columbia University.

STOMMEL, H. [1948]. "The Westward Intensification of Wind Driven Ocean Currents," *Trans. Amer. Geophys. Union* 29: 202–206.

STOMMEL, H. [1958]. "The Abyssal Circulation," *Deep Sea Res.* 5: 80–81.

STOMMEL, H., and ARONS, A. B. [1960]. "On the Abyssal Circulation of the World Ocean—II. An Idealized Model of Circulation Pattern and Amplitude in Oceanic Basins," *Deep Sea Res.* 6: 217–33.

STONEHOUSE, B. [1969]. "Environmental Temperatures of Tertiary Penguins," *Science* 163: 673–75.

STREETER, S. S. [1973]. "Bottom Waters and Benthonic Foraminifera in the North Atlantic—Glacial-Interglacial Contrasts," *Quat. Res.* 3: 131–41.

STUIVER, M., and QUAY, P. D. [1980]. "Changes in Atmospheric Carbon-14 Attributed to a Variable Sun," *Science* 207: 11–19.

SUESS, E. [1885]. *Das Antlitz der Erde,* 1. Prague: F.Tempsky.

SUESS, E. [1904–9]. *The Face of the Earth* (5 vol.). Oxford: Clarendon Press.

SUESS, H. E. [1980]. "The Radiocarbon Record in Tree-Rings of the Last 8,000 Years," *Radiocarbon J.* 22: 200–209.

SUGGATE, R. P. [1963]. "The Alpine Fault," *Trans. Roy. Soc. N. Z. Geol.,* 2: 105–129.

SUGIMURA, A., and UYEDA, S. [1973]. *Island Arcs: Japan and Its Environments.* Amsterdam: Elsevier.

SUGIMURA, A., MATSUDA, T., CHINZEI, K., and NAKAMURA, K. [1963]. "Quantitative Distribution of Late Cenozoic Volcanic Materials in Japan," *Bull. Volcanol.* 26: 125–40.

SUN, S. S., TATSUMOTO, M., and SCHILLING, J. G. [1975]. "Mantle Plume Mixing along the Reykjanes Ridge Axis: Lead Isotopic Evidence," *Science* 190: 143–47.

SUPKO, P. R., and PERCH-NIELSEN, K. [1977]. "General Synthesis of Central and South Atlantic Drilling Results, Leg 39, Deep Sea Drilling Project," in *Initial Reports of the Deep Sea Drilling Project,* vol. 39, pp. 1099–1131, ed. P. R. Supko, K. Perch-Nielsen, and others. Washington, D.C.: U.S. Government Printing Office.

SVERDRUP, H. U., JOHNSON, N. W., and FLEMING, R. H. [1942]. *The Oceans.* Englewood Cliffs, N.J.: Prentice-Hall, Inc.

SWALLOW, J. C. [1971]. "The Aries Current Measurement in the Western North Atlantic," *Phil. Trans. Series A.* 270: 451–60.

SWIFT, D. J. P. [1969]. "Outer Shelf Sedimentation: Processes and Products," in *The New Concepts of Continental Margin Sedimentation,* Lecture No. 5, American Geological Institute, Short Course Lecture Notes, DS 5, 26.

SWIFT, D. J. P. [1970]. "Quaternary Shelves and the Return to Grade," *Mar. Geol.* 8: 5–30.

SWIFT, D. J. P. [1974]. "Continental Shelf Sedimentation," in *Geology of Continental Margins,* pp. 117–135, ed. C. A. Burk and C. L. Drake. New York: Springer-Verlag.

SWIFT, D. J. P. [1975]. "Barrier-island Genesis: Evidence from the Central Atlantic Shelf, Eastern U.S.A.," *Sediment. Geol.* 14: 1–43.

SYKES, L. R. [1963]. "Seismicity of the South Pacific Ocean," *J. Geophys. Res.* 68: 5999.

SYKES, L. R. [1966]. "The seismicity and Deep Structure of Island Arcs," *J. Geophys. Res.* 71: 2781–3006.

SYKES, L. R. [1967]. "Mechanism of Earthquakes and Nature of Faulting on the Midocean Ridges," *J. Geophys. Res.* 72: 2131–53.

TALWANI, M. [1964]. "A Review of Marine Geophysics," *Mar. Geol.* 2: 29–80.

TALWANI, M., and ELDHOLM, O. [1973]. "The Boundary between Continental and Oceanic Basement at the Margin of Rifted Continents," *Nature* 241: 325–30.

TALWANI, M., and ELDHOLM, O. [1974]. "Margins of the Norwegian-Greenland Sea," in *The Geology of Continental Margins,* pp. 361–74, ed. C. A. Burk and C. L. Drake. New York: Springer-Verlag.

TALWANI, M., and ELDHOLM, O. [1977]. "Evolution of the Norwegian-Greenland Sea," *Geol. Soc. Am. Bull.* 88: 969–99.

TALWANI, M., HARRISON, C. G., and HAYES, D. E., ed. [1979]. "Deep Drilling Results in the Atlantic: Ocean Crust," *Maurice Ewing Series 2.* Washington, D.C.: American Geophysical Union.

TAPPAN, H. [1968]. "Primary Production, Isotopes, Extinctions and the Atmosphere," *Paleo. Paleo. Paleo.* 4: 187–210.

TAPPAN, H. [1980]. *The Paleobiology of Plant Protists.* San Francisco: W. H. Freeman and Company, Publishers.

TAPPAN, H., and LOEBLICH, A. R., JR. [1973]. "Evolution of the oceanic plankton," *Earth Sci. Rev.* 9: 207–40.

TARLING, D. H., and MITCHELL, J. G. [1976]. "Revised Cenozoic Polarity Time Scale," *Geology* 4: 133–36.

TARNEY, J., and others [1979]. "Nature of Mantle Heterogeneity in the North Atlantic: Evidence from Leg 49 Basalts," in *Maurice Ewing Series, 2,* pp. 285–301, ed. M. Talwani, C. G. Harrison, and D. E. Hayes. Washington, D.C.: American Geophysical Union.

TAYLOR, F. B. [1910]. "Bearing of the Tertiary Mountain Belts on the Origin of the Earth's Plan," *Geol. Soc. Am. Bull.* 21: 179–226.

TAYLOR, P. T., ZIETZ, I., and DENNIS, L. S. [1968]. "Geological Implications of Aeromagnetic Data for the Eastern Continental Margin of the United States," *Geophysics* 33: 755–80.

THIEDE, J. [1979A]. "History of the North Atlantic Ocean: Evolution of an Asymmetric Zonal Paleoenvironment in a Latitudinal Ocean Basin," in *Maurice Ewing Series, 3,* pp. 275–296, ed. M. Talwani, W. Hay, and W. B. F. Ryan. Washington, D.C.: American Geophysical Union.

THIEDE, J. [1979B]. "Paleogeography and Paleobathymetry of the Mesozoic and Cenozoic North Atlantic Ocean," *Geojournal* 3.3: 263–72.

THIEDE, J. [1980]. "Palaeo-oceanography, Margin Stratigraphy and Palaeophysiography of the Tertiary North Atlantic and Norwegian–Greenland Seas," *Phil. Trans. R. Soc. Land.* A., 294: 177–185.

THIEDE, J., and VAN ANDEL, Tj. H. [1977]. "The Paleoenvironment of Anaerobic Sediments on the Late Mesozoic South Atlantic Ocean," *Earth and Planet. Sci. Letts.* 33: 301–309.

THIERSTEIN, H. R., and BERGER, W. H. [1978]. "Injection Events in Ocean History," *Nature* 276: 461–466.

THORARINSSON, S. [1944]. "Tefrokronologiska studies Po Island," *Geografiska Ann.* 26: 1–217.

THUNELL, R. C. [1979]. "Pliocene-Pleistocene Paleotemperatures and Paleosalinity History of the Mediterranean Sea, Results from Deep Sea Drilling Project Sites 125 and 132," *Mar. Micropaleo.* 4: 173.

THUNELL, R. C., WILLIAMS, D. F., and KENNETT, J. P. [1977]. "Late Quaternary Paleoclimatology, Stratigraphy and Sapropel History in Eastern Mediterranean Deep-Sea Sediments," *Mar. Micropaleo.* 2: 371–88.

TISSOT, B. [1979]. "Effects on Prolific Petroleum Source Rocks and Major Coal Deposits Caused by Sea-Level Changes," *Nature* 277: 463–65.

TOKSOZ, M. N., and BIRD, P. [1977]. "Formation and Evolution of Marginal Basins and Continental Plateaus," in *Maurice Ewing Series, 1,* pp. 379–93, eds. M. Talwani and W. C. Pitman III. Washington, D.C.: American Geophysical Union.

TOLDERLUND, D. S., and BÉ, A. W. H. [1971]. "Seasonal Distributions of Planktonic Foraminifera in the Western North Atlantic," *Micropaleo.* 17: 297–329.

TUCHOLKE, B. E., and EWING, J. I. [1974]. "Bathymetry and Sediment Geometry of the Great Antilles Outer Ridge and Vicinity," *Geol. Soc. Am. Bull.* 85: 1789–1802.

TUCHOLKE, B. E., HOLLISTER, C. D., WEAVER, F. M., and VENNUM, W. R. [1976]. "Continental Rise and Abyssal Plain Sedimentation in the Southeast Pacific Basin, Leg. 35, Deep Sea Drilling Project," in *Initial Reports of the Deep Sea Drilling Project,* vol. 35, pp. 359–400, ed. C. D. Hollister, C. Craddock, and others. Washington, D.C.: U.S. Government Printing Office.

TUCHOLKE, B. E., and MOUNTAIN, G. S. [1979]. "Seismic Stratigraphy, Lithostratigraphy and Paleosedimentation Patterns in the North American Basin," in *Maurice Ewing Series, 3,* pp. 58–86, ed. M. Talwani, W. Hay, and W. B. F. Ryan. Washington, D.C.: American Geophysical Union.

TUCHOLKE, B., and VOGT, P., and others [1979]. *Initial Reports of the Deep Sea Drilling Project, Leg 43.* Washington, D.C.: U.S. Government Printing Office.

TURCOTTE, D. L., AHERN, J. L., and BIRD, T. M. [1980]. "The State of Stress at Continental Margins," *Tectonophysics* 42: 1–28.

TUREKIAN, K. K. [1964]. "The Geochemistry of the Atlantic Ocean Basin," *Trans. N. Y. Acad. Sci. Ser. II* 26: 312–30.

TUREKIAN, K. K. [1965]. "Some Aspects of the Geochemistry of Marine Sediments," in *Chemical Oceanography,* vol. 2, ed. J. P. Riley and G. Skirrow. New York: Academic Press.

UCHUPI, E. [1967]. "The Continental Margin South of Cape Hatteras, North Carolina: Shallow Structure," *Southeastern Geol.* 8: 155–77.

UCHUPI, E. [1968]. "Atlantic Continental Shelf and Slope of the United States—Physiology," *U.S. Geol. Survey Prof. Paper,* 529C.

UCHUPI, E. [1970]. "Atlantic Continental Shelf and Slope of the United States—Shallow Structure," *U.S. Geol. Survey Prof. Paper,* 529I.

UCHUPI, E., EMERY, K. O., BOWIN, C. O., and PHILLIPS, J. D. [1976]. "Continental Margin off Western Africa: Senegal to Portugal," *Am. Assoc. Petrol. Geol. Bull.* 60: 809–78.

UDDEN, J. A. [1914]. "Mechanical Composition of Clastic Sediments," *Geol. Soc. Am. Bull.* 25: 655–744.

UMBGROVE, J. H. F. [1947]. *The Pulse of the Earth.* The Hague: Martinus Nijhoff.

UNNI, C. K., and SCHILLING, J. G. [1978]. "Cl and Br Degassing by Volcanism along the Reykjanes Ridge and Iceland," *Nature* 272: 19–23.

UREY, H. C. [1947]. "The Thermodynamic Properties of Isotopic Substances," *J. of the Chem. Soc.* London (April), pp. 562–581.

UREY, H. C., LOWENSTAM, H. A., EPSTEIN, S., and McKINNEY, C. R. [1951]. "Measurements of Paleo-temperatures and Temperatures of the Upper Cretaceous of England, Denmark, and Southeastern U.S., *Geol. Soc. Am. Bull.* 62: 399–416.

USACHEV, P. I. [1949]. "The Micro-flora of Polar Ice," A. K. Nauk, U.S.S.R., *Institut Okeanologii, Trudy* 3: 216–259.

UYEDA, S. [1978]. *The New View of the Earth, Moving Continents and Moving Oceans.* San Francisco: W. H. Freeman and Company, Publishers.

UYEDA, S., and KANAMORI, H. [1979]. Back-Arc Opening and the Mode of Subduction. *J. Geophys. Res.* 84: 1049–61.

VACQUIER, V., RAFF, A. D., and WARREN, R. E. [1961]. "Horizontal Displacements in the Floor of the Northeastern Pacific Ocean," *Geol. Soc. Am. Bull.* 72: 1251–58.

VAIL, P. R., and HARDENBOL, J. [1979]. "Sea-Level Changes during the Tertiary," *Oceanus* 22: 71–79.

VAIL, P. R., MITCHUM, R. M., JR., and THOMPSON, S. III [1977]. "Global Cycles of Relative Changes of Sea Level," in *Seismic Stratigraphy—Applications to Hydrocarbon Exploration,* pp. 83–97, ed. C. E. Payton. American Association of Petroleum Geologists, Memoir 26.

VALENTINE, J. W. [1972]. "Global Tectonics and the Fossil Record." *J. Geol.* 80: 167–84.

VALENTINE, J. W. [1973]. *Evolutionary Paleoecology of the Marine Biosphere.* Englewood Cliffs, N.J.: Prentice-Hall, Inc.

VALENTINE, J. W., and MOORES, E. M. [1970]. "Plate Tectonics Regulation of Faunal Diversity and Sea Level: A Model," *Nature* (London) 228: 657–59.

VAN ANDEL, Tj. H. [1975]. "Mesozoic/Cenozoic Calcite Compensation Depth and the Global Distribution of Calcareous Sediments," *Earth and Planet. Sci. Letts.* 26: 187–95.

VAN ANDEL, Tj. H. [1981]. "Sediment Nomenclature and Sediment Classification during Phases I–III of the Deep Sea Drilling Project," in *Deep Sea Drilling Project Initial Reports Methodology Volume,* ed. G. R. Heath. (in press).

VAN ANDEL, Tj. H., and BALLARD, R. D. [1979]. "The Galapagos Rift at 86°W, 2, Volcanism Structure, and Evolution of the Rift Valley," *J. Geophys. Res.* 84: 5390–5406.

VAN ANDEL, Tj. H., HEATH, G. R., and MOORE, T. C., JR. [1975]. *Cenozoic Tectonics, Sedimentation, and Paleoceanography of the Central Equatorial Pacific.* Geol. Soc. Am. Memoir 143.

VAN ANDEL. Tj. H., HEATH, G. R., and MOORE, T. C., JR. [1976]. "Cenozoic History of the Central Equatorial Pacific: A Synthesis Based on Deep Sea Drilling Project Data," in *The Geophysics of the Pacific Ocean Basin and its Margin,* Geophysical Monograph 19, pp. 281–96. Washington, D.C.: American Geophysical Union.

VAN ANDEL, Tj. H., THIEDE, J., SCLATER J. G., and HAY, W. W. [1977]. "Depositional History of the South Atlantic Ocean during the Last 125 Million Years," *J. Geol.* 85: 651–98.

VAN ANDEL, Tj. H., and VEEVERS, J. J. [1967]. "Morphology and Sediments of the Timor Sea," *Bur. Min Resources, Geol. and Geophys. Bull.* 83: 173.

VAN COUVERING, J. A. and others [1976]. "The Terminal Miocene Event," *Mar. Micropaleo.* 1: 263–86.

VAN DER LINGEN, G. J., ed. [1977]. *Diagenesis of Deep-Sea Biogenic Sediments.* Stroudsburg, Pennsylvania: Dowden, Hutchinson and Ross, Inc.

VAN GORSEL, J. T., and TROELSTRA, S. R. [1981]. "Late Neogene Planktonic Foraminiferal Biostratigraphy and Climostratigraphy of the Solo River Section (Java, Indonesia)." *Mar. Micropaleo.* 6: 183–209.

VAN HINTE, J. E. [1976]. "A Cretaceous Time Scale," *Am. Assoc. Petrol. Geol. Bull.* 60: 498–516.

VAN VALKENBURG, S. D., and NORRIS, R. E. [1970]. "The Growth and Morphology of the Silicoflagellate *Dictyocha fibula* Ehrenberg, in Culture," *J. Phycol.* 6: 48–54.

VEEH, H. H., and VEEVERS, J. J. [1970]. "Sea Level at −175 m Off the Great Barrier Reef 13,600 to 17,000 Years Ago. *Nature* 226: 536–37.

VEEVERS, J. J. [1969]. "Paleogeography of the Timor Sea region," *Paleo. Paleo. Paleo.* 6: 125–40.

VEEVERS, J. J. [1972]. "Regional Site Surveys, Sites 259, 262, 263," in *Initial Reports of the Deep Sea Drilling Project,* pp. 561–66, ed. J. J. Veevers, J. R. Heirtzler, and others. Washington, D.C.: U.S. Government Printing Office.

VEEVERS, J. J., JONES, J. G., and TALENT, J. A. [1971]. "Indo-Australian Stratigraphy and the Configuration and Dispersal of Gondwanaland," *Nature* 229: 383–88.

VELLA, P. [1965]. "Sedimentary Cycles, Correlation, and Stratigraphic Classification," *Trans. R. Soc. N. Z. Geol.* 3: 1–9.

VENING MEINESZ, F. A. [1952]. "The Origin of Continents and Oceans," *Geol. Mijnbouw* 31: 373–84.

VENRICK, E. L. [1971]. "Recurrent Groups of Diatom Species in the North Pacific," *Ecology* 52: 614–25.

VERGNAUD-GRAZZINI, C., PIERRE, C., and LETOLLE, R. [1978]. "Paleoenvironment of the North-East Atlantic during the Cenozoic: Oxygen and Carbon Isotope Analysis at Deep Sea Drilling Project Sites 398, 400A and 401," *Oceanologica Acta* 1: 381–90.

VINCENT, E. [1975]. "Neogene Planktonic Foraminifera from the Central North Pacific, Leg 32, Deep Sea Drilling Project," in *Initial Reports of the Deep Sea Drilling Project,* vol. 32, pp. 765–791, eds. R. L. Larson, R. Moberly, and others. Washington, D.C.: U.S. Government Printing Office.

VINCENT, E., KILLINGLEY, J. S., and BERGER, W. H. [1980]. "The Magnetic Epoch-6 Carbon Shift, a Change in the Ocean's $^{13}C/^{12}C$ Ratio 6.2 Million Years Ago," *Mar. Micropaleo.* 5: 185–203.

VINE, F. J. [1966]. "Spreading of the Ocean Floor; New Evidence," *Science* 154: 1405–15.

VINE, F. J. [1968]. "Magnetic Anomalies Associated with Mid-Ocean Ridges," in *The History of the Earth's Crust,* pp. 73–89, ed. R. A. Phinney. Princeton, N.J.: Princeton University Press.

VINE, F. J., and MATTHEWS, D. H. [1963]. "Magnetic Anomalies over Oceanic Ridges," *Nature* 199: 947–49.

VINE, F. J., and WILSON, J. T. [1965]. "Magnetic Anomalies over a Young Oceanic Ridge off Vancouver Island," *Science* 150: 485–489.

VOGT, P. R. [1972A]. "The Faeroe-Iceland-Greenland Aseismic Ridge and the Western Boundary Undercurrent," *Nature* 239: 79–86.

VOGT, P. R. [1972B]. "Evidence for Global Synchronism in Mantle Plume Convection, and Possible Significance for Geology," *Nature* 240: 338–42.

VOGT, P. R. [1973]. "Subduction and Aseismic Ridges," *Nature* 241: 189–91.

VOGT, P. R. [1975]. "Changes in Geomagnetic Reversal Frequency at Times of Tectonic Change: Evidence for Coupling between Core and Upper Mantle Processes," *Earth and Planet. Sci. Letts.* 25: 313–21.

VOGT, P. R. [1979]. "Global Magmatic Episodes: New Evidence and Implications for the Steady-State Mid-oceanic Ridge," *Geology* 7: 93–98.

VON HUENE, R., and KULM, L. D. [1973]. "Tectonic Survey of Leg 18, in *Initial Reports of the Deep Sea Drilling Project,* vol. 18, pp. 961–976, ed. L. D. Kulm, R. Von Huene, and others. Washington, D.C.: U.S. Government Printing Office.

VON HUENE, R., LANGSETH, M., NASU, N., and OKADA, H. [1980]. "Summary, Japan Trench Transect," in *Initial Reports of the Deep Sea Drilling Project,* vol. 57, pp. 473–88, ed. R. von Huene, N. Nasu, and others. Washington, D.C.: U.S. Government Printing Office.

VON RAD, U., and ROSCH, H. [1974]. "Petrography and Diagenesis of Deep-Sea Cherts from the Central Atlantic," in *Pelagic Sediments: On Land and Under the Sea,* vol. 1, pp. 327–47, ed. K. J. Hsü and H. C. Jenkyns. Spec. Publs. Int. Assoc. Sediment.

WALCOTT, R. I. [1972]. "Gravity, Flexure and the Growth of Sedimentary Basins at a Continental Edge," *Geol. Soc. Am. Bull.* 83: 1845.

WALCOTT, R. I. [1978]. "Present Tectonics and Late Cenozoic Evolution of New Zealand," *Geophys. J. R. Astr. Soc.* 52: 137–64.

WALKER, G. P. L. [1975]. "Excess Spreading Axes and Spreading Rate in Iceland," *Nature* 255: 448–71.

WARNKE, D. A., and HANSEN, M. E. [1977]. "Sediments of Glacial Origin in the Area of Operations of Deep Sea Drilling Project Leg 38 (Norwegian-Greenland Seas): Preliminary Results from Sites 336 and 344," *Ber. Naturf. Ges. Freiburg i. Br.* 67: 371–92.

WARREN, B. A. [1970]. "General Circulation of the South Pacific," in *Scientific Exploration of the South Pacific,* pp. 33–49, ed. W. S. Wooster. Washington, D.C.: National Academy of Sciences.

WARREN, B. A. [1971]. "Antarctic Deep Water Contribution to the World Ocean," in *Research in the Antarctic,* publ. 93, pp. 630–43, ed. L. O. Quam. Washington, D.C.: American Association for the Advancement of Science.

WATKINS, N. D. [1972]. "A Review of the Development of the Geomagnetic Polarity Time Scale and a Discussion of Prospects for its Finer Definition," *Geol. Soc. Am. Bull.* 83: 551–74.

WATKINS, N. D., and BAKSI, A. K. [1974]. "Magnetostratigraphy and Oroclinal Folding of the Columbia River, Steens, and Owyhee Basalts in Oregon, Washington and Idaho," *Am. J. Sci.* 274: 148–89.

WATKINS, N. D., and HUANG, T. C. [1977]. "Tephras in Abyssal Sediments East of the North Island, New Zealand: Chronology, Paleowind Velocity, and Paleoexplosivity." *N. Z. J. Geol. Geophys.* 20: 179–98.

WATKINS, N. D., and KENNETT, J. P. [1972]. "Regional Sedimentary Disconformities and Upper Cenozoic Changes in Bottom Water Velocities between Australia and Antarctica," *Ant. Res. Ser.* 19: 273–93, American Geophys. Union.

WATTS, A. B. [1975]. "Marine Gravity," *Rev. of Geophys. and Space Physics* 13: 531–36.

WATTS, A. B., WEISSEL, J. K., and LARSON, R. L. [1977]. "Sea-Floor Spreading in Marginal Basins of the Western Pacific." *Tectonophysics* 37: 167–81.

WEGENER, A. [1912]. "Die entstehung der kontinente," *Geologische Rundschau.* 3: 276–292.

WEGENER, A. [1915]. *Die Entstehung der Kontinente und Ozeane.* Sammlung Vieweg, 23, Branunschweig.

WEGENER, A. [1924]. *The Origin of Continents and Oceans.* London: Dover.

WEISSEL, J. K. [1977]. "Evolution of the Lau Basin by the Growth of Small Plates," in *Maurice Ewing Series 1,* pp. 429–436, ed. M. Talwani and W. C. Pitman III. Washington, D.C.: American Geophysical Union.

WEISSEL, J. K., and HAYES, D. E. [1972]. "Magnetic Anomalies in the Southeast Indian Ocean." in *Antarctic Oceanology II: The Australian-New Zealand Sector,* Antarctic Research Series, vol. 19, pp. 165–96. ed. D. E. Hayes, pp. 165–96. Washington, D.C.: American Geophysical Union.

WEISSEL, J. K., and HAYES, D. E. [1974]. "The Australian-Antarctic Discordance: New Results and Implications," *J. Geophys. Res.* 79: 2579–87.

WELLMAN, H. W. [1970]. "Recent Fault Movements in New Zealand," in *Proceedings of the Royal Society, New Zealand International Symposium on Recent Crustal Movements and Assoc. Seismicity,* p. 36, Wellington, New Zealand.

WELLMAN, H. W. [1971]. "Reference Lines, Fault Classification, Transform Systems, and Ocean-Floor Spreading," *Tectonophysics* 12: 199–210.

WELLS, J. W. [1957]. "Coral Reefs," in *Geol. Soc. Am. Memoir 67,* ed. J. Hedgbeth, pp. 609–631.

WENTWORTH, C. K. [1922]. "A Scale of Grade and Class Terms for Clastic Sediments." *J. Geol.* 30: 377–392.

WERTENBAKER, W. [1974]. *The Floor of the Sea.* Boston: Little, Brown & Company.

WESER, O. E. [1973]. "Sediment classification," in *Initial Reports of the Deep Sea Drilling Project,* Leg. 18, pp. 9–13, ed. L. D. Kulm, R. von Huene, and others. Washington, D.C.: U.S. Government Printing Office.

WHITAKER, J. D. Mc D. [1974]. "Ancient Submarine Canyons and Fan Valleys." In *Modern and ancient geosynclinal sedimentation*, pp. 106-125, ed. R. H. Dott, Jr. and R. H. Shaver. SEPM Special Publication 19.

WHITE, W. M., SCHILLING, J. G., and HART, S. R. [1976]. "Evidence for the Azores Mantle Plume from Strontium Isotope Geochemistry of the Central North Atlantic." *Nature* 263: 659-63.

WHITMORE, F. C., EMERY, K. O., COOKE, H. B. S., and SWIFT, D. J. [1967]. "Elephant Teeth from the Atlantic Continental Shelf." *Science* 156: 1477-81.

WILLIAMS, H. and Mc BIRNEY, A. [1979]. *Volcanology.* San Francisco: Freeman, Cooper, and Co.

WILLIAMS, D. L., VON HERZEN, R. P., SCLATER, J. G., and ANDERSON, R. N. [1974]. "The Galapagos Spreading Center: Lithospheric Cooling and Hydrothermal Circulation." *Geophys. J. R. Astr. Soc.* 38: 587-608.

WILSON, J. L. [1974]. "Characteristics of Carbonate Platform Margins," *Am. Assoc. Petrol. Geol. Bull.* 58: 810-24.

WILSON, J. T. [1963]. "Hypothesis of Earth's Behavior," *Nature* 198: 925-29.

WILSON, J. T. [1963]. "Continental Drift." *Scientific American* 208: 86-100.

WILSON, J. T. [1965]. "A New Class of Faults and Their Bearing on Continental Drift." *Nature* 207: 343-47.

WILSON, J. T. [1968]. "Static or Mobile Earth: The Current Scientific Revolution." *Am. Philos. Soc. Proc.* 112: 309-20.

WILSON, J. T. [1973]. "Mantle Plumes and Plate Motions." *Tectonophysics* 19: 149-64.

WINDOM, H. L. [1976]. "Lithogenous Material in Marine Sediments," in *Chemical Oceanography,* (2nd ed) vol. 5, pp. 103-35, ed. J. P. Riley and G. Skirrow. New York: Academic Press.

WINTERER, E. L. [1973]. "Sedimentary Facies and Plate Tectonics of Equatorial Pacific," *Am. Assoc. Petrol. Geol. Bull.* 57: 265-82.

WISE, S. W., JR., and WEAVER, F. M. [1974]. "Chertification of oceanic sediments," in *Pelagic Sediments: On Land and Under the Sea,* pp. 301-326, ed. K. J. Hsü and H. C. Jenkyns. Special Publication International Association Sedimentology 1.

WOLFE, J. A. [1971]. "Tertiary Climatic Fluctuations and Methods of Analysis of Tertiary Floras." *Paleo. Paleo. Paleo.* 9: 27-57.

WOLFE, J. A. [1978]. "A Paleobotanical Interpretation of Tertiary Climates in the Northern Hemisphere." *Am. Sci.* 66: 694-703.

WOLFE, J. A., and HOPKINS, D. M. [1967]. "Climatic Changes Recorded by Tertiary Land Floras in Northwestern North America," in *Tertiary Correlations and Climatic Changes in the Pacific,* pp. 67-76, ed. K. Hatai. Pacific Science Congress 11th, Tokyo. Sendai, Japan: Sasaki Printing Co.

WOODRUFF, F., and DOUGLAS, R. G. [1981]. "Response of Deep Sea Benthic Foraminifera to Miocene Paleoclimatic Events, Deep Sea Drilling Project Site 289," *Cenop Symposium, Marine Micropal.* (in press).

WORNARDT, W. W., JR. [1969]. "Diatoms, Past, Present, Future," in *Proc. of the first Inter. Conf. on Planktonic Microfossils,* pp. 69-714, ed. P. Bronnimann and H. H. Renz. Leiden: Brill.

WORSLEY, T. R. [1974]. "The Cretaceous-Tertiary Boundary Event in the Ocean," in *Studies in Paleooceanography,* pp. 94-125, ed. W. W. Hay. SEPM Spec. Publ. 2

WORSLEY, T. R., and DAVIES, T. A. [1979]. "Sea Level Fluctuations and Deep-Sea Sedimentation Rates," *Science* 203: 455-456.

WRIGHT, L. D. [1978]. "River Deltas." In *Coastal Sedimentary Environments,* pp. 5-68, ed. R. A. Davis, Jr., New York: Springer-Verlag.

WUST, G., BROGMUS, W., and NOODT, E. N. [1954]. "Die Zonale Verteilung von Salzgehalt, Neiderschlag, Verdungstung, Temperatur und Dichte an der Oberflache der Ozèane." *Kieler Meeresforsch* 10: 137-61.

WYLLIE, P. J. [1971]. *The Dynamic Earth: Textbook in Geosciences.* New York: John Wiley and Sons, Inc.

YODER, H. S. [1976]. *Generation of Basaltic Magma.* Washington, D.C.: National Academy of Sciences.

YODER, H. S., JR., and TILLEY, C. E. [1962]. "Origin of Basalt Magmas: An Experimental Study of Natural and Synthetic Rock Systems." *J. Petro.* 3: 342-532.

YOUNG, R. A., and HOLLISTER, C. D. [1974]. "Quaternary Sedimentation in a Late Precambrian Shelf Sea, Victoria Island, Canadian Arctic Archipelago." *J. Sediment. Petrol.* 47: 943-53.

ZANEVELD, J. R. V., ROACH, D. M., and PAK, H. [1974]. "The Determination of the Index of Refraction Distribution of Oceanic Particulates." *J. Geophys. Res.* 79: 4091-95.

ZIEGLER, A. M., and others [1977]. "Silurian Continental Distributions, Paleogeography, Climatology, and Biogeography," *Tectonophysics* 40: 13-51.

ZIEGLER, A. M., and others [1979]. "Paleozoic Paleogeography." *Ann. Rev. Earth Planet. Sci.* 7: 473-502.

ZIEGLER, P. A. [1975]. "Geologic Evolution of North Sea and Its Tectonic Framework." *Am. Assoc. Petrol. Geol. Bull.* 59: 1073-97.

ZIEGLER, P. A. [1977]. "Geology and Hydrocarbon Provinces of the North Sea." *Geojournal* 1: 7-31.

ZONENSHAYN, L. P., and GORODNITSKIY, A. M. [1977]. "Paleozoic and Mesozoic Reconstructions of the Continents and Oceans, Article 1: Early and Middle Paleozoic Reconstructions." *Geotectonics* 11: 159-72.

Author Index

Abbott, D., 684–85, 687
Adams, C. G., 740–41
Airy, G. B., 20
Akiba, F., 740
Aldrich, L. T., 70
Almagor, G., 406–7
Alvarez, L. W., 712
Alvarez, W., 711
Ampferer, O., 107
Anderson, D. L., 17
Anderson, J. B., 445–48
Anderson, R. N., 231
Andrews, J. E., 187, 655
Andrews, P. B., 719
Andrews, P. B., 735
Anikouchine, W. A., 24, 27
Argand, E., 394
Arkell, W. J., 702
Arnold, Z. M., 553
Arons, A. B., 256
Arrhenius, G., 80, 503, 605, 691
Arthur, M. A., 490, 660, 663–65, 690, 699, 704–5, 710
Ashraf, A., 309
Atwater, T., 165, 181, 187–90, 373, 381–82, 391
Aumento, F., 227

Bacon, F., 106
Baker, G., 433
Baksi, A. K., 383
Ballard, R. D., 210. 218, 220, 350

Bally, A. W., 324, 329, 331–32, 334, 348–49, 353, 370
Bambach, R. K., 701
Bandy, O. L., 557, 559, 641, 738, 745
Banner, F. T., 66
Barazangi, M., 139, 141, 156, 368
Barker, P. F., 658, 672, 726, 729, 731
Barron, E. J., 111–12, 697–98, 701
Barron, J. A., 639, 738
Batten, R. L., 275
Baumgartner, T. R., 337–38
Bé, A. W. H., 457, 469, 541, 543, 547, 550, 562–66, 606, 611, 641
Beckinsale, R. D., 70
Behrendt, J. C., 344–45
Bender, M. L., 87, 610, 739
Benioff, H., 132, 156
Benson, R. H., 560–62, 722, 740
Berger, A. L., 647
Berger, W. H., 127, 397–98, 400–401, 457, 460, 464, 466–68, 470, 482, 543–44, 547, 571–72, 578, 581, 607, 611, 616, 619, 621, 627–29, 632, 665, 680, 683, 688–91, 704–5, 707, 711–12, 717, 723–24
Berggren, W. A., 67, 76, 91–92, 482, 598, 602, 637–38, 655–57, 664, 667–69, 677, 710, 728, 731, 740–43, 747
Berner, R. A., 470
Bernoulli, D., 663
Berry, W. B. A., 704
Bezrukov, P. L., 457
Bibee, L. D., 372–75

Biehler, S., 220
Biggs, R. B., 291–92
Bird, J. M., 178, 180, 350–51, 375, 390–92, 595
Biscaye, P. E., 428–29, 435, 503, 525–27, 621
Bjorklund, K. R., 579–80
Black, M., 564, 570
Blackwelder, B. W., 270
Blatt, H., 412
Bloom, A. L., 267, 269, 272
Blow, W. H., 66–67, 90
Bodvarsson, G., 212
Boersma, A., 712–13, 726
Bolli, H. M., 66, 91, 674
Boltovskoy, E., 538, 554–55
Bonatti, E., 216
Booth, J., 514
Boström, K., 492–93
Bott, M. H. P., 16, 39, 48–49, 207
Bouguer, P., 19
Bouma, A. H., 413
Bowen, D. Q., 273
Bowen, N. L., 221
Bowen, R., 703
Bradley, W. H., 6
Bradshaw, J. S., 543–45, 606
Bramlette, M. N., 6, 711
Bray, J. R., 648
Brenninkmeyer, B., 302
Brewster, N. A., 731, 733, 735, 739, 742
Briden, J. C., 324
Brisco, C. D., 423

Subject Index

AA lava, 218
Absolute age:
  of rocks, 53
Abyssal antidunes, 511
Abyssal barchans, 530–31
Abyssal bed forms, 528
Abyssal circulation:
  controls, 506
Abyssal cones, 419, 421–22
Abyssal hill provinces, 37–38
Abyssal landscape:
  characteristics, 528–29
Abyssal plains, 27, 37–38, 341, 423–25
  deposits on, 412
Abyssal zone, 258
  benthonic foraminifera, 559
Abyssopelagic, 258
Acantharia, 573
Accretionary basins:
  of forearc, 365
Accretionary borderland:
  New Zealand, 379–80
Accretionary prism, 357, 359, 362–64
  evolution, 361, 363–64
  New Zealand, 378–79
Accretionary trench systems, 357
Acme zones, 54, 64
Acoustic stratification, 510
Active margins, 29, 322–25, 353
Aegean arc, 390
Aegean Sea, 392, 394
Afar Depression, Ethiopia, 331
Africa, 3
  age of Atlantic opening, 332, 343

kaolinite, 431
  separation from South America,
  113
African plate, 132, 190, 195
African rift-valley system, 35
Agulhas Bank:
  evolution, 672
  phosphorite, 502
Agulhas Current, 241–42
Agulhas Plateau, 166
Air-gun method, 9, 42, 50
Alaska arc, 362
Alaska Current, 241–42
Albedo:
  Cretaceous, 698
  modern, 698
  paleoclimate changes, 647, 697–98
Aleutian abyssal plain, 464
Aleutian Arc, 151
  gravity, 153
Aleutian Current, 241
Aleutian-Kamchatka cusp, 164
Aleutian Trench, 39, 132
Alkali basalts:
  composition, 215
Alkali basalt series, 214
Alpine Fault, 187, 377
Alps, European:
  recent tectonic history, 384
Alteration:
  oceanic crust, 230
Amazon River, 25, 250, 288, 290, 422,
  427–28
American Plate, 176

Amphibolite facies:
  oceanic crust, 211
Amsterdam Islands, 200–201
Anaerobic sediments, 486–92
Anatolian Fault, 132
Andalusian stratotype, 740
Andaman-Nicobar-Mantawai islands,
  365
Andaman Sea Basin, 373
Andean Cordillera, 133
Andean episode:
  volcanism, 384
Andes:
  Cretaceous plutonism, 129
  recent tectonic history, 384
  volcanoes, 368
Andesite, 204
  formation, 157, 368–70
  volcanism, 156
Andesite line, 204, 369
Angola:
  diapiric dams, 337–38
Angola Basin, 468–69
Angular velocity:
  of plates, 138
  tectonic motion, 138
Anisotropy of magnetic susceptibility,
  514
Anoxia:
  oceanic, 236, 703
Anoxic basins, 619
Anoxic events, 703
Anoxic sediments, 486–92, 703–5
  Mesozoic, 663–64, 674–75

Bahama Banks, 346, 535
Bahama platform, 348
   paleoreconstructions, 191
Bahamas, 342, 524
   lagoons, 314
   shelf sediments, 312
Baja California:
   tectonics, 136
Balearic Basin:
   Mediterranean, 392-94
Baltimore Canyon, 416
Baltimore Canyon Trough, 344-46
Banda Arc, 390
Barbados, 272-73, 379
   dust, 433
   terraces, 272-73
   uplifted accretionary prism, 359
Barchans, 530
Bar-finger sands, 295
Barite, 503
Barnacles, 313-14
Barrier beaches, 305
Barrier islands, 289, 305-7
Barrier reefs, 316-18
Basalt, 18
   alkaline, 216-17
   alteration of, 230
   composition, 213-14, 369-70
   compositional variations, 226-29
   fractional crystallization, 226-29
   fractionation, 214
   glass composition, 217
   isotopic ratios, 223-26
   Layer 2, 210
   low temperature alteration, 478
   magma, 221
   magnetization, 232
   olivine-tholeiites, 216-17
   origin, 221-22
   quartz-tholeiites, 216-17
   sonic velocity, 209
Basaltic glass:
   and radiogenic dating, 70
Basin and range province:
   western United States, 189
Basin-to-basin fractionation, 457, 498,
   597, 689
Bathyal zone, 258
   benthonic foraminifera, 556-59
Bathypelagic, 258
Bauer Basin, Pacific, 494
Bay of Bengal:
   abyssal plain, 37
   cone, 422
   sediments, 194-95
Bay of Biscay:
   paleoreconstructions, 192
   western margin, 338-39
Bay of Islands:
   ophiolite, 211
Beaches, 295-305
   cusps, 302
   face, 299
   seasonal cycles, 304
   sediments, 299-300
Bengal cone, 422
Benguela Current, 241-42
   phosphorite, 502

Benioff zone, 132, 141, 143-44, 150,
   322, 356, 360, 363, 369
   heat flow, 146
   relation to trenches, 152
   volcanism, 156-57, 368-69
Benthic boundary layer, 506
Benthonic environments, 258
Benthonic foraminifera:
   abyssal water masses, 515-16
   Antarctic, 447-48
   bathymetric zones, 555-59
   biotopes, 555
   classification, 552-53
   depth biotopes, 554-59
   distribution, 554-59
   Eocene-Oligocene boundary, 723
   importance of, 551-52
   morphology, 552-53
   paleoceanographic methods, 607-8
   reworking, 415
   wall structure, 552-53
Bering Canyon, 419
Bering Sea, 26, 254, 372
   silica deposition, 476
Bering Straits, 656
Berms, 299
Bermuda platform:
   pteropod ooze, 460
Bikini Atoll, 316
Biochronology, 87
   correlations, 88
Biocoenosis, 258
Biofacies:
   term, 58
Biogenic sediments, 454-92
   definition, 455
   shallow water, 312-16
Biogeography, 596
   Neogene-Paleogene transition,
   733-34
   Oligocene oceans, 728
   Quaternary, 748
Biostratigraphy, 54, 62-68
   correlation, 602
   Neogene, 90
   term, 53
   zone, 62
Biotopes, 258
Bioturbation, 77, 80, 618-19
   microtektites, 453
Biozones, 62
   stratigraphy, 54
Birdfoot Delta, 294
Birnessite, 497
Bjorn ridge, 532, 534
Black Sea, 26
   anoxic sediments, 486-88
   Quaternary history, 487-88
Black shales, 490-91
   Cretaceous distribution, 704
   stratigraphic distribution, 706
Blake-Bahama basin, 344
Blake Bahama Outer Ridge, 42, 47, 423,
   534-35
Blake Plateau, 335, 342, 346, 348, 351,
   535
   basin, 344
   cross section, 347

Blake Spur anomaly, 345
Bonin-Mariana Islands, 39
Bonin Trench, 39
Bore, 298
Bosporus Straits:
   importance in Mediterranean circulation, 490
Bottom currents, 505-7
   velocities, 256, 506, 517
Bottom photographs, 507-10, 524,
   531
Bottomset beds, deltas, 295, 297
Bottom topography:
   effect on circulation, 255
Bottom waters:
   age, 236, 458
   circulation, 250, 255-57, 505-6
   density of, 251
   formation of, 250-55
   origin of, 235-36
   oxygen levels, 236
   paleocirculation, 607
   silica content, 475
   source areas, 615
Bouguer anomaly, 20
Bouma sequence, 413
Bouvet Islands:
   hot spot, 163
Box corer, 100, 102
Brahmaputra River, 288
Brazil Basin, 192, 468-69
Brazil Current, 241-42
Breaker zone, 298, 300-301
Brine Basin, Gulf of Mexico, 489
Broken ridge, 166, 196, 199, 612
Brown clay, 426
   global distribution, 457
Brunhes epoch, 73
   oxygen isotopic record, 85
Brunhes/Matuyama boundary:
   association with microtektites, 452
Bryozoans, 313-14
   limestones, 455

Cabot Strait, Newfoundland, 524
Caicos Outer Ridge, 533-35
Calabrian stage, 92
Calcalkaline, volcanic rocks, 157
Calcareous microfossils, 470-72, 538-
   73
   dissolution of, 470
Calcareous nannofossils, 564-73
   biogeography of, 567-69
   depth distribution, 569-70
   diagenesis, 472-73
   dissolution, 472, 571
   distribution, 567-69
   diversity, 567
   geologic range, 566
   importance for stratigraphy, 63
   living forms, 566
   modern diversity, 64
   Neogene zonation, 90
   paleobiogeography, 572-73
   Paleocene value, 572-73
   Paleogene zonation, 91
   sinking of, 483, 571

Glacio-eustatic change, 268
Glass:
    composition, 217
Global ash falls, 438
Global contraction, 105
    hypothesis, 106-7
Global expansion, 105
Global paleoceanographic evolution, 695-751
Global plate tectonics, 105, 131
Globigerina ooze, 460
Globigerinidae, 539
Globigerinoides datum, 93
Globorotaliidae, 539
Glomar Challenger, 94-98, 205, 338, 560
Gondwanaland, 106-7, 109, 111-12, 127, 181, 195-96, 323, 598-99, 656, 700-702
Gorda Escarpment, 36
Gorda Plate, 132
Gordon ridge, 117
Gorgonians, 507
Grab samplers, 101
Grain flow, 408-9
Grain size, velocities for erosion and transportation, 517
Grand Banks, 248, 345
    sediments, 314
    slump, 407, 411-12
GRAPE measurements, 615, 621
Gravitational anchor theory, 165
Gravitational compaction, 620
Gravitational force, 19
Gravity, 19
    acceleration of, 19
    structure of Aleutian Arc, 153
    transport of sediment, 405-15
Gravity anomalies, 19-20
    eastern North America, 326
    island arcs, 150, 152
    mid-ocean ridge, 145
Gravity corer, 100-101
Gravity sliding hypotheses, 171
Great Barrier Reef:
    Australia, 316
    lagoons, 314
Greater Antilles, 346
    outer ridge, 423, 533-35
Greater India, 195-96, 202
Greece:
    tectonics, 390
Greenland, 25
    ocean boundaries, 26
    paleoreconstruction, 192-93
Greenland Sea, 253-54
    bottom-water source, 251
Greenschist facies:
    oceanic crust, 211
Groins, 302
Ground water:
    sediment transport to oceans, 625
Group:
    stratigraphy, 54
Guadeloupe Island Chain, 160
Guayamas Basin, Gulf of California, 619
Guinea, 309
    Basin, 192

Guinea nose:
    paleoreconstructions, 190
Gulf Coast:
    barrier beaches, 305
Gulf of Aden, 34-35, 178
    age of opening, 332
    stage in life cycle, 179
    tectonic history, 199
Gulf of Alaska, 34
    sediments, 657
Gulf of California, 34-35
    age of opening, 332
    late Miocene isolation, 616
    organic content, 491
    silica deposition, 475-77
    tectonic history, 189
    tectonics, 135-36
    transform faults, 331
Gulf of Maine:
    tides, 310
Gulf of Mexico, 26
    continental margins, 344, 348-50
    early opening history, 190
    marginal sea shorelines, 265
    Quaternary climates, 639
    Quaternary melt-water influx, 616
    sea-level history, 269
    sediment dams, 336
    Sigsbee rise, 342
    tectonics, 349
Gulf of St. Lawrence:
    early history, 348
Gulf Stream, 237, 244, 248, 535
    association with sediment drifts, 533-34
    early flow, 348
    erosion, 348
    history, 351
    influence on glaciation, 745
    transport, 248
    velocity, 248
    volume, 242
    width, 248
Gulf Stream Current, 241-42
Gutenberg discontinuity, 16, 19
Guyana Basin:
    varves, 80
Guyots, 38, 115, 159
Gyres:
    anticyclonic, 242
    cyclonic, 242

Hadal zone, 258
Hadopelagic, 258
Half-life:
    40K, 70
    ^{14}C, 79
    radioisotopes, 68
    ^{230}Th, 79
Hatteras abyssal plain, 418, 424-25, 535
    turbidites, 424
Hatteras Canyon, 418, 424
Hatton ridge, 532, 534
Hawaiian-Emperor Chain, 38, 160, 163-66, 385-86
    Aleutian cusp, 170
    bend in chain, 164
    junction with Kuril, 170

Hawaiian hot spot, 161
Hawaiian Islands, 38, 159, 162, 184, 217
    age progression, 163-65
    hot spots, 163
    magnetic lineations, 182-83
    manganese nodules, 495
    surrounding moats, 167
    volcanic rocks, 218
    volcanism, 159
Heat flow, 17
    change with crustal ages, 145-46
    equation, 146
    marginal basin, 373
    mid-ocean ridge axis, 220
    ocean crust, 145
    ocean ridges, 231
Hekla:
    volcano, 438
Hellenides Mountains, 394
Hemipelagic sediments, 399, 426, 456
    classification, 401
Hermatypic corals, 313-14
Heteropods, 460, 562
Hiatuses, 608
    causes, 523
    Cenozoic record, 692-94
    deep sea, 521-23
    dissolution, 522-23
    global record, 692-94
    mapping, 513-14
    term, 59
High-alumina, volcanic rocks, 157
High-frequency echoes, 511
Hikurangi, Trough, 376
Himalayas, 26, 178, 194-95, 201, 355, 390-91, 422
    evolution, 686
    formation, 201-2
    recent tectonic history, 384
    stage in life cycle, 179
    Tethys Sea, 111
Holocene:
    varves, 81
Holocene-Pleistocene boundary, 80
Holocene transgression, 268, 270-71, 289-90, 297, 308, 315
Homo sapiens:
    evolution, 747
Horizon A, 45-47, 482, 667
Horizon β + B, 47
Horse latitudes, 241
Hot spots, 161-63
    lava production, 163
    volcanic history, 385
Huang Ho River, 288
Hudson Canyon, 308, 416, 418, 424
Hudson River, 422
Hunter Channel, 614, 681
Hunter fracture zone, 354-55
Huon Peninsula, New Guinea, 272
Hyaloclastites, 218, 399
Hydraulic piston corer, 77, 98-99, 102-3
Hydrogen sulfide:
    mid-ocean ridges, 231
Hydrosphere, 17
Hydrostatic/gas corer, 102
Hydrothermal activity:
    calcium input, 460

circulation, 230
  high temperature, 231-32
  metal-rich sediments, 493-95
  mid-ocean ridges, 146
Hyperbolic echoes, 511
Hypsographic curve, 26
Hypsometry, 23-26

Iberia:
  paleoreconstruction, 191-92
Iberian Portal, 740
Ice ages:
  late Pliocene development, 742-45
  Quaternary, 745-51
Icebergs, 443-44
Ice decay:
  orbital configuration, 648
Ice growth:
  orbital configuration, 648
Iceland, 217
  active volcanism, 35
  association with mid-Atlantic
    ridge, 35
  basalt composition changes,
    223-26
  glaciation, 743
  hot spot, 163
  ocean boundaries, 26
  paleomagnetic polarity sequence,
    77
  Reykjanes ridge, 119-20
  volcanism, 157-59, 437
Iceland-Faeroe ridge, 167, 534
  early history, 193
  evolution, 736-37
  subsidence, 730
Ice-rafted sediments:
  Antarctic Cenozoic, 730, 732
Ice sheets, 20, 82
  effect on oxygen isotopic com-
    position, 83-84
  effect on sea level, 268
  history, 85
  Northern Hemisphere, 272, 648
  oxygen isotopes, 82, 84
  sea-level change, 265
  volumes, 268
Ice shelves, 444
  dry base, 446-47, 449
  wet base, 446, 448-49
Igneous rocks, relation to ocean life
  cycles, 179
Ignimbrite, 157, 437
Illite, 427-28, 340
  distribution, 430
Inclination, 22
India:
  collision with Asia, 197, 201-2
  early spreading history, 112, 195
  evolution, 686
  Himalayan development, 111
  northward drift, 109
  paleoreconstructions, 111, 196-98
  poles of rotation, 181, 193
Indian Ocean:
  age, 128
  area, 24-25

aseismic ridges, 166
basalts, 217
bottom photos, 508-9
carbonate oozes, 462
Cenozoic history, 683-88
circulation, 241
continental rises, 30
depth, 25
early history, 599
erosion, 512
history of CCD, 689
margins, 156, 322
Mesozoic history, 683-84
modern sediment patterns, 686-
  88
paleobathymetry, 601
paleoceanographic evolution, 682-
  88
ridges, 193
salinity structure, 253
sedimentation rate change, 626
tectonic evolution, 193-202, 682-
  88
temperature structure, 252
trenches, 39
volume, 25
Indian Plate, 132, 195
  heat flow, 145
Indonesia, 25, 39
  arc, 151, 322
  archipelago, 21
  double arc, 151
  trenches, 39
  volcanism, 156, 369
  volcanoes, 437
Indonesian Seaway, 612
Indus River, 288
Inner core of earth, 14
Inshore, beach profile, 299
Instantaneous rotation vectors, 138
Instars:
  ostracods, 560
Interglacial episodes, 691
Interglacial-glacial cycle, 83
International Program for Ocean Drill-
  ing (IPOD), 94
International Stratigraphic Guide, 54,
  74
International Time Scale, 55
Interstadials:
  interglacials, 272
Intertropical convergence, 241
Interval-zones, 54, 65
Intramassif basins:
  of forearc, 365
Intraplate volcanism, 159
Intrusive rocks, 17
Ionian Trench, 390
Iridium:
  Cretaceous-Tertiary boundary, 712
Iron oxides, 492-95
Iron-rich basal sediments, 492-95
Island arc, 149-52
  cusps, 170
  episodic formation, 382
  terminology, 358
  typical cross section, 357
Island chains, 160-61
Island-arc continental margin, 29, 357

Isopachs:
  sediment distribution in West Pa-
    cific, 150
Isostasy, 20, 24
Isostatic anomalies, 20-21
Isostatic change, 269-70
Isostatic compensation, 20
Isostatic undercompensation, 20
Isotherms, 237
Isotopic stages, 84
Israel:
  slump features, 406-7
Italy:
  Pliocene-Pleistocene boundary, 89
  Calabrian stage, 92
Ivory Coast:
  microtektite strewnfield, 450-51,
    453

Jan Mayen ridge, 166
Japan, 7, 9, 62
  ash layers, 62
  magnetic lineations, 182-83
  Sea, 151
  volcanism, 156
Japan Arc, 354, 369, 383
Japan Current, 237
Japan Sea, 370, 373, 376
  basin, 371, 373
Japan Trench, 39, 132, 356
Jaramillo event, 73
Java Trench, 39, 132, 322
Joint Oceanographic Institute for Deep-
  Earth Sampling (JOIDES), 94
  sediment classification, 404
Juan de Fuca ridge, 117, 135, 188
Jurassic:
  age, 56
  coal deposits, 285
  Ferrar dolerites, 112
  global sea-level cycles, 280
  Gondwanaland, 112
  magnetic quiet zones, 124
  ocean history, 599
  oil deposits, 285
  oldest oceanic rocks, 127
  paleoreconstructions, 324, 701
  quiet zone, 344
  stages, 122
Juvenile water:
  geodetic, 266

Kaena event, 73
Kaikoura orogeny, 377
  New Zealand, 186
Kaolinite, 427-28, 431
  distribution, 429
Kaolinite-gibbsite-montmorillonite
  suite:
  clays, 431
Kapitean stage:
  New Zealand, 740
K-Ar dating technique, 72, 88
Kasten corer, 100, 102
Katabatic winds, 247
Kauai, Hawaii, 163
Kerguelen hot spot, 197, 200-201

Netherlands:
  barrier beaches, 305
Newark Paleomagnetic Interval, 327
New Brunswick, 524
New Caledonia, 186
New Caledonia Basin, 371
New England:
  continental rise, 47
  shelf sediments, 311
  shelves, 308
Newfoundland, 211
  outer ridge, 341
  paleoreconstructions, 190
New Guinea, 272–73
  collision zone, 355
  mountain formation, 114
  tectonism, 187
  terraces, 272–73
  uplift, 425
New Hebrides:
  arc, 364
New Hebrides Basin, 187, 370
New Jersey:
  continental margin, 346
New York:
  continental rise, 341–42
New Zealand, 3, 7, 9
  active arc, 354, 376–81
  active margin shelves, 264
  ash layers, 441
  Cenozoic position, 720
  Cretaceous plutonism, 129
  dust, 433
  geosyncline, 186
  hemipelagic sediments, 426
  late Miocene, 740
  papa, 426
  regional tectonic history, 185
Niger River, 25, 337
  delta, 294
Nihoa Island, 163
Nile Delta, 294
Nile River, 25
  Pliocene gorge, 741
Ninetyeast ridge, 166–67, 194–96, 256
  evolution, 685–86
  history, 199
NN-zones, 90
No-analogue situations:
  transfer function, 646
Nonaccretionary trench systems, 357
Nondipole field, 22
Nonvolcanic arc, 364
Norfolk Basin, 187
Norfolk ridge, 185–86
Norm, 214
  purpose, 216
Normal polarity, 22
  paleomagnetism, 71
  in tectonism, 357, 360
Normative basalt tetrahedron, 216
Normative classification, 214
North America, 3
  age of opening, 332
  Cretaceous plutonism, 129
  early separation, 192
  east coast continental margin,
    342–48

east coast sediments, 310–12
eastern margin, 326, 341, 345
microtektite strewn fields, 450–51,
  453
opening, 343
western margin tectonics, 187–89
North American Basin, 58
  deposits, 510
  rock sequence, 45
North American Plate, 132, 187–189,
  190
North American rise, 341
North Atlantic:
  abyssal plains, 37
  age of margins, 127
  age of opening, 127
  black shales, 663–64
  bottom-water flow path, 534
  bottom-water source, 251
  cable-laying, 6
  Cenozoic history, 665–69
  deep crustal drilling, 210
  depth–age relationship, 148
  dust distribution, 434
  dust storms, 432–33
  early carbonate and evaporites,
    660
  early evolution, 350–51
  early history, 190, 194
  early rifting, 329, 331
  geological cross section, 351
  history of CCD, 689
  ice-rafted sediments, 450
  margin evolution, 350–51
  margins, 325–26
  Mesozoic history, 660
  morphologic features, 27
  Neogene circulation, 668
  paleoceanographic evolution, 659–
    71
  Paleocene, 666–67
  Paleogene siliceous sedimenta-
    tion, 667
  Quaternary paleocirculation,
    604–5, 746
  sea-level history, 269–71
  sediment drifts, 532–35
  stratigraphy, 662
  tectonic evolution, 659–71
North Atlantic Current, 241
North Atlantic Deep Water, 247, 254,
  535
  benthonic foraminifera, 515–16
  effect on CCD history, 690
  evolution, 668, 737
Northeast Pacific:
  fracture zones, 36–37
  tectonic history, 187–89
North Equatorial Current, 241, 248
Northern Hemisphere:
  ice sheets, 89, 599, 648, 669, 743–
    45
  paleoreconstructions, 111
North Fiji Basin, 376
North magnetic pole, 21
North Pacific:
  depth–age relationship, 148
  ice-rafted sediments, 450

principal morphologic features, 28
silica deposition, 475
windblown quartz, 435
North Pacific Current, 241
North Pole, 21, 26, 71
North Sea, 26
  barrier beaches, 305
Northwest Atlantic:
  rock sequence, 58
Northwest Pacific:
  tectonics, 182
Norway:
  ahermatypic corals, 313
Norwegian Basin, 254
Norwegian–Greenland Sea:
  early history, 192
Norwegian Sea, 25, 35, 193, 253, 736
  age of opening, 332
  bottom-water source, 251
  ice-rafted sediments, 450
  Quaternary paleotemperature
    record, 644
Nothofagus, 727
Nova Scotia, 524
  continental margins, 344
  margins, 325
Nunivak event, 73
N-zones, 66–67, 90

Oahu, Hawaii, 161
Obduction, 389
Ocean basalts:
  alteration, 71
  radiometric dating, 71
  ridges, 133
Ocean basins, 23
  age, 147–49
  areas, 28
  floors, 27, 37
  history, 180
  sediment volume, 321
Ocean-bottom seismometers (OBS), 50
Ocean–continent boundary, 325
Ocean crust, 8, 24, 180, 204–32
  age, 120, 123–24, 126–27, 147–49,
    180
  age–depth relationship, 147–49
  alteration and weathering, 230–31
  biostratigraphic ages, 124
  cross section, 211
  deep-sea drilling, 210
  fracturing, 230
  lithologic units, 210
  magnetization, 232
  porosity, 209
  rock types, 213–218
  seismic reflection, 209
  seismic refraction, 47
  seismic velocity and age, 208
  structure, 206–13
  youthfulness, 124
Ocean floor:
  methods of study, 507
Oceanic islands:
  ages, 116
  value in sea-level studies, 269

Ripple marks (*cont.*)
photographs of, 508
velocities for formation, 517–18, 528
Rises, continental:
area, 28
distribution, 341
Rivers:
sediment transport, 288, 290, 624
silica influx, 475, 478–79
Rockall Plateau, 166, 329
western margins, 338–39
Rocard Island, 159
Rockall Bank, 534
Rock falls, 407
Romanche fracture zone, 169, 255, 469
Ross ice shelf, 252
evolution, 739
Ross Sea, 21, 250–51, 444
bottom-water formation, 252
Rotary drilling, 102

Sabellariid worms, 313
Sable Island Bank, 308
Sahara Desert:
Devonian polar ice cap, 107
sediment source, 432
St. Ann's Shoal:
Sierra Leone, 308
St. Lawrence River, 25, 422
St. Paul Islands, 200–201
hot spot, 163
Salinity:
surface water, 237
Salinity crisis:
effect on oceanic salt budget, 742
Mediterranean, 616, 738–42
Saltation:
sediment, 408
Salt diapirs, 334, 337–38
South Atlantic, 169
Salt-wedge estuary, 291–92
Samoan Passage, 257
San Andreas Fault, 132, 135, 143, 188, 355, 388
Sand:
beach transport, 303
velocity for erosion, 517
Sand waves, 516, 528, 530–32
San Lucas Fan, 421
Santa Barbara Basin, 619
anoxic sediments, 488
varves, 80–81
Santa Maria, Guatemala.
volcano, 443
São Paulo Plateau, 681
Sapropels, 486–87
formation, 489–90
Mediterranean, 489–90
Sargasso Sea, 462–63
Scandinavia:
isostatic rebound, 20–21
Scanning electron microscope (SEM), 472, 566
Scotia Arc, 155, 370
Scotia Sea, 151
Basin, 345, 373
Scour marks, 516–17

Scree deposits, 407
Scripps Canyon, California, 417
Scripps Institution of Oceanography, 7–8, 94, 511
Sea-floor sampling devices, 101–2
Sea-floor spreading, 104, 114–29, 180
marginal basins, 372
onset, 331–33
rate changes, 385
rates, 120, 149
sea-level change, 267, 282–84
Sea ice:
evolution, 657
formation, 251
related diatoms, 585
Sea level, 24
effect on sediment disposal, 289
global cycles, 280
history, 268–85
paleoclimatic effect, 699
relation to chert formation, 482–83
Sea-level change:
addition of juvenile water, 266
causes, 265–67, 282
CCD history, 690–91
effect on canyons, 418
effect on oceanic sediments, 625–26
effect on sedimentation, 267
glacial isostasy, 266
hydroisostatic deformation, 266
importance, 268
oceanic ridges, 129
Quaternary, 29
rates, 281–82
sedimentary cycles, 273–75
sedimento-eustatic, 266
tectonic-erosion, 266
tectono-eustatic, 266
Sea-level cycles:
economic importance, 285
Sea-level history:
effect on tectonism, 269
methods, 278
Quaternary, 268–73
Sea-level regression:
late Miocene, 739–40
Seamounts, 37–38, 159
chains, 38
Sea of Japan:
isolation, 740
late Miocene isolation, 616
Sea of Okhotsk:
silica deposition, 475
Seasonal thermocline, 235–36
Seawater:
oxygen isotopic composition, 82–83, 85
viscosity, 541
Sechron, 274
Secondary waves, 14
Sedimentary cycles, 273–75
economic importance, 285
Sedimentation rates:
changes through time, 625–27
erosion, 512
Sediment dams, 335–36
Sediment drifts, 532–33

deep sea, 519
North Atlantic, 534–35
Sediments:
extraterrestrial, 450–53
gravity flows, 405
mapping, 512–13
methods of classification, 399–400
rate of deposition, 464
relation to ocean life cycle, 179
subduction, 363
submarine canyons, 417
volumes in ocean, 321
wind and grain size, 413–18, 439
Sediment sampling methods, 93–103
Sediment transport:
rivers, 288
Sediment waves, 511
deep-sea, 530
Seismic first-motion studies, 134
Seismicity, 134, 139–43
Seismic margins, 29, 323
Seismic reflection, 9, 40–47, 276
causes of diagenesis, 473
oceanic crust, 209
reflection profiles, 510–12
seismic stratigraphy, 275
Seismic refraction, 40, 47–50
ocean crust, 206–10
Seismic stratigraphy, 275–82
Seismic velocities, 18
Seismic waves, 14, 17
Seismology, 14
Semidiurnal tides, 310
Series, 54
Seychelle Islands, 113
paleoreconstructions, 111–12
Shales, black, 490–91
Shear waves, 14
Sheeted-dike complex, 211–12, 228–29
Sheet flows, 218
Shelf:
break, 27, 29, 265
edge, 29
modern mud blanket, 310
relict sand blanket, 309
sediments, 308–16
topography, 307
zones, 258
Shelf provinces:
eastern North America, 310
Shield volcanoes, 217
Shikoku Basin, 371, 373
Shoreline:
length, 263
migration, 282
Short-headed, delta front fans, 421
Sial, 18
Siberian continental shelf, 35
Sicily:
Miocene–Pliocene boundary, 92
Sierra Leone rise, 169, 309
Sierra Nevada:
recent tectonic history, 384
Silica:
biogenic fixation by organisms, 478–79
Cenozoic interocean patterns, 622
cycle, 475, 478–79
diagenesis, 481

810 Subject Index

lapilli, 438
mechanisms, 368-70
oceanic ridges, 218-20
origin of cherts, 482
paleoclimatic effect, 648
patterns, 152-70
plate tectonics, 152-70
recent distribution, 154
seismicity, 156
tholeiitic, 204
Volcanoes, 131
distribution in West Pacific, 150
Pacific, 159
recent distribution, 154
volume of material produced, 437
Volcanogenic sediments, 436-42

Walvis Channel, 614
Walvis ridge, 159, 169, 469
early history, 192-93
evolution, 672, 674
Water depth zones, 259
Wave-generated currents, 300
Wave refraction, 298
Waves, 298, 300-302
climate, 305
setdown, 302
setup, 302
terminology, 298
wave height, 300-301
wave length, 297

Weathering:
oceanic crust, 230
Weddell Sea, 250-51, 253-54, 444, 446,
449
bottom-water formation, 252
Wegenerian revolution, 115
West Africa:
shelf, 308-9
West Antarctic ice sheet:
evolution, 738
West Australia:
adjacent ancient landmass, 195
western margin, 338-39
West Australian Current, 241-42
Western boundary currents, 242, 248,
256
nepheloid layer, 526
Western boundary undercurrents:
association with sediment drifts,
533-34
eastern North America, 423
West Florida:
continental shelf, 276
West Philippine Basin, 371, 373, 376
West-wind drift, 247
Whales:
evolution, 728-29, 747
Oligocene, 727
Wharton Basin, 197, 199
Whole-rock analyses, 70
Wilmington Canyon, 418
Wind:

ash distribution, 441-42
currents, 245
importance in circulation, 240-41
Windblown sediments, 432-36
Wisconsin Glaciation, 271-72
Worldwide Standardized Seismograph
Network (WWSSN), 139
Würm Glaciation, 272

Xenoliths, 18
X-ray radiography, 513

Yangtze River, 288
Yucatan:
shelf, 314
Yucatan Channel, 745
Yucatan Peninsula, 336
Yuryaku Seamount, 164

Zanclean stage, 92
Zeolite facies:
oceanic crust, 211
Zeolites, 503
Zhemchug Canyon, 419
Zonations:
calcareous nannofossils, 66-67
planktonic foraminifera, 66-67
Radiolaria, 67
Zooxanthellae, 313-14